Grundlehren der mathematischen Wissenschaften 310

A Series of Comprehensive Studies in Mathematics

Editors

M. Artin S.S. Chern J. Coates J.M. Fröhlich
H. Hironaka F. Hirzebruch L. Hörmander C.C. Moore
J.K. Moser M. Nagata W. Schmidt D.S. Scott
Ya.G. Sinai J. Tits M. Waldschmidt S. Watanabe

Managing Editors

M. Berger B. Eckmann S.R.S. Varadhan

Springer
Berlin
Heidelberg
New York
Barcelona
Budapest
Hong Kong
London
Milan
Paris
Santa Clara
Singapore
Tokyo

Mariano Giaquinta
Stefan Hildebrandt

Calculus of Variations I

The Lagrangian Formalism

With 73 Figures

Springer

Mariano Giaquinta
Università di Firenze, Dipartimento di Matematica Applicata "G. Sansone"
Via S. Marta 3, I-50139 Firenze, Italy

Stefan Hildebrandt
Universität Bonn, Mathematisches Institut
Wegelerstr. 10, D-53115 Bonn, Germany

Mathematics Subject Classification: 49-XX, 53-XX, 70-XX

ISBN 3-540-50625-X Springer-Verlag Berlin Heidelberg New York

Library of Congress Cataloging-in-Publication Data. Giaquinta, Mariano, 1947– . Calculus of variations/Mariano Giaquinta, Stefan Hildebrandt. p. cm. – (Grundlehren der mathematischen Wissenschaften; 310–311) Includes bibliographical references and indexes. Contents: 1. The Lagrangian formalism – 2. The Hamiltonian formalism. ISBN 3-540-50625-X (Berlin: v. 1). – ISBN 0-387-50625-X (New York: v. 1). – ISBN 3-540-57961-3 (Berlin: v. 2). – ISBN 0-387-57961-3 (New York: v. 2) 1. Calculus of variations. I. Hildebrandt, Stefan. II. Title. III. Series. QA315.G46 1996 515′.64 – dc20 96-20429

Printed in Germany

Cover design: Springer-Verlag, Design & Production

Typesetting: Asco Trade Typesetting Ltd., Hong Kong

SPIN: 10011540 41/3140/SPS – 5 4 3 2 1 0 – Printed on acid-free paper

Preface

This book describes the classical aspects of the variational calculus which are of interest to analysts, geometers and physicists alike. Volume 1 deals with the formal apparatus of the variational calculus and with nonparametric field theory, whereas Volume 2 treats parametric variational problems as well as Hamilton–Jacobi theory and the classical theory of partial differential equations of first order. In a subsequent treatise we shall describe developments arising from Hilbert's 19th and 20th problems, especially direct methods and regularity theory.

Of the classical variational calculus we have particularly emphasized the often neglected theory of inner variations, i.e. of variations of the independent variables, which is a source of useful information such as monotonicity formulas, conformality relations and conservation laws. The combined variation of dependent and independent variables leads to the general conservation laws of Emmy Noether, an important tool in exploiting symmetries. Other parts of this volume deal with Legendre–Jacobi theory and with field theories. In particular we give a detailed presentation of one-dimensional field theory for nonparametric and parametric integrals and its relations to Hamilton–Jacobi theory, geometrical optics and point mechanics. Moreover we discuss various ways of exploiting the notion of convexity in the calculus of variations, and field theory is certainly the most subtle method to make use of convexity. We also stress the usefulness of the concept of a null Lagrangian which plays an important role in several instances. In the final part we give an exposition of Hamilton–Jacobi theory and its connections with Lie's theory of contact transformations and Cauchy's integration theory of partial differential equations.

For better readability we have mostly worked with local coordinates, but the global point of view will always be conspicuous. Nevertheless we have at least once outlined the coordinate-free approach to manifolds, together with an outlook onto symplectic geometry.

Throughout this volume we have used the classical *indirect method* of the calculus of variations solving first Euler's equations and investigating thereafter which solutions are in fact minimizers (or maximizers). Only in Chapter 8 we have applied direct methods to solve minimum problems for parametric integrals. One of these methods is based on results of field theory, the other uses the concept of lower semicontinuity of functionals. *Direct methods* of the calculus of variations and, in particular, existence and regularity results for minimizers of multiple integrals will be subsequently presented in a separate treatise.

We have tried to write the present book in such a way that it can easily be read and used by any graduate student of mathematics and physics, and by nonexperts in the field. Therefore we have often repeated ideas and computations if they appear in a new context. This approach makes the reading occasionally somewhat repetitious, but the reader has the advantage to see how ideas evolve and grow. Moreover he will be able to study most parts of this book without reading all the others. This way a lecturer can comfortably use certain parts as text for a one-term course on the calculus of variations or as material for a reading seminar.

We have included a multitude of examples, some of them quite intricate, since examples are the true lifeblood of the calculus of variations. To study specific examples is often more useful and illustrative than to follow all ramifications of the general theory. Moreover the reader will often realize that even simple and time-honoured problems have certain peculiarities which make it impossible to directly apply general results.

In the *Scholia* we present supplementary results and discuss references to the literature. In addition we present historical comments. We have consulted the original sources whenever possible, but since we are no historians we might have more than once erred in our statements. Some background material as well as hints to developments not discussed in our book can also be found in the *Supplements*.

A last word concerns the size of our project. The reader may think that by writing two volumes about the classical aspects of the calculus of variations the authors should be able to give an adequate and complete presentation of this field. This is unfortunately not the case, partially because of the limited knowledge of the authors, and partially on account of the vast extent of the field. Thus the reader should not expect an encyclopedic presentation of the entire subject, but merely an introduction in one of the oldest, but nevertheless very lively areas of mathematics. We hope that our book will be of interest also to experts as we have included material not everywhere available. Also we have examined an extensive part of the classical theory and presented it from a modern point of view.

It is a great pleasure for us to thank friends, colleagues, and students who have read several parts of our manuscript, pointed out errors, gave us advice, and helped us by their criticism. In particular we are very grateful to Dieter Ameln, Gabriele Anzellotti, Ulrich Dierkes, Robert Finn, Karsten Große-Brauckmann, Anatoly Fomenko, Hermann Karcher, Helmut Kaul, Jerry Kazdan, Rolf Klötzler, Ernst Kuwert, Olga A. Ladyzhenskaya, Giuseppe Modica, Frank Morgan, Heiko von der Mosel, Nina N. Uraltseva, and Rüdiger Thiele. The latter also kindly supported us in reading the galley proofs. We are much indebted to Kathrin Rhode who helped us to prepare several of the examples. Especially we thank Gudrun Turowski who read most of our manuscript and corrected numerous mistakes. Klaus Steffen provided us with example [0] in 3,*1* and the regularity argument used in 3,*6* nr. 11. Without the patient and excellent typing and retyping of our manuscripts by Iris Pützer and Anke Thiedemann this book could not have been completed, and we appreciate their invaluable help as well as the patience of our Publisher and the constant and friendly encouragement by Dr. Joachim Heinze. Last but not least we would like to extend our thanks to *Consiglio Nazionale delle Ricerche*, to *Deutsche Forschungsgemeinschaft*, to *Sonderforschungsbereich* 256 *of Bonn University*, and to the *Alexander von Humboldt Foundation*, which have generously supported our collaboration.

Bonn and Firenze, February 14, 1994

Mariano Giaquinta
Stefan Hildebrandt

Contents of Calculus of Variations I and II

Calculus of Variations I: The Lagrangian Formalism

Calculus of Variations II: The Hamiltonian Formalism

Introduction

The Calculus of Variations is the art to find optimal solutions and to describe their essential properties. In daily life one has regularly to decide such questions as which solution of a problem is best or worst; which object has some property to a highest or lowest degree; what is the optimal strategy to reach some goal. For example one might ask what is the shortest way from one point to another, or the quickest connection of two points in a certain situation. The isoperimetric problem, already considered in antiquity, is another question of this kind. Here one has the task to find among all closed curves of a given length the one enclosing maximal area. The appeal of such optimum problems consists in the fact that, usually, they are easy to formulate and to understand, but much less easy to solve. For this reason the calculus of variations or, as it was called in earlier days, the isoperimetric method has been a thriving force in the development of analysis and geometry.

An ideal shared by most craftsmen, artists, engineers, and scientists is the principle of the economy of means: What you can do, you can do simply. This aesthetic concept also suggests the idea that nature proceeds in the simplest, the most efficient way. Newton wrote in his *Principia*: "*Nature does nothing in vain, and more is in vain when less will serve; for Nature is pleased with simplicity and affects not the pomp of superfluous causes.*" Thus it is not surprising that from the very beginning of modern science optimum principles were used to formulate the "laws of nature", be it that such principles particularly appeal to scientists striving toward unification and simplification of knowledge, or that they seem to reflect the preestablished harmony of our universe. Euler wrote in his *Methodus inveniendi* [2] from 1744, the first treatise on the calculus of variations: "*Because the shape of the whole universe is most perfect and, in fact, designed by the wisest creator, nothing in all of the world will occur in which no maximum or minimum rule is somehow shining forth.*" Our belief in the best of all possible worlds and its preestablished harmony claimed by Leibniz might now be shaken; yet there remains the fact that many if not all laws of nature can be given the form of an extremal principle.

The first known principle of this type is due to Heron from Alexandria (about 100 A.D.) who explained the law of reflection of light rays by the postulate that *light must always take the shortest path*. In 1662 Fermat succeeded in deriving the law of refraction of light from the hypothesis that *light always propagates in the quickest way from one point to another*. This assumption is now called *Fermat's principle.* It is one of the pillars on which geometric optics rests; the other one is *Huygens's principle* which was formulated about 15 years later. Further, in his letter to De la Chambre from January 1, 1662, Fermat motivated his principle by the following remark: "*La nature agit toujour par les voies les plus courtes.*" (Nature always acts in the shortest way.)

About 80 years later Maupertuis, by then President of the Prussian Academy of Sciences, resumed Fermat's idea and postulated his metaphysical principle of the *parsimonious universe,* which later became known as "*principle of least action*" or "*Maupertuis's principle*". He stated: *If there occurs some change in nature, the amount of action necessary for this change must be as small as possible.*

"Action" that nature is supposed to consume so thriftily is a quantity introduced by Leibniz which has the dimension "energy × time". It is exactly that quantity which, according to Planck's quantum principle (1900), comes in integer multiples of the elementary quantum h.

In the writings of Maupertuis the action principle remained somewhat vague and not very convincing, and by Voltaire's attacks it was mercilessly ridiculed. This might be one of the reasons why Lagrange founded his *Méchanique analitique* from 1788 on d'Alembert's principle and not on the least action principle, although he possessed a fairly general mathematical formulation of it already in 1760. Much later Hamilton and Jacobi formulated quite satisfactory versions of the action principle for point mechanics, and eventually Helmholtz raised it to the rank of the most general law of physics. In the first half of this century physicists seemed to prefer the formulation of natural laws in terms of space–time differential equations, but recently the principle of least action had a remarkable comeback as it easily lends itself to a global, coordinate-free setup of physical "field theories" and to symmetry considerations.

The development of the calculus of variations began briefly after the invention of the infinitesimal calculus. The first problem gaining international fame, known as "problem of quickest descent" or as "brachystochrone problem", was posed by Johann Bernoulli in 1696. He and his older brother Jakob Bernoulli are the true founders of the new field, although also Leibniz, Newton, Huygens and l'Hospital added important contributions. In the hands of Euler and Lagrange the calculus of variations became a flexible and efficient theory applicable to a multitude of physical and geometric problems. Lagrange invented the δ-calculus which he viewed to be a kind of "higher" infinitesimal calculus, and Euler showed that the δ-calculus can be reduced to the ordinary infinitesimal calculus. Euler also invented the multiplier method, and he was the first to treat variational problems with differential equations as subsidiary conditions. The development of the calculus of variations in the 18th century is described in the booklet by Woodhouse [1] from 1810 and in the first three chapters of H.H. Goldstine's historical treatise [1]. In this first period the variational calculus was essentially concerned with deriving necessary conditions such as Euler's equations which are to be satisfied by minimizers or maximizers of variational problems. Euler mostly treated variational problems for single integrals where the corresponding Euler equations are ordinary differential equations, which he solved in many cases by very skillful and intricate integration techniques. The spirit of this development is reflected in the first parts of this volume. To be fair with Euler's achievements we have to emphasize that he treated in [2] many more one-dimensional variational problems than the reader can find anywhere else including our book, some of which are quite involved even for a mathematician of today.

However, no *sufficient conditions* ensuring the minimum property of solutions of Euler's equations were given in this period, with the single exception of a paper by Johann Bernoulli from 1718 which remained unnoticed for about 200 years. This is to say, analysts were only concerned with determining solutions of Euler equations, that is, with stationary curves of one-dimensional variational problems, while it was more or less taken for granted that such stationary objects furnish a real extremum.

The sufficiency question was for the first time systematically tackled in Legendre's paper [1] from 1788. Here Legendre used the idea to study the second variation of a functional for deciding such questions. Legendre's paper contained some errors, pointed out by Lagrange in 1797, but his ideas proved to be fruitful when Jacobi resumed the question in 1837. In his short paper [1] he sketched an entire theory of the second variation including his celebrated theory of conjugate points, but all of his results were stated with essentially no proofs. It took a whole generation of mathematicians to fill in the details. We have described the basic features of the *Legendre–Jacobi theory* of the second variation in Chapters 4 and 5 of this volume.

Euler treated only a few variational problems involving multiple integrals. Lagrange derived the "Euler equations" for double integrals, i.e. the necessary differential equations to be satisfied by minimizers or maximizers. For example he stated the minimal surface equation which characterizes the stationary surface of the nonparametric area integral. However he did not indicate how one can obtain solutions of the minimal surface equation or of any other related Euler equation. Moreover neither he nor anyone else of his time was able to derive the *natural boundary conditions* to be satisfied by, say, minimizers of a double integral subject to free boundary conditions since the tool of "integration by parts" was not available. The first to successfully tackle two-dimensional variational problems with free boundaries was Gauss in his paper [3] from 1830 where he established a variational theory of capillary phenomena based on Johann Bernoulli's *principle of virtual work* from 1717. This principle states that in equilibrium no work is needed to achieve an infinitesimal displacement of a mechanical system. Using the concept of a potential energy which is thought to be attached to any state of a physical system, Bernoulli's principle can be replaced by the following hypothesis, *the principle of minimal energy*: The equilibrium states of a physical system are stationary states of its potential energy, and the stable equilibrium states minimize energy among all other "virtual" states which lie close-by.

For capillary surfaces not subject to any gravitational forces the potential energy is proportional to their surface area. This explains why the phenomenological theory of soap films is just the theory of surfaces of minimal area.

After Gauss free boundary problems were considered by Poisson, Ostrogradski, Delaunay, Sarrus, and Cauchy. In 1842 the French Academy proposed as topic for their great mathematical prize the problem to derive the natural boundary conditions which together with Euler's equations must be satisfied by minimizers and maximizers of free boundary value problems for multiple integrals. Four papers were sent in; the prize went to Sarrus with an honourable mentioning of Delaunay, and in 1861 Todhunter [1] held Sarrus's paper for "the most important original contribution to the calculus of variations which has been made during the present century". It is hard to believe that these formulas which can nowadays be derived in a few lines were so highly appreciated by the Academy, but we must realize that in those days integration by parts was not a fully developed tool. This example shows very well how the problems posed by the variational calculus forced analysts to develop new tools. Time and again we find similar examples in the history of this field.

In Chapters 1–4 we have presented all formal aspects of the calculus of variations including all necessary conditions. We have simultaneously treated extrema of single and multiple integrals as there is barely any difference in the degree of difficulty, at least as long as one sticks to variational problems involving only first order derivatives. The difference between one- and multidimensional problems is rarely visible in the formal aspect of the theory but becomes only perceptible when one really wants to construct solutions. This is due to the fact that the necessary conditions for one-dimensional integrals are ordinary differential equations, whereas the Euler equations for multiple integrals are partial differential equations. The problem to solve such equations under prescribed boundary conditions is a much more difficult task than the corresponding problem for ordinary differential equations; except for some special cases it was only solved in this century. As we need rather refined tools of analysis to tackle partial differential equations we deal here only with the formal aspects of the calculus of variations in full generality while existence questions are merely studied for one-dimensional variational problems. The existence and regularity theory of multiple variational integrals will be treated in a separate treatise.

Scheeffer and Weierstrass discovered that positivity of the second variation at a stationary curve is not enough to ensure that the curve furnishes a local minimum; in general one can only show that it is a *weak minimizer*. This means that the curve yields a minimum only in comparison to those curves whose tangents are not much different.

In 1879 Weierstrass discovered a method which enables one to establish a *strong minimum property* for solutions of Euler's equations, i.e. for stationary curves; this method has become known as *Weierstrass field theory*. In essence Weierstrass's method is a rather subtle convexity argument which uses two ingredients. First one employs a local convexity assumption on the integrand of the variational integral which is formulated by means of *Weierstrass's excess function*. Secondly, to make proper use of this assumption one has to embed the given stationary curve in a suitable field of such curves. This field embedding can be interpreted as an introduction of a particular system of normal coordinates which very much simplify the comparison of the given stationary curve with any neighbouring curve. In the plane it suffices to embed the given curve in an arbitrarily chosen field of stationary curves while in higher dimensions one has to embed the curve in a so-called *Mayer field*.

In Chapter 6 of this volume we shall describe Weierstrass field theory for nonparametric one-dimensional variational problems and the contributions of Mayer, Kneser, Hilbert and Carathéodory. The corresponding field theory for parametric integrals is presented in Chapter 8. There we have also a first glimpse at the so-called *direct method* of the calculus of variations. This is a way to establish directly the existence of minimizers by means of set-theoretic arguments; another treatise will entirely be devoted to this subject. In addition we sketch field theories for multiple integrals at the end of Chapters 6 and 7.

In chapter 7 we describe an important involutory transformation, which will be used to derive a dual picture of the Euler–Lagrange formalism and of field theory, called *canonical formalism*. In this description the dualism *ray versus wave* (or: particle–wave) becomes particularly transparent. The canonical formalism is a part of the *Hamilton–Jacobi theory*, of which we give a self-contained presentation in Chapter 9, together with a brief introduction to symplectic geometry. This theory has its roots in Hamilton's investigations on geometrical optics, in particular on systems of rays. Later Hamilton realized that his formalism is also suited to describe systems of point mechanics, and Jacobi developed this formalism further to an effective integration theory of ordinary and partial differential equations and to a theory of canonical mappings. The connection between *canonical* (or *symplectic*) *transformations* and *Lie's theory of contact transformations* is discussed in Chapter 10 where we also investigate the relations between the principles of Fermat and Huygens. Moreover we treat *Cauchy's method* of integrating partial differential equations of first order by the method of characteristics and illustrate the connection of this technique with Lie's theory.

The reader can use the detailed table of contents with its numerous catchwords as a guideline through the book; the detailed introductions preceding each chapter and also every section and subsection are meant to assist the reader in obtaining a quick orientation. A comprehensive *glimpse at the literature* on the Calculus of Variations is given at the end of Volume 2. Further references can be found in the Scholia to each chapter and in our bibliography. Moreover, important historical references are often contained in footnotes. As important examples are sometimes spread over several sections, we have added a *list of examples*, which the reader can also use to locate specific examples for which he is looking.

Contents of Calculus of Variations I
The Lagrangian Formalism

Part I. The First Variation and Necessary Conditions

4.2. Characterization of Null Lagrangians 55
(Structure of null Lagrangians. Exactly the Lagrangians of divergence form are null Lagrangians. The divergence and the Jacobian of a vector field as null Lagrangians.)

5. Variational Problems of Higher Order 59
(Euler equations. Equilibrium of thin plates. Gauss curvature. Gauss-Bonnet theorem. Curvature integrals for planar curves. Rotation number of a planar curve. Euler's area problem.)

6. Scholia 68

Chapter 2. Variational Problems with Subsidiary Conditions 87

1. Isoperimetric Problems 89
(The classical isoperimetric problem. The multiplier rule for isoperimetric problems. Eigenvalues of the vibrating string and of the vibrating membrane. Hypersurfaces of constant mean curvature. Catenaries.)

2. Mappings into Manifolds: Holonomic Constraints 97
(The multiplier rule for holonomic constraints. Harmonic mappings into hypersurfaces of $\mathbb{R}^{N+1}$. Shortest connection of two points on a surface in $\mathbb{R}^3$. Johann Bernoulli's theorem. Geodesics on a sphere. Hamiltons's principle and holonomic constraints. Pendulum equation.)

3. Nonholonomic Constraints 110
(Normal and abnormal extremals. The multiplier rule for one-dimensional problems with nonholonomic constraints. The heavy thread on a surface. Lagrange's formulation of Maupertuis's least action principle. Solenoidal vector fields.)

4. Constraints at the Boundary. Transversality 122
(Shortest distance in an isotropic medium. Dirichlet integral. Generalized Dirichlet integral. Christoffel symbols. Transversality and free transversality.)

5. Scholia 132

Chapter 3. General Variational Formulas 145

1. Inner Variations and Inner Extremals. Noether Equations 147
(Energy-momentum tensor. Noether's equations. Erdmann's equation and conservation of energy. Parameter invariant integrals: line and double integrals, multiple integrals. Jacobi's geometric version of the least action principle. Minimal surfaces.)

2. Strong Inner Variations, and Strong Inner Extremals 163
(Inner extremals of the generalized Dirichlet integral and conformality relations. H-surfaces.)

3. A General Variational Formula 172
(Fluid flow and continuity equation. Stationary, irrotational, isentropic flow of a compressible fluid.)

4. Emmy Noether's Theorem 182
(The n-body problem and Newton's law of gravitation. Equilibrium problems in elasticity. Conservation laws. Hamilton's principle in continuum mechanics. Killing equations.)

5. Transformation of the Euler Operator to New Coordinates 198
(Generalized Dirichlet integral. Laplace-Beltrami Operator. Harmonic mappings of Riemannian manifolds.)

6. Scholia 210

Part II. The Second Variation and Sufficient Conditions

Chapter 4. Second Variation, Excess Function, Convexity 217

1. Necessary Conditions for Relative Minima 220
   1.1. Weak and Strong Minimizers 221
   (Weak and strong neighbourhoods; weak and strong minimizers; the properties ($\mathcal{M}$) and ($\mathcal{M}'$). Necessary and sufficient conditions for a weak minimizer. Scheeffer's example.)
   1.2. Second Variation: Accessory Integral and Accessory Lagrangian 227
   (The accessory Lagrangian and the Jacobi operator.)
   1.3. Thc Lcgcndrc–Hadamard Condition 229
   (Necessary condition for weak minimizers. Ellipticity, strong ellipticity, and superellipticity.)
   1.4. The Weierstrass Excess Function $\mathscr{E}_F$ and Weierstrass's Necessary Condition 232
   (Necessary condition for strong minimizers.)
2. Sufficient Conditions for Relative Minima Based on Convexity Arguments 236
   2.1. A Sufficient Condition Based on Definiteness of the Second Variation 237
   (Convex integrals.)
   2.2. Convex Lagrangians 238
   (Dirichlet integral, area and length, weighted length.)
   2.3. The Method of Coordinate Transformations 242
   (Line element in polar coordinates. Carathéodory's example. Euler's treatment of the isoperimetric problem.)
   2.4. Application of Integral Inequalities 250
   (Stability via Sobolev's inequality.)
   2.5. Convexity Modulo Null Lagrangians 251
   (The H-surface functional.)
   2.6. Calibrators 254
3. Scholia 260

Chapter 5. Weak Minimizers and Jacobi Theory 264

1. Jacobi Theory: Necessary and Sufficient Conditions for Weak Minimizers Based on Eigenvalue Criteria for the Jacobi Operator 265
   1.1. Remarks on Weak Minimizers 265
   (Scheeffer's example: Positiveness of the second variation does not imply minimality.)
   1.2. Accessory Integral and Jacobi Operator 267
   (The Jacobi operator as linearization of Euler's operator and as Euler operator of the accessory integral. Jacobi equation and Jacobi fields.)

1.3. Necessary and Sufficient Eigenvalue Criteria for Weak Minima ... 271
(The role of the first eigenvalue of the Jacobi operator. Strict Legendre-Hadamard condition. Results from the eigenvalue theory for strongly elliptic systems. Conjugate values and conjugate points.)

2. Jacobi Theory for One-Dimensional Problems in One Unknown Function ... 276

2.1. The Lemmata of Legendre and Jacobi ... 276
(A sufficient condition for weak minimizers.)

2.2. Jacobi Fields and Conjugate Values ... 281
(Jacobi's function $\Delta(x, \xi)$. Sturm's oscillation theorem. Necessary and sufficient conditions expressed in terms of Jacobi fields and conjugate points.)

2.3. Geometric Interpretation of Conjugate Points ... 286
(Envelope of families of extremals. Fields of extremals and conjugate points. Embedding of a given extremal into a field of extremals. Conjugate points and complete solutions of Euler's equation.)

2.4. Examples ... 292
(Quadratic integrals. Sturm's comparison theorem. Conjugate points of geodesics. Parabolic orbits and Galileo's law. Minimal surfaces of revolution.)

3. Scholia ... 306

Chapter 6. Weierstrass Field Theory for One-Dimensional Integrals and Strong Minimizers ... 310

1. The Geometry of One-Dimensional Fields ... 312

1.1. Formal Preparations: Fields, Extremal Fields, Mayer Fields, and Mayer Bundles, Stigmatic Ray Bundles ... 313
(Definitions. The modified Euler equations. Mayer fields and their eikonals. Characterization of Mayer fields by Carathéodory's equations. The Beltrami form. Lagrange brackets. Stigmatic ray bundles and Mayer bundles.)

1.2. Carathéodory's Royal Road to Field Theory ... 327
(Null Lagrangian and Carathéodory equations. A sufficient condition for strong minimizers.)

1.3. Hilbert's Invariant Integral and the Weierstrass Formula. Optimal Fields. Kneser's Transversality Theorem ... 332
(Sufficient conditions for weak and strong minimizers. Weierstrass fields and optimal fields. The complete figure generated by a Mayer field: The field lines and the one-parameter family of transversal surfaces. Stigmatic fields and their value functions $\Sigma(x, e)$.)

2. Embedding of Extremals ... 350

2.1. Embedding of Regular Extremals into Mayer Fields ... 351
(The general case $N \geq 1$. Jacobi fields and pairs of conjugate values. Embedding of extremals by means of stigmatic fields.)

2.2. Jacobi's Envelope Theorem ... 356
(The case $N = 1$: First conjugate locus and envelope of a stigmatic bundle. Global embedding of extremals.)

2.3. Catenary and Brachystochrone 362
(Field theory for integrals of the kind $\int \omega(x, u)\sqrt{1 + (u')^2}\, dx$ corresponding to Riemannian metrics $ds = \omega(x, z)\sqrt{dx^2 + dz^2}$. Galilei parabolas. Minimal surfaces of revolution. Poincaré's model of the hyperbolic plane. Brachystochrone.)

2.4. Field-like Mayer Bundles, Focal Points and Caustics 372
(Conjugate base of Jacobi fields and its Mayer determinant $\Delta(x)$. The zeros of $\Delta(x)$ are isolated. Sufficient conditions for minimality of an extremal whose left endpoint freely varies on a prescribed hypersurface.)

3. Field Theory for Multiple Integrals in the Scalar Case: Lichtenstein's Theorem 384
(Fields for nonparametric hypersurfaces. Carathéodory equations. Hilbert's invariant integral. Embedding of extremals. Lichtenstein's theorem.)

4. Scholia 395

Supplement. Some Facts from Differential Geometry and Analysis 400
1. Euclidean Spaces 400
2. Some Function Classes 405
3. Vector and Covector Fields. Transformation Rules 408
4. Differential Forms 412
5. Curves in $\mathbb{R}^N$ 421
6. Mean Curvature and Gauss Curvature 425

A List of Examples 432

Bibliography 437

Subject Index 468

Contents of Calculus of Variations II
The Hamiltonian Formalism

Part III. Canonical Formalism and Parametric Variational Problems

Chapter 7. Legendre Transformation, Hamiltonian Systems, Convexity, Field Theories 3

1. Legendre Transformations 4
   1.1. Gradient Mappings and Legendre Transformations 5
   (Definitions. Involutory character of the Legendre transformation. Conjugate convex functions. Young's inequality. Support function. Clairaut's differential equation. Minimal surface equation. Compressible two-dimensional steady flow. Application of Legendre transformations to quadratic forms and convex bodies. Partial Legendre transformations.)
   1.2. Legendre Duality Between Phase and Cophase Space. Euler Equations and Hamilton Equations. Hamilton Tensor 18
   (Configuration space, phase space, cophase space, extended configuration (phase, cophase) space. Momenta. Hamiltonians. Energy-momentum tensor. Hamiltonian systems of canonical equations. Dual Noether equations. Free boundary conditions in canonical form. Canonical form of E. Noether's theorem, of Weierstrass's excess function and of transversality.)
2. Hamiltonian Formulation of the One-Dimensional Variational Calculus 26
   2.1. Canonical Equations and the Partial Differential Equation of Hamilton–Jacobi 26
   (Eulerian flows and Hamiltonian flows as prolongations of extremal bundles. Canonical description of Mayer fields. The 1-forms of Beltrami and Cartan. The Hamilton-Jacobi equation as canonical version of Carathéodory's equations. Lagrange brackets and Mayer bundles in canonical form.)
   2.2. Hamiltonian Flows and Their Eigentime Functions. Regular Mayer Flows and Lagrange Manifolds 33
   (The eigentime function of an r-parameter Hamiltonian flow. The Cauchy representation of the pull-back $h^*\kappa_H$ of the Cartan form κ_H with respect to an r-parameter Hamilton flow h by means of an eigentime function. Mayer flows, field-like Mayer bundles, and Lagrange manifolds.)
   2.3. Accessory Hamiltonians and the Canonical Form of the Jacobi Equation 41
   (The Legendre transform of the accessory Lagrangian is the accessory Hamiltonian, i.e. the quadratic part of the full Hamiltonian, and its canonical equations describe Jacobi fields. Expressions for the first and second variations.)

2.4. The Cauchy Problem for the Hamilton–Jacobi Equation 48
(Necessary and sufficient conditions for the local solvability of the Cauchy problem. The Hamilton-Jacobi equation. Extension to discontinuous media: refracted light bundles and the theorem of Malus.)

3. Convexity and Legendre Transformations 54

3.1. Convex Bodies and Convex Functions in $\mathbb{R}^n$ 55
(Basic properties of convex sets and convex bodies. Supporting hyperplanes. Convex hull. Lipschitz continuity of convex functions.)

3.2. Support Function, Distance Function, Polar Body 66
(Gauge functions. Distance function and support function. The support function of a convex body is the distance function of its polar body, and vice versa. The polarity map. Polar body and Legendre transform.)

3.3. Smooth and Nonsmooth Convex Functions. Fenchel Duality 75
(Characterization of smooth convex functions. Supporting hyperplanes and differentiability. Regularization of convex functions. Legendre-Fenchel transform.)

4. Field Theories for Multiple Integrals 94

4.1. DeDonder–Weyl's Field Theory 96
(Null Lagrangians of divergence type as calibrators. Weyl equations. Geodesic slope fields or Weyl fields, eikonal mappings. Beltrami form. Legendre transformation. Cartan form. DeDonder's partial differential equation. Extremals fitting a geodesic slope field. Solution of the local fitting problem.)

4.2. Carathéodory's Field Theory 106
(Carathéodory's involutory transformation, Carathéodory transform. Transversality. Carathéodory calibrator. Geodesic slope fields and their eikonal maps. Carathéodory equations. Vessiot-Carathéodory equation. Generalization of Kneser's transversality theorem. Solution of the local fitting problem for a given extremal.)

4.3. Lepage's General Field Theory 131
(The general Beltrami form. Lepage's formalism. Geodesic slope fields. Lepage calibrators.)

4.4. Pontryagin's Maximum Principle 136
(Calibrators and pseudonecessary optimality conditions. (I) One-dimensional variational problems with nonholonomic constraints: Lagrange multipliers. Pontryagin's function, Hamilton function, Pontryagin's maximum principle and canonical equations. (II) Pontryagin's maximum principle for multi-dimensional problems of optimal control.)

5. Scholia 146

Chapter 8. Parametric Variational Integrals 153

1. Necessary Conditions 154

1.1. Formulation of the Parametric Problem. Extremals and Weak Extremals 155
(Parametric Lagrangians. Parameter-invariant integrals. Riemannian metrics. Finsler metrics. Parametric extremals. Transversality of line elements. Eulerian covector field and Noether's equation. Gauss's equation. Jacobi's variational principle for the motion of a point mass in $\mathbb{R}^3$.)

1.2. Transition from Nonparametric to Parametric Problems and Vice Versa ... 166
(Nonparametric restrictions of parametric Lagrangians. Parametric extensions of nonparametric Lagrangians. Relations between parametric and nonparametric extremals.)

1.3. Weak Extremals, Discontinuous Solutions, Weierstrass–Erdmann Corner Conditions. Fermat's Principle and the Law of Refraction ... 171
(Weak D^1- and C^1-extremals. DuBois-Reymond's equation. Weierstrass-Erdmann corner conditions. Regularity theorem for weak D^1-extremals. Snellius's law of refraction and Fermat's principle.)

2. Canonical Formalism and the Parametric Legendre Condition ... 180

2.1. The Associated Quadratic Problem. Hamilton's Function and the Canonical Formalism ... 180
(The associated quadratic Lagrangian Q of a parametric Lagrangian F. Elliptic and nonsingular line elements. A natural Hamiltonian and the corresponding canonical formalism. Parametric form of Hamilton's canonical equations.)

2.2. Jacobi's Geometric Principle of Least Action ... 188
(The conservation of energy and Jacobi's least action principle: a geometric description of orbits.)

2.3. The Parametric Legendre Condition and Carathéodory's Hamiltonians ... 192
(The parametric Legendre condition or C-regularity. Carathéodory's canonical formalism.)

2.4. Indicatrix, Figuratrix, and Excess Function ... 201
(Indicatrix, figuratrix and canonical coordinates. Strong and semistrong line elements. Regularity of broken extremals. Geometric interpretation of the excess function.)

3. Field Theory for Parametric Integrals ... 213

3.1. Mayer Fields and their Eikonals ... 214
(Parametric fields and their direction fields. Equivalent fields. The parametric Carathéodory equations. Mayer fields and their eikonals. Hilbert's independent integral. Weierstrass's representation formula. Kneser's transversality theorem. The parametric Beltrami form. Normal fields of extremals and Mayer fields, Weierstrass fields, optimal fields, Mayer bundles of extremals.)

3.2. Canonical Description of Mayer Fields ... 227
(The parametric Cartan form. The parametric Hamilton-Jacobi equation or eikonal equation. One-parameter families of F-equidistant surfaces.)

3.3. Sufficient Conditions ... 229
(F- and Q-minimizers. Regular Q-minimizers are quasinormal. Conjugate values and conjugate points of F-extremals. F-extremals without conjugate points are local minimizers. Stigmatic bundles of quasinormal extremals and the exponential map of a parametric Lagrangian. F- and Q-Mayer fields. Wave fronts.)

3.4. Huygens's Principle ... 243
(Complete Figures. Duality between light rays and wave fronts. Huygens's envelope construction of wave fronts. F-distance function. Foliations by one-parameter families of F-equidistant surfaces and optimal fields.)

4. Existence of Minimizers ... 248
4.1. A Direct Method Based on Local Existence ... 248
(The distance function $d(P, P')$ related to F and its continuity and lower semicontinuity properties. Existence of global minimizers based on the local existence theory developed in *3.3*. Regularity of minimizers.)
4.2. Another Direct Method Using Lower Semicontinuity ... 254
(Minimizing sequences. An equivalent minimum problem. Compactness of minimizing sequences. Lower semicontinuity of the variational integral. A general existence theorem for obstacle problems. Regularity of minimizers. Existence of minimizing F-extremals. Inclusion principle.)
4.3. Surfaces of Revolution with Least Area ... 263
(Comparison of curves with the Goldschmidt polygon. Todhunter's ellipse. Comparison of catenaries and Goldschmidt polygons. Conclusive results.)
4.4. Geodesics on Compact Surfaces ... 270
(Existence and regularity of F-extremals wich minimize the arc length.)
5. Scholia ... 275

Part IV. Hamilton–Jacobi Theory and Partial Differential Equations of First Order

Chapter 9. Hamilton–Jacobi Theory and Canonical Transformations . 283

1. Vector Fields and 1-Parameter Flows ... 288
1.1. The Local Phase Flow of a Vector Field ... 290
(Trajectories, integral curves, maximal flows.)
1.2. Complete Vector Fields and One-Parameter Groups of Transformations ... 292
(Infinitesimal transformations.)
1.3. Lie's Symbol and the Pull-Back of a Vector Field ... 294
(The symbol of a vector field and its transformation law.)
1.4. Lie Brackets and Lie Derivatives of Vector Fields ... 298
(Commuting flows. Lie derivative. Jacobi identity.)
1.5. Equivalent Vector Fields ... 303
(Rectification of nonsingular vector fields.)
1.6. First Integrals ... 304
(Time-dependent and time-independent first integrals. Functionally independent first integrals. The motion in a central field. Kepler's problem. The two-body problem.)
1.7. Examples of First Integrals ... 314
(Lax pairs. Toda lattice.)
1.8. First-Order Differential Equations for Matrix-Valued Functions. Variational Equations. Volume Preserving Flows ... 317
(Liouville formula. Liouville theorem. Autonomous Hamiltonian flows are volume preserving.)
1.9. Flows on Manifolds ... 320
(Geodesics on S^2.)

2. Hamiltonian Systems ... 326
2.1. Canonical Equations and Hamilton–Jacobi Equations Revisited ... 327
(Mechanical systems. Action. Hamiltonian systems and Hamilton-Jacobi equation.)
2.2. Hamilton's Approach to Canonical Transformations ... 333
(Principal function and canonical transformations.)
2.3. Conservative Dynamical Systems. Ignorable Variables ... 336
(Cyclic variables. Routhian systems.)
2.4. The Poincaré–Cartan Integral. A Variational Principle for Hamiltonian Systems ... 340
(The Cartan form and the canonical variational principle.)
3. Canonical Transformations ... 343
3.1. Canonical Transformations and Their Symplectic Characterization ... 343
(Symplectic matrices. The harmonic oscillator. Poincaré's transformation. The Poincaré form and the symplectic form.)
3.2. Examples of Canonical Transformations. Hamilton Flows and One-Parameter Groups of Canonical Transformations ... 356
(Elementary canonical transformation. The transformations of Poincaré and Levi-Civita. Homogeneous canonical transformations.)
3.3. Jacobi's Integration Method for Hamiltonian Systems ... 366
(Complete solutions. Jacobi's theorem and its geometric interpretation. Harmonic oscillator. Brachystochrone. Canonical perturbations.)
3.4. Generation of Canonical Mappings by Eikonals ... 379
(Arbitrary functions generate canonical mappings.)
3.5. Special Dynamical Problems ... 384
(Liouville systems. A point mass attracted by two fixed centers. Addition theorem of Euler. Regularization of the three-body problem.)
3.6. Poisson Brackets ... 407
(Poisson brackets, fields, first integrals.)
3.7. Symplectic Manifolds ... 417
(Symplectic geometry. Darboux theorem. Symplectic maps. Exact symplectic maps. Lagrangian submanifolds.)
4. Scholia ... 433

Chapter 10. Partial Differential Equations of First Order and Contact Transformations ... 441

1. Partial Differential Equations of First Order ... 444
1.1. The Cauchy Problem and Its Solution by the Method of Characteristics ... 445
(Configuration space, base space, contact space. Contact elements and their support points and directions. Contact form, 1-graphs, strips. Integral manifolds, characteristic equations, characteristics, null (integral) characteristic, characteristic curve, characteristic base curve. Cauchy problem and its local solvability for noncharacteristic initial values: the characteristic flow and its first integral F, Cauchy's formulas.)

1.2. Lie's Characteristic Equations. Quasilinear Partial Differential Equations 463
(Lie's equations. First order linear and quasilinear equations, noncharacteristic initial values. First integrals of Cauchy's characteristic equations, Mayer brackets $[F, \Phi]$.)

1.3. Examples 468
(Homogeneous linear equations, inhomogeneous linear equations, Euler's equation for homogeneous functions. The reduced Hamilton-Jacobi equation $H(x, u_x) = E$. The eikonal equation $H(x, u_x) = 1$. Parallel surfaces. Congruences or ray systems, focal points. Monge cones, Monge lines, and focal curves, focal strips. Partial differential equations of first order and cone fields.)

1.4. The Cauchy Problem for the Hamilton–Jacobi Equation 479
(A discussion of the method of characteristics for the equation $S_t + H(t, x, S_x) = 0$. A detailed investigation of noncharacteristic initial values.)

2. Contact Transformations 485

2.1. Strips and Contact Transformations 486
(Strip equation, strips of maximal dimension (= Legendre manifolds), strips of type C_n^k, contact transformations, transformation of strips into strips, characterization of contact transformations. Examples: Contact transformations of Legendre, Euler, Ampère, dilations, prolongated point transformations.)

2.2. Special Contact Transformations and Canonical Mappings 496
(Contact transformations commuting with translations in z-direction and exact canonical transformations. Review of various characterizations of canonical mappings.)

2.3. Characterization of Contact Transformations 500
(Contact transformations of $\mathbb{R}^{2n+1}$ can be prolonged to special contact transformations of $\mathbb{R}^{2n+3}$, or to homogeneous canonical transformations of $\mathbb{R}^{2n+2}$. Connection between Poisson and Mayer brackets. Characterization of contact transformations.)

2.4. Contact Transformations and Directrix Equations 511
(The directrix equation for contact transformations of first type: $\Omega(x, z, \bar{x}, \bar{z}) = 0$. Involutions. Construction of contact transformations of the first type from an arbitrary directrix equation. Contact transformations of type r and the associated systems of directrix equations. Examples: Legendre's transformation, transformation by reciprocal polars, general duality transformation, pedal transformation, dilations, contact transformations commuting with all dilations, partial Legendre transformations, apsidal transformation, Fresnel surfaces and conical refraction. Differential equations and contact transformations of second order. Canonical prolongation of first-order to second-order contact transformations. Lie's G-K-transformation.)

2.5. One-Parameter Groups of Contact Transformations. Huygens Flows and Huygens Fields; Vessiot's Equation 541
(One-parameter flows of contact transformations and their characteristic Lie functions. Lie equations and Lie flows. Huygens flows are Lie flows generated by n-strips as initial values. Huygens fields as ray maps of Huygens flows. Vessiot's equation for the eikonal of a Huygens field.)

2.6. Huygens's Envelope Construction 557
(Propagation of wave fronts by Huygens's envelope construction: Huygens's principle. The indicatrix W and its Legendre transform F. Description of Huygens's principle by the Lie equations generated by F.)

3. The Fourfold Picture of Rays and Waves 565

3.1. Lie Equations and Herglotz Equations 566
(Description of Huygens's principle by Herglotz equations generated by the indicatrix function W. Description of Lie's equations and Herglotz's equations by variational principles. The characteristic equations $S_x = W_\xi/M$, $S_z = -1/M$ for the eikonal S and the directions D of a Huygens field.)

3.2. Hölder's Transformation 571
(The generating function F of a Hölder transformation $\mathscr{H}_F$ and its adjoint Φ. The Hölder transform H of F. Examples. The energy-momentum tensor $T = p \otimes F_p - F$. Local and global invertibility of $\mathscr{H}_F$. Transformation formulas. Connections between Hölder's transformation $\mathscr{H}_F$ and Legendre's transformation $\mathscr{L}_F$ generated by F: the commuting diagram and Haar's transformation $\mathscr{R}_F$. Examples.)

3.3. Connection Between Lie Equations and Hamiltonian Systems 587
(Hölder's transformation $\mathscr{H}_F$ together with the transformation $\theta \mapsto z$ of the independent variable generated by $\dot{z} = \Phi$ transforms Lie's equations into a Hamiltonian system $\dot{x} = H_y$, $\dot{y} = -H_x$. Vice versa, the Hölder transform $\mathscr{H}_H$ together with the "eigentime transformation" $z \mapsto \theta$ transforms any Hamiltonian system into a Lie system. Equivalence of Mayer flows and Huygens flows, and of Mayer fields and Huygens fields.)

3.4. Four Equivalent Descriptions of Rays and Waves. Fermat's and Huygens's Principles 595
(Under suitable assumptions, the four pictures of rays and waves due to Euler-Lagrange, Huygens-Lie, Hamilton, and Herglotz are equivalent. Correspondingly the two principles of Fermat and of Huygens are equivalent.)

4. Scholia 600

A List of Examples 605

A Glimpse at the Literature 610

Bibliography 615

Subject Index 646

Part I

The First Variation and Necessary Conditions

Chapter 1. The First Variation

In this chapter we shall develop the *formalism of the calculus of variations* in simple situations. After a brief review of the necessary and sufficient conditions for *extrema* of ordinary functions on $\mathbb{R}^n$, we investigate in Section 2 some of the basic necessary conditions that are to be satisfied by minimizers of *variational integrals*

$$\mathscr{F}(u) = \int_\Omega F(x, u(x), Du(x))\, dx. \tag{1}$$

Here Ω is an open domain of $\mathbb{R}^n$, $n \geq 1$, and $u(x)$ denotes a function defined in Ω which in general is a vector valued mapping of Ω into $\mathbb{R}^N$, $N \geq 1$, that is, $u(x) = (u^1(x), \ldots, u^N(x))$, and Du denotes the first derivative of u,

$$Du = (D_\alpha u^i), \quad i = 1, \ldots, N, \ \alpha = 1, \ldots, n.$$

Simple examples of variational integrals (1) are the *Dirichlet integral*

$$\mathscr{D}(u) = \tfrac{1}{2} \int_\Omega |Du|^2\, dx, \tag{2}$$

and the *nonparametric area integral*

$$\mathscr{A}(u) = \int_\Omega \sqrt{1 + |Du|^2}\, dx, \tag{3}$$

which yields the area of the graph of u over Ω. Both functionals (2) and (3) are defined for scalar functions (i.e., $N = 1$), but we shall see later that variational integrals involving vector valued functions appear in physics and differential geometry.

The most important *necessary condition* to be satisfied by any minimizer of a variational integral $\mathscr{F}$ is the vanishing of its *first variation* $\delta\mathscr{F}$. For instance, if u minimizes $\mathscr{F}$ with respect to *variations* $\delta u = \varphi$ which do not change the boundary values of u, we must have

$$\delta\mathscr{F}(u, \varphi) := \frac{d}{d\varepsilon} \mathscr{F}(u + \varepsilon\varphi)\bigg|_{\varepsilon=0} = 0 \tag{4}$$

for all φ with compact support in Ω. In the case of the functionals (2) and (3), equation (4) reads as

$$\int_\Omega D_\alpha u \, D_\alpha \varphi \, dx = 0 \tag{5}$$

and

$$\int_\Omega \frac{D_\alpha u}{\sqrt{1 + |Du|^2}} D_\alpha \varphi \, dx = 0, \tag{6}$$

respectively for all differentiable φ with compact support in Ω. (Here we use *Einstein's summation convention*: the index α is to be summed from 1 to n.)

By means of the simple, but basic, *fundamental lemma of the calculus of variations*, relation (4) yields, after an integration by parts (i.e., after applying the Gauss–Green theorem), the *Euler equations* as a necessary condition to be satisfied by minimizers of $\mathscr{F}$ which are of class C^2. This leads to the notion of *extremals* and to the Euler operator $L_F(u)$ associated with a *Lagrangian* $F(x, z, p)$. In the physical literature, the Euler operator is often denoted by $\frac{\delta F}{\delta u}$. For the integrals (2) and (3), the corresponding Euler equations are the *Laplace equation*

$$\Delta u := D_\alpha D_\alpha u = 0$$

and the *mean curvature equation*

$$D_\alpha \left\{ \frac{D_\alpha u}{\sqrt{1 + |Du|^2}} \right\} = 0,$$

respectively. Introducing the vector field

$$Tu = \frac{\text{grad } u}{\sqrt{1 + |\text{grad } u|^2}}$$

we can write the second equation as

$$\text{div } Tu = 0.$$

Already these two examples show the close connection between minimum problems for variational integrals and boundary value problems for partial differential equations.

If u is a minimizer of $\mathscr{F}$ with respect to variations which are allowed to change the boundary values of u, then the relation (4) holds for all differentiable φ and not only for those with compact support in Ω. Thus we obtain not only Euler's equations for u, but also the so-called *natural boundary conditions*, compare *2.4*. For minimizers u of the Dirichlet integral and of the area functional the natural boundary condition states that the normal derivative of u at the boundary of Ω has to vanish.

Several variants of the fundamental lemma of the calculus of variations will be proved in *2.3*. The proofs are carried out by means of the *mollifier technique*.

C^1-solutions u of (4) will be called *weak extremals*. If such a solution u is even of class C^2, an integration by parts will lead to Euler's equation $L_F(u) = 0$

the solutions of which are called *extremals*. However, weak extremals need not be of class C^2 as it will be seen from examples presented in *3.1*. There we shall also state a condition ensuring that weak extremals of one-dimensional variational problems are in fact of class C^2 and therefore solutions of Euler's equation. Analogous results for multidimensional problem are much more difficult to prove and will therefore be deferred to a separate treatise.

There are other complications for variational problems. We shall see that variational integrals, even if they are bounded from below, need not necessarily possess minimizers of class C^1 or of any other reasonable class. While we postpone the general discussion of existence of minimizers to another treatise, we shall at least exhibit a few examples of minimum problems without solutions in *3.2*. A brief discussion of the existence question for one-dimensional parametric problems is given in 8,*4*.

Finally in *3.3* we shall briefly investigate *broken extremals*, i.e. extremals which are only piecewise of class C^1.

In Section 4 we shall treat the so-called *null Lagrangians*. These are some kind of degenerate variational integrands F for which the corresponding Euler equations $L_F(u) = 0$ are satisfied by arbitrary functions u of class C^2. A very simple example of a null Lagrangian is provided by integrands of the type

$$F(Du) = a \cdot Du + b,$$

with constant coefficients a and b, for instance,

$$F(Du) = \operatorname{div} u.$$

Note that the integrals

$$\int_\Omega (a \cdot Du + b)\, dx, \qquad \int_\Omega \operatorname{div} u \, dx$$

depend only on the value of u on $\partial\Omega$. This, in fact, will be seen to be characteristic of null Lagrangians. In *4.2* we shall also derive a pointwise characterization of null Lagrangians.

Though defined by a *local* differential condition, null Lagrangians play a special role as a source of *global integral invariants*. For instance, the determinant $\det Du$ of the gradient Du of a mapping $u : \Omega \to \mathbb{R}^n$, $\Omega \subset \mathbb{R}^n$, is also a null Lagrangian. It enters in the *Gauss-Bonnet theorem* as well as in the *degree theory* of mappings.

Null Lagrangians will also play an important role in establishing *sufficient conditions* for minimizers via field theories.

Finally, *variational problems of higher order* will briefly be discussed in Section 5.

We shall continue our discussion of the basic formalism of the calculus of variations in Chapters 2 and 3. In Chapter 2 we shall treat variational problems with constraints by means of *Lagrange multipliers*, and in Chapter 3 we shall study *general variations* of variational integrals generated by variations of both the dependent and the independent variables. This will, in particular, lead us to the *Noether equation* for the *energy–momentum tensor* and to the important *theorem of Emmy Noether* concerning *conservation laws*. This theorem states that invariance properties of variational integrals with respect to some transformation group imply additional information on their extremals which are expressed in the form of conservation laws.

1. Critical Points of Functionals

Let $\Phi(\varepsilon)$ be a real valued function on the open interval $(-\varepsilon_0, \varepsilon_0)$, $\varepsilon_0 > 0$, which has a local extremum (i.e., maximum or minimum) at $\varepsilon = 0$. Then, if the derivative $\Phi'(0)$ of Φ at $\varepsilon = 0$ exists, we infer that

$$\Phi'(0) = 0\,.$$

If, in addition, Φ is of class C^2 on $(-\varepsilon_0, \varepsilon_0)$, we obtain from Taylor's formula

$$\Phi(\varepsilon) - \Phi(0) = \tfrac{1}{2}\Phi''(\delta\varepsilon)\varepsilon^2 \quad \text{for some } \delta \in (0, 1)$$

the inequality

$$\Phi''(0) \geq 0$$

if $\varepsilon = 0$ is a local (or "relative") minimum, and

$$\Phi''(0) \leq 0$$

if $\varepsilon = 0$ is a local maximum of Φ.

Thus *the conditions*

$$\Phi'(0) = 0 \quad \textit{and} \quad \Phi''(0) \geq 0$$

are necessary conditions for a local minimum of Φ *at* $\varepsilon = 0$. But they are not sufficient as we see from the function $\Phi(\varepsilon) = \varepsilon^3$. On the other hand, the

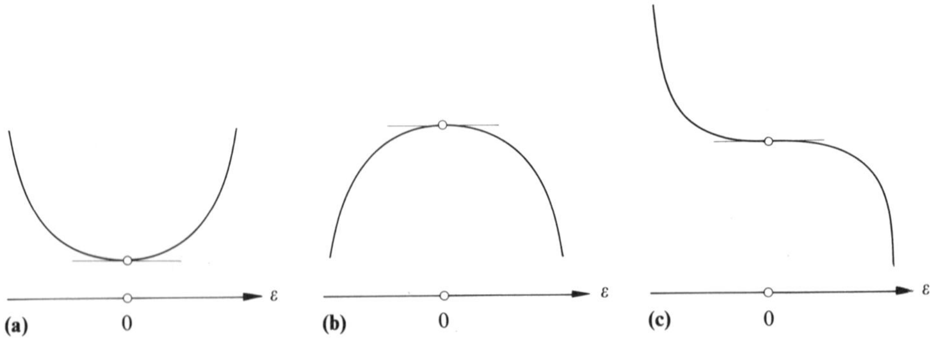

Fig. 1. The graph of a function $\Phi(\varepsilon)$. The point $\varepsilon = 0$ is **(a)** a minimizer, **(b)** a maximizer, **(c)** neither a minimizer nor a maximizer, but a stationary point of Φ.

conditions

$$\Phi'(0) = 0, \qquad \Phi''(0) > 0 \quad (\text{or } \Phi''(0) < 0)$$

are *sufficient*, but not necessary conditions for $\varepsilon = 0$ to be a strict local minimizer (or maximizer) of Φ, as is shown by the function $\Phi(\varepsilon) = \varepsilon^4$.

In other words, there is a slight gap between the necessary and the sufficient conditions. This phenomenon is quite typical also for the calculus of variations and will repeatedly be observed.

Let now $\mathscr{F}$ be a real valued function on some open set Ω of the Euclidean space $\mathbb{R}^n$ which is of class C^1. Then, for some $u_0 \in \Omega$ and $\zeta \in \mathbb{R}^n$ and for sufficiently small $\varepsilon_0 > 0$, the function

$$\Phi(\varepsilon) := \mathscr{F}(u_0 + \varepsilon\zeta), \quad -\varepsilon_0 < \varepsilon < \varepsilon_0,$$

is of class C^1. We call its derivative $\Phi'(0)$ at $\varepsilon = 0$ *the first variation of $\mathscr{F}$ at u_0 in direction of ζ* and write

$$\delta\mathscr{F}(u_0, \zeta) := \Phi'(0). \tag{1}$$

For $\mathscr{F} \in C^2(\Omega)$, we can also form the *second variation*

$$\delta^2\mathscr{F}(u_0, \zeta) := \Phi''(0) \tag{2}$$

of $\mathscr{F}$ at u_0 in direction of ζ. In general, if $\mathscr{F} \in C^m(\Omega)$, we define the *m-th variation of $\mathscr{F}$* by

$$\delta^m\mathscr{F}(u_0, \zeta) := \Phi^{(m)}(0).$$

If $\zeta = (\zeta^1, \ldots, \zeta^n)$, we compute that

$$\delta\mathscr{F}(u_0, \zeta) = \mathscr{F}_u(u_0)\cdot\zeta = \sum_{i=1}^n \mathscr{F}_{u^i}(u_0)\zeta^i$$

and

$$\delta^2\mathscr{F}(u_0, \zeta) = \mathscr{F}_{uu}(u_0)\cdot\zeta\zeta = \sum_{i,k=1}^n \mathscr{F}_{u^iu^k}(u_0)\zeta^i\zeta^k.$$

That is, the first variation $\delta\mathscr{F}(u_0, \zeta)$ is a linear form, and $\delta^2\mathscr{F}(u_0, \zeta)$ is a quadratic form with regard to ζ.

As *necessary condition for a local extremum of $\mathscr{F}$ at u_0* we obtain the relation

$$\delta\mathscr{F}(u_0, \zeta) = 0 \quad \text{for all } \zeta \in \mathbb{R}^n, \tag{3}$$

which is equivalent to

$$D\mathscr{F}(u_0) = 0,$$

where $D\mathscr{F}$ stands for the derivative of $\mathscr{F}$.

Every point u_0 which satisfies (3) is called a *critical* (or *stationary*) *point* of $\mathscr{F}$, and the number $\mathscr{F}(u_0)$ is said to be a *critical value* of $\mathscr{F}$.

Suppose now that $\mathscr{F}$ is of class $C^2(\Omega)$. *If $u_0 \in \Omega$ is a local minimizer of $\mathscr{F}$, then the second variation $\delta^2\mathscr{F}(u_0, \zeta)$ is a positive semi-definite quadratic form in*

ζ and, equivalently, the eigenvalues of the Hessian matrix $D^2\mathscr{F} = (\mathscr{F}_{u^iu^k})$ at u_0 are nonnegative.

Conversely, if the eigenvalues of $D^2\mathscr{F}(u_0)$ are positive, then the critical point $u_0 \in \Omega$ provides a strict local minimum of $\mathscr{F}$. In fact, strict convexity of $\mathscr{F}$ in some neighbourhood of u_0 serves the same purpose but is a slightly weaker condition as can be seen from the function $\mathscr{F}(u) = |u|^4$.

In other words, *the assumption*

$$\delta^2\mathscr{F}(u_0, \zeta) > 0 \quad (\text{or } <0) \tag{4}$$

for all $\zeta \in \mathbb{R}^n$ *with* $\zeta \neq 0$ *implies that a critical point* $u_0 \in \Omega$ *is a strict local minimizer* (*or maximizer*) *of* $\mathscr{F}$.

If at a critical point u_0 the Hessian matrix $D^2\mathscr{F}(u_0)$ has both positive and negative eigenvalues, then we infer from

$$\mathscr{F}(u_0 + \zeta) - \mathscr{F}(u_0) = \tfrac{1}{2}\delta^2\mathscr{F}(u_0, \zeta) + o(|\zeta|^2) \quad \text{as } \zeta \to 0 \tag{5}$$

that u_0 furnishes neither a maximum nor a minimum; in this case, u_0 is called a *saddle point* of $\mathscr{F}$.

A critical point u_0 of $\mathscr{F} \in C^2(\Omega)$ is said to be *nondegenerate* if $D^2\mathscr{F}(u_0)$ is nonsingular, i.e., if zero is no eigenvalue of $D^2\mathscr{F}(u_0)$.

The Taylor formula (5) implies that *nondegenerate critical points are isolated critical points*, and that the graph of $\mathscr{F}(u)$ in a sufficiently small neighbourhood of a nondegenerate critical point looks roughly the same as that of the nondegenerate quadratic form

$$\mathscr{Q}(\zeta) := \tfrac{1}{2}\delta^2\mathscr{F}(u_0, \zeta).$$

In fact, we can write

$$\mathscr{F}(u_0 + \zeta) - \mathscr{F}(u_0) = \sum_{i,k=1}^{n} a_{ik}(\zeta)\zeta^i\zeta^k,$$

$$a_{ik}(\zeta) := \int_0^1 (1 - t)\mathscr{F}_{u^iu^k}(u_0 + t\zeta)\, dt,$$

and it can be proved that *there is a diffeomorphism* $u = \varphi(v)$ *with* $u_0 = \varphi(0)$ *which maps some ball* $B = \{v: |v| < \rho\}$ *onto a neighbourhood* $\mathscr{U}$ *of* u_0 *such that*

$$\mathscr{F}(\varphi(v)) = \mathscr{F}(u_0) - \sum_{i=1}^{q} |v^i|^2 + \sum_{j=q+1}^{n} |v^j|^2 = \mathscr{F}(u_0) + \mathscr{Q}(v). \tag{6}$$

That is, up to composition with a diffeomorphism, $\mathscr{F}(u) - \mathscr{F}(u_0)$ is a quadratic form. This result is known as *Morse lemma*; see the Scholia for references.

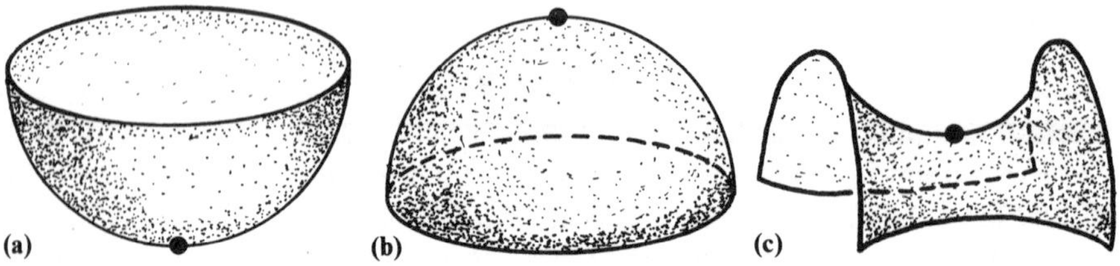

Fig. 2. **(a)** minimum, **(b)** maximum, **(c)** saddle.

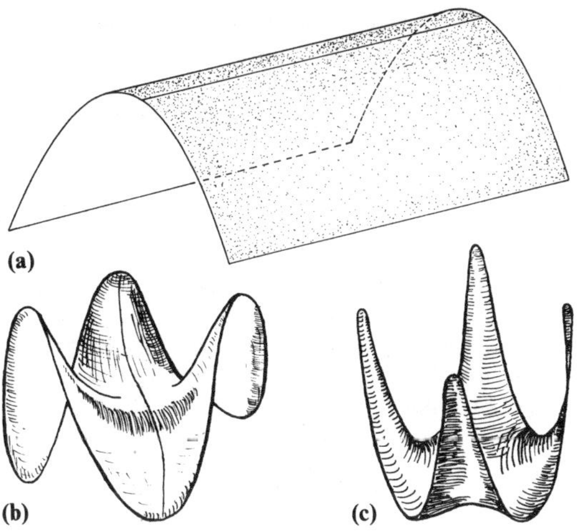

Fig. 3. Degenerate critical points. **(a)** Parabolic cylinder. **(b)** Monkey saddle, graph of $F(u, v) = u^3 - 3uv^2 = \operatorname{Re}\{w^3\}$, $w = u + iv$. **(c)** Graph of $F(u, v) = u^2v^2$.

The notions of *first and second variation* can easily be carried over to *functionals*

$$\mathscr{F} : V \to \mathbb{R}$$

which are defined on a subset V of a *linear space* X over $\mathbb{R}$. To this end, we choose some point $u_0 \in V$ and some vector $\zeta \in X$, and suppose that the segment $\{u \colon u = u_0 + \varepsilon\zeta, |\varepsilon| < \varepsilon_0\}$ is contained in V for some $\varepsilon_0 > 0$. Then we define the function $\Phi : (-\varepsilon_0, \varepsilon_0) \to \mathbb{R}$ by $\Phi(\varepsilon) := \mathscr{F}(u_0 + \varepsilon\zeta)$ and, if $\Phi'(0)$ exists, we once again introduce the *first variation* $\delta\mathscr{F}(u_0, \zeta)$ *of* $\mathscr{F}$ *at* u_0 *in direction of* ζ by

$$\delta\mathscr{F}(u_0, \zeta) = \Phi'(0).$$

Moreover, if $\Phi'(\varepsilon)$ exists for $|\varepsilon| < \varepsilon_0$ as well as $\Phi''(0)$, we define the *second variation* $\delta^2\mathscr{F}(u_0, \zeta)$ of $\mathscr{F}$ by

$$\delta^2\mathscr{F}(u_0, \zeta) = \Phi''(0).$$

Suppose now, that, as just defined, $\delta\mathscr{F}$ and $\delta^2\mathscr{F}$ exist at some $u_0 \in V$ for all directions ζ contained in some subspace Z of X. Secondly, assume that u_0 furnishes the minimum of the functional $\mathscr{F} : V \to \mathbb{R}$. Then we again conclude that

(7) $$\delta\mathscr{F}(u_0, \zeta) = 0, \quad \delta^2\mathscr{F}(u_0, \zeta) \geq 0 \qquad \text{for all } \zeta \in Z.$$

In the next section, we shall apply this observation to functionals $\mathscr{F} : V \to \mathbb{R}$ which are given by multiple integrals

(8) $$\mathscr{F}(u) = \int_\Omega F(x, u(x), Du(x))\, dx, \quad \Omega \subset \mathbb{R}^n,$$

and which are defined for functions $u : \Omega \to \mathbb{R}^N$ contained in some subset V of an appropriate function space X, say, $X \subset C^1(\Omega, \mathbb{R}^N)$.

Before we turn to such more concrete situations, we have to add a few remarks why we have introduced the first variation $\delta\mathscr{F}$ as fundamental notion of derivative. First of all, it is the oldest and time-honoured concept of a derivative of a functional, introduced by Larange and Euler, and therefore we keep the Lagrangian symbol δ. Secondly, and this is much more important, we prefer this notion as the fundamental one because it is so "weak". It can, for instance, happen that the first variation $\delta\mathscr{F}(u_0, \zeta)$ exists for "variations" ζ contained in some class Z which is smaller than X but still "sufficiently large" so that one can draw valuable conclusions on u_0 whereas other kinds of derivatives might not exist. Moreover, we need no norm on X in order to define $\delta\mathscr{F}$ and $\delta^2\mathscr{F}$.

To explain what we mean we will briefly discuss the two other principal notions of a derivative, the *Fréchet derivative* and the *Gâteaux derivative.*

Let X be a Banach space with the norm $|u|$ of its elements u, and denote by X^* the space of bounded linear functionals on X. Assume that $\mathscr{F} : V \to \mathbb{R}$ is a functional defined on some open subset V of X, and let u_0 be some point of V.

$\mathscr{F}$ is said to be *Fréchet differentiable* at u_0 if there exists some $\ell \in X^*$ *such that*

$$\mathscr{F}(u_0 + \zeta) = \mathscr{F}(u_0) + \ell(\zeta) + o(|\zeta|) \quad \text{as } |\zeta| \to 0$$

holds for all $\zeta \in X$ with $u_0 + \zeta \in V$. Then we set $D\mathscr{F}(u_0) = \mathscr{F}'(u_0) := \ell$ and call it the *Fréchet derivative of $\mathscr{F}$ at u_0*.

Furthermore $\mathscr{F}$ is called *Gâteaux differentiable* at u_0 if there exists some $\ell \in X^*$ such that

$$\mathscr{F}(u_0 + \varepsilon\zeta) = \mathscr{F}(u_0) + \varepsilon\ell(\zeta) + o(\varepsilon) \quad \text{as } \varepsilon \to 0$$

is satisfied for arbitrary $\zeta \in X$ and $|\varepsilon| < \varepsilon_0$ with sufficiently small $\varepsilon_0(\zeta) > 0$. One calls $d\mathscr{F}(u_0, \zeta) = \ell(\zeta)$ the *Gâteaux derivative of $\mathscr{F}$ at u_0*.

It is not hard to prove that Fréchet differentiability implies Gâteaux differentiability (and both derivatives coincide), whereas the converse does not hold. A simple counterexample is provided by the function $\mathscr{F} : \mathbb{R}^2 \to \mathbb{R}$, defined by

$$\mathscr{F}(0, 0) = 0, \qquad \mathscr{F}(u, v) = \left(\frac{uv^2}{u^2 + v^4}\right)^2 \quad \text{for } (u, v) \neq (0, 0).$$

Gâteaux differentiability at some point u_0 can be interpreted as the requirement that all directional derivatives at u_0 are to exist, and Fréchet differentiability at u_0 means the existence of the tangent plane to graph $\mathscr{F}$ at the point $(u_0, \mathscr{F}(u_0))$. The example shows that the existence of all directional derivatives does not secure the existence of the tangent plane.

Suppose now that we want to define the Fréchet or Gâteaux derivative for some functional of type (8), as given in physics or geometry. Then it is not a priori clear what its natural domain of

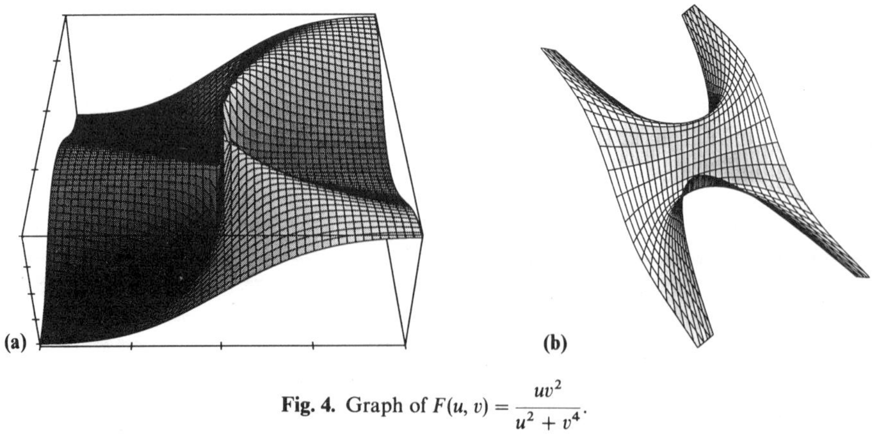

Fig. 4. Graph of $F(u, v) = \dfrac{uv^2}{u^2 + v^4}$.

definition X should be. It could, for example, be $C^1(\bar{\Omega}, \mathbb{R}^N)$, or the Sobolev space $H^{1,p}(\Omega, \mathbb{R}^N)$, $1 \le p \le \infty$, or the space $\mathrm{BV}(\Omega, \mathbb{R}^N)$ of functions with bounded variation. Thus the existence of $D\mathscr{F}(u_0)h$ and $d\mathscr{F}(u_0, h)$ depends on the chosen domain of definition X whereas the meaning and existence of $\delta\mathscr{F}$ is independent of the topology of X. On the other hand, it is easy to prove that

$$\delta\mathscr{F}(u_0, \zeta) = d\mathscr{F}(u_0, \zeta) = D\mathscr{F}(u_0)\zeta$$

holds for all $\zeta \in C^1(\bar{\Omega}, \mathbb{R}^N)$, provided that u and the integrand F are of class C^1. That is, all three derivatives coincide if we stay in the class C^1, as we shall do in Chapter 1. But it will become important to use $\delta\mathscr{F}(u_0, \zeta)$ when we turn to the existence and regularity theory for minimizers and stationary points of multiple integrals.

We finally remark that $\delta\mathscr{F}(u_0, \zeta)$ sometimes is denoted as "Gâteaux differential"; yet we prefer the classical name "first variation".

2. Vanishing First Variation and Necessary Conditions

Using the ideas of Section 1, we shall now derive necessary conditions which are to be satisfied by critical points of variational integrals. The principal condition is the vanishing of the first variation which, for smooth critical points, implies the *Euler equation* and, in case of free boundary values, also the *natural boundary condition.* We shall concentrate our attention to first order variational problems where the *Lagrangian* only involves the unknown function and its derivatives of first order. Higher order variational problems will briefly be touched at the end of this chapter.

2.1. The First Variation of Variational Integrals

In this section we shall consider functionals $\mathscr{F}$ of the type

$$\mathscr{F}(u) = \int_\Omega F(x, u(x), Du(x))\,dx, \tag{1}$$

which will be called *variational integrals.* We shall write $\mathscr{F}_\Omega(u)$ or $\mathscr{F}(u, \Omega)$ if we want to indicate the domain of integration Ω. The integrand $F(x, u, p)$ of such an integral $\mathscr{F}(u)$ will be denoted as *Lagrangian*, or *variational integrand*, or *Lagrange function.* The reader will have noticed that we denote the variational integrals that are associated to Lagrangians $F, G, \ldots$ by the script types $\mathscr{F}, \mathscr{G}, \ldots$ of the same letters. We will use this convention whenever it does not lead to some misinterpretation.

Since one has to distinguish between the Lagrangian $F(x, u, p)$ as function of the independent variables x, u, p, and the composed function $F(x, u(x), Du(x))$, it would be more precise to write $F(x, z, p)$ instead of $F(x, u, p)$. Yet the notation $F(x, u, p)$ and $F_u(x, u, p)$ for its partial derivative is more suggestive since it indicates the position of the unknown function u. Thus we shall use the notation $F(x, z, p)$ only if $F(x, u, p)$ might lead to some confusion. (For the same reason, many authors write $F(x, u, Du)$ instead of $F(x, z, p)$. Then, obviously, misinterpretations are close at hand, and notations like $\frac{\partial F}{\partial u^i_{x^\alpha}}$, $\frac{\partial^2 F}{\partial u^i_{x^\alpha} \partial u^k_{x^\beta}}$ instead of $F_{p^i_\alpha}$, $F_{p^i_\alpha p^k_\beta}$ are at least awkward if not misleading. We, therefore, shall avoid this way of writing which is quite common in the physical and also in the mathematical literature.)

In the following we shall generally suppose that Ω is a bounded open set in $\mathbb{R}^n$, and that $u : \bar{\Omega} \to \mathbb{R}^N$ is a function of class C^1. Moreover, we suppose that $F(x, u, p)$ is a real valued function of the variables (x, u, p) which is of class C^1 on some open set $\mathscr{U}$ of $\mathbb{R}^n \times \mathbb{R}^N \times \mathbb{R}^{nN}$ containing the 1-*graph* $\{(x, u(x), Du(x)) : x \in \bar{\Omega}\}$ of u. Then there is a number $\delta > 0$ such that the composed function $F(x, v(x), Dv(x))$ is defined for all $x \in \bar{\Omega}$ and for all $v \in C^1(\bar{\Omega}, \mathbb{R}^N)$ with $\|v - u\|_{C^1(\bar{\Omega})} < \delta$. Thus the integral

$$\mathscr{F}(v) = \int_\Omega F(x, v(x), Dv(x))\, dx$$

can be formed for all $v \in C^1(\bar{\Omega}, \mathbb{R}^N)$ with $\|v - u\|_{C^1(\bar{\Omega})} < \delta$. Consequently, the function

$$\Phi(\varepsilon) := \mathscr{F}(u + \varepsilon\varphi)$$

is defined for each $\varphi \in C^1(\bar{\Omega}, \mathbb{R}^N)$ and for $|\varepsilon| < \varepsilon_0$, where ε_0 is some positive number less than $\delta / \|\varphi\|_{C^1(\bar{\Omega})}$. Moreover, Φ is of class C^1 on $(-\varepsilon_0, \varepsilon_0)$, whence the *first variation* $\delta F(u, \varphi)$ *of* $\mathscr{F}$ *at* u *in direction of* φ is well defined by

$$\delta\mathscr{F}(u, \varphi) = \Phi'(0),$$

and a straight-forward computation yields

$$\delta\mathscr{F}(u, \varphi) = \int_\Omega \left\{ F_{u^i}(x, u, Du)\varphi^i + F_{p^i_\alpha}(x, u, Du)\varphi^i_{x^\alpha} \right\} dx. \tag{2}$$

Here we have used the *Einstein convention*: One has to sum over doubly appearing Greek indices from 1 to n, and over repeated Latin indices from 1 to N.

Introducing the notations $x = (x^\alpha)$, $u = (u^i)$, $p = (p^i_\alpha)$, and $p = (p_1, \ldots, p_n)$, $p_\alpha = (p^1_\alpha, \ldots, p^N_\alpha)$, $F_{p_\alpha} = (F_{p^1_\alpha}, \ldots, F_{p^N_\alpha})$, we can also write

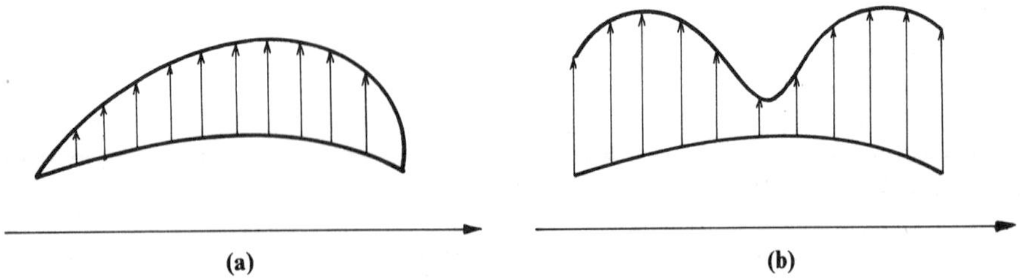

Fig. 5. Variation of a given curve by a vector field, **(a)** keeping the boundary values fixed, **(b)** with free boundary values.

$$\delta\mathscr{F}(u,\varphi) = \int_\Omega \left\{F_u(x,u,Du)\cdot\varphi + F_p(x,u,Du)\cdot D\varphi\right\} dx, \tag{2'}$$

and we note that, under our assumptions on F and $u(x)$, the first variation $\delta\mathscr{F}(u,\varphi)$ is a linear functional of $\varphi \in C^1(\bar{\Omega}, \mathbb{R}^N)$. Formula (2) suggests to introduce an expression $\delta F(u,\varphi)$ which is defined by

$$\delta F(u,\varphi)(x) := F_u(x,u(x),Du(x))\cdot\varphi(x) + F_p(x,u(x),Du(x))\cdot D\varphi(x) \tag{3}$$

for $x \in \bar{\Omega}$. We call it the *first variation of the Lagrangian F at u in direction of φ.* Then we can write

$$\delta\mathscr{F}(u,\varphi) = \int_\Omega \delta F(u,\varphi)\,dx. \tag{4}$$

We now want to show that the same formula holds with respect to a *general variation $\psi(x,\varepsilon)$ of $u(x)$ in direction of the vector field $\varphi(x)$*. This is a mapping $\psi: \bar{\Omega}\times[-\varepsilon_0,\varepsilon_0]\to\mathbb{R}^N$, $\varepsilon_0>0$, of class C^1 which satisfies

$$\text{(i)}\ \psi(x,0)=u(x), \qquad \text{(ii)}\ \left.\frac{\partial\psi}{\partial\varepsilon}(x,\varepsilon)\right|_{\varepsilon=0} = \varphi(x)$$

for all $x\in\bar{\Omega}$.

The variations previously considered were of the special kind $\psi(x,\varepsilon) = u(x)+\varepsilon\varphi(x)$, whereas general variations can be written as

$$\psi(x,\varepsilon) = u(x)+\varepsilon\varphi(x)+o(\varepsilon) \quad (\varepsilon\to 0)$$

on account of Taylor's formula.

Since ψ is of class C^1, it follows that $\|\psi(\cdot,\varepsilon)-u\|_{C^1(\bar{\Omega})}<\delta$ holds for sufficiently small values of $|\varepsilon|$. Then the chain rule yields the formula

$$\left.\frac{\partial}{\partial\varepsilon}F(x,\psi(x,\varepsilon),D_x\psi(x,\varepsilon))\right|_{\varepsilon=0} = \delta F(u,\varphi)(x), \tag{5}$$

whence

$$\left.\frac{d}{d\varepsilon}\mathscr{F}(\psi(\cdot,\varepsilon))\right|_{\varepsilon=0} = \delta\mathscr{F}(u,\varphi) = \int_\Omega \delta F(u,\varphi)\,dx. \tag{6}$$

A variation $\psi(x,\varepsilon)$ of $u(x)$ which is not subject to any boundary condition on $\partial\Omega$ will be called a

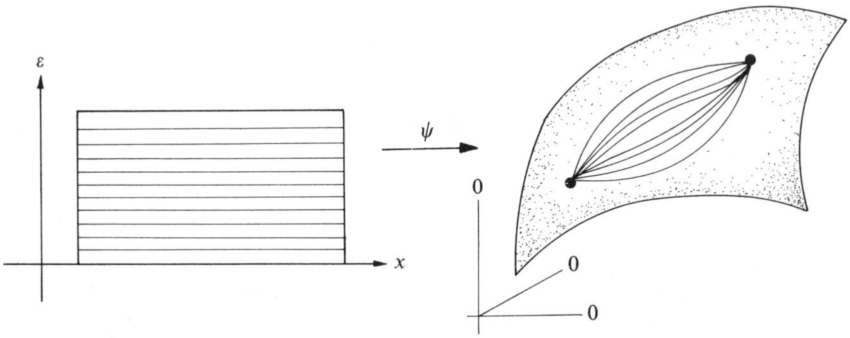

Fig. 6. General variations $\psi(x,\varepsilon)$ allow the variations to satisfy nonlinear constraints.

variation with free boundary values. If, however, the additional condition

$$\text{(iii)}\ \psi(x, \varepsilon) = u(x) \quad \text{for } x \in \partial\Omega \text{ and } |\varepsilon| < \varepsilon_0$$

is satisfied, we speak of a *variation with fixed boundary values*. Its *velocity vector* $\varphi(x) = \psi_\varepsilon(x, 0)$ fulfils $\varphi(x) = 0$ for $x \in \partial\Omega$.

Variations of $u(x)$ that are of the more general type $\psi(x, \varepsilon)$ have to be used if we consider variational problems where the comparison functions are subjected to nonlinear constraints. A typical example is the problem of shortest connection of two points on a surface S by curves which are restricted to lie on S (cf. Chapter 2).

We now consider the relation[1]

$$(7) \qquad \delta\mathscr{F}(u, \varphi) = 0 \quad \text{for all } \varphi \in C_c^\infty(\Omega, \mathbb{R}^N),$$

which, on account of (2), is equivalent to

$$(8) \quad \int_\Omega \left\{ F_{u^i}(x, u, Du)\varphi^i + F_{p_\alpha^i}(x, u, Du)D_\alpha\varphi^i \right\} dx = 0 \quad \text{for all } \varphi \in C_c^\infty(\Omega, \mathbb{R}^N).$$

A solution $u \in C^1(\Omega, \mathbb{R}^N)$ of (8) is called a *weak extremal of* $\mathscr{F}$, and equation (8) is denoted as *weak Euler equation* (or, more precisely, as Euler equation for *u in the weak form*).

We note that (8) follows from a rather weak minimum property of u. It suffices to assume that, for some $\delta_0 \in (0, \delta)$, the inequality

$$(9) \qquad \mathscr{F}(u) \le \mathscr{F}(u + \varphi)$$

holds for all $\varphi \in C_c^\infty(\Omega, \mathbb{R}^N)$ with $\|\varphi\|_{C^1(\bar\Omega)} < \delta_0$.

Therefore we give the following

Definition. *Let $\mathscr{C}$ be a subset of $C^1(\bar\Omega, \mathbb{R}^N)$. Then $u \in \mathscr{C}$ is called a* weak (relative) minimizer *of $\mathscr{F}$ in $\mathscr{C}$ if there is some $\delta_0 > 0$ such that $\mathscr{F}(u) \le \mathscr{F}(v)$ for all $v \in \mathscr{C}$ satisfying $\|v - u\|_{C^1(\Omega)} < \delta_0$.*

Note that (8) follows even from a weaker minimum property.

Proposition. *Suppose that, for every $x_0 \in \Omega$, there exists a number $r > 0$ with the following properties*:

(*i*) $B_r(x_0) \subset\subset \Omega$,

(*ii*) *for every $\varphi \in C_c^\infty(B_r(x_0), \mathbb{R}^N)$, there is a number $\varepsilon_0 > 0$ such that $\|\varepsilon\varphi\|_{C^1(\bar\Omega)} < \delta$ and $\mathscr{F}(u) \le \mathscr{F}(u + \varepsilon\varphi)$ hold for all ε with $|\varepsilon| < \varepsilon_0$. Then relation* (8) *is satisfied.*

---

[1] The reader may wonder why we operate with $C_c^\infty(\Omega)$ instead of $C_0^1(\bar\Omega)$. Here we follow the custom of the theory of distributions. It has the advantage that one need not distinguish between "test functions" for differential equations of different orders. Note that, under our assumptions, (7) implies that $\delta\mathscr{F}(u, \varphi) = 0$ holds for all $\varphi \in C_0^1(\bar\Omega, \mathbb{R}^N)$.

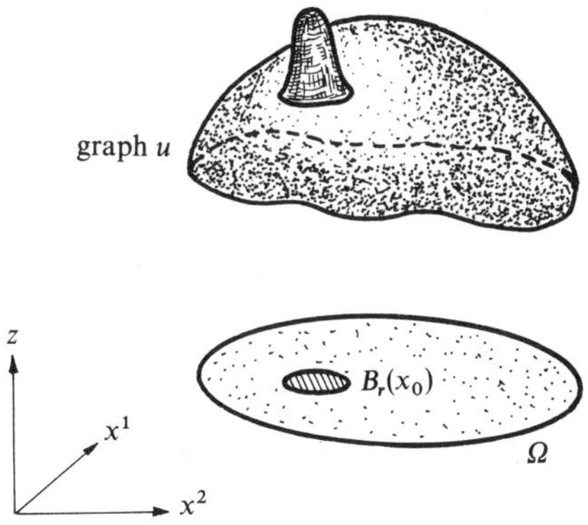

Fig. 7. Local perturbation of a graph u by a vector field $(0, \varphi(x))$ with supp $\varphi \subset B_r(x_0)$.

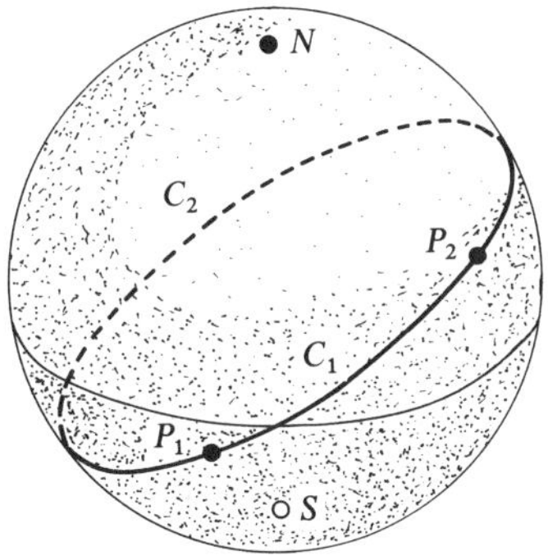

Fig. 8. If P_1 and P_2 are not antipodal points on the sphere, then there is exactly one great circle through P_1 and P_2 which is then decomposed by P_1, P_2 into a smaller arc C_1 and a larger arc C_2. The arc C_1 minimizes the arc length between P_1 and P_2 while C_2 is just a stationary arc.

Proof. By a suitable partition of unity

$$\eta_1 + \eta_2 + \cdots + \eta_k = 1, \quad \eta_i \in C_c^\infty(B_{r_i}(x_i)),$$

one can decompose $\varphi \in C_c^\infty(\Omega, \mathbb{R}^N)$ into functions $\varphi_i = \eta_i \varphi$ with small support whence $\delta\mathscr{F}(u, \varphi_i) = 0$ follows. If we sum these equations from 1 to k, we obtain $\delta\mathscr{F}(u, \varphi) = 0$. □

In other words, the weak Euler equations (8) already hold if the value of $\mathscr{F}$ is not diminished in case we put very small local bumps on u.

Yet it should be noted that there can exist weak extremals which have no weak minimum or maximum property at all ("saddles"). But we will show that a weak extremal satisfying the strict Legendre–Hadamard condition (some second derivative condition; see 5,*1.3*) possesses the following local minimum property:

For every $x_0 \in \Omega$, there exists a neighbourhood Ω' of x_0 with $\Omega' \subset\subset \Omega$ and a number $\varepsilon > 0$ such that $\mathscr{F}(v) \geq \mathscr{F}(u)$ holds for all v with $v - u \in C_c^1(\Omega', \mathbb{R}^N)$ and $\|v - u\|_{C^1(\bar{\Omega})} < \varepsilon$.

We also note that "local minimality" will in general not imply global minimality. This can be seen from the great circles on a sphere which are the extremals of the length functional. Sufficiently small arcs on a great circle provide a minimal connection of their endpoints whereas arcs containing a pair of antipodal points in their interior are not global minimizers.

2.2. The Fundamental Lemma of the Calculus of Variations, Euler's Equations, and the Euler Operator L_F

We will prove the following fundamental

Theorem. *Suppose that $F \in C^2(\mathcal{U})$, and that u is a weak extremal of $\mathcal{F}$, that is,*

$$\int_\Omega \{F_{u^i}(x, u, Du)\varphi^i + F_{p^i_\alpha}(x, u, Du)D_\alpha\varphi^i\}\,dx = 0 \quad \text{for all } \varphi \in C_c^\infty(\Omega, \mathbb{R}^N).$$

Let, moreover, u be of class $C^2(\Omega, \mathbb{R}^N)$. Then $u(x)$ satisfies

$$(1) \qquad D_\alpha F_{p^i_\alpha}(x, u(x), Du(x)) - F_{u^i}(x, u(x), Du(x)) = 0, \quad 1 \le i \le N,$$

for all $x \in \Omega$.

Actually, the same conclusion holds if, instead of $F \in C^2(\mathcal{U})$, we assume that $F_u \in C^0(\mathcal{U})$ and $F_p \in C^1(\mathcal{U})$.

Proof. By partial integration we obtain that

$$\int_\Omega \{F_{z^i}(x, u(x), Du(x)) - D_\alpha F_{p^i_\alpha}(x, u(x), Du(x))\}\,\varphi^i(x)\,dx = 0$$

holds for every test function $\varphi(x) = (\varphi^1(x), \ldots, \varphi^N(x))$ of class $C_c^\infty(\Omega, \mathbb{R}^N)$. In particular, if we fix some index i and choose $\varphi^j(x) = 0$ for $j \ne i$ and $\varphi^i = \eta$ for some arbitrary $\eta \in C_c^\infty(\Omega)$, we arrive at

$$\int_\Omega \{F_{z^i}(\ldots) - D_\alpha F_{p^i_\alpha}(\ldots)\}\eta(x)\,dx = 0.$$

Then, on account of the so-called *fundamental lemma of the calculus of variations*, which will be proved below, it follows that

$$F_{z^i}(\ldots) - D_\alpha F_{p^i_\alpha}(\ldots) = 0. \qquad \square$$

The Fundamental Lemma. *Let $f(x)$ be a continuous, real valued function on some open set Ω of $\mathbb{R}^n$, and suppose that*

$$(*) \qquad \int_\Omega f(x)\eta(x)\,dx \ge 0 \quad \textit{for all } \eta \in C_0^\infty(\Omega) \textit{ with } \eta \ge 0$$

or

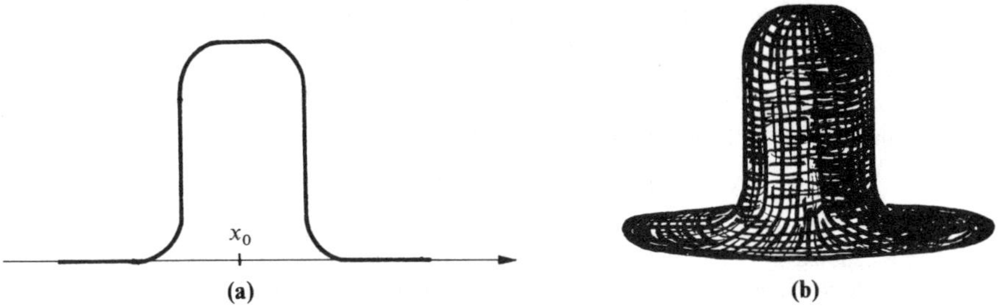

Fig. 9. The function η in the fundamental lemma: **(a)** crosscut through graph η, **(b)** graph η.

$$(**) \qquad \int_\Omega f(x)\eta(x)\,dx = 0 \quad \textit{for all } \eta \in C_c^\infty(\Omega)$$

holds. Then we have

$$f(x) \geq 0 \quad or \quad f(x) = 0,$$

respectively, for all $x \in \Omega$.

Proof. Suppose that $(*)$ is satisfied, and assume that there is a point $x_0 \in \Omega$ with $f(x_0) < 0$. Then we can find a number $\varepsilon > 0$ and a ball $B_r(x_0) \subset\subset \Omega$ such that $f(x) < -\varepsilon$ on $B_r(x_0)$. By means of the test function $\eta \in C_c^\infty(\Omega)$, defined by

$$\eta(x) := \begin{cases} \exp(-1/(r^2 - |x - x_0|^2)) & \text{for } x \in B_r(x_0), \\ 0 & \text{for } x \in \Omega - B_r(x_0), \end{cases}$$

we now arrive at the contradictory statement

$$0 \leq \int_\Omega f(x)\eta(x)\,dx = \int_{B_r(x_0)} f(x)\eta(x)\,dx < -\varepsilon \int_{B_r(x_0)} \eta(x)\,dx < 0.$$

That is, the relation must imply $f(x) \geq 0$ for all $x \in \Omega$. The second assertion is an immediate consequence of the first one. □

One calls (1) *the Euler equations* of the integral $\mathscr{F}(u) = \int_\Omega F(x, u, Du)\,dx$, and its C^2-solutions $u(x)$ are said to be the *extremals* of $\mathscr{F}$, or *F-extremals*. This does, of course, not imply that extremals are extremizers (i.e., maximizers or minimizers) of $\mathscr{F}$. In fact, even for functions of real variables, critical points may only be saddle points and not maximizers nor minimizers. The problem of finding the minimizers (extremals) of $\mathscr{F}$ in a class $\mathscr{C}$ will occasionally be denoted by

$$\text{“}\mathscr{F} \to \textit{min (stationary) in } \mathscr{C}\text{”}.$$

Quite often, and rightfully so, equations (1) are called the *Euler–Lagrange equations* of the corresponding variational integral $\mathscr{F}$. We have settled with the shorter notation, but to do justice to Lagrange who invented the formal δ-calculus (which, at first, he considered to be an infinitesimal calculus of "higher order" until Euler invented the method of embedding an extremum $u(x)$ in a family of variations $\psi(x, \varepsilon)$), we use the symbol L_F (or simply L) for the *Euler operator*

$$L_F(u) := F_u(\cdot, u(\cdot), Du(\cdot)) - D_\alpha F_{p_\alpha}(\cdot, u(\cdot), Du(\cdot)). \tag{2}$$

We can, for instance, interpret L as a mapping

$$L_F : C^s(\Omega, \mathbb{R}^N) \to C^{s-2}(\Omega, \mathbb{R}^N)$$

for $s \geq 2$. In the older literature, the Euler operator L_F is sometimes called the *Lagrangian derivative* (of F) and is denoted by $[F]$. In the physical literature, the expression $L_F(u)$ is called the *variational derivative of the functional* $\mathcal{F}$ *at* u, and one uses the notation $\delta\mathcal{F}/\delta u$ instead of $L_F(u)$. Thus we find formulas like

$$\delta\mathcal{F}(u, \varphi) = \int_\Omega \frac{\delta\mathcal{F}}{\delta u}\,\delta u\,dx \quad \text{for } \delta u := \varphi \in C_c^\infty(\Omega, \mathbb{R}^N).$$

In (1) and (2) as well as in similar formulas which will follow we have used a *convention on partial derivatives* that is to be kept in mind:

Partial derivatives attached as index mean that we first have to differentiate before a composition is made. For example, in order to form $F_{p_\alpha^i}(x, u(x), Du(x))$ one first has to compute the partial derivative $F_{p_\alpha^i}(x, u, p)$ and then to insert $u(x)$ for u and $Du(x)$ for p, whereas $D_\alpha F_{p_\beta^i}(x, u(x), Du(x))$ or $\frac{\partial}{\partial x^\alpha} F_{p_\beta^i}(x, u(x), Du(x))$ means that one has to take the partial derivative of the function $\Pi_i^\beta(x) := F_{p_\beta^i}(x, u(x), Du(x))$ with respect to the variable x^α. One has carefully to distinguish the expression $F_{p_\alpha^i x^\alpha}(x, u, p)$ from

$$\frac{\partial}{\partial x^\alpha} F_{p_\alpha^i}(x, u(x), Du(x))$$

$$= F_{p_\alpha^i x^\alpha}(x, u(x), Du(x)) + F_{p_\alpha^i u^k}(x, u(x), Du(x)) D_\alpha u^k + F_{p_\alpha^i p_\beta^k}(x, u(x), Du(x)) D_\alpha D_\beta u^k.$$

If no misunderstanding is possible, we shall as everyone else simply write

$$D_\alpha F_{p_\alpha^i} - F_{u^i} = 0, \quad i = 1, \ldots, N,$$

instead of (1), but the reader should keep in mind the correct interpretation. Using the notation (2), the Euler equations obtain the short form

$$L_F(u) = 0 \quad \text{in } \Omega.$$

Note that in older textbooks our expressions $F_{p x^\alpha}(x, u, Du)$ and $\frac{\partial}{\partial x^\alpha} F_p(x, u, Du)$ are often denoted by $\frac{\partial}{\partial x^\alpha} F_p$ and $\frac{d}{dx^\alpha} F_p$, respectively.

We shall now discuss a few examples.

[1] Consider the *Dirichlet integral* $\mathcal{D}(u)$ defined by

$$\mathcal{D}(u) = \tfrac{1}{2} \int_\Omega |Du|^2\,dx, \quad N = 1,$$

with $|Du|^2 = u_{x^\alpha} u_{x^\alpha}$. Its Lagrangian F is given by

$$F(p) = \tfrac{1}{2}|p|^2.$$

Then $F_{p^\alpha}(p) = p_\alpha$, and we obtain

$$\Delta u = 0 \quad \text{in } \Omega$$

as corresponding Euler equation, where

$$\Delta = D_1 D_1 + \cdots + D_n D_n$$

denotes the *Laplace operator* of $\mathbb{R}^n$. The solutions of $\Delta u = 0$ in Ω are called *harmonic functions* (in Ω), and the equation $\Delta u = 0$ is denoted as *Laplace equation* or *potential equation.*

In fact, the Newtonian potential

$$u(x) = \int_G \frac{\sigma(y)}{|x-y|^{n-2}}\,dy, \quad n \geq 3,$$

of some density function $\sigma \in L^\infty(G)$, defined on a bounded domain G of $\mathbb{R}^n$, satisfies

$$\Delta u = 0 \quad \text{in } \mathbb{R}^n - \bar{G}.$$

[2] The integral

$$\mathscr{F}(u) = \int_\Omega \{\tfrac{1}{2}|Du|^2 + f(x)u\}\,dx, \quad N = 1,$$

with the Lagrangian

$$F(x, z, p) = \tfrac{1}{2}|p|^2 + f(x)z$$

has the so-called *Poisson equation*

$$\Delta u = f(x) \quad \text{in } \Omega$$

as the corresponding Euler equation.

If $\sigma \in C^1(\bar{G})$, then the Newtonian potential $u(x) = \int_G |x-y|^{2-n}\sigma(y)\,dy$ satisfies the Poisson equation with the right-hand side $f(x) = -(n-2)\omega_n\sigma(x)$, where[2]

$$\omega_n = \frac{2\Gamma^n(1/2)}{\Gamma(n/2)} = \frac{2\pi^{n/2}}{\Gamma(n/2)}$$

is the surface area of the unit sphere in $\mathbb{R}^n$. Newtonian potentials play an important role as gravitational potentials and, in electrostatics, as Coulomb potentials.

[3] The nonlinear Poisson equation

$$\Delta u = f(u) \quad \text{in } \Omega$$

is the Euler equation of the integral

$$\int_\Omega \{\tfrac{1}{2}|Du|^2 + g(u)\}\,dx, \quad N = 1,$$

where g is a primitive function of f, i.e., $g'(z) = f(z)$. Similarly,

$$\int_\Omega \left\{\tfrac{1}{2}|Du|^2 - \frac{1}{p+1}|u|^{p+1} - g(u)\right\} dx, \quad p \geq 1,$$

leads to the Euler equation

$$-\Delta u = u|u|^{p-1} + f(u),$$

which is an important model equation in physics where it is assumed that

$$f(0) = 0, \quad \lim_{z\to\infty} \frac{f(z)}{|z|^p} = 0.$$

---

[2] For the computation of the surface area of the unit sphere in $\mathbb{R}^n$ in terms of Euler's Γ-function, $\Gamma(t) := \int_0^\infty x^{t-1}e^{-x}\,dx$, $t > 0$, compare e.g. Courant–John [1] or Fleming [1].

The case of the so-called limit exponent $p = \dfrac{n+2}{n-2}$ or $p+1 = \dfrac{2n}{n-2}$ is of particular interest.[3]

If $n = 4$, the limit exponent is $p = 3$ (or $p + 1 = 4$). For instance,

$$\int_\Omega \{\tfrac{1}{2}|Du|^2 - \tfrac{1}{4}|u|^4 + \tfrac{1}{2}\lambda |u|^2\}\, dx$$

has the Euler equation

$$-\Delta u = u^3 - \lambda u .$$

This very simple model appears in physics in connection with oversimplified versions of *Yang–Mills equations.*

4 Interpret $\mathbb{R}^4$ as space–time continuum of points $(x, t) \in \mathbb{R}^3 \times \mathbb{R}$ where x is the position vector in $\mathbb{R}^3$ and t denotes the time. Let Ω be some bounded domain in $\mathbb{R}^4$, and consider functions $u(x, t)$ of x and t. Then the integral

$$\int_\Omega \tfrac{1}{2}\{u_t^2 - |u_x|^2\}\, dx\, dt ,$$

with

$$u_x = (u_{x^1}, u_{x^2}, u_{x^3}) = Du$$

has the *wave equation*

$$\Box u = 0 \quad \text{in } \Omega$$

as Euler equation where $\Box$ denotes the *d'Alembert operator*

$$\Box = D_t D_t - D \cdot D , \quad D \cdot D = D_1 D_1 + D_2 D_2 + D_3 D_3 = \Delta .$$

Similarly,

$$\int_\Omega \tfrac{1}{2}\{u_t^2 - |u_x|^2 - m^2 u^2\}\, dx$$

leads to the so-called *Klein–Gordon equation*

$$\Box u + m^2 u = 0 , \quad \Box = D_t^2 - \Delta .$$

Adding the term $\dfrac{\lambda}{4} u^4 + ju$ to the Lagrangian $\tfrac{1}{2}\{u_t^2 - |u_x|^2 - m^2 u^2\}$, we arrive at the Euler equation

$$\Box u + m^2 u = \lambda u^3 + j .$$

5 The *area functional*

$$\mathscr{A}(u) = \int_\Omega \sqrt{1 + |Du|^2}\, dx$$

for hypersurfaces $z = u(x)$, $x \in \Omega \subset \mathbb{R}^n$, in $\mathbb{R}^{n+1}$ yields the *minimal surface equation*

$$\operatorname{div} Tu = 0 \text{ in } \Omega , \quad Tu := \frac{Du}{\sqrt{1 + |Du|^2}}$$

as Euler equation, which we can also write as

[3] This limit case is related to Sobolev's inequality and to the isoperimetric inequality.

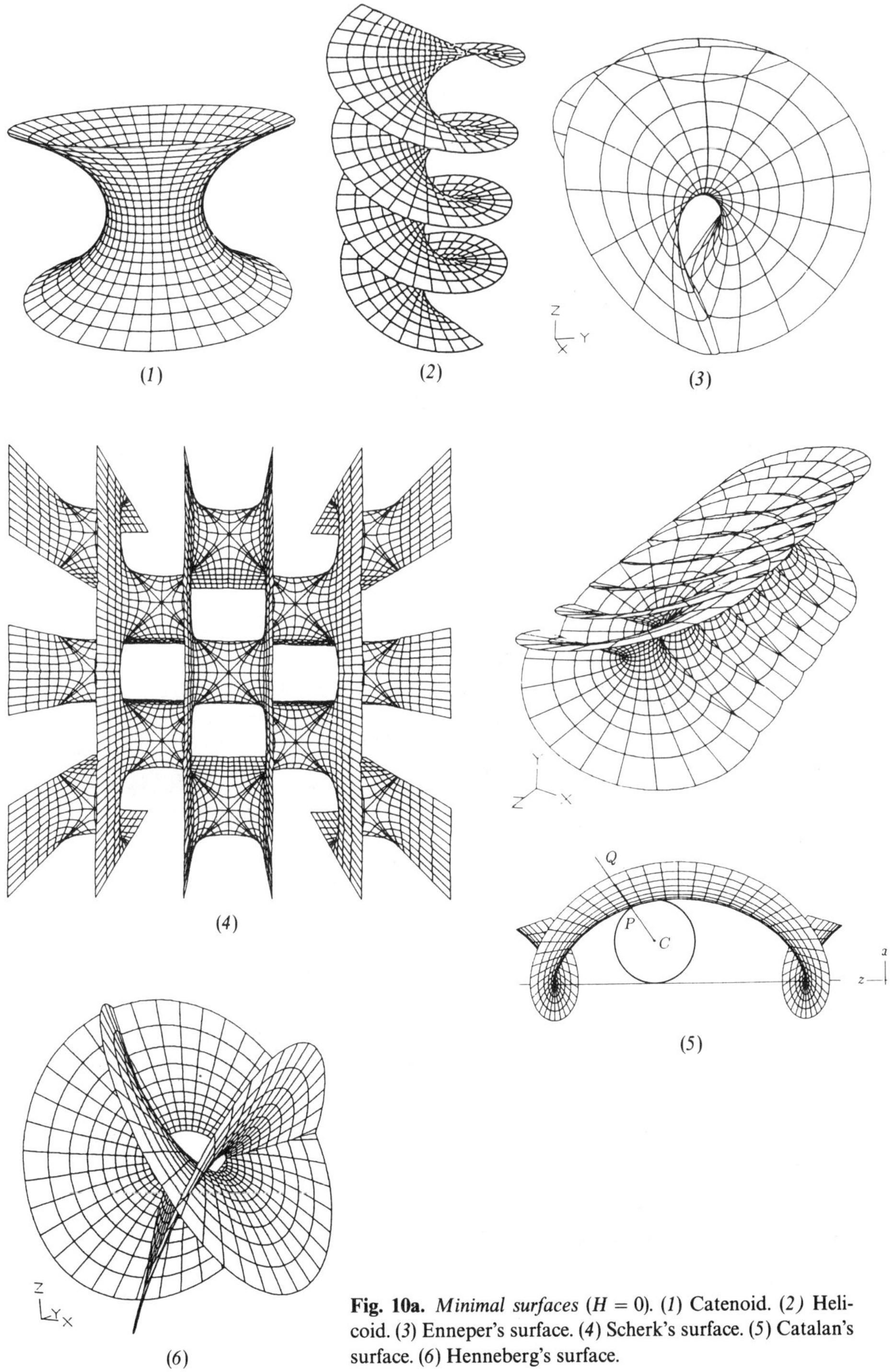

Fig. 10a. *Minimal surfaces* ($H = 0$). (*1*) Catenoid. (*2*) Helicoid. (*3*) Enneper's surface. (*4*) Scherk's surface. (*5*) Catalan's surface. (*6*) Henneberg's surface.

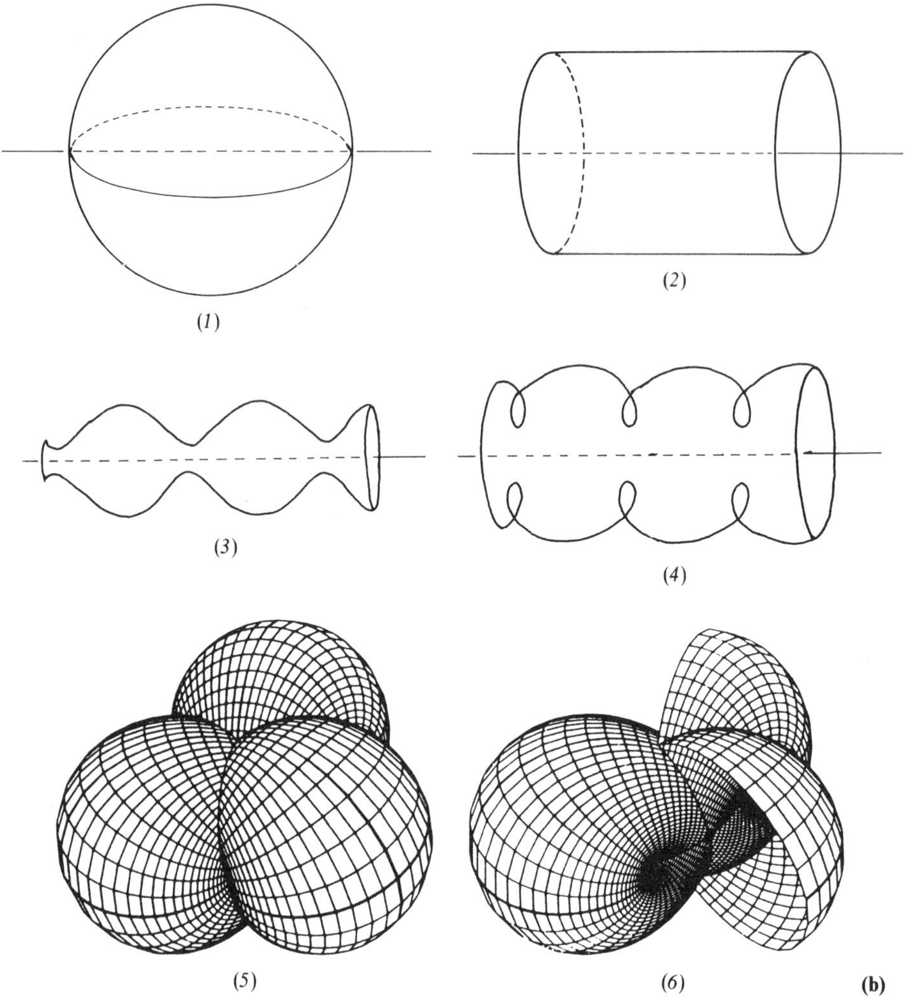

Fig. 10b. *Surfaces of constant mean curvature* $H \neq 0$. (*1*) Sphere. (*2*) Cylinder. (*3*) Unduloid. (*4*) Nodoid. (*5*) Wente's surface. (*6*) Part of Wente's surface.

The plane, the catenoid and the surfaces (*1*)–(*4*) are the only surfaces of revolution with constant mean curvature. Their meridians are obtained by the motion of the foci of conic sections rolling on a straight line.

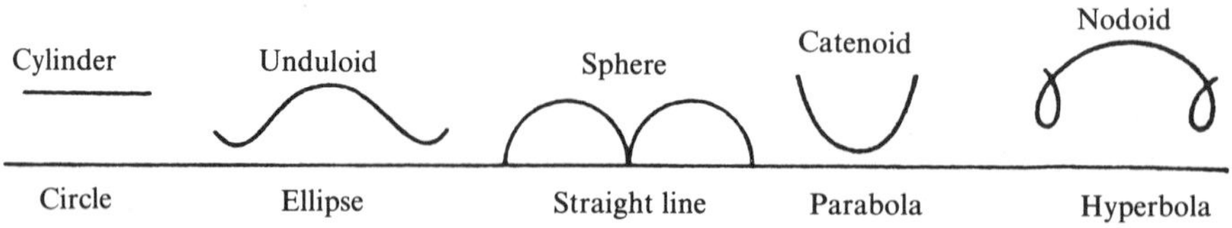

$$D_\alpha\left(\frac{D_\alpha u}{\sqrt{1+|Du|^2}}\right)=0.$$

The expression $H := \frac{1}{n}\operatorname{div} Tu$ is the mean curvature function[4] of graph(u). Thus the stationary points u of $\mathscr{A}(u)$ describe surfaces of zero mean curvature, which are called *minimal surfaces.*

Moreover, the extremals u of the functional

$$\mathscr{F}(u)=\int_\Omega \{\sqrt{1+|Du|^2}+Hu\},\quad H=H(x),$$

satisfy

$$\operatorname{div} Tu = nH$$

and, therefore, are describing surfaces of mean curvature H. More precisely, the graph of the function u in $\mathbb{R}^{n+1}$ is a surface that has the mean curvature $H(x)$ at the point $(x, u(x))$.

For $n = 1$,

$$\mathscr{L}(u)=\int_a^b \sqrt{1+u'^2}\,dx$$

is the length of the nonparametric curve $z = u(x)$, $a \le x \le b$. The corresponding Euler equation

$$\left(\frac{u'}{\sqrt{1+u'^2}}\right)'=0$$

implies that the extremals of $\mathscr{L}(u)$ are the straight lines $u(x) = \alpha x + \beta$, $a \le x \le b$, $\alpha, \beta \in \mathbb{R}$.

If u is a C^2-solution of the equation

$$\left(\frac{u'}{\sqrt{1+u'(x)^2}}\right)'=H,$$

$H = \text{const} \ne 0$, then the graph of u has to be a circular arc. In fact, the defining equation of u yields

$$\frac{u'}{\sqrt{1+u'^2}}=Hx+c$$

for some constant c, whence

$$u'^2=(Hx+c)^2(1+u'^2),$$

and therefore

$$u'=\pm\frac{Hx+c}{\sqrt{1-(Hx+c)^2}}.$$

This implies

$$u(x)=\mp\frac{1}{H}\sqrt{1-(Hx+c)^2}.$$

Setting $R := 1/H$ and $x_0 := -c/H$, we obtain

$$(x-x_0)^2+u(x)^2=R^2.$$

6 *Lagrangians of the kind $F(x, p)$, $n = 1$.* In this case, the Euler equation reduces to

$$\{F_p(x,u'(x))\}'=0,\quad x\in I,$$

[4] For the definition of the mean curvature, see the Supplement.

and can, therefore, be integrated. It follows that

$$F_p(x, u'(x)) = a$$

holds for some constant vector $a = (a^1, \dots, a^N) \in \mathbb{R}^N$. If $F_{pp} = (F_{p^i p^k})$ is nonsingular, we can – at least locally – solve this equation with respect to $u'(x)$ and obtain

$$u'(x) = g(x, a)$$

for some function $g(x, a)$. If $u(x_0) = b$, then

$$u(x) = b + \int_{x_0}^{x} g(t, a)\, dt .$$

In particular, for

$$F(x, p) = \omega(x)\sqrt{1 + |p|^2}, \quad \omega(x) > 0,$$

we arrive at

$$\frac{\omega(x)u'(x)}{\sqrt{1 + |u'(x)|^2}} = a, \quad x \in I,$$

whence $|a| \le \omega_0 = \inf_I \omega(x)$. It follows that

$$u(x) = b + \int_{x_0}^{x} \frac{a}{\sqrt{\omega^2(t) - |a|^2}}\, dt .$$

[7] *Conservation of energy and minimal surfaces of revolution.* If $n = 1$ and F is of the form $F(u, p)$, then the expression

$$p \cdot F_p(u, p) - F(u, p)$$

is a *first integral* for the solutions $u(x)$ of the Euler equation. This means that

$$u'(x) \cdot F_p(u(x), u'(x)) - F(u(x), u'(x)) \equiv h$$

for some constant h. In fact,

$$\frac{d}{dx}\left\{u' \cdot F_p(u, u') - F(u, u')\right\} = u'' \cdot F_p + u' \cdot \frac{d}{dx} F_p - F_u \cdot u' - F_p \cdot u''$$

$$= u' \cdot \left[\frac{d}{dx} F_p - F_u\right] = 0.$$

Because of its application in mechanics, the equation

$$u' \cdot F_p(u, u') - F(u, u') = h$$

is said to express the *conservation of energy* (cf. 2,2 [5] and 3,2 [1]).

The "conserved" expression $p \cdot F_p - F$ is not obvious to guess. In Chapter 3 we shall see how Noether's theorem yields this and a number of similar identities called "conservation laws".

Note that the conservation law $pF_p - F = \text{const}$ may have more solutions than the Euler equation. For instance, if $u(x) \equiv \text{const} =: c$ and $F(c, 0) = -h$, then $u(x) \equiv c$ satisfies the conservation law

$$u'F_p(u, u') - F(u, u') = h,$$

but it is a solution of the Euler equation only if $F_u(c, 0) = 0$. For example, if

$$F(u, p) = \sqrt{p^2 + \sin^2 u},$$

then every $u(x) \equiv \arcsin h$, $0 < h < 1$, satisfies the conservation law but is not a solution of the Euler equation. (Observe that, in spherical coordinates (r, φ, θ), curves on the unit sphere S^2, given by $\theta = \theta(\varphi)$, $\varphi_1 \le \varphi \le \varphi_2$, have the length

$$\int_{\varphi_1}^{\varphi_2} \sqrt{\dot\theta^2 + \sin^2\theta}\, d\varphi, \qquad \dot{} = \frac{d}{d\varphi}.)$$

If $N = 1$ and $F_{pp} \neq 0$, the conservation law

$$u'F_p(u, u') - F(u, u') = h$$

suffices to integrate Euler equation. In fact, if $p \neq 0$, then the implicit function theorem yields that

$$u' = \psi(u, h) \quad \text{or} \quad dx = \frac{du}{\psi(u, h)},$$

whence

$$x = x_0 + \int_{u_0}^{u} \frac{dz}{\psi(z, h)} = f(u, u_0, h).$$

As an application, we consider *surfaces of revolution minimizing area* among all rotationally symmetric surfaces having the same boundary. Let $z = u(x)$, $a \le x \le b$, be the meridian of such a surface S; we assume that $u(x) > 0$ on $[a, b]$. Then S will be generated by rotating the graph of u about the x-axis, and the area $\mathscr{A}(u)$ of S is given by

$$\mathscr{A}(u) = 2\pi \int_a^b u\sqrt{1 + u'^2}\, dx.$$

More generally, we consider functionals of the type

$$\int_a^b \omega(u)\sqrt{1 + (u')^2}\, dx$$

with the Lagrangian

$$F(z, p) = \omega(z)\sqrt{1 + p^2}.$$

Then every extremal $u(x)$, $a \le x \le b$, will satisfy

$$\omega(u) = -h\sqrt{1 + (u')^2} \quad \text{on } [a, b].$$

This shows that $h \neq 0$ if $\omega(u(x)) \neq 0$ on $[a, b]$, and that $h < 0$ if $\omega(u(x)) > 0$ on $[a, b]$. Moreover, we conclude that either $\omega(u(x)) \equiv 0$, or $\omega(u(x)) \neq 0$ for all $x \in [a, b]$.

Let

$$\kappa = \left(\frac{u'}{\sqrt{1 + u'^2}}\right)' = \frac{u''}{(1 + u'^2)^{3/2}}$$

be the curvature of graph u. Since

$$L_F(u) = \frac{\omega_u(u)}{\sqrt{1 + u'^2}} - \omega(u)\left(\frac{u'}{\sqrt{1 + u'^2}}\right)',$$

the Euler equation can be written as

$$\omega(u)\sqrt{1 + u'^2}\,\kappa = \omega_u(u).$$

Thus any extremal $u(x)$ which is of class C^2 on $[a, b]$ and satisfies $\omega_u(u(x)) = 0$ for some x must also satisfy either $\omega(u(x)) = 0$, or $\kappa(x) = 0$, or both. In other words, the assumption $\omega_u(u(x)) \neq 0$ for all $x \in (a, b)$ implies that $\kappa(x) \neq 0$ and $\omega(u(x)) \neq 0$. Hence every extremal $u(x)$ with $\omega_u(u(x)) \neq 0$ is free of inflection points and, therefore, is either convex or concave, and *singular extremals*, characterized by $\omega(u(x)) \equiv 0$, are excluded. The energy integral then implies that every nonsingular extremal $u(x)$ will satisfy the equation

$$(u')^2 = \frac{1}{h^2}\{\omega(u)^2 - h^2\}.$$

In the special case of *surfaces of revolution*, the Lagrangian is

$$F(u, p) = 2\pi u\sqrt{1 + p^2},$$

with $\omega(u) = 2\pi u$, $\omega_u(u) = 2\pi$. Hence there exist no singular extremals, and the energy constant h of every extremal $u(x) > 0$ satisfies

$$0 < -h \leq 2\pi \min\{u(x): a \leq x \leq b\}.$$

Let us, instead of h, introduce the new constant $c = -\frac{h}{2\pi}$. Then $u(x) \geq c$, and we may introduce a new function $v(x)$ by

$$u(x) = c \cosh v(x).$$

Since

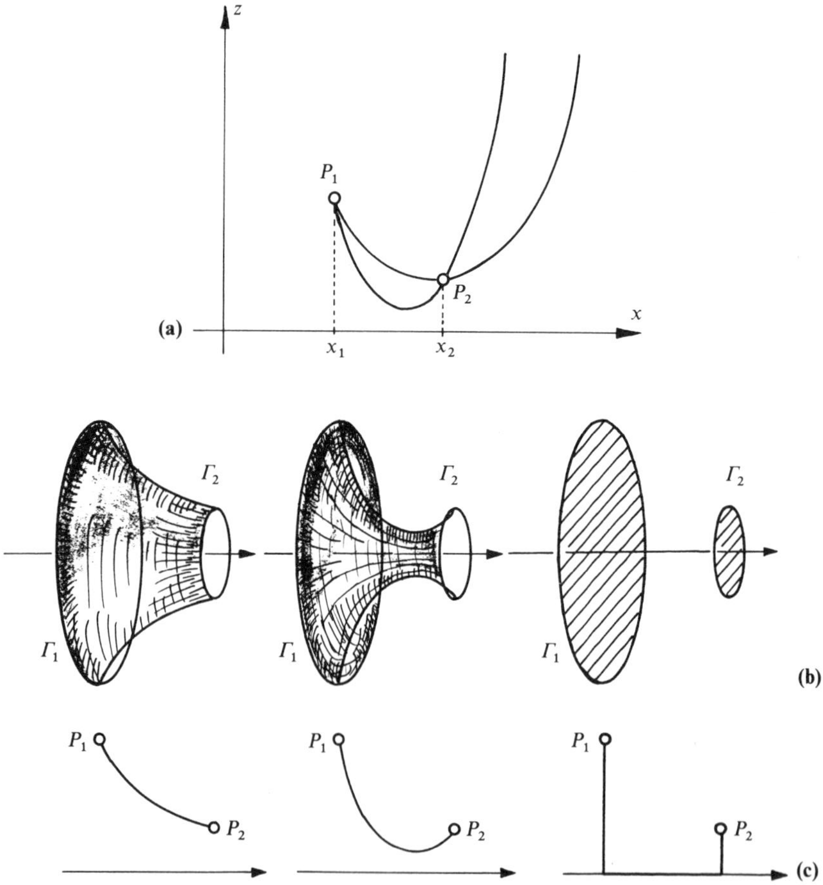

Fig. 11. (a) Two points P_1 and P_2 in the upper halfplane can be joined by two, one, or no catenary arc. Correspondingly the two circles Γ_1 and Γ_2 generated by P_1 and P_2 span two, one, or no catenoid. **(b)** Suppose that P_1, P_2 are connected by two catenary arcs, and let G be the Goldschmidt polygon between P_1 and P_2. **(c)** Then Γ_1 and Γ_2 span two catenoids and the minimal surface consisting of the two disks in Γ_1 and Γ_2.

$$(cu')^2 = u^2 - c^2,$$

we infer that

$$(cv')^2 = 1,$$

whence

$$v(x) = \pm\frac{x - x_0}{c},$$

and therefore

$$u(x) = c\cosh\frac{x - x_0}{c}.$$

This 2-parameter family of real analytic curves describes all possible extremals of the functional $\mathscr{A}(u)$. They are called *catenaries* or *chain lines* because they turn out to be equilibrium positions of a chain (or rather of an infinitely thin, inextensible heavy thread with a uniform mass density), cf. *2,1* [5]. Correspondingly, the surfaces of revolution with catenaries of the type $u(x) = c\cosh\frac{x - x_0}{c}$ as meridians are called *catenoids.* It turns out that they are surfaces of mean curvature zero, and therefore it can be proved that they furnish a stationary value of the area functional not only in the class of rotationally symmetric curves but among all surfaces with fixed, rotationally symmetric boundaries.

We finally want to remind the reader that the weak form (8) in *2.1* of the Euler equations only requires that $u \in C^1(\Omega, \mathbb{R}^N)$, whereas relation (1) of this subsection, at least in the classical interpretation, needs that u be of class $C^2(\Omega, \mathbb{R}^N)$. This fact has been assumed in the statement of the theorem, and it will turn out that it is not automatically satisfied. We shall see in Section 3 that *there exist variational integrals $\mathscr{F}$ with weak extremals of class C^1 that are not of class C^2. In other words, there are weak extremals (or: weak solutions of the Euler equations) which are not extremals, that is, classical solutions of the Euler equation.*

2.3. Mollifiers. Variants of the Fundamental Lemma

The results of this section are not needed for the understanding of a large part of this volume and can, therefore, be skipped at the first reading. We want to develop several variants of the fundamental lemma which are important for the discussion of the so-called weak forms of differential equations. Particularly simple proofs can be given by means of *mollifiers* or *smoothing operators.*

We start by defining mollifiers on $L^p(\mathbb{R}^n)$, $1 \le p < \infty$; the general case of mollifiers on $L^p(\Omega)$, $\Omega \subset \mathbb{R}^n$, will be reduced to this special situation. In the following, we shall write

$$\int f(x)\,dx := \int_{\mathbb{R}^n} f(x)\,dx$$

and

$$\langle f, g\rangle := \int f(x)g(x)\,dx, \qquad \|f\|_{L^p} := \left\{\int |f(x)|^p\,dx\right\}^{1/p}$$

for functions $f \in L^p$, $g \in L^q$ with $\frac{1}{p} + \frac{1}{q} = 1$.

A function $k \in C^\infty(\mathbb{R}^n)$ will be called a *smoothing kernel* if it satisfies the four conditions[5]

$$k(x) = k(-x), \qquad \int k(x)\,dx = 1,$$

$$k(x) \geq 0 \quad \text{on } \mathbb{R}^n, \qquad k(x) = 0 \quad \text{for } |x| \geq 1.$$

An example is furnished by the function

$$k(x) = (1/K_0)\cdot k_0(x), \quad K_0 = \int k_0(x)\,dx,$$

where

$$k_0(x) = \begin{cases} \exp\dfrac{1}{|x|^2 - 1} & \text{for } |x| < 1, \\ 0 & \text{for } |x| \geq 1. \end{cases}$$

For a given smoothing kernel k we define the family of kernels $k_\varepsilon(x)$, $\varepsilon > 0$, by

$$k_\varepsilon(x) = \varepsilon^{-n} k\left(\frac{x}{\varepsilon}\right).$$

We have $k_\varepsilon \in C_c^\infty(\mathbb{R}^n)$, $\operatorname{supp} k_\varepsilon = \overline{B}_\varepsilon(0)$, and $\int k_\varepsilon(x)\,dx = 1$. Correspondingly, a family of *mollifiers* or *smoothing operators* S_ε is defined to be the convolution operators $S_\varepsilon u = k_\varepsilon * u$, $\varepsilon > 0$, that is,

$$(S_\varepsilon u)(x) = \int k_\varepsilon(x - y)u(y)\,dy = \int \varepsilon^{-n} k\left(\frac{x - y}{\varepsilon}\right)u(y)\,dy.$$

Note that these integrals have only to be extended over the ball $B(x, \varepsilon) = \{y \in \mathbb{R}^n : |x - y| < \varepsilon\}$. Introducing new coordinates z by $y = x - \varepsilon z$, we obtain

$$(S_\varepsilon u)(x) = \int k(z)u(x - \varepsilon z)\,dz,$$

where the integral has only to be carried out over $B(0, 1)$.

It is an easy exercise to prove that, for every $u \in L^p(\Omega)$, $1 \leq p \leq \infty$, the mollified function $S_\varepsilon u$ is of class $C^\infty(\mathbb{R}^n)$.

Lemma 1. (*i*) *For every $p \in [1, \infty]$, the mollifier S_ε defines a bounded linear operator on $L^p(\mathbb{R}^n)$ which satisfies $\|S_\varepsilon u\|_{L^p} \leq \|u\|_{L^p}$ for all $u \in L^p(\mathbb{R}^n)$.*

(*ii*) *If $u \in C_c^0(\mathbb{R}^n)$, then* $\lim_{\varepsilon\to 0} \|u - S_\varepsilon u\|_{C^0(\mathbb{R}^n)} = 0$.

(*iii*) *If $u \in L^p(\mathbb{R}^n)$, $1 \leq p < \infty$, then* $\lim_{\varepsilon\to 0} \|u - S_\varepsilon u\|_{L^p} = 0$.

---

[5] The assumption $k(x) = k(-x)$ guaranteeing property (iv) of Lemma 1 below is often omitted.

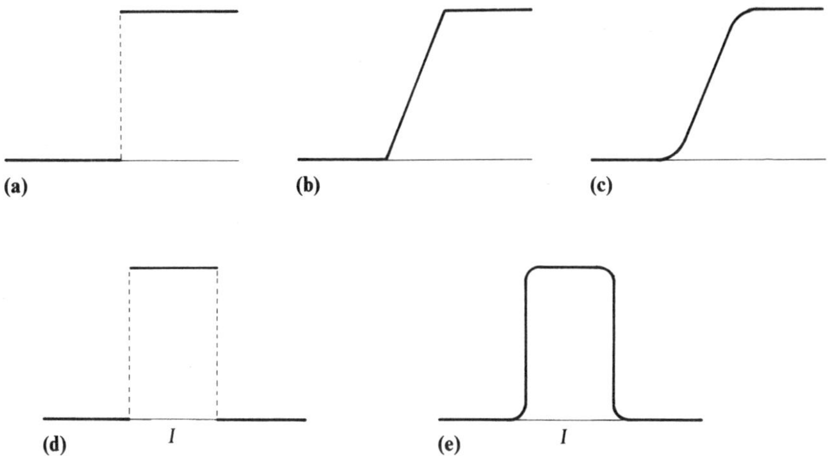

Fig. 12. **(a)** A Heaviside function with a jump discontinuity. **(b)** A Heaviside function mollified with the characteristic function of an interval. **(c)** Mollification of a Heaviside function with a C^∞-kernel. **(c)** The characteristic function of an interval I. **(e)** A smooth kernel k.

(*iv*) *If* $u \in L^p(\mathbb{R}^n)$ *and* $v \in L^q(\mathbb{R}^n)$, $\frac{1}{p} + \frac{1}{q} = 1$, *then* $\langle S_\varepsilon u, v\rangle = \langle u, S_\varepsilon v\rangle$

(*v*) *If* $\operatorname{supp} u \subset \Omega$, *then* $\operatorname{supp} S_\varepsilon u \subset \Omega_\varepsilon$, *where* $\Omega_\varepsilon := \{x \in \mathbb{R}^n : \operatorname{dist}(x, \bar{\Omega}) < \varepsilon\}$.

(*vi*) *For every* $u \in C^\infty(\mathbb{R}^n)$, *we have* $D_\alpha S_\varepsilon u = S_\varepsilon D_\alpha u$, *in other words, differentiation and mollification commute.*

Proof. (i) The linearity of the operator S_ε is obvious. Suppose now that $u \in L^p(\mathbb{R}^n)$. If $p = 1$, then

$$\|S_\varepsilon u\|_{L^1} = \int |(S_\varepsilon u)(x)|\,dx \le \iint k_\varepsilon(x-y)|u(y)|\,dy\,dx,$$

and by Fubini's theorem and $\int k_\varepsilon(x-y)\,dy = 1$, we obtain

$$\|S_\varepsilon u\|_{L^1} \le \int |u(y)|\cdot\left\{\int k_\varepsilon(x-y)\,dx\right\}dy = \int |u(y)|\,dy = \|u\|_{L^1}.$$

If $p = \infty$, we conclude from

$$|(S_\varepsilon u)(x)| \le \int k_\varepsilon(x-y)|u(y)|\,dy \le \|u\|_{L^\infty}\cdot\int k_\varepsilon(x-y)\,dy = \|u\|_{L^\infty}$$

that $\|S_\varepsilon u\|_{L^\infty} \le \|u\|_{L^\infty}$.

Finally, if $1 < p < \infty$ and $q = \frac{p}{p-1}$, Hölder's inequality yields

$$
\begin{aligned}
|(S_\varepsilon u)(x)|^p &= \left| \int k_\varepsilon(x-y)u(y)\,dy \right|^p \\
&= \left| \int k_\varepsilon^{1-1/p}(x-y)k_\varepsilon^{1/p}(x-y)u(y)\,dy \right|^p \\
&\le \left(\int k_\varepsilon(x-y)\,dy \right)^{p-1} \int k_\varepsilon(x-y)|u(y)|^p\,dy \\
&= (S_\varepsilon |u|^p)(x),
\end{aligned}
$$

whence

$$
\begin{aligned}
\|S_\varepsilon u\|_{L^p}^p &= \int |(S_\varepsilon u)(x)|^p\,dx \le \iint k_\varepsilon(x-y)|u(y)|^p\,dy\,dx \\
&= \int |u(y)|^p \left\{ \int k_\varepsilon(x-y)\,dx \right\} dy = \int |u(y)|^p\,dy = \|u\|_{L^p}^p .
\end{aligned}
$$

(ii) If $u \in C_c^0(\mathbb{R}^n)$, there exists a compact set $K' \subset \mathbb{R}^n$ such that $u(x) = 0$ for $x \notin K'$. This implies the existence of a compact set K with $K' \subset K$ such that $(S_\varepsilon u)(x) = 0$ for $x \notin K$ and $0 < \varepsilon < 1$, whence

$$
\|u - S_\varepsilon u\|_{C^0(\mathbb{R}^n)} = \|u - S_\varepsilon u\|_{C^0(K)} \quad \text{for } 0 < \varepsilon < 1 .
$$

On the other hand, we infer for $x \in K$ that

$$
\begin{aligned}
|u(x) - (S_\varepsilon u)(x)| &= \left| \int k_\varepsilon(x-y)\{u(y) - u(x)\}\,dy \right| \\
&\le \int k_\varepsilon(x-y)|u(y) - u(x)|\,dy \\
&\le \sup\{|u(y) - u(x)| \colon y \in K \cap B_\varepsilon(x)\} .
\end{aligned}
$$

Since u is uniformly continuous on K, it follows that $\lim_{\varepsilon\to 0} \|u - S_\varepsilon u\|_{C^0(\mathbb{R}^n)} = 0$.

(iii) Suppose that $u \in L^p(\mathbb{R}^n)$, $1 < p < \infty$, and let v be an arbitrary function in $C_c^\infty(\mathbb{R}^n)$. Then, by (i),

$$
\begin{aligned}
\|u - S_\varepsilon u\|_{L^p} &\le \|u - v\|_{L^p} + \|v - S_\varepsilon v\|_{L^p} + \|S_\varepsilon(u - v)\|_{L^p} \\
&\le 2\|u - v\|_{L^p} + \|v - S_\varepsilon v\|_{L^p} .
\end{aligned}
$$

It is an immediate consequence of Lusin's theorem that $C_c^0(\mathbb{R}^n)$ is a dense subset of $L^p(\mathbb{R}^n)$. Hence, for a given $\delta > 0$, there is $v \in C_c^0(\mathbb{R}^n)$ such that $\|u - v\|_{L^p} < \delta/4$. As in (ii), we can find a compact set K such that $v(x) = 0$ and $(S_\varepsilon v)(x) = 0$ for $0 < \varepsilon < 1$ and $x \notin K$. By virtue of (ii), we infer that

$$
\|v - S_\varepsilon v\|_{L^p} = \|v - S_\varepsilon v\|_{L^p(K)} \le (\text{meas } K)^{1/p} \cdot \|v - S_\varepsilon v\|_{C^0(\mathbb{R}^n)} < \delta/2
$$

if $0 < \varepsilon < \varepsilon_0$, for some $\varepsilon_0 > 0$. Altogether,

$$\|u - S_\varepsilon u\|_{L^p} < \delta \quad \text{if } 0 < \varepsilon < \varepsilon_0 .$$

(iv) If $u \in L^p$, $v \in L^q$, and $\frac{1}{p} + \frac{1}{q} = 1$, then we obtain $S_\varepsilon u \in L^p$, $S_\varepsilon v \in L^q$, $u \cdot S_\varepsilon v \in L^1$, $S_\varepsilon u \cdot v \in L^1$, $u \cdot v \in L^1$, and furthermore, by Fubini's theorem,

$$\begin{aligned}
\langle u, S_\varepsilon v\rangle &= \int u(x)\left(\int k_\varepsilon(x-y)v(y)\,dy\right)dx \\
&= \iint k_\varepsilon(x-y)u(x)v(y)\,dx\,dy \\
&= \int v(y)\left(\int k_\varepsilon(y-x)u(x)\,dx\right)dy = \langle S_\varepsilon u, v\rangle .
\end{aligned}$$

(v) is an obvious consequence of the definition of S_ε.

(vi) Let u be of class $C^\infty(\mathbb{R}^n)$. Then we have

$$\begin{aligned}
(D_\alpha S_\varepsilon u)(x) &= \frac{\partial}{\partial x^\alpha}\int k_\varepsilon(x-y)u(y)\,dy = \int_{B(x,\varepsilon)} \frac{\partial}{\partial x^\alpha} k_\varepsilon(x-y)u(y)\,dy \\
&= -\int_{B(x,\varepsilon)} \frac{\partial}{\partial y^\alpha} k_\varepsilon(x-y)u(y)\,dy = \int_{B(x,\varepsilon)} k_\varepsilon(x-y)\frac{\partial}{\partial y^\alpha}u(y)\,dy \\
&= \int k_\varepsilon(x-y)(D_\alpha u)(y)\,dy = (S_\varepsilon D_\alpha u)(x).
\end{aligned}$$

□

The relation (iii) implies that $C^\infty(\mathbb{R}^n)$ is dense in $L^p(\mathbb{R}^n)$ if $1 \le p < \infty$. Again usin's theorem, we can infer that $C_c^0(\Omega)$ is dense in $L^p(\Omega)$, provided that $\mathbb{R}^n$ and $1 \le p < \infty$. The set $C_c^0(\Omega)$ can be embedded into $C_c^0(\mathbb{R}^n)$ by setting $= 0$ for $x \notin \Omega$ and $u \in C_c^0(\Omega)$. Then we infer from (ii) and (v) the following

Corollary. *Let Ω be an open set of $\mathbb{R}^n$ and $1 \le p < \infty$. Then $C_0^\infty(\Omega)$ is dense in $L^p(\Omega)$ (with respect to the norm of $L^p(\Omega)$).*

Lemma 2. (*i*) *Let $\Omega' \subset\subset \Omega$ and $1 \le p \le \infty$. Then there is an $\varepsilon_0 > 0$ such that $\|S_\varepsilon u\|_{L^p(\Omega')} \le \|u\|_{L^p(\Omega)}$ for $0 < \varepsilon < \varepsilon_0$ and $u \in L^p(\Omega)$.*

(ii) *If $\Omega' \subset\subset \Omega$, $u \in L^p(\Omega)$, and $1 \le p < \infty$, then*

$$\lim_{\varepsilon \to 0} \|u - S_\varepsilon u\|_{L^p(\Omega')} = 0 .$$

(iii) *If $u \in L^1(\Omega)$, $\Omega \subset \mathbb{R}^n$, and $\varphi \in C_c^\infty(\Omega)$, then*

$$\langle S_\varepsilon u, \varphi\rangle = \langle u, S_\varepsilon \varphi\rangle \quad \textit{for } 0 < \varepsilon < \varepsilon_0$$

and for sufficiently small $\varepsilon_0 > 0$.

Proof. This lemma follows immediately from Lemma 1 if we embeded $L^p(\Omega)$ into $L^p(\mathbb{R}^n)$ by setting $u(x) = 0$ for $x \notin \Omega$, and $u \in L^p(\Omega)$. □

Lemma 3 (General form of the fundamental lemma). *Suppose that f is of class $L^1(\Omega)$ and satisfies*

$$(*) \qquad \int_\Omega f(x)\eta(x)\,dx \geq 0 \quad \textit{for all } \eta \in C_c^\infty(\Omega) \textit{ with } \eta \geq 0$$

or

$$(**) \qquad \int_\Omega f(x)\eta(x)\,dx = 0 \quad \textit{for all } \eta \in C_c^\infty(\Omega).$$

Then we obtain $f(x) \geq 0$ or $f(x) = 0$, respectively a.e. on Ω.

Proof. Suppose that $(*)$ holds, and choose an open set Ω' with $\Omega' \subset\subset \Omega$. Then there is a number $\varepsilon_0 > 0$ such that $\Omega'_\varepsilon \subset\subset \Omega$ for $0 < \varepsilon < \varepsilon_0$. Hence, we infer from $(*)$ that $(S_\varepsilon f)(x) = \int k_\varepsilon(x - y) f(y)\,dy \geq 0$ for all $x \in \Omega'$.

On the other hand, the relation

$$\|f - S_\varepsilon f\|_{L^1} \to 0 \quad \text{as } \varepsilon \to 0$$

implies the existence of a sequence $\{\varepsilon_k\}$ with $0 < \varepsilon_k < \varepsilon_0$ and $\varepsilon_k \to 0$ as $k \to \infty$, such that $(S_{\varepsilon_k} f)(x) \to f(x)$ holds a.e. on $\mathbb{R}^n$, whence we conclude that $f(x) \geq 0$ is satisfied a.e. on Ω'. Since Ω' is an arbitrary open set with $\Omega' \subset\subset \Omega$, we finally obtain that $f(x) \geq 0$ a.e. on Ω.

The second assertion of the lemma follows readily from the first one. □

The next result is a variant of the fundamental lemma which is often called *Du Bois–Reymond's lemma.*

Lemma 4. *Suppose that Ω is a domain in $\mathbb{R}^n$ and that $f \in L^1(\Omega)$ satisfies*

$$\int_\Omega f(x) D_\alpha \eta(x)\,dx = 0, \quad 1 \leq \alpha \leq n,$$

for all $\eta \in C_c^\infty(\Omega)$. Then $f(x)$ coincides with a constant function a.e. on Ω.

Proof. As in the proof of Lemma 3, there is a subdomain $\Omega' \subset\subset \Omega$ and a number $\varepsilon_0 > 0$ such that $\Omega'_\varepsilon \subset\subset \Omega$ and $S_\varepsilon \eta \in C_c^\infty(\Omega)$, provided that $0 < \varepsilon < \varepsilon_0$ and $\eta \in C_c^\infty(\Omega')$. We again set $f(x) = 0$ for $x \notin \Omega$. For every $\eta \in C_c^\infty(\Omega')$ and $0 < \varepsilon < \varepsilon_0$ we then infer

$$\begin{aligned} 0 &= \int_\Omega f(x)(D_\alpha S_\varepsilon \eta)(x)\,dx = \langle f, D_\alpha S_\varepsilon \eta\rangle = \langle f, S_\varepsilon D_\alpha \eta\rangle = \langle S_\varepsilon f, D_\alpha \eta\rangle \\ &= -\langle D_\alpha S_\varepsilon f, \eta\rangle, \end{aligned}$$

whence, by virtue of the fundamental lemma, we may conclude that $D_\alpha S_\varepsilon f = 0$, $\alpha = 1, \ldots, n$, holds on Ω'. Thus there exists a number $c(\varepsilon)$ such that $(S_\varepsilon f)(x) = c(\varepsilon)$ is satisfied a.e. on Ω' if $0 < \varepsilon < \varepsilon_0$.

Since $\lim_{\varepsilon \to 0} \|f - S_\varepsilon f\|_{L^1} = 0$, there is a sequence $\{\varepsilon_k\}$ with $0 < \varepsilon_k < \varepsilon_0$ and $\lim_{k \to \infty} \varepsilon_k = 0$ such that $(S_{\varepsilon_k} f)(x) \to f(x)$ a.e. on Ω', as $k \to \infty$. It follows that $c = \lim_{k \to \infty} c(\varepsilon_k)$ exists, and that $f(x) = c$ holds a.e. on Ω'. Since Ω' is an arbitrary subdomain of the domain Ω, we finally conclude that $f(x) = \text{const}$ a.e. on Ω. □

Another variant of the fundamental lemma is provided by

Lemma 5. *Let Ω be an open set in $\mathbb{R}^n$, $f, g \in L^1(\Omega)$, and suppose that*

$$(*) \qquad \int_\Omega f(x)\varphi(x)\,dx = 0$$

holds for all $\varphi \in C_c^\infty(\Omega)$ subject to the subsidiary condition

$$(**) \qquad \int_\Omega g(x)\varphi(x)\,dx = 0\,.$$

Then there exists a number $\lambda \in \mathbb{R}$ such that

$$f(x) = \lambda g(x) \quad \text{a.e. on } \Omega\,.$$

In particular, if meas $\Omega < \infty$, *and if f is a function of class $L^1(\Omega)$ satisfying $(*)$ for all functions $\varphi \in C_c^\infty(\Omega)$ with mean value zero, then $f(x)$ coincides a.e. in Ω with a constant function.*

Proof. For $f, g \in L^2(\Omega)$, we have the following geometric interpretation of the assertion: Suppose that f is orthogonal in $L^2(\Omega)$ to all functions φ which are perpendicular to g and of class $C_c^\infty(\Omega)$. (As $C_c^\infty(\Omega)$ is dense in $L^2(\Omega)$, this means that f is orthogonal to all functions which, in turn, are orthogonal to g.) Then it is claimed that f is a multiple of g, i.e., $f = \lambda g$ for some $\lambda \in \mathbb{R}$. This interpretation suggests also the proof that even works in the general case.

If $(**)$ holds for all $\varphi \in C_c^\infty(\Omega)$, then the assertion follows at once from Lemma 3. Hence we can assume that there is some $\psi \in C_c^\infty(\Omega)$ such that

$$\int_\Omega g(x)\psi(x)\,dx \neq 0$$

and, multiplying ψ by some suitable constant, we have that there is some $\psi \in C_c^\infty(\Omega)$ with

$$\langle g, \psi \rangle = 1\,.$$

Now we write any $\eta \in C_c^\infty(\Omega)$ in the form

$$\eta = \langle g, \eta \rangle \psi + \{\eta - \langle g, \eta \rangle \psi\}\,,$$

this way decomposing η into a sum $\varphi_0 + \varphi$ for two $C_c^\infty(\Omega)$-functions $\varphi_0 := \langle g, \eta\rangle\psi$ and $\varphi := \eta - \langle g, \eta\rangle\psi$, the second of which is orthogonal to g, i.e.

$$\langle g, \varphi\rangle = 0.$$

By assumption, we obtain

$$\langle f, \varphi\rangle = 0.$$

Setting $\lambda := \langle f, \psi\rangle$, we can write this relation in the form

$$\langle f - \lambda g, \eta\rangle = 0,$$

which is satisfies for all $\eta \in C_c^\infty(\Omega)$. As λ is a fixed number independent of η, we can apply Lemma 3 to $f - \lambda g$, thus obtaining $f - \lambda g = 0$, or $f = \lambda g$. This proves the first part of the assertion. The second part follows from the first by choosing $g(x) \equiv 1$. □

2.4. Natural Boundary Conditions

In addition to the assumptions of *2.2*, let us furthermore suppose that $\partial\Omega$ is a manifold of class C^1, $u \in C^1(\bar{\Omega}, \mathbb{R}^N)$, and that instead of *2,1*, (8), the function u is even satisfying the stronger relation

$$(1) \qquad \delta\mathscr{F}(u, \varphi) = 0 \quad \text{for all } \varphi \in C^1(\bar{\Omega}, \mathbb{R}^N).$$

The last equation is, for instance, a consequence of the minimum property

$$(2) \qquad \mathscr{F}(u) \le \mathscr{F}(u + \varphi) \quad \text{for all } \varphi \in C^1(\bar{\Omega}, \mathbb{R}^N) \text{ with } \|\varphi\|_{C^1(\bar{\Omega})} < \delta_0,$$

which is supposed to hold for some $\delta_0 > 0$.

Proposition. *If* $\partial\Omega \in C^1$, $F \in C^2(\mathscr{U})$, $u \in C^1(\bar{\Omega}, \mathbb{R}^N) \cap C^2(\Omega, \mathbb{R}^N)$, *and if* $\delta\mathscr{F}(u, \varphi) = 0$ *for all* $\varphi \in C^1(\bar{\Omega}, \mathbb{R}^N)$, *then* u *is an extremal of* $\mathscr{F}$ *which, on* $\partial\Omega$, *satisfies the "natural boundary conditions"*

$$(3) \qquad \nu_\alpha F_{p_\alpha^i}(x, u, Du) = 0, \quad i = 1, \ldots, N.$$

Here $\nu(x) = (\nu_1(x), \ldots, \nu_n(x))$ *denotes the exterior normal of* $\partial\Omega$ *at the point* $x \in \partial\Omega$.

Remark. It will be apparent from the proof that, instead of the assumption $F \in C^2(\mathscr{U})$, it suffices to assume that the partial derivatives F_u and F_p exist and that $F_u \in C^0(\mathscr{U})$, $F_p \in C^1(\mathscr{U})$.

Proof. (i) We first prove assertion (3) under the slightly stronger assumption that $u(x)$ is of class $C^2(\bar{\Omega}, \mathbb{R}^N)$. Then a partial integration implies that

$$(4) \quad \delta\mathscr{F}(u, \varphi) = \int_\Omega L_F(u) \cdot \varphi \, dx + \int_{\partial\Omega} \nu_\alpha(x) F_{p_\alpha^i}(x, u(x), Du(x)) \varphi^i(x) \, d\mathscr{H}^{n-1}(x).$$

Here $d\mathscr{H}^{n-1}$ stands for the $(n-1)$-dimensional area element of $\partial\Omega$. By the Theorem in *2.2*, we know already that $L_F(u) = 0$ on Ω, whence

(5) $$\delta\mathscr{F}(u, \varphi) = \int_{\partial\Omega} \nu_\alpha(x) F_{p_\alpha^i}(x, u(x), Du(x))\varphi^i(x)\, d\mathscr{H}^{n-1}(x)$$

for all $\varphi \in C^1(\bar{\Omega}, \mathbb{R}^N)$.

By an obvious generalization of the "fundamental lemma" we infer from (1) and (5) the relations

$$\nu_\alpha(x) F_{p_\alpha^i}(x, u(x), Du(x)) = 0 \quad \text{for all } x \in \partial\Omega,\ 1 \le i \le N,$$

which, in vectorial form, can be written as

(6) $$\nu(x) \cdot F_p(x, u(x), Du(x)) = 0 \quad \text{on } \partial\Omega.$$

(ii) If only $u \in C^1(\bar{\Omega}, \mathbb{R}^N) \cap C^2(\Omega, \mathbb{R}^N)$ is assumed then we are not allowed to perform the partial integration on Ω, but we have to restrict ourselves to a subdomain $\Omega' \subset\subset \Omega$ with $\partial\Omega' \in C^1$. We then obtain

(7) $$\delta\mathscr{F}_{\Omega'}(u, \varphi) = \int_{\partial\Omega'} \nu_\alpha(x) F_{p_\alpha^i}(x, u(x), Du(x))\varphi^i(x)\, d\mathscr{H}^{n-1}(x)$$

for all $\varphi \in C^1(\bar{\Omega}, \mathbb{R}^N)$.

Now we choose for Ω' a sequence of smoothly bounded subdomains $\Omega_k \subset\subset \Omega$ with $\Omega_k \nearrow \Omega$ such that $\partial\Omega_k$ converges in C^1 to $\partial\Omega$ as $k \to \infty$.[6] Then we conclude that $\text{meas}(\Omega - \Omega_k) \to 0$ whence

$$\delta\mathscr{F}_{\Omega_k}(u, \varphi) \to \delta\mathscr{F}(u, \varphi) \quad \text{as } k \to \infty$$

follows for each test function $\varphi \in C^1(\bar{\Omega}, \mathbb{R}^N)$. In conjunction with (7) we arrive at (5) from where the proof proceeds as in (i), and the Proposition is proved. □

We see from the minimum property (2) that free (or natural) boundary conditions occur whenever we have a minimum within a class of functions, the boundary values of which are not fixed. For vector-valued extremals we often meet problems where some of the components are fixed whereas others are allowed to move freely. Then only the latter ones will satisfy natural boundary conditions. We shall later discuss relevant results for important special problems.

Let us consider some examples.

[1] The natural boundary condition associated with the Dirichlet integral

$$\mathscr{D}(u) = \frac{1}{2}\int_\Omega |Du|^2\, dx$$

---

[6] This means that, *locally*, the graphs representing $\partial\Omega_k$ converge in C^1 to the graphs of $\partial\Omega$.

is the so-called *Neumann boundary condition*

$$\frac{\partial u}{\partial \nu} = 0 \quad \text{on } \partial\Omega,$$

where ν denotes the exterior normal to $\partial\Omega$.

Clearly, this is also the natural boundary condition for every variational integral of the kind

$$\int_\Omega \{\tfrac{1}{2}|Du|^2 + g(x, u)\}\, dx .$$

[2] The natural boundary condition of the area functional

$$\mathscr{A}(u) = \int_\Omega \sqrt{1 + |Du|^2}\, dx$$

as well as of every functional of the type

$$\int_\Omega \{\sqrt{1 + |Du|^2} + g(x, u)\}\, dx$$

is given by

$$\nu \cdot Tu = 0 \text{ on } \partial\Omega, \quad Tu := \frac{Du}{\sqrt{1 + |Du|^2}},$$

which has a remarkable geometric interpretation.

Let α be the angle between the normal

$$\boldsymbol{n} = \frac{1}{\sqrt{1 + |Du|^2}}(Du, -1)$$

of the hypersurface $(x, u(x))$, $x \in \bar{\Omega}$, and the normal $\mathbf{v} = (\nu, 0)$ to the cylinder $\partial\Omega \times \mathbb{R}$ in $\mathbb{R}^{n+1}$. Then

$$\cos \alpha = \mathbf{v} \cdot \boldsymbol{n} = \nu \cdot Tu,$$

and the free boundary condition $\nu \cdot Tu = 0$ says that the minimal surface $(x, u(x))$ which satisfies $u \in C^1(\bar{\Omega}) \cap C^2(\Omega)$,

$$\delta\mathscr{A}(u, \varphi) = 0 \quad \text{for all } \varphi \in C^1(\bar{\Omega}),$$

intersects the cylinder $\partial\Omega \times \mathbb{R}$ perpendicularly.

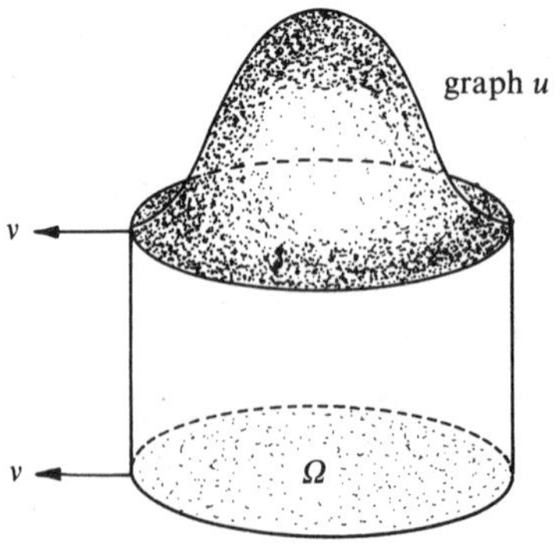

Fig. 13. The Neumann condition.

3. Remarks on the Existence and Regularity of Minimizers

In this section we have collected some simple but instructive remarks about the existence and regularity problem of minimizers, mainly for one-dimensional variational problems. A more refined discussion leading to the so-called *direct methods of the calculus of variations and* to the *regularity theory for weak extremals of multiple integrals* will be given in another treatise.

We begin in *3.1* by presenting some examples of variational integrals that have weak C^1-extremals which are not of class C^2. Then we shall show that weak extremals of *elliptic Lagrangians* for one-dimensional integrals are necessarily of class C^2. Our proof of this regularity result is based on some reasoning of Du Bois–Reymond and Hilbert using an *integrated form of the Euler equation.*

Actually, it is not a priori clear whether C^1, C^2 or some other function space is the natural setting where a first-order variational problem is to be solved. In fact it is part of the problem to define the *class of admissible functions* where the functional is to be minimized. In general it is not true that every "reasonable problem" (whatever is meant by "reasonable") has a solution, and if there exists a solution, one cannot take it for granted that this solution is smooth: there are minimizers which do not satisfy Euler's equation. In *3.2* we shall discuss a few examples of variational problems without solutions.

Finally in *3.3* we shall consider *broken extremals* which are continuous, piecewise smooth functions that satisfy Euler's equations except for finitely many points. At such *corners* a broken extremal must satisfy the *Weierstrass–Erdmann corner conditions.*

3.1. Weak Extremals Which Do Not Satisfy Euler's Equation. A Regularity Theorem for One-Dimensional Variational Problems

The following examples show that there are weak extremals and even weak minimizers which are not smooth solutions of Euler's equations.

[1] A very trivial example is given by the functional

$$\mathscr{F}(v) = \int_{-1}^{1} [Dv(x) - 2|x|]^2 \, dx, \quad n = N = 1,$$

which is minimized by the functions $u(x) = x|x| + \text{const}$ that are of class C^1 on $[-1, 1]$ but not in C^2 on $(-1, 1)$. Thus $u(x)$ will satisfy (8) in *2.1* but not (1).

However, this example is not quite satisfactory since $F_p \notin C^1$. Therefore we will present two more refined examples of Lagrangians that are real analytic (and even polynomial), and for which there exist solutions $u \in C^1(\bar{\Omega}, \mathbb{R}^N)$ of (8) in *2.1* that are not of class $C^2(\Omega, \mathbb{R}^N)$. Thus, even for nicely behaving Lagrangians, weak extremals need not be "classical" extremals.

[2] Let $n = N = 1$ and $F(x, u, p) = u^2(2x - p)^2$. Then

$$u(x) := \begin{cases} 0 & \text{for } -1 \leq x \leq 0, \\ x^2 & \text{for } 0 \leq x \leq 1 \end{cases}$$

is a function of class C^1 which is the (uniquely) determined minimizer of

$$\mathscr{F}(v) = \int_{-1}^{1} v^2(x)(2x - Dv(x))^2\, dx,$$

within the class of functions $v \in C^1([-1, 1])$ which satisfy the boundary conditions $v(-1) = 0$ and $v(1) = 1$. Obviously, $u(x)$ is not of class C^2.

[3] Let $n = 2$, $N = 1$, $\Omega = \{(x, y): x^2 + y^2 < 1\}$, and

$$\mathscr{F}(u) = \tfrac{1}{2}\int_\Omega (u_x^2 - u_y^2)\, dx\, dy.$$

Formally, the Euler equation of this integral is the wave equation

$$u_{xx} - u_{yy} = 0.$$

Consider now a real valued function $\psi(t)$ of class $C^1(\mathbb{R})$ which is nowhere twice differentiable. Then $u(x, y) := \psi(x - y)$ is of class $C^1(\Omega)$, but the second partial derivatives of u exist nowhere on Ω. Because of $u_x = -u_y$ we obtain $u_x\varphi_x - u_y\varphi_y = u_x\varphi_y - u_y\varphi_x$ for each $\varphi \in C_c^\infty(\Omega)$, and therefore

$$\begin{aligned} \delta\mathscr{F}(u, \varphi) &= \int_\Omega (u_x\varphi_x - u_y\varphi_y)\, dx\, dy = \int_\Omega (u_x\varphi_y - u_y\varphi_x)\, dx\, dy \\ &= \int_\Omega u(\varphi_{yx} - \varphi_{xy})\, dx\, dy = 0. \end{aligned}$$

Note that the Hessian $\begin{pmatrix} F_{pp} & F_{qp} \\ F_{qp} & F_{qq} \end{pmatrix} = \begin{pmatrix} 1 & 0 \\ 0 & -1 \end{pmatrix}$ of the integrand $F(q, p) = p^2 - q^2$ in this example is indefinite, whereas the Hessian $F_{pp}(x, u, p) = 2u^2$ of the integrand F in [2] is positive semidefinite.

The situation is even more desperate than indicated by the previous two examples. For instance, there exist real analytic Lagrangians F such that the corresponding variational integral $\mathscr{F}$ with respect to fixed boundary values is minimized by Lipschitz functions, but by no C^1-function, and a fortiori by no C^2-function. This phenomenon was discovered by Euler, and he viewed it as rather paradoxical.[7]

---

[7](E735) *De insigni paradoxo quod in analysi maximorum et minimorum occurit*, Mem. Acad. Sci. St. Pétersbourg **3** (1811), and Euler [1] Ser. I, vol. 25, 286–292. The paper was written in 1779.

Here and in the future, the letter E, as in (E735), refers to Eneström's catalogue of Euler's papers; see: Jahresberichte der DMV, Ergänzungsband IV, 1910–1913.

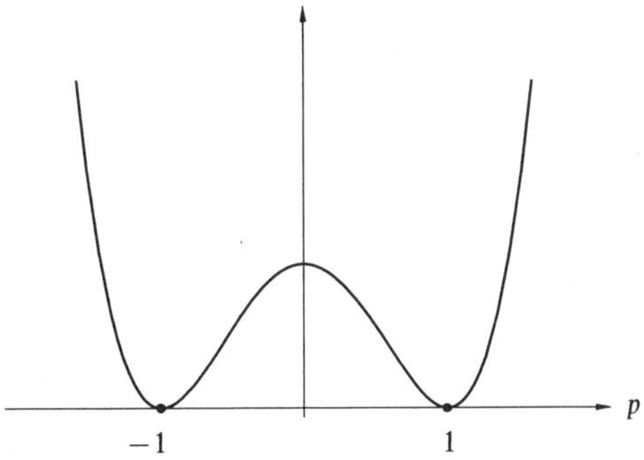

Fig. 14. $F(p) = (p^2 - 1)^2$.

4 *Euler's paradox*. Consider, for instance, the Lagrangian

$$F(p) = (p^2 - 1)^2.$$

It is easy to see that the infimum of the corresponding variational integral

$$\mathscr{F}(u) = \int_0^1 (u'^2 - 1)^2 \, dx$$

on the class $\mathscr{C}$ of functions $u \in C^1([0, 1])$ satisfying $u(0) = 0$ and $u(1) = 0$ is zero, and that this infimum cannot be attained within $\mathscr{C}$. However, there are uncountably many zig-zag functions $u(x)$ on $I = [0, 1]$ fulfilling the boundary conditions $u(0) = 0$ and $u(1) = 0$. In fact, any Lipschitz function $u(x)$ satisfying $u'(x) = \pm 1$ a.e. on I furnishes the minimal value zero for $\mathscr{F}$.

This example shows that the spaces C^2 or C^1 are by no means the natural classes where every variational problem is to be tackled. In fact, it is part of the problem to formulate its setting and to find out what should be a suitable class of "admissible functions" where $\mathscr{F}$ is to be minimized. Problems with "broken minimizers" are by no means an artificial matter but appear naturally in applications. For instance, sailing against the wind, the sailor has to tack, that is, he has to follow a zig-zag line by switching between two most favorable angles in order to use the wind in the best way. An amusing discussion of this and related problems can be found in L.C. Young's treatise [1], pp. 155–160.

5 Let us conclude our discussion by considering the Lagrangian

$$F(z, p) = (1 + z^2)\{1 + (p^2 - 1)^2\}$$

with the corresponding variational integral

$$\mathscr{F}(u) = \int_0^1 (1 + u^2)\{1 + (u'^2 - 1)^2\} \, dx.$$

Subjecting the admissible functions u to the boundary conditions $u(0) = 0$ and $u(1) = 0$, we easily verify that the infimum of $\mathscr{F}$ is one among C^1-functions, or Lipschitz functions, or even among absolutely continuous functions. However, the infimum is attained for no such function since

$\mathscr{F}(u) = 1$ would both imply $u(x) = 0$ and $u'(x) = \pm 1$ a.e. on $[0, 1]$ which is impossible. In other words, if one insists in obtaining a minimizer, one has to enlarge the class of admissible "functions" even further by admitting "infinitesimal" zig-zags. For this approach we refer the reader to the discussion in L.C. Young [1].

Now we turn to the proof of a regularity theorem for extremals of one-dimensional variational problems the basic ideas of which are due to Du Bois–Reymond and Hilbert. Consider a variational integral

$$(1) \qquad \mathscr{F}(u) := \int_a^b F(x, u(x), u'(x))\, dx$$

whose Lagrangian $F(x, u, p)$ is a C^1-function on $I \times \mathbb{R} \times \mathbb{R}^N$ where $I := [a, b]$. The functional $\mathscr{F}$ is defined on the class $\mathrm{Lip}(I, \mathbb{R}^N)$ of Lipschitz functions $u: I \to \mathbb{R}^N$ which satisfy a Lipschitz condition

$$|u(x) - u(y)| \le L|x - y| \quad \text{for all } x, y \in I,$$

where the constant L may depend on u.

Definition. *A function* $u \in \mathrm{Lip}(I, \mathbb{R}^N)$ *is said to be a* weak Lipschitz-extremal *if it satisfies*

$$(2) \quad \int_a^b [F_u(x, u, u')\cdot \varphi + F_p(x, u, u')\cdot \varphi']\, dx = 0 \quad \textit{for all } \varphi \in C_c^\infty((a, b), \mathbb{R}^N).$$

Remark 1. This definition is an extension of the notion of a weak extremal introduced in *2.1* where we required that a weak extremal of $\mathscr{F}$ be of class C^1. There are various other generalizations of this notion. In this volume we want to agree upon that a weak extremal is always of class C^1. However, if we consider solutions u of (2) which are not C^1 but of a larger class $\mathscr{C}$ such as $D^1(I, \mathbb{R}^N)$ or $\mathrm{Lip}(I, \mathbb{R}^N)$, we call them *weak* $\mathscr{C}$*-extremals*.

Remark 2. If u is a (local) minimizer of $\mathscr{F}$ in the class of all functions $v \in \mathrm{Lip}(I, \mathbb{R}^N)$ with prescribed boundary values $v(a)$ and $v(b)$, one proves as in Section 2 that (2) holds true.

An integration by parts yields

$$\int_a^b F_u(x, u, u')\cdot \varphi\, dx = -\int_a^b \left[\int_a^x F_u(t, u(t), u'(t))\, dt\right]\cdot \varphi'(x)\, dx$$

for all $\varphi \in C_c^\infty((a, b), \mathbb{R}^N)$, and therefore we can write (2) in the equivalent form

$$(3) \qquad \int_a^b \left[F_p(x, u, u') - \int_a^x F_u(t, u(t), u'(t))\, dt\right]\cdot \varphi'(x)\, dx = 0$$

$$\text{for all } \varphi \in C_c^\infty((a, b), \mathbb{R}^N).$$

By Du Bois–Reymond's lemma (cf. *2.3*, Lemma 4) we deduce from (3) the following result.

Proposition 1. *For any weak Lipschitz-extremal u of $\mathcal{F}$ there exists a constant vector $c \in \mathbb{R}^N$ such that*

$$F_p(x, u(x), u'(x)) = c + \int_a^x F_u(t, u(t), u'(t))\, dt \tag{4}$$

holds true for almost all $x \in (a, b)$.

Since $F_u(t, u, p)$ is a continuous function of the $2N + 1$ variables $(t, u, p) \in I \times \mathbb{R}^N \times \mathbb{R}^N$, we infer that $F_u(t, u(t), u'(t))$ is a bounded measurable function of the variable $t \in I$. Therefore the right-hand side of (4) is an *absolutely continuous* function of $x \in I$, and we obtain

Proposition 2. *Any weak Lipschitz-extremal u of $\mathcal{F}$ satisfies the Euler equation*

$$\frac{d}{dx} F_p(x, u(x), u'(x)) = F_u(x, u(x), u'(x)) \quad \text{a.e. on } I. \tag{5}$$

Remark 3. Note that in general we are not allowed to write the left-hand side of (5) in the differentiated form

$$\frac{d}{dx} F_p(x, u(x), u'(x)) = F_{px}(x, u(x), u'(x)) + F_{pu}(x, u(x), u'(x)) \cdot u'(x) + F_{pp}(x, u(x), u'(x)) \cdot u''(x).$$

We denote equation (4) as *Du Bois–Reymond's equation* or as *integrated form of Euler's equation.*

We shall use Du Bois–Reymond's equation to prove regularity of weak Lipschitz-extremals provided that the Lagrangian F satisfies a suitable *ellipticity condition.* We shall proceed in two steps.

Proposition 3. *Let u be a weak extremal of $\mathcal{F}$, and suppose that F_p is of class C^1 on $I \times \mathbb{R}^N \times \mathbb{R}^N$ and that the Hessian matrix $F_{pp}(x, u(x), u'(x))$ is invertible for all $x \in (a, b)$. Then $u(x)$ is of class C^2 on (a, b).*

Proof. By definition a weak extremal u is of class C^1. Hence the right-hand side

$$\pi(x) := c + \int_a^x F_u(t, u(t), u'(t))\, dt \tag{6}$$

of (4) is of class $C^1(I)$, and the left-hand side $F_p(x, u(x), u'(x))$ is continuous on I whence we obtain

$$F_p(x, u(x), u'(x)) - \pi(x) = 0 \quad \text{for all } x \in I. \tag{7}$$

Now we define a mapping $G: I \times \mathbb{R}^N \to \mathbb{R}^N$ by

$$G(x, p) := F_p(x, u(x), p) - \pi(x), \tag{8}$$

which is of class $C^1(I \times \mathbb{R}^N, \mathbb{R}^N)$ and satisfies

$$\det G_p(x, u'(x)) \neq 0 \quad \text{for all } x \in (a, b).$$

Moreover we infer from (7) that

$$G(x, u'(x)) = 0 \quad \text{for all } x \in I.$$

Then the implicit function theorem implies that $u'(x)$ is of class C^1 on (a, b). □

Proposition 4. *Assume that F_p is of class C^1 on $I \times \mathbb{R}^N \times \mathbb{R}^N$ and that $F_{pp}(x, z, p)$ is positive definite on $I \times \mathbb{R}^N \times \mathbb{R}^N$. Then every weak Lipschitz-extremal $u: I \to \mathbb{R}^N$ of $\mathscr{F}$ is necessarily of class $C^2(I, \mathbb{R}^N)$.*

Proof. Consider a mapping $\Phi: (x, z, p) \to (x, z, q)$ of $I \times \mathbb{R}^N \times \mathbb{R}^N$ into itself which is defined by

$$\Phi(x, z, p) := (x, z, F_p(x, z, p)). \tag{9}$$

We will show in 7,*1* that Φ yields a C^1-diffeomorphism of $I \times \mathbb{R}^N \times \mathbb{R}^N$ onto its image $\mathscr{U} := \Phi(I \times \mathbb{R}^N \times \mathbb{R}^N)$. Denote by $\Psi: \mathscr{U} \to I \times \mathbb{R}^N \times \mathbb{R}^N$ the inverse of Φ. We infer from (4) that

$$\Phi(x, u(x), u'(x)) = (x, u(x), \pi(x)) \quad \text{a.e. on } I, \tag{10}$$

where $\pi(x)$ is defined by (6). Set $\sigma(x) := (x, u(x), u'(x))$, $e(x) := (x, u(x), \pi(x))$, $\Psi(e(x)) =: (x, u(x), v(x))$. Then (10) is equivalent to $e(x) = \Phi(\sigma(x))$ a.e. on I whence

$$\sigma(x) = \Psi(e(x)) \quad \text{a.e. on } I, \tag{11}$$

that is,

$$u'(x) = v(x) \quad \text{a.e. on } I. \tag{12}$$

Since $u \in \mathrm{Lip}(I, \mathbb{R}^N)$, there is a null set $\mathscr{N}$ in I such that

$$|u'(x)| \leq k \quad \text{for } x \in I - \mathscr{N}, \tag{13}$$

and because of $e(x) = \Phi(\sigma(x))$ a.e. on I we can assume that

$$e(x) = \Phi(\sigma(x)) \quad \text{for } x \in I - \mathscr{N}. \tag{14}$$

Set $\mathscr{K} := \{\Phi(x, z, p): (x, z, p) \in \mathbb{R} \times \mathbb{R}^N \times \mathbb{R}^N \text{ and } |p| \leq k\}$. Then (13) implies that

$$\Phi(\sigma(x)) \in \mathscr{K} \quad \text{for } x \in I - \mathscr{N}, \tag{15}$$

and on account of (14) it follows that

$$e(x) \in \mathscr{K} \quad \text{for } x \in I - \mathscr{N}. \tag{16}$$

Since $e(x)$ is continuous on I and $\mathscr{K}$ is closed, we infer from (16) that

$$e(x) \in \mathscr{K} \quad \text{for all } x \in I, \tag{17}$$

and therefore $e(I)$ is contained in $\mathscr{U}$, the domain of Ψ. By virtue of $\Psi \in C^1$ we see that $\Psi \circ e$ is continuous, and so $v(x)$ is continuous on I. On the other hand, by (12) the absolutely continuous function $u(x)$ can be represented in the form

$$u(x) = u(a) + \int_a^x v(t)\, dt.$$

Consequently u is of class C^1 on I, and Proposition 3 yields that $u(x)$ is even of class C^2 on (a, b). In fact, (10) implies that $\pi \in C^1(I, \mathbb{R}^N)$, and then relation (11) yields $u' \in C^1(I, \mathbb{R}^N)$, i.e. $u \in C^2(I, \mathbb{R}^N)$. □

Remark. A careful inspection of our reasoning shows that we can weaken our assumptions in several respects. For instance, the function $\pi(x)$ is already absolutely continuous if $F_u(x, u(x), u'(x))$ is of class L^1 on (a, b). On the other hand relation (17) is then more difficult to achieve, as for some points $x \in I$ the images $e(x)$ might belong to $\partial\mathscr{U}$ but not to $\mathscr{U}$, if we do not assume that $u \in \mathrm{Lip}(I, \mathbb{R}^N)$, and so we could not form $\Psi(e(x))$. Thus for general weak extremals a more subtle discussion is needed to generalize the preceding reasoning.

3.2. Remarks on the Existence of Minimizers

Following *3.1* [4] we have noted that there is no reason why minimizers should always be smooth. Moreover variational problems might have no solutions at all, even if we enlarge the class of "admissible functions" considerably by replacing C^2 or C^1 by D^1 or by the class of Lipschitz functions. In this subsection we shall exhibit further examples which will shed more light on those phenomena.

[1] *Weierstrass's example.* In his criticism of the so-called *Dirichlet's principle*, Weierstrass pointed out that it is by no means clear why a functional like Dirichlet's integral $\frac{1}{2}\int |Du|^2\, dx$ should have a minimizer in the class of functions $u : \bar{\Omega} \to \mathbb{R}$ with prescribed boundary values $u|_{\partial\Omega}$, although it is bounded from below. The existence of a finite lower bound does in general not guarantee the existence of a minimum. We shall discuss Dirichlet's principle, its historical context, and its mathematical implications in another treatise. Here we just want to show Weierstrass's example. Consider the variational integral

$$\mathscr{F}(u) := \int_{-1}^{1} x^2 \left(\frac{du}{dx}\right)^2 dx,$$

which is to be minimized in the class

$$\mathscr{C} := \{u \in C^1([-1, 1]) \colon u(-1) = -1,\, u(1) = 1\}.$$

Clearly we have $\inf_{\mathscr{C}} \mathscr{F} \geq 0$, and by inserting the function $u_\varepsilon \in \mathscr{C}$, defined by

$$u_\varepsilon(x) := \frac{\operatorname{arc\,tg} \frac{x}{\varepsilon}}{\operatorname{arc\,tg} \frac{1}{\varepsilon}}, \qquad |x| \leq 1,\ \varepsilon > 0,$$

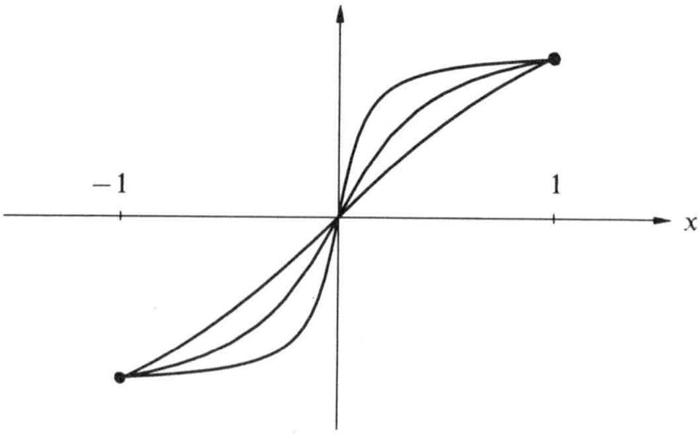

Fig. 15. The Weierstrass example.

we obtain $\mathscr{F}(u_\varepsilon) \to 0$ as $\varepsilon \to +0$. Therefore we have

$$\inf_{\mathscr{C}} \mathscr{F} = 0.$$

However, there is no function $u \in \mathscr{C}$ such that $\mathscr{F}(u) = 0$ since this relation would imply $u'(x) \equiv 0$, i.e. $u(x) \equiv$ const, and this is impossible because of the boundary condition.

In order to find a minimizer we would have to admit discontinuous functions as admissible functions such as $u_0(x) := \operatorname{sign} x$. Then we would have $\mathscr{F}(u_0) = 0$.

[2] Suppose we want to minimize

$$\mathscr{F}(u) := \int_0^1 (1 + |u'|^2)^{1/4}\, dx$$

in the class $\mathscr{C}$ of Lipschitz functions $u : [0, 1] \to \mathbb{R}$ satisfying $u(0) = 0$ and $u(1) = 1$. By considering $\mathscr{F}(u_\delta)$ for the admissible functions

$$u_\delta(x) := \begin{Bmatrix} 0 & \text{for } 0 \le x \le 1 - \delta \\ 1 + \delta^{-1}\cdot(x - 1) & \text{for } 1 - \delta \le x \le 1 \end{Bmatrix}, \quad \delta > 0,$$

we obtain

$$\begin{aligned} \mathscr{F}(u_\delta) &= \int_0^{1-\delta} dx + \int_{1-\delta}^1 (1 + \delta^{-2})^{1/4}\, dx \\ &= (1 - \delta) + \delta^{1/2}(1 + \delta^2)^{1/4} \to 1 \quad \text{as } \delta \to 0, \end{aligned}$$

whence $\inf_{\mathscr{C}} \mathscr{F} = 1$. But obviously there is no element $u \in \mathscr{C}$ such that $\mathscr{F}(u) = 1$ as this would imply $u'(x) \equiv 0$.

[3] Suppose we want to find a graph of minimal area over the unit disk B of $\mathbb{R}^2$ which lies above the point $P_0 := (0, 0, 1)$ in $\mathbb{R}^2 \times \mathbb{R}$. By considering circular cones above the discs $B_\varepsilon = \{(x, y)\colon x^2 + y^2 < \varepsilon^2\}$ with vertices in P_0 and adding the annuli $\{(x, y)\colon \varepsilon^2 < x^2 + y^2 < 1\}$ we obtain admissible graphs whose areas tend to π, the area of the disc B. Therefore the infimum of the area functional in the class $\mathscr{C}$ of admissible graphs is π and clearly there is no function $u \in C^1(\bar{B})$ with $u|_{\partial B} = 0$ such that $\int_B \sqrt{1 + |\nabla u|^2}\, dx = \pi$ and $u(0, 0) \ge 1$. The infimum of the area is assumed by a "singular object" consisting of the disc B and a "hair" H raised in the center 0 of B. Another interpretation of the phenomenon is that the infimum is not assumed in $\mathscr{C}$ but by the disc B, i.e. the functional neglects the constraint $u(0) = 1$ if one wants to minimize it. Physically this corresponds to the fact that one cannot push up a soap film by a needle; for this purpose one needs a "blunt" object.

[4] Consider the functional

$$\mathscr{F}(u) := \int_0^1 \frac{1}{1+|u'|^2}\,dx$$

on the class $\mathscr{C} := \{u \in C^1([0,1]): u(0) = 0, u(1) = 1\}$. Obviously $0 < \mathscr{F}(u) < 1$, and it is not difficult to see that $\inf_{\mathscr{C}} \mathscr{F} = 0$ and $\sup_{\mathscr{C}} \mathscr{F} = 1$. In fact near each such function we can find $u, v \in \mathscr{C}$ such that $\mathscr{F}(u) < \varepsilon$ and $1 - \varepsilon < \mathscr{F}(v)$, where ε is an arbitrarily prescribed positive number. However neither $\inf_{\mathscr{C}} \mathscr{F}$ nor $\sup_{\mathscr{C}} \mathscr{F}$ can be realized by $\mathscr{F}(u)$ for some $u \in \mathscr{C}$.

[5] *Nonparametric surfaces of prescribed mean curvature.* In 2.2 [5] we have seen that a nonparametric surface $u(x)$, $x \in \Omega$, of prescribed mean curvature H has to satisfy the equation

(1) $$\operatorname{div} Tu = nH, \quad \text{where } Tu := \frac{Du}{\sqrt{1+|Du|^2}}$$

and n is the dimension of Ω. The equation $\operatorname{div} Tu = nH$ is the Euler equation of the functional

(2) $$\int_\Omega (\sqrt{1+|Du|^2} + nH)\,dx.$$

For the sake of simplicity we assume that H is constant. We infer from (1) that

(3) $$-\int_\Omega Tu \cdot D\varphi\,dx = nH \int_\Omega \varphi\,dx$$

for all $\varphi \in C_c^1(\Omega)$. Moreover, for each ball B_R of radius R contained in Ω there exists a sequence of spherically symmetric functions φ on B_R which converge in $L^2(\Omega)$ to the characteristic function of B_R. Since $|Tu| < 1$ we then infer from (3) that

(4) $$n|H||B_R| < |\partial B_R|.$$

Thus we infer for $\Omega = B_R$ that a solution u of (1) in B_R can only exist if the *necessary condition*

(5) $$|H| < 1/R$$

holds true.

[6] *Parametric surfaces of prescribed mean curvature.* A similar argument works for a two-dimensional surface $u: B \to \mathbb{R}^3$ of constant mean curvature which is defined on a disc $B \subset \mathbb{R}^2$. Suppose that the surface $u(x, y) = (u^1(x, y), u^2(x, y), u^3(x, y))$ is given in conformal parameters x, y, i.e. we assume

(6) $$|u_x|^2 = |u_y|^2, \quad u_x \cdot u_y = 0.$$

Then u has to satisfy the equation

(7) $$\Delta u = 2H u_x \wedge u_y,$$

see 3,2 [4]. Integration by parts yields

(8) $$\int_B \Delta u\,dx\,dy = \int_{\partial B} \frac{\partial u}{\partial n}\,d\mathscr{H}^1,$$

where n is the exterior normal to ∂B. On the other hand we have

$$\int_B \Delta u\, dx\, dy = 2H \int_B u_x \wedge u_y\, dx\, dy \tag{9}$$

$$= H \int_B \frac{\partial}{\partial x}(u \wedge u_y)\, dx\, dy + H \int_B \frac{\partial}{\partial y}(u_x \wedge u)\, dx\, dy$$

$$= H \int_{\partial B} \{(u \wedge u_y)\, dy - (u_x \wedge u)\, dx\}$$

$$= H \int_{\partial B} u \wedge du.$$

If $u|_{\partial B}$ defines a topological mapping of ∂B onto a closed Jordan curve Γ, we have

$$A(\Gamma) := \int_\Gamma u \wedge du = \int_{\partial B} u \wedge du, \tag{10}$$

where the number $A(\Gamma)$ depends only on the curve Γ and not on its representation $u|_{\partial B}$. We infer from (8)–(10) that

$$A(\Gamma) \cdot H = \int_{\partial B} \frac{\partial u}{\partial n}\, d\mathscr{H}^1. \tag{11}$$

Let t be a tangential unit vector field along ∂B. Then we have $\left|\frac{\partial u}{\partial n}\right| = \left|\frac{\partial u}{\partial t}\right|$ on ∂B whence

$$\left|\int_{\partial B} \frac{\partial u}{\partial n}\, d\mathscr{H}^1\right| \le \int_{\partial B} \left|\frac{\partial u}{\partial t}\right| d\mathscr{H}^1 = L(\Gamma), \tag{12}$$

where $L(\Gamma)$ is the length of Γ. We obtain from (11) and (12) the condition

$$|A(\Gamma)|\,|H| \le L(\Gamma), \tag{13}$$

which is necessary for a solution $u : B \to \mathbb{R}^3$ of (6), (7) mapping ∂B topologically onto a Jordan curve Γ. If Γ is a circle of radius R, then $L(\Gamma) = 2\pi R$ and $A(\Gamma) = 2\pi R^2$; hence the necessary condition (13) reduces in this case to $|H| R < 1$ as to be expected from the preceding example.

7 *Capillary surfaces in a tube with a corner.* In outer space, where gravity is zero, a nonparametric surface $u : \Omega \to \mathbb{R}$ above a convex domain Ω in $\mathbb{R}^2$ describing the equilibrium surface of a liquid in the cylindrical tube $\Omega \times \mathbb{R}$ has to satisfy the equations

$$\begin{aligned} \operatorname{div} Tu &= 2H \quad \text{in } \Omega, \\ \nu \cdot Tu &= \cos\gamma \quad \text{on } \partial\Omega, \end{aligned} \tag{14}$$

where H is a positive constant, ν the exterior unit normal to $\partial\Omega$, and γ is a constant angle, $0 \le \gamma \le \pi/2$. Note that γ is the angle between the liquid surface and the boundary $\partial\Omega \times \mathbb{R}$ of the capillary tube.

We remark that equations (14) are the Euler equation and the free boundary condition (see *2.4*) of the variational integral

$$\mathscr{E}(u) := \int_\Omega \sqrt{1 + Du^2}\, dx\, dy + 2H \int_\Omega u\, dx\, dy - (\cos\gamma) \int_{\partial\Omega} u\, d\mathscr{H}^1.$$

Set $\Omega^\ell := \{(x, y) : x + iy = re^{i\varphi},\ 0 < r < \ell,\ 0 < \varphi < 2\alpha < \pi\}$ and suppose that there is some $\ell > 0$ such that $\Omega^\ell = \Omega \cap B_\ell(0)$. Let Ω^* be the triangle with the corners 0, $P_1 = (\ell, 0)$, and $P_2 = (\ell \cos 2\alpha, \ell \sin 2\alpha)$. Then $2 \operatorname{meas} \Omega^* = \ell^2 \sin 2\alpha$, and the length $h = \overline{P_1 P_2}$ of the linear segment Γ bounded by P_1 and P_2 is $h = 2\ell \sin\alpha$. Let $0 < \rho \ll \ell$ and set $\Omega^*_\rho := \Omega^* \cap B_\rho(0)$, $C_\rho := \Omega^* \cap \partial B_\rho(0)$, cf. Fig. 16. Furthermore let $u \in C^1(\overline{\Omega} - \{0\}) \cap C^2(\Omega)$ be a solution of (14). Then we obtain

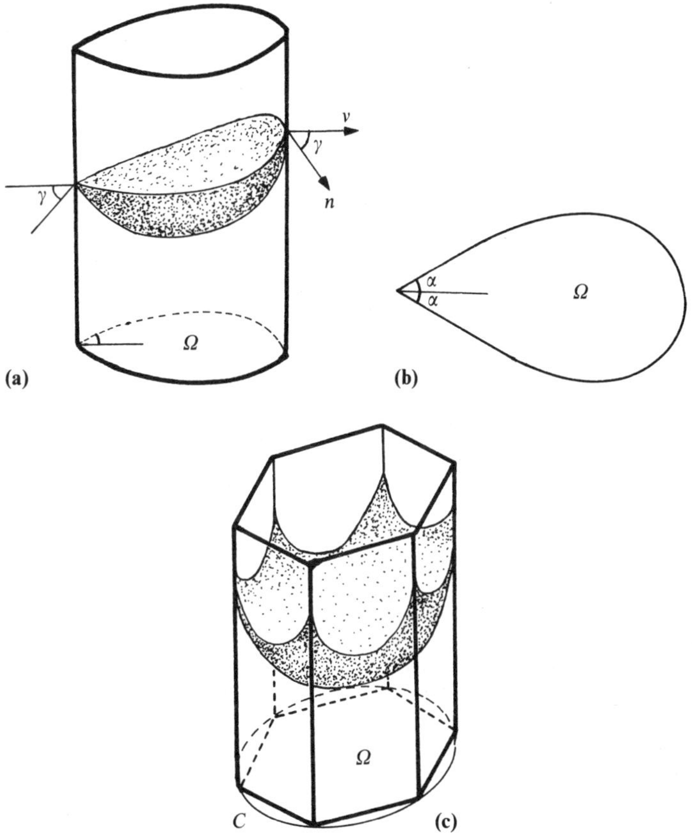

Fig. 16. (a) A solution of the capillary problem [7] may exist if $\alpha + \gamma \geq \pi/2$, but cannot exist if $\alpha + \gamma < \pi/2$ where **(b)** Ω is a domain with a corner of half the opening α. **(c)** Let Ω be a regular hexagon inscribed in the circle C. Then any lower hemisphere whose equator lies above C provides a solution for the case $\alpha = \pi/3$ and $\alpha + \gamma = \pi/2$.

$$2H \text{ meas } \Omega_\rho^* = \int_{\Omega_\rho^*} \text{div } T \, dx \, dy = \int_{\partial\Omega_\rho^*} (\nu \cdot Tu) \, d\mathscr{H}^1$$

$$= 2(\ell - \rho) \cos \gamma + \int_{C_\rho} (\nu \cdot Tu) \, d\mathscr{H}^1 + \int_\Gamma (\nu \cdot Tu) \, d\mathscr{H}^1,$$

and $\nu \cdot Tu < 1$. As $\rho \to +0$ we infer

$$2H \text{ meas } \Omega^* = 2\ell \cos \gamma + \int_\Gamma (\nu \cdot Tu) \, d\mathscr{H}^1,$$

whence

$$H\ell^2 \sin 2\alpha > 2\ell(\cos \gamma - \sin \alpha).$$

Letting $\ell \to +0$ we arrive at $\cos \gamma - \sin \alpha \leq 0$, and $\sin \alpha = \cos(\pi/2 - \alpha)$, $0 < \alpha < \pi/2$, imply that

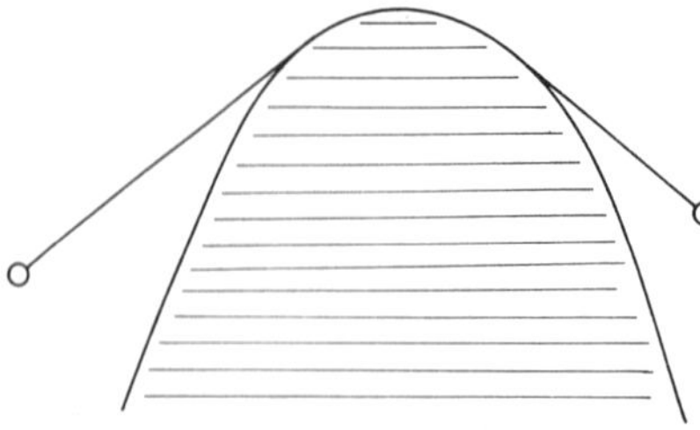

Fig. 17.

Fig. 18.

$\cos \gamma \le \cos(\pi/2 - \alpha)$, i.e. $\alpha + \gamma \ge \pi/2$. Thus we conclude that there is no solution of (14) in a domain Ω of the opening 2α if $\alpha + \gamma < \pi/2$. Note that for $\alpha + \gamma = \pi/2$ the problem can have a solution which is part of a spherical cap.

Further examples of this nature can be found in 4,*2.3* [1] [2].

Finally we want to show by way of example that constraints might cause a loss of regularity.

[8] An *obstacle problem.* Suppose we want to join two points P_1 and P_2 in the plane by a curve staying above an obstacle $\mathscr{K}$ (see Fig. 17). If the obstacle is bounded by a circular or parabolic arc, it is geometrically obvious that the connecting curve of least length consists of three parts $\Gamma', \Gamma, \Gamma''$, the tangential connections Γ' and Γ'' from P_1 and P_2 to the obstacle, and the connection Γ of the two tangent points Q_1 and Q_2 along $\partial\mathscr{K}$. Clearly the curve $\Gamma' \cup \Gamma \cup \Gamma''$ is of class C^1, but not of class C^2 since the curvatures of Γ' and Γ'' are zero while the curvature of Γ is non zero. In fact it is easy to see that the shortest connection of P_1 and P_2 above the circular or parabolic obstacle has a Lipschitz continuous tangent which is not of class C^1. If the obstacle is allowed to have corners, the shortest connection might even have a discontinuous tangent (see Fig. 18). A similar discussion can be carried out for the Dirichlet integral instead of the arc-length functional.

3.3. Broken Extremals

In this subsection we shall derive some necessary conditions for minimizers of a variational integral

$$\mathscr{F}(u) := \int_a^b F(x, u(x), u'(x))\, dx \tag{1}$$

in the class D^1 of *piecewise smooth functions* $u: [a, b] \to \mathbb{R}^N$. Recall that $u \in D^1([a, b], \mathbb{R}^N)$ if $u \in C^0([a, b], \mathbb{F}^N)$ and if there is a decomposition $a = t_0 < t_1 < t_2 < \cdots < t_k < t_{k+1} = b$ of $[a, b]$ into finitely many intervals $I_j = [t_{j-1}, t_j]$, $1 \le j \le k + 1$, such that $u|_{I_j}$ is of class C^1. Such a function u is differentiable except for finitely many points $t_1, \ldots, t_k$ where at least the one-sided derivatives $u'(t_j - 0)$ and $u'(t_j + 0)$ exist. If $u'(t_j - 0) \ne u'(t_j + 0)$, $1 \le j \le k$, we call t_j a *corner of u.*

In the following we assume that the Lagrangian $F(x, z, p)$ of the functional $\mathscr{F}$ in (1) is of class C^1 on $[a, b] \times \mathbb{R}^N \times \mathbb{R}^N$ (or on a neighbourhood $\mathscr{U}$ of the 1-graph of the minimizer that will be investigated). We are now going to prove the *Weierstrass–Erdmann corner conditions* which state that for any D^1-minimizer u of $\mathscr{F}$ the expressions

$$\pi(x) := F_p(x, u(x), u'(x)) \tag{2}$$

and

$$\varphi(x) := \pi(x) \cdot u'(x) - F(x, u(x), u'(x)) \tag{3}$$

satisfy

$$\pi(x - 0) = \pi(x + 0), \qquad \varphi(x - 0) = \varphi(x + 0) \tag{4}$$

and can therefore be viewed as continuous functions of $x \in [a, b]$. Let us formulate this result as

Proposition 1. *Let $I = [a, b]$, $u \in D^1(I, \mathbb{R}^N)$, and suppose that $\mathscr{F}(u) \le \mathscr{F}(v)$ for all $v \in \mathscr{C}_r$ and some $r > 0$ where $\mathscr{C}_r := \{v \in D^1(I, \mathbb{R}^N)\colon v(a) = u(a),\ v(b) = u(b),\ \sup_I |u - v| < r\}$. Let $x_0 \in (a, b)$ be a corner point of u, and set $z_0 := u(x_0)$, $p_0^+ := u'(x_0 + 0)$, $p_0^- := u'(x_0 - 0)$. Then we obtain the two corner conditions*

$$F_p(x_0, z_0, p_0^-) = F_p(x_0, z_0, p_0^+) \tag{I}$$

and

$$F(x_0, z_0, p_0^-) - p_0^- \cdot F_p(x_0, z_0, p_0^-) = F(x_0, z_0, p_0^+) - p_0^+ \cdot F_p(x_0, z_0, p_0^+). \tag{II}$$

Proof. (i) The first formula is an immediate consequence of formula (4) in *3.1*, Proposition 1. In fact, (I) is true for any weak D^1-extremal (and even for any weak Lipschitz-extremal) of $\mathscr{F}$.

(ii) While (I) followed from "variations of the dependent variables z", condition (II) will be derived by means of "variations of the independent variables x". Therefore (II) should be treated in Chapter 3 where we systematically investigate variations of general type. But since the computations for the case at hand are very simple we shall immediately derive the variational formula that is needed. The basic idea is to use *inner variations*, i.e. variations of the independent variables.

Let λ be an arbitrary function of class $C_c^\infty(I)$ and set $\tau_\varepsilon(x) := x + \varepsilon\lambda(x)$, $x \in I$, $|\varepsilon| \le \varepsilon_0$. For $0 < \varepsilon_0 \ll 1$ we obtain a smooth 1-parameter family of diffeomorphisms τ_ε of I onto itself such that

$$\tau_\varepsilon(a) = a, \qquad \tau_\varepsilon(b) = b, \qquad \tau_0(x) = x \quad \text{for all } x \in I,$$

and $\tau'_\varepsilon(x) > 0$, and we can assume that the functions $z_\varepsilon := u \circ \tau_\varepsilon^{-1}$ are contained in $\mathscr{C}_r$ whence

$$\frac{d}{d\varepsilon}\mathscr{F}(z_\varepsilon)|_{\varepsilon=0} = 0 \tag{5}$$

by the minimality of u. Since

$$\frac{d}{dx}z_\varepsilon(x) = u'(\tau_\varepsilon^{-1}(x))\frac{d}{dx}\tau_\varepsilon^{-1}(x)$$

and

$$\frac{d}{dx}\tau_\varepsilon^{-1}(x) = \frac{1}{1+\varepsilon\lambda'(\tau_\varepsilon^{-1}(x))},$$

we obtain by a change of variables that

$$\mathscr{F}(z_\varepsilon) = \int_a^b F\left(x+\varepsilon\lambda(x), u(x), \frac{u'(x)}{1+\varepsilon\lambda'(x)}\right)(1+\varepsilon\lambda'(x))\,dx. \tag{6}$$

We infer from (5) and (6) that

$$\int_a^b \{\lambda F_x(x,u,u') + \lambda'[F(x,u,u') - u'\cdot F_p(x,u,u')]\}\,dx = 0 \tag{7}$$

for all $\lambda \in C_c^\infty(I)$. By Du Bois–Reymond's lemma (see *2.3*) we deduce the existence of a constant c^* such that *Erdmann's equation*

$$\begin{aligned} &F(x,u(x),u'(x)) - u'(x)\cdot F_p(x,u(x),u'(x)) \\ &\quad = c^* + \int_a^x F_x(t,u(t),u'(t))\,dt \end{aligned} \tag{8}$$

is satisfied for all $x \in I$. Since the right-hand side of (8) is an absolutely continuous function, condition (II) is an immediate consequence of (8). □

Let us note that the reasoning above can also be applied to Lipschitz minimizers of $\mathscr{F}$. Thus we obtain

Proposition 2. *If $u \in \mathrm{Lip}(I, \mathbb{R}^N)$, $I = [a, b]$, is a minimizer of $\mathscr{F}$ among Lipschitz functions with the same boundary values as u, we obtain Du Bois–Reymond's equation*

$$F_p(x,u(x),u'(x)) = c + \int_a^x F_u(t,u(t),u'(t))\,dt \tag{9}$$

as well as equations (8) *for suitable constants $c \in \mathbb{R}^N$ and $c^* \in \mathbb{R}$. In particular if $F_x(x,z,p) \equiv 0$, then*

$$F(x,u(x),u'(x)) - u'(x)\cdot F_p(x,u(x),u'(x)) \equiv c,$$

i.e. the function $F - p\cdot F_p$ is a first integral on all Lipschitz minimizers of $\mathscr{F}$.

The last statement of Proposition 2 generalizes the *law of conservation of energy* formulated in *2.2* [7] to weak minimizers of $\mathscr{F}$ or, more generally, to solutions u of (5).

4. Null Lagrangians

Let $F(x, z, p)$ be a C^2-variational integrand which is defined on some open subset $\mathscr{U}$ of $\mathbb{R}^n \times \mathbb{R}^N \times \mathbb{R}^{nN}$. Denote the 1-graph (or contact graph) of some mapping $u \in C^1(\bar{\Omega}, \mathbb{R}^N)$ by

(1) $$\text{1-graph } u := \{(x, u(x), Du(x)) \colon x \in \bar{\Omega}\}.$$

Definition. *We call F a* null Lagrangian *if $L_F(u) = 0$ holds for all $u \in C^2(\bar{\Omega}, \mathbb{R}^N)$ with* 1-graph $u \subset \mathscr{U}$.

A simple example of a null Lagrangian is furnished by integrands of the type

(2) $$F(p) = a \cdot p + b = a_i^\alpha p_\alpha^i + b$$

with constant coefficients a_i^α and b. If $n = N$, $a_i^\alpha = \delta_i^\alpha$ and $b = 0$, we obtain the important null Lagrangian

(3) $$F(p) = p_1^1 + p_2^2 + \cdots + p_n^n = \operatorname{trace} p,$$

for which

$$F(Du) = \operatorname{div} u$$

if $u = (u^1, \ldots, u^n)$. Another interesting null Lagrangian is furnished by

(4) $$F(p) = \det p.$$

A direct verification of this fact may seem difficult, but we can easily prove the claim by the following reasoning: For any mapping $u \in C^2(\bar{\Omega}, \mathbb{R}^n)$, we have

$$\begin{aligned} F(Du)\, dx^1 \wedge dx^2 \wedge \cdots \wedge dx^n &= (\det Du)\, dx^1 \wedge \cdots \wedge dx^n \\ &= du^1 \wedge du^2 \wedge \cdots \wedge du^n \\ &= d(u^1 \wedge du^2 \wedge \cdots \wedge du^n). \end{aligned}$$

Hence, if $\bar{\Omega}$ is compact and if $\partial\Omega$ is sufficiently smooth, we obtain by means of Stokes's theorem that

$$\mathscr{F}(u) = \int_\Omega F(Du)\, dx = \int_{\partial\Omega} u^1 \wedge du^2 \wedge \cdots \wedge du^n.$$

Hence $\mathscr{F}(u)$ depends only on the boundary values of u and Du along $\partial\Omega$ and is therefore invariant on variations

$$v_\varepsilon(x) := u(x) + \varepsilon\varphi(x), \quad |\varepsilon| \ll 1,$$

with $\operatorname{supp} \varphi \subset \Omega$. Hence u must be an extremal of $\mathscr{F}$. If $\bar{\Omega}$ is not compact or $\partial\Omega$ not smooth, we can apply the reasoning to smoothly bounded subdomains $\Omega' \subset\subset \Omega$. Thus we obtain that $F(p) = \det p$ is a null Lagrangian.

Two other null Lagrangians are obtained for $n = 1$, $N = 2$ in the following way.

Let G be some domain in $\mathbb{R}^2$, and let $u(x, y)$, $v(x, y)$ be two functions of class $C^1(G)$ satisfying the Cauchy–Riemann equations

$$u_x = v_y, \qquad u_y = -v_x \tag{5}$$

in G. Define the two Lagrangians

$$F(x, y, p, q) := u(x, y)p - v(x, y)q \tag{6}$$

and

$$H(x, y, p, q) := u(x, y)q - v(x, y)p \tag{7}$$

on $G \times \mathbb{R}^2$, and set

$$\mathscr{F}(c) := \int_a^b F(x, y, \dot{x}, \dot{y})\, dt, \tag{8}$$

$$\mathscr{H}(c) := \int_a^b H(x, y, \dot{x}, \dot{y})\, dt \tag{9}$$

for curves $c(t) = (x(t), y(t))$, $a \le t \le b$, of class C^1 whose trace is contained in G. A straight-forward computation shows that both F and H are null Lagrangians. An immediate consequence of this fact is Cauchy's integral theorem as we shall see in *4.1* [1].

Null Lagrangians are important for several reasons. First of all, they are used to define *Hilbert invariant integrals* which play a major role in the derivation of sufficient conditions for strong relative minima via *field theories* (cf. Chapters 6 and 7). Secondly, null Lagrangians give rise to define *integral invariants.* Such invariants belong to the most interesting objects of differential geometry and topology. One important example is the *degree of a mapping*, another one the *winding number* of a closed path which is closely related to Cauchy's integral theorem. A third and somewhat more sophisticated example is provided by the *curvatura integra* $\int_M K\, dA$ of an orientable, closed, two-dimensional manifold M which only depends on the Euler characteristic of M. This is the content of the celebrated *Gauss–Bonnet theorem* that we shall briefly discuss in 5 [4]. In this case we have to consider a second order variational problem $\int_\Omega F(x, u, Du, D^2u)\, dx$.

Let us finally mention that variational integrals corresponding to null Lagrangians have remarkable lower semicontinuity properties.

4.1. Basic Properties of Null Lagrangians

The following result is fairly obvious:

Proposition 1. *Suppose that $\mathscr{F}(u) = \mathscr{F}(v)$ holds for all $u, v \in C^1(\bar{\Omega}, \mathbb{R}^N)$ satisfying $u = v$ on $\partial\Omega$ and* 1-graph u, 1-graph $v \subset \mathscr{U}$. *Then F is a null Lagrangian.*

Proof. Let u be an arbitrary function of class $C^1(\bar{\Omega}, \mathbb{R}^N) \cap C^2(\Omega, \mathbb{R}^N)$ whose 1-graph is contained in $\mathscr{U}$. Then, for any $\varphi \in C_c^\infty(\Omega, \mathbb{R}^N)$, we infer that

$$\mathscr{F}(u + \varepsilon\varphi) = \mathscr{F}(u) \quad \text{for } 0 \le |\varepsilon| \ll 1,$$

whence

$$\delta\mathscr{F}(u, \varphi) = 0$$

and therefore $L_F(u, \varphi) = 0$ if $u \in C^2(\Omega, \mathbb{R}^N)$. □

Now we prove a converse of Proposition 1.

Proposition 2. *Let $F(x, z, p)$ be a null Lagrangian which is defined on $\bar{\Omega} \times G \times \mathbb{R}^{nN}$ where G is a simply connected domain in $\mathbb{R}^N$. Then the corresponding variational integral $\mathscr{F}$ satisfies $\mathscr{F}(u) = \mathscr{F}(v)$ for all $u, v \in C^1(\bar{\Omega}, \mathbb{R}^N)$ satisfying $u = v$ on $\partial\Omega$, and whose 1-graphs are contained in $\bar{\Omega} \times G \times \mathbb{R}^{nN}$.*

Proof. We shall see in 3,5 that $L_F(u)$ transforms like a vector field if we change the dependent variables. Hence the property of F to be a null Lagrangian remains unaltered if we subject the dependent variables to a diffeomorphism, and we conclude that it is sufficient to prove the assertion for G being an open ball. Then the stated result follows from the next proposition. □

Proposition 3. *Let $F(x, z, p)$ be a null Lagrangian which is defined on a normal domain*

$$\mathscr{N} = \{(x, z) \colon x \in \bar{\Omega},\ a^i(x) < z^i < b^i(x),\ i = 1, \dots, N\},$$

where $a^i(x)$, $b^i(x)$ are the components of two functions $a, b \in C^0(\bar{\Omega}, \mathbb{R}^N)$ satisfying $a^i(x) < b^i(x)$ on $\bar{\Omega}$. Then the variational integral $\mathscr{F}$ associated with F satisfies $\mathscr{F}(u) = \mathscr{F}(v)$ for all $u, v \in C^1(\bar{\Omega}, \mathbb{R}^N)$ such that $u = v$ on $\partial\Omega$ and graph u, graph $v \subset \mathscr{N}$.

Proof. (i) Let us first assume that u and v are both of class $C^2(\bar{\Omega}, \mathbb{R}^N)$. Then we introduce the C^2-family of mappings $\psi(\varepsilon) : \bar{\Omega} \to \mathbb{R}^N$ defined by

$$\psi(x, \varepsilon) := u(x) + \varepsilon\varphi(x), \quad \varphi(x) := v(x) - u(x).$$

Note that $\varphi(x) = 0$ for $x \in \partial\Omega$. Since F is a null Lagrangian, we have $L_F(\psi(\varepsilon)) = 0$ on $\bar{\Omega}$, whence we infer by partial integration

$$\delta\mathscr{F}(\psi(\varepsilon), \varphi) = \int_\Omega L_F(\psi(\varepsilon)) \cdot \varphi \, dx = 0.$$

Therefore,

$$\mathscr{F}(v) - \mathscr{F}(u) = \int_0^1 \frac{d}{d\varepsilon}\mathscr{F}(\psi(\varepsilon)) \, d\varepsilon = \int_0^1 \delta\mathscr{F}(\psi(\varepsilon), \varphi) \, d\varepsilon = 0.$$

(ii) Now we drop the additional assumption $u, v \in C^2(\bar{\Omega}, \mathbb{R}^N)$. We choose two sequences $\{u_k\}$ and $\{v_k\}$ of functions $u_k, v_k \in C^2(\bar{\Omega}, \mathbb{R}^N)$ with graphs contained in $\mathcal{N}$ such that

$$\lim_{k\to\infty} \|u - u_k\|_{C^1(\bar{\Omega})} = 0 \quad \text{and} \quad \lim_{k\to\infty} \|v - v_k\|_{C^1(\bar{\Omega})} = 0$$

holds. Then we set

$$\varphi_k := v_k - u_k \quad \text{and} \quad \psi_k(x, \varepsilon) := (1-\varepsilon)u_k(x) + \varepsilon v_k(x) = u_k(x) + \varepsilon\varphi_k(x),$$

where $x \in \bar{\Omega}$ and $0 \le \varepsilon \le 1$. It follows that

$$\lim_{k\to\infty} \|\varphi_k\|_{C^0(\partial\Omega)} = 0 \quad \text{and} \quad \operatorname{graph} \psi_k(\varepsilon) \subset \mathcal{N} \quad \text{for } 0 \le \varepsilon \le 1.$$

By virtue of formula (5) in *2.4*, we have

$$\delta\mathcal{F}(\psi_k(\varepsilon), \varphi_k) = \int_{\partial\Omega} \nu_\alpha F_{p_\alpha^i}(x, \psi_k(\varepsilon), D\psi_k(\varepsilon))\varphi_k^i \, d\mathcal{H}^{n-1},$$

and therefore

$$\mathcal{F}(v_k) - \mathcal{F}(u_k) = \int_0^1 \frac{d}{d\varepsilon}\mathcal{F}(\psi_k(\varepsilon))\, d\varepsilon = \int_0^1 \delta\mathcal{F}(\psi_k(\varepsilon), \varphi_k)\, d\varepsilon$$

$$= \int_0^1 \int_{\partial\Omega} \nu_\alpha F_{p_\alpha^i}(x, \psi_k(\varepsilon), D\psi_k(\varepsilon))\varphi_k^i \, d\mathcal{H}^{n-1}\, d\varepsilon.$$

The left hand side converges to $\mathcal{F}(v) - \mathcal{F}(u)$ as $k \to \infty$, whereas the right-hand side tends to zero, since $\|\psi_k(\varepsilon)\|_{C^1(\bar{\Omega})}$ is uniformly bounded with respect to k and ε, and $\varphi_k|_{\partial\Omega}$ tends uniformly to zero. Consequently, $\mathcal{F}(u) - \mathcal{F}(v) = 0$. □

[1] *Cauchy's integral theorem.* Let us apply Proposition 2 to the case $n = 1$, $N = 2$ and to the two variational integrals $\mathcal{F}(c)$ and $\mathcal{H}(c)$ defined by (6)–(9) in the introduction to *4* where G is a simply connected domain in $\mathbb{R}^2$, and $u(x, y)$, $v(x, y)$ are C^1-solutions of the Cauchy–Riemann equations

$$u_x = v_y, \qquad u_y = -v_x$$

in G. Then we infer that both $\mathcal{F}(c)$ and $\mathcal{H}(c)$ depend only on the endpoints $c(a)$ and $c(b)$ of the arbitrary C^1-path $c: [a, b] \to G$. Hence, for any holomorphic function

$$f(z) = u(x, y) + iv(x, y)$$

of $z = x + iy \in G$, the complex line integral

$$\int_c f(z)\, dz = \int_b^a f(c(t))\dot{c}(t)\, dt = \mathcal{F}(c) + i\mathcal{H}(c)$$

depends only on the endpoints of c. That is, the complex line integral $\int_c f(z)\,dz$ vanishes for any closed C^1-curve c contained in G and for any function $f(z)$ that is holomorphic in G. This is just *Cauchy's integral theorem*, and our previous reasoning in the proof of Proposition 2 yields a perfectly elementary proof of this basic result of complex function theory as we have not to use the two-dimensional Gauss–Green theorem. Moreover this example shows that it is essential in Proposition 2 to assume G to be simply connected. For multiply connected domains G we might obtain periods.

4.2. Characterization of Null Lagrangians

This section can be omitted at the first reading. The ideas and results that are here presented will be relevant for the Weierstrass field theory. We exhibit further examples of null Lagrangians, and we shall give necessary and sufficient conditions characterizing null Lagrangians.

Let G be a domain in $\mathbb{R}^n \times \mathbb{R}^N$, and denote by $F(x, z, p)$ a Lagrange function which is defined for $(x, z, p) \in G \times \mathbb{R}^{nN}$. We suppose that F and F_z are of class C^0, whereas F_p is assumed to be of class C^1.

Let (x, z) be an arbitrary point in G, $p = (p_\alpha^i)$ an arbitrary point in $\mathbb{R}^{nN}$, and $q = (q_{\alpha\beta}^k)$ an arbitrary point in $\mathbb{R}^{n^2N}$ which satisfies $q_{\alpha\beta}^k = q_{\beta\alpha}^k$. Then we can choose a function $u \in C^2(\Omega, \mathbb{R}^N)$ with $\operatorname{graph}(u) \subset G$, such that

$$u^i(x) = z^i, \quad u_{x^\alpha}^i(x) = p_\alpha^i, \quad u_{x^\alpha x^\beta}^k(x) = q_{\alpha\beta}^k$$

holds. For instance, we may take the function

$$u^i(y) = z^i + p_\alpha^i(y^\alpha - x^\alpha) + \tfrac{1}{2} q_{\alpha\beta}^i (y^\alpha - x^\alpha)(y^\beta - x^\beta).$$

If F is a null Lagrangian, we have $L_F(u) = 0$ whence, at the point x, in particular it follows that

$$F_{z^i}(x, z, p) - F_{p_\alpha^i x^\alpha}(x, z, p) - F_{p_\alpha^i z^k}(x, z, p) p_\alpha^k - F_{p_\alpha^i p_\beta^k}(x, z, p) q_{\alpha\beta}^k = 0$$

holds for $i = 1, \ldots, N$. These equations are clearly equivalent to the two sets of relations

$$(1) \qquad F_{z^i}(x, z, p) - F_{p_\alpha^i x^\alpha}(x, z, p) - F_{p_\alpha^i z^k}(x, z, p) p_\alpha^k = 0$$

and

$$(2) \qquad F_{p_\alpha^i p_\beta^k}(x, z, p) q_{\alpha\beta}^k = 0.$$

Since $q_{\alpha\beta}^k = q_{\beta\alpha}^k$, equation (2) in turn is equivalent to

$$(3) \qquad F_{p_\alpha^i p_\beta^k} + F_{p_\beta^i p_\alpha^k} = 0.$$

If $n = 1$ or $N = 1$, this implies that $F(x, z, p)$ is a linear function of p for each pair $(x, z) \in G$, that is, F can be written as

$$(4) \qquad F(x, z, p) = A(x, z) + B_i^\alpha(x, z) p_\alpha^i,$$

with coefficients A, $A_z \in C^0(G)$ and $B_i^\alpha \in C^1(G)$.

Suppose now that F is of the type (4), whether $\min\{n, N\} = 1$ holds or not. Then (1) turns out to be equivalent to

$$\{B_{i,x^\alpha}^\alpha(x, z) - A_{z^i}(x, z)\} + [B_{i,z^k}^\alpha(x, z) - B_{k,z^i}^\alpha(x, z)] p_\alpha^k = 0,$$

which means that

$$(5) \qquad A_{z^i} = B_{i,x^\alpha}^\alpha \quad \text{and} \quad B_{k,z^i}^\alpha = B_{i,z^k}^\alpha$$

is satisfied on G.

Conversely, if F is of the form (4) and fulfills (5), one easily checks that F is a null Lagrangian. Thus we have proved

Proposition 1. *A variational integrand $F(x, z, p)$, $(x, z, p) \in G \times \mathbb{R}^{nN}$, is a null Lagrangian if it is of the form*

$$(4) \qquad F(x, z, p) = A(x, z) + B_i^\alpha(x, z) p_\alpha^i,$$

where the coefficients A and B_i^α satisfy

$$(5) \qquad A_{z^i} = B_{i,x^\alpha}^\alpha \quad \textit{and} \quad B_{k,z^i}^\alpha = B_{i,z^k}^\alpha$$

on G. If $n = 1$ or $N = 1$, these conditions are not only sufficient but also necessary for F to be a null Lagrangian.

The matter is more complicated if $m := \min\{n, N\} > 1$. Before we turn to this case, we shall derive another condition which is equivalent to (4) and (5). To this end, we slightly change the meaning of G. Suppose that G is a set in $\mathbb{R}^n \times \mathbb{R}^N$ which is of the form

(6) $$G = \{(x, z)\colon x \in \bar{\Omega}, z \in B(x)\}.$$

Here Ω is a domain in $\mathbb{R}^n$, $\partial\Omega \in C^1$; $B(x) \subset \mathbb{R}^N$ is, for every $x \in \Omega$, a simply connected domain in $\mathbb{R}^N$, and $\{(x, z)\colon x \in \Omega, z \in B(x)\}$ is supposed to be a domain in $\mathbb{R}^n \times \mathbb{R}^N$. We furthermore assume that $F \in C^1$ and $F_p \in C^2$ on G.

Proposition 2. *The Lagrange function F can be written in the form* (4), (5) *if and only if there exist functions $S^\alpha(x, z)$, $\alpha = 1, \ldots, n$, of class $C^1(G)$ with $S^\alpha_{x^\alpha} \in C^1(G)$ and $S^\alpha_z \in C^2(G)$ such that*

(7) $$F(x, z, p) = S^\alpha_{x^\alpha}(x, z) + S^\alpha_{z^i}(x, z)p^i_\alpha$$

holds for all $(x, z, p) \in G \times \mathbb{R}^{nN}$.

In conjunction with Proposition 1, we then obtain

Corollary. *If F is a null Lagrangian and if either $n = 1$ or $N = 1$, then F is of the divergence from* (7). *Conversely, every variational integrand of divergence form is a null Lagrangian.*

Moreover, we infer from (7) that

(8) $$F(x, u(x), Du(x)) = D_\alpha S^\alpha(x, u(x))$$

holds for every $u \in C^1(\bar{\Omega}, \mathbb{R}^N)$ with $\operatorname{graph}(u) \subset G$, and therefore

(9) $$\mathscr{F}(u) = \int_\Omega F(x, u, Du)\, dx = \int_{\partial\Omega} v_\alpha(x) S^\alpha(x, u(x))\, d\mathscr{H}^{n-1}(x),$$

which implies the invariance property $\mathscr{F}(u) = \mathscr{F}(v)$ if $u|_{\partial\Omega} = v|_{\partial\Omega}$, as stated in Propositions 2 and 3 of *4.1*.

Proof of Proposition 2. Clearly, (7) implies (4), and (5). Suppose now conversely that (4) and (5) hold. The integrability conditions $B^\alpha_{k,z^i} = B^\alpha_{i,z^k}$ imply the existence of functions $\Sigma^\alpha(x, z)$ with $\Sigma^\alpha_{z^i} = B^\alpha_i$ on G, and $B^\alpha_i = F_{p^i_\alpha} \in C^2$ yields that Σ^α and $\Sigma^\alpha_{z^i}$ are of class C^2.

Now we infer from $A_{z^i} = B^\alpha_{i,x^\alpha}$ that the divergence $\operatorname{div}_x \Sigma = \Sigma^\alpha_{x^\alpha}$ of the vector field $\Sigma = (\Sigma^1, \ldots, \Sigma^n)$ satisfies $A_{z^i} = (\operatorname{div}_x \Sigma)_{z^i}$ or, equivalently,

$$\operatorname{grad}_z A = \operatorname{grad}_z \operatorname{div}_x \Sigma.$$

Hence there exists a function $f(x)$ on Ω such that

$$A(x, z) = (\operatorname{div}_x \Sigma)(x, z) + f(x)$$

for all $(x, z) \in G$. We note that $\operatorname{div}_x \Sigma \in C^1(G)$ and also $A = F|_{p=0} \in C^1(G)$, whence $f \in C^1(\bar{\Omega})$. Then there is a vector field $v(x) = (v^1(x), \ldots, v^n(x))$ of class $C^1(\bar{\Omega}, \mathbb{R}^n)$ such that $v^\alpha_{x^\alpha} = \operatorname{div}_x v = f$ holds. If we set $S^\alpha(x, z) = \Sigma^\alpha(x, z) + v^\alpha(x)$, it follows that $S^\alpha_{x^\alpha} = A$ and $S^\alpha_{z^i} = B^\alpha_i$, which implies (7). □

Addendum. A similar reasoning yields: *If F satisfies* (4) *and* (5) *with merely $A, B^\alpha_i \in C^1(G)$, then there exist functions $S^1, \ldots, S^n \in C^0(G)$ with $S^\alpha_{z^i} \in C^1(G)$, $S^1_{x^1}, \ldots, S^n_{x^n} \in C^0(G)$ such that $S^\alpha_{x^\alpha} = A$, $S^\alpha_{z^i} = B^\alpha_i$, and therefore* (8) *holds true.*

We now return to the original assumptions on F and Ω, and investigate the structure of a null Lagrangian F in the general case $m := \min\{n, N\} > 1$. As we have already seen, F is a null Lagrangian if and only if it satisfies both (1) and (2). In particular, we obtain

(10) $$F_{p^i_\alpha p^k_\beta} = 0 \quad \text{if either } \alpha = \beta \text{ or } i = k.$$

From this we first infer that, for fixed $(x, z) \in G$, the function $F(x, z, p)$ is a polynomial with respect to p, and a closer inspection[8] yields that the degree of this polynomial is at most m. Thus we can write F in the form

(11) $$F(x, z, p) = A(x, z) + \sum_{\nu=1}^{m} \pi^\nu(x, z, p),$$

where, for fixed $(x, z) \in G$, the term $\pi^\nu(x, z, p)$ is a form of degree ν with respect to $p = (p^i_\alpha)$. Then (3) implies that

(12) $$\overset{\nu}{\pi}_{p^i_\alpha p^k_\beta} + \overset{\nu}{\pi}_{p^i_\beta p^k_\alpha} = 0.$$

Let us write $\pi^\nu(x, z, p)$ as

$$\pi^\nu(x, z, p) = B^{\alpha_1\alpha_2\ldots\alpha_\nu}_{i_1 i_2\ldots i_\nu}(x, z) p^{i_1}_{\alpha_1} p^{i_2}_{\alpha_2} \ldots p^{i_\nu}_{\alpha_\nu}.$$

Without loss of generality we may assume that the coefficients $B^{\alpha_1\ldots\alpha_\nu}_{i_1\ldots i_\nu}$ are symmetric with respect to the columns $\begin{pmatrix}\alpha_\ell\\ i_\ell\end{pmatrix}$, i.e., they remain unchanged if we permutate two columns $\begin{pmatrix}\alpha_\ell\\ i_\ell\end{pmatrix}$ and $\begin{pmatrix}\alpha_k\\ i_k\end{pmatrix}$ with each other. Then equation (12) is completely described by

$$B^{\alpha\beta\ldots}_{ik\ldots} + B^{\beta\alpha\ldots}_{ik\ldots} = 0$$

if $\nu \geq 2$, from which we infer that the coefficients $B^{\alpha\beta\ldots}_{ik\ldots}$ are skew symmetric in α, β as well as in i, k. By a standard reasoning used in multilinear algebra we obtain that π^ν is of the form

(13) $$\pi^\nu(x, z, p) = \sum_{|A|=|J|=\nu} B^A_J(x, z) P^J_A,$$

where the sum is to be extended over all multiindices $A = (\alpha_1, \alpha_2, \ldots, \alpha_\nu)$, $J = (i_1, \ldots, i_\nu)$ of length ν with $\alpha_1 < \alpha_2 < \cdots < \alpha_\nu$ and $i_1 < i_2 < \cdots < i_\nu$, and P^J_A denotes the determinant

(14) $$P^J_A = \begin{vmatrix} p^{\alpha_1}_{i_1}, p^{\alpha_1}_{i_2}, \ldots, p^{\alpha_1}_{i_\nu} \\ p^{\alpha_2}_{i_1}, p^{\alpha_2}_{i_2}, \ldots, p^{\alpha_2}_{i_\nu} \\ \vdots \\ p^{\alpha_\nu}_{i_1}, p^{\alpha_\nu}_{i_2}, \ldots, p^{\alpha_\nu}_{i_\nu} \end{vmatrix}.$$

In other words, $\pi^\nu(x, z, p)$ is a linear combination of the ν-dimensional minors of the matrix $p = (p^i_\alpha)$, with coefficients depending on x and z. If, moreover, F is independent of x and z, equations (1) are trivially satisfied. This yields

Proposition 3. *A null Lagrangian necessarily is an affine linear combination* $A(x, z) + \sum_{1 \leq |A|=|J| \leq m} B^A_J(x, z) P^J_A$ *of the subdeterminants of the matrix* $p = (p^j_\alpha)$ *with coefficients* A, B^A_J *depending on* x *and* z. *Conversely, every variational integrand* $F(p)$ *which is of the type*

$$F(p) = A + \sum B^A_J P^J_A,$$

with coefficients A, B^A_J *independent of* x *and* z, *is a null Lagrangian.*

Let us furthermore state the characterization of the general null Lagrangian:

Proposition 4. *A variational integrand* $F(x, z, p)$ *is a null Lagrangian if and only if it is of the form* (11) *where the summands* $\pi^\nu(x, z, p)$ *are of the type* (13) *and satisfy*

[8] For instance, if we choose $\alpha = \beta$ and $i = k$, (10) implies that F is a linear function of p^i_α. From here we may proceed by induction.

$$A_{z^i} - \pi^1_{p^i_\alpha x^\alpha} = 0$$

(15)
$$\pi^\nu_{z^i} - \pi^{\nu+1}_{p^i_\alpha x^\alpha} - \pi^\nu_{p^i_\alpha z^k} p^k_\alpha = 0 \quad (1 \le \nu \le m-1).$$

$$\pi^m_{z^i} - \pi^m_{p^i_\alpha z^k} p^k_\alpha = 0$$

Note that in particular a variational integrand $F(p)$, independent of x and z, is a null Lagrangian if and only if it is a linear combination of the minors of the matrix $p = (p^i_\alpha)$.

Concerning a further discussion of general null Lagrangian that exploits (15), the reader is referred to H. Rund [4], [7].

Finally we shall derive a handy condition that is both necessary and sufficient for F to be a null Lagrangian. For the proof, we have to assume that the boundary $\partial\Omega$ is of class C^1. We shall also add two examples that will show how this condition can be used. We can view the following result as a generalization of the Corollary of Proposition 2.

Proposition 5. *Suppose that G is of type* (6).

(i) *If there are functions $\omega^\alpha(x, z, p)$ of class $C^1(\bar{G} \times \mathbb{R}^{nN})$ such that*

(16)
$$F(x, u(x), Du(x)) = D_\alpha \omega^\alpha(x, u(x), Du(x))$$

holds on $\bar{\Omega}$ for every $u \in C^2(\bar{\Omega}, \mathbb{R}^N)$, then F is a null Lagrangian.

(ii) *Conversely, if F is a null Lagrangian of class C^2, and if the sets $B(x)$ appearing in the definition of* (6) *are starshaped with respect to some fixed point z_0, then there exist functions $\omega^\alpha(x, z, p)$ of class $C^1(\bar{G} \times \mathbb{R}^{nN})$ such that* (16) *holds on $\bar{\Omega}$ for every $u \in C^2(\bar{\Omega}, \mathbb{R}^N)$.*

Proof. (i) Integrating (16) over Ω and applying a partial integration, we infer that

(17)
$$\mathscr{F}(u) = \int_{\partial\Omega} \nu_\alpha(x)\omega^\alpha(x, u(x), Du(x))\, d\mathscr{H}^{n-1}(x)$$

holds for every $u \in C^2(\bar{\Omega}, \mathbb{R}^N)$, whence we obtain that

$$\mathscr{F}(u + \varepsilon\varphi) = \mathscr{F}(u)$$

is satisfied for each $\varphi \in C^\infty_c(\Omega, \mathbb{R}^N)$ and $|\varepsilon| \ll 1$. This implies

$$\delta\mathscr{F}(u, \varphi) = 0 \quad \text{for all } \varphi \in C^\infty_c(\Omega, \mathbb{R}^N)$$

and, consequently, $L_F(u) = 0$ for arbitrary $u \in C^2(\bar{\Omega}, \mathbb{R}^N)$.

(ii) Without loss of generality we can assume that all sets $B(x)$ are starshaped with respect to $z_0 = 0 \in \mathbb{R}^N$. Then we can write

$$\begin{aligned} &F(x, u(x), Du(x)) - F(x, 0, 0) \\ &= \int_0^1 \frac{d}{d\varepsilon} F(x, \varepsilon u, \varepsilon Du)\, d\varepsilon \\ &= \int_0^1 \delta F(\varepsilon u, u)\, d\varepsilon \\ &= \int_0^1 L_F(\varepsilon u)\cdot u\, d\varepsilon + \int_0^1 \operatorname{div}\{u \cdot F_p(x, \varepsilon u, \varepsilon Du)\}\, d\varepsilon. \end{aligned}$$

Since F is a null Lagrangian, we arrive at

$$F(x, u(x), Du(x)) = F(x, 0, 0) + \operatorname{div} \int_0^1 u(x)\cdot F_p(x, \varepsilon u(x), \varepsilon Du(x))\, d\varepsilon.$$

We can choose a vector field $v(x)$ of class $C^1(\bar{\Omega}, \mathbb{R}^n)$ such that

$$F(x, 0, 0) = \operatorname{div} v(x)$$

holds. Setting

$$\omega^\alpha(x, z, p) := v^\alpha(x) + \int_0^1 z^i F_{p^i_\alpha}(x, \varepsilon z, \varepsilon p)\, d\varepsilon,$$

we obtain that (16) is satisfied for every $u \in C^2(\bar{\Omega}, \mathbb{R}^N)$.

Hence we have proved that the existence of a C^1-vector field $\omega = (\omega^1, \ldots, \omega^n)$ satisfying

$$F(x, u(x), Du(x)) = \operatorname{div}\{\omega(x, u(x), Du(x))\}$$

for each $u \in C^2(\bar{\Omega}, \mathbb{R}^N)$ is both necessary and sufficient for F to be a null Lagrangian. □

Remark. Formula (17) seems to imply that we have to assume $u = v$ on $\partial\Omega$ as well as $Du = Dv$ on $\partial\Omega$ in order to prove $\mathscr{F}(u) = \mathscr{F}(v)$. Yet the Propositions 2 and 3 of *4.1* show that, for a null Lagrangian F, we have $\mathscr{F}(u) = \mathscr{F}(v)$ if $u, v \in C^1(\bar{\Omega}, \mathbb{R})$, $u = v$ on $\partial\Omega$, and $\operatorname{graph}(u) \subset G$, $\operatorname{graph}(v) \subset G$. Then formula (17) yields

(18) $$\int_{\partial\Omega} \nu_\alpha(x)\omega^\alpha(x, u(x), Du(x))\, d\mathscr{H}^{n-1}(x) = \int_{\partial\Omega} \nu_\alpha(x)\omega^\alpha(x, v(x), Dv(x))\, d\mathscr{H}^{n-1}(x).$$

Note that, at the right-hand side of (16), second derivatives of u will appear, but not at the left-hand side.

Two special cases. (i) If $\omega^\alpha(x, z, p) = S^\alpha(x, z)$, then the null Lagrangian F described by (16) is of the special form considered in (7) and (8). This expression is used in Weyl's field theory (cf. Supplement 3).

(ii) Suppose that there are functions $S^\alpha(x, z)$ on G such that

(19) $$F(x, u(x), Du(x)) = \det\{D_\beta S^\alpha(x, u(x))\}$$

or equivalently,

$$F(x, u(x), Du(x))\, dx^1 \wedge \cdots \wedge dx^n = dS^1(x, u(x)) \wedge \cdots \wedge dS^n(x, u(x)).$$

Then F possesses the divergence structure (16) since

(20) $$d[S^1 \wedge dS^2 \wedge \cdots \wedge dS^n] = dS^1 \wedge dS^2 \wedge \cdots \wedge dS^n.$$

This null Lagrangian F, which can be written as

(21) $$F(x, z, p) = \det\{S^\alpha_{x^\beta}(x, z) + S^\alpha_{z^i}(x, z)p^i_\beta\},$$

is used in Carathéodory's field theory for multiple integrals.

The examples (16) and (19) generalize the two null Lagrangians $F(Du) = \operatorname{div} u$ and $F(Du) = \det Du$ considered in the introduction to this section.

5. Variational Problems of Higher Order

In a similar way as in *2.1* and *2.2* we are led to Euler equations for higher order variational integrals

(1) $$\mathscr{F}(u) = \int_\Omega F(x, u(x), Du(x), D^2u(x), \ldots, D^m u(x))\, dx.$$

Under suitable differentiability assumptions on $u(x)$, $\varphi(x)$, and the Lagrangian

$F(x, u, p, q, r, \ldots, s)$, we obtain as first variation of $\mathscr{F}$ at u in direction of φ the formula

(2) $$\delta\mathscr{F}(u, \varphi) = \int_\Omega \{F_u(\ldots)\cdot\varphi + F_p(\ldots)\cdot D\varphi + F_q(\ldots)\cdot D^2\varphi + \cdots + F_s(\ldots)\cdot D^m\varphi\}\, dx,$$

where $(\ldots)$ stands for $(x, u(x), Du(x), \ldots, D^m u(x))$. Thus the Euler equations have the form

(3) $$F_u - DF_p + D^2 F_q - \cdots + (-1)^m D^m F_s = 0,$$

where $F_u, F_p, F_q, \ldots, F_s$ stand for the functions $F_u(x, u(x), Du(x), \ldots, D^m u(x))$, etc., of x.

Let us, particularly, consider variational integrals of second order:

$$\mathscr{F}(u) = \int_\Omega F(x, u(x), Du(x), D^2u(x))\, dx.$$

The associated Euler equations are

(4) $$F_{u^i}(\ldots) - D_\alpha F_{p^i_\alpha}(\ldots) + D_\alpha D_\beta F_{q^i_{\alpha\beta}}(\ldots) = 0,$$

with $(\ldots) = (x, u(x), Du(x), D^2u(x))$.

We now consider a few examples.

1 The equilibrium of thin plates leads to the study of variational integrals

$$\mathscr{I}(u) = \tfrac{1}{2}\int_\Omega |\Delta u|^2\, dx,$$

with the first variation

$$\delta\mathscr{I}(u, \varphi) = \int_\Omega \Delta u\, \Delta\varphi\, dx.$$

This yields the so-called *plate equation* or *biharmonic equation*

$$\Delta\Delta u = 0 \quad \text{in } \Omega$$

as corresponding Euler equation.

2 The Euler equation of

$$\int_\Omega \{\tfrac{1}{2}|\Delta u|^2 + F(x, u, Du)\}\, dx$$

is given by

$$\Delta\Delta u + L_F(u) = 0.$$

where $L_F(u)$ denotes the Euler operator of $F(x, z, p)$.

Null Lagrangians of higher order $F(x, z, p, q, \ldots, s)$ will analogously be defined as variational integrands whose Euler operator

$$L_F(u) := F_u - DF_p + D^2 F_q - \cdots + (-1)^m D^m F_s$$

annihilates all functions u of class C^{2m}.

We can state a necessary and sufficient condition for null Lagrangians of higher order which is the analogue of Proposition 5 in 4.2 and is proved in the same way.

A variational integrand $F(x, z, p, q, \ldots)$ is a null Lagrangian if and only if there are functions $\omega^\alpha(x, z, p, q, \ldots)$, $1 \le \alpha \le n$ such that

$$F(x, u(x), Du(x), \ldots, D^m u(x)) = D_\alpha \omega^\alpha(x, u(x), Du(x), \ldots, D^{2m-1}u(x)) \tag{5}$$

holds on $\bar{\Omega}$ for every $u \in C^{2m}(\bar{\Omega}; \mathbb{R}^N)$.

3 Let $N = 1$, $n = 2$, $x = x^1$, $y = x^2$, and

$$F(D^2 u) = \det D^2 u = u_{xx}u_{yy} - u_{xy}^2 .$$

Then F is a null Lagrangian since we can write

$$\begin{aligned} F(D^2 u) &= (u_x u_{yy})_x - (u_x u_{xy})_y = -(u_y u_{xy})_x + (u_y u_{xx})_y \\ &= -\tfrac{1}{2}\{(u_x)^2_{yy} - 2(u_x u_y)_{xy} + (u_y)^2_{xx}\} . \end{aligned} \tag{6}$$

The same holds for $N = 1$, $n \ge 2$, and $F(D^2 u) = \det D^2 u$, on account of the relation

$$\begin{aligned} (\det D^2 u)\cdot dx^1 \wedge \cdots \wedge dx^n &= du_{x^1} \wedge du_{x^2} \wedge \cdots \wedge du_{x^n} \\ &= d\{(u_{x^1})\cdot du_{x^2} \wedge \cdots \wedge du_{x^n}\} . \end{aligned}$$

4 A celebrated theorem by Gauss and Bonnet states that the total curvature (= curvatura integra) $\int_M K\, dA$ of an arbitrary closed and orientable, two-dimensional submanifold M of $\mathbb{R}^3$ is a topological invariant. Precisely speaking, we have

$$\int_M K\, dA = 4\pi(1 - p) = 2\pi\chi(M) . \tag{7}$$

Here K denotes the Gauss curvature of M, dA its area element, and $\int_M K\, dA$ the total curvature of M, whereas p is the genus of M and $\chi(M)$ its Euler characteristic, i.e., $\chi(M) = 2(1 - p)$; cf. e.g. Alexandrov–Hopf [1], Seifert–Threlfall [1], Blaschke [5], Dubrovin–Fomenko–Novikov [1]. We want to sketch a proof for surfaces of the topological type of the sphere, i.e., for $p = 0$, by using null Lagrangians. In this case we have to prove

$$\textstyle\int_M K\, dA = 4\pi . \tag{8}$$

Let us locally represent M as graph of some function $z = u(x, y)$ above some domain Ω in the x, y-plane, and suppose that $u \in C^2(\bar{\Omega})$. Then the area element dA of M can be written as

$$dA = \sqrt{1 + u_x^2 + u_y^2}\, dx\, dy$$

and its Gauss curvature is given by[9]

$$K = \frac{u_{xx}u_{yy} - u_{xy}^2}{(1 + u_x^2 + u_y^2)^2} .$$

Hence we obtain

$$K\, dA = F(Du, D^2 u)\, dx\, dy \tag{9}$$

[9] See Supplement, nr. 6.

if we introduce the Lagrangian

$$F(Du, D^2u) := \frac{u_{xx}u_{yy} - u_{xy}^2}{(1 + u_x^2 + u_y^2)^{3/2}}. \tag{10}$$

Let

$$\mathscr{F}(u) := \int_\Omega F(Du, D^2u)\, dx\, dy \tag{11}$$

be the variational integral associated with F. Then we have

$$\int_{M \cap \mathscr{U}} K\, dA = \mathscr{F}(u), \tag{12}$$

where $\mathscr{U}$ is a sufficiently small ball in $\mathbb{R}^3$ where $M \cap \mathscr{U}$ is given by $M \cap \mathscr{U} = \text{graph } u$.

We claim that there is some 1-form

$$\omega = a(Du, D^2u)\, dy - b(Du, D^2u)\, dx \tag{13}$$

such that

$$K\, dA = F(Du, D^2u)\, dx \wedge dy = d\omega, \tag{14}$$

or equivalently, that

$$F(Du, D^2u) = \frac{\partial}{\partial x} a(Du, D^2u) + \frac{\partial}{\partial y} b(Du, D^2u). \tag{15}$$

Suppose that (14) or (15) would be verified. Then it follows from our previous discussion that F is a null Lagrangian, and that $\mathscr{F}(u)$ remains fixed among all deformations of u keeping the boundary values of u fixed. In particular we have

$$\mathscr{F}(u) = \mathscr{F}(v)$$

for all $v \in C^2(\bar{\Omega})$ satisfying $u - v = 0$ on $\Omega - \Omega'$ where $\Omega' \subset\subset \Omega$. By means of (12) we infer that $\int_M K\, dA$ remains unchanged by local deformations of M (i.e., by deformations of M which are localized to a sufficiently small neighbourhood of some point of M). Repeating such local deformations, we can deform M via a family of manifolds M' into a 2-sphere S such that

$$\int_M K\, dA = \int_{M'} K\, dA = \int_S K\, dA.$$

If $S = S_R(0)$, then $\int_S dA = 4\pi R^2$ and $K = R^{-2}$ whence $\int_S K\, dA = 4\pi$, and formula (8) would be proved.

It remains to verify (15) for suitable functions a and b. Since a and b are certainly not uniquely determined by F if they exist at all, we should have a great choice among possible candidates. Hence any computation will be somewhat artificial if it is not guided by geometric intuition; the following verification of (15) is merely a clever guesswork.

Consider the functions

$$a := \frac{-u_{xy}u_y}{(1 + u_x^2)\sqrt{1 + u_x^2 + u_y^2}}, \qquad b := \frac{u_{xx}u_y}{(1 + u_x^2)\sqrt{1 + u_x^2 + u_y^2}}.$$

Then we compute that

$$\frac{\partial}{\partial x} a = \frac{1}{(1 + u_x^2 + u_y^2)^{3/2}(1 + u_x^2)^2} \cdot a^*,$$

where

$$a^* = [(u_{xxy}u_y + u_{xy}^2)(1 + u_x^2) - 2u_x u_y u_{xx} u_{xy}](1 + u_x^2 + u_y^2)$$
$$- (1 + u_x^2)(u_x u_y u_{xx} u_{xy} + u_y^2 u_{xy}^2),$$

and

$$\frac{\partial}{\partial x} b = \frac{1}{(1 + u_x^2 + u_y^2)^{3/2}(1 + u_x^2)^2} \cdot b^*,$$

$$b^* = [(u_{xxy}u_y + u_{xx}u_{yy})(1 + u_x^2) - 2u_x u_y u_{xx} u_{xy}](1 + u_x^2 + u_y^2) - (1 + u_x^2)(u_x u_y u_{xx} u_{xy} + u_y^2 u_{xx} u_{yy}).$$

It follows that $F = \frac{\partial}{\partial x} a + \frac{\partial}{\partial y} b$, and thus (15) is verified.

Generalizations of (7) have been proved by H. Hopf [1], Allendoerfer–Weil [1], Chern [1], and many others. Further references can be found in Kobayashi–Nomizu [1], Vol. 2, notes 20 and 21, pp. 358–364.

[5] *Curvature integrals for planar curves.* Let us consider a smooth immersed curve $\mathscr{C}$ in $\mathbb{R}^2$, i.e. we assume that $\mathscr{C}$ has a parametric representation $c(t)$, $t \in [a, b]$, satisfying $\dot{c}(t) \neq 0$. Let $\{\mathfrak{t}(t), \mathfrak{n}(t)\}$ be the moving frame along $c(t)$ consisting of the tangent vector $\mathfrak{t} := \dot{c}/|\dot{c}|$ and the (unit) normal $\mathfrak{n}$ which is positively oriented with respect to $\mathfrak{t}$, that is,

$$\det(\mathfrak{t}, \mathfrak{n}) = 1. \tag{16}$$

The curvature function $\kappa = \kappa_c$ of $\mathscr{C}$ is defined by the formula

$$\dot{\mathfrak{t}} = |\dot{c}| \kappa \mathfrak{n} \tag{17}$$

and the line element is given by $ds = |\dot{c}(t)|\, dt$. Moreover let $f(r)$ be a smooth real-valued function on $\mathbb{R}$. Then we define the curvature integral $\mathscr{F}(\mathscr{C})$ by the formula

$$\mathscr{F}(\mathscr{C}) := \int_{\mathscr{C}} f(\kappa)\, ds = \int_a^b f(\kappa_c)|\dot{c}|\, dt, \tag{18}$$

which is independent of the parameter representation $c : [a, b] \to \mathbb{R}^2$ that we have chosen. Let $\varphi : [a, b] \to \mathbb{R}^2$ be a smooth vector field along c with compact support in (a, b), and consider the family of curves $\mathscr{C}_\varepsilon$ which have the parameter representations $c_\varepsilon(t)$ given by

$$c_\varepsilon(t) := c(t) + \varepsilon\varphi(t), \quad a \le t \le b. \tag{19}$$

Suppose that $\mathscr{F}$ is stationary at $\mathscr{C}$ with respect to all such variations, i.e. the first variation

$$\delta\mathscr{F}(\mathscr{C}, \varphi) := \frac{d}{d\varepsilon}\mathscr{F}(\mathscr{C}_\varepsilon)\Big|_{\varepsilon=0} \tag{20}$$

of $\mathscr{F}$ at $\mathscr{C}$ vanishes,

$$\delta\mathscr{F}(\mathscr{C}, \varphi) = 0. \tag{21}$$

It is both more convenient and more instructive to derive Euler's equations directly from (21) using the orthonormal frame $\{\mathfrak{t}, \mathfrak{n}\}$ than to apply the general formulas refering to Euclidean coordinates. In fact we obtain

$$\frac{d}{d\varepsilon}\mathscr{F}(c_\varepsilon) = \int_a^b \left\{ f'(\kappa_{c_\varepsilon})\left(\frac{d}{d\varepsilon}\kappa_{c_\varepsilon}\right)|\dot{c}_\varepsilon| + f(\kappa_{c_\varepsilon})\frac{d}{d\varepsilon}|\dot{c}_\varepsilon| \right\} dt,$$

where

$$\kappa_{c_\varepsilon} = |\dot{c}_\varepsilon|^{-3}[\dot{c}_\varepsilon, \ddot{c}_\varepsilon],$$

and $[\alpha, \beta]$ stands for $\det(\alpha, \beta)$ if $\alpha, \beta \in \mathbb{R}^2$.

From now on we assume that $\mathscr{C}$ is parametrized with respect to the arc length, i.e. $|\dot{c}(t)| \equiv 1$. Then we have

$$\dot{c} = \mathfrak{t}, \quad \ddot{c} = \dot{\mathfrak{t}} = \kappa\mathfrak{n}, \quad \dot{\mathfrak{n}} = -\kappa\mathfrak{t} = -\kappa\dot{c} \tag{22}$$

and

(23) $$\kappa = [\dot{c}, \ddot{c}] \quad (\text{where } \kappa = \kappa_c).$$

It follows that

$$\frac{d}{d\varepsilon}\kappa_\varepsilon\bigg|_{\varepsilon=0} = [\dot{c}, \ddot{\varphi}] + [\dot{\varphi}, \ddot{c}] - 3\kappa\dot{c}\cdot\dot{\varphi}$$

and

$$\frac{d}{d\varepsilon}|\dot{c} + \varepsilon\dot{\varphi}|\bigg|_{\varepsilon=0} = \dot{c}\cdot\dot{\varphi}.$$

Therefore we obtain

(24) $$\delta\mathscr{F}(\mathscr{C}, \varphi) = \int_a^b \{f'(\kappa)([\dot{c}, \ddot{\varphi}] + [\dot{\varphi}, \ddot{c}] - 3\kappa\dot{c}\cdot\dot{\varphi}) + f(\kappa)\dot{c}\cdot\dot{\varphi}\}\, dt.$$

If we choose *tangential variations* φ as $\varphi = \zeta\mathfrak{t}$, $\zeta \in C_c^\infty((a, b))$, then a straight-forward computation shows on account of (22), (23) and $[\mathfrak{t}, \mathfrak{n}] = 1$ that

$$\delta\mathscr{F}(\mathscr{C}, \varphi) = \int_a^b \{f'(\kappa)\dot{\kappa}\zeta + f(\kappa)\dot{\zeta}\}\, dt,$$

and an integration by parts yields

$$\delta\mathscr{F}(\mathscr{C}, \varphi) = \int_a^b \{f'(\kappa)\dot{\kappa}\zeta - f'(\kappa)\dot{\kappa}\zeta\}\, dt = 0,$$

for any choice of ζ. Consequently tangential variations lead to no relation at all for a stationary curve $\mathscr{C}$. This was to be expected from the parameter invariance of the integral (18) and from the fact that for tangential φ the representation $c_\varepsilon(t)$ is just a reparametrization of $c(t)$ apart from terms of higher order with respect to ε which do not influence the value of $\delta\mathscr{F}$.

On the other hand, for *normal variations* $\varphi = \zeta\mathfrak{n}$, $\zeta \in C_c^\infty((a, b))$ a brief computations yields

$$\delta\mathscr{F}(\mathscr{C}, \varphi) = \int_a^b \{f'(\kappa)\ddot{\zeta} + (f'(\kappa)\kappa^2 - f(\kappa)\kappa)\zeta\}\, dt,$$

whence

(25) $$\delta\mathscr{F}(\mathscr{C}, \varphi) = \int_a^b \left\{\frac{d^2}{dt^2}f'(\kappa) + f'(\kappa)\kappa^2 - f(\kappa)\kappa\right\}\zeta\, dt,$$

and therefore

(26) $$\delta\mathscr{F}(\mathscr{C}, \zeta\mathfrak{n}) = \int_a^b \{f'''(\kappa)\dot{\kappa}^2 + f''(\kappa)\ddot{\kappa} + f'(\kappa)\kappa^2 - f(\kappa)\kappa\}\zeta\, dt.$$

Then the fundamental lemma leads to the Euler equation

(27) $$f''(\kappa)\ddot{\kappa} + f'''(\kappa)\dot{\kappa}^2 + \kappa\{\kappa f'(\kappa) - f(\kappa)\} = 0$$

for $\kappa(t)$. Now $\mathfrak{t}$ is obtained by integrating

(28) $$\begin{pmatrix}\dot{\mathfrak{t}}\\ \dot{\mathfrak{n}}\end{pmatrix} = \begin{pmatrix}0 & \kappa\\ -\kappa & 0\end{pmatrix}\begin{pmatrix}\mathfrak{t}\\ \mathfrak{n}\end{pmatrix},$$

and finally the extremals of (18) are determined by integrating $\dot{c} = \mathfrak{t}$.

Note that $\kappa(t) \equiv 0$ is always a solution of (27), and *therefore all straight lines are extremals of the integral* $\mathscr{F}$, *for any choice of* f.

If $\kappa(t) \neq 0$ we can obtain $c(t)$ from $\kappa(t)$ in a very convenient way by introducing the inclination angle $\omega(t)$ of $\mathscr{C}$ at $c(t)$ with respect to a given axis, say, the x-axis. This angle is given by

(29)
$$\omega(t) = \alpha + \int_a^t \kappa(t)\,dt, \quad \alpha := \arc\tan\frac{\dot y(a)}{\dot x(a)}$$

if $c(t) = (x(t), y(t))$. In fact, $[\mathfrak{t}, \mathfrak{n}] = 1$ implies that $\kappa = [\mathfrak{t}, \kappa\mathfrak{n}] = [\dot c, \ddot c] = \dot x\ddot y - \dot y\ddot x$ whence

$$\omega(t) = \alpha + \int_a^t \frac{\dot x\ddot y - \dot y\ddot x}{\dot x^2 + \dot y^2}\,dt = \alpha + \arc\tan\frac{\dot y(t)}{\dot x(t)}\bigg|_a^t,$$

that is,

$$\omega(t) = \arc\tan\frac{\dot y(t)}{\dot x(t)}$$

and therefore $\frac{dy}{dx} = \tan\omega$, as we have claimed. This implies

(30)
$$\dot x(t) = \cos\omega(t), \qquad \dot y(t) = \sin\omega(t),$$

and consequently $c(t) = (x(t), y(t))$ with $c(t_0) = (x_0, y_0)$ is given by

(31)
$$x(t) = x_0 + \int_a^t \cos\omega(t)\,dt,$$
$$y(t) = y_0 + \int_a^t \sin\omega(t)\,dt,$$

and the inclination angle $\omega(t)$ is obtained from the curvature function $\kappa(t)$ by (29).

Of particular interest is the integral

(32)
$$\mathcal{F}(\mathscr{C}) = \int_a^b (\kappa^2 + \lambda)\,ds, \quad \kappa = \kappa_c,$$

where λ is supposed to be a real constant. This functional plays an important role in nonlinear elasticity, since it is connected with the study of *elastic lines* as we shall point out in 2,5. The Euler equation (27) has now the form

(33)
$$2\ddot\kappa + \kappa^3 - \lambda\kappa = 0.$$

Multiplying this equation by $\dot\kappa$ we infer that there is a constant μ such that

(34)
$$\dot\kappa^2 + \tfrac{1}{4}\kappa^4 - \tfrac{1}{2}\lambda\kappa^2 = \mu,$$

whence we obtain $t = t(\kappa)$ as an elliptic integral:

(35)
$$t = \int \frac{d\kappa}{\sqrt{\mu + \frac{1}{2}\lambda\kappa^2 - \frac{1}{4}\kappa^4}}.$$

Thus the solution $\kappa(t)$ of (33) can be described by elliptic functions, and the extremals $\mathscr{C}$ of (32) can be represented in the form (31) where $\omega(t)$ is given by (29).

We finally note that (33) has the solutions $\kappa(t) \equiv 0$ and $\kappa(t) \equiv \sqrt{\lambda}$ (if $\lambda \ge 0$). Hence *straight lines as well as circles of radius* $1/\sqrt{\lambda}$ *are extremals of the integral* (32).

[6] *Rotation number of a closed curve in* $\mathbb{R}^2$. For a curve $\mathscr{C}$ in $\mathbb{R}^2$ represented by a smooth immersion $c: [a, b] \to \mathbb{R}^2$ we consider the integral

(36)
$$\mathcal{F}(\mathscr{C}) = \int_a^b \kappa_c|\dot c|\,dt = \int_{\mathscr{C}} \kappa\,ds,$$

where $\kappa = \kappa_c$ is the curvature function of $\mathscr{C}$ with respect to the representation c. Then the computations of [5], applied to $f(\kappa) = \kappa$, show that the Euler equation corresponding to $\mathcal{F}$ reduces the identity $0 = 0$ and that

$$\delta\mathcal{F}(\mathscr{C}, \varphi) = 0 \quad \text{for all } \varphi \in C_c^\infty((a, b), \mathbb{R}^2).$$

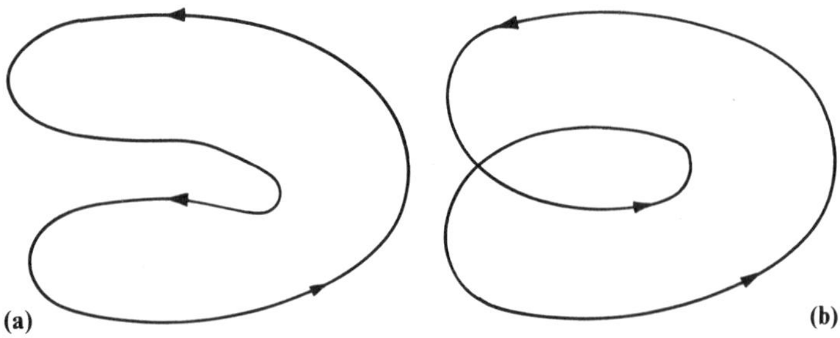

Fig. 19. The rotation number $\nu = \frac{1}{2\pi}\int_C \kappa\, ds$. **(a)** $\nu = 1$; **(b)** $\nu = 2$.

Hence, by writing $\mathscr{F}(\mathscr{C})$ in the form

$$\mathscr{F}(\mathscr{C}) = \int_a^b F(\dot{c}(t), \ddot{c}(t))\, dt,$$

we infer that *the integrand $F(\dot{c}, \ddot{c})$ is a null Lagrangian.* Thus the integral $\int_{\mathscr{C}} \kappa\, ds$ plays a similar role for closed curves in $\mathbb{R}^2$ as the integral $\int_M K\, dA$ for a closed surface M in $\mathbb{R}^3$.

Recall that $\omega(s) = \alpha + \int_a^s \kappa(s)\, ds$ is the inclination angle between the tangent $t(s) = \dot{c}(s)$ and the x-axis provided that α is a suitably chosen constant. Then we infer that for a closed smooth convex curve $\mathscr{C}$ we have $\omega(b) - \omega(a) = \pm 2\pi$, that is,

(37) $$\int_{\mathscr{C}} \kappa\, ds = \pm 2\pi.$$

More generally, a suitable "cutting device" yields that for any closed curve $\mathscr{C}$ which is smooth and regular there is an integer ν such that

(38) $$\int_{\mathscr{C}} \kappa\, ds = 2\pi\nu.$$

The number ν is called the *rotation number of the curve $\mathscr{C}$.*

The rotation number of a closed Jordan curve is ± 1, and $\nu = 1$ ($\nu = -1$) if $\mathscr{C}$ winds in the positive (negative) sense about its interior G.

[7] *Euler's area problem* consists in minimizing the area of the domain swept out by the curvature radius ρ of a curve $\mathscr{C}$ connecting two given points P_1 and P_2 in $\mathbb{R}^2$. If $\mathscr{C}$ is represented by a smooth immersion $c : [a, b] \to \mathbb{R}^2$ with the curvature $\kappa = \kappa_c$ and the curvature radius $\rho = 1/\kappa$, then the area in question is given by

(39) $$\mathscr{F}(\mathscr{C}) = \int_{\mathscr{C}} \rho\, ds = \int_{\mathscr{C}} \frac{ds}{\kappa} = \int_a^b \kappa_c^{-1} |\dot{c}|\, dt,$$

where ds denotes the line element of $\mathscr{C}$. The extremals of the integral $\mathscr{F}(\mathscr{C}) = \int_{\mathscr{C}} f(\kappa)\, ds$ are characterized by the Euler equation

$$\frac{d^2}{dt^2}\{f'(\kappa)\} + \kappa^2 f'(\kappa) - \kappa f(\kappa) = 0$$

provided that $|\dot{c}(t)| = 1$, i.e. $t = s$. In case of the integral $\int \rho\, ds$ this equation reduces to

(40) $$(\rho^2)^{\cdot\cdot} + 2 = 0,$$

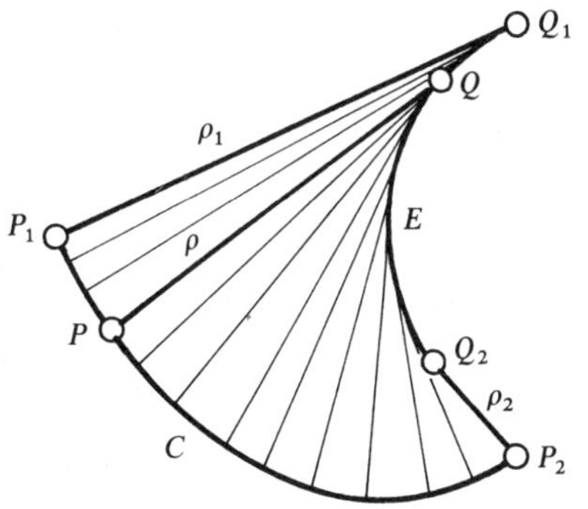

Fig. 20. Euler's area problem. Note that for a convex arc C with strictly increasing curvature the curvature radius ρ sweeps out a domain G bounded by $C = P_1P_2$, its evolute $E = Q_1Q_2$, and the connecting straight segments P_1Q_1 and P_2Q_2.

whence

$$\rho^2 = \beta^2 - (t - \alpha)^2, \quad \beta > 0,$$

with suitable constants α, β, i.e.

$$\rho(t) = \pm\sqrt{\beta^2 - (t - \alpha)^2}, \qquad |\dot{c}(t)| = 1. \tag{41}$$

The corresponding curves are cycloidal arcs. In fact, the cycloid generated by a wheel of radius R rolling on the upper side of the x-axis is described by

$$\begin{aligned} x &= R(\varphi - \sin\varphi) + x_0, \\ y &= R(1 - \cos\varphi) = 2R\sin^2\frac{\varphi}{2}, \end{aligned} \qquad 0 \le \varphi \le 2\pi,$$

if we consider the full festoon connecting two consecutive cusps. The function of the arc length is given by

$$s = 4R\left(1 - \cos\frac{\varphi}{2}\right) = 8R\sin^2\frac{\varphi}{4},$$

and the inclination angle of the tangent, ω, is related to the turning angle φ by $\omega = \frac{1}{2}(\pi - \varphi)$ whence $d\omega = -\frac{1}{2}\,d\varphi$ and $ds = 2R\sin\frac{\varphi}{2}\,d\varphi$. From

$$\frac{1}{\rho} = \kappa = \frac{d\omega}{ds},$$

we then infer that

$$\rho = -4R\sin\tfrac{1}{2}\varphi,$$

whence

$$\rho(s) = -\sqrt{(4R)^2 - (s - 4R)^2} \quad \text{for } |s| \le 4R,$$

which corresponds to the cycloidal arc described by $|\varphi| \le \pi$, and $\rho(4R - t) = \rho(4R + t)$. Therefore

$$\rho(t + 4R) = -\sqrt{(4R)^2 - t^2} \quad \text{if } |t| \le 4R,$$

and $s = 4R + t$ with $-4R \le t \le 4R$ corresponds to the cycloidal arc described by $0 \le \varphi \le 2\pi$, and we infer at once that the solutions of $(\rho^2)^{\cdot\cdot} + 2 = 0$ are cycloidal arcs.

6. Scholia

Section 1

1. Today, it seems to be a most natural idea to view integrals of the kind

$$\mathscr{F}(u) = \int_\Omega F(x, u(x), Du(x))\, dx$$

as *functions of functions u*, and it is difficult to imagine that this has been a revolutionary idea in 1887 when Volterra [2], [3] studied *functions of curves*. When one is looking at the history of the calculus of variations, it becomes even less understandable why this idea should have been so great a novelty because it should have been so obvious to anyone working in the field. Yet one has to realize that even the general notion of a function of one or several real variables only obtained its final form in the last century. Thus it took some time until the functional analytic point of view was integrated into the common mathematical education. The functional analytic program, extensively formulated by Hadamard (see [5], p. 2260), is very anti-Kronecker.[10]

The calculus of variations certainly is the oldest and most important root of functional analysis. Hadamard wrote in [5]: *Le Calcul des Variations est, pour les opérations fonctionelles, ce que le Calcul differentielle est pour les fonctions.*[11]

Nevertheless many other fields have contributed to the origin of functional analysis; we in particular mention the creation of set theory by Cantor, the development of the theory of integral equations by Volterra, Fredholm, Hilbert, and F. Riesz, and the study of the theory of topological spaces by Fréchet, Hausdorff, and Banach. It is most interesting to read Hadamard's address [6] to the International Mathematical Congress at Bologna in 1928.

2. The notion of a *functional* has been introduced by Hadamard in his treatise on the Calculus of variations (cf. [4], p. 282), whereas the notions of *functional equations* and *functional operations* seem to be much older; cf. the survey paper by Pincherle [2]. Nowadays one sees little reason to distinguish between functions of a real variable and functions of more general objects. Often the terms *functions* and *mappings* are synonymously used. Nevertheless, also the old name *functional* is kept for real (or complex) valued functions defined on some general set.

3. A theory of differentiation of functionals was inaugurated by Volterra and, more abstractly, by Fréchet and Gâteaux, see also P. Levy [1]. The notion of a *first variation* of some functional is essentially due to Euler (1771); cf. Euler [6]. The symbol $\delta\mathscr{F}$ was introduced by Lagrange (1755) (see Lagrange [13]).

4. A proof of the Morse lemma can be found in Abraham–Marsden [1], pp. 175–176; Milnor [1], pp. 6–8; Zeidler [1] I, p. 110, IV, p. 560; Guillemin–Sternberg [1], pp. 16–19.

The normal form (6) implies that *nondegenerate critical points are isolated critical points* and, moreover, that a C^3-function behaves in a sufficiently small neighborhood of a nondegenerate critical point like a nondegenerate quadratic form Q, the index q of which is determined by the number of negative eigenvalues of the Hessian matrix $D^2\mathscr{F}(u_0)$. If, however, u_0 is a degenerate critical point of $\mathscr{F}$, the behaviour of $\mathscr{F}$ in the neighborhood of u_0 is much more complicated, as we see from the following examples:

---

[10] Cf. for example H.M. Edwards, *An appreciation of Kronecker*, The Mathematical Intelligencer **9**, Nr. 1, 28–35 (1987).

[11] "*The calculus of variations is for functionals what the differential calculus is for functions.*"

[1] The graph of $\mathscr{F}(u, v) = u^2$ describes a parabolic cylinder, and all points of the u-axis are nonisolated, degenerate critical points.

[2] The graph of $\mathscr{F}(u, v) = u^2 v^2$ is a nonconvex surface over the u, v-plane, and all points of the u-axis and the v-axis are nonisolated, degenerate critical points.

[3] The function $\mathscr{F}(u, v) = e^{-1/r} \sin^2\left(\frac{1}{r}\right)$, $r = \sqrt{u^2 + v^2}$, possesses the origin as a degenerate, nonisolated critical point, at which the circles with radii $r = 1/(n\pi)$, $n = 1, 2, \ldots$, cluster. These circles consist of degenerate, nonisolated critical points.

[4] The function $\mathscr{F}(u, v) = u^3 - 3uv^2$, which is the real part of the holomorphic function $(u + iv)^3$, has the origin as degenerate but isolated critical point. Its graph is a monkey saddle that by the u, v-plane is intersected into six simply connected parts, three of which lie above the plane, the other ones below.

5. Sufficient conditions for the first variation $\delta\mathscr{F}(u_0, \zeta)$ of some general functional $\mathscr{F}$ to be linear with respect to ζ can be found in Vainberg [1], pp. 37–40. It, however, is an easy exercise to verify that $\delta\mathscr{F}(u_0, \zeta)$ is positively homogeneous of first order with respect to ζ.

Section 2

1. The modern history of the calculus of variations starts with Johann Bernoulli's challenge to solve the brachystochrone problem (1696) that was mainly directed against his elder brother Jacob. The two Bernoullis and Johann's pupil Euler are the founders of the variational calculus. The methods of the early period were completed and masterly described in Euler's celebrated treatise "*Methodus inveniendi lineas curvas maximi minimive proprietate gaudentes sive solutio problematis isoperimetrici latissimo sensu accepti*" from 1744. During this period and, actually, at least until the end of the eighteenth century, mathematicians used the name *method of the isoperimetric problem* instead of *calculus of variations*, which was coined by Euler only in 1756, after he had learned of Lagrange's δ-calculus. Also later Euler used the notation "calculus of variations" but rarely; mostly he spoke of shortest lines, of isoperimetric problems, of curves upon which a maximal property is imposed, of brachystochrones. As Carathéodory has stated: "*He had no particular name for the discipline which he had developed with such great success, and which really he had founded anew.*"

The first paper from another author than Euler carrying the notation "calculus of variations" in its title is Legendre's great mémoire[12] from 1786.

The early history of the calculus of variations is a well beaten track which we do not want to follow once again. We will instead refer the reader to several excellent introductions, in particular to the four surveys by Carathéodory, given in:

(1) *The beginning of research in the calculus of variations* ([12]; see *Schriften* [16], Vol. 2, pp. 93–107).

(2) *Basel und der Beginn der Variationsrechnung* ([14]; see *Schriften* [16], Vol. 2, pp. 108–128).

(3) *Einführung in Eulers Arbeiten über Variationsrechnung* ([15]; see *Schriften* [16], Vol. 5, pp. 107–174, cf. also: Euler, *Opera Omnia* I 24, pp. VIII–LXIII).

(4) *Geometrische Optik* (cf. [11], pp. 1–15).

---

[12] *Memoire sur la manière de distinguer les maxima des minima dans le Calcul des Variations*, Mémoires de l'Acad. roy. des Sciences (1786), 1788, 7–37.

A very detailed survey of the history of the one-dimensional calculus of variations has been given by H.H. Goldstine [1], who reviewed the main lines of development from the time of Fermat (1662) until the beginning of this century. There one can also find a comprehensive bibliography.

A more popular historical account putting the origin of the calculus of variations into perspective to the general development of science can be found in Hildebrandt–Tromba [1].

A selection of some of the papers by Johann and Jacob Bernoulli, Euler, Lagrange, Legendre, and Jacobi, with annotations by P. Stäckel, can be found (in a German translation) in Ostwald's Klassiker der exakten Wissenschaften Nr. 46 (1894) and Nr. 47 (1894).

An introduction to the early achievements of the calculus of variations is given in the treatises by Woodhouse [1], Lagrange [9], and Todhunter [1]. The contributions of Gauss to the calculus of variations are discussed by Bolza in his paper "Gauss und die Variationsrechnung" (cf. Gauss, Werke [1], Vol. 10, Bolza's treatise [1] and the monograph by Funk [1]).

2. The notation *Lagrangian* (or *Lagrange function*) for a variational integrand $F(x, z, p)$ has been coined in the physical literature; originally, it was only used for integrands of the type $F = T - V$, where T and V denote the kinetic and potential energy of some physical system. Carathéodory [10], p. 190, uses the notation *Grundfunktion* ("basic function").

3. Euler's differential equation was first stated by Euler in his *Methodus inveniendi* [2], Chapter 2, nr. 21. Quite often, one speaks of Lagrange's differential equation, or of the Euler–Lagrange-equations. Yet Lagrange himself attributes this equation to Euler: "*Cette équation est celle qu'Euler a trouvée le premier*" (Oeuvres [12], Vol. 10, p. 397).[13] Euler derived this equation in a geometric way by considering any variational integral as sum of infinitesimals, to which the rules of ordinary calculus are applied. Although Euler's derivation is not quite rigorous, at least not by our present-day standard, his approach is perfectly sound and can easily be made accurate (cf. A. Kneser [4]).

The name *extremal* for C^2-solutions of the Euler equations has been introduced by A. Kneser [3]. Tonelli has denoted *weak extremals as extremaloids* [1], but we prefer the first terminology, because it corresponds to the expression *weak solution* of a partial differential equation which is nowadays used.

4. With Lagrange's letter of August 12, 1755 to Euler, the development of the variational calculus took a new turn.[14] In this letter, Lagrange explained the δ-symbolism which quickly leads to Euler's equation. Euler was very impressed by the new method, and already one year later he lectured about it at the Berlin Academy. In the "registres" of the academy, the protocol nr. 441 about the meeting on September 16, 1756, reads: "*Mr. Euler a lû Elementa calculi variationum.*" At this and the preceding meeting (September 9, 1756), Euler for the first time lectured about Lagrange's δ-calculus; but his notes were published[15] much later, because Euler wanted to respect Lagrange's right of priority. Lagrange published his method in two papers, which appeared in 1762 and 1772, respectively.[16] There Lagrange considered the first variation of the area functional and derived the *minimal surface* equation.

Lagrange and also Euler first believed that the calculus of variations was an entirely new calculus, so to say a differential calculus on a higher level. In his "Institutionum calculi integralis",

[13] "*That is the equation that Euler was the first to discover.*"

[14] Lagrange, Oeuvres [12], Vol. 14, pp. 138–144.

[15] (E296) *Elementa calculi variationum*, Novi comment. acad. sci. Petrop. **10** (1764), 1766, 51–93; (E297) *Analytica explicatio methodi maximorum et minimorum*, Novi comment. acad. sci. Petrop. **10** (1764), 1766, 94–134. Cf. also: Opera omnia, Ser. I, Vol. 25.

[16] *Essai d'une nouvelle méthode pour déterminer les maxima et les minima des formules intégrales indéfinies*, Miscellanea Taurinensia **2** (1760/61), 1762, 173–195; *Sur la méthode des variations*, Misc. Taur. **4** (1766/69), 1771, 163–187.

Euler once again represented the δ-calculus, this time also admitting variations of the independent variable.[17] Moreover, he derived the first variation of general double integrals.

About 1771, Euler discovered the artifice, by which the variational calculus can be reduced to well-known methods of the differential calculus.[18] It consists in embedding a given extremal $u(x)$ in a family $\psi(x, \varepsilon)$ of mappings such that the relations

$$\psi(x, 0) = u(x) \quad \text{and} \quad \frac{\partial \psi}{\partial \varepsilon}(x, 0) = \varphi(x)$$

hold for an arbitrary vector field $\varphi(x)$. Introducing $\Phi(\varepsilon) := \mathscr{F}(\psi(\varepsilon))$, one obtains $\Phi'(0) = 0$ if u is a minimizer (or maximizer) of $\mathscr{F}$. This is nowadays the standard method to derive the Euler equation; we have presented it in Subsection 2. Euler's technique in combination with Lagrange's symbolism is so flexible and powerful, that the name "calculus of variations" is truly justified. Since the middle of the nineteenth century, this is the generally accepted designation for the field involving extremum problems related to simple or multiple integrals.

5. For some time it was customary to use the symbols

$$\frac{d}{dx^\alpha} G(x, u(x), Du(x)) \quad \text{and} \quad \frac{\partial}{\partial x^\alpha} G(x, u(x), Du(x))$$

instead of our notations

$$D_\alpha G(x, u(x), Du(x)) \quad \text{and} \quad G_{x^\alpha}(x, u(x), Du(x)),$$

respectively. In other words, $\frac{\partial}{\partial x^\alpha} G$ stands for the partial derivative G_{x^α}, whereas $\frac{d}{dx^\alpha} G$ denotes the total derivative $G_{x^\alpha} + G_{z^i} u^i_{x^\alpha} + G_{p^i_\beta} u^i_{x^\alpha x^\beta}$.

However, the notation used by us is presently preferred by the majority of authors. The reader is to be reminded not to confuse our notation $D_\alpha G$ (or $\frac{\partial}{\partial x^\alpha} G$) with that of older authors when he is consulting the old literature.

6. Lagrange[19] discovered that the function

$$u(x) = \int_G \frac{\gamma(y)}{|x - y|} \, dy, \quad G \subset \mathbb{R}^3,$$

has the property that, for any point $x \in \mathbb{R}^3 - \bar{G}$, the vector grad $u(x)$ describes the gravitational force, exerted by a mass distribution on G of density $\gamma(y)$ upon a unit mass concentrated at x. Green (1828) and Gauss (1838) suggested the names *potential function* and *potential*, respectively, for the function u. Apparently Laplace[20] first stated that, in cartesian coordinates x^1, x^2, x^3, the potential function satisfies the equation

$$D_1^2 u + D_2^2 u + D_3^2 u = 0$$

in $\mathbb{R}^3 - \bar{G}$ which, therefore, is called *Laplace equation* or else *potential equation*. For the differential

---

[17] (E385) *Institutionum calculi integralis*, Vol. 2, appendix: *De calculo variationum*, Petersburg 1770; cf. Euler [5].

[18] (E420) *Methodus nova et facilis calculum variationis tractandi*, Novi Comment. Petropolis **16** (1771), 35–70.

[19] Par. sav. étr. 7 (1773), cf. Oeuvres **6**, p. 349.

[20] Par. Hist. 1787 (1789), p. 252; cf. Oeuvres **11**, p. 278. In Par. Hist. 1782 (1785), p. 135 (Oeuvres **10**, p. 302), Laplace had already given the formula with respect to polar coordinates.

expression on the left-hand side, Murphy introduced[21] the symbol Δu. The notation "*harmonic functions*" for the solutions of $\Delta u = 0$ goes back to Thomson and Tait (Treatise on natural philosophy 1867).

On G, Newton's potential satisfies *Poisson's equation*

$$\Delta u = -4\pi\gamma .$$

For a sphere G of constant density γ, this was discovered by Poisson,[22] the first correct proof is due to Gauss[23] under the assumption that $\gamma \in C^1(G)$. In his thesis (Tübingen, 1882), O. Hölder verified this equation under the weaker assumption $\gamma \in C^\alpha(G)$, $0 < \alpha < 1$. Hölder's result is the root of the celebrated Korn–Lichtenstein–Schauder estimates.

The potential equation $\Delta\phi = 0$ appeared for the first time in a paper by Euler[24] in 1761. Euler observed that the velocity $v(x)$ of an incompressible fluid satisfies

$$\operatorname{div} v = 0 .$$

Introducing the velocity potential $\phi(x)$ for an irrotational fluid motion by

$$v = \operatorname{grad} \phi ,$$

Euler obtained the equation

$$\Delta\phi = 0 ,$$

for which he then tried to find polynomial solutions, in default of a general method to construct solutions.

7. The first to give a general solution $u(x, t)$ of the one-dimensional wave equation was d'Alembert.[25] He proved that every solution $u(x, t)$ has to be of the form

$$u(x, t) = f(x + t) + g(x - t) .$$

In the sequel there arose a vivid dispute between Euler and d'Alembert about the nature of the arbitrary functions f and g. D'Alembert insisted that they had to be analytic, whereas Euler wanted to admit nonanalytic terms f and g, this way breaking with a sacred *dogma of Leibniz* that everything in nature should be representable by analytic expressions. Later Daniel Bernoulli and Lagrange participated in the discussion, but the question remained unresolved for more than fifty years. Only in the nineteenth century, Euler's opinion was accepted as the correct one. This fundamental dispute has been described by Riemann,[26] and in great detail by Truesdell.[27] We also refer to the recent treatise by Bottazini [1].

---

[21] According to H. Burkhardt and W.F. Meyer, *Potentialtheorie*, Enzyklopädie der Mathemat. Wissenschaften IIA7b, p. 468. See also: R. Murphy, *Elementary principles of the theory of electricity, heat and molecular actions*, **1**, Cambridge 1823, p. 93.

[22] N. Bull. philom. **3** (1813), p. 388.

[23] *Allgemeine Lehrsätze in Beziehung auf die im verkehrten Verhältnisse des Quadrats der Entfernung wirkenden Anziehungs- und Abstoßungskräfte*, Result. aus d. Beob. des magn. Vereins im Jahre 1839, Leipzig 1840 (cf. Werke [1], Vol. 5, pp. 206–211).

[24] (E258) *Principia motus fluidorum*, Novi comm. acad. sci. Petrop. **6** (1756/57), 1761, 271–311. (Cf. also: Opera omnia, Ser. II, Vol. 12, 133–168.) According to Jacobi, a paper with the title "*De motu fluidorum in genere*" has been read to the Berlin Academy on August 31, 1752.

[25] Mémoires de l'Académie de Berlin (1747), p. 214.

[26] *Ueber die Darstellbarkeit einer Funktion durch eine trigonometrische Reihe*, Habilitationsschrift, Göttingen 1854 (Göttinger Abh. **13** (1867)); cf. also Riemann's Werke, second ed., XII, 227–271.

[27] *The rational mechanics of flexible or elastic bodies* 1638–1788, in: Euler's Opera omnia, Ser. II, Vol. XI_2.

8. Whereas Lagrange in his paper of 1762 (Misc. Taur. **2**) had not been able to determine a nonplanar solution of the minimal surface equation, Meusnier in 1776 discovered that certain catenaries generate surfaces of revolution which are minimal surfaces.[28] Actually, these solutions are already contained among the solutions of two isoperimetric problems found by Euler in 1744.[29]

9. The method, presented in Section 2, to derive Euler's equations from the relation $\delta\mathscr{F}(u, \varphi) = 0$, by performing a partial integration and then applying the fundamental lemma, is due to Lagrange. For Lagrange, the fundamental lemma was self-evident, and so it was for all other mathematicians for about a century. According to Huke [1] the first rigorous proof was given by Sarrus in his prize-essay [1] from 1848. The present-day proof was essentially found by P. Du Bois-Reymond [1] in 1879. These ideas eventually led to the theory of distributions created by L. Schwartz [1]. A historical survey of the contributions to the fundamental lemma was given by A. Huke [1].

10. Already Euler discovered that, for Lagrangians of the type $F(z, p)$, the expression $F - p \cdot F_p$ forms a first integral of the Euler equations (cf. Methodus inveniendi [2], Chapter 2, nr. 30).

11. The various refinements and variants of the fundamental lemma presented in *2.3* have a long history, part of which is described in Bolza's treatise [3]. The proofs are considerably simplified by using *mollifiers* (or *smoothing operators*). This important tool has been developed by Friedrichs [2], [3], and earlier by Leray [1].

12. The derivation of the *natural boundary condition* proved to be a formidable task which could not be carried out by Euler, Lagrange, and the other great mathematicians of the eighteenth century, mainly because the tool of integration by parts for multiple integrals was not available to them. The first to derive a free boundary condition for a double integral was Gauss (1829) in his paper on equilibrium configurations of fluids subject to surface tension.[30]

Soon after the publication of Gauss's work (1830), there appeared several papers contributing to the variational calculus of multiple integrals: As the most influential one, we quote the paper by Poisson (1831).[31] According to Bolza [1], the succeeding publications by Ostrogradski (1836), Delaunay (1843), Cauchy (1844), Sarrus (1846) are more or less influenced by Poisson; yet it is unknown if Poisson had known Gauss's paper when he wrote his mémoire, since he does not quote it. On the other hand, Poisson cited Gauss in his "capillary paper" published in 1831.[32] For a detailed survey of the development of the variational calculus for multiple integrals and, connected with it, of the rule of partial integration, we refer the reader to Bolza [1] and Todhunter [1]. In any case, there is no doubt that the variational calculus caused considerable problems to the mathematicians of the last century, although their difficulties are quite incomprehensible to us. For instance, the prize mémoire of Sarrus (*Recherches sur le calcul des variations*, 1846) extends over some 127

---

[28] *Mémoire sur la courbure des surfaces*, Paris, Mémoires de Mathématique et de Physique (de savans etrangers) de l'Académie **10** (1785), 447–550 (lu 1776).

[29] *Methodus inveniendi lineas curvas* [2], Chapter 5, nrs. 44 and 47 (pp. 194–198); cf. also Opera omnia, Ser. I, Vol. 24, pp. 182 and 185 (E65).

[30] *Principia generalia theoriae figurae fluidorum in statu aequilibrii*, communicated to the Göttinger Ges. der Wiss. Sept. 28, 1829; appeared 1830, and also in Göttinger Abh. **7**, 39–88 (1832); cf. Werke **5**, 29–77.

[31] *Mémoire sur le calcul des variations*, Mém. Acad. roy. Sci. **12**, 223–331 (1833), communicated Nov. 10, 1831.

[32] See Bolza [1], pp. 42–44, and Todhunter [1], p. 54, nr. 84, remark (3).

pages. That even Jacobi was impressed by the difficulties caused by the variational calculus for multiple integrals, can be seen from his following remarks:[33]

"Lately, the most distinguished mathematicians such as Poisson and Gauss were concerned with the determination of the [first] variation of the double integral, which is causing infinite trouble because of the arbitrary functions. Nevertheless one is led to them by quite ordinary problems, for instance, by the problem to determine among all surfaces spanning a skew quadrilateral in space the one with smallest surface area. It is not known to me, that someone had thought of investigating the second variation of such double integrals. Also I myself have, despite much effort, merely realized that this topic belongs to the most difficult ones."

13. Newton's variational problem. Actually the first genuine problem of the calculus of variations in modern times was formulated by Newton in 1685, namely, to determine the shape of a body offering *least resistance* while moving with constant velocity through a homogeneous fluid. Suppose that the body has a maximal cross section $\Omega \subset \mathbb{R}^2$ orthogonal to the velocity vector and that the front surface of the body is described as a graph of a function $u : \Omega \to \mathbb{R}$. Then the law of friction postulated by Newton led him to the expression

$$\mathscr{F}(u) = c \int_\Omega \frac{dx}{1 + |Du|^2}, \quad c = \text{const},$$

as formula for the resistance of the body. If we assume that Ω is a disk $B_R(0)$ of radius R and that the body is rotationally symmetric, i.e. $u(x) = z(r)$, $r = |x|$, then by using polar coordinates r, θ about 0 we obtain $dx = r\,dr\,d\theta$, and $|Du(x)|^2 = z'(r)^2$, $' = d/dr$, and therefore $\mathscr{F}(u)$ can be written as

$$\mathscr{G}(z) = 2\pi c \int_0^R \frac{r}{1 + z'(r)^2}\,dr.$$

This functional has often been discussed in the literature, and there has been a still continuing discussion in how far Newton's resistance formula is physically relevant. We refer the reader to H.H. Goldstine [1], P. Funk [1], K. Weierstrass [1], Vol. 7, and to the recent paper by Buttazzo/Ferone/Kawohl [1] where also the minimization problem for the general integral

$$\int \frac{dx}{1 + |Du|^2}$$

is treated. Concerning Newton's original work we refer to Newton's *Papers* [2], Vol. VI, pp. 456–480, and to the *Principia* [1] (cf. Scholium to Proposition XXXIV).

---

[33] Jacobi's *Variations-Rechnung*, handwritten notes of lectures held by Jacobi (1837/38). The notes were taken by Rosenhain. The original text is the following:

Es haben sich in der neuesten Zeit die ausgezeichnetsten Mathematiker wie Poisson und Gauss mit der Auffindung der Variation des Doppelintegrals beschäftigt, die wegen der willkürlichen Funktionen unendliche Schwierigkeiten macht. Dennoch wird man durch ganz gewöhnliche Aufgaben darauf geführt, z.B. durch das Problem: unter allen Oberflächen, die durch ein schiefes Viereck im Raum gelegt werden können, diejenige anzugeben, welche den kleinsten Flächeninhalt hat. Es ist mir nicht bekannt, daß schon irgend jemand daran gedacht hätte, die zweite Variation solcher Doppelintegrale zu untersuchen; auch habe ich, trotz vieler Mühe, nur erkannt, daß der Gegenstand zu den allerschwierigsten gehört."

Section 3

1. The papers of Du Bois-Reymond [1], [2] were an important step in the development of the direct methods of the calculus of variations. The results of this section are built on his work and on extensions due to Hilbert, which he presented in his Göttingen lectures (summer term 1899). Hilbert's regularity proof is described in a paper by Whittemore [1]. The most general regularity results for weak extremals of one-dimensional variational problems can be found in the work of Tonelli.

2. References concerning the *history of Dirichlet's principle* can be found in Monna [1].

3. Example [6] in 1,*3.2* is due to Heinz [3]. To make this example completely rigorous, one needs boundary regularity results for H-surfaces, the most general of which are due to Heinz [4].

4. Example [7] in 1,*3.2* was given by Concus and Finn. References to the literature and a survey of related results can be found in Finn's monograph [1].

5. Apparently variational problems with an obstacle condition were first considered by Gauss and Steiner; cf. Bolza [1], pp. 81 Steiner [1] and [2], Vol. 2, pp. 177–308, and Weierstrass [2]. Concerning Weierstrass's results on obstacle problems we refer to the Scholia of Chapter 8 (see 8,5, no. 3).

6. The corner conditions for weak extremals of nonparametric variational problems were first published by Erdmann [1] in 1877. Weierstrass derived the corner conditions for weak extremals of parametric integrals already in 1865 and presented them in his lectures at Berlin (see also Chapter 8).

Section 4

1. Null Lagrangians can already be found in Euler's work.[34] Integrals of null Lagrangians play an essential role as invariant integrals (or independent integrals) in field theories. This was first discovered by Beltrami [2] in 1868 and rediscovered by Hilbert [1], [3]. The development of these methods for multiple integrals was proposed by Hilbert as problem 23 in his celebrated lecture, held at the International Mathematical Congress in Paris (1900); cf. [3]. Hilbert's program was in part carried out by Carathéodory, H. Weyl, and others; a survey of these results is given by the monographs of Funk [1], Rund [4], and Klötzler [4]. We shall describe some of the main results in Chapter 7.

2. Concerning the relation between null Lagrangians and *characteristic classes*, we refer the reader to Dubrovin–Fomenko–Novikov [1], Vol. 1, pp. 452–454.

3. The results on the characterization of null Lagrangians have in part been taken from Rund [4]. An approach to null Lagrangians via differential forms is described in Funk [1]. Various results and examples can be found in Olver [1].

---

[34] *Methodus inveniendi* (E65), Chapter 1, nr. 34, and Chapter 2, nr. 50.

Section 5

1. Curvature integrals for space curves. Here we generalize 1,5 $\boxed{5}$ to smooth curves $\mathscr{C}$ immersed in $\mathbb{R}^3$. Let any $\mathscr{C}$ be described by a parametric representation $c(t)$, $t \in [a, b]$, $\dot{c}(t) \neq 0$. For $\alpha, \beta, \gamma \in \mathbb{R}^3$ we denote the vector product of α and β by $\alpha \wedge \beta$, and $[\alpha, \beta, \gamma]$ is the spatial product

$$(1) \qquad [\alpha, \beta, \gamma] := \det(\alpha, \beta, \gamma) = (\alpha \wedge \beta)\cdot\gamma.$$

Then the curvature $\kappa(t)$ and the torsion $\tau(t)$ of the curve $\mathscr{C}$ with the representation $c(t)$ are given by

$$(2) \qquad \kappa := \frac{|\dot{c} \wedge \ddot{c}|}{|\dot{c}|^3}, \qquad \tau := \frac{[\dot{c}, \ddot{c}, \dddot{c}]}{|\dot{c} \wedge \ddot{c}|^2},$$

respectively, and $ds = |\dot{c}(t)|\, dt$ is the line element of $\mathscr{C}$. We write κ_c and τ_c instead of κ and τ if we want to indicate the dependence of c. Moreover let $f(u, v)$ be a smooth real-valued function on $\mathbb{R}^2$. Then we define the curvature integral $\mathscr{F}(\mathscr{C})$ by

$$(3) \qquad \mathscr{F}(\mathscr{C}) := \int_{\mathscr{C}} f(\kappa, \tau)\, ds = \int_a^b f(\kappa_c, \tau_c)|\dot{c}|\, dt.$$

Let $\varphi : [a, b] \to \mathbb{R}^2$ be a smooth vector field along c with compact support in (a, b), and consider the family of curves $\mathscr{C}_\varepsilon$ given by the representation

$$c_\varepsilon(t) = c(t) + \varepsilon\varphi(t), \quad a \le t \le b.$$

The first variation of $\mathscr{F}$ at $\mathscr{C}$ in direction of φ is defined by

$$\delta\mathscr{F}(\mathscr{C}, \varphi) := \frac{d}{d\varepsilon}\mathscr{F}(\mathscr{C}_\varepsilon)\Big|_{\varepsilon=0},$$

and this expression vanishes if $\mathscr{C}$ is an extremal. We now want to derive the Euler equations. To this end we assume that $\mathscr{C}$ is an extremal given by $c : [a, b] \to \mathbb{R}^3$ with $|\dot{c}| = 1$. Let $\{\mathfrak{t}, \mathfrak{n}, \mathfrak{b}\}$ be the moving frame along $\mathscr{C}$ satisfying Frenet's formulas

$$(4) \qquad \dot{\mathfrak{t}} = \kappa\mathfrak{n}, \qquad \dot{\mathfrak{n}} = -\kappa\mathfrak{t} + \tau\mathfrak{b}, \qquad \dot{\mathfrak{b}} = -\tau\mathfrak{n}.$$

Then we have

$$(5) \qquad \delta\mathscr{F}(\mathscr{C}, \varphi) = \int_a^b \left\{ f'(\kappa, \tau)\frac{d}{d\varepsilon}\kappa_{c_\varepsilon}\Big|_{\varepsilon=0} + f^*(\kappa, \tau)\frac{d}{d\varepsilon}\tau_{c_\varepsilon}\Big|_{\varepsilon=0} + f(\kappa, \tau)\frac{d}{d\varepsilon}|\dot{c} + \varepsilon\dot{\varphi}|\Big|_{\varepsilon=0} \right\} dt,$$

where $f' = f_u$, $f'' = f_{uu}$, $f^* = f_v$, $f^{**} = f_{vv}$, $f'^* = f_{uv}, \ldots$, etc.

Let $\zeta \in C_c^\infty((a, b))$. As in the planar case, tangential variations $\varphi = \zeta\mathfrak{t}$ lead to no equation, whereas normal variations $\varphi = \zeta\mathfrak{n}$ imply the Euler equation of fourth order

$$(6) \qquad \frac{d^2}{dt^2}\left\{f'(\kappa, \tau) + 2\frac{\tau}{\kappa}f^*(\kappa, \tau)\right\} - \frac{d}{dt}\left\{f^*(\kappa, \tau)\left(\frac{3\dot{\tau}}{\kappa} - \frac{2\dot{\kappa}\tau}{\kappa^2}\right)\right\}$$
$$+ f'(\kappa, \tau)(\kappa^2 - \tau^2) + f^*(\kappa, \tau)\left\{\frac{\ddot{\tau}}{\kappa} - \frac{\dot{\kappa}\dot{\tau}}{\kappa^2} + 2\tau\kappa\right\} - \kappa f = 0,$$

and binormal variations $\varphi = \zeta\mathfrak{b}$ yield the Euler equation of sixth order

$$(7) \qquad \frac{d^3}{dt^3}\left\{\frac{1}{\kappa}f^*(\kappa, \tau)\right\} + \frac{d^2}{dt^2}\left\{\frac{\dot{\kappa}}{\kappa^2}f^*(\kappa, \tau)\right\} - \frac{d}{dt}\left\{2f'(\kappa, \tau)\tau + f^*(\kappa, \tau)\left(\frac{\tau^2}{\kappa} - \kappa\right)\right\}$$
$$+ \dot{\tau}f'(\kappa, \tau) + f^*(\kappa, \tau)\left(\frac{2\tau\dot{\tau}}{\kappa} - \frac{\dot{\kappa}\tau^2}{\kappa^2}\right) = 0.$$

Then the curvature κ and torsion τ of an extremal $\mathscr{C}$ of (3) can be obtained by integrating (6) and (7), and thereafter $\mathscr{C}$ is to be determined by integrating equation (28) in Section 5, or alternatively from formulas (29) and (31) in Section 5.

The variational problem $\int_{\mathscr{C}} f(\kappa, \tau)\, ds \to$ extr. for curves $\mathscr{C}$ in $\mathbb{R}^3$ was treated by Irrgang [1] as a Lagrange multiplier problem (see Chapter 2) by adding the equation $\mathfrak{t} = \dot{c}$ and the Frenet equations

$$\mathfrak{t}' - \kappa\mathfrak{n} = 0, \qquad \mathfrak{n}' + \kappa\mathfrak{t} - \tau\mathfrak{b} = 0, \qquad \mathfrak{b}' + \tau\mathfrak{n} = 0$$

as subsidiary condition. He derived conditions which guarantee that an extremal yields a semistrong minimum or maximum (i.e. an extremum with respect to small variations of c and $\mathfrak{t}$). He in particular treated the variational problem for the affine length where $f(\kappa, \tau) = (\kappa^2\tau)^{1/6}$. By transforming the extremal $c(s)$ to the parameter of the affine arc length, σ, via $c(s) = X(\sigma)$ and $d\sigma = (\kappa^2\tau)^{1/6}\, ds$, the Euler equations appear in the form

$$\frac{d^4 X}{d\sigma^4} = 0.$$

Then Irrgang [1], pp. 397–399, concluded that all extremals can be obtained from the cubic parabola

$$x = \sigma, \qquad y = \tfrac{1}{2}\sigma^2, \qquad z = \tfrac{1}{6}\sigma^3$$

by means of volume preserving affine transformations which lead to a 10-parameter family of cubic parabolas in $\mathbb{R}^3$. Other examples studied by Irrgang are integrands of the form $f(\kappa, \tau) = \varphi(\kappa/\tau)$ and also Lagrangians which are either linear in (κ, τ) or a function of τ alone, e.g. $f(\tau) = \sqrt{\tau}$. In the case $f = f(\tau)$ one obtains the natural equations of an arbitrary extremal in the form

$$\kappa = \phi(\tau), \qquad s - s_0 = \int_{s_0}^{\tau} \psi(\tau)\, d\tau,$$

where ϕ and ψ are algebraic expressions in f and its derivatives. Thus we can determine all extremals by quadratures.

If f does not depend on the variable v, i.e. $f^* = 0$, equations (6) and (7) reduce to the system

$$\frac{d^2}{dt^2} f'(\kappa) + f'(\kappa)(\kappa^2 - \tau^2) - \kappa f(\kappa) = 0, \qquad \frac{d}{dt}\{2f'(\kappa)\tau\} - \dot{\tau} f'(\kappa) = 0, \tag{8}$$

which can be written as

$$f''(\kappa)\ddot{\kappa} + f'''(\kappa)\dot{\kappa}^2 + f'(\kappa)(\kappa^2 - \tau^2) - \kappa f(\kappa) = 0, \qquad 2f''(\kappa)\dot{\kappa}\tau + f'(\kappa)\dot{\tau} = 0. \tag{9}$$

For planar curves we have $\tau = 0$, and then (9) reducdes to the equation

$$f''(\kappa)\ddot{\kappa} + f'''(\kappa)\dot{\kappa}^2 + f'(\kappa)\kappa^2 - f(\kappa)\kappa = 0 \tag{10}$$

derived in 5 [5].

For space curves the second equation of (8) or (9) can be written as

$$2\tau \frac{d}{dt} f'(\kappa) + f'(\kappa)\dot{\tau} = 0, \tag{11}$$

whence

$$\frac{d\tau}{\tau} = -2\frac{df'(\kappa)}{f'(\kappa)}.$$

Hence there is a constant v such that

$$\tau = \frac{v}{f'(\kappa)^2}. \tag{12}$$

Inserting (12) in (8) we obtain a second order equation for κ. However, by an observation of Radon, we can reduce this integration problem to a much simpler problem, in fact, to a mere quadrature. To this end we introduce the functions $\alpha(t)$, $\beta(t)$, $\gamma(t)$ by

$$\alpha := \kappa f'(\kappa) - f(\kappa), \qquad \beta := \frac{d}{dt} f'(\kappa), \qquad \gamma := \frac{v}{f'(\kappa)}. \tag{13}$$

Using the identity $(f(\kappa))^{\cdot} = f'(\kappa)\dot{\kappa}$ in conjunction with (8_1) and (12) we obtain the equation

$$\begin{bmatrix} \dot{\alpha} \\ \dot{\beta} \\ \dot{\gamma} \end{bmatrix} = \begin{bmatrix} 0 & \kappa & 0 \\ -\kappa & 0 & \tau \\ 0 & -\tau & 0 \end{bmatrix} \begin{bmatrix} \alpha \\ \beta \\ \gamma \end{bmatrix}, \tag{14}$$

which is of the same form as Frenet's formulae. Introducing $\xi := (\alpha, \beta, \gamma)$ we can write (14) as

$$\dot{\xi} = A\xi, \tag{14'}$$

where A is the skew-symmetric matrix of Frenet's formulae. Thus

$$\frac{d}{dt}|\xi|^2 = \langle \dot{\xi}, \xi \rangle + \langle \xi, \dot{\xi} \rangle = \langle A\xi, \xi \rangle + \langle \xi, A\xi \rangle$$

$$= \langle A\xi, \xi \rangle - \langle A\xi, \xi \rangle = 0,$$

and therefore $|\xi(t)|^2 \equiv \text{const} =: c$. We therefore have

$$[\kappa f'(\kappa) - f(\kappa)]^2 + \left[\frac{d}{dt} f'(\kappa)\right]^2 + \left[\frac{\nu}{f'(\kappa)}\right]^2 = c. \tag{15}$$

Now we pass from $\kappa(t)$ to the inverse function $t(\kappa)$ whose derivative can be obtained from (15) in the form

$$\frac{dt}{d\kappa} = g(\kappa),$$

where $g(\kappa)$ is a function of the independent variable κ. A quadrature leads to $t(\kappa)$, which yields $\kappa(t)$, and from (12) we obtain $\tau(t)$. Thus we have the natural equations $\kappa = \kappa(s)$, $\tau = \tau(s)$ of all extremals $c(s)$ if we recall that $t = s$ is the parameter of the arc length because of $|\dot{c}| = 1$. In fact, even the extremals of $\int f(\kappa)\, ds$ in $\mathbb{R}^3$ can be found by quadratures alone; cf. Blaschke [5], Section 27.

For the integral

$$\mathscr{F}(\mathscr{C}) = \int_a^b (\kappa^2 + \lambda)\, ds, \quad \lambda = \text{const},$$

we obtain the Euler equations

$$2\ddot{\kappa} + \kappa^3 - \lambda\kappa - 2\kappa\tau^2 = 0, \qquad 2\dot{\kappa}\tau + \kappa\dot{\tau} = 0.$$

If $\tau \not\equiv 0$, then $\kappa \not\equiv 0$, and then the second equation yields $\kappa^4\tau^2 = \nu^2$ with a suitable constant $\nu \neq 0$, and then the first equation above can be transformed into

$$2\ddot{\kappa} + \kappa^3 - \lambda\kappa - 2\nu^2\kappa^{-3} = 0,$$

which has the integral

$$4\dot{\kappa}^2 + \kappa^4 - 2\lambda\kappa^2 + 4\nu^2\kappa^{-2} = \mu,$$

with a suitable constant μ. Integrating by separation of variables yields $t = t(\kappa)$, and the inverse function $\kappa = \kappa(t)$ is the desired curvature function. Finally τ is obtained from $\tau(t) = \pm\nu\kappa^{-2}(t)$.

For $N = 2$ the extremals of $\int (\kappa^2 + \lambda)\, ds$ were already determined by Euler; for $N = 3$ there exist detailed investigations by Langer and Singer as well as by Bryant and Griffith (2,5 nr. 16).

2. The extremals of $\int \rho\, ds$ were first described by Euler in his paper E56 (*Curvarum maximi minimive proprietate gaudentium inventio nova et facilis*) which appeared in the Commentarii of St. Petersburg in 1736; cf. Opera omnia, Ser. I, Vol. 25. Euler treated the problem again in his paper E99 from 1738 and in his book *Methodus inveniendi* from 1744 (cf. Euler [2], Cap. II, Section 51, Ex. II, pp. 654–666). We also refer to the comments of Carathéodory [16], Vol. 5, Nr. 43, pp. 143–144, concerning the question what are the actual minimizers. Clearly the integral $\int_{\mathscr{C}} \rho\, ds$ is the area between a curve $\mathscr{C}$, its evolute, and the normals of $\mathscr{C}$ in its endpoints. By means of "field theory", the minimum question is discussed in A. Kneser [1], Sections 53–54.

3. *Parameter-invariant integrals.* Consider a Lagrangian $F(z, p, q)$ on $\mathbb{R}^N \times \mathbb{R}^N \times \mathbb{R}^N$ (or some open subset thereof), and a variational integral of the kind

$$\mathscr{F}(c) := \int_a^b F(c(t), \dot{c}(t), \ddot{c}(t))\, dt \tag{16}$$

defined on smooth curves $c:[a, b] \to \mathbb{R}^N$. We suppose that $\mathscr{F}(c)$ is parameter-invariant, i.e. $\mathscr{F}(c \circ \tau) = \mathscr{F}(c)$ for any smooth orientation-preserving parameter transformation τ. If $\{\tau_\varepsilon\}_{|\varepsilon|<\varepsilon_0}$ is a smooth 1-parameter family of such transformations, we obtain

$$\frac{d}{d\varepsilon}\mathscr{F}(c \circ \tau_\varepsilon) = 0 \quad \text{for } |\varepsilon| < \varepsilon_0$$

and for any choice of c. Let $\tau_\varepsilon(t) = \tau(t, \varepsilon) := t + \varepsilon\zeta(t)$ where $\zeta(t)$ is an arbitrarily chosen function of class $C_c^\infty(a, b)$. Then, for $|\varepsilon| \ll 1$, $\{\tau_\varepsilon\}$ is an admissible family of parameter transformations such that $\tau(t, 0) = t$, $\tau'(t, 0) = \zeta(t)$, $' = d/d\varepsilon$ whence we derive the Taylor expansion

$$c(\tau(t, \varepsilon)) = c(t) + \varepsilon\dot{c}(t)\zeta(t) + O(\varepsilon^2).$$

This implies

$$0 = \frac{d}{d\varepsilon}\mathscr{F}(c \circ \tau_\varepsilon)\bigg|_{\varepsilon=0} = \delta\mathscr{F}(c, \varphi), \quad \varphi := \zeta\dot{c}, \tag{17}$$

and we note that $\varphi = \zeta\dot{c}$ is a tangential variation. We can write this equation in the form

$$0 = \int_a^b \{F_z(e)\cdot\zeta\dot{c} + F_p(e)\cdot(\zeta\dot{c})^{\cdot} + F_q(e)\cdot(\zeta\dot{c})^{\cdot\cdot}\}\, dt, \tag{18}$$

where $e := (c, \dot{c}, \ddot{c})$. Performing suitable integrations by part, this equation can be transformed into

$$\int_a^b \frac{d}{dt}\{F(e) - F_p(e)\cdot\dot{c} - F_q(e)\cdot\ddot{c} + [F_q(e)]^{\cdot}\cdot\dot{c}\}\zeta\, dt = 0. \tag{19}$$

By Du Bois–Reymond's lemma we infer $\{\dots\} = \text{const}$, and it is easy to see that this constant is zero. For this purpose we note that $G(z)F(z, p, q)$ leads to an invariant integral for any choice of G provided that $F(z, p, q)$ generates an invariant integral. Then we choose G in such a way that $G(z) \equiv 1$ on a compact set of $\mathbb{R}^N$ containing $c([a, b])$ and $G(z) \equiv 0$ for $|z| \gg 1$. Extending c to a smooth map $\mathbb{R} \to \mathbb{R}^N$ with $|c(t)| \to \infty$, the assertion follows at once. Thus we have

$$F(e) = \left[F_p(e) - \frac{d}{dt}F_q(e)\right]\cdot\dot{c} + F_q(e)\cdot\ddot{c} \tag{20}$$

for any smooth curve c and $e = (c, \dot{c}, \ddot{c})$.

Suppose now that the Lagrangian F does not depend on z, i.e. $F(p, q)$ is a function of p, q alone. Then the Euler equations of

$$\mathscr{F}(c) = \int_a^b F(\dot{c}(t), \ddot{c}(t))\, dt$$

reduce to

$$-\frac{d}{dt}F_p(\dot{c}, \ddot{c}) + \frac{d^2}{dt^2}F_q(\dot{c}, \ddot{c}) = 0.$$

Hence there is a constant vector $A \in \mathbb{R}^2$ such that

$$F_p(\dot{c}, \ddot{c}) - \frac{d}{dt}F_q(\dot{c}, \ddot{c}) = A. \tag{21}$$

In connection with (20) we see that *any extremal c satisfies*

$$F(\dot c, \ddot c) = A\cdot \dot c + F_q(\dot c, \ddot c)\cdot \ddot c. \tag{22}$$

In case of the integral $\int \rho\, ds = \int F\, dt$ the Lagrangian F has the form

$$F(p, q) = \frac{|p|^4}{[p, q]}, \qquad [p, q] := \det(p, q), \tag{23}$$

whence $-F(p, q) = F_q(p, q)\cdot q$. Equation (22) therefore becomes

$$2F(\dot c, \ddot c) = A\cdot \dot c. \tag{24}$$

Set

$$A := (a, b), \qquad r := |A| = \sqrt{a^2 + b^2}, \qquad U := \frac{1}{r}\begin{pmatrix} b & a \\ -a & b\end{pmatrix},$$

where U is a rotation about the angle γ, $\tan\gamma = a/b$. Rotating the Cartesian coordinates by this angle, equation (24) is transformed into

$$2F(\dot c, \ddot c) = r\dot c_2 \tag{25}$$

since F is rotationally invariant on account of (23). If we choose a nonparametric representation $c(x) = (x, y(x))$ for c, equation (25) becomes

$$2(1 + y'^2)^2/y'' = ry', \quad ' = \frac{d}{dx}, \tag{26}$$

whence we obtain

$$y'(x) = \sqrt{\frac{x + k}{(r/4) - (x + k)}} \tag{27}$$

for some constant k. Let $R := r/8$ and introduce the variable φ by

$$x = R(1 - \cos\varphi) + x_0. \tag{28}$$

Then we infer from (27) that

$$y = R(\varphi - \sin\varphi) + y_0, \tag{29}$$

and thus we obtain that the extremals $\mathscr{C}$ of $\int \rho\, ds$ have the parameter representation

$$c(\varphi) = (x_0 + R(1 - \cos\varphi), y_0 + R(\varphi - \sin\varphi)), \tag{30}$$

i.e. the extremals are cycloids, as we had already seen in 5 [7].

Similarly we can use equation (22) to determine the extremals of $\int(\kappa^2 + \lambda)\, ds$ (cf. also 5 [5]). We are then led to the relation

$$\kappa^2 = \lambda - \frac{A\cdot \dot c}{|\dot c|}, \tag{31}$$

with some constant vector $A = (a, b) \in \mathbb{R}^2$. If we now introduce $\omega(s)$ and α as in 5 [5], we have

$$\kappa = \frac{d\omega}{ds}, \qquad \frac{dc}{ds} = \left(\frac{dx}{ds}, \frac{dy}{ds}\right) = (\cos(\omega + \alpha), \sin(\omega + \alpha)),$$

whence

$$\kappa^2 = \left(\frac{d\omega}{ds}\right)^2 = g(\omega), \quad g(\omega) := \lambda - a\cos(\omega + \alpha) - b\sin(\omega + \alpha) \tag{32}$$

and

$$d\omega = \sqrt{g(\omega)}\, ds. \tag{33}$$

Therefore,

(34) $$x(\omega) = x_0 + \int_\alpha^{\omega+\alpha} \frac{\cos\omega}{\sqrt{g(\omega)}}\,d\omega, \qquad y(\omega) = y_0 + \int_\alpha^{\omega+\alpha} \frac{\sin\omega}{\sqrt{g(\omega)}}\,d\omega.$$

Differentiating (32) with respect to s, a brief computation shows that the extremals of $\int(\kappa^2 + \lambda)\,ds$ satisfy

(35) $$2\kappa = [A, c] + \text{const}.$$

4. Radon's variational problem. In his thesis [1] Radon has investigated the extremals of functionals of the kind

(36) $$\mathscr{F}(c) = \int_a^b f(c, \omega, \kappa)\,ds,$$

where $c(s) = (x(s), y(s))$ denotes the representation of an immersed curve $\mathscr{C}$ in $\mathbb{R}^2$ with respect to its arc length s, $\omega(s)$ is the angle between the x-axis and the tangent, and $\kappa(s)$ is the curvature of $c(s)$, i.e.

$$|\dot c(s)| = 1, \quad \dot c(s) = (\cos\omega(s), \sin\omega(s)), \quad \dot\omega = \kappa = [\dot c, \ddot c].$$

The integral (36) is just another form of the general parameter-invariant integral discussed in nr. 3,

(37) $$\mathscr{F}(c) = \int_a^b F(c, \dot c, \ddot c)\,ds, \quad |\dot c| = 1.$$

In fact, by the reasoning given in 8,*1.1* one easily proves that the parameter invariance of $\mathscr{F}$ is equivalent to the homogeneity relation

$$F(x, y, k\dot x, k\dot y, k^2\ddot x + l\dot x, k^2\ddot y + l\dot y) = kF(x, y, \dot x, \dot y, \ddot x, \ddot y)$$

for all $k > 0$ and each $l \in \mathbb{R}$, where $c(s) = (x(s), y(s))$, etc. Set $v = |\dot c| = \sqrt{\dot x^2 + \dot y^2}$, $k = 1/v$, $l = -(\dot x\ddot x + \dot y\ddot y)/v^4$, $\dot x/v = \cos\omega$, $\dot y/v = \sin\omega$, $\dot x\ddot y - \dot y\ddot x = v^3\kappa$, then $ds = v\,dt$ and $\kappa = d\omega/ds$. Hence we obtain $F(c, \dot c, \ddot c) = F(x, y, \cos\omega, \sin\omega, -\kappa\sin\omega, \kappa\cos\omega)|\dot c|$, and therefore, if we write F as

(38) $$F(c, \dot c, \ddot c) = f(c, \omega, \kappa)|\dot c|,$$

we see that (37) can be expressed in the form (36). Radon derived *sufficient conditions* for an extremal of (36) to be a *semistrong minimizer* of (36) with respect to given boundary conditions (i.e. to furnish an extremum with respect to variations which are close by in the x, y, ω-space). It is of geometric interest to formulate Euler's equation also for (36). Thus, let the first variation of (36) vanish, i.e.

(39) $$0 = \int_a^b (f_x\delta x + f_y\delta y + f_\omega\delta\omega + f_\kappa\delta\kappa + f\delta v)\,dt \quad (|\dot c| = 1),$$

$v := |\dot c|$, where the sign δ means that we have to replace $c(t) = (x(t), y(t))$ by $\underline{c}(t) = (\underline{x}(t), \underline{y}(t)) = (x(t) + \varepsilon\xi(t), y(t) + \varepsilon\eta(t))$, then differentiate with respect to ε, and finally set $\varepsilon = 0$. Clearly we have $\delta x = \xi$, $\delta y = \eta$.

(i) Let $\delta x = \xi \in C_c^\infty(a, b)$, $\delta y = \eta = 0$. From

$$\underline{\omega} = \operatorname{arc\,ctg}\frac{\dot x + \varepsilon\dot\xi}{\dot y},$$

we then infer

$$\delta\omega = -\dot y\dot\xi = -\dot\xi\sin\omega,$$

and

$$\underline{\kappa} = \frac{d\underline{\omega}}{ds}\frac{1}{|\dot{\underline{c}}|}$$

yields

$$\delta\kappa = -\ddot\xi\sin\omega - 2\kappa\dot\xi\cos\omega.$$

Finally,

$$\delta v = \dot{\xi} \cos \omega .$$

Then it follows from (39) that

$$\int_a^b \left\{ f_x - \frac{d}{ds}\left[(f - \kappa f_\kappa) \cos \omega - \left(f_\omega - \frac{d}{ds} f_\kappa \right) \sin \omega \right] \right\} \xi \, ds = 0. \tag{40}$$

(ii) If $\delta x = 0$, $\delta y = \eta \in C_c^\infty((a, b))$ we similarly infer $\delta\omega = \dot{\eta} \cos \omega$, $\delta\kappa = \ddot{\eta} \cos \omega - 2\kappa\dot{\eta} \sin \omega$, $\delta v = \dot{\eta} \sin \theta$, and therefore

$$\int_a^b \left\{ f_y - \frac{d}{ds}\left[(f - \kappa f_\kappa) \sin \omega + \left(f_\omega - \frac{d}{ds} f_\kappa \right) \cos \omega \right] \right\} \eta \, ds = 0. \tag{41}$$

Set

$$\begin{aligned} P &:= (f - \kappa f_\kappa) \cos \omega - \left(f_\omega - \frac{d}{ds} f_\kappa \right) \sin \omega , \\ Q &:= (f - \kappa f_\kappa) \sin \omega + \left(f_\omega - \frac{d}{ds} f_\kappa \right) \cos \omega , \\ R &:= f_x \sin \omega - f_y \cos \omega + \kappa (f - \kappa f_\kappa) + \frac{d}{ds}\left(f_\omega - \frac{d}{ds} f_\kappa \right). \end{aligned} \tag{42}$$

Then we obtain from (40) and (41) the two Euler equations

$$f_x - \frac{d}{ds} P = 0, \qquad f_y - \frac{d}{ds} Q = 0. \tag{43}$$

Moreover we have the general identities

$$f = P \cos \omega + Q \sin \omega + \kappa f_\kappa \tag{44}$$

and

$$R \sin \omega = f_x - \frac{d}{ds} P, \qquad R \cos \omega = f_y - \frac{d}{ds} Q, \tag{45}$$

which hold for arbitrary functions x, y, ω, κ. We infer from (45) that the two equations (43) are linearly dependent, and that they are equivalent to the single equation $R = 0$, i.e. to

$$f_x \sin \omega - f_y \cos \omega + \kappa (f - \kappa f_\kappa) + \frac{d}{ds}\left(f_\omega - \frac{d}{ds} f_\kappa \right) = 0. \tag{46}$$

In a similar way we can derive the second variation of $\mathscr{F}$ and develop the theory of conjugate points for $\mathscr{F}$; see Chapter 5 and Radon [1], Sections II–IV.

If f is independent of x and y, equations (43) yield the two integrals

$$P = \text{const}, \qquad Q = \text{const}, \tag{47}$$

and the identity (44) in conjunction with (47) leads to the third integral

$$f - \kappa f_\kappa = \alpha \cos(\omega - \beta), \tag{48}$$

with two suitable constants α and β.

Radon has explicitly treated a number of specific cases such as $f(\kappa) = \kappa^c$, in particular $f(\kappa) = 1/\kappa$, which appears in Euler's problem (see nrs. 2 and 3).

5. *The general curvature integral for surfaces in* $\mathbb{R}^3$. Let $X(u, v)$, $(u, v) \in \bar{B} \subset \mathbb{R}^2$, be a regular smooth parametric surface in $\mathbb{R}^3$; we assume at least $X \in C^4(\bar{B}, \mathbb{R}^3)$. To simplify notation we write $w = (u, v)$ and $u^1 = u$, $u^2 = v$. Let $g_{\alpha\beta} = X_{u^\alpha} \cdot X_{u^\beta}$ be the coefficients of the first fundamental form of

X, and set

$$g = \det(g_{\alpha\beta}), \qquad (g^{\alpha\beta}) = (g_{\alpha\beta})^{-1},$$

that is,

$$g^{11} = g_{22}/g, \qquad g^{12} = g^{21} = -g_{12}/g, \qquad g^{22} = g_{11}/g.$$

Moreover, denote by $\mathcal{N} = g^{-1/2} X_u \wedge X_v$ the surface normal (or Gauss map) of X, and let $b_{\alpha\beta}$ be the coefficients of the second fundamental form of X,

$$b_{\alpha\beta} = \langle \mathcal{N}, X_{u^\alpha u^\beta} \rangle = -\langle \mathcal{N}_{u^\alpha}, X_{u^\beta} \rangle.$$

Then the Gauss curvature K and the mean curvature H of X are given by

$$K = \frac{b}{g}, \qquad H = \tfrac{1}{2} b_{\alpha\beta} g^{\alpha\beta},$$

where

$$b = \det(b_{\alpha\beta}).$$

Furthermore, let

$$b^\alpha_\beta = g^{\alpha\gamma} b_{\gamma\beta}, \qquad (h^{\alpha\beta}) = (b_{\alpha\beta})^{-1},$$

$$\Gamma^\gamma_{\alpha\beta} = \tfrac{1}{2} g^{\gamma\delta}[g_{\alpha\delta,\beta} + g_{\delta\beta,\alpha} - g_{\alpha\beta,\delta}].$$

Then we have

$$X_{u^\alpha u^\beta} = \Gamma^\gamma_{\alpha\beta} X_{u^\gamma} + b_{\alpha\beta} \mathcal{N}.$$

Consider now a Lagrangian f which depends on two variables, and introduce the corresponding *curvature integral* $\mathcal{F}$ by

(1) $$\mathcal{F}(X) := \int_X f(K, H)\, dA = \int_B f(K, H) \sqrt{g}\, du^1\, du^2$$

of X which is parameter-invariant, that is,

$$\mathcal{F}(X) = \mathcal{F}(X \circ \tau)$$

for every diffeomorphism $\tau : \bar{B}^* \to \bar{B}$ with $\det D\tau > 0$. Therefore, tangential variations of X will not change $\mathcal{F}(X)$, and so the two tangential Euler equations will trivially be satisfied. We can expect only one nontrivial Euler equation which is obtained by "normal variations".

Consider now an extremal X of $\mathcal{F}$ and a family of normal variations $\underline{X}(w, \varepsilon)$ of $X(w)$ defined by

(2) $$\underline{X}(w, \varepsilon) = X(w) + \varepsilon \varphi(w) \mathcal{N}(w), \quad w \in \bar{B},\ |\varepsilon| \ll 1,$$

where $\varphi \in C_c^\infty(B)$.

Then we obtain

(3) $$0 = \frac{d}{d\varepsilon} \mathcal{F}(\underline{X}(\cdot, \varepsilon)) \bigg|_{\varepsilon=0} =: \delta\mathcal{F}(X, \varphi),$$

where

(4) $$\delta\mathcal{F}(X, \varphi) = \int_B [f_K \delta K + f_H \delta H + (f \delta\sqrt{g})/\sqrt{g}] \sqrt{g}\, du^1\, du^2.$$

In order to compute δK, δH, $\delta\sqrt{g}$ where $\delta K = \left(\frac{d}{d\varepsilon}\underline{K}\right)\bigg|_{\varepsilon=0}$, $\underline{K}$ = Gauss curvature of $\underline{X}$, etc., we expand $\underline{K}(w, \varepsilon)$, $\underline{H}(w, \varepsilon)$ and $\sqrt{\underline{g}(w, \varepsilon)}$ in terms of ε and indicate terms of second and higher order in ε by $+ \cdots$. To this end we note that

$$\underline{X}_{u^\alpha} = X_{u^\alpha} + \varepsilon[\varphi\mathcal{N}_{u^\alpha} + \varphi_{u^\alpha}\mathcal{N}],$$
$$\underline{X}_{u^\alpha u^\beta} = X_{u^\alpha u^\beta} + \varepsilon[\varphi\mathcal{N}_{u^\alpha u^\beta} + \varphi_{u^\alpha}\mathcal{N}_{u^\beta} + \varphi_{u^\beta}\mathcal{N}_{u^\alpha} + \varphi_{u^\alpha u^\beta}\mathcal{N}].$$

Then we first obtain

$$\underline{g}_{\alpha\beta} = g_{\alpha\beta} - 2\varepsilon\varphi b_{\alpha\beta} + \cdots,$$

whence

$$\underline{g} = \det(\underline{g}_{\alpha\beta}) = g - 2\varepsilon\varphi g b_{\alpha\beta} g^{\alpha\beta} + \cdots$$

and

$$\sqrt{\underline{g}} = \sqrt{g}[1 - 2\varepsilon\varphi H + \cdots].$$

Therefore

(5) $$(\delta\sqrt{g})/\sqrt{g} = -2H\varphi.$$

Secondly we find that

$$\underline{b}_{\alpha\beta} = b_{\alpha\beta} + \varepsilon[\varphi(Kg_{\alpha\beta} - 2Hb_{\alpha\beta}) + \varphi_{\alpha\beta}] + \cdots,$$

where

(6) $$\varphi_{\alpha\beta} := \varphi_{u^\alpha u^\beta} - \Gamma^\gamma_{\alpha\beta}\varphi_{u^\gamma}$$

are the covariant second derivatives of φ. Let us introduce the second-order operators Δ and $\Box$ by

(7) $$\Delta\varphi := g^{\alpha\beta}\varphi_{\alpha\beta}, \qquad \Box\varphi := Kh^{\alpha\beta}\varphi_{\alpha\beta}.$$

Then we infer from the above expansions that

$$\underline{K} = K + \varepsilon[2HK\varphi + \Box\varphi]] + \cdots,$$
$$\underline{H} = H + \varepsilon[(2H^2 - K)\varphi + \tfrac{1}{2}\Delta\varphi] + \cdots.$$

A straight-forward but somewhat tedious computation shows that $\Delta\varphi = g^{\alpha\beta}[\varphi_{u^\alpha u^\beta} - \Gamma^\gamma_{\alpha\beta}\varphi_{u^\gamma}]$ can be written as

(8) $$\Delta\varphi = \frac{1}{\sqrt{g}} D_\alpha[\sqrt{g} g^{\alpha\beta} D_\beta\varphi],$$

where $D_\alpha = \dfrac{\partial}{\partial u^\alpha}$, that is, Δ is the *Laplace–Beltrami operator* associated with the metric tensor $(g_{\alpha\beta})$. Similarly $\Box\varphi$ can be written as

(9) $$\Box\varphi = \frac{1}{\sqrt{g}} D_\alpha[\sqrt{g} K h^{\alpha\beta} D_\beta\varphi].$$

The above expansions of $\underline{K}$ and $\underline{H}$ yield

(10) $$\delta K = 2HK\varphi + \Box\varphi, \qquad \delta H = (2H^2 - K)\varphi + \tfrac{1}{2}\Delta\varphi.$$

If we now insert (5) and (10) in (4) and take (3) into account, an integration by parts yields

$$\int_B \varphi[\Box f_K + \tfrac{1}{2}\Delta f_H + 2HKf_K + (2H^2 - K)f_H - 2fH]\, dA = 0$$

for all $\varphi \in C_c^\infty(B)$, and then the fundamental lemma leads to the following *Euler equation for the curvature integral* (4):

(11) $$\Box f_K + \tfrac{1}{2}\Delta f_H + 2HKf_K + (2H^2 - K)f_H - 2fH = 0.$$

Let us consider now some important examples which are special cases of (11).

$\boxed{1}$ *Minimal surfaces* are defined as extremals of the area functional

$$\mathscr{A}(X) = \int_B \sqrt{g}\, du\, dv = \int_B |X_u \wedge X_v|\, du\, dv. \tag{12}$$

Clearly, $\mathscr{A}(X)$ is the special case $f = 1$ of the integral (4), and thus the Euler equation reduces to

$$H = 0. \tag{13}$$

Thus we obtain the by now well-known result that minimal surfaces are just the surfaces of zero mean curvature.

$\boxed{2}$ *Willmore surfaces* are the extremals of the functional

$$\mathscr{W}(X) = \int_X H^2\, dA = \int_B H^2 \sqrt{g}\, du\, dv. \tag{14}$$

According to (11), applied to $f(H) = H^2$, they are the solutions of the Euler equation

$$\Delta H + 2H(H^2 - K) = 0. \tag{15}$$

$\boxed{3}$ *The total curvature*

$$\mathscr{K}(X) = \int_X K\, dA = \int_B K\sqrt{g}\, du\, dv \tag{16}$$

of the surface X leads to the Euler equation $0 = 0$, which shows that $K\sqrt{g}$ is a null Lagrangian, and that $\mathscr{K}(X)$ is an invariant integral. From here one is again led to the Gauss–Bonnet theorem (cf. 5 $\boxed{4}$).

$\boxed{4}$ The integral $\int f(K)\, dA$ has the Euler equation

$$\Box f'(K) + 2H[Kf'(K) - f(K)] = 0. \tag{17}$$

$\boxed{5}$ The Euler equation of $\int f(H)\, dA$ is given by

$$\tfrac{1}{2}\Delta f'(H) + (2H^2 - K)f'(H) - 2Hf(H) = 0. \tag{18}$$

6. The Euler equations of general second- and higher-order curvature integrals for submanifolds of Riemannian manifolds were derived e.g. by Rund [9].

Of particular importance is *Einstein's gravitational potential*

$$\mathscr{G}(g) := \int_M R(g) d\,\mathrm{vol}(g) \tag{19}$$

considered as functional of a Riemannian (or Minkowskian) metric g on an n-dimensional manifold M, $n \geqq 3$, which on M defines a line element ds which locally is given by

$$ds^2 = g_{ik}\, dx^i\, dx^k$$

if g_{ik} are the local components of g. The Euler equations of (19), which should by rights be a fourth order equation, actually is a second order equation, the so-called *Einstein field equation*,

$$\mathrm{Ric} = \tfrac{1}{2} R g, \tag{20}$$

where $R = \mathrm{trace}_g\, \mathrm{Ric}$ is the trace of the Ricci curvature Ric. Locally (20) means

$$R_{ik} = \tfrac{1}{2} R g_{ik}, \tag{20$'$}$$

whence

$$R_i^k = g^{jk} R_{ij} = \tfrac{1}{2} R g_{ij} g^{jk} = \tfrac{1}{2} R \delta_i^k$$

and therefore

$$R = R^i_i = \tfrac{1}{2} R \delta^i_i = \frac{n}{2} R ;$$

hence, for $n \geqq 3$, $R = 0$. We thus conclude that "in vacuum" the Einstein field equations are equivalent to

(21) $$\mathrm{Ric}(g) = 0 \quad \text{or} \quad R_{ij} = 0 .$$

The special structure of

$$R \, d \,\mathrm{vol} = g^{kl} R^i_{ikl} \sqrt{|\det g|} \, dx$$

implies that the field equations are of second and not of fourth order, see e.g. Dubrovin–Fomenko–Novikov [1], Vol. 1, pp. 390–397 for the derivation of (21).

In the presence of matter the Einstein field equations become

(22) $$R_{ik} - \tfrac{1}{2} R g_{ik} = -\lambda T_{ik} ,$$

$\lambda = \mathrm{const}$; they are the Euler equations of an action integral

(23) $$\mathscr{A} = \mathscr{G} + \lambda \mathscr{L} = \int_M (R + \lambda L) \, d \,\mathrm{vol} ,$$

where L is a suitably chosen energy density related to the energy–momentum tensor (T_{ik}).

A similar phenomenon is observed for the Monge–Ampère equation which can be obtained as Euler equation of a second-order variational problem.

Chapter 2. Variational Problems with Subsidiary Conditions

In this chapter we shall derive necessary conditions for extremizers (i.e. maximizers or minimizers) of variational integrals which are subject to various kinds of constraints. In particular we shall treat the following three cases.

(i) *Isoperimetric constraints.* Here some integral is held to be constant, say

$$\int_\Omega G(x, u, Du)\, dx = \text{const}.$$

For example, we may look for minimizers of the area functional

$$\int_\Omega \sqrt{1 + |Du|^2}\, dx$$

among graphs $\{(x, u(x)) : x \in \Omega\}$ of functions $u : \Omega \to \mathbb{R}$ over some domain Ω which have prescribed boundary values on $\partial\Omega$ and enclose a prescribed volume

$$\int_\Omega |u|\, dx = V,$$

with the x-plane. This is the classical *isoperimetric problem* in its nonparametric formulation. Its dual problem provides another example of a variational problem with isoperimetric constraints: Among all graphs with prescribed boundaries on $\partial\Omega$ and with prescribed surface area, one has to find the one enclosing maximum volume. In other words, for given $A > \operatorname{meas} \Omega$, we are to maximize

$$\int_\Omega |u|\, dx$$

among all graphs satisfying

$$\int_\Omega \sqrt{1 + |Du|^2}\, dx = A$$

and such that u is prescribed on $\partial\Omega$.

(ii) *Holonomic constraints.* These are subsidiary conditions of the kind

$$G(x, u(x)) = 0.$$

The most important conditions of this type are side conditions of the form

$$G(u(x)) = 0.$$

Constraining the competing functions of a variational problem by such a condition means that we are looking for extremizers of a given functional $\int_\Omega F(x, u, Du)\, dx$ among a class of mappings $u: \Omega \to \mathbb{R}^N$ whose values are confined to some submanifold M of $\mathbb{R}^N$ which is described by the equation $G(z) = 0$.

A typical problem with holonomic constraints is to minimize the length of curves confined to a sphere, say, to S^{N-1}, which join two given points of this manifold. The constraint in this case is

$$|u|^2 = 1.$$

Another example is provided by the so-called problem of *energy-minimizing harmonic maps* into a sphere (or, more generally, into a Riemannian manifold). This amounts to minimizing the Dirichlet integral

$$\frac{1}{2}\int_\Omega |Du|^2\, dx, \quad \Omega \subset \mathbb{R}^n,$$

among mappings $u: \Omega \to S^{N-1}$ into an $(N-1)$-dimensional sphere S^{N-1} of $\mathbb{R}^N$ (or into a Riemannian manifold), with prescribed boundary values on $\partial\Omega$.

(iii) *Nonholonomic constraints.* These are subsidiary conditions of the type

$$G(x, u(x), Du(x)) = 0.$$

Most problems of optimal control theory fall into this category. For instance, sending a spaceship from the Earth to another planet, one may wish to minimize its flying time or its fuel consumption. However, the satellite is not completely free to move but will be subject to gravitational influences by the Sun and its planets. According to the Newtonian laws of mechanics, these influences can be expressed by differential equations, i.e., by nonholonomic constraints.

In all of these cases, the derivation of necessary conditions and especially of analogues of the Euler equations is reduced to the procedure of 1,*2* using the artifice of *Lagrange multipliers.* Actually, multipliers were invented and diversely applied by Euler long before Lagrange. Yet they are firmly attached to Lagrange's name because of his applications of multipliers in mechanics, where they are used to compute constraining forces in an elegant way.

We recall what Lagrange multipliers are for ordinary functions. Let f be a smooth real-valued function on $\mathbb{R}^n$ and let S be the level set of a smooth function g defined on $\mathbb{R}^n$, that is

$$S = \{x \in \mathbb{R}^n : g(x) = c\}.$$

If f restricted to S attains an extremum at some point x_0 of S where $Dg(x_0) \neq 0$, then there is a real number, called Lagrange multiplier, such that

$$Df(x_0) = \lambda Dg(x_0),$$

i.e. the gradient of f has no tangential component along S. In other words, for a suitable number λ, the function $f + \lambda g$ has an unconstrained critical point at x_0.

For isoperimetric side conditions, multipliers turn out to be constants, while they are functions of the independent variables in the case of holonomic and nonholonomic conditions.

The existence and smoothness of Lagrange multipliers can in many cases be established in an elementary way within the setting of smooth functions, provided that certain natural assumptions are satisfied. The reader should, however, be aware of the fact that, in specific cases, authors often do not check whether these assumptions are actually fulfilled, and our book is no exception in this respect. Thus the reader should carefully examine every single example and establish the existence of multipliers. Unfortunately this seems to be a highly nontrivial matter for nonholonomic constraints. For multiple integrals with nonholonomic side conditions, Lagrange multipliers are merely a formal tool, the applicability of which is not yet well established. For one-dimensional problems, the situation is much more satisfactory, but even in this case it is fairly cumbersome to establish the existence and regularity of multiplier functions. We shall briefly discuss *Lagrange problems*, that is, one-dimensonal variational problems with nonholonomic constraints, in Section 3. The modern synopsis and extension of the multiplier rule for Lagrange problems and related questions is the celebrated *Pontryagin maximum principle* which will be touched in Chapter 7.

In the last section we shall investigate constraints at the boundary. A typical example is provided by admissible functions whose boundary values are confined to certain preassigned manifolds called *supporting surfaces*. It will then turn out that extremizers with respect to such constraints have to satisfy a so-called *transversality condition* at their boundaries. That is, extremizers meet their supporting surfaces in a well-defined way which is determined by the structure of the Lagrangian. In many cases transversality means orthogonality.

The notion of transversality will also play an important role in the one-dimensonal field theory and in the Hamilton–Jacobi theory.

1. Isoperimetric Problems

In this section we shall derive necessary conditions for solutions of *isoperimetric problems*, that is, for local extrema $u \in C^1(\bar{\Omega}, \mathbb{R}^N)$ of variational integrals

$$\mathscr{F}(u) := \int_\Omega F(x, u, Du)\, dx,$$

which, besides boundary conditions on $\partial\Omega$, are subject to a *subsidiary condition* of the kind

$$\mathscr{G}(u) = c,$$

with some constant c, where $\mathscr{G}(u)$ is an integral of the form

(1) $$\mathcal{G}(u) = \int_\Omega G(x, u, Du)\, dx$$

or

(2) $$\mathcal{G}(u) = \int_\Sigma G(x, u, Du)\, d\mathcal{H}^{n-1},$$

respectively, where Σ is a subset of $\partial\Omega$ with a positive Hausdorff[1] measure $\mathcal{H}^{n-1}(\Sigma)$.

The name for this class of problems is derived from the classical isoperimetric problem (isos = equal, perimetron = circumference or perimeter) which will be discussed in Example [1] at the end of this section. It is the question to determine all bounded planar regions of maximal area with a boundary contour of prescribed length.

To have a clear cut situation, we formulate the following *assumptions*:

Let Ω be a bounded domain in $\mathbb{R}^n$, $u \in C^1(\bar{\Omega}, \mathbb{R}^N)$, and let $\mathcal{U}$ be an open set in $\mathbb{R}^n \times \mathbb{R}^N \times \mathbb{R}^{nN}$ containing $\{(x, u(x), Du(x)) : x \in \bar{\Omega}\}$. Suppose also that $F(x, z, p)$ and $G(x, z, p)$ are Lagrangians of class $C^2(\mathcal{U})$. Define $\mathcal{G}$ by (1) and set

$$c := \mathcal{G}(u).$$

We finally assume that $\mathcal{S}$ is a class of mappings $v \in C^1(\bar{\Omega}, \mathbb{R}^N)$ such that, *for every v in $\mathcal{S}$ and for every pair of functions $\varphi, \psi \in C_c^\infty(\Omega, \mathbb{R}^N)$, there exist numbers $\varepsilon_0 > 0$, $t_0 > 0$ such that $v + \varepsilon\varphi + t\psi \in \mathcal{S}$ for $|\varepsilon| < \varepsilon_0$, $|t| < t_0$.* We say in this case that $\mathcal{S}$ has the *variational property* $(\mathcal{V})$.

Remark. If a set $\mathcal{S}$ of admissible functions has the property $(\mathcal{V})$, then $\mathcal{S}$ is in a weak sense open. More precisely, any $v \in \mathcal{S}$ can be varied in all "smooth" directions φ with compact support in Ω.

Theorem 1. *Suppose that u furnishes a weak minimum (or maximum) of the functional $\mathcal{F}$ in the class $\mathcal{S}_c := \mathcal{S} \cap \{v : \mathcal{G}(v) = c\}$. Then there exists a real number λ, called the Lagrange multiplier, such that*

(3) $$\delta\mathcal{F}(u, \varphi) + \lambda\delta\mathcal{G}(u, \varphi) = 0 \quad \text{for all } \varphi \in C_c^\infty(\Omega, \mathbb{R}^N)$$

holds, provided that $\delta\mathcal{G}(u, \varphi)$ does not vanish for all $\varphi \in C_c^\infty(\Omega, \mathbb{R}^N)$. If we also assume that $u \in C^2(\Omega, \mathbb{R}^N)$, then the Euler equations

(4) $$D_\alpha\{F_{p_\alpha^i} + \lambda G_{p_\alpha^i}\} - \{F_{z^i} + \lambda G_{z^i}\} = 0, \quad 1 \le i \le N,$$

are satisfied on Ω.

Proof. By assumption, we can find a function $\psi \in C_c^\infty(\Omega, \mathbb{R}^N)$ such that

---

[1] For open sets Σ on smooth boundaries $\partial\Omega$, $\mathcal{H}^{n-1}(\Sigma)$ can be understood as the usual surface area of Σ.

$\delta\mathscr{G}(u, \psi) = 1$. With this function and with an arbitrary $\varphi \in C_c^\infty(\Omega, \mathbb{R}^N)$, we define the functions

$$\Phi(\varepsilon, t) := \mathscr{F}(u + \varepsilon\varphi + t\psi), \qquad \Psi(\varepsilon, t) := \mathscr{G}(u + \varepsilon\varphi + t\psi)$$

for $(\varepsilon, t) \in [-\varepsilon_0, \varepsilon_0] \times [-t_0, t_0] =: Q$, and, for sufficiently small numbers $\varepsilon_0 > 0$ and $t_0 > 0$, we obtain

$$\Phi(\varepsilon, t) \geq \Phi(0, 0) \quad (\text{or } \leq \Phi(0, 0))$$

for all $(\varepsilon, t) \in Q$ with $\Psi(\varepsilon, t) = c$. Since $\Psi_t(0, 0) = \delta\mathscr{G}(u, \psi) = 1$, we may apply the standard Lagrange multiplier theorem. Thus we infer the existence of a number $\lambda \in \mathbb{R}$ such that the function $\Phi(\varepsilon, t) + \lambda\Psi(\varepsilon, t)$ has $(\varepsilon, t) = (0, 0)$ as a critical point. Consequently

$$\Phi_\varepsilon(0, 0) + \lambda\Psi_\varepsilon(0, 0) = 0,$$

$$\Phi_t(0, 0) + \lambda\Psi_t(0, 0) = 0$$

or

$$\delta\mathscr{F}(u, \varphi) + \lambda\delta\mathscr{G}(u, \varphi) = 0,$$

$$\delta\mathscr{F}(u, \psi) + \lambda\delta\mathscr{G}(u, \psi) = 0.$$

The second equation yields

$$\lambda = -\delta\mathscr{F}(u, \psi), \tag{5}$$

and we see that the value of λ is independent of the chosen variation $\varphi \in C_c^\infty(\Omega, \mathbb{R}^N)$. Hence the first relation gives the desired equation (3).

The rest of the assertion follows as in 1,*2.2*. □

The afore-stated result can easily be generalized to r subsidiary conditions of integral type. Suppose that $G_k(x, z, p)$, $1 \leq k \leq r$, are Lagrangians of class $C^1(\mathscr{U})$, and set

$$\mathscr{G}_k(v) := \int_\Omega G_k(x, v, Dv)\, dx.$$

Let $c = (c_1, \ldots, c_r)$, where $c_k := \mathscr{G}_k(u)$, and set

$$\mathscr{S}_c := \mathscr{S} \cap \{v\colon \mathscr{G}_k(v) = c_k,\ k = 1, \ldots, r\}.$$

Here $\mathscr{S}$ once again denotes a class of mappings with the variational property $(\mathscr{V})$.

Theorem 2. *Suppose that u furnishes a weak extremum of $\mathscr{F}$ in the class $\mathscr{S}_c$, and assume that there are functions $\psi_1, \ldots, \psi_r \in C_c^\infty(\Omega, \mathbb{R}^N)$ such that the matrix*

$$A = (a_{kl}), \qquad a_{kl} := \delta\mathscr{G}_k(u, \psi_l)$$

has maximal rank r. Then there exist numbers $\lambda_1, \ldots, \lambda_r \in \mathbb{R}$ such that the functional

$$\mathscr{F}^* := \mathscr{F} + \lambda_1 \mathscr{G}_1 + \lambda_2 \mathscr{G}_2 + \cdots + \lambda_r \mathscr{G}_r$$

satisfies

$$\delta\mathscr{F}^*(u, \varphi) = 0 \quad \textit{for all } \varphi \in C_c^\infty(\Omega, \mathbb{R}^N). \tag{6}$$

Proof. Let $t = (t^1, t^2, \ldots, t^r)$, and set

$$\Phi(\varepsilon, t) = \mathscr{F}(u + \varepsilon\varphi + t^1\psi_1 + \cdots + t^r\psi_r),$$

$$\Psi_k(\varepsilon, t) := \mathscr{G}_k(u + \varepsilon\varphi + t^1\psi_1 + \cdots + t^r\psi_r).$$

Note that $\Phi, \Psi_k \in C^1$, $A = \left(\frac{\partial \Psi_k}{\partial t^l}(0, 0)\right)$, and that

$$\Phi(\varepsilon, t) \geq \Phi(0, 0) \quad (\text{or } \leq \Phi(0, 0))$$

for $|\varepsilon| < \varepsilon_0$, $|t| < t_0$, with $0 < \varepsilon_0, t_0 \ll 1$, and

$$\Psi_k(\varepsilon, t) = c_k, \quad 1 \leq k \leq r.$$

On account of the classical multiplier theorem, there exist numbers $\lambda_1, \ldots, \lambda_r \in \mathbb{R}$ such that

$$\Phi_\varepsilon(0, 0) + \sum_{k=1}^{r} \lambda_k \Psi_{k,\varepsilon}(0, 0) = 0,$$

$$\Phi_{t^m}(0, 0) + \sum_{k=1}^{r} \lambda_k \Psi_{k,t^m}(0, 0) = 0, \quad 1 \leq m \leq r.$$

From the second set of equations, $\lambda_1, \ldots, \lambda_r$ can be computed and turn out to be independent of the choice of $\varphi \in C_c^\infty(\Omega, \mathbb{R}^N)$. Then the first equation implies relation (6), and the theorem is proved. □

Similarly, if

$$\mathscr{G}(v) = \int_\Sigma G(x, v, Dv)\, d\mathscr{H}^{n-1}, \quad \Sigma \subset \partial\Omega, \tag{7}$$

the existence of a function $\psi \in C_c^1(\Sigma, \mathbb{R}^N)$ with

$$\delta\mathscr{G}(u, \psi) = \int_\Sigma \{G_z(x, u, Du)\cdot\psi + G_p(x, u, Du)\cdot D\psi\}\, d\mathscr{H}^{n-1} \neq 0$$

implies that a weak relative extremizer u of the functional $\mathscr{F}$ on the class

$$\mathscr{S}_c := \{v \in C^1(\bar{\Omega}, \mathbb{R}^N) \colon \mathscr{G}(v) = c \text{ and } v = u \text{ on } \partial\Omega - \Sigma\},$$

with $c := \mathscr{G}(u)$ yields the existence of a Lagrange multiplier $\lambda \in \mathbb{R}$ such that

$$\delta\mathscr{F}(u, \varphi) + \lambda\delta\mathscr{G}(u, \varphi) = 0 \quad \text{for all } \varphi \in C_c^1(\Omega \cup \Sigma, \mathbb{R}^N).$$

If $u \in C^2(\bar{\Omega}, \mathbb{R}^N)$, we can infer

$$(8) \qquad 0 = \int_\Omega L_F(u)\cdot\varphi\,dx + \int_\Sigma \nu_\alpha F_{p^i_\alpha}(x, u, Du)\varphi^i\,d\mathscr{H}^{n-1}$$
$$+ \int_\Sigma \lambda\{G_{z^i}(x, u, Du)\varphi^i + G_{p^i_\alpha}(x, u, Du)D_\alpha\varphi^i\}\,d\mathscr{H}^{n-1}.$$

Suppose that $G = G(x, z)$, i.e., that G does not depend on p. Then we conclude that

$$(8') \qquad L_F(u) = 0 \quad \text{in } \Omega$$

and

$$(8'') \qquad \nu_\alpha F_{p^i_\alpha}(x, u, Du) + \lambda G_{z^i}(x, u) = 0 \quad \text{on } \Sigma$$

hold, where $\nu = (\nu_1, \ldots, \nu_n)$ denotes the exterior normal of $\partial\Omega$.

Let us now consider some examples.

[1] *The classical isoperimetric problem.* Suppose that $\mathscr{C}$ is a closed Jordan curve of length L whose interior domain has maximal area among all interiors of closed simple curves with length L. We assume that $\mathscr{C}$ is a regular curve of class C^2 and want to show that *$\mathscr{C}$ is a circle.*

A simple reasoning shows that the interior Ω of $\mathscr{C}$ is a convex set (cf. Fig. 1). Then the assertion is a consequence of the following reasoning.

Let $z = f(x)$, $a_1 \le x \le a_2$, be a fixed curve which connects two points $P_1 = (a_1, b_1)$ and $P_2 = (a_2, b_2)$ in the x, z-plane, and consider the class $\mathscr{S}$ of C^1-curves $z = v(x)$, $a_1 \le x \le a_2$, with $v(a_1) = b_1$, $v(a_2) = b_2$ which satisfy $f(x) < v(x)$ on (a_1, a_2). Then the class $\mathscr{S}$ satisfies condition $(\mathscr{V})$. Let $\mathscr{S}_c$ be the class of functions $v \in \mathscr{S}$ satisfying

$$\mathscr{L}(v) := \int_{a_1}^{a_2} \sqrt{1 + v'^2}\,dx = c$$

for some fixed number c larger than $|P_1 - P_2|$.

For $v \in \mathscr{S}$, we denote by

$$\mathscr{A}(v) = \int_{a_1}^{a_2} \{v(x) - f(x)\}\,dx,$$

the area between the graphs of f and v.

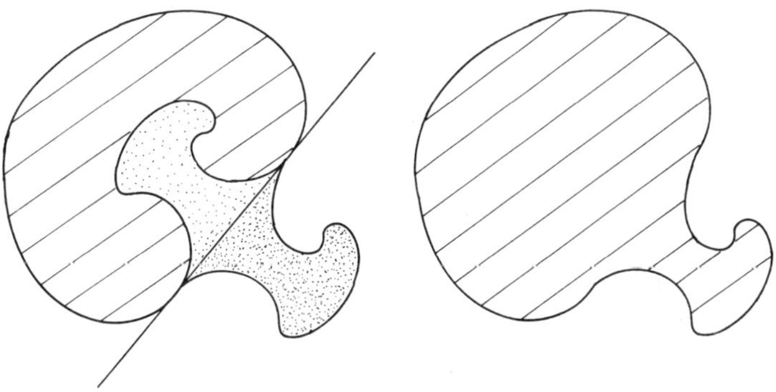

Fig. 1. *The solution of the isoperimetric problem must be convex.* Otherwise one could find a domain a larger area which has the same perimeter.

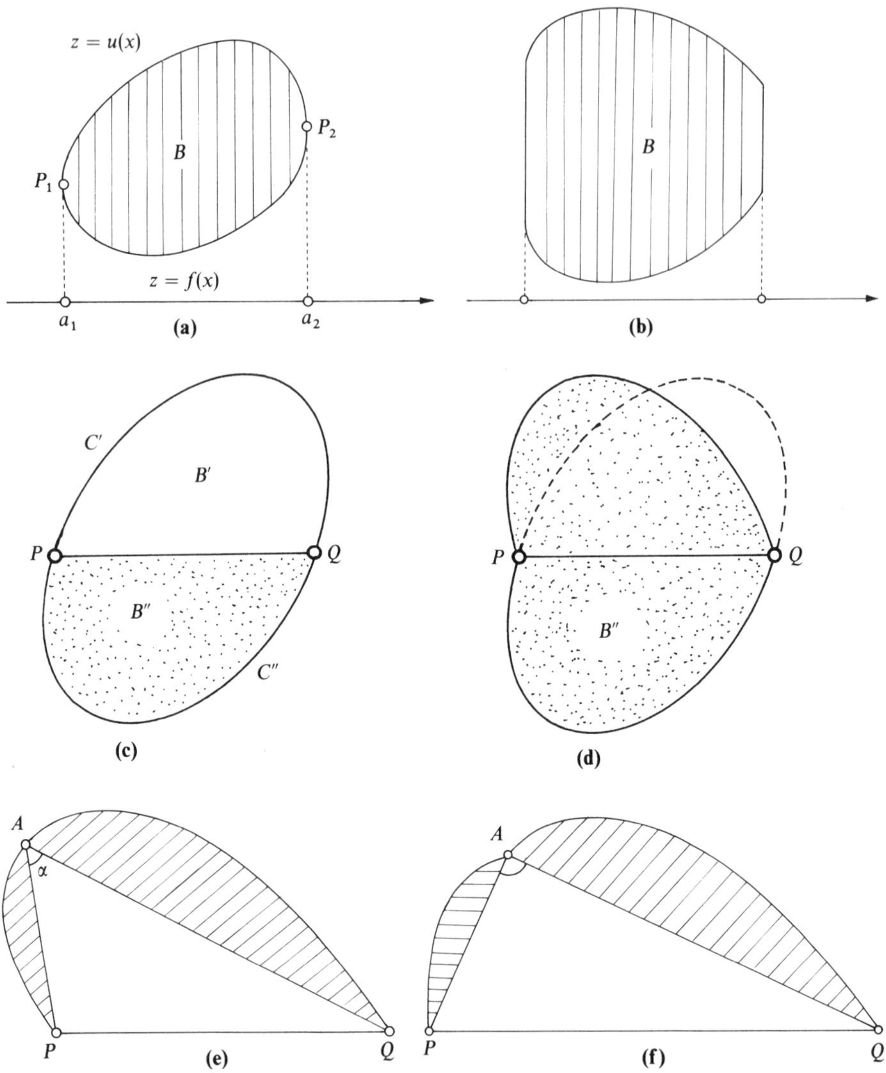

Fig. 2. Two proofs of the *isoperimetric property of the circle* based on the assumption that the isoperimetric problem has a solution. **(a)** A first proof using a Lagrange multiplier as described in [1]; **(b)** This proof also functions if B is not assumed to be strictly convex. **(c–f)** A second proof due to Steiner: **(c)** Let B be a solution of the isoperimetric problem; we can assume B to be convex. Then we choose two points P, Q on the boundary C of B which divide C in two arcs C', C'' of equal length. The chord PQ decomposes B in two convex parts B' and B'' of equal perimeter. **(d)** B', B'' must have equal area. In fact, if e.g. *area* B' < *area* B'', then the union of B'' and its mirror image with respect to the axis L through P, Q has the same perimeter as B, but a larger area. **(e)** Now we show that B' (and analogously B'') is a semidisk. Otherwise there would exist a point A on C'' such that the angle $\alpha = \sphericalangle PAQ$ were not a right one. Imagine that the two shaded lunulae could be moved like a pair of scissors by installing a hinge at A. **(f)** By changing α to a right angle we enlarge the area of the triangle PAQ. Adding the lunulae we obtain a new domain whose union with its mirror image in the axis L has the same perimeter as B but a large area, a contradiction.

Suppose now that $u \in \mathscr{S}_c$ is a C^2-maximizer of $\mathscr{A}$ in $\mathscr{S}_c$ which is not an affine function, that is, a straight line (and this is a consequence of the assumption $|P_1 - P_2| < c$). Then there exists some $\varphi \in C_c^\infty((a_1, a_2))$ such that $\delta\mathscr{L}(u, \varphi) \neq 0$ holds. Hence there is a Lagrange multiplier $\lambda \in \mathbb{R}$ such that u is an extremal of $\mathscr{A} + \lambda\mathscr{L}$ or, equivalently, of

$$\int_{a_1}^{a_2} (u + \lambda\sqrt{1 + u'^2})\, dx.$$

Hence u is a solution of the Euler equation

$$\left(\frac{u'}{\sqrt{1 + u'^2}}\right)' = \frac{1}{\lambda},$$

and we infer that the graph of u has constant curvature $1/\lambda$ and must therefore be a circular arc, cf. also 1,*2.2* [5]. Note that, in the present discussion, we have assumed the existence of a C^2-solution of the isoperimetric problem. It is another matter how this existence can be established. We also refer the reader to 4,*2.3* [3].

In the following examples we shall omit the easy proof that the sufficient condition of Theorem 1 for the existence of a Lagrange multiplier is satisfied. We will only write down the formal equations.

[2] *Eigenvalue problem of the vibrating string.* Let $u(x)$, $a \leq x \leq b$, be the minimizer of the integral

$$\int_a^b |v'(x)|^2\, dx$$

in the class of functions $v \in C^1([a, b])$ which satisfy $v(a) = 0$, $v(b) = 0$ as well as

$$\int_a^b |v(x)|^2\, dx = 1.$$

Then u is a weak extremal of

$$\int_a^b \{|v'(x)|^2 - \lambda|v(x)|^2\}\, dx$$

and, if u is of class C^2, we arrive at the Euler equation

$$u'' + \lambda u = 0.$$

Multiplying by u and integrating over $[a, b]$, we, after a partial integration, obtain

$$\int_a^b |u'|^2\, dx = \lambda.$$

Thus the Lagrange multiplier λ turns out to be an eigenvalue of the *eigenvalue problem of the vibrating string*

$$u'' + \lambda u = 0, \quad u(a) = 0, \qquad u(b) = 0,$$

and, in fact, it is the smallest eigenvalue.

Similarly, the other eigenvalues of this problem can be interpreted as Lagrange multipliers of the integral $\int_a^b |v'|^2\, dx$ on certain subspaces of $C_0^1([a, b])$.

[3] *Eigenvalue problem of the vibrating membrane.* Consider a bounded domain Ω of $\mathbb{R}^n$, and let $u(x)$ be the minimum of the integral

$$\int_\Omega |Dv|^2\, dx$$

in the class of functions $v \in C^1(\overline{\Omega})$ which satisfy

$$v(x) = 0 \text{ on } \partial\Omega \quad \text{and} \quad \int_\Omega |v(x)|^2 \, dx = 1 .$$

Then u is a weak extremal of

$$\int_\Omega \{|Dv|^2 - \lambda |v|^2\} \, dx ,$$

and if u is of class $C^2(\Omega)$, it satisfies the Euler equation

$$\Delta u + \lambda u = 0 \quad \text{in } \Omega$$

whence

$$\int_\Omega |Du|^2 \, dx = \lambda .$$

Hence the Lagrange multiplier λ turns out to be the smallest eigenvalue of the *eigenvalue problem of the vibrating membrane.*

Similarly the other eigenvalues of the membrane can be interpreted as Lagrange multipliers of the Dirichlet integral with respect to the subsidiary condition $\int_\Omega |u|^2 \, dx = 1$.

[4] *Hypersurfaces of constant mean curvature.* Let $u(x)$, $x \in \bar{\Omega}$, be a surface which is defined on some domain Ω of $\mathbb{R}^n$ and has prescribed boundary values f on $\partial\Omega$. We assume that u minimizes area

$$\mathscr{A}(u) = \int_\Omega \sqrt{1 + |Du|^2} \, dx$$

among all C^1-surfaces with the prescribed boundary values f on $\partial\Omega$, for which the volume functional

$$\mathscr{V}(u) = \int_\Omega u(x) \, dx ,$$

is kept fixed. Then $u(x)$ is an extremal of the functional

$$\mathscr{A}(u) + \lambda \mathscr{V}(u) = \int_\Omega \{\sqrt{1 + |Du|^2} + \lambda u\} \, dx$$

and, therefore, satisfies the Euler equation

$$\operatorname{div} Tu = nH ,$$

with $H = \lambda/n$ and $Tu = (1 + |Du|^2)^{-1/2} Du$. This means that $(x, u(x))$ is a hypersurface of constant mean curvature H.

[5] *Catenary or chain line.* Let us now look at the equilibrium position of an inextensible heavy thread of constant density, say, density 1. Such a thread is an idealized physical system which is thought to be an infinitely thin thread. Another interpretation of such a thread is that of a very thin chain with very small links. Let the chain be fixed at two points $P_1 = (a_1, b_1)$ and $P_2 = (a_2, b_2)$ in the x, z-plane. If the position of the chain is described by $z = u(x)$, $a_1 \le x \le a_2$, this imposes the boundary conditions $u(a_1) = b_1$ and $u(a_2) = b_2$. Since the chain is assumed to be inextensible, its length c is prescribed. Thus we have the isoperimetric side condition

$$\int_{a_1}^{a_2} \sqrt{1 + u'^2} \, dx = c .$$

A stable equilibrium position must have a center of gravity which is as low as possible. If we assume that gravity acts in direction of the negative z-axis, this means that a stable equilibrium position will minimize the quotient

$$\int_{a_1}^{a_2} u\sqrt{1+u'^2}\,dx \bigg/ \int_{a_1}^{a_2} \sqrt{1+u'^2}\,dx.$$

Since the denominator is constant, it will be a minimum of

$$\int_{a_1}^{a_2} u\sqrt{1+u'^2}\,dx$$

under the subsidiary condition

$$\int_{a_1}^{a_2} \sqrt{1+u'^2}\,dx = c$$

and the boundary conditions $u(a_1) = b_1$, $u(a_2) = b_2$. As we know, a C^2-minimum $z = u(x)$ will be an extremal of the variational integral

$$\int_{a_1}^{a_2} (u+\lambda)\sqrt{1+u'^2}\,dx$$

for some appropriate number $\lambda \in \mathbb{R}$, the Lagrange multiplier of the problem. Introducing

$$v(x) := u(x) + \lambda,$$

we obtain an extremal of the afore-considered variational integral

$$\int_{a_1}^{a_2} v\sqrt{1+v'^2}\,dx;$$

cf. 1,*2.2* [7]. Thus the meridian curves $v(x) = a\cosh\dfrac{x-x_0}{a}$ of surfaces of revolution which furnish a stationary value of the area functional, and the equilibrium positions of a chain, the so-called[2] *catenaries* are, up to a translation in direction of the z-axis, solutions of the same Euler equation, that is, both problems are in some sense equivalent.

2. Mappings into Manifolds: Holonomic Constraints

In this section we shall discuss local extrema and, more generally, stationary points (extremals) of variational integrals $\mathcal{F}(u) = \int_\Omega F(x, u, Du)\,dx$ which, in addition to appropriate boundary conditions, have to satisfy a subsidiary condition of the kind

$$(1) \qquad G(x, u(x)) = 0,$$

where $G(x, z)$ denotes a scalar or vector valued function of the variables $(x, z) \in \mathbb{R}^n \times \mathbb{R}^N$. Conditions of the type (1) are called *holonomic* constraints, whereas subsidiary conditions of the more general form

$$(2) \qquad G(x, u(x), Du(x)) = 0$$

[2] The name *catenary* is derived from Latin *catena, chain.*

are denoted as *nonholonomic* (holos = integer; the meaning of holonomic is "integrable").

Note that every holonomic constraint (1) can, by differentiation, be transformed into the nonholonomic constraint

$$G_x(x, u(x)) + G_z(x, u(x)) \cdot Du(x) = 0$$

from which (1) can be regained by integration, whereas a constraint

$$A(x, u(x)) + B(x, u(x)) \cdot Du(x) = 0$$

can in general not be brought to the form (1). Strictly speaking, the nonholonomic or nonintegrable constraints are those which cannot be brought to the form (1) by integration.

If $n = 1$ and if x is interpreted as a time parameter t, a time-dependent condition (1) or (2) is called *rheonomic* (= flowing) whereas a time-independent condition is said to be *scleronomic* (= rigid). In other words, the condition $G(u, \dot{u}) = 0$ is scleronomic, whereas $G(t, u, \dot{u}) = 0$ is called rheonomic.

If we conceive $u \in C^1(\bar{\Omega}, \mathbb{R}^N)$ as a mapping from $\bar{\Omega}$ into $\mathbb{R}^N$, and if we interpret the equations

$$G(z) = 0$$

for a given function $G = (G^1, \ldots, G^r)$, $1 \leq r \leq N - 1$, as the defining equations of an $(N - r)$-dimensional submanifold M in $\mathbb{R}^N$, then the problem to minimize $\mathcal{F}$ under the subsidiary condition

$$G(u(x)) = 0 \tag{3}$$

can be seen as a minimization problem for mappings $u : \bar{\Omega} \to M$. For instance, the question of shortest distance of two points P_1 and P_2 on a sphere of radius R leads to the problem of finding the minimum of the integral

$$\int_{t_1}^{t_2} |\dot{u}(t)| \, dt$$

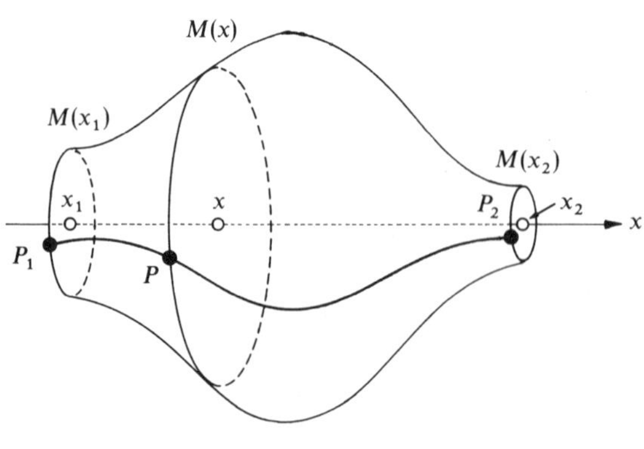

Fig. 3. The curve graph u lies on a surface $S = \{(x, z) : z \in M(x)\}$ where $M(x)$ is a manifold in $\mathbb{R}^2$ given by $M(x) = \{z \in \mathbb{R}^2 : G(x, z) = 0\}$.

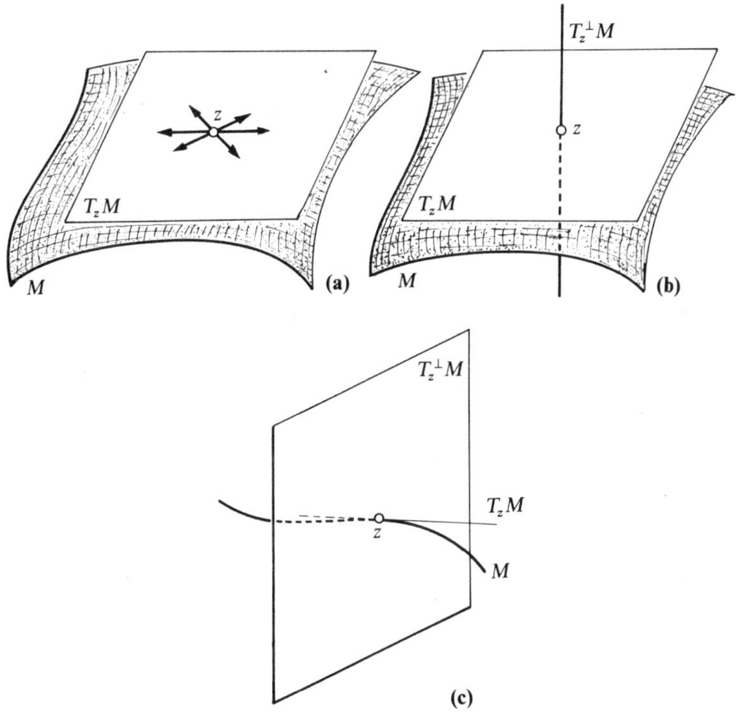

Fig. 4. **(a)** The tangent space T_zM of a manifold M at z. **(b)** The normal space $T_z^{\perp}M$ of surface M in $\mathbb{R}^3$. **(c)** The normal space $T_z^{\perp}M$ of a curve M in $\mathbb{R}^3$.

among all mappings $u \in C^1([t_1, t_2], \mathbb{R}^3)$ with $u(t_1) = P_1$ and $u(t_2) = P_2$ which are subject to the constraint $|u(t)| = R$. As it is well known, the solutions are subarcs of great circles on $S_R^2 = \{z \in \mathbb{R}^3: |z| = R\}$; cf. Example 4 below.

If the subsidiary condition is of the more general form (1),

$$G(x, u(x)) = 0,$$

then we can conceive the admissible mappings $u: \Omega \to \mathbb{R}^N$ as functions which map $x \in \Omega$ into a submanifold $M(x)$ of $\mathbb{R}^N$ described by

$$M(x) = \{z \in \mathbb{R}^N: G(x, z) = 0\}.$$

In this case, the appropriate geometrical picture of admissible mappings is that of sections of a suitable fibre bundle.

Another geometrical interpretation is obtained if, for a given function $G(x, z) = (G^1(x, z), \ldots, G^r(x, z))$, we see the relations

$$G^1(x, z) = 0, \ldots, G^r(x, z) = 0$$

as the defining equations of an $(n + N - r)$-dimensional submanifold M of $\mathbb{R}^n \times \mathbb{R}^N$. Then mappings u subject to (1) can be understood as functions, the graph of which is contained in M.

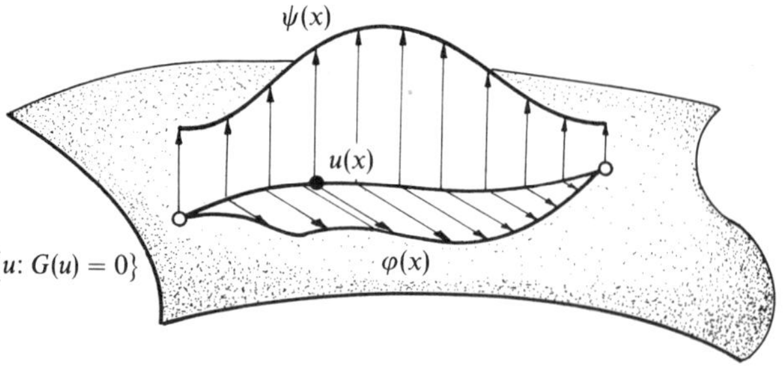

Fig. 5. A vector field ψ along u which is perpendicular to all vector fields φ along u which are tangent to M.

Holonomic side conditions can fairly easily be treated, whereas nonholonomic conditions are much more troublesome, even in the case $n = 1$. For this reason, we shall presently restrict ourselves to the investigation of holonomic conditions and defer the investigation of nonholonomic ones to the next section.

Suppose now that Ω is a bounded domain in $\mathbb{R}^n$, and that $u \in C^1(\bar{\Omega}, \mathbb{R}^N)$ satisfies

$$G(x, u(x)) = 0 \quad \text{for all } x \in \bar{\Omega}.$$

For the sake of simplicity, we assume that $G(x, z)$ is of class $C^2(\mathbb{R}^n \times \mathbb{R}^N, \mathbb{R}^r)$, $1 \le r \le N - 1$, and we also suppose that the $r \times N$-matrix function $G_z = (G^i_{z^k})$ has maximal rank r in all points of the set

$$\mathscr{M} := \{(x, z): x \in \bar{\Omega}, G(x, z) = 0\}.$$

Then, for every $x \in \bar{\Omega}$, the set

$$M(x) := \{z \in \mathbb{R}^N: G(x, z) = 0\}$$

is an $(N - r)$-dimensional manifold in $\mathbb{R}^N$, with normal vector fields $G^1_z(x, z), \ldots, G^r_z(x, z)$ of class C^1.

Denote by $\Pi(x, z)$ the orthogonal projection of $\mathbb{R}^N$ onto the tangent space $T_z M(x)$ of $M(x)$ at z; Π depends continuously on x and z.

Definition. *For a mapping $u: \bar{\Omega} \to \mathbb{R}^N$ with $u(x) \in M(x)$, a function $\varphi \in C^1(\bar{\Omega}, \mathbb{R}^N)$ is said to be a* tangential vector field along u *if $\varphi(x) \in T_{u(x)} M(x)$ holds for all $x \in \bar{\Omega}$.*

Then we can state the following version of the fundamental lemma.

Lemma 1. *Let ψ be a mapping of class $C^0(\Omega, \mathbb{R}^N)$ and suppose that*

$$\int_\Omega \psi(x) \cdot \varphi(x)\, dx = 0 \tag{4}$$

is satisfied for all tangential vector fields $\varphi \in C_c^1(\Omega, \mathbb{R}^N)$ *along* u. *Then we obtain* $\Pi(x, u)\psi = 0$ *or, more precisely,*

$$\Pi(x, u(x))\, \psi(x) = 0 \quad \textit{for all } x \in \Omega. \tag{5}$$

Proof. Fix some $x_0 \in \Omega$, set $z_0 = u(x_0)$, and introduce an orthonormal frame $\tau_1, \ldots, \tau_{N-r}, \nu_1, \ldots, \nu_r$ defined for (x, z) in a sufficiently small neighbourhood $\mathcal{N}_0$ of the point (x_0, z_0) in $\mathcal{M}$, where $\tau_1, \ldots, \tau_{N-r}$ are tangential to $M(x)$. We can assume that the vectors $\tau_i(x, z)$, $\nu_k(x, z)$ depend in a C^1-way on $(x, z) \in \mathcal{N}_0$. Choose some ball $B = B_R(x_0)$ with $\{(x, u(x)) \colon x \in B\} \subset \mathcal{N}_0$, and some arbitrary functions $\varphi^1(x), \ldots, \varphi^{N-r}(x)$ of class $C_c^1(B)$. Then

$$\varphi(x) := \varphi^1(x)\tau_1(x, u(x)) + \cdots + \varphi^{N-r}(x)\tau_{N-r}(x, u(x))$$

is a tangential vector field of class $C_c^1(B, \mathbb{R}^N)$ along the mapping u. Moreover, there are functions $a^i(x)$, $b^k(x)$ of class $C^0(B)$ such that

$$\psi(x) = \sum_{i=1}^{N-r} a^i(x)\tau_i(x, u(x)) + \sum_{k=1}^{r} b^k(x)\nu_k(x, u(x))$$

holds for $x \in B$. Then we infer from (4) that

$$\int_\Omega \{a^1(x)\varphi^1(x) + \cdots + a^{N-r}(x)\varphi^{N-r}(x)\}\, dx = 0$$

is fulfilled for arbitrary $\varphi^i(x) \in C_c^1(B)$, $1 \le i \le N - r$. Now the fundamental lemma yields that

$$a^i(x) = 0 \quad \text{on } B,\ 1 \le i \le N - r,$$

which is equivalent to

$$\Pi(x, u(x))\, \psi(x) = 0$$

for all $x \in B$, and in particular for $x = x_0$. Since x_0 was chosen to be an arbitrary point of Ω, the assertion (5) is proved. □

Consider the class $\mathscr{C}$ of mappings $v \in C^1(\bar{\Omega}, \mathbb{R}^N)$ whose 1-graph $\{(x, v(x), Dv(x)) \colon x \in \bar{\Omega}\}$ is contained in the domain of definition of F, and which satisfy the holonomic constraint

$$G(x, v(x)) = 0 \quad \text{for all } x \in \bar{\Omega}.$$

Suppose, moreover, that u is contained in $\mathscr{C}$ and furnishes a weak relative minimum of $\mathscr{F}$ in $\mathscr{C}$, that is,

$$\mathscr{F}(u) \le \mathscr{F}(v)$$

for all $v \in \mathscr{C}$ with $v = u$ on $\partial\Omega$ and $\|v - u\|_{C^1(\bar{\Omega})} < \delta$ for some $\delta > 0$.

Let $\varphi \in C_c^1(\Omega, \mathbb{R}^N)$ be an arbitrary vector field which is tangential along u. By an elementary construction, we can find a mapping $\psi : \bar{\Omega} \times [-t_0, t_0] \to \mathbb{R}^N$, $t_0 > 0$, of class C^1, such that

$$\psi(x, t) = u(x) + t\varphi(x) + o(t) \quad \text{as } t \to 0$$

holds as well as

$$\psi(x, t) = u(x) \quad \text{for } (x, t) \in \partial\Omega \times [-t_0, t_0]$$

and

$$\psi(\cdot, t) \in \mathscr{C} \quad \text{for every } t \in [-t_0, t_0].$$

Therefore, the function

$$\Phi(t) := \mathscr{F}(\psi(\cdot, t))$$

satisfies

$$\Phi(t) \geq \Phi(0)$$

for sufficiently small values of $|t|$, whence

$$\Phi'(0) = 0,$$

that is,

$$\delta\mathscr{F}(u, \varphi) = 0.$$

Thus we have proved

Lemma 2. *If $u \in \mathscr{C}$ is a weak relative minimum of $\mathscr{F}$ in the class $\mathscr{C}$ defined by the holonomic constraint $G(x, v(x)) = 0$, then the first variation $\delta\mathscr{F}(u, \varphi)$ vanishes for all vector fields $\varphi \in C_c^1(\Omega, \mathbb{R}^N)$ which are tangential along u.*

This leads us to the following

Definition. *A mapping $u \in C^1(\bar{\Omega}, \mathbb{R}^N)$ subject to some holonomic constraint $G(x, u(x)) = 0$ is said to be a* weak constrained extremal *if $\delta\mathscr{F}(u, \varphi) = 0$ holds for all $\varphi \in C_c^1(\Omega, \mathbb{R}^N)$ which are tangential along u.*

Thus we can say that *every constrained weak relative minimum is a weak constrained extremal.*

If u is a weak constrained extremal of class $C^2(\Omega, \mathbb{R}^N)$, we obtain

$$0 = \delta\mathscr{F}(u, \varphi) = \int_\Omega L_F(u) \cdot \varphi \, dx$$

for all $\varphi \in C_c^1(\Omega, \mathbb{R}^N)$ which are tangential along u. By Lemma 1, we infer that

$$\Pi(x, u) L_F(u) = 0. \tag{6}$$

In other words, *$L_F(u)(x)$ is perpendicular to the tangent space $T_{u(x)}M(x)$, for all $x \in \Omega$.*

Therefore $L_F(u)(x)$ has to be a linear combination of the r normal vectors $G_z^1(x, u(x)), \ldots, G_z^r(x, u(x))$. Thus there exist functions $\mu_1(x), \ldots, \mu_r(x)$, $x \in \Omega$, such that

$$L_F(u) = \mu_1 G_z^1(\cdot, u) + \mu_2 G_z^2(\cdot, u) + \cdots + \mu_r G_z^r(\cdot, u) \tag{7}$$

is satisfied. If $u \in C^{s+2}(\Omega, \mathbb{R}^N)$, and if G is of class C^{s+1}, we may infer from (7) that $\mu_j \in C^s(\Omega)$, $j = 1, \ldots, r$. Set

$$\lambda_j(x) := -\mu_j(x)$$

and

$$F^*(x, z, p) := F(x, z, p) + \sum_{j=1}^{r} \lambda_j(x) G^j(x, z). \tag{8}$$

Then equation (7) is obviously equivalent to

$$L_{F^*}(u) = 0 \tag{9}$$

or to

$$D_\alpha F^*_{p^i_\alpha}(x, u(x), Du(x)) - F^*_{z^i}(x, u(x), Du(x)) = 0. \tag{9'}$$

Thus we have proved:

Theorem. *If u is a weak constrained extremal of $\mathscr{F}$ with respect to the holonomic conditions $G^j(x, u(x)) = 0$, $1 \le j \le r$, and if u is of class $C^2(\Omega, \mathbb{R}^N)$, then there exist functions $\lambda_1, \ldots, \lambda_r \in C^0(\Omega)$ such that u is an extremal of the Lagrangian*

$$F^*(x, z, p) := F(x, z, p) + \sum_{j=1}^{r} \lambda_j(x) G^j(x, z).$$

In other words, the multiplier rule derived in Section 1 for isoperimetric constraints also holds for holonomic constraints, with the modification that the Lagrange multipliers λ_j now are functions of $x \in \Omega$ and not only constant numbers.

In the next section we shall, under suitable assumptions, formulate the multiplier rule as well for nonholonomic constraints. Before that we will discuss a few examples of holonomic constraints.

[1] *Harmonic mappings into hypersurfaces of* $\mathbb{R}^{N+1}$. Let

$$\mathscr{D}(u) = \frac{1}{2}\int_\Omega |Du|^2\, dx, \qquad |Du|^2 = u^i_{x^\alpha} u^i_{x^\alpha},$$

be the Dirichlet integral of C^1-mappings $u : \bar{\Omega} \to \mathbb{R}^{N+1}$.

Then the Euler equations for $u = (u^1, \ldots, u^{N+1})$ can be written as

$$\Delta u = 0.$$

If we instead consider mappings subject to the constraint $|u| = 1$, i.e., mappings $u : \Omega \to S^N$ of $\Omega \subset \mathbb{R}^n$ into the unit sphere $S^N = \{z \in \mathbb{R}^{N+1} : |z| = 1\}$, then the constrained Euler equations are

$$-\Delta u = \mu(x) u.$$

In order to determine the multiplier $\mu(x)$, we first notice that $|u| = 1$ implies

$$\mu = -u \cdot \Delta u .$$

On the other hand, by differentiating $|u|^2 = 1$ twice, it follows that

$$u \cdot \Delta u + |Du|^2 = 0,$$

whence

$$\mu = |Du|^2,$$

and therefore

$$-\Delta u = u|Du|^2.$$

If we replace the constraint $|u| = 1$ by the scalar constraint $G(u) = \text{const.}$, we arrive at the constrained Euler equation

$$-\Delta u = \mu G_u(u) \quad \text{with } \mu = \frac{G_{u^i u^k}(u) u^i_{x^\alpha} u^k_{x^\alpha}}{|G_u(u)|^2} .$$

[2] *Shortest connection of two points on a surface in* $\mathbb{R}^3$. Consider mappings $u: I \to \mathbb{R}^3$ of an interval $I = [a, b]$ into $\mathbb{R}^3$ which are subject to some constraint

$$G(u) = c,$$

where G is a scalar function on $\mathbb{R}^3$. We interpret these mappings as parametric representations of curves that lie on a surface S described by $\{z \in \mathbb{R}^3 : G(z) = c\}$. The regular C^2-extremals of the length functional

$$\mathscr{L}(u) = \int_a^b |\dot{u}(t)| \, dt,$$

subjected to the holonomic constraints $G(u) = c$, are extremals of the functional

$$\int_a^b \{|\dot{u}| + \lambda G(u)\} \, dt$$

for some multiplier $\lambda(t)$, and thus we obtain the constrained Euler equation

$$\frac{d}{dt}\left\{\frac{\dot{u}}{|\dot{u}|}\right\} = \lambda G_z(u).$$

Let us introduce the parameter of the arc length s by $ds = |\dot{u}| \, dt$, and set $v(s) = u(t(s))$ and $\mu(s) = \lambda(t(s)) \dfrac{dt}{ds}$. Then we arrive at

$$\frac{d^2}{ds^2} v = \mu G_z(v).$$

By Frenet's formula,[3] we have $\dfrac{d^2 v}{ds^2} = \kappa \mathbf{v}$ where $\mathbf{v}$ is the principal normal of the shortest line $v(s)$. Thus $\mathbf{v}$ and the surface normal $G_z(v)$ are collinear, and we obtain

Johann Bernoulli's theorem. *At each point P of a shortest line, its osculating plane intersects the tangent plane of S perpendicularly.*

[3] Cf. the Supplement.

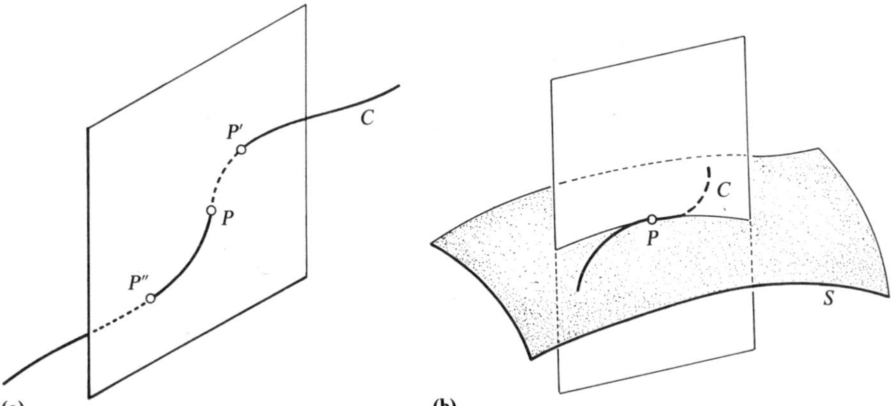

(a) **(b)**

Fig. 6. (a) The *osculating plane E* of a curve C at some of its points, P, is the limit of the plane E^* containing P and two other points P' and P'' on C tending to P.

Fig. 6. (b) Johann Bernoulli's theorem: The osculating plane $E(P)$ of a geodesic C of S intersects S perpendicularly at P.

A constrained extremal $u(t)$, $a \le t \le b$, of the length functional $\mathscr{L}(u)$ with regard to the subsidiary condition $G(u) = c$ is called a *geodesic* on the surface $S = \{z \in \mathbb{R}^3: G(z) = c\}$ if $u(t)$ is parametrized in proportion to the arc length, that is, if

$$|\dot{u}(t)| \equiv \text{const} \neq 0$$

holds. By the same reasoning as before, we infer that

$$\ddot{u} = \lambda G_z(u)$$

holds for some function $\lambda(t)$.

Conversely, every nonconstant C^2-solution $u(t)$, $a \le t \le b$, of

$$(*) \qquad \ddot{u} = \lambda G_z(u) \quad \text{and} \quad G(u) = c$$

is necessarily a geodesic on S, since $G(u) = c$ implies $\dot{u} \cdot G_z(u) = 0$, whence $0 = \dot{u} \cdot \ddot{u} = \frac{1}{2}\frac{d}{dt}|\dot{u}|^2$, and therefore $|\dot{u}(t)| \equiv \text{const} \neq 0$ on $[a, b]$.

On the other hand, the constrained extremals $u(t)$ of the one-dimensional Dirichlet integral

$$\frac{1}{2}\int_a^b |\dot{u}|^2 \, dt,$$

with regard to the subsidiary condition $G(u) = c$ coincide with the solutions of $(*)$. *Thus, we can also define geodesics on S as nonconstant harmonic mappings of some interval into the surface S. Moreover, at each point of a geodesic u, the osculating plane is perpendicular to the surface S.*

The subsidiary condition $|\dot{u}| = \text{const}$ is a nonholonomic constraint which selects the geodesics among all extremals of the length and relates them to the extremals of the one-dimensional Dirichlet integral.

[3] *Geodesics on a hypersurface in* $\mathbb{R}^{n+1}$. Let $S = \{z \in \mathbb{R}^{n+1}: G(z) = c\}$ be a regular hypersurface in $\mathbb{R}^{n+1}$. Then *geodesics on S* are defined as S-constrained extremals of $\frac{1}{2}\int |\dot{u}|^2 \, dt$, or as S-constrained extremals of $\int |\dot{u}| \, dt$, subject to the holonomic constraint $G(u) = c$, and they can be characterized by the equations

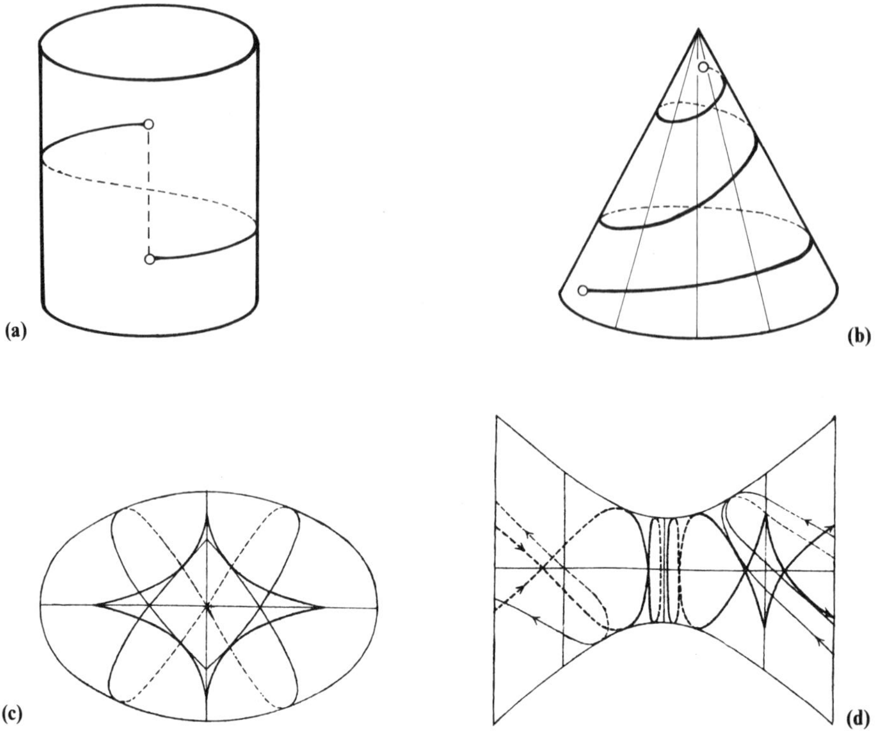

Fig. 7. Geodesics on **(a)** a cylinder, **(b)** a cone, **(c)** an ellipsoid, **(d)** a hyperboloid.

$$\ddot{u} = \lambda G_z(z) \quad \text{for suitable } \lambda(t) \text{ and } G(u) = c.$$

We include in the definition of a geodesic $u(t)$, $a \le t \le b$, that $u(t) \ne \text{const}$; or $|\dot{u}(t)| \ne 0$. By the reasoning of [2], it follows that $|\dot{u}(t)| \equiv \text{const}$, and thus we may, without loss of generality, assume that $|\dot{u}(t)| \equiv 1$.

[4] *Geodesics on a sphere.* Let $u(t)$, $a \le t \le b$, be a geodesic on the sphere

$$S = \{z \in \mathbb{R}^{n+1}: |z| = R\}$$

of radius $R > 0$ in $\mathbb{R}^{n+1}$. We shall prove:

Every geodesic on a sphere S has to be part of a great circle on S. Conversely, every great circle on S is a geodesic.

In fact, if $u(t)$ is a geodesic on S, then we can assume that $|\dot{u}(t)| \equiv 1$, i.e., t is the parameter of the arc length. Therefore, the curvature function $\kappa(t)$ of the geodesic $u(t)$ is given by

$$\kappa(t) = |\ddot{u}(t)|.$$

By [3], there exists a function $\lambda(t)$ such that

$$\ddot{u} = \lambda \frac{u}{R}.$$

On the other hand, by differentiating $|u|^2 = R^2$ twice with respect to t, we obtain

$$\ddot{u} \cdot u = -|\dot{u}|^2 = -1.$$

Thus it follows that $\lambda = -1/R$, and therefore $\kappa(t) \equiv 1/R$. Hence $u(t)$ is a circular arc of radius R on S and, therefore, part of a great circle. The converse can be proved by the same formulas.

[5] *Hamilton's principle and holonomic constraints.* Consider ℓ points $M_1, \ldots, M_\ell$ in $\mathbb{R}^3$ where the masses $m_1, \ldots, m_\ell$ are concentrated, $m_j > 0$, and suppose that the point masses M_j move in time under the influence of forces F_j. Denote by $X_j(t) = (x_j(t), y_j(t), z_j(t))$ the position vector of M_j at the time t with regard to some fixed cartesian system of coordinates x, y, z in $\mathbb{R}^3$.

We suppose that the ℓ forces $F_1, \ldots, F_\ell$ are time-independent and depend only on the positions $X_1, \ldots, X_\ell$ of $M_1, \ldots, M_\ell$:

$$F_j = F_j(X), \qquad X = (X_1, \ldots, X_\ell).$$

The motion of the ℓ point masses M_j is ruled by *Newton's lex secunda* (= second law) which, in Euler's formula, is written as[4]

$$m_j \ddot{X}_j = F_j(X), \quad 1 \le j \le \ell.$$

We assume that the forces $F = (F_1, \ldots, F_\ell)$ are *conservative*, that is, there exists a function $V(X)$ on $\mathbb{R}^{3\ell}$ such that

$$F_j = -\operatorname{grad}_{X_j} V = -(V_{x_j}, V_{y_j}, V_{z_j}) \quad \text{or} \quad F = -\operatorname{grad}_X V.$$

We now picture the ℓ point masses in $\mathbb{R}^3$ as a single point in $\mathbb{R}^{3\ell}$ with the position vector X, the motion of which is given by the Newton system of ordinary differential equations.

This system can be viewed as the Euler equations of the functional

$$\mathscr{S}(X) = \int_{t_1}^{t_2} L(X(t), \dot{X}(t))\, dt,$$

which is called the *action integral* and has the variational integrand

$$L(X, Y) := T(Y) - V(X),$$

where

$$T(Y) := \sum_{j=1}^{\ell} \frac{1}{2} m_j |Y_j|^2 .$$

One calls $T(\dot{X})$ the *kinetic energy*, $V(X)$ the *potential energy*, $L = T - V$ the *free energy*, and $E = T + V$ the *total energy* of the system $\{M_1, \ldots, M_\ell\}$ at the time t.

The observation that the Euler equations for $\mathscr{S}(X)$ coincide with Newton's law of motion

$$m_j \ddot{X}_j = -\operatorname{grad}_{X_j} V$$

is called *Hamilton's principle*, or *principle of least action* (although it rather should be denoted as "principle of stationary action").

It is shown in mechanics that Hamilton's principle also gives the correct equations of motion if the system of point masses is subject to holonomic constraints

$$G^1(t, X) = 0, \ldots, G^r(t, X) = 0.$$

In other words, there are functions $\lambda_1(t), \ldots, \lambda_r(t)$ such that the correct motion is described by

$$m_j \ddot{X}_j = -\operatorname{grad}_{X_j} V + \sum_{k=1}^{r} \lambda_k \operatorname{grad}_{X_j} G^k .$$

The second term on the right-hand side is interpreted as forces exerted by the constraints on the mass points.

[4] No summation with respect to repeated indices is assumed in this example.

Note that the constraints $G^j = 0$ yield r restrictions for the 3ℓ variables X. Thus, if the assumptions of the implicit function theorem are satisfied, we have exactly $N = 3\ell - r$ degrees of freedom and can remove r superfluous variables. Correspondingly, the variables

$$X = (\ldots, X_j, \ldots) = (\ldots, x_j, y_j, z_j, \ldots)$$

can – at least locally – be represented in the form

$$X = \phi(q),$$

with appropriate new variables $q = (q^1, \ldots, q^N)$ which are no longer subject to constraints. In other words, the $q^1, \ldots, q^N$ are *free variables*. Transforming the action integral $\mathscr{S}$ to the new functions $q(t)$, $\dot{q}(t)$, we obtain

$$\mathscr{S}(q) = \int_{t_1}^{t_2} \wedge(q, \dot{q})\, dt,$$

with a new Lagrangian $\wedge(q, \dot{q})$. *Hamilton's principle then states that the Euler equations*

$$\frac{d}{dt}\wedge_{\dot{q}^k}(q, \dot{q}) - \wedge_{q^k}(q, \dot{q}) = 0, \quad k = 1, \ldots, N,$$

yield the constrained equations of motion. In mechanics, these equations are usually called the *general Lagrange equations* or *Lagrange equations of second kind.*[5]

Since $L(X, Y) = T(Y) - V(X)$ does not depend on the independent variable t, the *law of conservation of energy*, proved in 1,*2.2* [7] yields that the expression

$$E := Y \cdot L_Y - L$$

is a first integral of the extremals of L. (Another proof of this fact based on E. Noether's theorem will be given in 3,*4* [1]) A brief calculation gives

$$E(X, Y) = T(Y) + V(X).$$

Thus the total energy $E(X(t), \dot{X}(t))$ is constant in time for every motion $X(t)$ which is an extremal of the action integral $\int L\, dt$. This fact justifies the notion "conservation of energy" introduced in 1,*2.2* [7]. Since also the Lagrangian $\wedge(q, \dot{q})$ of the holonomically constrained system $\{M_1, \ldots, M_\ell\}$ does not explicitly depend on t, we infer as well that

$$\dot{q}^k \wedge_{\dot{q}^k}(q, \dot{q}) - \wedge(q, \dot{q}) = \text{const}$$

is satisfied for every motion $q(t)$ governed by the Hamilton principle with the action integral

$$\mathscr{S} = \int_{t_2}^{t_2} \wedge(q, \dot{q})\, dt.$$

This remark shows that Hamilton's principle is more than just a casual observation. First of all, it provides us with a convenient tool to transform Newton's equations to new coordinates which may be better suited for the purpose of integration and which conveniently take the constraining forces into account. This will become particularly apparent in the so-called Hamilton–Jacobi theory

---

[5] Heinrich Hertz believed that Hamilton's principle does not yield the correct equations of motion in case of nonholonomic constraints. He was corrected by Otto Hölder [2] who in 1896 showed that Hamilton's principle can also be justified for nonholonomic side conditions, for instance for the motion of a sled, of skates, or of a rolling ball on a plane.

For a different opinion concerning the validity of Hamilton's principle with regard to mechanical systems subjected to nonholonomic constraints, see Prange [2], pp. 564–565, 568, 585–586, and Rund [4], pp. 352–358, in particular: p. 358. There one can also find further references to the literature concerning Hamilton's principle.

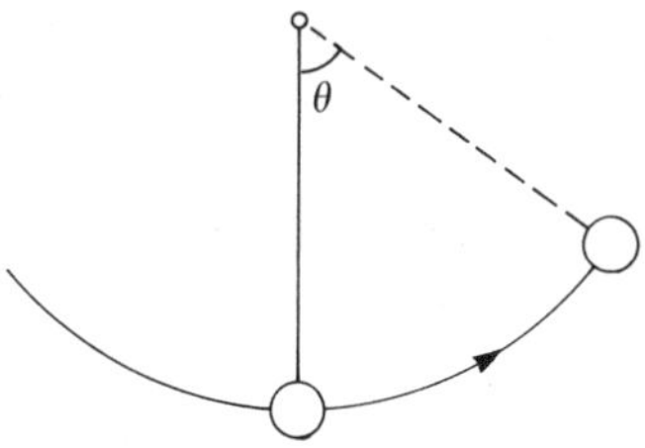

Fig. 8. The pendulum: A constrained motion on a circular orbit.

which will be developed in the third and fourth part of this treatise. Moreover, Hamilton's principle can be carried over to continuum mechanics. As was proved by von Helmholtz, all basic equations of classical physics can be derived from a variational principle. The same holds for general relativity and for quantum mechanics, as was shown by Hilbert and Schrödinger respectively. Also the work of Feyman was inspired by variational principles.

In present-day physics, Hamilton's principle is used as a comfortable tool to formulate the *fundamental equations of motion* for general field theories that are constructed as models of the physical reality. In this context Emmy Noether's theorem (see 3,*4*) yields conservation laws implied by the symmetries of the field.

6 *Pendulum equation.* In order to demonstrate Hamilton's principle for a simple constrained system, we consider a particle M which is forced to lie on a fixed circle of radius R in a vertical plane, where the z-axis is directed vertically and is pointing in the direction opposite to the gravitation. Introducing polar coordinates r, θ around the center of the circle, we obtain $r(t) = R$ for the motion of M constrained to the circle. In order to derive the equation governing the motion $\theta = \theta(t)$ we note that, on account of $x = R \sin \theta$, $z = R(1 - \cos \theta)$, we obtain

$$T = \frac{1}{2} m R^2 \dot{\theta}^2$$

for the kinetic energy of M, and

$$V = mgz = mgR(1 - \cos \theta)$$

for its potential energy. Thus the action integral is given by

$$\int_{t_1}^{t_2} \left\{ \frac{1}{2} m R^2 \dot{\theta}^2 - mgR(1 - \cos \theta) \right\} dt,$$

whence we arrive at the "*pendulum equation*"

$$R\ddot{\theta} + g \sin \theta = 0$$

as Lagrangian equation of motion. The associated linearized equation about the stable equilibrium solution $\theta = 0$ is

$$R\ddot{\theta} + g\theta = 0,$$

which formally coincides with the equation of a *harmonic oscillator*:

$$m\ddot{x} + kx = 0.$$

This is the equation of motion for a particle constrained to lie on a given straight line, say, the x-axis, and subject to a force with the potential energy $V(x) = \frac{1}{2}kx^2$, k some positive constant. A spring which, according to *Hooke's law*, exerts a force $F(x) = -kx$ provides an example of a simple mechanical system with the potential energy $V(x) = \frac{1}{2}kx^2$.

3. Nonholonomic Constraints

Now we shall discuss necessary conditions for local extrema of variational integrals $\mathscr{F}(u) = \int_\Omega F(x, u, Du)\,dx$ which, besides boundary conditions, are subject to nonholonomic constraints

(1) $$G(x, u(x), Du(x)) = 0.$$

We note that every variational problem of higher order can be reduced to a variational problem of first order with some linear nonholonomic constraint. For instance, a second order variational integral

$$\int_\Omega F(x, u, Du, D^2u)\,dx$$

can be written as

$$\int_\Omega F(x, u, Du, Dv)\,dx$$

if we introduce $v(x)$ by the equation

$$v(x) - Du(x) = 0,$$

which is then to be considered as a nonholonomic constraint $G(w, Dw) = 0$ for the function $w(x) := (u(x), v(x))$.

If $n = 1$ and $\Omega = (a, b)$, we can also subordinate an isoperimetric condition

(2) $$\int_a^b H(x, u(x), Du(x))\,dx = c$$

by introducing the supplementary function

$$v(x) := \int_a^x H(t, u(t), Du(t))\,dt,$$

which has to satisfy the boundary conditions

$$v(a) = 0, \qquad v(b) = c,$$

and is linked to $u(x)$ by the constraint

(3) $$Dv(x) - H(x, u(x), Du(x)) = 0.$$

Then we consider $\int_a^b F(x, u, Du)\,dx$ as variational integral for the function $w(x) = (u(x), v(x))$ subject to the constraint (3). The special feature of this variational problem is that the component v of w does not explicitly appear in the variational integral.

Consider a Lagrangian $F(x, z, p)$ and a function $G(x, z, p) = (G^1(x, z, p), \ldots, G^r(x, z, p))$ of class C^1 on $\mathbb{R}^n \times \mathbb{R}^N \times \mathbb{R}^{nN}$. Moreover, let Ω be a bounded domain in $\mathbb{R}^n$, $u \in C^1(\bar{\Omega}, \mathbb{R}^N)$, and assume that (1) holds on Ω and that u is a weak minimizer of

$$\mathscr{F}(v) = \int_\Omega F(x, v, Dv)\, dx$$

in the class $\mathscr{C}$ of admissible functions $v \in C^1(\bar{\Omega}, \mathbb{R}^N)$ which satisfy $v|_{\partial\Omega} = u|_{\partial\Omega}$ as well as

$$G(x, v(x), Dv(x)) = 0. \tag{4}$$

Hence there is a number $\delta > 0$ such that

$$\mathscr{F}(u) \le \mathscr{F}(v)$$

for all $v \in \mathscr{C}$ with $\|v - u\|_{C^1(\bar{\Omega})} < \delta$.

A mapping $\psi(x, \varepsilon)$ of $\bar{\Omega} \times [-\varepsilon_0, \varepsilon_0]$, $\varepsilon_0 > 0$, into $\mathbb{R}^N$ of class C^1 is said to be an *admissible variation of* $u(x)$ if the following three conditions are satisfied:

(i) $\psi(x, 0) = u(x)$ for $x \in \bar{\Omega}$;

(ii) the infinitesimal variation $\varphi(x) := \dfrac{\partial\psi}{\partial\varepsilon}(x, 0)$ is of class $C_c^1(\Omega, \mathbb{R}^N)$,

(iii) $\psi(\cdot, \varepsilon) \in \mathscr{C}$ for all $\varepsilon \in [-\varepsilon_0, \varepsilon_0]$.

We infer from (i) and (ii) that

$$\psi(x, \varepsilon) = u(x) + \varepsilon\varphi(x) + o(\varepsilon) \quad \text{as } \varepsilon \to 0 \tag{5}$$

holds.

However, not every minimizer u possesses an admissible variation. For instance, the integral $\mathscr{F}(u) = \int_0^1 (|\dot{u}^1|^2 + |\dot{u}^2 - 1|^2)\, dt$ has $u_0(t) = (0, t)$ as minimizer in the set $\mathscr{C}$ of functions $u(t) = (u^1(t), u^2(t))$ which are of class $C^1([0, 1], \mathbb{R}^2)$ and satisfy $u^1(0) = u^2(0) = u^1(1) = 0$, $u^2(1) = 1$, and $\dot{u}^2 - \sqrt{1 + |\dot{u}^1|^2} = 0$. It can easily be seen that there is no admissible variation of u_0 in the class $\mathscr{C}$. In fact, the slope $\dot{\psi}^2$ of the second component ψ^2 of an admissible variation ψ is at least one. As we have to satisfy the boundary conditions $\psi^2(0) = 0$, $\psi^2(1) = 1$, we obtain $\dot{\psi}^2(t) \equiv 1$ which implies $\psi(t) \equiv u(t)$.

Extremizers of this type are called *rigid*. It is plausible that no conditions can be derived for rigid extremizers.

The phenomenon of rigid extremizers shows that it is not a trivial matter to construct admissible variations. If this is possible we define the "linearization" A of the differential operator G at u by forming a variation (5) and setting

$$(A\varphi)(x) = \frac{\partial}{\partial\varepsilon} G(x, \psi(x, \varepsilon), D_x\psi(x, \varepsilon))\Big|_{\varepsilon=0}. \tag{6}$$

Then we obtain

$$A\varphi = a(x)D\varphi + b(x)\varphi, \tag{7}$$

with

$$a(x) := G_p(x, u(x), Du(x)), \qquad b(x) := G_z(x, u(x), Du(x))$$

and obviously, we have the following

Lemma 1. *If $\psi(x, \varepsilon)$ is an admissible variation of $u(x)$, then the infinitesimal variation $\varphi = \psi_\varepsilon(\cdot, 0)$ satisfies*

$$A\varphi = 0 \tag{8}$$

and

$$\delta\mathscr{F}(u, \varphi) = 0. \tag{9}$$

One can try to use this Lemma for obtaining a *generalization of the multiplier theorem to nonholonomic constraints*,[6] which roughly states the following:

Under suitable assumptions on u and $\mathscr{C}$ which involve the linearization A of the constraining operator G, there exist Lagrange multipliers $\lambda_1(x), \ldots, \lambda_r(x)$ with $\lambda = (\lambda_1, \ldots, \lambda_r) \in L^1(\Omega, \mathbb{R}^r)$ such that

$$\int_\Omega \{(F_z + \lambda \cdot G_z) \cdot \varphi + (F_p + \lambda \cdot G_p) \cdot D\varphi\}\, dx = 0 \tag{10}$$

hold for all $\varphi \in C_c^1(\Omega, \mathbb{R}^N)$. Here we have set

$$\lambda \cdot G_z = \sum_{j=1}^r \lambda_j G_z^j, \qquad \lambda \cdot G_p = \sum_{j=1}^r \lambda_j G_p^j,$$

where the arguments of F_z, G_z^j, etc., are to be taken as $(x, u(x), Du(x))$.

Under suitable smoothness assumptions on u and λ, one then could infer from (10) the Euler equations

$$D_\alpha(F_{p_\alpha^i} + \lambda \cdot G_{p_\alpha^i}) - (F_{z^i} + \lambda G_{z^i}) = 0, \quad 1 \le i \le N. \tag{11}$$

We remark, however, that it appears to be difficult to verify Klötzler's assumptions for the existence of multipliers in concrete situations.

One-dimensional variational problems with differential equations as subsidiary conditions are often called *Lagrange problems*. It seems appropriate to use the same notation for multidimensional variational problems with nonholonomic constraints.

The theory of one-dimensional Lagrange problems is well developed, and *optimal control theory* can be considered as its natural extension. Parts of it may be formulated in the C^1-framework and are, at least in principle, as elementary as the previous discussion of isoperimetric or holonomic side conditions. It is, in fact, customary to reduce one-dimensional isoperimetric variational problems to Lagrange problems and to consider these as the general class of variational problems which have to be treated, compare the Scholia. We shall now briefly discuss one-dimensional Lagrange-problems. For a more detailed account we refer to Bolza [3], Bliss [5], Carathéodory [10], [16], Prange [2], Funk [1], L.C. Young [1], Hestenes [5], Rund [4], Sagan [1] and Cesari [1].

---

[6] For such an approach compare e.g. Klötzler [4], pp. 72–78.

For multiple integrals the investigation of Lagrange problems is much less complete. In particular there is no elementary theory that guarantees the existence of multipliers within the C^1-setting in such a simple way as in the case of isoperimetric or holonomic constraints.

Let us now verify Lagrange's multiplier rule for minimizers of single integrals. To this end we consider a mapping $u \in C^2([a, b], \mathbb{R}^N)$ which minimizes the variational integral

$$\mathscr{F}(u) = \int_a^b F(x, u(x), u'(x))\,dx$$

under the nonholonomic constraints

$$G^\beta(x, u(x), u'(x)) = 0, \quad \beta = 1, \ldots, r, \tag{12}$$

where $r < N$. We assume that the Lagrangian $F(x, z, p)$ and the constraints $G^\beta(x, z, p)$ are of class C^3, and that the r functions G^β are independent, i.e. that the Jacobian matrix $(G^\beta_{p^i})$ has rank r. Without loss of generality we can then assume that

$$\frac{\partial(G^1, \ldots, G^r)}{\partial(p^1, \ldots, p^r)} \neq 0. \tag{13}$$

Set

$$v(x) = (v^1(x), \ldots, v^r(x)), \quad v^\alpha(x) := u^\alpha(x), \ 1 \le \alpha \le r,$$

$$w(x) = (w^1(x), \ldots, w^m(x)), \qquad w^i(x) := u^{r+i}(x), \ 1 \le i \le m,$$

where $m := N - r$. By (12) and (13) we can express $v'(x)$ in terms of x, $v(x)$, $w(x)$ and $w'(x)$. Now we define an $(r + 1)$-parameter variation $\mathscr{W}(x, \tau, \varepsilon)$ of $w(x)$ by

$$\mathscr{W}(x, \tau, \varepsilon) := w(x) + \tau\varphi(x) + \varepsilon^1\eta_1(x) + \cdots + \varepsilon^r\eta_r(x),$$

where $\tau \in \mathbb{R}$, $\varepsilon = (\varepsilon^1, \ldots, \varepsilon^r) \in \mathbb{R}^r$, and $\varphi, \eta_1, \ldots, \eta_r$ are arbitrarily chosen functions of class $C^2([a, b], \mathbb{R}^m)$ vanishing for $x = a$ and $x = b$. Obviously we have

$$\begin{gathered} \mathscr{W}(x, 0, 0) = w(x) \quad \text{for all } a \le x \le b; \\ \mathscr{W}(a, \tau, \varepsilon) = w(a), \qquad \mathscr{W}(b, \tau, \varepsilon) = w(b) \quad \text{for all } \tau, \varepsilon; \\ \frac{\partial}{\partial \tau}\mathscr{W} = \varphi, \qquad \frac{\partial}{\partial \varepsilon^\beta}\mathscr{W} = \eta_\beta. \end{gathered} \tag{14}$$

Now we fix the parameters τ, ε and consider the system of r ordinary differential equations

$$G^\beta(x, y, \mathscr{W}, y', \mathscr{W}') = 0, \quad 1 \le \beta \le r, \tag{15}$$

in the r unknown functions $y(x) = (y_1(x), \ldots, y_r(x))$. For $(\tau, \varepsilon) = (0, 0)$ we have the solution $y = v$. Then we infer from well-known results on ordinary differential equations that there is a number $\delta > 0$ such that for every pair (τ, ε) with $|\tau|, |\varepsilon^1|, \ldots, |\varepsilon^r| < \delta$ there is a solution $y(x) = \mathscr{V}(x, \tau, \varepsilon)$ of (15) which is continuous in (x, τ, ε) together with its first derivatives and its second derivatives of the form $\mathscr{V}_{x\tau}$ and $\mathscr{V}_{x\varepsilon^\alpha}$, and which satisfies the "initial conditions"

$$\mathscr{V}(a, \tau, \varepsilon) = v(a) \quad \text{for } |\tau|, |\varepsilon^1|, \ldots, |\varepsilon^r| < \delta. \tag{16}$$

Moreover the unique solvability of the Cauchy problem for (15) implies

$$\mathscr{V}(x, 0, 0) = v(x). \tag{17}$$

In order to obtain an "admissible variation" $\mathscr{U}(x, \tau)$ of $u(x)$ satisfying the boundary conditions

$$\mathscr{U}(a, \tau) = u(a), \qquad \mathscr{U}(b, \tau) = u(b), \tag{18}$$

we try to solve the equation

$$\mathscr{V}(b, \tau, \varepsilon) = v(b) \tag{19}$$

in terms of ε. Note that (19) furnishes a system of r scalar equations for the unknowns $\varepsilon^1, \dots, \varepsilon^r$. If $\varepsilon = \varepsilon(\tau)$ were a solution of (19), then we obtain from (14_2), (16) and (19) that

$$\mathscr{U}(x, \tau) := (\mathscr{V}(x, \tau), \mathscr{W}(x, \tau)) \tag{20}$$

satisfies (18), and (14_1) and (17) imply

$$\mathscr{U}(x, 0) = u(x) \quad \text{for } a \le x \le b \tag{21}$$

since $\varepsilon(\tau)|_{\tau=0} = 0$.

Let us now discuss equation (19). At this stage we only know that *the function* $\Phi(\tau, \varepsilon)$, *defined by*

$$\begin{gathered} \Phi(\tau, \varepsilon) := \int_a^b F(x, Z(x, \tau, \varepsilon), Z'(x, \tau, \varepsilon))\, dx, \\ Z(x, \tau, \varepsilon) = (\mathscr{V}(x, \tau, \varepsilon), \mathscr{W}(x, \tau, \varepsilon)) \end{gathered} \tag{22}$$

has a minimum at $(\tau, \varepsilon) = (0, 0)$ *with respect to the constraint* (19). Then, by the standard multiplier rule, there are real numbers $\ell_0, \ell_1, \dots, \ell_r$, not all vanishing, such that $(\tau, \varepsilon) = (0, 0)$ is a critical point of the function

$$\Psi(\tau, \varepsilon) := \ell_0 \Phi(\tau, \varepsilon) + \sum_{\alpha=1}^{r} \ell_\alpha \mathscr{V}^\alpha(b, \tau, \varepsilon), \tag{23}$$

i.e. we have

$$\Psi_\tau(0, 0) = 0, \qquad \Psi_{\varepsilon^\beta}(0, 0) = 0, \quad 1 \le \beta \le r. \tag{24}$$

The constants $\ell_0, \ell_1, \dots, \ell_r$ are only determined up to a common factor (different from zero). We distinguish between the *principal case* $\ell_0 \ne 0$ and the *exceptional case* $\ell_0 = 0$. In the principal case we are allowed to assume $\ell_0 = 1$. If we have

$$\det \mathscr{V}_\varepsilon(b, 0, 0) \ne 0, \tag{25}$$

then it follows that

$$\operatorname{rank}(\mathscr{V}_\tau(b, 0, 0), \mathscr{V}_{\varepsilon^1}(b, 0, 0), \mathscr{V}_{\varepsilon^r}(b, 0, 0)) = r \tag{26}$$

and we, therefore, are in the principal case $\ell_0 \ne 0$. Now we note that equations (24) can be written as

$$\begin{gathered} \ell_0 \Phi_\tau(0, 0) + \sum_{\alpha=1}^{r} \ell_\alpha \mathscr{V}_\tau^\alpha(b, 0, 0) = 0, \\ \ell_0 \Phi_{\varepsilon^\beta}(0, 0) + \sum_{\alpha=1}^{r} \ell_\alpha \mathscr{V}_{\varepsilon^\beta}^\alpha(b, 0, 0) = 0. \end{gathered} \tag{27}$$

Recall now that the variation $Z(x, \tau, \varepsilon) = (\mathscr{V}(x, \tau, \varepsilon), \mathscr{W}(x, \tau, \varepsilon))$ of $u(x) = (v(x), w(x))$ is a solution of

$$G^\alpha(x, Z(x, \tau, \varepsilon), Z'(x, \tau, \varepsilon)) = 0, \quad 1 \le \alpha \le r, \tag{28}$$

where $' = \frac{d}{dx}$. Differentiating these relations with respect to τ and ε^β respectively and evaluating the resulting equations at $(\tau, \varepsilon) = (0, 0)$, we obtain

$$\begin{gathered} \bar{G}_z^\alpha \cdot \zeta + \bar{G}_p^\alpha \cdot \zeta' = 0, \\ \bar{G}_z^\alpha \cdot \psi_\beta + \bar{G}_p^\alpha \cdot \psi_\beta' = 0, \end{gathered} \tag{29}$$

where we have set

$$\begin{gathered} \bar{G}_z^\alpha(x) := G_z^\alpha(x, u(x), u'(x)), \qquad \bar{G}_p^\alpha(x) := G_p^\alpha(x, u(x), u'(x)), \\ \zeta(x) := Z_\tau(x, 0, 0) = (\mathscr{V}_\tau(x, 0, 0), \varphi(x)), \\ \psi_\beta(x) := Z_{\varepsilon^\beta}(x, 0, 0) = (\mathscr{V}_{\varepsilon^\beta}(x, 0, 0), \eta_\beta(x)). \end{gathered} \tag{30}$$

Furthermore we infer from (14) and (16) that

$$\begin{aligned} &\zeta(a) = 0, \quad \psi_1(a) = 0, \ldots, \psi_r(a) = 0, \\ &\zeta^\alpha(b) = 0, \quad \psi_1^\alpha(b) = 0, \ldots, \psi_r^\alpha(b) = 0 \ \text{ for } \alpha = r+1, \ldots, N. \end{aligned} \tag{31}$$

Now we turn to the construction of the Lagrange multiplier functions $\lambda_1(x), \ldots, \lambda_r(x)$. The idea is to use arbitrary displacements $\varphi(x)$ and suitably chosen fixed displacements $\eta_1(x), \ldots, \eta_r(x)$. Let us first investigate how the numbers $\ell_0, \ell_1, \ldots, \ell_r$ depend on the choice of $\varphi, \eta_1, \ldots, \eta_r$. We infer from (29)–(31) that the function $(\zeta^1(x), \ldots, \zeta^r(x))$ vanishes for $x = a$ and satisfies an inhomogeneous linear system of the kind

$$\frac{\partial \zeta^\alpha}{dx} + c_\beta^\alpha(x)\zeta^\beta = f^\alpha(x), \quad 1 \le \alpha \le r,$$

since (13) holds true. The functions $f^\alpha(x)$ are completely determined by $u(x)$ and $\varphi(x)$ and do not depend on the choice of $\eta_1, \ldots, \eta_r$. Therefore ζ is determined by the choice of φ and is independent of the selection of $\eta_1, \ldots, \eta_r$. Similarly one verifies that ψ_β depends only on the choice of η_β and not on φ and η_α, $\alpha \ne \beta$. This implies that also $\Phi_{\varepsilon\beta}(0,0)$ is independent of φ and η_α, $\alpha \ne \beta$, but is solely depending on the choice of η_β. Furthermore, equation (27_2) is equivalent to

$$\ell_0 \Phi_{\varepsilon\beta}(0,0) + \sum_{\alpha=1}^r \ell_\alpha \psi_\beta^\alpha(b) = 0, \quad 1 \le \beta \le r. \tag{32}$$

Introducing the functions $\psi^\alpha(x)$ by

$$\psi^\alpha(x) := (\psi_1^\alpha(x), \ldots, \psi_r^\alpha(x)) = Z_\varepsilon^\alpha(x, 0, 0), \tag{33}$$

equation (32) can be written as

$$\ell_0 \mathring{\Phi}_\varepsilon + \ell_1 \psi^1(b) + \cdots + \ell_r \psi^r(b) = 0, \quad \mathring{\Phi}_\varepsilon := \Phi_\varepsilon(0,0). \tag{34}$$

Set

$$\begin{aligned} B = B(\eta_1, \ldots, \eta_r) &:= (\mathring{\Phi}_\varepsilon, \psi^1(b), \ldots, \psi^r(b)), \\ k &:= \operatorname{rank} B(\eta_1, \ldots, \eta_r). \end{aligned} \tag{35}$$

Case I. Suppose that there is a system of C^2-functions $\eta_1, \eta_2, \ldots, \eta_r : [a, b] \to \mathbb{R}^m$ vanishing at the endpoints $x = a, b$ such that $k = r$.

(Ia) Suppose that $\det(\psi^1(b), \ldots, \psi^r(b)) \ne 0$, i.e. that (25) holds true. Then we are in the principal case whence $\ell_0 = 1$, and $(\ell_1, \ldots, \ell_r)$ is the uniquely determined solution of

$$\mathring{\Phi}_\varepsilon + \ell_1 \psi^1(b) + \cdots + \ell_r \psi^r(b) = 0.$$

Then $\ell_0, \ldots, \ell_r$ depend only on the choice of $\eta_1, \ldots, \eta_r$ and are independent of the choice of φ.

(Ib) Let $\det(\psi^1(b), \ldots, \psi^r(b)) = 0$. Then we can assume that $\psi^r(b)$ is a linear combination of $\psi^1(b), \ldots, \psi^{r-1}(b)$ and $\det(\mathring{\Phi}_\varepsilon, \psi^1(b), \ldots, \psi^{r-1}(b)) \ne 0$. Equation (34) now implies that $\ell_0 = 0$, i.e. we are in the exceptional case. Now we can choose $\ell_r = 1$, and then $\ell_1, \ldots, \ell_{r-1}$ can uniquely be determined from (34). Thus we may again assume that $\ell_0, \ell_1, \ldots, \ell_r$ are independent of the choice of φ.

Case II. Suppose that $k < r$ for all admissible $\eta_1, \ldots, \eta_r$ and let s be the maximum of $k(\eta_1, \ldots, \eta_r)$, $s < r$. Then we choose a system of admissible $\eta_1, \ldots, \eta_r$ for which $k = s$. Then we can assume that

$$\operatorname{rank}(\mathring{\Phi}_{\varepsilon\beta}, \psi_\beta^1(b), \ldots, \psi_\beta^r(b))_{1 \le \beta \le s} = s,$$

and consequently we can express s of the numbers $\ell_0, \ldots, \ell_r$ in terms of $r + 1 - s$ of the other ℓ_α by resolving the first s equations of (32). Since ψ_β^α is independent of φ and η_γ for $\gamma \ne \beta$, we infer that the coefficients of the linear functions which express the first kind of ℓ's in terms of the remaining ℓ's are independent of $\varphi, \eta_{s+1}, \ldots, \eta_r$. We use this observation to pass to a better choice of the numbers $\ell_0, \ldots, \ell_r$. First we choose an admissible φ, and then we replace all the functions $\eta_{s+1}, \ldots, \eta_r$ by φ. Since we obtain $k(\eta_1, \ldots, \eta_s, \varphi, \ldots, \varphi) = s$, we can apply the above reasoning to the new η-system $\eta_1, \ldots, \eta_s, \varphi, \ldots, \varphi$. Next we choose all ℓ's of the second kind as one. Then we obtain ℓ's of the first which depend on $\eta_1, \ldots, \eta_s$ but are independent of φ, and such that $(\ell_0, \ell_1, \ldots, \ell_r)$ satisfies equations

(32). However, for $\beta = s+1, \dots, r$ these equations are the same as (27_1) on account of our special choice of $\eta_{s+1}, \dots, \eta_r$. Hence we have proved that there is a vector $(\ell_0, \ell_1, \dots, \ell_r) \neq 0$ in $\mathbb{R}^{r+1}$ which is a solution of (27) and does not depend on the special choice of φ. We finally note that in case II we can even choose the ℓ_α in such a way that $\ell_0 = 0$ and $(\ell_1, \ell_2, \dots, \ell_r) \neq 0$.

As the upshot of our discussion we can state:

Lemma 2. *One can always find functions $\eta_1, \dots, \eta_r$ of class $C^2([a,b], \mathbb{R}^m)$, vanishing at $x = a$ and $x = b$, and a constant vector $(\ell_0, \ell_1, \dots, \ell_r) \in \mathbb{R}^{r+1} - \{0\}$ such that for any $\varphi \in C^2([a,b], \mathbb{R}^m)$, $\varphi(a) = 0$, $\varphi(b) = 0$, equations (27) are satisfied, i.e.*

$$\text{(36)} \qquad \begin{gathered} \ell_0 \Phi_\tau(0,0) + \sum_{\alpha=1}^r \ell_\alpha \zeta^\alpha(b) = 0, \\ \ell_0 \Phi_{\varepsilon\beta}(0,0) + \sum_{\alpha=1}^r \ell_\alpha \psi_\beta^\alpha(b) = 0, \quad 1 \le \beta \le r. \end{gathered}$$

Moreover, if $\det(\psi_\beta^\alpha(b))_{1 \le \alpha, \beta \le r} = 0$, we can assume that $\ell_0 = 0$ whereas $\det(\psi_\beta^\alpha(b))_{1 \le \alpha, \beta \le r} \neq 0$ implies that $\ell_0 \neq 0$.

Now we choose arbitrary C^1-functions $\lambda_1(x), \dots, \lambda_r(x)$ and set

$$\text{(37)} \qquad F^*(x, z, p) := \ell_0 F(x, z, p) + \sum_{\alpha=1}^r \lambda_\alpha G^\alpha(x, z, p).$$

Multiplying (29_1) by λ_α, summing with respect to α from 1 to r, integrating with respect to x, and adding the result to (36_1), we arrive at the relation

$$\text{(38)} \qquad \int_a^b [\bar{F}_z^* \cdot \zeta + \bar{F}_p^* \cdot \zeta'] \, dx + \sum_{\alpha=1}^r \ell_\alpha \zeta^\alpha(b) = 0.$$

Here the functions $\bar{F}_z^*$ and $\bar{F}_p^*$ are defined as

$$\text{(39)} \qquad \bar{F}_z^*(x) := F_z^*(x, u(x), u'(x)), \qquad \bar{F}_p^*(x) := F_p^*(x, u(x), u'(x)).$$

An integration of (38) by parts implies that

$$\text{(40)} \qquad \int_a^b \left[\bar{F}_z^* - \frac{d}{dx} \bar{F}_p^* \right] \cdot \zeta \, dx + \sum_{\alpha=1}^r \zeta^\alpha(b) \cdot \{\ell_\alpha + F_{p_\alpha}^*(b, u(b), u'(b))\} = 0$$

if we take (31) into account.

Now we want to fix the functions $\lambda_1, \dots, \lambda_r$. We are going to choose them as solution of the Cauchy problem

$$\text{(41)} \qquad \bar{F}_{z\beta}^* - \frac{d}{dx} \bar{F}_{p\beta}^* = 0, \quad \ell_\beta + \bar{F}_{p\beta}^*(b) = 0, \qquad 1 \le \beta \le r.$$

Note that the differential equations in (41) are a system of linear ordinary differential equations of first order for $\lambda_1, \dots, \lambda_r$, given by

$$\text{(42)} \qquad \ell_0 \left(\bar{F}_{z\beta} - \frac{d}{dx} \bar{F}_{p\beta} \right) + \sum_{\alpha=1}^r \lambda_\alpha \left(\bar{G}_{z\beta}^\alpha - \frac{d}{dx} \bar{G}_\beta^\alpha \right) - \sum_{\alpha=1}^r \lambda'_\alpha \bar{G}_{p\beta}^\alpha = 0,$$

and the matrix of the coefficients of λ'_α in (42) is invertible since (13) implies that

$$\text{(43)} \qquad \det(G_{p\beta}^\alpha) \neq 0.$$

Furthermore, the initial conditions (41) are just

$$\text{(44)} \qquad \sum_{\alpha=1}^r \lambda_\alpha(b) \bar{G}_{p\beta}^\alpha(b) = \ell_\beta - \ell_0 \bar{F}_{p\beta}(b).$$

Since $(\bar{G}_\beta^\alpha(b))$ is invertible, we can first determine the initial values $\lambda_1(b), \dots, \lambda_r(b)$ from (44) and then the functions $\lambda_1(x), \dots, \lambda_r(x)$, $a \le x \le b$, from (42).

Thus the Cauchy problem (41) *has a* C^1*-solution* $(\lambda_1, \lambda_2, \ldots, \lambda_r)$ *which is independent of the choice of* φ since the numbers $\ell_0, \ell_1, \ldots, \ell_r$ are independent of the choice of φ as we have noted before.

If the functions $\lambda_1, \ldots, \lambda_r(x)$ satisfy (41), equation (38) becomes

$$\int_a^b \sum_{\beta=r+1}^{N} \zeta^\beta \left[\bar{F}^*_{z^\beta} - \frac{d}{dx} \bar{F}^*_{p^\beta} \right] dx = 0. \tag{45}$$

Since $\zeta^{\beta+r}(x) = \varphi^\beta(x)$, $1 \le \beta \le m$, $m = N - r$, it follows that for all $\varphi \in C^2([a, b], \mathbb{R}^m)$, $\varphi(a) = \varphi(b) = 0$, we have

$$\int_a^b \sum_{\beta=1}^{m} \varphi^\beta \left[\bar{F}^*_{z^{\beta+r}} - \frac{d}{dx} \bar{F}^*_{p^{\beta+r}} \right] dx = 0. \tag{46}$$

The fundamental lemma yields

$$\bar{F}^*_{z^\beta} - \frac{d}{dx} \bar{F}^*_{p^\beta} = 0, \quad r + 1 \le \beta \le N. \tag{47}$$

Combining equations (41) and (47), we arrive at

$$\bar{F}^*_z - \frac{d}{dx} \bar{F}^*_p = 0. \tag{48}$$

Thus we have proved

Theorem 1 (Lagrange multiplier rule I). *Let* $u \in C^2([a, b], \mathbb{R}^N)$ *be a minimizer of the functional*

$$\mathscr{F}(u) = \int_a^b F(x, u, u')\, dx$$

under the nonholonomic constraints

$$G^\beta(x, u, u') = 0, \quad 1 \le \beta \le r, \quad (r < N),$$

satisfying (13) *in a neighbourhood of* 1-graph(u), *and suppose that* $F, G^1, \ldots, G^r$ *are of class* C^3. *Then there exist a constant* ℓ_0 *(which can be taken as zero or one) and functions* $\lambda_1(x), \ldots, \lambda_r(x)$ *of class* $C^1([a, b])$ *such that u is an extremal of the unconstrained variational integral*

$$\mathscr{F}^*(u) = \int_a^b F^*(x, u, u')\, dx,$$

with the Lagrangian

$$F^*(x, z, p) := \ell_0 F(x, z, p) + \sum_{\alpha=1}^{r} \lambda_\alpha(x) G^\alpha(x, z, p).$$

Clearly the principal case $\ell_0 \ne 0$ is of particular importance since for $\ell_0 = 0$ the Lagrangian F does not appear in the Euler equation (48). Let us now discuss how the exceptional case $\ell_0 = 0$ can be excluded.

If $\ell_0 = 0$, equation (48) reduces to

$$\sum_{\alpha=1}^{r} \left\{ \lambda_\alpha \bar{G}^\alpha_{z^i} - \frac{d}{dx}(\lambda_\alpha \bar{G}^\alpha_{p^i}) \right\} = 0, \quad 1 \le i \le N. \tag{49}$$

Since $(\ell_1, \ell_2, \ldots, \ell_r) \ne 0$, we deduce from (43) and (44) that not all $\lambda_\alpha(b)$ are zero. Then it follows from (42) that not all $\lambda_\alpha(x)$ are identically zero, and we obtain

Lemma 3. *If* $\ell_0 = 0$, *then the homogeneous system* (44) *has a nontrivial solution* $\lambda(x) := (\lambda_1(x), \ldots, \lambda_r(x))$.

This result motivates the following terminology due to Hahn.

Definition. *A minimizer of* $\mathcal{F}$ *under the constraints* $G^\beta(x, u, u') = 0$, $\beta = 1, \dots, r$, *is said to be* abnormal *if there is a nontrivial* C^1*-solution* $\lambda = (\lambda_1, \dots, \lambda_r)$ *of* (49); *otherwise u is called* normal.

Then we can state Lemma 3 as follows.

Lemma 3′. *If* $\ell_0 = 0$, *then u is an abnormal minimizer.*

Moreover we obtain the following important result due to Hahn and Bolza.

Theorem 2. *If u is a normal minimizer, then we can choose a system of functions* $\eta_1, \dots, \eta_r \in C^2([a, b], \mathbb{R}^m)$, *vanishing at* $x = a$ *and* $x = b$, *such that*

$$\det(\psi^\alpha_\beta(b))_{1 \le \alpha, \beta \le r} \ne 0. \tag{50}$$

Then the corresponding constants $\ell_0, \ell_1, \dots, \ell_r$ *satisfy* $\ell_0 \ne 0$, *i.e. we are in the principal case.*

Proof. From the preceding discussion of the cases (Ia), (Ib), (II) and from Lemma 3′ we infer that for a normal minimizer u there must exist an admissible system $\eta_1, \dots, \eta_r$ such that (50) holds true, and that $\ell_0 \ne 0$. □

From Theorems 1 and 2 we derive

Theorem 3 (Lagrange multiplier rule II). *If u is a normal* C^2*-minimizer of* $\mathcal{F}$ *under the constraints* $G^\beta(x, u, u') = 0$, $\beta = 1, \dots, r$, $(r < N)$, *then u is an* F^**-extremal for the Lagrangian*

$$F^*(x, z, p) = F(x, z, p) + \sum_{\alpha=1}^{r} \lambda_\alpha G^\alpha(x, z, p)$$

and the multipliers $\lambda_1, \dots, \lambda_r$ *are uniquely determined by u.*

Proof. It remains to be seen that $\lambda_1, \dots, \lambda_r$ are uniquely determined by u. In fact, suppose that $\bar\lambda_1, \dots, \bar\lambda_r$ were a second set of multipliers. Then u were an extremal of both $F + \sum_{\alpha=1}^r \lambda_\alpha G^\alpha$ and $F + \sum_{\alpha=1}^r \bar\lambda_\alpha G^\alpha$. Subtracting the corresponding Euler equations we would obtain

$$\sum_{\alpha=1}^{r} \left\{ \mu_\alpha \bar G^\alpha_{z^i} - \frac{d}{dx}(\mu_\alpha \bar G^\alpha_{p^i}) \right\} = 0, \quad 1 \le i \le N.$$

Since u is normal, we would obtain $(\mu_1, \dots, \mu_r) = 0$, i.e. $\lambda_1 = \bar\lambda_1, \dots, \mu_1 = \bar\mu_r$. □

In the following examples, we give some "formal appplications" of the multiplier rule to nonholonomic constraints, without actually proving that the rule can rightfully be used. Neither we shall investigate whether the multipliers are smooth functions. Concerning these questions, we refer the reader to the literature quoted before.

[1] Let $F(x, z, p, q)$ be the integrand of the second order variation integral

$$\mathcal{F}(u) = = \int_\Omega F(x, u, Du, D^2u)\, dx,$$

which, by introducing $v := Du$, will be transformed into

$$\mathcal{F}(u) = \int_\Omega F(x, u, Du, Dv)\, dx,$$

where (u, v) is subject to the nonholonomic constraint

$$v - Du = 0.$$

Then the multiplier rule requires to form the Euler equations of the functional

$$\int \{F(x, u, Du, Dv) + \lambda[v - Du]\}\, dx,$$

which amount to

$$F_u - DF_p + D\lambda = 0, \qquad \lambda - DF_q = 0,$$

whence the well-known equations

$$F_u - DF_p + D^2 F_q = 0$$

can be inferred.

[2] *Equilibrium conditions for a heavy thread which lies on a given surface.*[7] Let P_1 and P_2 be two points on a regular surface M in $\mathbb{R}^3$ described by

$$G(z) = 0.$$

Consider an (infinitely thin and inextensible) thread of length ℓ whose endpoints are attached to P_1 and P_2, and which is constrained to lie on M. We assume that the virtual positions of the thread are given by a parameter representation

$$z = u(s) = (u^1(s), u^2(s), u^3(s)), \quad 0 \le s \le \ell,$$

by means of the arc length s, that is,

$$|\dot{u}(s)|^2 = 1.$$

Then u has to satisfy

$$G(u(s)) = 0, \qquad u(0) = P_1, \qquad u(\ell) = P_2.$$

In equilibrium, the center of gravity of the thread has the lowest possible position above the ground. That is, if the 3-axis points in the opposite direction of gravity, then the integral

$$\int_0^\ell u^3\, dx$$

will be minimized by the equilibrium position. Applying the multiplier rule, we find that the first variation of

$$\int_0^\ell \{u^3 + \mu G(u) + \lambda[|\dot{u}|^2 - 1]\}\, ds$$

has to vanish in equilibrium whence we obtain the Euler equations

$$e_3 + \mu G_z(u) - 2\dot{\lambda}\dot{u} - 2\lambda\ddot{u} = 0, \quad e_3 := (0, 0, 1).$$

Moreover, the conditions $|\dot{u}|^2 = 1$ and $G(u) = 0$ imply

$$\dot{u}\cdot\ddot{u} = 0 \quad \text{and} \quad G_z(u)\cdot\dot{u} = 0,$$

whence, multiplying the Euler equation by $\dot{u}$ and S respectively, we obtain

$$\dot{u}^3 - 2\dot{\lambda} = 0$$

[7] Cf. also *1* [5], where the unconstrained problem is treated.

and

$$S^3 - 2\lambda \ddot{u} \cdot S = 0,$$

where $S := \dot{u} \wedge \mathcal{N}$ is the side normal of $u(s)$ and $\mathcal{N} := \dfrac{G_z(u)}{|G_z(u)|}$ is the surface normal of M along the curve $u(s)$. The first equation yields the first integral

$$u^3 - 2\lambda = c,$$

with some constant c.

Furthermore, we have

$$\ddot{u} = \kappa \mathbf{v},$$

where κ is the curvature of u, while $\mathbf{v}$ denotes its principal normal. As usual,

$$\kappa_g = \kappa \mathbf{v} \cdot S = \ddot{u} \cdot S$$

stands for the geodesic curvature of the curve $u(s)$, and we infer from the second equation the relation

$$S^3 - 2\lambda \kappa_g = 0,$$

and therefore

$$\dot{u}^2 G_{z^3}(u) - \dot{u}^3 G_{z^2}(u) = [u^3 - c] |G_z(u)| \kappa_g$$

is another relation for u.

3 *The principle of least action as stated by Lagrange.* We consider ℓ point masses m_j, $1 \le j \le \ell$, with the position vectors $X_j(t)$ and the potential energy $V(X)$, $X = (X_1, \ldots, X_\ell)$. The kinetic energy of the motion $X(t)$ is given by

$$T = \sum_{j=1}^{\ell} \frac{m_j}{2} |Y_j|^2, \quad Y_j = \dot{X}_j.$$

The virtual (or possible) motions $X(t)$, $t_1 \le t \le t_2$, from an initial point P_1 to an endpoint P_2 are subject to the following conditions:

(i) $$X(t_1) = P_1 \quad \text{and} \quad X(t_2) = P_2;$$

(ii) $T(\dot{X}) + V(X) = h$, where h denotes some fixed number;

(iii) $$G^j(X) = 0, \quad 1 \le j \le r < N := 3\ell.$$

The numbers t_1 and t_2, however, are not kept fixed; that is, we are allowed to vary the dependent variables $X = (X_1, \ldots, X_\ell)$ as well as the independent variable t. Condition (ii) is suggested by the fact that, for conservative systems, the total energy $E = T + V$ is a first integral of the Euler equations, cf. 2 5. Then we can state

Maupertuis's principle of least action in the formulation of Lagrange: *The actual motion of a system of ℓ points masses is a minimizer (or at least a stationary point) of the action integral*

$$\mathcal{A}(X) = \int_{t_1}^{t_2} T(\dot{X})\, dt$$

in the class of admissible functions defined by (i)–(iii).

By means of the multiplier rule we can now see that *Hamilton's principle and Maupertuis principle (as formulated by Lagrange) are equivalent.*

In fact, applying the multiplier rule to the Lagrangian

$$F = T + \lambda \cdot [T + V - h] + \sum_{k=1}^{r} \lambda_k G^k,$$

we obtain the Euler equations

$$\frac{d}{dt}\{(1+\lambda)m_j \dot{X}_j\} = \lambda \operatorname{grad}_{X_j} V + \sum_{k=1}^{r} \lambda_k \operatorname{grad}_{X_j} G^k.$$

By the "energy theorem" stated in 1,*2.2* [7] 3,*4.1* [1] and 3,*4.4* [1] we obtain that $F - \sum_{j=1}^{\ell} Y_j \cdot F_{Y_j}$ is a first integral of the actual motion $X(t)$. Because of $\sum_{j=1}^{\ell} Y_j \cdot F_{Y_j} = 2(1+\lambda)T$ and of the constraints (ii) and (iii), we obtain that

$$(1 + 2\lambda)T(\dot{X}) = c$$

for some constant c which, by the boundary condition (6) of 3,*2*, has to be zero:

$$(1 + 2\lambda)T(\dot{X}) = 0.$$

(Note that this boundary condition is a consequence of the assumption that t_1 and t_2 can be varied. If t_1 and t_2 are not allowed to change, then c can be any number.)

If $\dot{X} \neq 0$, we have $T(\dot{X}) > 0$, and therefore $1 + 2\lambda = 0$, or $\lambda = -1/2$. Then the Euler equations become

$$m_j \ddot{X}_j = -\operatorname{grad}_{X_j} V + \sum_{k=1}^{r} 2\lambda_k \operatorname{grad}_{X_j} G^k$$

and, according to *2* [5] these are the correct equations describing the constrained motion.

Since we clearly may reverse this reasoning, *the equivalence of Hamilton's and Maupertuis's principles* is proved.

We have, however, to admit that this equivalence has only formally been verified, that is, we have not checked whether the multiplier rule can really be used; this is left to the reader.

[4] *Solenoidal vector fields.* Consider a variational integral

$$\mathscr{F}(u) = \int_\Omega F(x, u, Du)\, dx$$

for vector fields $u \in C^2(\bar{\Omega}, \mathbb{R}^n)$, $\Omega \subset \mathbb{R}^n$, which are subject to the constraint

$$\operatorname{div} u = 0.$$

Such vector fields are called *solenoidal* (or tubular). Applying the Lagrange multiplier rule to the integrand

$$F(x, u, Du) + \lambda \operatorname{div} u,$$

we obtain the Euler equation

$$L_F(u) = \operatorname{grad} \lambda.$$

If $n = 3$, and Ω is simply connected, we have the equivalent equation

$$\operatorname{curl} L_F(u) = 0.$$

For

$$F(u) = \tfrac{1}{2}|u|^2,$$

the Euler equation reduces to

$$u = \operatorname{grad} \lambda \quad \text{with } \Delta\lambda = 0$$

since $0 = \operatorname{div} u = \Delta\lambda$. That is, extremals are gradients of harmonic functions.

4. Constraints at the Boundary. Transversality

In this section we shall derive free boundary conditions for weak extrema of variational integrals

$$\mathscr{F}(u) = \int_\Omega F(x, u, Du)\, dx,$$

which on $\partial\Omega$ are subject to nonlinear constraints. We restrict ourselves to holonomic constraints

$$G(x, u(x)) = 0 \quad \text{for } x \in \partial\Omega.$$

Suppose that Ω is a bounded domain in $\mathbb{R}^n$ with $\partial\Omega \in C^1$, and let $G(x, z) = (G^1(x, z), \dots, G^r(x, z))$, $1 \le r \le N - 1$, be of class C^2 on $\mathbb{R}^n \times \mathbb{R}^N$. We also assume that $G_z = (G^i_{z^k})$ has maximal rank r at all points of the set $\{(x, z): x \in \partial\Omega, G(x, z) = 0\}$. Then, for every $x \in \partial\Omega$, the set

$$M(x) := \{z \in \mathbb{R}^N: G(x, z) = 0\}$$

is an $(N - r)$-dimensional manifold in $\mathbb{R}^N$ with the normal vector fields $G^k_z(x, z)$, $1 \le k \le r$, of class C^1. As in Section 2, $\Pi(x, z)$ denotes the orthogonal projection of $\mathbb{R}^N$ onto the tangent space $T_z M(x)$ of $M(x)$ at z. For the following we assume that u is at least of class $C^1(\bar{\Omega}, \mathbb{R}^N)$.

Definition. *A mapping $\varphi \in C^1(\bar{\Omega}, \mathbb{R}^N)$ is said to be a* tangential vector field *along $u|_{\partial\Omega}$ if $\varphi(x) \in T_{u(x)} M(x)$ holds for all $x \in \partial\Omega$.*

Similarly to the corresponding result in Section 2, we can prove:

Lemma 1. *Let $\psi \in C^0(\partial\Omega, \mathbb{R}^N)$ satisfy*

$$\int_{\partial\Omega} \psi(x) \cdot \varphi(x)\, d\mathscr{H}^{n-1}(x) = 0$$

for all $\varphi \in C^1(\bar{\Omega}, \mathbb{R}^N)$ which are tangential along $u|_{\partial\Omega}$. Then it follows that

$$\Pi(x, u(x))\psi(x) = 0 \quad \textit{for all } x \in \partial\Omega.$$

Consider the class $\mathscr{C}$ of mappings $v \in C^1(\bar{\Omega}, \mathbb{R}^N)$ such that F is defined on $\{(x, v(x), Dv(x)): x \in \bar{\Omega}\}$, and which satisfy

$$G(x, v(x)) = 0 \quad \text{for all } x \in \partial\Omega.$$

Assume that u is a weak minimizer of $\mathscr{F}$ in $\mathscr{C}$, that is,

$$\mathscr{F}(u) \le \mathscr{F}(v)$$

holds for all $v \in \mathscr{C}$ with $\|u - v\|_{C^1(\bar{\Omega})} < \delta$, where δ is some sufficiently small positive constant. Then we prove as in Section 2 that

(1) $\delta\mathscr{F}(u, \varphi) = 0$ for all $\varphi \in C^1(\bar{\Omega}, \mathbb{R}^N)$ which are tangential along $u|_{\partial\Omega}$.

We call $u \in C^1(\bar{\Omega}, \mathbb{R}^N)$ a *stationary point of* $\mathscr{F}$ *in* $\mathscr{C}$ if (1) is satisfied. Then it follows that *every weak minimizer u of* $\mathscr{F}$ *in* $\mathscr{C}$ *also is a stationary point of* $\mathscr{F}$ *in* $\mathscr{C}$.

If u is a stationary point of class $C^2(\Omega, \mathbb{R}^N)$, we in particular have

$$\delta\mathscr{F}(u, \varphi) = 0 \quad \text{for all } \varphi \in C_c^\infty(\Omega, \mathbb{R}^N),$$

and therefore

$$L_F(u) = 0 \quad \text{on } \Omega.$$

Then by the same reasoning as in the proof of the Proposition in 1,*2.4*, we infer from (1) that

$$\int_{\partial\Omega} \nu_\alpha(x) F_{p_\alpha^i}(x, u(x), Du(x)) \varphi^i(x)\, d\mathscr{H}^{n-1}(x) = 0 \tag{2}$$

holds for all $\varphi \in C^1(\bar{\Omega}, \mathbb{R}^N)$ that are tangential along $u|_{\partial\Omega}$, and Lemma 1 yields

$$\Pi(x, u(x))\nu(x) \cdot F_p(x, u(x), Du(x)) = 0, \tag{3}$$

where $\nu \cdot F_p = (\nu_\alpha F_{p_\alpha^1}, \dots, \nu_\alpha F_{p_\alpha^n})$ is the product of the exterior normal $\nu = (\nu_1, \dots, \nu_n)$ on $\partial\Omega$ with $F_p = (F_{p_\alpha^i})$. Collecting these results, we arrive at the following

Theorem 1. *If* $u \in C^1(\bar{\Omega}, \mathbb{R}^N) \cap C^2(\Omega, \mathbb{R}^N)$ *is a stationary point of* $\mathscr{F}$ *with respect to the subsidiary condition*

$$G(x, u(x)) = 0 \quad \text{or} \quad u(x) \in M(x) \ \textit{for } x \in \partial\Omega,$$

then u satisfies the Euler equation

$$L_F(u) = 0 \quad \textit{on } \Omega$$

as well as the nonlinear boundary condition

$$(\nu \cdot F_p)(x) \perp T_{u(x)}M(x) \quad \textit{on } \partial\Omega, \tag{4}$$

that is, the (co-)vector $Z(x) = (Z_1(x), \dots, Z_N(x))$ *with the components*

$$Z_i(x) := \nu_\alpha(x) F_{p_\alpha^i}(x, u(x), Du(x))$$

is perpendicular to the surface $M(x)$ *for every* $x \in \partial\Omega$.

The boundary condition (4) will be refered to as the *transversality condition*, or we shall say that the stationary point u intersects $M(x)$ *transversally* at every $x \in \partial\Omega$.

It is not difficult to see that the reasoning can be reversed, and we obtain the following

Theorem 1′. *An extremal of* $\mathscr{F}$ *is a stationary point of* $\mathscr{F}$ *in the class* $\mathscr{C} = \{v \in C^1(\bar{\Omega}, \mathbb{R}^N)\colon G(x, v(x)) = 0 \text{ on } \partial\Omega\}$ *if and only if u intersects* $M(x) := \{z \in \mathbb{R}^N\colon G(x, z) = 0\}$ *transversally at every* $x \in \partial\Omega$.

If $r = 1$, then $M(x) = \{z \in \mathbb{R}^N\colon G(x, z) = 0\}$ describes a hypersurface in $\mathbb{R}^N$ with the normal $G_z(x, z)$. Thus transversality of u implies the existence of a function $\lambda(x)$, $x \in \partial\Omega$, such that

$$\nu_\alpha(x) F_{p_\alpha^i}(x, u(x), Du(x)) = \lambda(x) G_{z^i}(x, u(x))$$

holds, or briefly

$$\nu_\alpha F_{p_\alpha} = \lambda G_z .$$

If $n = N - 1$, this is equivalent to

$$\det(F_{p_1}, F_{p_2}, \ldots, F_{p_n}, G_z) = 0 .$$

In most applications, the constraint will be *scleronomic*, that is, of the type $G(u(x)) = 0$ on $\partial\Omega$. In this case we call

$$M := \{z \in \mathbb{R}^N\colon G(z) = 0\},$$

the *supporting surface* of the boundary values of the admissible functions.

Theorem 1″. *An extremal of $\mathcal{F}$ will be a stationary point of $\mathcal{F}$ in the class $\mathcal{C}$ of admissible functions with boundary values on the supporting surface M if and only if it is intersecting M transversally at all points of $\partial\Omega$.*

For $n = 1$ and $\Omega = (a, b)$, transversality of u with respect to M means that the covector $F_p(x, u(x), u'(x))$ is orthogonal to M for $x = a$ and $x = b$. More precisely,

$$F_p(x, u(x), u'(x)) \perp T_{u(x)} M \quad \text{for } x = a, b .$$

If $u(x)$ is kept fixed at $x = a$ but can freely move on M at $x = b$, we obtain that $F_p(b, u(b), u'(b))$ is orthogonal to M at the point $u(b)$.

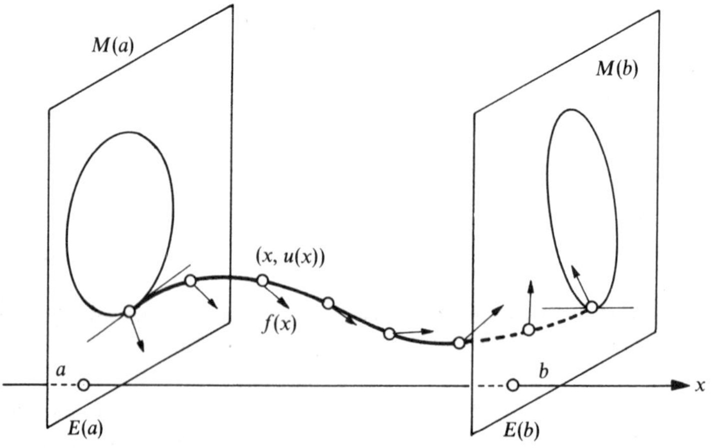

Fig. 9. The vector field $f(x) := \nu(x) \cdot F_p(x, u(x), Du(x))$ is perpendicular to $M(x)$ at $u(x)$ for $x = a, b$.

We consider some examples.

1 *Shortest distance in an isotropic medium.* Let $\omega(z) > 0$ be of class $C^1(\mathbb{R}^N)$, it will be considered as weight function (or "density") for measuring the *weighted length* $\mathscr{L}(u)$ of some path

$$z = u(t), \quad t_1 \le t \le t_2,$$

of class C^1. We set

$$\mathscr{L}(u) := \int_{t_1}^{t_2} \omega(u)|\dot{u}|\,dt.$$

The Euler equations are equivalent to

$$\frac{d}{dt}\left\{\omega(u)\frac{\dot{u}}{|\dot{u}|}\right\} = \omega_z(u)|\dot{u}|.$$

If the extremal u is parametrized by the arc length, i.e., if $|\dot{u}| = 1$, then we set $\frac{d}{dt} = {}'$ and obtain

$$\{\omega(u)u'\}' = \omega_z(u)$$

as equations equivalent to the Euler equations. From

$$F(z, p) = \omega(z)|p|,$$

we infer that

$$F_p(z, p) = \omega(z)\frac{p}{|p|}.$$

Hence transversality is in this case equivalent to orthogonality. In particular, the shortest connection u between some fixed point P of $\mathbb{R}^N$ and some manifold M in $\mathbb{R}^N$ has to meet M at a right angle

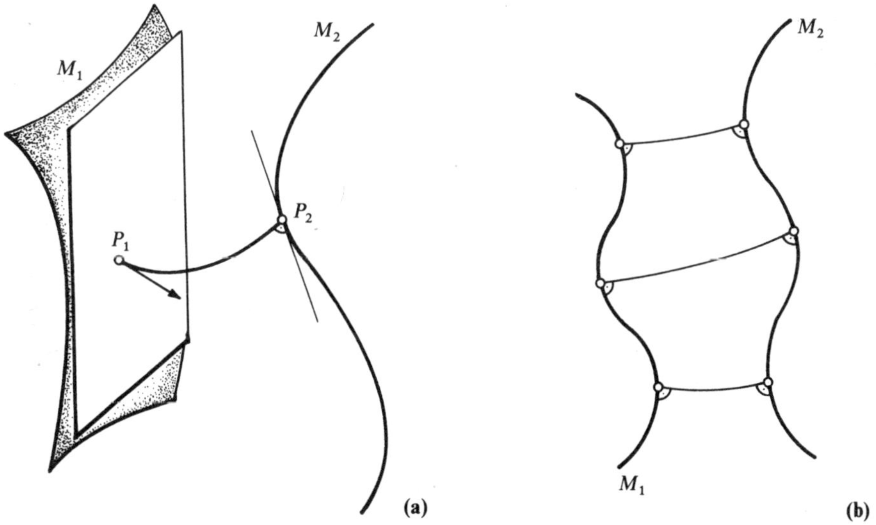

Fig. 10. (a) Free transversality of extremals with endpoints on a surface and a curve. **(b)** Locally shortest (or largest) connections of two manifolds M_1 and M_2 meet these manifolds perpendicularly.

whatever the choice of ω may be. Similarly the shortest connection between two submanifolds M_1 and M_2 hits both M_1 and M_2 perpendicularly. The same results hold for "stationary curves" of the length functional $\mathscr{L}$ whose endpoints are constrained to lie on certain supporting surfaces in $\mathbb{R}^N$.

[2] *Dirichlet integral.* Consider the Lagrangian

$$F(p) = \tfrac{1}{2}|p|^2.$$

The associated variational functional is the Dirichlet integral

$$\mathscr{D}(u) = \frac{1}{2}\int_\Omega |Du|^2\,dx.$$

Since $F_{p_\alpha} = p_\alpha$, we obtain

$$v_\alpha F_{p_\alpha}(Du) = v_\alpha D_\alpha u = \frac{\partial u}{\partial v}.$$

Transversality of $u \in C^1(\bar{\Omega}, \mathbb{R}^N)$ on $\partial\Omega$ with respect to some supporting surface M of the boundary values $u|_{\partial\Omega}$ therefore means that $\frac{\partial u}{\partial v}(x)$ is perpendicular to all tangent vectors a of M at the point $u(x)$. Hence u is transversal to M for all points of $\partial\Omega$ if it intersects M in its boundary values $u(x)$, $x \in \partial\Omega$, at a right angle.

[3] *Generalized Dirichlet integral.* In the previous example, the Dirichlet integral is defined with respect to the Euclidean line element

$$ds^2 = \delta_{ik}\,dz^i\,dz^k,$$

where δ_{ik} denotes the Kronecker symbol. Now we want to define a similar integral corresponding to a general Riemannian line element

$$ds^2 = g_{ik}(z)\,dz^i\,dz^k,$$

where $G(z) = (g_{ik}(z))$ denotes a symmetric, positive definite $N \times N$-matrix function on $\mathbb{R}^N$ which is of class C^2. The generalized Dirichlet integral $\mathscr{D}(u)$ with respect to the Riemannian line element ds^2 is, for $u \in C^1(\bar{\Omega}, \mathbb{R}^N)$, defined by

$$\mathscr{D}(u) = \frac{1}{2}\int_\Omega g_{ik}(u) D_\alpha u^i D_\alpha u^k\,dx$$

or, equivalently, by

$$\mathscr{D}(u) = \frac{1}{2}\int_\Omega G(u) D_\alpha u \cdot D_\alpha u\,dx.$$

As we shall see in 3,5, the ordinary Dirichlet integral from [2] will be transformed into this generalized Dirichlet integral if we subject the cartesian coordinates of $\mathbb{R}^N$ to a nonlinear coordinate transformation, since such a change of coordinates transforms the line element $ds^2 = \delta_{ik}\,dz^i\,dz^k$ (with respect to the old cartesian coordinates $z^1, \ldots, z^N$) into a line element $ds^2 = g_{ik}(z)\,dz^i\,dz^k$ (with respect to the new curvilinear coordinates which are again denoted by $z^1, \ldots, z^N$).

More generally, the generalized Dirichlet integral $\mathscr{D}(u) = \frac{1}{2}\int_\Omega g_{ik}(u) D_\alpha u^i D_\alpha u^k\,dx$ can even be defined for an arbitrary symmetric $N \times N$-matrix function $G(z) = (g_{ik}(z))$ on $\mathbb{R}^N$ which is of class C^1.

Here G may even be indefinite as in the case of the so-called *Lorentz metric*

$$ds^2 = (dz^1)^2 + (dz^2)^2 + (dz^3)^2 - (dz^4)^2,$$

where

$$G = \begin{bmatrix} 1 & & & 0 \\ & 1 & & \\ & & 1 & \\ 0 & & & -1 \end{bmatrix}.$$

This metric plays an important role in the *theory of special relativity*. Similarly, the metrics in the theory of general relativity are indefinite.

The Euler equations of the generalized Dirichlet integral can be brought into the form

$$D_\alpha\{g_{ik}(u)D_\alpha u^i\} - \tfrac{1}{2}g_{ij,k}(u)D_\alpha u^i D_\alpha u^j = 0,$$

where we have set

$$g_{ij,k}(z) := \frac{\partial}{\partial z^k} g_{ij}(z).$$

Equivalently we may write

$$g_{ik}(u)\Delta u^i + \{g_{ik,j}(u) - \tfrac{1}{2}g_{ij,k}(u)\}D_\alpha u^i D_\alpha u^j = 0.$$

If we introduce the *Christoffel symbols of the first kind*[8]

$$\Gamma_{ikj}(z) := \tfrac{1}{2}\{g_{kj,i} - g_{ij,k} + g_{ik,j}\}(z),$$

we can also write

$$g_{ik}(u)\Delta u^i + \Gamma_{ikj}(u)D_\alpha u^i D_\alpha u^j = 0.$$

We can transform these equations into a more elegant form if the matrix (g_{ik}) is invertible. Denote by (g^{ik}) the inverse matrix to (g_{ik}), that is,

$$G = (g_{ik}), \qquad G^{-1} = (g^{ik}),$$

and let us introduce the expressions

$$\Gamma_{ij}^l(z) := g^{lk}(z)\Gamma_{ikj}(z),$$

which are denoted as *Christoffel symbols of the second kind.* The Euler equations can then be transformed into the equivalent form

$$\Delta u^k + \Gamma_{ij}^k(u)D_\alpha u^i D_\alpha u^j = 0, \quad 1 \le k \le N.$$

Moreover, we derive from

$$F(z, p) = \tfrac{1}{2}g_{ik}(z)p_\alpha^i p_\alpha^k$$

that

$$F_{p_\alpha^i}(u, Du) = g_{ik}(u)u_{x^\alpha}^k.$$

If $\nu = (\nu_1, \ldots, \nu_n)$ denotes the exterior normal on $\partial\Omega$, we arrive at

$$\nu_\alpha F_{p_\alpha^i}(u, Du) = g_{ik}(u)\frac{\partial u^k}{\partial \nu}.$$

Suppose now that the extremal u is transversal to the supporting manifold M at $x \in \partial\Omega$. Then, for every tangent vector $a \in T_{u(x)}M$, we have that $a \perp \nu \cdot F_p(u, Du)$, or equivalently, if $a = (a^1, \ldots, a^N)$, that

$$a^i \nu_\alpha F_{p_\alpha^i}(u, Du) = 0$$

or

$$g_{ik}(u)a^i\frac{\partial u^k}{\partial \nu} = 0.$$

---

[8] Note that the definition of Γ_{ikj} slightly varies in the literature. Some authors write Γ_{ijk} for the expression which we have denoted by Γ_{ikj}.

We now want to show that this relation can be interpreted as a generalized orthogonality condition. To this end we introduce the "generalized inner product"

$$\langle a, b\rangle := G(u(x))\, a\cdot b = g_{ik}(u(x))\, a^i b^k$$

for two vectors $a, b \in \mathbb{R}^N$. If $x \in \partial\Omega$, then transversality of the extremal u and the supporting manifold M can be written as

$$\left\langle a, \frac{\partial u}{\partial v}(x)\right\rangle = 0 \quad \text{for all } a \in T_{u(x)}M\,.$$

In other words, transverality means orthogonality with respect to the "scalar product" $\langle\cdot,\cdot\rangle$ on the tangent space $T_{u(x)}\mathbb{R}^N$ which is to be identified with $\mathbb{R}^N$. In the case of a positive definite matrix $G(z)$, this scalar product defines the so-called *Riemannian metric.*

For $n = 1$, the Euler equations reduce to

$$\frac{d^2}{dt^2}u^k + \Gamma^k_{ij}(u)\frac{du^i}{dt}\frac{du^j}{dt} = 0$$

if the independent variable is written as t (instead of x). The solutions of these equations are denoted as *geodesics* with respect to the line element $ds^2 = g_{ik}(z)\, dz^i\, dz^k$.

There is a second kind of *transversality condition* which was introduced by Kneser; we shall call it *free transversality*, because this condition is suggested by integrals

$$\mathscr{F}(u, \Omega) = \int_\Omega F(x, u(x), Du(x))\, dx$$

for which the parameter domain Ω is not fixed but is left free. To put it in other words, free transversality is a natural boundary condition resulting for mappings $u: \bar{\Omega} \to \mathbb{R}^N$ which are stationary for $\mathscr{F}(u, \Omega)$, where the boundary of Ω may freely move.

Methodically, free transversality is a topic that should be treated in Chapter 3, because there we shall consider changes of the parameter domain as well as "variations of the independent variables". However, it may be helpful to immediately compare the two notions of transversality.

We shall restrict our consideration to single integrals

$$\int_a^b F(x, u(x), u'(x))\, dx$$

and to mappings $u: [a, b] \to \mathbb{R}^N$ with fixed right endpoint $P_2 = (b, u(b))$, whereas the left endpoint $P_1 = (a, u(a))$ may freely move on a C^1-submanifold $\mathscr{M}$ of $\mathbb{R} \times \mathbb{R}^N$, given by some scalar equation

$$G(x, z) = 0.$$

Note that this viewpoint differs very much from the one previously assumed. There the constraint $G(x, z) = 0$ was interpreted as condition for z, keeping the boundary point x of $\partial\Omega$ fixed. The set $M(x) = \{z \in \mathbb{R}^N\colon G(x, z) = 0\}$ was viewed as $(N-1)$-dimensional manifold in $\mathbb{R}^N$ describing the set of permitted boundary values $u(x)$ for some $x \in \partial\Omega$ and any admissible function u. Presently the left endpoint a of $\Omega = (a, b)$ is allowed to move, and therefore the endpoint $(a, u(a)) = P_1$ of the curve $z = u(x)$, $a \le x \le b$, can slide on the N-dimensional manifold $\mathscr{M}$ in $\mathbb{R}^{N+1}$.

Definition. *We say that the curve $z = u(x)$, $a \le x \le b$, and the hypersurface $\mathscr{M} = \{(x, z) \in \mathbb{R}^{N+1}: G(x, z) = 0\}$ are* freely transversal *to each other at $P = (a, u(a))$ with respect to the Lagrangian F if the following holds:*

(4′)
(i) $G(a, u(a)) = 0$, *that is:* $P \in \mathscr{M}$;
(ii) *the vector* $\mathscr{N}(a) := (F - u' \cdot F_p, F_p)|_{x=a}$ *is orthogonal to the tangent space* $T_P\mathscr{M}$ *of* $\mathscr{M}$ *at* P.

(Here the arguments of F and F_p are $(x, u(x), u'(x))$.)

More general, let $(x, z, p) \in \mathbb{R} \times \mathbb{R}^N \times \mathbb{R}^N$ be some *line element* in $\mathbb{R}^{N+1}$ with the supporting point (x, z) and the direction $(1, p)$, and let $(x, z, v) \in \mathbb{R} \times \mathbb{R}^N \times \mathbb{R}^{N+1}$ be some *hypersurface element* with the same supporting point (x, z) and the normal direction v describing the position of the element in $\mathbb{R}^{N+1}$. Then (x, z, p) and (x, z, v) are said to be *freely transversal to each other* if and only if v and $\mathscr{N}$ are collinear, where $\mathscr{N}$ is defined by

$$\mathscr{N} := (F(x, z, p) - p \cdot F_p(x, z, p), \quad F_p(x, z, p)).$$

The notion of free transversality is justified by the following

Theorem 2. *If $F \in C^2$, and if $u \in C^2([a, b], \mathbb{R}^N)$ is a minimizer (or a stationary point) of*

$$\mathscr{F}(u) = \int_a^b F(x, u(x), u'(x))\, dx,$$

where b and $u(b)$ are fixed, whereas $(a, u(a))$ can freely move on a regular hypersurface $\mathscr{M}$ of $\mathscr{R}^{N+1}$ defined by some scalar equation $G(x, z) = 0$ then the curve $z = u(x)$ and the hypersurface $\mathscr{M}$ are freely transversal to each other at the left endpoint $P_1 = (a, u(a))$ of the curve.

Proof. First we embed $z = u(x)$, $a \le x \le b$, into a smooth family of curves

$$z = \psi(x, \varepsilon), \quad \alpha(\varepsilon) \le x \le b, \quad |\varepsilon| < \varepsilon_0,$$

satisfying

$$\alpha(0) = a, \qquad \psi(x, 0) = u(x), \qquad \psi_\varepsilon(x, 0) = \varphi(x),$$
$$G(\alpha(\varepsilon), \psi(\alpha(\varepsilon), \varepsilon)) = 0, \quad \psi(b, \varepsilon) = u(b).$$

Set

$$\phi(\varepsilon) := \int_{\alpha(\varepsilon)}^b F(x, \psi(x, \varepsilon), \psi'(x, \varepsilon))\, dx,$$

where $'$ denotes the partial derivative with respect to x. Since u is assumed to be a minimizer or at least a stationary point of $\mathscr{F}$ with respect to the boundary conditions described in the theorem, we have

$$\frac{d\phi}{d\varepsilon}(0) = 0.$$

On the other hand, we infer

$$\begin{aligned}\frac{d\phi}{d\varepsilon}(0) &= -F(a, u(a), u'(a))\cdot\alpha_\varepsilon(0)\\ &\quad+\int_a^b \{F_z(x, u, u')\cdot\varphi + F_p(x, u, u')\cdot\varphi'\}\,dx\\ &= -F(a, u(a), u'(a))\cdot\alpha_\varepsilon(0)\\ &\quad+\int_a^b L_F(u)\cdot\varphi\,dx + [\varphi\cdot F_p(x, u, u')]_a^b.\end{aligned}$$

Moreover, we have $L_F(u) = 0$ and $\varphi(b) = 0$, whence

$$0 = F\cdot\alpha_\varepsilon(0) + F_p\cdot\varphi(a);$$

here the arguments of F and F_p are to be taken as $a, u(a), u'(a)$. This equation can be rewritten as

$$0 = (F - u'(a)\cdot F_p)\cdot\alpha_\varepsilon(0) + F_p\cdot(u'(a)\alpha_\varepsilon(0) + \varphi(a)).$$

Moreover, the relation

$$G(\alpha(\varepsilon), \psi(\alpha(\varepsilon), \varepsilon)) = 0$$

yields

$$0 = G_x(a, u(a))\alpha_\varepsilon(0) + G_z(a, u(a))\cdot(u'(a)\alpha_\varepsilon(0) + \varphi(a)),$$

that is, $(\alpha_\varepsilon(0), u'(a)\alpha_\varepsilon(0) + \varphi(a)) \in T_P\mathcal{M}$, $P = (a, u(a))$. By a reasoning that has repeatedly been used, we can assume that $(\alpha_\varepsilon(0), u'(a)\alpha_\varepsilon(0) + \varphi(a))$ is an arbitrarily given tangent vector to M at P, and thus we infer that $(F - u'(a)\cdot F_p, F_p)$ is orthogonal to $T_P\mathcal{M}$. □

Remark 1. From the second "proof" of Theorem 2 sketched below we shall see that *transversality* and *free transversality* are very closely related to each other.

Remark 2. *If F is of the type $F(x, z, p) = \omega(x, z)\sqrt{1 + |p|^2}$, $\omega > 0$, then free transversality of $z = u(x)$ and $\mathcal{M}$ at $x = a$ means that the curve $z = u(x)$ meets the hypersurface $\mathcal{M}$ perpendicularly at $P_1 = (a, u(a))$.*

In fact,

$$(F - p\cdot F_p, F_p) = \frac{\omega(x, z)}{\sqrt{1 + |p|^2}}\cdot(1, p),$$

from where the statement follows at once.

Remark 3. Similarly we can define free transversality for multiple integrals. This concept plays a particular role in Carathéodory's field theory. We refer the reader to Funk [1], pp. 382–401, and to our presentation in 7,4.2.

We sketch a second proof of Theorem 2 which, although being not quite complete, is rather interesting and can be made perfectly rigorous. The idea is to free the variable x from its distinguished role and to give all variables $x, z^1, \ldots, z^N$ equal rights, which corresponds to the situation that $\mathcal{M}$ is a submanifold in the x,z-space. Then one can play back free transversality to ordinary transverality. More precisely, *introducing a parametric integral $\mathcal{H}(c)$ coinciding with the nonparametric integral $\mathcal{F}(c)$ on nonparametric curves $c(x) = (x, u(x))$ we see that free transversality for $\mathcal{F}$ is just the transversality condition with respect to $\mathcal{H}$.*

To carry out this idea, we introduce new variables

$$\xi = (\xi^0, \xi^1, \ldots, \xi^N), \qquad \pi = (\pi^0, \pi^1, \ldots, \pi^N)$$

and a function

$$H = H(\xi, \pi),$$

which will be connected with x, z, p, and $F(x, z, p)$ by the relations

$$\xi^0 = x, \quad \xi^1 = z^1, \ldots, \xi^N = z^N,$$

$$p^i = \pi^i/\pi^0 \quad \text{for } 1 \le i \le N,$$

$$H(\xi, \pi) := \pi^0 F(\xi^0, \xi^1, \ \ldots\ , \ \xi^N, \pi^1/\pi^0, \ldots, \pi^N/\pi^0).$$

Let us introduce the new functional

$$\mathcal{H}(c) := \int_{t_1}^{t_2} H(c, \dot{c})\, dt$$

for parametric curves $c(t)$, $t_1 \le t \le t_2$, in $\mathbb{R}^{N+1}$. This integral is invariant with respect to orientation preserving parameter transformations σ, i.e. $\mathcal{H}(c) = \mathcal{H}(c \circ \sigma)$, and the integrand $H(\xi, \pi)$ of $\mathcal{H}$ does not depend of the independent variable t. Moreover, for every nonparametric curve $z = u(x)$, $a \le x \le b$, we have

$$\mathcal{F}(u) = \mathcal{H}(c)$$

if we set

$$t = x, \quad c(t) = (x, u(x)), \qquad t_1 = a, \qquad t_2 = b$$

or more precisely,

$$\int_a^b F(x, u(x), u'(x))\, dx = \int_{t_1}^{t_2} H(c(t), \dot{c}(t))\, dt.$$

Suppose now that, for the given minimizing or stationary nonparametric curve u of $\mathcal{F}$, the corresponding curve $c(t) = (t, u(t))$ is minimizing or stationary for $\mathcal{H}$ among all C^1-curves $\tilde{c}(t)$, $t_1 \le t \le t_2$, such that $\frac{d}{dt}\tilde{c}(t) \ne 0$ and

$$P_1 = \tilde{c}(t_1) \in \mathcal{M}, \qquad P_2 = \tilde{c}(t_2) = \text{fixed}.$$

Since $\mathcal{H}$ is parameter-invariant, we can equivalently interpret this property as minimality or stationary of $\mathcal{H}$ with respect to curves for which the endpoints of the parameter interval $[t_1, t_2]$ are fixed. According to Theorem 1, this is equivalent to the property that $c(t)$ intersects $\mathcal{M}$ transversally in the ordinary sense at $t = t_1$ with respect to the Lagrangian H. In other words, we obtain that

$$H_\pi(c(t_1), \dot{c}(t_1)) \perp T_{P_1}\mathcal{M}, \quad P_1 = c(t_1) \in \mathcal{M}$$

holds. On the other hand, we infer from the defining relations that

$$H_{\pi^0}(\xi, \pi) = F(x, z, p) - p \cdot F_p(x, z, p),$$

$$H_{\pi^i}(\xi, \pi) = F_{p^i}(x, z, p), \quad 1 \le i \le N,$$

is satisfied. Then we conclude from $H_\pi(c(t_1), \dot{c}(t_1)) \perp T_{P_1}\mathcal{M}$ that the nonparametric curve $z = u(x)$ and the hypersurface $\mathcal{M}$ are freely transversal to each other.

We have not proved that $c(t)$ is a minimizer or a stationary point for $\mathscr{H}$, if $u(x)$ is one for $\mathscr{F}$. We shall discuss this question in some detail in 8,*1.2*. In particular we shall prove that $c(t) = (t, u(t))$ is an extremal of $\mathscr{H}$ if u is an extremal of $\mathscr{F}$. However, we shall see in 8,*1.2* that nonparametric minimizers need not be parametric ones.[9]

5. Scholia

1. Already in his two papers on isoperimetric problems from 1732 and 1736, respectively, Euler introduced *multipliers* (which are nowadays called *Lagrange multipliers*).[10] But these publications contain several serious mistakes and erroneous methods, which show that Euler had not yet mastered these questions.[11] Only in his *Methodus inveniendi* (E65) from 1744, he obtained a basically sound theory (apart from the fact that Euler did not think of extremals which neither are maximizers nor minimizers). In Chapter 6 of this treatise, he developed the multiplier theory for isoperimetric side conditions, proving the existence of a constant multiplier, and in Chapter 5, he treated problems with several subsidiary conditions of isoperimetric type. For instance, he proved that, for two variational integrals $\mathscr{F}$, $\mathscr{G}$ and two constants λ, μ, the functional $\lambda\mathscr{F} + \mu\mathscr{G}$ yields the same extremals as the isoperimetric problem where $\mathscr{F}$ is to be given an extremum, keeping $\mathscr{G}$ fixed.

In Chapter 3 of the *Methodus inveniendi*, Euler dealt with variational integrals

$$\int_a^b F(x, u, v, v', \ldots, v^{(k)})\,dx$$

subject to nonholonomic constraints of the kind

$$u' - G(x, u, v, v', \ldots, v^{(k)}) = 0,$$

and he proved the existence of a multiplier function. Although Euler did not advance to the most general type of Lagrange problem, his analysis of the constraint problem is an achievement of first class that Carathéodory called "*eine Spitzenleistung, wie sie auch einem Euler nicht allzu oft geglückt ist.*"[12]

Later on, Lagrange has applied his δ-method to variational problems with nonholonomic constraints, thus obtaining a new approach to the multiplier rule. His name became attached to the multiplier rule by the applications to mechanics that he derived in his *Méchanique analitique*[13] in 1788 (see Lagrange [1], Sect. 4, §1). Moreover, he treated general nonholonomic conditions; yet his proof contained two serious gaps which were only filled by A. Mayer (1886), A. Kneser (1900), and Hilbert (1906). For a brief historical account, we refer to Bolza [3], pp. 566–569, and to Goldstine [1], pp. 148–150.

The general Lagrange problem for simple integrals is carefully treated in: Bolza [3], Chapters 11 and 12; Carathéodory [10], Chapter 18; Hestenes [5], Chapter 6; Rund [4], Chapter 5; L.C. Young [1], Vol. 2; Cesari [1]. The last three monographs also contain a detailed bibliography and

---

[9] See also Bolza [3], pp. 198–201.

[10] (E27) *Problematis isoperimetrici in latissimo sensu accepti solutio generalis*, Comm. acad. sci. Petrop. *6* (1732/3), 1738, 123–155; (E56) *Curvarum maximi minimive proprietate gaudentium inventio nova et facilis*, Comm. acad. sci. Petrop. *8* (1736), 1741, 147–158; cf. also: Opera omnia, Ser. I. vol. 25.

[11] Carathéodory, Schriften [16], Vol. 5, pp. 130–135.

[12] "*a major accomplishment that even an Euler did not achieve too often*".

[13] This is no orthographic error; only the second edition was called *Mécanique analytique*. The first volume of this edition appeared in 1811, the second one in 1815, two years after Lagrange's death.

describe the relations to optimization theory. Our treatment of the multiplier rule for the Lagrange problem is essentially taken from Bolza [3].

For multiple integrals, the state of art is much less satisfactory; some results can be found in Klötzler [4] and in Bittner [1].

2. The first proof of the isoperimetric property of the circle appears in the commentary of Theon to Ptolemy's *Almagest* and in the collected works of Pappus. The author of the proof is Zenodoros, who must have lived sometime between Archimedes (died 212 B.C.) and Pappus (about A.D. 340), since he quotes Archimedes and is cited by Pappus. Concerning the history of the isoperimetric problem, we refer to Gericke [1] and Blaschke [2]. The first complete proof of the isoperimetric property of the circle was provided by Weierstrass in his lectures at Berlin University.

3. The problem of the vibrating string was already considered by the Pythagoreans. In modern times, experimental results were found by Mersenne (1625) and Galileo (1638). Theoretical results were obtained by Taylor (1713), Johann Bernoulli (1727), Daniel Bernoulli, Euler, d'Alembert, and Lagrange. We refer to Truesdell [1], [2]. The variational approach to eigenvalue problems was particularly stressed in the treatise of Courant–Hilbert [1–4], which had great influence on the development of mathematical physics and quantum mechanics. Also Rayleigh's celebrated treatise [1] on *The theory of sound* and the papers by W. Ritz [1] were very influential.

4. Hypersurfaces of constant mean curvature not only appear as soap bubbles, but also as capillary surfaces without gravity in equilibrium; cf. Finn [1], in particular Chapter 6.

5. The problem to determine the shape of a freely hanging rope (or chain) was posed by Jacob Bernoulli in 1690; the solution was found by Leibniz, Huygens, and Johann Bernoulli in 1691. Earlier, this problem had been considered by Galileo who wrote that the equilibrium configuration of a rope has the shape of a parabola. The notation *catenary* (Latin: catenaria) seems to be due to Leibniz. The treatment of the equilibrium configuration of a chain as an isoperimetric problem is due to Euler (cf. *Methodus inveniendi*, Chapter 5, nrs. 47, 73, 74).

Recently it was disputed whether or not Galileo thought the catenary to be a parabola. The matter seems clear if we hear what Salviati says in the second day of Galileo's *Discorsi*:[14]

"*Fix two nails at a wall at equal height above the horizontal, and at a distance from each other which is twice the length of the rectangle where the semiparabola is to be constructed. At both nails a fine chain is suspended which is so long that its lowest point is at a distance equal to the length of the rectangle below the horizontal through the nails: This chain has the shape of a parabola. If we puncture the line formed by the chain onto the wall, we have described a complete parabola.*"

However, the historian M. Fierz [1], p. 63, pointed out that, on the fourth day of the *Discorsi*, Salviati gives a somewhat different description of the catenary:

"*We feel amazement and joy, if the little or tightly stretched rope approaches the parabolic form, and the similarity is such that if, in a plane, you draw a parabola, and you consider it invertedly, the vertex below and the base parallel to the horizontal, and you suspend a chain at the endpoints of the base of the parabola, then the chain is more or less curving and will attach itself to the said parabola,*

[14] *Discorsi e dimostrazioni matematiche*, Elsevir, Leiden, 1638; cf. *giornata seconda*, p. 146: *Ferminsi ad alto due chiodi in un parete equidistanti all' Orizonte, e trà di loro lontani il doppio della larghezza del rettangolo, su 'l quale vogliamo notare la semiparabola, e da questi due chiodi penda una catenella sottile, e tanto lunga, che la sua sacca si stenda quanta è la lunghezza del Prisma: questa catenella si piega in figura Parabolica. Si che andando punteggiando sopra 'l muro la strada, che vi fà essa catenella, haremo descritto un' intera Parabola.*

and the attachment is the more precise, the less the parabola is curved, that is, the more it is stretched; in parabolas with an elevation [slope] of 45 degrees, the chain covers the parabola almost perfectly."[15]

In the opinion of Fierz, this second statement of Galileo shows clearly, that he was well aware of the fact that the parabola is merely an approximation for the catenary. Fierz goes even further by pointing out that, for $|x| < 1$, the parabola $y = 1 + \frac{1}{2}x^2$ is a good approximation to $y = \cosh x$. He could be right, but at least this reasoning seems a bit farfetched; in any case it does not explain why, in the second dialogue, Galileo described the chain as a "complete parabola" ("intera parabola"). Of course, he could have discovered later that the parabola is merely an approximation to the true catenary and he might have forgotten to erase what he had written in the first place. This, in fact, is the speculation of A. Herzig and I. Szabó [1], pp. 53–54. As the present authors are no historians of mathematics, they have to ask the reader to form his own opinion.

6. The notion of a *harmonic mapping* was introduced by S. Bochner [1] in 1940, and in full generality by Fuller [1]. Three surveys of the development till 1988 were given by Eells and Lemaire [1], [2], [3]. Harmonic mappings have become an important tool in differential geometry.

7. The first published paper on shortest lines on a surface is due to Euler.[16] However, it now seems to be certain that Johann Bernoulli possessed the "*law of the osculating plane*" in 1698.[17]

8. The modern theory of geodesics in Riemannian manifolds began with Riemann's lecture on occasion of his Habilitationskolloquium June 10, 1854: "*Über die Hypothesen, welche der Geometrie zugrunde liegen.*"[18]

9. The history of the principle of least action has often been described. Yet the matter is still controversial, and there seems to be no general agreement who invented the principle, Leibniz, Euler, or Maupertuis. A popular account of the controversy between Maupertuis and Euler on the one hand, and S. König/Voltaire on the other is given in Hildebrandt–Tromba [1]. A more profound discussion can be found in Carathéodory [16], Vol. 5, pp. 160–165, Fleckenstein [1], and F. Klein [3], Vol. 1, pp. 191–207; see also Pulte [1]. We mention that the first mathematical treatment of the action principle was given by Euler (1744) in the *Additamentum* II of his *Methodus invendiendi.*

Moreover, there are different mathematical versions of the least action principle which sometimes are emphatically distinguished from each other. We do not want to participate in these controversies and just refer the reader to Klein's opinion ([3], Vol. 1, pp. 192–193),[19] and to Prange [2], pp. 565–566 and footnote 153 on p. 607.

---

[15] "*Recandovi insieme maraviglia, e diletto, che la corda così tesa, e poco, ò molto tirata, si piega in linee, le quali assai si avvicinano alle paraboliche, e la similitudine è tanta che se voi segnerete in una superficie piana, & eretta all' Orizonte una linea parabolica, e tenendola inversa, cioè col vertice in giù, e con la base parallela all' Orizonte, facendo pendere una catenella sostenuta nelle estremità della base della segnata parabola, vedrete allentando più, ò meno la detta catenuzza incurvarsi, e adattarsi alla medesima parabola; e tale adattemento tanto più esser preciso, quanto la segnata parabola sar à men' curva, cioé più distesa; si che nelle parabole descritte con elevazioni sotto à i grad.* 45 *la catenella camina quasi ad unguem sopra la parabola*". Cf. *Discorsi, giornata quarta*, p. 284.

[16] (E9) *De linea brevissima in superficie quacunque duo quaelibet puncta jungente*, Comm. acad. sci. Petrop. **3** (1728) 1732, 110–124; cf. also: Opera omnia, Ser. I, Vol. 25, 1–12.

[17] Johann Bernoulli's Opera omnia, Vol. 4, p. 108, Nr. 166.

[18] Werke [3], second ed., nr. 13, p. 272.

[19] "*The Lagrange equations arise immediately from the variational problem $\delta \int L dt = 0$ (the limits of the integral kept fixed). Remarkably, this idea appears in Lagrange only between the lines; therefore we find the strange fact that this relation – mainly by the influence of Jacobi – is generally known in Germany, and therefore also in France, as "Hamilton's principle", whereas in England no one will understand this terminology. There one denotes this equation by the correct though unintuitive name*

We only mention that there are essentially three different versions of the action principle:

(i) Hamilton's form, as described in 2 [5]. Here one has to consider a variational problem on a fixed time interval.

(ii) Lagrange's form, as presented in 3 [3]. This is a variational problem on a time interval with free endpoints, with the nonholonomic constraint $T + V = h$. Here a variation of the independent variable is required in order to prove the desired result; cf. 3,2 [1].

(iii) Jacobi's geometrical version. A special case of it is discussed in 3,*1* [2]. It leads to a first order differential equation due to Gauss, describing the geometrical shape of the orbits. The general form of Jacobi's geometrical action principle will be considered in Chapter 8.

---

of the "principle of stationary action". ... The widely-known "principle of least action" is another, by Lagrange preferred concept which he found at the beginning of his studies in 1759. In the 18th century this principle drew great interest, in particular from philosophical quarters. ... Here Maupertuis is especially to be mentioned.

The form of this principle follows from the first one by a combination of the equations

$$\left.\begin{aligned} L &= T - \mathcal{U} \\ h &= T + \mathcal{U} \end{aligned}\right\} \quad 2T = L + h .$$

Hence we obtain

$$\int L\, dt = \int 2T\, dt - h(t_1 - t_0)$$

and the resulting variational problem becomes $\delta \int 2T\, dt = 0$ *.... But the integral* $\int 2T\, dt$ *is what has long been called the action whence the name of the principle is derived .... Recently the principle obtained another, important form by Jacobi who altogether eliminated the time t. He wrote* $T = h - \mathcal{U} = \sqrt{T(h - \mathcal{U})}$ *and attained*

$$\delta \int \sqrt{(h - \mathcal{U}) \sum a_{\alpha\beta}\, dq_\alpha\, dq_\beta} = 0 .$$

In this form the principle, of course, determines solely the trajectory and not the time in which it is traversed.

If in this "Jacobi principle" one wishes just as Riemann to view $\sqrt{\sum a_{\alpha\beta}\, dq_\alpha\, dq_\beta} = \sqrt{ds^2} = ds$ *as the line element in some n-dimensional space, then one finally arrives at*

$$\delta \int \sqrt{h - \mathcal{U}}\, ds = \delta \int v\, ds = 0 ,$$

where v now means the speed of a moving point in an n-dimensional space, which represents a lucid geometric counterpart to the mechanical problem.

The original remarks of Klein read as follows:

Die Lagrangesche Gleichungen erwachsen unmittelbar aus dem Variationsproblem $\delta \int L\, dt = 0$ *(bei festgehaltenen Grenzen). Merkwürdigerweise steht dieser Ansatz bei Lagrange nur zwischen den Zeilen; es konnte sich daher die merkwürdige Tatsache entwickeln, daß diese Beziehung in Deutschland – hauptsächlich durch Jacobis Wirken – und dadurch auch in Frankreich allgemein als "Hamiltonsches Prinzip" bezeichnet wird, während in England niemand diese Ausdrucksweise versteht; dort benennt man die Gleichung vielmehr mit einem korrekten aber unanschaulichen Namen als "Prinzip der stationären Wirkung". ... Das vielverbreitete "Prinzip der kleinsten Wirkung" ist eine andere, von Lagrange bevorzugte Fassung, die er bei Beginn seiner Studien 1759 vorfand. Im 18. Jahrhundert wandte man diesem Prinzip besonders von philosophischer Seite lebhaftes Interesse zu. ... Besonders Maupertuis wäre hier zu nennen. Die Form des Prinzips ergibt sich aus der ersten durch Kombination der Gleichungen*

10. The name "*solenoidal vector field*" was coined by Maxwell.

11. The notion of *transversality* (in our terminology: *free transversality*) was introduced by A. Kneser [3]. It is, of course, very dangerous to change a well established terminology. On the other hand, it seemed to be useful to have a name for the property 2,4, (4). We chose the compromise to denote both relations 2,4, (4) and (4′) as transversality relations. Only if some confusion is possible, we will remind the reader which kind of transversality relation is meant.

12. The problems of Lagrange, Mayer, and Bolza. The problem of Bolza with variable end-points is usually considered as the most general variational problem that involves only simple integrals. Any Lagrange problem and any Mayer problem can be reduced to a Bolza problem; in fact, all three problems are equivalent. During the first half of this century, a quite complete theory of necessary and sufficient conditions was developed for such questions.

A slightly more special form of these problems is the optimal control problem which during the fifties moved into the center of interest when engineers tried to solve the problem of controlling a system, governed by a set of differential equations, in such a way that a given performance index is to be minimized or maximized. The main difference to Bolza's problem is that, in optimal control problems, the differential equations usually appear in a normal form and that the competing functions lie in a closed set instead of an open one. The theory of optimal control soon went off on a different road, and the necessary conditions for such problems were eventually formulated by means of the *Pontryagin maximum principle*; see 7,4.4 for a brief discussion.

Basically we shall not deal with optimal control problems in our treatise; we refer the reader to the extensive literature on this field, e.g. Bliss [5], Young [1], Hestenes [5], Cesari [1] and Fleming-Rishel [1] (not quoting any of the engineering-type books). Here we shall content ourselves to just formulate Bolza, Mayer and Lagrange problems.

$$\left.\begin{aligned} L &= T - \mathcal{U} \\ h &= T + \mathcal{U} \end{aligned}\right\} \quad 2T = L + h.$$

Es wird also

$$\int L\,dt = \int 2T\,dt - h(t_1 - t_0),$$

und als Variationsproblem ergibt sich $\delta \int 2T\,dt = 0 \ldots$ *. Das Integral* $\int 2T\,dt$ *ist nun das, was man seit alters her als "actio" = Wirkung bezeichnet, woher sich der Name des Prinzips ... erklärt. ... In neuerer Zeit gewann das Prinzip wiederum eine andere, bedeutungsvolle Gestalt durch Jacobi, der die Zeit t vollends eliminierte. Er schrieb* $T = h - \mathcal{U} = \sqrt{T(h - \mathcal{U})}$ *und erhielt*

$$\delta \int \sqrt{(h - \mathcal{U}) \sum a_{\alpha\beta}\, dq_\alpha\, dq_\beta} = 0.$$

In dieser Form legt das Prinzip natürlich nur die Bahnkurve fest, nicht die Zeit, in der sie durchlaufen wird.

Will man in diesem "Jacobischen Prinzip" nach Riemannscher Weise $\sqrt{\sum a_{\alpha\beta}\, dq_\alpha\, dq_\beta} = \sqrt{ds^2} = ds$ *als Bogenelement auffassen in einem n-dimensionalen Raum, so ergibt sich schließlich*

$$\delta \int \sqrt{h - \mathcal{U}}\, ds = \delta \int v\, ds = 0,$$

wo v nun die Geschwindigkeit eines im n-dimensionalen Raum bewegten Punktes bedeutet, der für das mechanische Problem ein übersichtliches geometrisches Gegenbild darstellt.

In the *problem of Bolza*, one has to minimize a functional of the kind

$$\mathcal{J}(x_1, x_2, u) := G(x_1, u(x_1), x_2, u(x_2)) + \int_{x_1}^{x_2} F(x, u, u')\, dx$$

among all curves $u : [x_1, x_2] \to \mathbb{R}^N$ satisfying both a set of differential equations

$$\phi_\alpha(x, u(x), u'(x)) = 0, \quad \alpha = 1, \dots, m,$$

and a set of boundary conditions of the form

$$\psi_\beta(x_1, u(x_1), x_2, u(x_2)) = 0, \quad \beta = 1, \dots, p,$$

where $m < N$ and $p < 2N + 2$.

The special case $F = 0$ is denoted as a *Mayer problem*, whereas the case $G = 0$ is a *Lagrange problem* with variable endpoints.

By definition, Mayer and Lagrange problems are special cases of Bolza problems, but it is easy to see that any Bolza problem can be reduced to either a Mayer problem or a Lagrange problem. In fact, the Bolza problem considered above is equivalent to the task to find the minimizer of the functional

$$\mathcal{M}(x_1, x_2, U) := G(x_1, u(x_1), x_2, u(x_2)) + v(x_2)$$

among all curves $U(x) = (u(x), v(x))$, $x_1 \le x \le x_2$, satisfying

$$\phi_\alpha(x, u, u') = 0, \qquad v' - F(x, u, u') = 0,$$

and

$$\psi_\beta(x_1, u(x_1), x_2, u(x_2)) = 0, \quad v(x_1) = 0,$$

and this is a Mayer problem.

Furthermore, the Bolza problem above is also equivalent to the Lagrange problem which consists in minimizing the functional

$$\int_{x_1}^{x_2} [v(x) + F(x, u(x), u'(x))]\, dx$$

among all curves $U = (u, v)$ satisfying

$$\phi_\alpha(x, u, u') = 0, \quad v'(x) = 0,$$

$$\psi_\beta(x_1, u(x_1), x_2, u(x_2)) = 0,$$

$$v(x_1)(x_2 - x_1) - G(x_1, u(x_1), x_2, u(x_2)) = 0.$$

13. Optimal control problems. Suppose that $x(t) = (x^1(t), \dots, x^N(t))$ characterizes the state of a physical system described by a system of differential equations

$$\dot{x}^i(t) = f^i(x(t), u(t)), \quad 1 \le i \le N,$$

or equivalently

$$\dot{x}(t) = f(x(t), u(t)),$$

where $u(t) = (u^1(t), \dots, u^k(t))$ are parameter functions. We interpret $u(t)$ as *control functions*. The class of *admissible controls* will be specified in advance. For every initial value $x_0 = x(t_0)$ and any control function $u(t)$, the system $\dot{x} = f(x, u)$ has a unique solution which is called trajectory. The quadruple

$$U := \{u, t_0, t_1, x_0\}$$

is called a *control process*. We fix a function $F(x, u)$; to every control process U with the corresponding trajectory $x(t)$, $t_1 \le t \le t_2$, we assign the number

$$\mathscr{F}(U) := \int_{t_0}^{t_1} F(x(t), u(t))\, dt.$$

Then the control process U is said to be *optimal* if

$$\mathscr{F}(U) \le \mathscr{F}(U^*)$$

holds true for any other control process $U^* = \{u^*, t_0, t_1, x_0\}$ satisfying $x^*(t_1) = x(t_1)$, where $x^*(t)$ denotes the trajectory corresponding to u^*.

Actually, one is usually interested in a somewhat more general situation. One allows processes $U^* = \{u^*, t_0, t_1^*, x_0\}$ where t_1^* may vary but the trajectory $x^*(t)$ is subject to $x^*(t_1^*) = x(t_1)$, in addition to $x^*(t_0) = x_0$. An important special case is $F \equiv 1$ where the functional $\mathscr{F}(U)$ reduces to $\int_{t_0}^{t_1} dt = t_1 - t_0$ which represents the time between the two positions $x_0 = x(t_0)$ and $x_1 = x(t_1)$. In this particular case, optimality means minimal time to go from x_0 to x_1.

Note that every minimizing problem for the functional $\int F(t, x(t), \dot{x}(t))\, dt$ can be viewed as an optimal control problem: Minimize $\int F(t, x(t), u(t))\, dt$ for all controls $u(t)$ and the corresponding trajectories $x(t)$ given by

$$\dot{x}(t) = u(t).$$

Further information and references to optimal control problems can be found in L.C. Young [1] and L. Cesari [1].

14. Geodesics on a surface of revolution. The classical **theorem of Clairaut** states: *Let $\mathscr{C}$ be a geodesic on a smooth regular surface of revolution S. Then at any point P of $\mathscr{C}$ the radius $r(P)$ of the circle of latitude at P multiplied by the sine of the inclination angle $\alpha(P)$ of $\mathscr{C}$ with respect to the meridian through P is a constant.*

In fact, a surface of revolution, S, can locally be described either by the parametric representation

$$x = r\cos\theta, \quad y = r\sin\theta, \quad z = f(r), \qquad r_1 < r < r_2,\ 0 \le \theta \le 2\pi, \tag{1}$$

where $f(r)$ is a smooth function of r and $0 < r_1 < r_2 < \infty$, or by

$$x = g(z)\cos\theta, \quad y = g(z)\sin\theta, \quad z = z, \quad z_1 < z < z_2,\ 0 \le \theta \le 2\pi, \tag{1'}$$

where $g(z)$ is a smooth function of z such that $g(z) > 0$.

In the first case the arc length of curves $\mathscr{C}$ on S described by $r = r(\theta)$, $\theta_1 \le \theta \le \theta_2$, is given by

$$L(\mathscr{C}) = \int_{\theta_1}^{\theta_2} \sqrt{r^2 + [1 + f'(r)^2]\dot{r}^2}\, d\theta, \tag{2}$$

while in the second case the length of curves $\mathscr{C}$ on S described by $z = z(\theta)$, $\theta_1 \le \theta \le \theta_2$, is to be computed from

$$L(\mathscr{C}) = \int_{\theta_1}^{\theta_2} \sqrt{g(z)^2 + [1 + g'(z)^2]\dot{z}^2}\, d\theta, \tag{2'}$$

where $\cdot = d/d\theta$, $f' = df/dr$, $g' = dg/dz$. According to 2.2 [7], the extremals of (2) of the form $r = r(\theta)$ satisfy

$$\frac{r^2}{\sqrt{r^2 + [1 + f'(r)^2]r^2}} = \text{const}, \tag{3}$$

whereas the extremals $r = z(\theta)$ of (2′) are solutions of

$$\frac{g(z)^2}{\sqrt{g(z)^2 + [1 + g'(z)^2]\dot{z}^2}} = \text{const}. \tag{3'}$$

In the first case we introduce the arc-length function $s(\theta)$ by

$$\dot{s} = \sqrt{r^2 + [1 + f'(r)^2]\dot{r}^2}, \quad s(\theta_1) = 0, \tag{4}$$

and then we write the extremal $r = r(\theta)$, $\theta_1 \le \theta \le \theta_2$, as

$$\theta = \theta(s), \quad r = r(s), \qquad 0 \le s \le L,$$

which leads to the representation

$$X(s) = (r(s)\cos\theta(s), r(s)\sin\theta(s), f(r(s))).$$

Its unit tangent vector $\dot{X}$ is given by

$$\frac{dX}{ds} = (\cos\theta, \sin\theta, f'(r))\frac{dr}{ds} + (-\sin\theta, \cos\theta, 0)r\frac{d\theta}{ds}.$$

A circle of latitude $c(\theta) = (r\cos\theta, r\sin\theta, f(r))$ satisfies

$$\dot{c} = (-r\sin\theta, r\cos\theta, 0), \quad |\dot{c}| = r.$$

Hence we obtain

(5) $$\frac{dX}{ds}(s)\cdot\frac{dc}{d\theta}(\theta) = r^2(s)\frac{d\theta}{ds}(s)$$

if $\theta = \theta(s)$ and $r = r(s)$. On the other hand we have

(6) $$\frac{dX}{ds}(s)\cdot\frac{dc}{d\theta}(\theta) = r(s)\cos\beta(\theta),$$

where $\beta(\theta)$ denotes the inclination angle of the geodesic $\mathscr{C}$ with respect to the circle of latitude c. Furthermore, on account of (4) the conservation law (3) can be written as

(7) $$r^2\frac{d\theta}{ds} = \text{const}.$$

Then we infer from (5)–(7) that each extremal $r = r(\theta)$ with the inclination angle $\beta(\theta)$ with respect to its circle of latitude satisfies

(8) $$r\cos\beta = \text{const} \quad (:= k).$$

The corresponding inclination angle $\alpha = \alpha(\theta)$ of $X(s)$ for $s = s(\theta)$ with respect to its meridian

$$m(r) = (r\cos\theta, r\sin\theta, f(r)), \quad r_1 < r < r_2,$$

at $r = r(\theta)$ is related to $\beta(\theta)$ by $\alpha = \pi/2 - \beta$, whence relation (8) becomes

(9) $$r\sin\alpha = k.$$

In case 2 we derive the same relation for all extremals of the form $z = z(\theta)$ on account of the conservation law (3′), introducing $s(\theta)$ by

(4′) $$\dot{s} = \sqrt{g(z)^2 + [1 + g'(z)^2]\dot{z}^2}, \quad s(\theta_1) = 0.$$

Now the radius r in (9) is to be taken as $g(z(\theta))$, and the meridians are described by

$$m(z) = (g(z)\cos\theta, g(z)\sin\theta, z), \quad z_1 < z < z_2.$$

As we have noted in *2.2* [7], the conservation laws (3) and (3′) respectively may have more solutions than the Euler equations of (2) or (2′). Hence Clairaut's equation (9) is a necessary but not sufficient condition for the extremals of (2) or (2′). A straight-forward computation shows the following:

(i) A solution $r = r(\theta)$, $\theta_1 \le \theta \le \theta_2$, of $r(\theta)\sin\alpha(\theta) \equiv \text{const}$ defines a geodesic arc on a part (1) of S provided that $\dot{r}(\theta) \not\equiv 0$ on any open subinterval of $[\theta_1, \theta_2]$. However, no circle (i.e. no solution $r(\theta) \equiv \text{const}$) can be a geodesic on a part (1) of S.

(ii) Consider a smooth function $z = z(\theta)$, $\theta_1 \le \theta \le \theta_2$, such that $r(\theta) := g(z(\theta))$ is a solution of $r(\theta)\sin\alpha(\theta) \equiv \text{const}$. Then $z(\theta)$ defines a geodesic on a part (1′) of S if $\dot{z}(\theta) \not\equiv 0$ on any open subinterval of $[\theta_1, \theta_2]$. A circle defined by $z(\theta) \equiv z_0$, i.e. $r(\theta) \equiv g(z_0)$, is a geodesic on a part (1′) of S if and only if $g'(z_0) = 0$.

The latter result means that *all latitude circles of extremal radius are closed geodesics on S.*

Because of $r > 0$ on S we infer that the constant $k = 0$ in (9) corresponds to the meridians. Thus we infer that *all meridians of S are geodesics.* Suppose now that $k \neq 0$, say, $k > 0$. Then (7) implies $d\theta/ds > 0$, and (9) yields $r \geq k > 0$ on the entire geodesic.

These observations allow us to discuss the global behaviour of geodesics on S. To simplify matters we consider a geodesic $X(s) = (x(s), y(s), z(s))$ which is defined for all $s \geq s_0$ and dives into the halfspace $\mathscr{H}_0 := \{z \geq z_0\}$ for $s = s_0$, i.e. $z(s_0) = z_0$ and $(dz/ds)(s_0) > 0$. We assume that $S \cap \mathscr{H}_0$ can be represented in the form

$$x = g(z) \cos \theta, \quad y = g(z) \sin \theta, \quad z = z, \qquad z_0 \leq z < \bar{z} \leq \infty, \quad 0 \leq \theta \leq 2\pi,$$

where $g(z)$ is a smooth function satisfying $g(z) > 0$, and $\lim_{z \to \bar{z}-0} g(z) = \infty$ if $\bar{z} < \infty$. Denote the polar coordinates of $X(s)$ by $r(s)$, $\theta(s)$, $z(s)$, $r(s) = g(z(s))$, and let

$$k := \inf\{r(s) \sin \alpha(s) : s_0 \leq s < \infty\} > 0 .$$

Then we distinguish two cases:

(i) If $r(s) > k$ for all $s \geq s_0$, then $0 < \sin \alpha(s) < 1$ and therefore $0 < \alpha(s) < \pi/2$ for all $s \geq s_0$, which means that $dz/ds > 0$ on $[s_0, \infty)$, i.e. $z(s)$ is strictly increasing. Thus the geodesic "never turns back". Let

$$z_1 := \lim_{s \to \infty} z(s) .$$

If $z_1 = \bar{z}$, the geodesic disappears towards infinity. On the other hand, if $z_1 < \bar{z}$ and $r_1 := g(z_1)$, it follows that $z(s) \to z_1$ and $r(s) \to r_1$ as $s \to \infty$, and $z(s) < z(s') < z_1$ for all $s, s' \in (s_0, \infty)$ satisfying $s < s'$. That is, the geodesic $X(s)$ is a spiralling curve tending to limit circle $C(r_1) := \{(x, y, z): x^2 + y^2 = r_1^2, r = z_1\}$ as $s \to \infty$ without ever reaching it. More precisely we infer from $|dX/ds| = 1$ that

$$\left(\frac{dz}{ds}\right)^2 = [1 + g'(z)^2]^{-1} \cdot \left\{1 - r^2 \left(\frac{d\theta}{ds}\right)^2\right\}, \tag{10}$$

and $r^2\, d\theta/ds = k$ yields

$$\frac{d\theta}{ds}(s) \to \frac{k}{r_1^2} \quad \text{as } s \to \infty, \quad \text{where } r_1 \geq k .$$

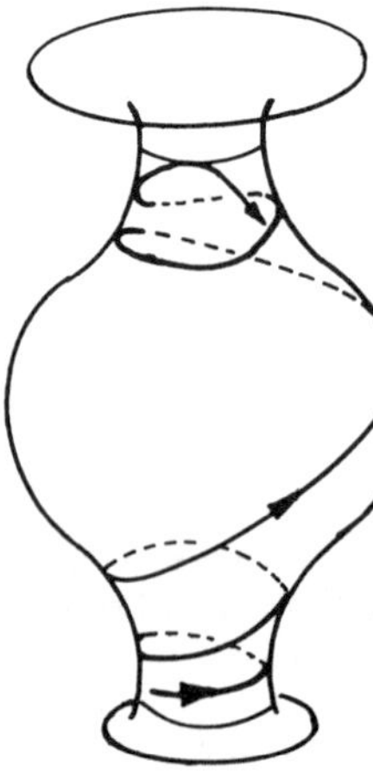

Fig. 11. Geodesics on a surface of revolution.

Thus we obtain

$$\lim_{s\to\infty} \frac{dz}{ds}(s) = \sqrt{\frac{1-(k/r_1)^2}{1+g'(z_1)^2}}.$$

Since $z_1 = \lim_{s\to\infty} z(s) < \infty$ it follows that $\lim_{s\to\infty} \frac{dz}{ds}(s) = 0$, and therefore $k = r_1$ and $\lim_{s\to\infty}\left(\frac{d\theta}{ds}(s)\right)^2 = 1/r_1^2$.

(ii) Suppose now that there is some $s_1 \in (s_0, \infty)$ such that $r(s_1) = k$ and $r(s) > k$ for $s_0 \le s < s_1$. Then we infer from $r^2\, d\theta/ds = k$ that $(d\theta/ds)(s) \to 1/k$ as $s \to s_1$. On account of (10) we then obtain

$$\lim_{s\to s_1} \frac{dz}{ds}(s) = 0 \quad \text{and} \quad \lim_{s\to s_1} \frac{dr}{ds}(s) = 0,$$

and $r \sin\alpha = k$ implies

$$\lim_{s\to s_1} \alpha(s) = \pi/2,$$

i.e. the geodesic $X(s)$ is tangent to the latitude circle C of S passing through the point $P_1 := X(s_1)$. Note that C cannot be a closed geodesic of S because in this case C and $X(s)$ have to coincide, which is impossible. Consequently we have $g'(z) \neq 0$ for $z_1 := z(s_1)$, according to our previous discussion. Furthermore, since $r(s) = g(z(s)) > k$ for $s_0 \le s < s_1$ and $r(s) \to k$ as $s \to s_1$ we infer that $g'(z_1) < 0$ whence $k = g(z_1) > g(z)$ for $z_1 < z < z_1 + \varepsilon$ and $0 < \varepsilon \ll 1$. Since $k = \inf\{r(s)\colon s \ge s_0\}$ we see that the geodesic cannot cross the circle C but has to bend back, i.e. $X(s)$ is reflected and returns to the region at the left of C. Then $X(s)$ may either disappear towards infinity in direction of the negative z-axis, or else it will be reflected by another circle C', etc. If $X(s)$, $-\infty < s < \infty$, is a nonclosed geodesic oscillating between two latitude circles C and C', it can be proved that $X(s)$ lies dense in the part of S bounded by C and C'. It is quite remarkable that one can obtain so much information on the global behaviour of geodesics on a surface of revolution without any detailed computation, just by inspecting Clairaut's equation (9) or its equivalent (7). It seems that (9) was essentially already known to Jacob Bernoulli.

15. Geodesics on a cylinder or a cone. If S is a circular cylinder, then the discussion of nr. *14* shows that the latitude circles are the closed geodesics on S, and that the meridians (i.e. the generating lines of S) are geodesics. All the other geodesics have a constant angle of inclination α with respect to the meridians, $0 < |\alpha| < \pi/2$. These curves are the *helices* on S, which may be right- or left-handed.

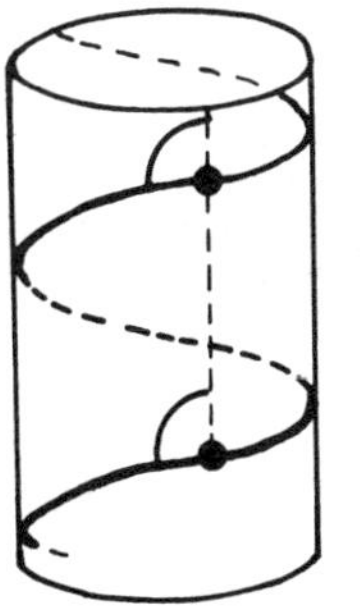

Fig. 12. Geodesics on a cylinder and Clairaut's theorem.

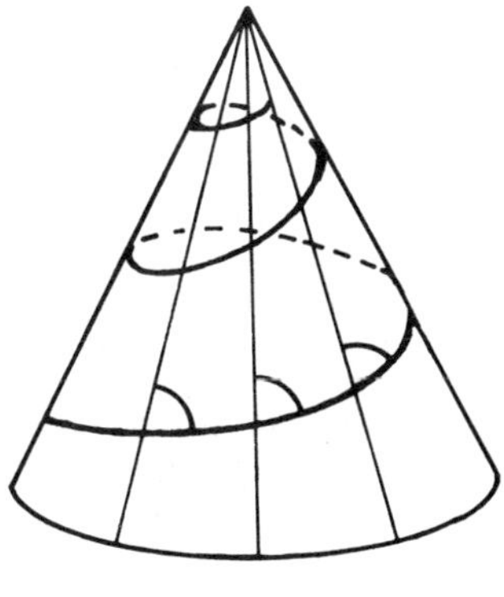

Fig. 13. Geodesics on a cone and Clairaut's theorem.

On a circular cone S there are no closed geodesics, but all meridians (= generating lines of S) are geodesics. Any other geodesic comes from infinity, reaches some latitude circle C of radius $r_0 > 0$ where it turns back and returns to infinity.

16. Euler's treatment of the elastic line. The study of elastic materials started with Galilei (*Discorsi*, 1638) and Hooke (*Lectures de potentia restitutiva, or of spring explaining the power of springing bodies*, 1678), and the latter formulated the basic law of linear elasticity: *ut tensio sic vis.* The bending of elastic beams was treated by Mariotte (*Traité du mouvement des eaux et des autres fluides*, 1686) and Leibniz (*Acta Eruditorum*, 1684), and Jacob Bernoulli was the first to describe the form of a clamped elastic strip in equilibrium (*Acta Eruditorum*, 1694). For a penetrating historical survey we refer to Truesdell, *The rational mechanics of flexible or elastic bodies* (in: Euler, *Opera omnia*, ser. II, Vol. 11_2). Following a suggestion by Daniel Bernoulli, Euler gave the first treatment of elastic lines by means of the calculus of variations in the *Additamentum I* to his *Methodus inveniendi* (1744, pp. 245–310), which carries the title *De curvis elasticis.* Euler characterized the equilibrium position of an elastic line by the following variational principle: *Among all curves of equal length, joining two points where they have prescribed tangents, to determine that which minimizes the value of the expression* $\int ds/\rho^2$.

In other words, Euler interpreted an ideal elastic line as an inextensible curve $\mathscr{C}$ with a "potential energy" of $\int \kappa^2\, ds$, κ being the curvature function of $\mathscr{C}$, whose positions of (stable) equilibrium are characterized by the minima of the potential energy, i.e. by Johann Bernoulli's *principle of virtual work.*[20] Thus the problem of the elastic line leads to the isoperimetric problem

$$\int_{\mathscr{C}} \kappa^2\, ds \to \min \quad \text{with} \quad \int_{\mathscr{C}} ds = L,$$

where any representation $c : [a, b] \to \mathbb{R}^2$ of $\mathscr{C}$ has to satisfy the boundary conditions

$$c(a) = P_1, \qquad c(b) = P_2, \qquad \frac{\dot c(a)}{|\dot c(a)|} = \mathfrak{t}_1, \qquad \frac{\dot c(b)}{|\dot c(b)|} = \mathfrak{t}_2 .$$

By using a (constant) multiplier λ we are led to the task to determine the extremals of the functional

$$\int_{\mathscr{C}} (\kappa^2 + \lambda)\, ds .$$

This problem we have treated in 1,5 $\boxed{5}$, and in 1,6. Euler approached the problem by writing the integral $\int \kappa^2\, ds$ in the nonparametric form

$$\int \frac{y''^2}{(1 + y'^2)^{5/2}}\, dx,$$

where $\mathscr{C}$ is described by $y = y(x)$ and the subsidiary condition $\int ds = L$ becomes

$$\int \sqrt{1 + y'^2}\, dx = L .$$

He described the solutions in terms of elliptic integrals which he computed by series expansions. This way Euler found nine species of elastic lines in the plane, four of which are depicted in Fig. 14.

There is an extensive literature on the *linea elastica.* We only mention the thesis of Born [1], Göttingen 1909, where the stability of the elastic line is investigated. Further references can be found in Funk's treatise [1] and in the literature on elasticity theory. In $\mathbb{R}^3$ and, more generally, in spaces of constant curvature the extremals of the integral $\int_{\mathscr{C}} (\kappa^2 + \lambda)\, ds$ were in detail discussed by Bryant–Griffiths [1] and by Langer–Singer [1, 2]. The latter authors proved that there exist infinitely many

[20] This principle was apparently first stated in a letter by Johann Bernoulli to Varignon, written January 26, 1717. It was published in Varignon's *Nouvelle Mécanique*, Vol. 2, p. 174, in 1725.

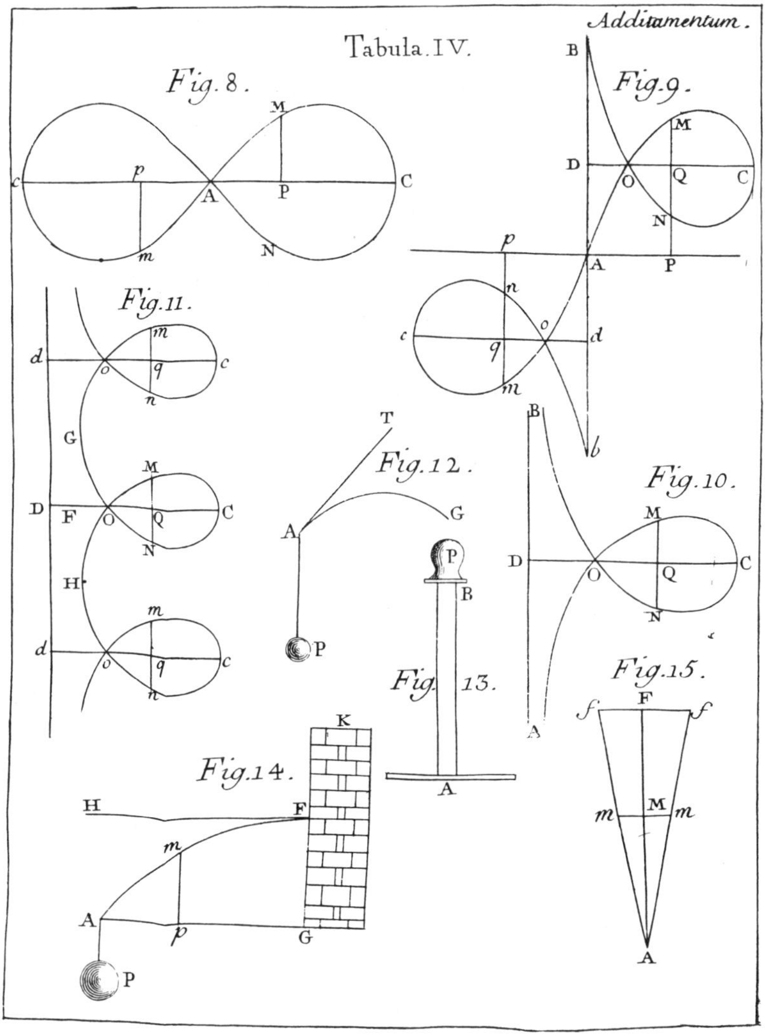

Fig. 14. Elasticae as depicted by Euler, *Methodus inveniendi*, table IV.

closed embedded elasticae in $\mathbb{R}^3$ which are knotted and lie on embedded tori of revolution; the knot types are torus knots.

Another approach to the Euler equation of the elastic line is to formulate the variational principle as a Lagrange problem (cf. Section 3, and also nr. *12* of these Scholia). Introducing the angle $\omega(s)$ by

$$\omega(s) = \alpha + \int_0^s \kappa \, ds,$$

the elastic energy of a line $c(s) = (x(s), y(s))$, $|\dot{c}(s)| = 1$, becomes

$$\int_0^L \dot{\omega}(s)^2 \, ds,$$

and the coordinate functions of $c(s)$ are related to $\omega(s)$ by the two nonholonomic subsidiary conditions

$$\dot{x}(s) - \cos \omega(s) = 0, \qquad \dot{y}(s) - \sin \omega(s) = 0.$$

Thus we are led to the variational problem

$$\int_0^L \{\dot{\omega}^2 + \lambda_1(\dot{x} - \cos \omega) + \lambda_2(\dot{y} - \sin \omega)\}\, ds \to \text{stat}.$$

for the unknown functions $\omega(s)$, $x(s)$, $y(s)$, and the Euler equations of this functional are

$$\ddot{\omega} - 2\lambda_1 \sin \omega + 2\lambda_2 \cos \omega = 0,$$

$$\dot{\lambda}_1 = 0, \qquad \dot{\lambda}_2 = 0.$$

Hence the multiplier functions $\lambda_1(s)$ and $\lambda_2(s)$ are constants, and by a suitable rotation of the coordinate system we can achieve that $\lambda_2 = 0$. Introducing $a^2 := -2\lambda_1$, the Euler equation becomes

$$\ddot{\omega} + a^2 \sin \omega = 0,$$

which is just the pendulum equation. We obtain

$$\sin \frac{\omega(s)}{2} = k \operatorname{sn}(as)$$

where $k = \sin \frac{\gamma}{2}$ and γ is the maximum of $\omega(s)$ and sn u is Jacobi's doubly periodic function sin am u. Introducing $\ell := L/2$, we have

$$K := \int_0^{\pi/2} \frac{d\varphi}{\sqrt{1 - k^2 \sin^2 \varphi}} = a\ell.$$

Then we can express $c(s)$ in terms of sn and of Jacobi's theta-function $\vartheta_2(v)$, c.f. e.g. Dörrie [1], pp. 521–522.

17. Delaunay's variational problem from 1842 is the task to connect two given line elements of $\mathbb{R}^3$ by a longest or shortest space curve of constant curvature. This very interesting problem is another Lagrange problem whose Euler equations were first integrated by Weierstrass in 1884 in terms of elliptic functions (cf. *Werke* [1], Vol. 3, pp. 183–218). The true minimum or maximum problem was studied by H.A. Schwarz (unpublished results) and Venske [1] in a thesis inspired by Schwarz (Göttingen, 1891). A careful discussion of these results and a historical survey can be found in Carathéodory [16], Vol. 2, pp. 12–39. Moreover, in his book [10] Carathéodory presented a treatment of Delaunay's problem via Hamiltonian formalism (cf. pp. 382–388 and 400) which had been found by Josepha v. Schwarz [1].

The more general problem of *finding extrema of the length functional among curves of bounded curvature* was studied by H.A. Schwarz, E. Schmidt, and Carathéodory; cf. Blaschke [5], Vol. 1, Sections 31–32, and Carathéodory [16], Vol. 2, pp. 65–92.

Chapter 3. General Variational Formulas

In the first two chapters we have investigated the rate of change of variational integrals $\mathscr{F}$ with respect to variations of the dependent variables. In particular, we have derived the necessary conditions

$$\delta\mathscr{F}(u, \varphi) = 0 \quad \text{for all } \varphi \in C_c^\infty(\Omega, \mathbb{R}^N)$$

and

$$L_F(u) = 0$$

for extremizers u of $\mathscr{F}$, and suitable modifications of these conditions were shown to hold for constrained extremizers.

There are as good reasons to consider the variation of $\mathscr{F}$ with respect to changes of the independent variables. In Section 1, we shall consider one-parameter families of diffeomorphisms $y \to x = \xi(y, \varepsilon)$ from Ω onto itself which reduce to the identity on Ω for $\varepsilon = 0$. For a minimizer u of

$$\mathscr{F}(u) = \int_\Omega F(x, u(x), Du(x))\, dx,$$

we infer that

$$\frac{d}{d\varepsilon}\int_\Omega F(y, u(\xi(y, \varepsilon)), D_y u(\xi(y, \varepsilon)))\, dy\bigg|_{\varepsilon=0} = 0.$$

This leads to a new type of necessary conditions for extremizers which, in integrated form, are written as

$$\partial\mathscr{F}(u, \lambda) = 0,$$

λ being the velocity field of $\xi(\cdot, \varepsilon)$ at $\varepsilon = 0$, that is,

$$\lambda(y) = \frac{\partial}{\partial\varepsilon}\xi(y, \varepsilon)\bigg|_{\varepsilon=0}.$$

Since every vector field $\lambda \in C_c^\infty(\Omega, \mathbb{R}^n)$ generates a family of diffeomorphisms $\xi(y, \varepsilon) := y + \varepsilon\lambda(y)$ from Ω onto itself, it follows that

$$(1) \qquad \partial\mathscr{F}(u, \lambda) = 0 \quad \text{for all } \lambda \in C_c^\infty(\Omega, \mathbb{R}^n).$$

The linear functional $\partial\mathscr{F}$ is called the *inner variation* of $\mathscr{F}$; it is to be seen as the counterpart of the first variation $\delta\mathscr{F}$. Correspondingly we define the *inner extremals* of $\mathscr{F}$ as counterparts to the extremals by requiring (1). Applying a partial integration to (1), we infer by means of the fundamental lemma the system of partial differential equations

$$D_\beta T_\alpha^\beta + F_{x^\alpha} = 0 \tag{2}$$

for u, denoted as *Noether equations*. Here

$$T_\alpha^\beta = F_{p_\beta^i} u_{x^\alpha}^i - F\delta_\alpha^\beta$$

is the so-called *Hamilton tensor* (or: *energy–momentum tensor*). The Noether equations (2) are the counterpart to the Euler equations $L_F(u) = 0$ that follow from

$$\delta\mathscr{F}(u, \varphi) = 0 \quad \text{for all } \varphi \in C_c^\infty(\Omega, \mathbb{R}^N). \tag{3}$$

In Section 2 we shall consider variations of $\mathscr{F}$ with respect to general changes $\xi(\cdot, \varepsilon): \Omega_\varepsilon \to \Omega$ of the independent variables. The corresponding "extremals" will be called *strong inner extremals*. As we shall see, they are solutions to the Noether equations (2) satisfying the free boundary conditions

$$T_\alpha^\beta \nu_\beta = 0 \quad \text{on } \partial\Omega, \; 1 \le \alpha \le n, \tag{4}$$

where $\nu = (\nu_1, \ldots, \nu_n)$ is the exterior normal to $\partial\Omega$. This corresponds to the natural boundary condition

$$F_{p_\alpha^i} \nu_\alpha = 0 \quad \text{on } \partial\Omega, \; 1 \le i \le N, \tag{5}$$

that was derived from

$$\delta\mathscr{F}(u, \varphi) = 0 \quad \text{for all } \varphi \in C^1(\bar{\Omega}, \mathbb{R}^N).$$

We shall also see how the property of being a strong inner extremal is related for $n = 2$ to the so-called *conformality relations* which form the basis for the study of two-dimensional minimal surfaces by means of harmonic functions.

In Section 3 we investigate the variational behaviour of integrals $\mathscr{F}(u, \Omega)$ if both the mapping u and the independent variables are varied. In particular, also the domain of integration Ω will be modified. The resulting *variational formula* for $\mathscr{F}$ has various interesting applications, the most important of which is *Emmy Noether's theorem* which is discussed in Section 4. It states that the extremals of some variational integral will satisfy certain *conservation laws* if the integral remains invariant under the action of a continuous transformation group. In fact, *infinitesimal invariance* suffices to establish such conservation laws, and it is irrelevant that the variation is generated by a group; the reasoning works in the same way for very general types of variations.

Finally, in Section 5, we shall investigate the transformation behaviour of the Euler operator if both the dependent and the independent variables are transformed in a general way. In particular we obtain that the Euler expression $L_F(u)$ can be viewed as a covector field along the mapping $u: \Omega \to \mathbb{R}^N$.

1. Inner Variations and Inner Extremals. Noether Equations

Consider a variational integral

$$\mathscr{F}(u) = \int_\Omega F(x, u(x), Du(x))\, dx,$$

the Lagrangian $F(x, z, p)$ of which is supposed to be of class C^1 on $\bar{\Omega} \times \mathbb{R}^N \times \mathbb{R}^{nN}$, $\Omega \subset \mathbb{R}^n$. In Chapter 1 we have seen that for every weak minimizer $u \in C^1(\bar{\Omega}, \mathbb{R}^N)$ of $\mathscr{F}$ (with respect to fixed boundary conditions), the first variation of $\mathscr{F}$ at u in direction of every C^∞-function with compact support in Ω has to vanish, i.e.,

(1) $$\delta\mathscr{F}(u, \varphi) = 0 \quad \text{for all } \varphi \in C_c^\infty(\Omega, \mathbb{R}^N).$$

This was proved by embedding the minimizer $u(x)$ into a family of comparison functions $\psi(x, \varepsilon)$ with

$$\psi(x, 0) = u(x), \quad \frac{\partial\psi}{\partial\varepsilon}(x, 0) = \varphi(x) \qquad \text{for } x \in \Omega,$$

$$\psi(x, \varepsilon) = u(x) \quad \text{for } x \in \partial\Omega \text{ and } |\varepsilon| < \varepsilon_0,$$

and by noticing that the function $\Phi(\varepsilon) := \mathscr{F}(\psi(\varepsilon))$ satisfies $\Phi(\varepsilon) \geq \Phi(0)$ for $|\varepsilon| < \varepsilon_0$, whence $\Phi'(0) = 0$ follows; but this is equivalent to relation (1). In other words, condition (1) follows from a suitable *variation of the dependent variables.*

We now want to show that a similar relation can be obtained from *variations of the independent variables.* To make this precise, we choose an arbitrary vector field $\lambda \in C_c^\infty(\Omega, \mathbb{R}^n)$ with components $\lambda^1, \ldots, \lambda^n$. Then we construct a family $\xi(\varepsilon) : \bar{\Omega} \to \bar{\Omega}$ of diffeomorphisms $x = \xi(y, \varepsilon)$ of $\bar{\Omega}$ onto itself depending smoothly on the parameter $\varepsilon \in (-\varepsilon_0, \varepsilon_0)$, $\varepsilon_0 > 0$, which satisfy

(2) $$\begin{gathered}\xi(x, 0) = x, \quad \frac{\partial}{\partial\varepsilon}\xi(x, 0) = \lambda(x) \quad \text{for } x \in \bar{\Omega},\\ \xi(x, \varepsilon) = x \quad \text{for } x \in \partial\Omega \text{ and } |\varepsilon| < \varepsilon_0,\end{gathered}$$

or, equivalently,

(3) $$\xi(y, \varepsilon) = y + \varepsilon\lambda(y) + o(\varepsilon) \quad \text{for } y \in \bar{\Omega}, \ \xi(y, \varepsilon) = y \text{ for } y \in \partial\Omega.$$

Such a family is, for example, given by

(3′) $$\xi(y, \varepsilon) := y + \varepsilon\lambda(y), \quad y \in \bar{\Omega}, \ |\varepsilon| < \varepsilon_0,$$

if $\varepsilon_0 > 0$ is a sufficiently small number. In fact, there is a number $K > 0$ such that

$$|\lambda(y_1) - \lambda(y_2)| \leq K|y_1 - y_2| \quad \text{for all } y_1, y_2 \in \bar{\Omega},$$

whence

$$|\xi(y_1, \varepsilon) - \xi(y_2, \varepsilon)| \geq |y_1 - y_2| - |\varepsilon||\lambda(y_1) - \lambda(y_2)|$$
$$\geq (1 - \varepsilon K)|y_1 - y_2| \geq \tfrac{1}{2}|y_1 - y_2|$$

for all $y_1, y_2 \in \bar{\Omega}$, provided that $|\varepsilon| \leq \varepsilon_0 \leq (2K)^{-1}$. In conjunction with the implicit function theorem we infer that $\xi(\cdot, \varepsilon)$ is a C^∞-diffeomorphism if $|\varepsilon| \leq \varepsilon_0$, and $\lambda \in C_c^\infty(\Omega, \mathbb{R}^n)$ implies that $\xi(y, \varepsilon) = y$ for all y sufficiently close to $\partial\Omega$. Therefore we have

$$\xi(\Omega, \varepsilon) = \Omega \quad \text{for } |\varepsilon| \ll 1$$

and we infer that $\xi(\cdot, \varepsilon)$ furnishes a C^∞-diffeomorphism of Ω onto itself. The inverse mapping $\eta(\varepsilon)$ of (3) or (3') can be written as

$$\eta(x, \varepsilon) = x - \varepsilon\lambda(x) + o(\varepsilon), \tag{4}$$

whence

$$\eta_x(x, \varepsilon) = I - \varepsilon\lambda_x(x) + o(\varepsilon),$$

that is

$$\frac{\partial \eta^\alpha}{\partial x^\beta}(x, \varepsilon) = \delta_\beta^\alpha - \varepsilon\lambda_{x^\beta}^\alpha(x) + o(\varepsilon),$$

and in particular

$$\frac{\partial}{\partial \varepsilon} \det\left(\frac{\partial \eta^\alpha}{\partial x^\beta}\right)\Bigg|_{\varepsilon=0} = -\lambda_{x^\alpha}^\alpha = -\operatorname{div} \lambda.$$

Actually, this result can be rephrased as follows: *If $A(t)$ is a square-matrix valued function of class C^1 such that $A(t_0) = I$, then*

$$(\det A)'(t_0) = (\operatorname{trace} A')(t_0), \quad \textit{where } ' = \frac{d}{dt}.$$

Now we consider the family of comparison functions $v(\varepsilon) := u \circ \xi(\varepsilon)$ which are given by

$$v(y, \varepsilon) = u(\xi(y, \varepsilon)) = u(y + \varepsilon\lambda(y) + o(\varepsilon)) \tag{5}$$

for $y \in \bar{\Omega}$ and $|\varepsilon| < \varepsilon_0$. Then we define

$$\Phi(\varepsilon) := \mathscr{F}(v(\varepsilon)) = \int_\Omega F(y, v(y, \varepsilon), Dv(y, \varepsilon))\, dy.$$

If $u(x)$ is a weak minimizer of $\mathscr{F}$ (with respect to fixed boundary values), it follows that

$$\Phi(\varepsilon) \geq \Phi(0) \quad \text{for } |\varepsilon| < \varepsilon_0$$

and therefore

$$\Phi'(0) = 0.$$

Definition 1. *The expression*

$$\partial\mathcal{F}(u, \lambda) := \Phi'(0) = \frac{d}{d\varepsilon}\mathcal{F}(u \circ \xi(\varepsilon))\bigg|_{\varepsilon=0}$$

will be called the (first) inner variation *of the functional $\mathcal{F}$ at u in direction of the vector field $\lambda \in C_c^\infty(\Omega, \mathbb{R}^n)$.*

Since we shall not consider any inner variations of higher order, we will simply speak of *the inner variation of $\mathcal{F}$.*

Lemma 1. *For $u \in C^1(\bar{\Omega}, \mathbb{R}^N)$ and $\lambda \in C_c^\infty(\Omega, \mathbb{R}^n)$, the inner variation of $\mathcal{F}$ is given by*

(6) $$\partial\mathcal{F}(u, \lambda) = \int_\Omega (F_{p^i_\beta} u^i_{x^\alpha}\lambda^\alpha_{x^\beta} - F\lambda^\alpha_{x^\alpha} - F_{x^\alpha}\lambda^\alpha)\, dx,$$

where the arguments of F, F_{x^α}, $F_{p^i_\beta}$ are to be taken as $(x, u(x), Du(x))$.

Proof. Because of $v_y(y, \varepsilon) = u_x(\xi(y, \varepsilon)) \cdot \xi_y(y, \varepsilon)$ (or $v^i_{y^\alpha} = u^i_{x^\beta}(\xi)\xi^\beta_{y^\alpha}$), we obtain that

$$\Phi(\varepsilon) = \int_\Omega F(y, u(\xi(y, \varepsilon)), Du(\xi(y, \varepsilon)) \cdot D\xi(y, \varepsilon))\, dy.$$

Note that $\det D\eta(x, \varepsilon)$ is close to 1 for $|\varepsilon| \ll 1$. The transformation theorem for multiple integrals yields

$$\Phi(\varepsilon) = \int_\Omega F(\eta(x, \varepsilon), u(x), Du(x) \cdot \xi_y(\eta(x, \varepsilon), \varepsilon)) \det D\eta(x, \varepsilon)\, dx.$$

On account of

$$\frac{\partial\eta}{\partial\varepsilon}(x, 0) = -\lambda(x), \qquad \frac{\partial}{\partial\varepsilon}\det \eta_x\bigg|_{\varepsilon=0} = -\lambda^\alpha_{x^\alpha},$$

$$\xi_y(x, 0) = 1, \qquad \frac{\partial}{\partial\varepsilon}\{\xi_y(\eta(x, \varepsilon), \varepsilon)\}\bigg|_{\varepsilon=0} = D\lambda(x),$$

we infer that

$$\Phi'(0) = \int_\Omega \{F_{x^\alpha} \cdot (-\lambda^\alpha) + F_{p^i_\beta} u^i_{x^\alpha}\lambda^\alpha_{x^\beta} + F \cdot (-\lambda^\alpha_{x^\alpha})\}\, dx,$$

where F_{x^α}, $F_{p^i_\beta}$ and F have the argument $(x, u(x), Du(x))$. □

In analogy to formulas (3) and (4) of 1,*2.1* we introduce the expression $\partial F(u, \lambda)$ by

(7) $$\partial F(u, \lambda)(x) := F_{p^i_\beta} u^i_{x^\alpha}(x)\lambda^\alpha_{x^\beta}(x) - F\lambda^\alpha_{x^\alpha}(x) - F_{x^\alpha}\lambda^\alpha(x),$$

where the arguments of F, F_{x^α} and $F_{p^i_\beta}$ are to be taken as $(x, u(x), Du(x))$. Then we can write

$$\partial\mathscr{F}(u, \lambda) = \int_\Omega \partial F(u, \lambda)\, dx.$$

Let us introduce the *Hamilton tensor* (or *energy–momentum tensor*)

$$T(x, z, p) = (T^\beta_\alpha(x, z, p))$$

by

$$T^\beta_\alpha := p^i_\alpha F_{p^i_\beta} - \delta^\beta_\alpha F, \tag{8}$$

where the arguments of T^β_α, $F_{p^i_\beta}$ and F are to be taken as $(x, z, p) \in \bar{\Omega} \times \mathbb{R}^N \times \mathbb{R}^{nN}$. Then we can write formulas (6) and (7) as

$$\partial\mathscr{F}(u, \lambda) = \int_\Omega [T^\beta_\alpha(x, u, Du)\lambda^\alpha_{x^\beta} - F_{x^\alpha}(x, u, Du)\lambda^\alpha]\, dx \tag{6'}$$

and

$$\partial F(u, \lambda) = T^\beta_\alpha(\cdot, u, Du)\lambda^\alpha_{x^\beta} - F_{x^\alpha}(\cdot, u, Du)\lambda^\alpha. \tag{7'}$$

Definition 2. *A mapping $u \in C^1(\bar{\Omega}, \mathbb{R}^N)$ is said to be an* inner extremal *of $\mathscr{F}$ if*

$$\partial\mathscr{F}(u, \lambda) = 0 \tag{9}$$

holds for all vector fields $\lambda \in C^\infty_c(\Omega, \mathbb{R}^n)$.

Obviously, the notion of an *inner extremal* of $\mathscr{F}$ is analogously coined to that of a *weak extremal* which, by definition, is a mapping u of class $C^1(\bar{\Omega}, \mathbb{R}^N)$ satisfying

$$\delta\mathscr{F}(u, \varphi) = 0 \quad \text{for all } \varphi \in C^\infty_c(\Omega, \mathbb{R}^N), \tag{9'}$$

whereas *extremals* u of $\mathscr{F}$ are, by definition, of class $C^1(\bar{\Omega}, \mathbb{R}^N) \cap C^2(\Omega, \mathbb{R}^N)$ and satisfy (9′) or, equivalently, $L_F(u) = 0$ in Ω.

An inspection shows that the notions $\delta\mathscr{F}$ and $\partial\mathscr{F}$ are quite differently defined. For instance, the derivatives F_{x^α} appear in $\partial\mathscr{F}$ but not in $\delta\mathscr{F}$, whereas the derivatives F_{u^i} arise in $\delta\mathscr{F}$ but not in $\partial\mathscr{F}$. Hence δ and ∂ will have different ranges of applicability if we try to lower the regularity assumptions on F. Correspondingly, weak extremals and inner extremals are very different objects which, nevertheless, are closely connected. Before we discuss this connection, we want to derive from (9) a system of partial differential equations characterizing every inner extremal of class C^2.

Theorem 1. *Suppose that $F \in C^2$ (or at least $F_p \in C^1$). Then every inner extremal $u \in C^2(\Omega, \mathbb{R}^N)$ of the functional $\mathscr{F}$ satisfies the equations*

$$D_\beta T^\beta_\alpha(x, u, Du) + F_{x^\alpha}(x, u, Du) = 0, \tag{10}$$

which will be called "Noether equations".[1] *Conversely, every C^2-solution u of* (10) *is an inner extremal of $\mathscr{F}$.*

Proof. Any inner extremal u of $\mathscr{F}$ satisfies

$$\int_\Omega (T_\alpha^\beta(x, u, Du) D_\beta \lambda^\alpha - F_{x^\alpha}(x, u, Du)\lambda^\alpha)\, dx = 0$$

for all $\lambda \in C_c^\infty(\Omega, \mathbb{R}^n)$. If u is of class $C^2(\Omega, \mathbb{R}^N)$, we obtain by a partial integration that

$$-\int_\Omega [D_\beta T_\alpha^\beta(x, u, Du) + F_{x^\alpha}(x, u, Du)]\lambda^\alpha\, dx = 0,$$

and the fundamental lemma yields (10). The converse is proved by reversing the reasoning. □

The Noether equations (10) are the counterpart to the Euler equations

$$D_\alpha F_{p_\alpha^i}(x, u, Du) - F_{z^i}(x, u, Du) = 0, \quad 1 \le i \le N,$$

which follow from (9′), whereas the Noether equations are a consequence of (9).

We can give the Noether equations a different form using the following lemma.

Lemma 2. *If $F \in C^2$ (or at least $F_p \in C^1$) and $u \in C^2(\Omega, \mathbb{R}^N)$, then*

$$\partial\mathscr{F}(u, \lambda) = -\int_\Omega \{L_F(u) \cdot Du\} \cdot \lambda\, dx$$

holds for all $\lambda \in C_c^\infty(\Omega, \mathbb{R}^n)$ or, in coordinates

$$\partial\mathscr{F}(u, \lambda) = -\int_\Omega \{D_\beta F_{p_\beta^i} - F_{z^i}\} u_{x^\alpha}^i \lambda^\alpha\, dx.$$

Proof. Integrating by parts, we conclude that

$$\begin{aligned} -\partial\mathscr{F}(u, \lambda) &= \int_\Omega \{F_{x^\alpha}\lambda^\alpha + F\lambda_{x^\alpha}^\alpha - F_{p_\beta^i} u_{x^\alpha}^i \lambda_{x^\beta}^\alpha\}\, dx \\ &= \int_\Omega \{F_{x^\alpha} - D_\alpha F + D_\beta(F_{p_\beta^i} u_{x^\alpha}^i)\}\lambda^\alpha\, dx \\ &= \int_\Omega \{D_\beta F_{p_\beta^i} - F_{z^i}\} u_{x^\alpha}^i \lambda^\alpha\, dx. \end{aligned}$$ □

By the fundamental lemma, we infer:

[1] In honour of Emmy Noether; cf. 3, *3* and 3, *4*.

Proposition 1. *A mapping u of class $C^1(\bar{\Omega}, \mathbb{R}^N) \cap C^2(\Omega, \mathbb{R}^N)$ is an inner extremal of $\mathcal{F}$ if and only if it satisfies in Ω the equations*

$$L_F(u) \cdot D_\alpha u = 0, \quad 1 \le \alpha \le n, \tag{11}$$

which are equivalent to

$$\{D_\beta F_{p^i_\beta} - F_{z^i}\} u^i_{x^\alpha} = 0, \quad 1 \le \alpha \le n. \tag{11'}$$

It follows from Theorem 1 and Proposition 1 that *the Noether equations*

$$D_\beta T^\beta_\alpha + F_{x^\alpha} = 0, \quad 1 \le \alpha \le n,$$

can be written in the equivalent form

$$L_F(u) \cdot D_\alpha u = 0, \quad 1 \le \alpha \le n.$$

Hence the Noether equations mean that the Euler expression $L_F(u)$ is perpendicular to the vectors $u_{x^1}, u_{x^2}, \ldots, u_{x^n}$. If $1 \le n \le N$, we obtain by $z = u(x)$ a parameter representation of some n-dimensional surface $\mathcal{S}$ in $\mathbb{R}^N$. Then, if u is an inner extremal of class C^2, Noether's equations express the fact that $L_F(u)$ is a vector field orthogonal to $\mathcal{S}$, i.e., $L_F(u) \perp T_u\mathcal{S}$.

Another consequence of Lemma 2 is the following result

Proposition 2. *Every extremal is also an inner extremal.*

The preceding result does not imply that any weak extremal is necessarily an inner extremal. The problem is caused by the fact that, for the proof of Lemma 2, one needs u to be of class C^2. There are, however, weak extremals which are not of class C^2 (see, for instance, 1,*3.1* [2] and [3]). In this case it is not clear what can be done, and one has to expect difficulties as we can already see from the following example concerning Lipschitz continuous solutions u of the two relations

$$\delta\mathcal{F}(u, \varphi) = 0 \quad \text{for all } \varphi \in C_c^\infty(\Omega, \mathbb{R}^N) \tag{*}$$

and

$$\partial\mathcal{F}(u, \lambda) = 0 \quad \text{for all } \lambda \in C_c^\infty(\Omega, \mathbb{R}^n) \tag{**}$$

for the variational integral

$$\mathcal{F}(u) = \int_0^1 (\dot{u}^2 - 1)^2 \, dt,$$

with the Lagrangian

$$F(p) = (p^2 - 1)^2.$$

According to Du Bois–Reymond's lemma (see 1,*2.3*, Lemma 4), the first equation (*) is equivalent to

$$(\dot{u}^2 - 1)\dot{u} = \text{const a.e.} \quad \text{on } [0, 1].$$

Consequently every Lipschitz function u with the property that $\dot{u}$ only takes the values 1, 0, or -1 is a solution of (*). On the other hand, equation (**) is equivalent to

$$(\dot{u}^2 - 1)^2 - 4(\dot{u}^2 - 1)\dot{u}^2 = \text{const a.e.} \quad \text{on } [0, 1]$$

or to

$$(\dot{u}^2 - 1)(3\dot{u}^2 + 1) = \text{const a.e.} \quad \text{on } [0, 1].$$

Hence (**) has only those Lipschitz functions u with $\dot{u}(t) = \pm 1$ or 0 as solutions which satisfy either $\dot{u}(t) \equiv 0$ or $\dot{u}(t) = \pm 1$. That is, there are Lipschitz continuous solutions of (*) which do not satisfy (**).

Now we want to show that, indeed, *weak extremals need not be inner extremals.* This can be seen from the following example suggested by K. Steffen.

[0] Set $n = 1$ and $N \geq 2$. According to H. Whitney [1] there is a C^1-function $F(p)$ on $\mathbb{R}^N$ and a continuous nonconstant curve $t \mapsto v(t) \in \mathbb{R}^N$, $0 \leq t \leq 1$, such that $F_p(v(t)) \equiv 0$ but $F(v(t)) \not\equiv \text{const}$.

Let us define the functional $\mathscr{F}$ by

$$\mathscr{F}(u) := \int_0^1 F(u'(x))\,dx.$$

Then

$$u(x) := \text{const} + \int_0^x v(t)\,dt, \qquad 0 \leq x \leq 1,$$

is a weak extremal of $\mathscr{F}$ since

$$\int_0^1 F_p(u'(x)) \cdot \varphi(x)\,dx = 0 \qquad \text{for all } \varphi \in C_c^\infty((0, 1), \mathbb{R}^N),$$

but u is not an inner extremal since the integral

$$\partial\mathscr{F}(u, \lambda) = \int_0^1 [F_p(v) \cdot v - F(v)]\lambda'\,dx = -\int_0^1 F(v)\lambda'\,dx$$

does not vanish for all $\lambda \in C_c^\infty((0, 1))$, as $F(v(t)) \not\equiv \text{const}$.

Yet one can show that every weak[2] extremal u satisfying the strict Legendre–Hadamard condition

$$\tfrac{1}{2}F_{p_\alpha^i p_\beta^k}(x, u(x), Du(x))\xi^i\xi^k\eta_\alpha\eta_\beta \geq \lambda|\xi|^2|\eta|^2$$

for $x \in \bar{\Omega}$ and some constant $\lambda > 0$, must necessarily be of class $C^2(\Omega, \mathbb{R}^N)$ if $F(x, z, p)$ is smooth, say, of class C^3. That is, every weak extremal is an extremal and, therefore, also an inner extremal if it satisfies the strict Legendre-Hadamard condition.

The "regularity theorem" used in this context is not easy to prove. Thus we wish to note that there is a much simpler way to verify that *every weak extremal satisfying the strict Legendre–Hadamard condition is necessarily an inner extremal.* It is, in fact, a fairly easy matter to show that any weak extremal fulfilling the Legendre-Hadamard condition is a locally weak minimizer, cf. 5,*1.3*, Theorem 4. Then the assertion follows from the linearity of $\partial\mathscr{F}(u, \cdot)$, a suitable decomposition $\lambda = \lambda_1 + \cdots + \lambda_k$, and from

Proposition 3. *Any weak minimizer $u \in C^1(\bar{\Omega}, \mathbb{R}^N)$ is both a weak extremal and an inner extremal.*

Proof. This result is a consequence of what we have said at the beginning of this section. □

Analogously we have: *Weak minimizers of a minimum problem with holonomic constraints are inner extremals. The same is true for weak minimizers of isoperimetric problems if the constraining side conditions are parameter-invariant.*

---

[2] Note that, by definition, weak extremals are of class C^1. Later we shall admit weak extremals which are much less regular. Then the regularity theorem does not always hold.

Leaving the question undecided whether any weak extremal of $\mathscr{F}$ will necessarily be an inner extremal, we consider the converse problem: Is an inner extremal of $\mathscr{F}$ also a weak extremal? This is easily seen to be false. For instance, every constant function u will be an inner extremal of the functional

$$\int_\Omega G(u)\,dx,$$

whereas $u = \text{const}$ will in general not be a weak extremal. More generally, any function $u \in C^1(\bar{\Omega}, \mathbb{R}^N)$, $n < N$, is always an inner extremal of $\int_\Omega G(u, Du)\,dx$ if the integrand $G(z, p)$ is positive for $p \neq 0$ and statisfies $G(z, p \cdot a) = \det a G(z, p)$ for all a with $\det a > 0$; cf. Propositions 3 and 4 of this section.

A somewhat more interesting counterexample is obtained in the following way: Consider a variational integral of the kind

$$\mathscr{F}(u) = \int_a^b F(u, \dot{u})\,dt,$$

where $n = 1$, $N \geq 1$. By Du Bois–Reymond's lemma, any inner extremal of $\mathscr{F}$ is characterized by the "conservation of energy"-law

$$\dot{u} \cdot F_p(u, \dot{u}) - F(u, \dot{u}) = \text{const}$$

(see 1,*2.2* [7], and the [1] below). If $N > 1$, this single first order equation will clearly not be equivalent to the N Euler equations if F is a general Lagrangian. However, even for $N = 1$, this conservation law may have more solutions than the Euler equation $\frac{d}{dt} F_p(u, \dot{u}) - F_z(u, \dot{u}) = 0$ as we have seen in 1,*2.2* [7].

Let us now discuss some more examples.

[1] *One-dimensional problems: Erdmann's equation and conservation of energy.* Let $n = 1$, $\Omega = (a, b) \subset \mathbb{R}$, $u \in C^1([a, b], \mathbb{R}^N)$, and $u' = \frac{du}{dx}$. Then u is an inner extremal of $\mathscr{F}(u) = \int_a^b F(x, u, u')\,dx$ if and only if

$$\int_a^b \{F_x \cdot \lambda + (F - u' \cdot F_p)\lambda'\}\,dx = 0$$

holds for all $\lambda \in C_c^\infty((a, b))$. Set $f(x) := F_x(x, u(x), u'(x))$ and write $f(x) = \frac{d}{dx} \int_a^x f(t)\,dt$. Then (cf. 1,*3.1*)

$$\int_a^b F_x \cdot \lambda\,dx = \int_a^b \left(\frac{d}{dx} \int_a^x f(t)\,dt\right) \lambda(x)\,dx = -\int_a^b \left(\int_a^x f(t)\,dt\right) \lambda'(x)\,dx,$$

whence

$$\int_a^b \left\{-\int_a^x f(t)\,dt + (F - u' \cdot F_p)\right\} \lambda'\,dx = 0.$$

On account of Du Bois–Reymond's lemma (cf. 1,*2.3*, Lemma 4), there is a constant c such that *Erdmann's equation*

$$(F - u' \cdot F_p)(x) = c + \int_a^x F_x\,dx \tag{12}$$

is satisfied for every inner extremal u.

If F does not depend on x, i.e., if $F = F(z, p)$, then $F_x = 0$, and we obtain the *energy theorem*:

$$F(x, u, u') - u' \cdot F_p(x, u, u') = c \quad \text{on } (a, b). \tag{13}$$

This was already proved in 1,*2.2* [7]; see also 1,*3.3*.

[2] *Parameter-invariant line integrals.* Let $n = 1$ and consider Lagrangians of type $F(z, p)$ which are positively homogeneous of first degree with respect to p, i.e.,

$$F(z, kp) = k \cdot F(z, p) \quad \text{for } k > 0. \tag{14}$$

We assume that $F(z, p)$ is continuous on $\mathbb{R}^N \times \mathbb{R}^N$, and of class c^2 for $p \neq 0$. A simple example of such an integrand is provided by

$$F(z, p) = \omega(z)|p|.$$

If $z = c(t)$, $t_1 \le t \le t_2$, denotes a curve in $\mathbb{R}^N$ of class $C^1([t_1, t_2], \mathbb{R}^N)$ which is regular (i.e., $|\dot{c}| \neq 0$), then *the integral*

$$\mathscr{F}(c) = \int_{t_1}^{t_2} F(c, \dot{c})\, dt$$

is parameter-invariant. That is, if $t = \sigma(\tau)$ describes a diffeomorphism of $[t_1, t_2]$ onto itself with $\dfrac{d\sigma}{d\tau}(\tau) > 0$, then it follows that

$$\mathscr{F}(c \circ \sigma) = \mathscr{F}(c).$$

Thus every regular C^2-curve $c(t)$ has to be an inner extremal of $\mathscr{F}$ and will, therefore, satisfy Noether's equation

$$\dot{c} \cdot L_F(c) = 0.$$

Hence, for every C^2-curve $c(t)$, the Euler expression $L_F(c)$ is always orthogonal to the tangent vector $\dot{c}$. Therefore the Euler system

$$L_F(c) = 0$$

of a parameter-invariant integral consists of at most $N - 1$ (instead of N) independent equations, or in other words, we can reduce the system of Euler equations to be satisfied by an extremal $c: [t_1, t_2] \to \mathbb{R}^N$, to a system of $N - 1$ equations. If, in particular, $N = 2$, an extremal of a parameter-invariant line integral is characterized by a single equation. This equation will now be computed.

We write $c(t) = (x(t), y(t))$, $t_1 \le t \le t_2$, for curves in $\mathbb{R}^2$, and $F(x, y, p, q)$ for the Lagrangian of which we suppose that

$$F(x, y, kp, kq) = kF(x, y, p, q) \quad \text{for } k > 0. \tag{14'}$$

Suppose that $c(t)$ is of class C^2 and regular, i.e.,

$$\dot{x}(t)^2 + \dot{y}(t)^2 \neq 0.$$

Then $L_F(c)$ is orthogonal to $\dot{c} = (\dot{x}, \dot{y})$, or, equivalently, collinear with the normal $(\dot{y}, -\dot{x})$. Hence there is a function $T_F(x, y, \dot{x}, \dot{y}, \ddot{x}, \ddot{y})$ such that

$$F_x - \frac{d}{dt}F_p = T_F \dot{y}, \qquad F_y - \frac{d}{dt}F_q = -T_F \dot{x}. \tag{15}$$

Thus the Euler equations are equivalent to the single equation

$$T_F = 0. \tag{16}$$

Let us compute the expression T_F. First we note that (15) can be written as

$$\begin{aligned} F_x - \{F_{px}\dot{x} + F_{py}\dot{y} + F_{pp}\ddot{x} + F_{pq}\ddot{y}\} &= T_F \dot{y}, \\ F_y - \{F_{qx}\dot{x} + F_{qy}\dot{y} + F_{qp}\ddot{x} + F_{qq}\ddot{y}\} &= -T_F \dot{x}. \end{aligned} \tag{17}$$

Note that F_x, F_y and F_p, F_q are positively homogeneous of order 1 or 0, respectively, with regard to p, q whence

$$\begin{aligned} F_x &= F_{xp}\dot{x} + F_{xq}\dot{y}, \qquad & F_y &= F_{yp}\dot{x} + F_{yq}\dot{y}, \\ 0 &= F_{pp}\dot{x} + F_{pq}\dot{y}, \qquad & 0 &= F_{qp}\dot{x} + F_{qq}\dot{y}. \end{aligned} \tag{17*}$$

The two equations in the second line of (17*) yield

$$(F_{pp}, F_{pq}, F_{qq}) \sim (\dot{y}^2, -\dot{x}\dot{y}, \dot{x}^2).$$

Hence there exists a function $F_1 = F_1(x, y, p, q)$ such that, for $p = \dot{x}$, $q = \dot{y}$, we have

$$(18) \qquad F_{pp} = \dot{y}^2 F_1, \qquad F_{pq} = -\dot{x}\dot{y}F_1, \qquad F_{qq} = \dot{x}^2 F_1.$$

Now we replace F_x and F_y in (17) according to the two equations in the first line of (17*), and apply (18).

Then it follows that

$$(19) \qquad T_F = F_{xq} - F_{yp} + F_1\{\dot{x}\ddot{y} - \dot{y}\ddot{x}\}.$$

We conclude that the Euler equations

$$F_x - \frac{d}{dt}F_p = 0, \qquad F_y - \frac{d}{dt}F_q = 0$$

are equivalent to the single equation

$$(20) \qquad F_{xq} - F_{yp} + F_1\{\dot{x}\ddot{y} - \dot{y}\ddot{x}\} = 0,$$

which, by introducing the curvature

$$\kappa = \frac{\dot{x}\ddot{y} - \dot{y}\ddot{x}}{(\dot{x}^2 + \dot{y}^2)^{3/2}},$$

can also be written as

$$(21) \qquad F_1 \cdot \kappa \cdot (\dot{x}^2 + \dot{y}^2)^{3/2} = F_{yp} - F_{xq}.$$

This equation was derived by Weierstrass.[3]

If F is a two-dimensional Riemann line element, i.e.,

$$F(x, y, p, q) = \sqrt{e(x, y)p^2 + 2f(x, y)pq + g(x, y)q^2},$$

with $eg - f^2 > 0$, then a straight-forward computation yields[4] that

$$(22) \qquad T_F(c, \dot{c}, \ddot{c}) = \kappa_g\sqrt{e(c)g(c) - f^2(c)},$$

where κ_g denotes the geodesic curvature of the curve $c(t)$ with respect to the line element

$$ds^2 = e\,dx^2 + 2f\,dx\,dy + g\,dy^2.$$

Hence geodesics are curves of geodesic curvature zero. For geodesics on a curved surface, this is also an immediate consequence of Johann Bernoulli's theorem (cf. 2,2 [2]).

Let us consider the particular example

$$F(x, y, p, q) = \omega(x, y)\sqrt{p^2 + q^2},$$

with $\omega(x, y) > 0$. The Euler equations to be satisfied by an extremal $c(t) = (x(t), y(t))$ are

$$\frac{d}{dt}\left\{\omega(c)\frac{\dot{c}}{|\dot{c}|}\right\} - \operatorname{grad}\omega(c)|\dot{c}| = 0.$$

Introducing the tangent vector $\mathfrak{t}(t) = \dfrac{\dot{c}(t)}{|\dot{c}(t)|}$, the normal $\mathfrak{n}(t)$, the curvature $\kappa(t)$, and the curvature radius $\rho(t)$ along $c(t)$, we have

$$\kappa = \frac{1}{\rho} \quad \text{and} \quad \frac{d\mathfrak{t}}{dt} = \frac{|\dot{c}|}{\rho}\mathfrak{n},$$

[3] It is throughout used in his celebrated lecture notes [1] Vol. 7; cf., in particular, pp. 107–108.

[4] See the Supplement for the definition of κ_g. Compare also Bolza [3], p. 210.

and the normal part of the Euler equation can be written as

$$\frac{\omega(c)}{\rho} = \operatorname{grad}\omega(c)\cdot\mathfrak{n} = \frac{\partial\omega}{\partial\mathfrak{n}}(c).$$

Thus equation (16) can be written in the equivalent form

$$\frac{1}{\rho} = \left(\frac{\partial}{\partial\mathfrak{n}}\log\omega\right)(c) \tag{16'}$$

due to Gauss, and the expression T_F is given by

$$T_F(c, \dot{c}, \ddot{c}) = \kappa\omega(c) - \frac{\partial\omega}{\partial\mathfrak{n}}(c).$$

Consider now an arbitrary C^2-motion $c(t)$ with speed $v(t) := |\dot{c}(t)| > 0$. From

$$\dot{\mathfrak{t}} = \frac{v}{\rho}\mathfrak{n}, \quad \ddot{c} = (v\mathfrak{t})^{\cdot} = \dot{v}\mathfrak{t} + v\dot{\mathfrak{t}},$$

we conclude that

$$\ddot{c} = \dot{v}\mathfrak{t} + \frac{v^2}{\rho}\mathfrak{n}.$$

If $c(t)$ describes the motion of a point mass m in some conservative field with the potential energy $V(x, y)$, Newton's equations state

$$m\ddot{c} = -\operatorname{grad} V(c) = -\frac{\partial V}{\partial\mathfrak{t}}(c)\,\mathfrak{t} - \frac{\partial V}{\partial\mathfrak{n}}(c)\,\mathfrak{n},$$

which is equivalent to the system of equations

$$m\dot{v} = -\frac{\partial V}{\partial\mathfrak{t}}(c), \qquad \frac{mv^2}{\rho} = -\frac{\partial V}{\partial\mathfrak{n}}(c). \tag{$*$}$$

Introduce the arc-length function $s = s(t)$ along $c(t)$ by $\dot{s} = v$. Then the first of the two equations $(*)$ yields

$$m\dot{s}\ddot{s} = -\frac{\partial V}{\partial\mathfrak{t}}(c)\dot{s} = -\operatorname{grad} V(c)\cdot\dot{c} = -\frac{dV(c)}{dt}$$

and this implies the conservation of total energy:

$$\tfrac{1}{2}m\dot{s}^2 + V(c) = h$$

for some constant h. Conversely, by differentiating this equation, we can return to the first equation of $(*)$ if $\dot{s} \neq 0$. In other words, the differential equation

$$\dot{s} = \sqrt{\frac{2}{m}\{h - V(c)\}}$$

describes the motion in time along the orbit of the point mass, provided that $\dot{s} > 0$.

The orbit of the point mass, i.e., its trace in $\mathbb{R}^2$ described during the motion, is characterized by the second equation of $(*)$. Introducing the function

$$\omega(x, y) := \sqrt{2\{h - V(x, y)\}},$$

we can equivalently write this equation as

$$\frac{1}{\rho} = \frac{\partial}{\partial\mathfrak{n}}\log\omega(c),$$

which is the Euler equation of the parameter-invariant integral

$$\mathscr{F}(c) = \int_{t_1}^{t_2} \omega(c)|\dot{c}|\,dt.$$

Thus we infer the following result:

Let $c(t)$ be some motion in $\mathbb{R}^2$ with $|\dot{c}(t)| > 0$ which is ruled by Newton's equations

$$m\ddot{c} = -\operatorname{grad} V(c).$$

This motion can equivalently be described by the following procedure:
Determine an extremal of the variational integral $\int \omega(c)|\dot{c}|\,dt$, $\omega(c) = \sqrt{2(h - V(c))}$, which starts at the same point as the actual motion c. Clearly this extremal will in general not be the actual motion; only its trace will coincide with that of c. Next we choose a representation $\gamma(s)$ of the extremal by the arc length parameter s, i.e., $\left|\frac{d\gamma}{ds}\right| = 1$. Then the correct motion $c(t)$ is obtained from $\gamma(s)$ by the formula $c(t) = \gamma(s(t))$ where the inverse $t(s)$ of the function $s(t)$ is to be determined from the formula

$$\frac{dt}{ds} = \frac{\sqrt{m}}{\omega(\gamma)}$$

together with the proper initial condition for $t(s)$.

This is *Jacobi's geometric version of the least action principle.* There is an analogue for the motion of ℓ point masses in $\mathbb{R}^3$ which we shall prove later.

If the graph of $c(t)$ is a nonparametric curve over the x-axis described by the function $z = u(x)$, we can introduce the representation

$$\gamma(x) = (x, u(x)),$$

and the function $x(t)$ is determined as inverse of $t(x)$ which is solution of the differential equation

$$\frac{dt}{dx} = \frac{\sqrt{m(1 + u'^2)}}{\omega(\gamma)}.$$

Then the actual motion is obtained as $c(t) = \gamma(x(t))$.

[3] *Parameter invariant double integrals. Minimal surfaces.* Suppose that $n = 2$ and $N = 3$, and set

$$z = (z^1, z^2, z^3), \qquad p = (p^1, p^2, p^3), \qquad q = (q^1, q^2, q^3).$$

We consider Lagrangians $F(z, p, q)$ which are independent of the two independent variables x, y and have the special form

$$F(z, p, q) = G(z, p \wedge q), \tag{23}$$

where

$$p \wedge q = (p^2q^3 - p^3q^2, p^3q^1 - p^1q^3, p^1q^2 - p^2q^1)$$

is the vector product of p, q, and $G(z, X)$ is continuous on $\mathbb{R}^3 \times \mathbb{R}^3$, of class C^2 for $X \neq 0$, and positively homogeneous of first degree with respect to $X = (X^1, X^2, X^3) \in \mathbb{R}^3$, i.e.,

$$G(z, k \cdot X) = k \cdot G(z, X) \quad \text{for } k > 0. \tag{24}$$

Consider smooth mappings $u : \bar{\Omega} \to \mathbb{R}^3$, $\Omega \subset \mathbb{R}^2$, which are viewed as parameter representations of surfaces

$$\{z \in \mathbb{R}^3 : z = u(x, y), (x, y) \in \Omega\},$$

in $\mathbb{R}^3$. Then we define the integral

$$\mathscr{F}(u) = \int_\Omega F(u, u_x, u_y)\,dx\,dy = \int_\Omega G(u, u_x \wedge u_y)\,dx\,dy, \tag{25}$$

which is parameter invariant. In fact, for every diffeomorphism $\sigma : \bar{\Omega} \to \bar{\Omega}$ of $\bar{\Omega}$ onto itself with positive Jacobian, we have

$$\mathscr{F}(u \circ \sigma) = \mathscr{F}(u).$$

Consequently, every mapping u of class C^2 is an inner extremal and must, therefore, satisfy the two Noether equations

$$L_F(u) \cdot u_x = 0, \qquad L_F(u) \cdot u_y = 0.$$

In other words, the two vectors $L_F(u)$ and $u_x \wedge u_y$ are collinear. Hence there exists a real-valued function $T_F(u, Du, D^2 u)$ such that

$$L_F(u) = T_F(u, Du, D^2 u) \cdot (u_x \wedge u_y) \tag{26}$$

holds for every $u \in C^2(\Omega, \mathbb{R}^3)$, and we infer that the system

$$L_F(u) = 0$$

consisting of three scalar Euler equations is equivalent to the single scalar equation

$$T_F(u, Du, D^2 u) = 0. \tag{27}$$

The function T_F can be computed by a similar reasoning as in [2]. We shall, however, restrict ourselves to the special case of the Lagrangian

$$F(p, q) = G(p \wedge q) = |p \wedge q| = \sqrt{|p|^2 |q|^2 - (p \cdot q)^2},$$

whose variational integral is the area functional

$$\mathscr{A}(u) = \int_\Omega |u_x \wedge u_y| \, dx \, dy.$$

We compute that

$$F_p(p, q) = q \wedge G_X(p \wedge q), \qquad F_q(p, q) = -p \wedge G_X(p \wedge q), \quad G_X(X) = \frac{X}{|X|}.$$

Introducing the surface normal

$$\xi = |u_x \wedge u_y|^{-1} \cdot (u_x \wedge u_y),$$

we obtain

$$L_F(u) = -D_x F_p - D_y F_q = (u_x \wedge \xi)_y - (u_y \wedge \xi)_x,$$

whence

$$L_F(u) = u_x \wedge \xi_y - u_y \wedge \xi_x.$$

Comparing this expression with $u_x \wedge u_y$, we see that both terms transform in the same way if we change variables. In fact, substituting x, y by the new independent variables α, β such that the corresponding Jacobian

$$\mathscr{J} = \frac{\partial(x, y)}{\partial(\alpha, \beta)}$$

is positive, both $L_F(u)$ and $u_x \wedge u_y$ are multiplied by $\mathscr{J}$. Thus the expression T_F is invariant with respect to admissible changes of the independent variables, and hence we can simplify the computation of T_F introducing suitable variables, and this can be done locally.

Since u is a regular surface, we can assume that it is locally the graph of some smooth function, say,

$$u(x, y) = (x, y, \psi(x, y)).$$

Then a straight-forward computation shows that

$$T_F = -\frac{(1+\psi_y^2)\psi_{xx} - 2\psi_x\psi_y\psi_{xy} + (1+\psi_x^2)\psi_{yy}}{1+\psi_x^2+\psi_y^2},$$

whence it follows that $-T_F$ is twice the *mean curvature* H of the surface u:

(28) $$T_F = -2H.$$

Therefore, (26) takes the form

(26′) $$L_F(u) = -2H\cdot(u_x \wedge u_y).$$

In other words, the three Euler equations $L_F(u) = 0$ are equivalent to the single scalar equation

$$H = 0.$$

That is, *the extremals of the area functional $\mathscr{A}(u)$ are given by surfaces of zero mean curvature.* These surfaces are called *minimal surfaces.* (Cf. also 1,*2.2* [5].) Note that here "minimal" does not mean "minimal area", but only "stationary area".

[4] *Parameter invariant multiple integrals.* Let us briefly consider the generalization of [2] and [3] to higher dimensions. To this end, we assume that $1 \le n < N$, and consider Lagrangians of the form

(29) $$F(z, p) = G(z, p_1 \wedge p_2 \wedge \cdots \wedge p_n),$$

$z \in \mathbb{R}^N$, $p = (p_\alpha^i) \in \mathbb{R}^{nN}$, where $G(z, P)$ is a continuous function of $(z, P) \in \mathbb{R}^N \times \mathbb{R}^m$, $m = \binom{N}{n}$, which is of class C^2 for $P \ne 0$ and positively homogeneous of the first degree, i.e.,

(30) $$G(z, k\cdot P) = k\cdot G(z, P) \quad \text{for all } k > 0,$$

and $p_1 \wedge p_2 \wedge \cdots \wedge p_n$ denotes the exterior product P of the vectors $p_1, \ldots, p_n \in \mathbb{R}^N$, $p_\alpha = (p_\alpha^i)$, which is given by

$$P = (P^J), \quad J = (i_1, \ldots, i_n), \; i_1 < i_2 < \cdots < i_n,$$
$$P^J = \det(p_\beta^{i_\alpha}) = \det(p^{i_1}, p^{i_2}, \ldots, p^{i_n}), \quad p^{i_\alpha} = (p_\beta^{i_\alpha}).$$

Then, for any $u \in C^1(\bar{\Omega}, \mathbb{R}^N)$ which is regular (i.e., rank $Du = n$), the integral

(31) $$\mathscr{F}(u) = \int_\Omega F(u, Du)\,dx = \int_\Omega G(u, D_1u \wedge D_2u \wedge \cdots \wedge D_nu)\,dx$$

is invariant with respect to diffeomorphisms $\sigma: \bar{\Omega} \to \bar{\Omega}$ of $\bar{\Omega}$ onto itself with positive Jacobian $J = \det D\sigma$, that is, $\mathscr{F}(u \circ \sigma) = \mathscr{F}(u)$.

Let $G(z) = (g_{ik}(z))$ be a positive definite, symmetric matrix function on $\mathbb{R}^N$. Then

$$ds^2 = g_{ik}(z)\,dz^i\,dz^k$$

defines a Riemannian line element on $\mathbb{R}^N$, and the variational integral

(32) $$\mathscr{F}(u) = \int_\Omega \sqrt{\det\{g_{ik}(u)D_\alpha u^i D_\beta u^k\}}\,dx$$

is the area of the n-dimensional surface in $\mathbb{R}^N$, given by a parametric representation $z = u(x)$, $x \in \bar{\Omega}$, with respect to this line element. The variational integrand

$$F(z, p) = \sqrt{\det\{g_{ik}(z)p_\alpha^i p_\beta^k\}}$$

furnishes an example of a Lagrangian that can be expressed in the form (29).

From the relation $\mathscr{F}(u \circ \sigma) = \mathscr{F}(u)$ we infer that each C^2-mapping $u(x)$ is an inner extremal and will, therefore, satisfy the n Noether equations

$$L_F(u)\cdot u_{x^\alpha} = 0, \quad \alpha = 1, \ldots, n.$$

Let $X_I: \bar{\Omega} \to \mathbb{R}^N$, $I = 1, \ldots, N-n$, be independent vector fields along u which satisfy

$$X_I \cdot D_\alpha u = 0 \quad \text{and} \quad X_I \cdot X_K = 0 \quad \text{for } I \neq K$$

($\alpha = 1, \ldots, n$; $I, K = 1, \ldots, N - n$). Then there exist functions $T_F^1, \ldots, T_F^{N-n}$ such that

$$L_F(u) = T_F^I \cdot X_I \tag{33}$$

holds. Therefore, the system of Euler equations is equivalent to the system of $N - n$ equations

$$T_F^1 = 0, \ldots, T_F^{N-n} = 0. \tag{34}$$

Since we can construct the X_I solely in terms of $u_{x^1}, \ldots, u_{x^n}$, we can achieve that the functions T_F^I only depend on u, Du, and D^2u.

Let us set

$$\mathscr{F}(u, \Omega) := \int_\Omega F(u, Du)\, dx. \tag{35}$$

Then conditions (29) and (30) imply that

$$\mathscr{F}(u, \Omega) = \mathscr{F}(u \circ \sigma, \Omega^*) \tag{36}$$

holds for every $\Omega \subset \mathbb{R}^n$, $u \in C^1(\bar{\Omega}, \mathbb{R}^N)$, *and each diffeomorphism* σ *of* $\bar{\Omega}^*$ *onto* $\bar{\Omega}$.

Proposition 4. *The integral* $\mathscr{F}(u, \Omega)$ *possesses the general invariance property* (36) *if and only if its integrand* F *satisfies*

$$F(z, p \cdot a) = (\det a) \cdot F(z, p) \tag{37}$$

for all $z \in \mathbb{R}^N$, $p \in \mathbb{R}^{nN}$, *and every* $n \times n$-*matrix* $a = (a_\beta^\alpha)$ *with* $\det a > 0$.

Proof. Suppose that (36) holds, that is,

$$\int_\Omega F(u, Du)\, dx = \int_{\Omega^*} F(v, Dv)\, dy,$$

where $v = u \circ \sigma$. For given z, p, and a, we choose u, σ, x_0 and Ω in such a way that $x_0 \in \Omega$, $u(x_0) = z$, $Du(x_0) = p$, and $D\sigma(y_0) = a$, where $y_0 = \sigma^{-1}(x_0)$ and $\Omega^* = \sigma^{-1}(\Omega)$. Applying the transformation theorem to the left-hand side of the last equation and substituting $v = u(\sigma)$, $Dv = Du(\sigma) \cdot D\sigma$ on the right, we arrive at

$$\int_{\Omega^*} F(u(\sigma), Du(\sigma)) \cdot \det D\sigma\, dy = \int_{\Omega^*} F(u(\sigma), Du(\sigma) \cdot D\sigma)\, dy.$$

Now we choose Ω^* to be a ball $B_r(y_0)$, divide by meas $B_r(y_0)$, and let r tend to zero. Then we conclude that (37) holds.

Conversely, we may derive (36) from (37) by reversing the previous reasoning. □

Corollary. *The homogeneity relation* (14) *is both necessary and sufficient that* F *is the Lagrangian of a one-dimensional variational integral which is invariant in the sense of* (36).

Finally we shall derive a condition which is equivalent to (37) and generalizes Euler's well-known relation for homogeneous functions.

Proposition 5. *Suppose that* $F(z, p) > 0$ *holds if* $p \neq 0$. *Then the identity* (37) *is satisfied if and only if*

$$F_{p_\beta^i}(z, p) p_\alpha^i = \delta_\alpha^\beta F(z, p) \tag{38}$$

is true for all $(z, p) \in \mathbb{R}^N \times \mathbb{R}^{nN}$ *with* $p \neq 0$.

Proof. (i) We note that, for $a = (a_\beta^\alpha)$,

$$\frac{\partial}{\partial a_\beta^\alpha} \det a = A_\alpha^\beta,$$

where A_α^β is the algebraic cofactor in det a. By differentiating the identity (37) with respect to a_β^α, we obtain that

$$F_{p_\beta^i}(z, a\cdot p)p_\alpha^i = F(z, p)A_\alpha^\beta$$

provided that $p \neq 0$ and det $a > 0$. This implies (38) if we choose $a_\beta^\alpha = \delta_\beta^\alpha$.

(ii) Conversely, let (38) be satisfied. We choose $p = q\cdot a$, $q = (q_\alpha^i)$, $a = (a_\beta^\alpha)$, that is, $p_\beta^i = q_\alpha^i a_\beta^\alpha$, and assume $q \neq 0$, det $a > 0$, and for the moment also $q\cdot a \neq 0$ (this is, as we shall see, a consequence of $q \neq 0$ and det $a > 0$). Then (38) yields

$$F_{p_\beta^i}(z, q\cdot a)q_\gamma^i a_\alpha^\gamma = F(z, q\cdot a)\delta_\alpha^\beta .$$

If we multiply both sides by A_ν^α and take the equation $a_\alpha^\gamma A_\nu^\alpha = \delta_\nu^\gamma\cdot \det a$ into account, it follows that

$$F_{p_\beta^i}(z, q\cdot a)q_\nu^i \det a = F(z, q\cdot a)A_\nu^\beta .$$

Introducing

$$G(a) := F(z, q\cdot a), \qquad H(a) := \det a ,$$

we infer that

$$G_{a_\beta^\nu}(a)H(a) = G(a)H_{a_\beta^\nu}(a). \tag{39}$$

The functions $G(a)$ and $H(a)$ will be considered on the open set $\mathscr{P} := \{a\colon \det a > 0\}$, where $H(a)$ is a polynomial and therefore of class C^1, while $G(a)$ is of class C^1 only on the open set $\mathscr{P} - \mathscr{P}_0$, $\mathscr{P}_0 := \{a \in \mathscr{P}\colon G(a) = 0\}$.

Let $p \neq 0$, $e = (\delta_\beta^\alpha)$, and $a = (a_\beta^\alpha) \in \mathscr{P}$. Since $\mathscr{P}$ is connected, there exists some C^1-path $\sigma : [0, 1] \to \mathscr{P}$ such that $\sigma(0) = e$ and $\sigma(1) = a$ holds. Let us first assume that $\sigma(t) \in \mathscr{P} - \mathscr{P}_0$ for all $t \in [0, 1]$. Then, replacing a in (39) by $\sigma(t)$ and multiplying by $\dot\sigma_\nu^\beta(t)$, we obtain for

$$\varphi(t) := G(\sigma(t)), \qquad \psi(t) := H(\sigma(t)),$$

the equation

$$\dot\varphi(t)\psi(t) = \varphi(t)\dot\psi(t),$$

whence

$$\varphi(t)/\psi(t) = \varphi(0)/\psi(0).$$

Since $\psi(0) = H(e) = \det e = 1$, we arrive at

$$\varphi(t) = \psi(t)\varphi(0),$$

which for $t = 1$ yields the desired equation

$$F(z, q\cdot a) = \det a\cdot F(z, q).$$

Finally, if there would exist a parameter value $t \in [0, 1]$ such that $\sigma(t) \in \mathscr{P}_0$, then there had to be a smallest value $\tau \in [0, 1]$ such that $\sigma(\tau) \in \mathscr{P}_0$. Since $\sigma(0) = e \notin \mathscr{P}_0$, we infer that $\tau > 0$ and that $\sigma(t) \notin \mathscr{P}_0$ for $0 \le t < \tau$. Consequently, we obtain

$$\varphi(t) = \psi(t)\cdot\varphi(0) \quad \text{for } 0 \le t < \tau,$$

and, by continuity, it follows that

$$\varphi(\tau) = \psi(\tau)\cdot\varphi(0).$$

From $\varphi(\tau) = 0$ and $\varphi(0) = F(z, q) \neq 0$, we would obtain that $\psi(t) = H(\sigma(\tau)) = 0$, which would be a contradiction to $\sigma(\tau) \in \mathscr{P}$ and $H > 0$ on $\mathscr{P}$. This completes the proof of the second part of the proposition. □

Note that condition (38) is equivalent to the identity

$$T_\alpha^\beta(z, p) = 0 \quad \text{for all } (z, p) \in \mathbb{R}^N \times \mathbb{R}^{nN} \text{ with } p \neq 0. \tag{38'}$$

2. Strong Inner Variations, and Strong Inner Extremals

In this section we want to consider variations of the independent variables which vary the domain of definition Ω of the mappings $u: \Omega \to \mathbb{R}^N$ in consideration. This will lead us to the notion of *strong inner extremals*. Any such extremal that is of class C^2 will not only satisfy the Noether equations but also a *free boundary condition*. The most interesting example of this section concerns the strong inner extremals of the (generalized) Dirichlet integral which are seen to satisfy the so-called *conformality relations*.

For the sake of simplicity we shall throughout assume that $F(x, u, p)$ is a Lagrangian of class C^2 on $\mathbb{R}^n \times \mathbb{R}^N \times \mathbb{R}^{nN}$.

In the previous section we have considered variations $x = \xi(y, \varepsilon)$ of the independent variables x by families of diffeomorphisms defined on the same parameter domain Ω as x. Now we want to admit that the domains of definition of $\xi(\varepsilon)$ may change as ε varies.

As in Section 1, we start with a prescribed vector field $\lambda: \bar{\Omega} \to \mathbb{R}^n$, $\Omega \subset \mathbb{R}$, which now is assumed to be of class C^1. Then we want to construct a family $\xi(\varepsilon): \bar{\Omega}_\varepsilon^* \to \bar{\Omega}$ of diffeomorphisms of $\bar{\Omega}_\varepsilon^*$ onto $\bar{\Omega}$, depending smoothly on the parameter $\varepsilon \in (-\varepsilon_0, \varepsilon_0)$, $\varepsilon_0 > 0$, which satisfy

$$(1) \qquad \Omega_0^* = \Omega, \quad \xi(x, 0) = x, \quad \frac{\partial}{\partial \varepsilon}\xi(x, \varepsilon)\bigg|_{\varepsilon=0} = \lambda(x) \qquad \text{for } x \in \bar{\Omega}.$$

Corresponding to Section 1, (3), we want to write $\xi(y, \varepsilon)$ in the form

$$(2) \qquad \xi(y, \varepsilon) = y + \varepsilon\lambda(y) + o(\varepsilon) \quad \text{for } y \in \bar{\Omega}_\varepsilon^*.$$

There is a conceptual difficulty caused by the fact that $\lambda(y)$ originally is only defined on $\bar{\Omega}$, whereas in (2) $\lambda(y)$ should be defined on $\bar{\Omega}_\varepsilon^*$. This difficulty will be overcome by interpreting the assumption "$\lambda \in C^1(\bar{\Omega}, \mathbb{R}^n)$" in the following sense: *There exists a function* $\tilde{\lambda} \in C_c^1(\mathbb{R}^n, \mathbb{R}^n)$ *such that* $\lambda = \tilde{\lambda}|_{\bar{\Omega}}$.

In other words, we shall operate with vector fields λ of class $C^1(\mathbb{R}^n, \mathbb{R}^n)$, the support of which is not necessarily contained in $\bar{\Omega}$. We shall assume that $\operatorname{int} \bar{\Omega} = \Omega$ and $\partial\Omega \in C^1$ to ensure that the class $C^1(\bar{\Omega}, \mathbb{R}^n)$ comprises all functions $\lambda \in C^1(\Omega, \mathbb{R}^n)$ such that λ and $D\lambda$ can continuously be extended to $\bar{\Omega}$.

Also, if we take λ as starting point, it is not a priori clear how the domains Ω_ε^* are to be chosen. Thus, in order to construct $\xi(y, \varepsilon)$ satisfying (2), we begin by defining the inverse mappings $\eta(\varepsilon) = \xi^{-1}(\varepsilon)$. We choose mappings $\eta(\varepsilon): \bar{\Omega} \to \mathbb{R}^n$ of the type

$$\eta(x, \varepsilon) = x - \varepsilon\lambda(x) + o(\varepsilon), \quad x \in \bar{\Omega},$$

and set

$$\Omega_\varepsilon^* := \eta(\Omega, \varepsilon).$$

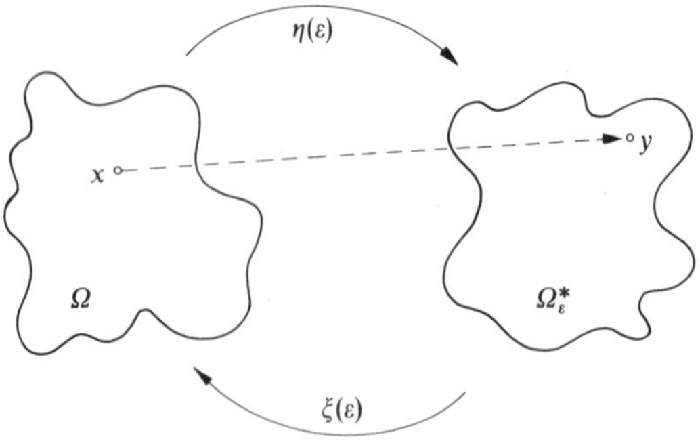

Fig. 1. The mappings $\xi(\varepsilon)$ and $\eta(\varepsilon)$.

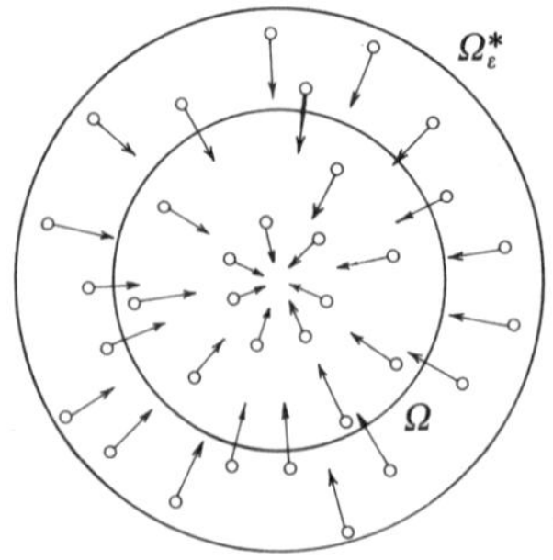

Fig. 2. A contraction $\xi(\varepsilon)\colon \bar{\Omega}_\varepsilon^* \to \bar{\Omega}$ by a radial vector field.

For instance, we can simply take

$$\eta(x, \varepsilon) := x - \varepsilon\lambda(x), \quad x \in \bar{\Omega},$$

if λ is an arbitrarily given function of class $C^1(\bar{\Omega}, \mathbb{R}^n)$.

We claim that the family $\eta(\varepsilon)$, $|\varepsilon| < \varepsilon_0$, can be chosen in such a way that the following holds:

Lemma. *For any* $\lambda \in C_c^1(\mathbb{R}^n, \mathbb{R}^n)$, *there is a number* $\varepsilon_0 > 0$ *and a function* $\eta(x, \varepsilon)$ *defined on* $\mathbb{R}^n \times (-\varepsilon_0, \varepsilon_0)$ *such that the following is satisfied*:

(i) *The function* $\eta(x, \varepsilon)$ *is of class* C^1 *and satisfies*

$$\eta(x, \varepsilon) = x - \varepsilon\lambda(x) + o(\varepsilon)$$

for any $x \in \Omega$; *in particular*

$$\frac{\partial \eta}{\partial \varepsilon}(x, 0) = -\lambda(x) \quad \textit{for any } x \in \mathbb{R}^n.$$

(ii) *For any* $\varepsilon \in (-\varepsilon_0, \varepsilon_0)$, *the mapping* $\eta(\cdot, \varepsilon)$ *furnishes a* C^1*-diffeomorphism of* $\mathbb{R}^n$ *onto itself.*

(iii) *The inverse* $\xi(\cdot, \varepsilon)$ *of* $\eta(\cdot, \varepsilon)$ *defines a mapping* $\xi(y, \varepsilon)$ *of class* C^1 *on* $\mathbb{R}^n \times (-\varepsilon_0, \varepsilon_0)$ *which satisfies*

$$\xi(y, \varepsilon) = y + \varepsilon\lambda(y) + o(\varepsilon)$$

for any $y \in \mathbb{R}^n$.

(iv) *Restricting the region of definition of* $\eta(\varepsilon) = \eta(\cdot, \varepsilon)$ *and* $\xi(\varepsilon) = \xi(\cdot, \varepsilon)$ *to* $\bar{\Omega}$ *and* $\bar{\Omega}_\varepsilon^*$, *respectively, where* $\Omega_\varepsilon^* := \eta(\Omega, \varepsilon)$, *the mapping* $\eta(\varepsilon)$ *is a* C^1*-diffeomorphism of* $\bar{\Omega}$ *onto* $\bar{\Omega}_\varepsilon^*$ *with the inverse* $\xi(\varepsilon)$, *and we have*

$$\Omega_0^* = \Omega, \quad \xi(x, 0) = x, \quad \frac{\partial}{\partial \varepsilon}\xi(x, \varepsilon)\bigg|_{\varepsilon=0} = \lambda(x) \qquad \textit{for } x \in \bar{\Omega}.$$

We shall call such a mapping $\xi(y, \varepsilon)$ a *strong variation of the independent variables with the infinitesimal generator* λ. We also speak of an *admissible family of parameter transformations.*

Proof. It suffices to consider the mappings

$$\eta(x, \varepsilon) = x - \varepsilon\lambda(x) \quad \text{for } x \in \Omega_0,$$

and, without loss of generality, we can assume that the vector field $\lambda \in C^1(\mathbb{R}^n, \mathbb{R}^n)$ satisfies a uniform Lipschitz condition on $\mathbb{R}^n$, that is,

$$|\lambda(x_1) - \lambda(x_2)| \le K|x_1 - x_2|$$

holds for some $K > 0$ and all $x_1, x_2 \in \mathbb{R}^n$. Then we conclude that

$$|\eta(x_1, \varepsilon) - \eta(x_2, \varepsilon)| \ge (1 - |\varepsilon|K)|x_1 - x_2| \ge \tfrac{1}{2}|x_1 - x_2|$$

holds for all $x_1, x_2 \in \mathbb{R}^n$ and for $|\varepsilon| < \varepsilon_0$ where ε_0 is a sufficiently small number satisfying $0 < \varepsilon_0 < (2K)^{-1}$. Thus the mapping $\eta(\cdot, \varepsilon): \mathbb{R}^n \to \mathbb{R}^n$ is injective provided that $|\varepsilon| < \varepsilon_0$, and the inverse-mapping theorem implies that $\eta(\cdot, \varepsilon)$ is a C^1-diffeomorphism of $\mathbb{R}^n$ onto itself. It is easy to see that the inverse mapping $\xi(\cdot, \varepsilon)$ can be written as

$$\xi(y, \varepsilon) = y + \varepsilon\lambda(y) + o(\varepsilon)$$

for any $y \in \mathbb{R}^n$ whence the other assertions follow without difficulty. □

For a given $u \in C^1(\bar{\Omega}, \mathbb{R}^N)$ we now consider a family of comparison functions $v(\varepsilon) := u \circ \xi(\varepsilon)$ which are defined by

$$v(y, \varepsilon) = u(\xi(y, \varepsilon)) \quad \text{for } y \in \bar{\Omega}_\varepsilon^* \text{ and } |\varepsilon| < \varepsilon_0 .$$

Here $\xi(y, \varepsilon)$ is a strong variation of the independent variables with the infinitesimal generator λ. We call $v(y, \varepsilon)$, $y \in \bar{\Omega}_\varepsilon^*$, $|\varepsilon| < \varepsilon_0$, a *strong inner variation of the function* $u(x)$, $x \in \bar{\Omega}$, *in direction of* λ.

Correspondingly, we form

$$\Phi(\varepsilon) := \mathscr{F}(v(\varepsilon), \Omega_\varepsilon^*) = \int_{\Omega_\varepsilon^*} F(y, v(y, \varepsilon), Dv(y, \varepsilon))\, dy$$

for $|\varepsilon| < \varepsilon_0$, $D = D_y$.

As in Section 1, the expression

$$\text{(3)} \qquad \partial\mathscr{F}(u, \lambda) := \Phi'(0) = \frac{d}{d\varepsilon}\mathscr{F}(v(\varepsilon), \Omega_\varepsilon^*)\bigg|_{\varepsilon=0}$$

will be called the *inner variation of* $\mathcal{F}$ *at* u *in direction of the vector field* $\lambda \in C^1(\bar{\Omega}, \mathbb{R}^n)$. The only difference to the previous section is that, instead of $\lambda \in C_c^\infty(\Omega, \mathbb{R}^n)$, we now allow variations λ of class $C^1(\bar{\Omega}, \mathbb{R}^n)$.

As in Lemma 1 of Section 1 we see that $\Phi'(0)$ exists and that $\partial\mathcal{F}(u, \lambda)$ is given by

$$\partial\mathcal{F}(u, \lambda) = \int_\Omega \{T_\alpha^\beta(x, u, Du)\lambda_{x^\beta}^\alpha - F_{x^\alpha}(x, u, Du)\lambda^\alpha\}\, dx. \tag{4}$$

where $T = (T_\alpha^\beta)$ denotes the Hamilton tensor which was defined as

$$T_\alpha^\beta(x, z, p) := p_\alpha^i F_{p_\beta^i}(x, z, p) - \delta_\alpha^\beta F(x, z, p).$$

In analogy to Section 1 we formulate

Definition 1. *A mapping* $u : \bar{\Omega} \to \mathbb{R}^N$ *of class* C^1 *is called a* strong inner extremal *of* $\mathcal{F}$ *if it satisfies*

$$\partial\mathcal{F}(u, \lambda) = 0 \quad \text{for all } \lambda \in C^1(\bar{\Omega}, \mathbb{R}^n). \tag{5}$$

According to our discussion, the following definition is equivalent:

Definition 1′. *A function* $u \in C^1(\bar{\Omega}, \mathbb{R}^N)$ *is a* strong inner extremal *of* $\mathcal{F}$ *if and only if, for every strong inner variation* $v(\varepsilon) = u \circ \xi(\varepsilon)$, *defined by an admissible family of parameter transformations* $\xi(\varepsilon) : \bar{\Omega}_\varepsilon^* \to \bar{\Omega}$, *and for* $\Phi(\varepsilon) := \mathcal{F}(v(\varepsilon), \bar{\Omega}_\varepsilon^*)$, *the condition* $\Phi'(0) = 0$ *is satisfied.*

Clearly, *every strong inner extremal is also an inner extremal*, but the converse is not true.

Proposition 1. *Any strong inner extremal* u *of class* $C^1(\bar{\Omega}, \mathbb{R}^N) \cap C^2(\Omega, \mathbb{R}^N)$ *satisfies both the Noether equations*

$$D_\beta T_\alpha^\beta(x, u, Du) + F_{x^\alpha}(x, u, Du) = 0, \quad 1 \le \alpha \le n, \tag{6}$$

on Ω *as well as the boundary conditions*

$$\nu_\beta T_\alpha^\beta(x, u, Du) = 0, \quad 1 \le \alpha \le n, \tag{6′}$$

on $\partial\Omega$ *where* $\nu = (\nu_1, \ldots, \nu_n)$ *is the exterior normal to* $\partial\Omega$.

Proof. By (4) and (5), we have

$$0 = \partial\mathcal{F}(u, \lambda) = \int_\Omega (T_\alpha^\beta \lambda_{x^\beta}^\alpha - F_{x^\alpha}\lambda^\alpha)\, dx$$

for any $\lambda \in C^1(\bar{\Omega}, \mathbb{R}^n)$. Applying a partial integration, we infer that

$$-\int_{\Omega} (D_\beta T_\alpha^\beta + F_{x^\alpha})\lambda^\alpha \, dx + \int_{\partial\Omega} \nu_\beta T_\alpha^\beta \lambda^\alpha \, d\mathscr{H}^{n-1} = 0$$

for all $\lambda \in C^1(\bar{\Omega}, \mathbb{R}^n)$. By means of the fundamental lemma, we arrive at the equations (6) and (6'). □

We now discuss some examples.

1 Let us once again consider 1-dimensional integrals

$$\mathscr{F}(u) = \int_a^b F(x, u, u') \, dx.$$

As we know from *1* 1, any inner extremal u of $\mathscr{F}$ satisfies Erdmann's equation

$$(F - u' \cdot F_p)(x) = c + \int_a^x F_x \, dx$$

for some constant c. If u is a strong inner extremal of $\mathscr{F}$, it also satisfies the boundary condition (6) which reduces to

$$(F - u' \cdot F_p)(x) = 0 \quad \text{for } x = a \text{ and for } x = b.$$

Thus Erdmann's equation obtains the special form

$$(F - u' \cdot F_p)(x) = \int_a^x F_x \, dx. \tag{7}$$

If $F_x = 0$, we arrive at the special form

$$F(x, u, u') - u' \cdot F_p(x, u, u') = 0 \quad \text{on } [a, b] \tag{8}$$

of the *energy theorem.*

This already follows if $(F - u' \cdot F_p)(x) = 0$ is satisfied for only one of the two points $x = a, b$ which, in turn, is a consequence of "$\partial\mathscr{F}(u, \lambda) = 0$ for all $\lambda \in C^1([a, b])$ with $\lambda(b) = 0$, or else $\lambda(a) = 0$."

2 Let $\mathscr{F}(u) = \int_\Omega F(u, Du) \, dx$ be an invariant n-dimensional integral with

$$F(z, p) > 0 \quad \text{if } p \neq 0.$$

In Proposition 5 of Section 1, we have proved the identity

$$\delta_\alpha^\beta F(z, p) - F_{p_\beta^i}(z, p) p_\alpha^i = 0 \quad \text{for } p \neq 0.$$

Thus the boundary condition (6') is trivially satisfied for every $u \in C^1(\bar{\Omega}, \mathbb{R}^N)$. This expresses the fact that every smooth u is not only an inner extremal, but even a strong inner extremal of an arbitrary invariant integral $\int_\Omega F(u, Du) \, dx$; see *1*, (36).

3 *Inner extremals of the generalized Dirichlet integral.* For $n = 2$ and $N \geq 2$, we consider the integrand

$$F(z, p) = g_{ik}(z) p_\alpha^i p_\alpha^k,$$

where the coefficients g_{ik} are assumed to be symmetric: $g_{ik}(z) = g_{ki}(z)$. Then the corresponding functional

(9) $$\mathscr{F}(u) = \int_\Omega g_{ik}(u) D_\alpha u^i D_\alpha u^k \, dx$$

is called the *generalized Dirichlet integral*[5] of some function $u \in C^1(\bar{\Omega}, \mathbb{R}^N)$ (with respect to the possibly indefinite matrix g_{ik}).

We set

$$p_\alpha^i(x) := u_{x^\alpha}^i(x), \quad p_\alpha = (p_\alpha^1, \ldots, p_\alpha^N), \quad \langle p_\alpha, p_\beta \rangle := g_{ik}(u) p_\alpha^i p_\beta^k.$$

Since

$$F_{x^\alpha} = 0, \quad F_{p_\beta^i}(u, Du) = 2 g_{ik}(u) p_\beta^k,$$

it follows that

$$F_{p_\beta^i}(u, Du) u_{x^\alpha}^i = 2 \langle p_\alpha, p_\beta \rangle.$$

Thus the Hamilton tensor $T(x, u, Du)$ with the components

(10) $$T_\alpha^\beta = F_{p_\beta^i} u_{x^\alpha}^i - \delta_\alpha^\beta F$$

is of the form

(11) $$\begin{bmatrix} T_1^1, & T_2^1 \\ T_1^2, & T_2^2 \end{bmatrix} = \begin{bmatrix} a, & b \\ b, & -a \end{bmatrix},$$

where we have set

(11′) $$a := \langle p_1, p_1 \rangle - \langle p_2, p_2 \rangle, \qquad b := 2 \langle p_1, p_2 \rangle.$$

Then it follows from (4) that

(12) $$\partial \mathscr{F}(u, \lambda) = \int_\Omega T_\alpha^\beta \lambda_{x^\beta}^\alpha \, dx$$

for every $\lambda \in C^1(\bar{\Omega}, \mathbb{R}^n)$ whence

(13) $$\partial \mathscr{F}(u, \lambda) = \int_\Omega \{a[\lambda_{x^1}^1 - \lambda_{x^2}^2] + b[\lambda_{x^2}^1 + \lambda_{x^1}^2]\} \, dx.$$

From this formula, we infer the following results:

Proposition 2. *If $u \in C^1(\bar{\Omega}, \mathbb{R}^N)$, $\Omega \subset \mathbb{R}^2$, is an inner extremal of the generalized Dirichlet integral* (9), *then*

$$f(\zeta) := a(x^1, x^2) - ib(x^1, x^2)$$

is a holomorphic function of the complex variable $\zeta = x^1 + ix^2 \in \Omega \subset \mathbb{C}$.

Proof. Assume first that $u \in C^2(\Omega, \mathbb{R}^N)$. Then it follows from (13) that, for $\lambda \in C_c^\infty(\Omega, \mathbb{R}^2)$, we have

(14) $$0 = \int_\Omega \{a[\lambda_{x^1}^1 - \lambda_{x^2}^2] + b[\lambda_{x^2}^1 + \lambda_{x^1}^2]\} \, dx,$$

and a partial integration yields

[5] Often, one adds the factor $\frac{1}{2}$, cf. 2,4 [3]. We have dropped it to simplify the following formulas.

$$0 = \int_\Omega \{-\lambda^1(a_{x^1} + b_{x^2}) + \lambda^2(a_{x^2} - b_{x^1})\}\, dx .$$

On account of the fundamental lemma, we obtain the Cauchy–Riemann equations

$$a_{x^1} = -b_{x^2}, \qquad a_{x^2} = b_{x^1},$$

and therefore $f(\zeta)$ is holomorphic in Ω.

If we drop the additional assumption $u \in C^2$ we can still proceed in this way by choosing an arbitrary $\mu \in C_c^\infty(\Omega, \mathbb{R}^2)$ and a mollifier S_ε with $0 < \varepsilon \ll 1$, and by inserting $\lambda = S_\varepsilon \mu$ as test function in (14). Then we get instead of the upper equation the relation

$$0 = \int_\Omega \{-\mu^1[(S_\varepsilon a)_{x^1} + (S_\varepsilon b)_{x^2}] + \mu^2[(S_\varepsilon a)_{x^2} - (S_\varepsilon b)_{x^1}]\}\, dx,$$

from which we conclude that the function

$$S_\varepsilon f = S_\varepsilon a - i S_\varepsilon b$$

is holomorphic in $\Omega'_\varepsilon := \{\zeta \in \Omega\colon \operatorname{dist}(\zeta, \partial\Omega) > \varepsilon\}$. Since $S_\varepsilon f$ converges uniformly to f, the representation formula of holomorphic functions by the Cauchy integral implies that also f is holomorphic in Ω. □

Proposition 3. *If $u \in C^1(\bar{\Omega}, \mathbb{R}^N)$, $\Omega \subset \mathbb{R}^2$, is a strong inner extremal of the generalized Dirichlet integral* (9), *then u satisfies the conformality relations*

$$(15) \qquad g_{ik}(u) u^i_{x^1} u^k_{x^1} = g_{ik}(u) u^i_{x^2} u^k_{x^2}, \qquad g_{ik}(u) u^i_{x^1} u^k_{x^2} = 0$$

on $\bar{\Omega}$.

Proof. If $u \in C^2(\Omega, \mathbb{R}^N)$, we infer from (6′) that $\nu_\beta T^\beta_\alpha = 0$ holds on $\partial\Omega$. Consider some $x_0 \in \partial\Omega$ where $\nu(x_0) = (1, 0)$. Then we get $T^1_1(x_0) = 0$ and $T^1_2(x_0) = 0$, or $a(x_0) = 0$ and $b(x_0) = 0$. Since the integral (9) is invariant with respect to rotations σ of the x^1, x^2-plane, also $u \circ \sigma$ are strong inner extremals of (9). Thus we can conclude $a = 0$ and $b = 0$ on $\partial\Omega$. Since $a = \operatorname{Re} f$ and $b = \operatorname{Im} f$ are harmonic on Ω and continuous on $\bar{\Omega}$, the maximum principle[6] for harmonic functions implies that $a(x) \equiv 0$ and $b(x) \equiv 0$ on $\bar{\Omega}$, and (15) is proved.

If we drop the additional assumption $u \in C^2$, Proposition 1 is no longer applicable. Then we proceed in a completely different way:

Choose arbitrary functions $\rho, \sigma \in C_c^\infty(\Omega)$, and determine functions $h, k \in C^2(\bar{\Omega})$ with

$$(16) \qquad \Delta h = \rho, \qquad \Delta k = \sigma \quad \text{on } \Omega .$$

This is certainly possible if $\partial\Omega$ is sufficiently smooth, say, $\Omega \in C^3$. Choosing

$$\lambda^1 := h_{x^1} + k_{x^2}, \qquad \lambda^2 := -h_{x^2} + k_{x^1},$$

we obtain

$$\lambda^1_{x^1} - \lambda^2_{x^2} = \rho, \qquad \lambda^1_{x^2} + \lambda^2_{x^1} = \sigma,$$

and therefore

$$(17) \qquad \int_\Omega \{a\rho + b\sigma\}\, dx = 0 \quad \text{for all } \rho, \sigma \in C_c^\infty(\Omega).$$

---

[6] Cf. e.g. Courant–Hilbert [1–4].

By virtue of the fundamental lemma, we infer that

$$a = 0 \text{ and } b = 0 \quad \text{on } \bar{\Omega}.$$

This reasoning also works if only $\partial\Omega \in C^1$ holds. In this case, we still can find solutions h, k of (16) which are of class $C^2(\Omega)$ and satisfy

$$\int_\Omega |Dh|^2\,dx < \infty \quad \text{and} \quad \int |Dk|^2\,dx < \infty.$$

By an approximation argument, we can again verify relation (17), and the assertion is proved in full generality. □

If $g = (g_{ik})$ is positive definite, we can interpret

$$\langle p_\alpha, p_\beta \rangle_g := g_{ik}(z) p_\alpha^i p_\beta^k$$

as a scalar product on the tangent space $T_z\mathbb{R}^N$ to $\mathbb{R}^N$ at z (which is isomorphic to $\mathbb{R}^N$), and

$$\|p\|_g^2 := (\|p_1\|_g^2 + \|p_2\|_g^2)^{1/2},$$

with

$$\|p_1\|_g := \langle p_1, p_1 \rangle_g^{1/2}, \qquad \|p_2\|_g := \langle p_2, p_2 \rangle_g^{1/2}$$

can be interpreted as "norms". Accordingly, the generalized Dirichlet integral can be written as

$$\int_\Omega \|Du\|_g^2\,dx,$$

and the conformality relations (15) take the form

$$(15') \qquad \|p_1\|_g^2 = \|p_2\|_g^2, \qquad \langle p_1, p_2 \rangle_g = 0.$$

Remark 1. We can consider a third type of variations given by $x = \xi(y, \varepsilon)$, $y \in \bar{\Omega}$, $|\varepsilon| < \varepsilon_0$. Here all transformations $\xi(\varepsilon)$ are defined on the same region $\bar{\Omega}$, but we do no longer assume that $\xi(y, \varepsilon) = y$ holds for $y \in \partial\Omega$ and $|\varepsilon| < \varepsilon_0$. In other words, the functions $\xi(y, \varepsilon)$ will be of the type

$$\xi(y, \varepsilon) = y + \varepsilon\lambda(y) + o(\varepsilon), \quad y \in \bar{\Omega},$$

and $\xi(\varepsilon) = \xi(\cdot, \varepsilon)$ is assumed to be a diffeomorphism of $\bar{\Omega}$ onto itself, if $|\varepsilon| < \varepsilon_0$.

Clearly these transformations are stronger than those of Section *1* because they do not reduce to the identity on $\partial\Omega$ but they are weaker than the variations of Section *2* since $\bar{\Omega}$ is mapped onto itself. For this intermediate type of variations, the infinitesimal generators $\lambda(y)$ are tangent fields along $\partial\Omega$. Hence, by introducing

$$\tau = (\tau_1, \tau_2), \qquad \tau_\alpha := \nu_\beta T_\alpha^\beta = \nu_\beta\{F_{p_\beta^i} u_{x^\alpha}^i - \delta_\alpha^\beta F\},$$

we infer from

$$\frac{d}{d\varepsilon} \mathscr{F}(u \circ \xi(\varepsilon))\Big|_{\varepsilon=0} = 0$$

for all "medium type" variations $\xi(\varepsilon)$ that the Noether equations $L_F(u) \cdot u_{x^\alpha} = 0$ hold and that τ is normal along $\partial\Omega$, i.e.,

$$(18) \qquad \tau = R\nu$$

holds along $\partial\Omega$ for some scalar function R, or equivalently

$$(18') \qquad \nu_\beta T_\alpha^\beta = R\nu_\alpha, \quad \alpha = 1, 2.$$

Suppose that $\partial\Omega$ contains a straight arc Σ parallel to the x^2-axis. Then we have $\nu = (\pm 1, 0)$ on Σ whence $T_2^1 = 0$ or $b = 0$ on Σ, where $b = \langle p_1, p_2 \rangle$. Although this does not suffice to conclude $b = 0$ on $\bar{\Omega}$, we can make this idea effective if Ω is the unit disk centered at the origin by using polar coordinates.

First we note that $f(\zeta) = \langle u_\zeta, u_\zeta \rangle$ and $g(\zeta) := \zeta^2 f(\zeta)$ are holomorphic on Ω, and that

$$rD_r - iD_\theta = \zeta \cdot (D_1 - iD_2) = 2\zeta \frac{\partial}{\partial \zeta},$$

where r, θ are polar coordinates on Ω defined by $\zeta = re^{i\theta}$. Introducing $\mathscr{U}(r, \theta) := u(r\cos\theta, r\sin\theta)$, we obtain

$$\begin{aligned} r^2\langle \mathscr{U}_r, \mathscr{U}_r \rangle - \langle \mathscr{U}_\theta, \mathscr{U}_\theta \rangle - 2ir\langle \mathscr{U}_r, \mathscr{U}_\theta \rangle \\ = 4\zeta^2 \langle u_\zeta, u_\zeta \rangle = \zeta^2 f(\zeta) = g(\zeta). \end{aligned}$$

Moreover, the boundary condition (18) yields $\langle \mathscr{U}_r, \mathscr{U}_\theta \rangle = 0$ on $\partial\Omega$, whence $\operatorname{Im} g(\zeta) = 0$ on $\partial\Omega$, and therefore $g(\zeta) \equiv \text{const}$ on $\bar{\Omega}$. It follows that $f(\zeta) \equiv 0$ on $\bar{\Omega}$.

Since the generalized Dirichlet integral is invariant with respect to conformal mappings of the ζ-plane the same conclusion holds for every simply connected bounded domain Ω of $\mathbb{R}^2$, if we take the Riemann mapping theorem into account. Thus we have found:

Proposition 3′. *If* $\left.\frac{d}{d\varepsilon}\mathscr{F}(u \circ \xi(\varepsilon))\right|_{\varepsilon=0} = 0$ *holds for all medium type variations* $\xi(\varepsilon)$ *of the independent variables, and if* Ω *is a simply connected bounded domain of* $\mathbb{R}^2$ *with* $\partial\Omega \in C^1$, *then the conformality conditions* (15) *are satisfied in* $\bar{\Omega}$.

Remark 2. This result is not anymore true for multiply connected domains as we have used that all simply connected domains are of the same conformal type. Our reasoning actually shows that the conformality relations (15) in the simply connected domain Ω are equivalent to the nonlinear boundary condition $\langle \mathscr{U}_r, \mathscr{U}_\theta \rangle = 0$ on $\partial\Omega$.

Inner extremals and, even more so, strong inner extremals play an important role in the theory of conformal mappings and in differential geometry. To give some idea, we consider surfaces

$$u(x) = (u^1(x), u^2(x), u^3(x)), \quad x \in \bar{\Omega},$$

in $\mathbb{R}^3$ which are defined on bounded parameter domains $\Omega \subset \mathbb{R}^2$, $x = (x^1, x^2)$.

4 We want to investigate variational integrals

$$\mathscr{F}(u) = \int_\Omega \{|Du|^2 + Q(u)\cdot(u_{x^1} \wedge u_{x^2})\}\, dx^1\, dx^2, \tag{19}$$

with the Lagrangians

$$F(z, p) = |p_1|^2 + |p_2|^2 + Q(z)\cdot(p_1 \wedge p_2),$$

where $Q = (Q^1, Q^2, Q^3)$ is a function of class $C^1(\mathbb{R}^3, \mathbb{R}^3)$. If $Q(u) = \frac{1}{3}u$, then $\int_\Omega Q(u)\cdot(u_{x^1} \wedge u_{x^2})\, dx^1\, dx^2$ is the algebraic volume of the cone from the vertex zero to the surface $u(x)$.

A simple computation yields

$$F_{p_1}(z, p) = 2p_1 - Q(z) \wedge p_2, \qquad F_{p_2}(z, p) = 2p_2 + Q(z) \wedge p_1.$$

Let us introduce the function $H \in C^0(\mathbb{R}^3)$ by

$$H = \tfrac{1}{4}\operatorname{div} Q = \tfrac{1}{4}Q^i_{z^i}. \tag{20}$$

Then the Euler equation

$$F_u - D_1 F_{p_1} - D_2 F_{p_2} = 0$$

turns out to be equivalent to

(21) $$\Delta u = 2H(u)u_{x^1} \wedge u_{x^2}.$$

By Proposition 2, the function

(22) $$f(\zeta) = |u_{x^1}|^2 - |u_{x^2}|^2 - 2i\, u_{x^1} \cdot u_{x^2}$$

is holomorphic in $\zeta = x^1 + ix^2$ for every inner extremal u of $\int_\Omega |Du|^2\,dx$, and $f(\zeta) \equiv 0$ if u is a strong inner extremal of the Dirichlet integral. On the other hand, $\int_\Omega Q(u)\cdot(u_{x^1} \wedge u_{x^2})\,dx$ is an invariant integral (cf. *1* [3]). Thus $u \in C^1(\bar{\Omega}, \mathbb{R}^3)$ is a (strong) inner extremal of $\mathscr{F}(u)$ if and only if it is a (strong) inner extremal of the Dirichlet integral, and we can draw the following two conclusions:

(i) *If $u \in C^1(\bar{\Omega}, \mathbb{R}^3) \cap C^2(\Omega, \mathbb{R}^3)$ is both an extremal and a strong inner extremal of $\mathscr{F}$, then it satisfies the Euler equations* (21) *as well as the conformality relations*

$$|u_{x^1}|^2 = |u_{x^2}|^2, \quad u_{x^1} \cdot u_{x^2} = 0.$$

Moreover, the surface $z = u(x)$ has mean curvature $\mathscr{H}(x) = H(u(x))$ at all regular points $x \in \Omega$.

In fact, if we set

$$e = |u_{x^1}|^2, \qquad f = u_{x^1} \cdot u_{x^2}, \qquad g = |u_{x^2}|^2,$$

$$l = X \cdot u_{x^1x^1}, \qquad m = X \cdot u_{x^1x^2}, \qquad n = X \cdot u_{x^2x^2},$$

where $X = |u_{x^1} \wedge u_{x^2}|^{-1} \cdot (u_{x^1} \wedge u_{x^2})$ is the surface normal of u at a regular point $x \in \Omega$, we obtain for $\mathscr{H}(x) := H(u(x))$ that

$$\mathscr{H} = \frac{\Delta u \cdot X}{2|u_{x^1} \wedge u_{x^2}|} = \frac{X \cdot u_{x^1x^1} + X \cdot u_{x^2x^2}}{2\sqrt{eg - f^2}} = \frac{en - 2fm + gl}{2(eg - f^2)}.$$

(ii) *If u is an inner extremal of $\mathscr{F}$, then $f(\zeta)$ defined by* (22) *is a holomorphic function of ζ.*

On the other hand, if u is a C^2-solution of the equation

$$\Delta u = 2H(u)u_{x^1} \wedge u_{x^2},$$

where $H(z)$ is a function of class C^1, we can determine some C^1-vector field $Q = (Q^1, Q^2, Q^3)$ such that $Q^i_{z^i} = 4H$. It follows u is an extremal of (19) and, therefore, u is an inner extremal. Consequently $f(\zeta)$ is a holomorphic function. Thus we have found:

(iii) *For every C^2-solution u of* (21), *$H \in C^1$, the associated function $f(\zeta) = |u_{x^1}|^2 - |u_{x^2}|^2 - 2iu_{x^1} \cdot u_{x^2}$ is holomorphic.*

3. A General Variational Formula

So far we have considered various kinds of (first) variations of a multiple integral which led us to the notions of extremals, inner extremals, and strong inner extremals. In these variations, either the dependent or the independent variables were altered. Now one might wish to unify these two variational aspects by both varying the dependent and the independent variables simultaneously, and to derive a corresponding variational formula. This idea will be carried out in the next Section. Presently we want to take a different point of view. Instead of fixing a domain Ω and a mapping $u : \Omega \to \mathbb{R}^N$ which then both will be varied, we consider a family of functions and a family of domains as well as a family of Lagrangians depending on the same parameter. To see the right picture, the reader should think of this parameter as a time parameter; thus we shall temporarily write t instead of ε. Considering functions $u(x, t)$ we now can interpret them as functions of space x and time t instead of viewing them as families of mappings $u(\cdot, t)$. Correspondingly we consider a given family of domains Ω_t^* varying smoothly in time t, and the Lagrangian $F(x, z, p, t)$ might be time-dependent. We then want to investigate how the function

$$\Phi(t) := \int_{\Omega_t^*} F(x, u(x, t), u_x(x, t), t)\, dt$$

changes in time t. For instance, in fluid mechanics one is led to study the total mass $M(t)$ or the total energy $E(t)$ of a moving fluid contained in varying domains Ω_t^*. (Two instructive examples can be found in [1] and [2] below). We shall compute an expression for $\dot{\Phi}(0)$ which will be called *general variational formula*; it is described in the Proposition below. In the next section this formula will be used to derive the variation of the usual variational integral

$$\mathscr{F}(u) = \int_\Omega F(x, u, Du)\, dx,$$

with respect to simultaneous variations of the dependent and the independent variables. This result will in turn be used to derive Emmy Noether's conservation laws which are to be satisfied by minimizers or stationary points of integrals that are invariant (or at least infinitesimally invariant) with respect to continuous transformation groups. In the next section we shall state these conservation laws as *Emmy Noether's theorem.*

From now on we shall return to our previous notation in order to be able to compare the results of this section with those of Sections *1* and *2*; that is, we shall denote the variational parameter once again by ε instead by t.

Since the emphasis will rest on the formal aspects, we shall assume here and in the next section that all appearing functions are as smooth as necessary. We shall refrain from formulating precise differentiability conditions.

Let us begin by considering a family of domains Ω_ε^*, $|\varepsilon| < \varepsilon_0$, which are "slight variations" of a fixed domain Ω in $\mathbb{R}^n$. More precisely, we consider a smooth family of diffeomorphisms $\eta(\cdot, \varepsilon) : \bar{\Omega} \to \bar{\Omega}_\varepsilon^*$, $|\varepsilon| < \varepsilon_0$, given by

$$y = \eta(x, \varepsilon), \qquad x \in \bar{\Omega}, \qquad |\varepsilon| < \varepsilon_0,$$

where Ω and $\Omega_\varepsilon^* = \eta(\Omega, \varepsilon)$ are domains in $\mathbb{R}^n$. We assume that $\eta(x, 0)$ is the identity map, that $\Omega_0^* = \Omega$. By introducing the *infinitesimal generator*

$$(1) \qquad \mu(x) := \frac{\partial \eta}{\partial \varepsilon}(x, 0), \quad x \in \bar{\Omega},$$

of the family $\eta(x, \varepsilon)$, we can write

$$(2) \qquad \eta(x, \varepsilon) = x + \varepsilon\mu(x) + o(\varepsilon) \quad \text{as } \varepsilon \to 0.$$

Let, secondly, $F(y, z, p, \varepsilon)$ be a family of Lagrangians which are defined for

$(y, z, p, \varepsilon) \in \mathbb{R}^n \times \mathbb{R}^N \times \mathbb{R}^{nN} \times (-\varepsilon_0, \varepsilon_0)$. Then, for any smooth map $z = v(y)$, $y \in \bar{\Omega}_\varepsilon^*$, $|\varepsilon| < \varepsilon_0$, the integral

(3) $$\mathscr{F}(v, \varepsilon, \Omega_\varepsilon^*) := \int_{\Omega_\varepsilon^*} F(y, v(y), Dv(y), \varepsilon)\, dy$$

is defined.

Next we choose a smooth family of mappings $z = v(y, \varepsilon)$, $y \in \bar{\Omega}_\varepsilon^*$, $|\varepsilon| < \varepsilon_0$, with values in $\mathbb{R}^N$, and set

(4) $$u(x) := v(x, 0), \qquad \varphi(x) := \frac{\partial v}{\partial \varepsilon}(x, 0)$$

as well as

(5) $$\Phi(\varepsilon) := \mathscr{F}(v(\varepsilon), \varepsilon, \Omega_\varepsilon^*).$$

In particular, we have

$$\Phi(0) = \mathscr{F}(u, 0, \Omega) = \int_\Omega F(x, u(x), Du(x), 0)\, dx.$$

Lemma. *We have*

(6) $$\Phi'(0) = \int_\Omega \{F_\varepsilon(\dots) + F_{z^i}(\dots)\varphi^i + F_{p_\alpha^i}(\dots)\varphi^i_{x^\alpha} + D_\alpha[F(\dots)\mu^\alpha]\}\, dx,$$

where $(\dots)$ *stands for* $(x, u(x), Du(x), 0)$, *and* $D_\alpha = \dfrac{\partial}{\partial x^\alpha}$.

Proof. Formula (2) implies that

$$\eta^\alpha_{x^\beta}(x, \varepsilon) = \delta^\alpha_\beta + \varepsilon\mu^\alpha_{x^\beta}(x) + o(\varepsilon).$$

Hence $\det(\eta^\alpha_{x^\beta}(x, 0)) = 1$, and therefore

$$\det(\eta^\alpha_{x^\beta}(x, \varepsilon)) > 0 \quad \text{for } x \in \bar{\Omega} \text{ and } |\varepsilon| \ll 1.$$

In addition we get

$$\frac{\partial}{\partial \varepsilon} \det(\eta^\alpha_{x^\beta}(x, \varepsilon))\bigg|_{\varepsilon=0} = \mu^\alpha_{x^\alpha}(x) = \operatorname{div} \mu(x).$$

On account of the transformation rule, we obtain

$$\Phi(\varepsilon) = \int_\Omega F(\eta(x, \varepsilon), v(\eta(x, \varepsilon), \varepsilon), D_y v(\eta(x, \varepsilon), \varepsilon), \varepsilon) \det D_x\eta(x, \varepsilon)\, dx.$$

Denote the integrand of the right-hand side by $f(x, \varepsilon)$. Then it follows that

$$\Phi(\varepsilon) = \int_\Omega f(x, \varepsilon)\, dx$$

and

$$\Phi'(\varepsilon) = \int_\Omega \frac{\partial f}{\partial \varepsilon}(x, \varepsilon)\, dx.$$

The chain rule yields

$$\frac{\partial f}{\partial \varepsilon}(x, 0) = F_{z^i}(\dots)\varphi^i(x) + F_{p^i_\alpha}(\dots)\varphi^i_{x^\alpha}(x) + F_\varepsilon(\dots)$$
$$+ \{D_\alpha F(\dots)\}\mu^\alpha(x) + F(\dots)\mu^\alpha_{x^\alpha},$$

where $(\dots) = (x, u(x), Du(x), 0)$. Noticing that

$$\{D_\alpha F\}\mu^\alpha + F\mu^\alpha_{x^\alpha} = D_\alpha[F\mu^\alpha] = \operatorname{div}[F\mu],$$

the desired formula follows at once. □

We now restrict our attention to the special case of a Lagrangian F that is independent of the parameter ε, i.e., $F = F(x, z, p)$. Then we have

(7) $$\Phi(\varepsilon) = \mathscr{F}(v(\varepsilon), \Omega^*_\varepsilon) = \int_{\Omega^*_\varepsilon} F(y, v(y, \varepsilon), D_y v(y, \varepsilon))\, dy$$

and in particular

$$\Phi(0) = \mathscr{F}(u, \Omega) = \int_\Omega F(x, u(x), Du(x))\, dx.$$

Definition. *The* general variation $\delta^*\mathscr{F}(u; \varphi, \mu)$ *of the functional* $\mathscr{F}(u) = \int_\Omega F(x, u, Du)\, dx$ *in direction of* (φ, μ) *is defined by*

(8) $$\delta^*\mathscr{F}(u; \varphi, \mu) := \Phi'(0).$$

Here φ, μ are the infinitesimal generators of the families $v(y, \varepsilon)$ and $\eta(x, \varepsilon)$, defined by

$$\mu(x) := \frac{\partial \eta}{\partial \varepsilon}(x, 0), \quad \varphi(x) := \frac{\partial v}{\partial \varepsilon}(x, 0), \qquad x \in \bar{\Omega}.$$

By means of the expression

$$\delta\mathscr{F}(u, \varphi) = \int_\Omega \{F_{z^i}(x, u, Du)\varphi^i + F_{p^i_\alpha}(x, u, Du)\varphi^i_{x^\alpha}\}\, dx$$

the result of the lemma can be stated in the following way, where the second formula follows from the first by an integration by parts:

Proposition. *The general variation* $\delta^*\mathscr{F}(u; \varphi, \mu)$, *defined by* (8), *is given by*

(9) $$\delta^*\mathscr{F}(u; \varphi, \mu) = \delta\mathscr{F}(u, \varphi) + \int_\Omega \operatorname{div}[F(x, u, Du)\mu]\, dx$$

or by

$$\delta^*\mathscr{F}(u; \varphi, \mu) = \int_\Omega L_F(u)\cdot\varphi\, dx + \int_\Omega \operatorname{div}[F(x, u, Du)\mu + F_p(x, u, Du)\cdot\varphi]\, dx. \tag{10}$$

Equivalently, we can write

$$\delta^*\mathscr{F}(u; \varphi, \mu) = \delta\mathscr{F}(u, \varphi) + \int_\Omega D_\alpha[F\mu^\alpha]\, dx \tag{9'}$$

and

$$\delta^*\mathscr{F}(u; \varphi, \mu) = \int_\Omega L_F(u)\cdot\varphi\, dx + \int_\Omega D_\alpha[F\mu^\alpha + F_{p^i_\alpha}\varphi^i]\, dx. \tag{10'}$$

Remark 1. If $\eta(x, \varepsilon) = x$ for $|\varepsilon| < \varepsilon_0$, i.e., if $\eta(\varepsilon)$ is the identity map from Ω onto $\Omega = \Omega^*_\varepsilon$ for all ε with $|\varepsilon| < \varepsilon_0$, then $\mu(x) \equiv 0$, and we see that

$$\delta^*\mathscr{F}(u; \varphi, 0) = \delta\mathscr{F}(u, \varphi).$$

Remark 2. On the other hand, if the family $v(y, \varepsilon)$, $y \in \bar\Omega^*_\varepsilon$, has been obtained from a given function $u(x)$, $x \in \bar\Omega$, by setting $v(y, \varepsilon) := u(\xi(y, \varepsilon))$, where $\xi(y, \varepsilon)$, $y \in \bar\Omega^*_\varepsilon$, describes the inverse $\xi(\varepsilon): \bar\Omega^*_\varepsilon \to \bar\Omega$ of the mapping $\eta(\varepsilon): \bar\Omega \to \bar\Omega^*_\varepsilon$, then

$$\mu = -\lambda,$$

where $\lambda(y) := \frac{\partial\xi}{\partial\varepsilon}(y, 0)$ denotes the infinitesimal generator of the family $\xi(\varepsilon)$. From this we conclude that

$$\varphi(x) = \frac{\partial v}{\partial\varepsilon}(x, 0) = Du(x)\cdot\lambda(x) = (u^i_{x^\alpha}(x)\lambda^\alpha(x)),$$

and the Proposition yields

$$\delta^*\mathscr{F}(u; \varphi, \mu) = \delta\mathscr{F}(u, Du\cdot\lambda) - \int_\Omega \operatorname{div}[F\lambda]\, dx$$

or

$$\delta^*\mathscr{F}(u; \varphi, \mu) = \int_\Omega [L_F(u)\cdot(Du\cdot\lambda) - \operatorname{div}\{F\lambda - F_p\cdot(Du\cdot\lambda)\}]\, dx.$$

We notice that

$$\delta\mathscr{F}(u, Du\cdot\lambda) - \int_\Omega \operatorname{div}[F\lambda]\, dx$$

$$= \int_\Omega \{F_{z^i}u^i_{x^\alpha}\lambda^\alpha + F_{p^i_\beta}D_\beta(u^i_{x^\alpha}\lambda^\alpha) - F\lambda^\alpha_{x^\alpha} - (D_\alpha F)\lambda^\alpha\}\, dx$$

$$= \int_\Omega (F_{z^i} u^i_{x^\alpha} \lambda^\alpha + F_{p^i_\beta} u^i_{x^\alpha x^\beta} \lambda^\alpha + F_{p^i_\beta} u^i_{x^\alpha} \lambda^\alpha_{x^\beta} - F \lambda^\alpha_{x^\alpha} - F_{x^\alpha} \lambda^\alpha$$

$$- F_{z^i} u^i_{x^\alpha} \lambda^\alpha - F_{p^i_\beta} u^i_{x^\alpha x^\beta} \lambda^\alpha)\, dx$$

$$= \int_\Omega (-F_{x^\alpha} \lambda^\alpha - F \lambda^\alpha_{x^\alpha} + F_{p^i_\beta} u^i_{x^\alpha} \lambda^\alpha_{x^\beta})\, dx.$$

In other words, the inner variation of $\mathscr{F}$ at u in direction of λ is given by

$$\partial \mathscr{F}(u, \lambda) = \delta^* \mathscr{F}(u; Du \cdot \lambda, -\lambda). \tag{11}$$

We could use this formula as an alternative for defining $\partial \mathscr{F}$ for C^2-functions via $\delta^* \mathscr{F}$, and an approximation procedure would then allow us to define $\partial \mathscr{F}$ for C^1-functions, and to establish the original formula

$$\partial \mathscr{F}(u, \lambda) = \int_\Omega (-F \lambda^\alpha_{x^\alpha} + [F_{p^i_\beta} u^i_{x^\alpha} - \delta^\beta_\alpha F] \lambda^\alpha)\, dx.$$

Remark 3. Often one uses symbols[7] like δu, δx, $\delta^* u$ which are defined by

$$\delta u = \varphi, \qquad \delta x = \mu, \qquad \delta^* u = \delta u + Du \cdot \delta x.$$

Then formula (10′) can be written as

$$\delta^* \mathscr{F}(u; \delta u, \delta x) = \int_\Omega L_F(u) \cdot \delta u\, dx + \int_\Omega \operatorname{div}\{F\, \delta x + F_p\, \delta u\}\, dx. \tag{12}$$

Remark 4. Let us carry over the general variational formulas (9) and (10) to variational integrals

$$\mathscr{F}(u, \Omega) = \int_\Omega F(x, u, Du, D^2 u, \ldots, D^m u)\, dx$$

of m-th order, with Lagrangians

$$F(x, z, p, q, \ldots, s),$$

where $z = (z^i)$, $p = (p^i_\alpha)$, $q = (q^i_{\alpha\beta}), \ldots, s = (s^i_{\alpha_1 \ldots \alpha_m})$. As before, we consider families $\eta(\varepsilon) : \bar{\Omega} \to \bar{\Omega}^*_\varepsilon$ and $v(\varepsilon) : \bar{\Omega}^*_\varepsilon \to \mathbb{R}^N$ with $\Omega = \Omega^*_0$, $\eta(0) = id_{\bar{\Omega}}$, $v(0) = u$, $\mu(x) = \frac{\partial \eta}{\partial \varepsilon}(x, 0)$, $\varphi(x) = \frac{\partial v}{\partial \varepsilon}(x, 0)$, and set

$$\delta^* \mathscr{F}(u; \varphi, \mu) := \Phi'(0),$$

where $\Phi(\varepsilon)$ is defined by

$$\Phi(\varepsilon) = \mathscr{F}(v(\varepsilon), \Omega^*_\varepsilon) = \int_{\Omega^*_\varepsilon} F(y, v(y, \varepsilon), Dv(y, \varepsilon), \ldots, D^m v(y, \varepsilon))\, dy.$$

---

[7] To spare the reader any confusion, we have to remark that in the literature one often finds the notations $\bar{\delta} u = \varphi$ or $\delta^* u = \varphi$ for the "variation of u with respect to fixed arguments", and $\delta u = \bar{\delta} u + Du \cdot \delta x$ or $\delta u = \delta^* u + Du \cdot \delta x$ for the "total" variation of u. Similarly, one finds $\delta \mathscr{F} = \Phi'(0)$ instead of the notation (8). Our notation has the advantage that it integrates the previous definition of $\delta \mathscr{F}$.

It is easy to see that formula (9) remains unaltered:

$$\delta^*\mathscr{F}(u;\varphi,\mu) = \delta\mathscr{F}(u,\varphi) + \int_\Omega \operatorname{div}[F\mu]\,dx, \tag{13}$$

with $F = F(x, u, Du, \ldots, D^m u)$.

The analogue of (10) will be more complicated. We note that $\delta F(u, \varphi)$ can be written as

$$\delta F(u,\varphi) = L_F(u)\cdot\varphi + \operatorname{div} S_F(u,\varphi), \tag{14}$$

where $S_F = (S_F^\alpha)$ is an $\mathbb{R}^n$-valued differential operator with components $S_F^1, \ldots, S_F^n$. The operator $S_F(u, \varphi)$ can easily be computed by a symbolic computation. Recall that

$$\delta F(u,\varphi) = F_z\cdot\varphi + F_p\cdot D\varphi + F_q\cdot D^2\varphi + \cdots + F_s\cdot D^m\varphi, \tag{15}$$

where the arguments of $F_z, F_p, \ldots, F_s$ have to be taken as $(x, u, Du, \ldots, D^m u)$. Consider the term

$$F_r\cdot D^k\varphi = F_{r^i_{\alpha_1\ldots\alpha_k}} D_{\alpha_1}D_{\alpha_2}\ldots D_{\alpha_k}\varphi^i$$

involving the k-th order derivatives of φ. We shall prove that

$$F_r\cdot D^k\varphi = (-1)^k D^k F\cdot\varphi + \operatorname{div} T_k, \quad 1 \le k \le m, \tag{16}$$

for some differential expression $T_k(u, \varphi)$. In fact

$$\begin{aligned} F_r\cdot D^k\varphi &= D[F_r\cdot D^{k-1}\varphi] - DF_r\cdot D^{k-1}\varphi \\ &= D[F_r\cdot D^{k-1}\varphi - DF_r\cdot D^{k-2}\varphi] + D^2F_r\cdot D^{k-2}\varphi \\ &= D[F_r\cdot D^{k-1}\varphi - DF_r\cdot D^{k-2}\varphi + D^2F_r\cdot D^{k-3}\varphi] + D^3F_r\cdot D^{k-3}\varphi \\ &= \cdots, \end{aligned}$$

and finally

$$F_r\cdot D^k\varphi = (-1)^k D^k F_r\cdot\varphi + D[F_r\cdot D^{k-1}\varphi - DF_r\cdot D^{k-2}\varphi + \cdots + (-1)^{k-1}D^{k-1}F_r\cdot\varphi],$$

and (16) is established. From

$$L_F(u) = F_z - DF_p + D^2F_q - \cdots + (-1)^m D^m F_s$$

and from the relations (15) and (16) we then infer that (14) holds if we choose

$$S_F = T_1 + T_2 + \cdots + T_m.$$

We also observe that $S_F(u, \varphi)$ depends on $x, u, Du, \ldots, D^{2m-1}u, \varphi, D\varphi, \ldots, D^{m-1}\varphi$; moreover, $S_F(u, \varphi)$ depends linearly on φ, i.e.,

$$S_F(u, \lambda_1\varphi_1 + \lambda_2\varphi_2) = \lambda_1 S_F(u,\varphi_1) + \lambda_2 S_F(u,\varphi_2)$$

for $\lambda_1, \lambda_2 \in \mathbb{R}$ and two generators φ_1, φ_2.

The formulas (13) and (14) then imply the following analogue of (10):

$$\delta^*\mathscr{F}(u;\varphi,\mu) = \int_\Omega \{L_F(u)\cdot\varphi + \operatorname{div}[F\mu + S_F(u,\varphi)]\}\,dx. \tag{17}$$

Note that

$$\operatorname{div}[F\mu + S_F(u,\varphi)] = D_\alpha[F(x, u, Du, \ldots, D^m u)\mu^\alpha + S_F^\alpha(u,\varphi)].$$

Clearly, the operator $S_F(u, \varphi)$ appearing in (14) is not uniquely defined because we can always add a null divergence to S_F without violating equation (14).

The principal application of our variational formulas (10) will be *Emmy Noether's theorem* treated in the next section. Before we turn to this topic we shall discuss two examples from the theory of fluid motion, the *continuity equation*, and the equations describing the *flow of a compressible, stationary, irrotational fluid*. In particular the continuity equation, expressing the "conservation of mass", is the prototype of a variety of conservation laws which play an important role in continuum physics. However, the following examples are not essential for the understanding of the next section. Thus the reader may as well immediately turn to Section 4 and skip the rest of this section.

[1] *Fluid flow and continuity equation.* Let G be a domain in $\mathbb{R}^3$ and I be an open interval in $\mathbb{R}$. We consider a smooth vector field $v(x, t)$ as a function of the space variable $x \in G$ and of the time $t \in I$.

This vector field $v = (v^1, v^2, v^3)$ is interpreted as the *velocity field of a fluid flow* in G. Such a flow is given by a mapping $y = f(x, t)$ which describes the position y of a particle at the time t that, at the time t_0, was at the locus x. This mapping is determined as the unique solution of the initial value problem

$$(*) \qquad \dot{y} = v(y, t), \qquad y(t_0) = x.$$

That is, f must satisfy the relations

$$\frac{\partial f}{\partial t}(x, t) = v(f(x, t), t), \qquad f(x, t_0) = x.$$

For some time $t_0 \in I$, we fix a domain $\Omega \subset\subset G$ and consider its images $\Omega_t = f(\Omega, t)$ under the mappings $y = f(x, t)$, $x \in \Omega$, with varying time.

Furthermore we assume the existence of a smooth real-valued function $\rho(x, t)$ on $G \times I$ which is describing the *mass density* of the flow. Then the function

$$M(t) := \int_{\Omega_t} \rho(y, t)\, dy$$

describes the *total mass* of the fluid which, at the time t, is contained in Ω_t.

One of the basic assumptions of continuum mechanics is the *law of conservation of mass* stating that, for arbitrary choice of Ω, the mass $M(t)$ is preserved with varying t, i.e.,

$$M(t) \equiv \text{const}.$$

We want to show that this integral condition is equivalent to a differential condition:

The law ofconservation of mass is equivalent to the partial differential equation of first order

$$\frac{\partial \rho}{\partial t} + \operatorname{div}\{\rho v\} = 0.$$

This relation is called *continuity equation.*

In fact, $M(t) \equiv \text{const}$ implies $\dot{M}(t) \equiv 0$. If we set $\varepsilon = t - t_0$, $F(y, \varepsilon) = \rho(y, t_0 + \varepsilon)$, $\eta(x, \varepsilon) = f(x, t_0 + \varepsilon)$, and $\Omega_\varepsilon^* = \Omega_{t_0+\varepsilon}$, then

$$\mu(x) = \eta_\varepsilon(x, 0) = f_t(x, t_0) = v(f(x, t_0), t_0) = v(x, t_0),$$

and we can identify $\dot{M}(t_0)$ with $\Phi'(0)$, where $\Phi(\varepsilon)$ is given by (5). Since $F_z = 0$, $F_p = 0$, it follows that

$$\frac{dM}{dt}(t_0) = \int_\Omega [\rho_t + \operatorname{div}\{\rho v\}]_{t=t_0}\, dx.$$

Because of $\dot{M}(t_0) = 0$, the integral vanishes for arbitrary choice of Ω whence

$$[\rho_t + \operatorname{div}\{\rho v\}]_{t=t_0} = 0 \quad \text{on } G.$$

If we change t_0, only the initial condition in $(*)$ varies, and we obtain a different representation of the flow to which the same reasoning can be applied. Hence the continuity equation must hold for all $t_0 \in I$.

Conversely, the continuity equation yields $\dot{M}(t) \equiv 0$ for any choice of $\Omega \subset\subset G$, and therefore $M(t) \equiv \text{const}$.

Following Stokes, we define the *material derivative* $\frac{D}{Dt}$ of the flow determined by the velocity field $v(x, t)$ as

$$\frac{D}{Dt} = \frac{\partial}{\partial t} + v^\alpha(x, t)\frac{\partial}{\partial x^\alpha} = \frac{\partial}{\partial t} + v(x, t) \cdot D .$$

Then the continuity equation can be written as

$$\frac{D\rho}{Dt} + \rho \operatorname{div} v = 0 .$$

A volume preserving flow is called *incompressible*. For such a flow, the volume $V(t) = \int_{\Omega_t} dy$ is time-invariant for any choice of Ω. Therefore, *incompressible flows are characterized by the equation*

$$\operatorname{div} v = 0 .$$

In other words, *a fluid motion is incompressible if and only if its velocity field $v(x, t)$ is divergence free.*

The *continuity equation of an incompressible fluid motion* then reduces to

$$\frac{D}{Dt}\rho = 0$$

or, equivalently, to

$$\rho_t + v \cdot \operatorname{grad} \rho = 0 .$$

[2] *Stationary, irrotational, isentropic flow of a compressible fluid.* Consider a fluid moving in some domain Ω of $\mathbb{R}^n$, $n = 2$ or 3. Let $\rho(x, t)$ be its mass density and $v(x, t)$ its velocity field. Assuming conservation of mass, we infer from [1] that

$$\rho_t + \operatorname{div}\{\rho v\} = 0 .$$

For *stationary* (or *steady*) *motions*, v and ρ do not depend on t whence the continuity equation reduces to

$$\operatorname{div}\{\rho v\} = 0 .$$

A steady flow is called *irrotational* if

$$\operatorname{curl} v = 0 .$$

Thus a stationary irrotational flow is characterized by the equations

$$\operatorname{div}\{\rho v\} = 0, \qquad \operatorname{curl} v = 0,$$

and for *incompressible* flows of this type, we have

$$\operatorname{div} v = 0, \qquad \operatorname{curl} v = 0 .$$

We mention that, by introducing the 1-form $\omega = \omega_\alpha\, dx^\alpha$, $\omega_\alpha := \delta_{\alpha\beta} v^\beta$, on G, these equations can be written as

$$d\omega = 0, \qquad \delta\omega = 0,$$

where d is the exterior derivative on forms and $\delta = -*d*$ the coderivative, $*$ being the $*$-operator on the Euclidean domain G; cf. the Supplement. *Therefore the velocity fields v of stationary, irrotational, and incompressible fluid flows are corresponding to harmonic 1-forms ω on G.*

On simply connected subdomains G' of G we can introduce the Helmholtz *velocity potential* ϕ by

$$v = \operatorname{grad} \phi$$

(or $\omega = d\phi$). In terms of ϕ, we describe the steady irrotational flow in G' by

$$\operatorname{div}\{\rho \operatorname{grad} \phi\} = 0.$$

If the flow is incompressible, this equation reduces to

$$\Delta\phi = 0 \quad \text{in } G'.$$

Let us introduce the function $Q(x)$ by

$$Q := |v|^2 = |\operatorname{grad} \phi|^2 \quad (= |\omega|^2).$$

We now assume that the fluid is a *polytropic gas*, and that the flow is *isentropic*.

The first assumption means that the *equation of state* of the gas takes the special form

$$p = A(S)\rho^\gamma, \quad \gamma > 1,$$

where p is the pressure, γ the ratio of the specific heats, and $A(S)$ is a positive function of the specific entropy S.

The assumption of an isentropic flow means

$$S = S(x, t) \equiv \text{const},$$

whence the equation of state is

$$p = A\rho^\gamma, \quad \gamma > 1,$$

with some positive constant A.

Bernoulli's law for a steady, irrotational and isentropic flow of a polytropic gas states that

$$\tfrac{1}{2}Q + \frac{\gamma}{\gamma - 1}\frac{p}{\rho} = h \quad (= \text{const}).$$

The function

$$c := \sqrt{\frac{dp}{d\rho}} = \sqrt{A\gamma\rho^{\gamma-1}}$$

is called the *local speed of sound* of the gas.

The steady flow is said to be *subsonic*, *sonic*, or *supersonic* if the *Mach number*

$$M = Q/c$$

is less than, equal to, or greater than one.

Introduce the local speed of sound c_0 and the density ρ_0 at points of stagnation (where $Q = 0$). Then we can write Bernoulli's law as

$$\frac{\gamma - 1}{2}Q + c^2 = c_0^2,$$

where $c_0 = \sqrt{(\gamma - 1)h}$ and also $c_0 = \sqrt{A\gamma\rho_0^{\gamma-1}}$. By choosing suitable units, we may arrange that $c_0 = 1$ and $\rho_0 = 1$ whence $A\gamma = 1$ and $c^2 = \rho^{\gamma-1}$, as well as

$$\frac{\gamma - 1}{2}Q + \rho^{\gamma-1} = 1.$$

Thus we can express the density ρ as a function of the velocity square. Although we should introduce a new symbol for $\rho(Q)$ in order to distinguish it from the function $\rho(x) = \rho(Q(x))$, we keep the notation $\rho(Q)$, as it is customary in fluid mechanics. It follows that

$$\rho = \left\{1 - \frac{\gamma - 1}{2} Q\right\}^{1/(\gamma-1)} \quad \text{and} \quad M^2 = \frac{Q}{1 - \frac{\gamma - 1}{2} Q},$$

and $M = 1$ corresponds to the velocity $Q_* = \dfrac{2}{\gamma + 1}$. We than compute that

$$M^2 = -\frac{2Q}{\rho}\frac{d\rho}{dQ}$$

and

$$1 - M^2 = \frac{1}{\rho}\left\{\rho + 2Q\frac{d\rho}{dQ}\right\} = \frac{1}{\rho^2}\frac{d}{dQ}\{Q\rho^2(Q)\}.$$

That is, the Mach number M is less than one if and only if

$$\frac{d}{dQ}\{Q\rho^2(Q)\} > 0$$

holds.

Let us introduce the functions

$$H(s) := \int_0^s \rho(\tau)\, d\tau$$

and

$$F(p) := H(|p|^2).$$

Then a straight-forward computation shows that the equation

$$\operatorname{div}\{\rho(|\operatorname{grad}\phi|^2)\operatorname{grad}\phi\} = 0$$

or

$$\operatorname{div}\{\rho(|D\phi|^2)D\phi\} = 0$$

is the Euler equation of the variational integral

$$\mathscr{F}(\phi) := \int_{G'} F(D\phi)\, dx.$$

This allows us to treat isentropic, steady irrotational flows of a gas as a variational problem. More generally, the equations

$$d\omega = 0, \quad \delta\{\rho(|\omega|^2)\omega\} = 0$$

are equivalent to the Euler equation of the functional

$$\mathscr{F}(\omega) = \int_G H(|\omega|^2)\, dx$$

if we vary over a given cohomology class of 1-forms ω which is represented by a particular 1-form ω_0 with $d\omega_0 = 0$ and with prescribed periods $\int_\gamma \omega_0$.

4. Emmy Noether's Theorem

The principal aim of this section is to derive Emmy Noether's theorem which states that invariance of a variational integral with respect to some one-parameter variation of the dependent or the independent variables (or both) implies a *conservation law* for any of its extremals. As an important application of this result we obtain conservation laws for extremals of integrals that are invariant under the action of some Lie group.

The starting point of our discussion are mappings $(x, z) \mapsto (x^*, z^*)$ of the type

$$x^* = Y(x, z, \varepsilon), \quad z^* = W(x, z, \varepsilon) \tag{1}$$

depending smoothly on some parameter ε, $|\varepsilon| < \varepsilon_0$, which for $\varepsilon = 0$ reduce to the identity map $(x, z) \to (x, z)$, that is,

$$x = Y(x, z, 0),$$
$$z = W(x, z, 0)$$

for any choice of x and z. For the sake of simplicity we assume that the mapping (1) is defined for all $(x, z) \in \mathbb{R}^n \times \mathbb{R}^N$ and is sufficiently smooth; its values (x^*, z^*) are supposed to be in $\mathbb{R}^n \times \mathbb{R}^N$. Since we only want to describe the formal aspects of the variational technique, we refrain from formulating sharp assumptions which allow us to carry out all computations. The reader can figure out without major difficulties what assumptions are really needed.

We define the **infinitesimal generator** (μ, ω) of the mapping (1) by

$$\mu(x, z) := \frac{\partial Y}{\partial \varepsilon}(x, z, 0), \quad \omega(x, z) := \frac{\partial W}{\partial \varepsilon}(x, z, 0). \tag{2}$$

To attain clear formulas, we shall in the sequel write y for x^* and w for z^*, but in applications we shall use the $*$-notation because it immediately indicates what are the original and the transformed variables, respectively.

For any smooth function $u : \bar{\Omega} \to \mathbb{R}^N$, $\Omega \subset \mathbb{R}^n$, we now apply (1) to $z = u(x)$, thus obtaining the variations

$$\eta(x, \varepsilon) := Y(x, u(x), \varepsilon), \quad w(x, \varepsilon) := W(x, u(x), \varepsilon), \tag{3}$$

$x \in \bar{\Omega}$, $|\varepsilon| < \varepsilon_0$, of the independent variables x and of the dependent variables $z = u(x)$. These variations have the *infinitesimal generators*

$$\bar{\mu}(x) := \frac{\partial \eta}{\partial \varepsilon}(x, 0) = \mu(x, u(x)), \quad \bar{\omega}(x) := \frac{\partial w}{\partial \varepsilon}(x, 0) = \omega(x, u(x)). \tag{4}$$

Since $\eta(x, 0) = x$, it follows that

$$\eta(x, \varepsilon) = x + \varepsilon \bar{\mu}(x) + o(\varepsilon) \quad \text{as } \varepsilon \to 0. \tag{5}$$

Set $\Omega_\varepsilon^* := \eta(\Omega, \varepsilon)$, and let us write $\eta(\varepsilon) = \eta(\cdot, \varepsilon)|_{\bar{\Omega}}$. Because of (5), the mappings $\eta(\varepsilon) : \bar{\Omega} \to \bar{\Omega}_\varepsilon^*$ are diffeomorphisms of $\bar{\Omega}$ onto $\bar{\Omega}_\varepsilon^*$ if $|\varepsilon| < \varepsilon_0$ provided that $\varepsilon_0 > 0$ is sufficiently small. We can view these diffeomorphisms as perturbations of the identity map $x \to x$ of $\bar{\Omega}$ onto itself. By the reasoning used in Section 2 we can assume that $\bar{\mu}$ can be extended to a smooth function on $\mathbb{R}^n$ that has compact support. Thus the inverse $\xi(\varepsilon) : \bar{\Omega}_\varepsilon^* \to \bar{\Omega}$ of $\eta(\varepsilon)$ can be written as

$$\xi(y, \varepsilon) = y - \varepsilon\bar{\mu}(y) + o(\varepsilon) \tag{5'}$$

for any $y \in \bar{\Omega}_\varepsilon^*$.

Consider now the family of functions $w(x, \varepsilon)$, $x \in \bar{\Omega}$, which is defined by (3). By means of the transformation

$$x = \xi(y, \varepsilon), \quad y \in \Omega_\varepsilon^*,$$

we introduce a new family of functions $v : \bar{\Omega}_\varepsilon^* \to \mathbb{R}^N$, setting

$$v(y, \varepsilon) := w(\xi(y, \varepsilon), \varepsilon), \quad y \in \bar{\Omega}_\varepsilon^*. \tag{6}$$

Because of $\Omega_0^* = \Omega$ and $v(y, 0) = w(\xi(y, 0), 0)$ for any $y \in \bar{\Omega}$ we have $u = v(\cdot, 0)$. Hence we can view $v(\cdot, \varepsilon)$, $|\varepsilon| < \varepsilon_0$, as a *variation of* u which has the first variation

$$\varphi(x) := \frac{\partial v}{\partial \varepsilon}(x, 0), \quad x \in \bar{\Omega}. \tag{7}$$

Consider now a Lagrangian $F(x, z, p)$ defined for all $(x, z, p) \in \mathbb{R}^n \times \mathbb{R}^N \times \mathbb{R}^{nN}$ which is at least of class C^2.

Then, for any smooth mapping $u : \bar{\Omega} \to \mathbb{R}^N$ of some bounded domain $\Omega \subset \mathbb{R}^n$, we define the integrals

$$\mathscr{F}(u, \Omega) := \int_\Omega F(x, u(x), Du(x))\, dx \tag{8}$$

and correspondingly

$$\mathscr{F}(v(\varepsilon), \Omega_\varepsilon^*) := \int_{\Omega_\varepsilon^*} F(y, v(y, \varepsilon), v_y(y, \varepsilon))\, dy \tag{9}$$

for the variations $v(\varepsilon) := v(\cdot, \varepsilon)$ of u, where we have set $v_y = D_y v = (v^i_{y^\alpha})$.

Definition. *The functional* $\mathscr{F}(u, \Omega)$ *is said to be* invariant *with respect to the transformation* (1) *if, for any domain* $\Omega \subset \mathbb{R}^n$ *and for any smooth mapping* $u : \bar{\Omega} \to \mathbb{R}^N$, *we have*

$$\mathscr{F}(v(\varepsilon), \Omega_\varepsilon^*) = \mathscr{F}(u, \Omega) \quad \textit{for all } \varepsilon \in (-\varepsilon_0, \varepsilon_0),$$

where the variation $v(\varepsilon)$ *of* u *is defined by* (3), (5′), *and* (6). *Furthermore, we call* $\mathscr{F}(u, \Omega)$ infinitesimally invariant *with respect to* (1) *if we have*

$$\left.\frac{d}{d\varepsilon}\mathscr{F}(v(\varepsilon), \Omega_\varepsilon^*)\right|_{\varepsilon=0} = 0. \tag{10}$$

Evidently, invariance implies infinitesimal invariance. If (1) defines a one-parameter group of mappings then both notions are equivalent.

Now we can formulate

Emmy Noether's theorem. *Suppose that the functional $\mathscr{F}(u, \Omega)$ is invariant or at least infinitesimally invariant with respect to a family of transformations* (1) *which have the infinitesimal generators $\mu = (\mu^1, \dots, \mu^n)$ and $\omega = (\omega^1, \dots, \omega^N)$. Then every extremal $u \in C^2(\Omega, \mathbb{R}^N)$ of $\mathscr{F}(u, \Omega)$ satisfies the "conservation law"*

$$D_\alpha(F_{p^i_\alpha}\omega^i - T^\alpha_\beta \mu^\beta) = 0 \tag{11}$$

or briefly,

$$\operatorname{div}(F_p \cdot \omega - T \cdot \mu) = 0. \tag{11'}$$

Equivalently we can write

$$D_\alpha\{F\mu^\alpha + F_{p^i_\alpha}(\omega^i - u^i_{x^\beta}\mu^\beta)\} = 0 \tag{12}$$

or

$$\operatorname{div}\{F\mu + F_p \cdot (\omega - Du \cdot \mu)\} = 0. \tag{12'}$$

Note that here and in the following, the functions μ and ω actually mean $\mu(x, u(x))$ and $\omega(x, u(x))$ respectively, just as $F_{p^i_\alpha}$ stands for $F_{p^i_\alpha}(x, u(x), Du(x))$, etc. In other words, we sloppily write μ and ω instead of $\bar{\mu}$ and $\bar{\omega}$ (as defined by (4)).

Remark 1. If every solution u of a differential equation $G(x, u, Du, D^2u, \dots) = 0$ *satisfies the relation*

$$\operatorname{div} S(x, u, Du, D^2u, \dots) = 0, \quad \text{or } D_\alpha S^\alpha = 0, \tag{$*$}$$

then one calls $(*)$ a *conservation law* for the differential equation $G = 0$.

The key to Noether's theorem are the variational formulas (9′) and (10′) of the previous Section 3. In our context they imply the following:

Proposition 1. *Let $v(\varepsilon)$, $|\varepsilon| < \varepsilon_0$, be the variation of an arbitrary smooth function $u: \bar{\Omega} \to \mathbb{R}^N$ by means of the transformation* (1) *with the infinitesimal generator (μ, ω), i.e., $v(y, \varepsilon) = W(\xi(y, \varepsilon), \varepsilon)$. Then the derivative $\Phi'(0)$ of $\Phi(\varepsilon) := \mathscr{F}(v_\varepsilon, \Omega^*_\varepsilon)$ is given by*

$$\Phi'(0) = \delta^* \mathscr{F}(u; \varphi, \mu), \tag{13}$$

where

$$\varphi := \omega - Du \cdot \mu = (\omega^1 - u^1_{x^\alpha}\mu^\alpha, \dots, \omega^N - u^N_{x^\alpha}\mu^\alpha),$$

and $\delta^* \mathscr{F}(u; \varphi, \mu)$ is the "general variation" of the functional $\mathscr{F}(u) := \mathscr{F}(u, \Omega)$. Equivalent expressions for $\Phi'(0)$ are given by the formulas

$$\begin{aligned} \Phi'(0) &= \delta\mathscr{F}(u, \varphi) + \int_\Omega \operatorname{div}[F(x, u, Du)\mu]\, dx \\ &= \int_\Omega L_F(u) \cdot \varphi\, dx + \int_\Omega \operatorname{div}\{F\mu + F_p \cdot \varphi\}\, dx \end{aligned} \tag{14}$$

and

$$\Phi'(0) = \delta\mathcal{F}(u, \omega) - \partial\mathcal{F}(u, \mu). \tag{15}$$

Proof. From $y = \xi(y, 0)$ and $v(y, \varepsilon) = w(\xi(y, \varepsilon), \varepsilon)$ we infer

$$\frac{\partial \xi^\beta}{\partial y^\alpha}(y, 0) = \delta_\alpha^\beta, \qquad \frac{\partial v^i}{\partial y^\alpha}(y, 0) = \frac{\partial w^i}{\partial x^\beta}(y, 0)\frac{\partial \xi^\beta}{\partial y^\alpha}(y, 0)$$

for $y \in \bar{\Omega}$, whence

$$v^i_{y^\alpha}(x, 0) = w^i_{x^\alpha}(x, 0) \quad \text{for } x \in \bar{\Omega}.$$

Since $w(x, 0) = u(x)$ on $\bar{\Omega}$, we arrive at

$$v^i_{y^\alpha}(x, 0) = u^i_{x^\alpha}(x) \quad \text{for } x \in \bar{\Omega}.$$

On the other hand, by differentiating

$$w(x, \varepsilon) = v(\eta(x, \varepsilon), \varepsilon)$$

with respect to ε and setting $\varepsilon = 0$, it follows that

$$w^i_\varepsilon(x, 0) = v^i_{y^\alpha}(x, 0)\eta^\alpha_\varepsilon(x, 0) + v^i_\varepsilon(x, 0).$$

Hence the first variation $\varphi = v_\varepsilon(\cdot, 0)$ satisfies

$$\varphi^i = \omega^i - u^i_{x^\alpha}\mu^\alpha, \quad 1 \le i \le N,$$

or

$$\varphi = \omega - Du \cdot \mu.$$

Applying formulas (7)–(10) of Section 3, we then obtain relations (13) and (14). Thus we have

$$\Phi'(0) = \int_\Omega \{F_{u^i}[\omega^i - u^i_{x^\beta}\mu^\beta] + F_{p^i_\alpha} D_\alpha[\omega^i - u^i_{x^\beta}\mu^\beta] + \mu^\alpha D_\alpha F + F D_\alpha \mu^\alpha\}\, dx,$$

where the arguments of F, F_{u^i}, $F_{p^i_\alpha}$ are $(x, u(x), Du(x))$. A straight-forward computation yields

$$\begin{aligned}\{\dots\} &= F_{x^\alpha}\mu^\alpha + F_{u^i}\omega^i + F_{p^i_\alpha}[\omega^i_{x^\alpha} - u^i_{x^\beta}\mu^\beta_{x^\alpha}] + F\mu^\alpha_{x^\alpha} \\ &= (F_{u^i}\omega^i + F_{p^i_\alpha}\omega^i_{x^\alpha}) - (T^\alpha_\beta \mu^\beta_{x^\alpha} - F_{x^\alpha}\mu^\alpha) \\ &= \delta F(u, \omega) - \partial F(u, \mu),\end{aligned}$$

whence we obtain (15). □

A first consequence of Proposition 1 is the following result:

Theorem 1 (E. Noether's identities). *Suppose that the functional $\mathcal{F}(u, \Omega)$ is infinitesimally invariant with respect to the transformation (1) which has the infinitesimal generator (μ, ω). Then, for any smooth mapping $u : \bar{\Omega} \to \mathbb{R}^N$, it follows that*

(16) $$L_F(u)\cdot\omega + (D_\beta T_\alpha^\beta + F_{x^\alpha})\mu^\alpha + D_\alpha[F_{p_\alpha^i}\omega^i - T_\beta^\alpha\mu^\beta] = 0$$

and

(17) $$L_F(u)\cdot\varphi + D_\alpha\{F\mu^\alpha + F_{p_\alpha^i}\varphi^i\} = 0,$$

where $\varphi = (\varphi^1, \ldots, \varphi^N)$ *and* $\varphi^i = \omega^i - u^i_{x^\beta}\mu^\beta$. *The last equation can be written as*

(17′) $$L_F(u)\cdot\varphi + \operatorname{div}\{F\mu + F_p\cdot\varphi\} = 0$$

or as

(17″) $$L_F(u)\cdot\varphi = D_\alpha\{T_\beta^\alpha\mu^\beta - F_{p_\alpha^i}\omega^i\}.$$

In brief, the expression $L_F(u)\cdot\varphi$ *with* $\varphi = \omega - Du\cdot\mu$ *is a divergence.*

Proof. By Proposition 1, infinitesimal invariance of $\mathscr{F}(u, \Omega)$ with respect to (1) implies that

(18) $$\int_\Omega \{L_F(u)\cdot\varphi + D_\alpha[F\mu^\alpha + F_{p_\alpha^i}\varphi^i]\}\,dx = 0$$

and

$$\int_\Omega \{(F_{u^i}\omega^i + F_{p_\alpha^i}\omega^i_{x^\alpha}) + (F_{x^\alpha}\mu^\alpha + T_\beta^\alpha\mu^\beta_{x^\alpha})\}\,dx = 0.$$

By a partial integration, the last equation yields

(19) $$\int_\Omega \{L_F(u)\cdot\omega + (D_\beta T_\alpha^\beta + F_{x^\alpha})\mu^\alpha + D_\alpha[F_{p_\alpha^i}\omega^i - T_\beta^\alpha\mu^\beta]\}\,dx = 0.$$

Now we fix u and choose Ω as a ball $B_r(x_0)$. Dividing both sides of (18) and (19) by meas $B_r(x_0)$ and letting r tend to zero, we arrive at the identities (16) and (17). □

Proof of E. Noether's theorem. Let u be an extremal for the Lagrangian F, i.e., we have the Euler equations

$$L_F(u) = 0.$$

By the results of Section 1, the Euler equations imply the Noether equations

$$D_\beta T_\alpha^\beta + F_{x^\alpha} = 0.$$

Then (12) follows from (17), and (11) is a consequence of (16). By the way, note that (11) can also be inferred from (17″), without using the Noether equation. □

If we replace the assumption of infinitesimal invariance in Theorem 1, i.e., of $\Phi'(0) = 0$ by the weaker supposition that

$$\Phi'(0) = \int_\Omega \operatorname{div}\{\ldots\}\,dx$$

holds for some suitable expression $\{\dots\}$, then the main conclusion of this theorem remains essentially unchanged. Precisely speaking, we obtain the following stronger versions of E. Noether's identities and of E. Noether's theorem:

Theorem 2. *If the integral $\mathcal{F}(u) = \int_\Omega F(x, u, Du)\,dx$ is infinitesimally invariant with respect to the family of transformations* (1) *modulo some divergence, that is, if*

$$\Phi'(0) + \int_\Omega \operatorname{div} A(u, \omega, \mu)\,dx = 0$$

holds for some expression $Q = (A^1, A^2, \dots, A^n)$, then we obtain the identity

$$-L_F(u)\cdot(\omega - Du\cdot\mu) = \operatorname{div}\{A(u, \omega, \mu) + F\mu + F_p\cdot(\omega - Du\cdot\mu)\}.$$

Therefore, every extremal u of $\mathcal{F}$ satisfies

$$\operatorname{div}\{A(u, \omega, \mu) + F\mu + F_p\cdot(\omega - Du\cdot\mu)\} = 0.$$

Remark 2. Our previous discussion remains virtually unchanged if we replace the transformation $(x, z) \to (x^*, z^*)$ given by (1) by a more general transformation of the kind

$$x^* = Y(x, z, p, \varepsilon),$$
$$z^* = W(x, z, p, \varepsilon)$$

satisfying

$$x = Y(x, z, p, 0),$$
$$z = W(x, z, p, 0)$$

for any $(x, z, p) \in \mathbb{R}^n \times \mathbb{R}^N \times \mathbb{R}^{nN}$, or, even more generally, by some mapping

$$x^* = Y(x, z, p, q, \dots, \varepsilon),$$
$$z^* = W(x, z, p, q, \dots, \varepsilon),$$

which for $\varepsilon = 0$ reduces to the identity map with respect to x, z. Then, for any smooth mapping $u: \Omega \to \mathbb{R}^N$, the formulas (3) will be replaced by

$$\eta(x, \varepsilon) := Y(x, u(x), Du(x), \varepsilon),$$
$$w(x, \varepsilon) := W(x, u(x), Du(x), \varepsilon)$$

or by

$$\eta(x, \varepsilon) := Y(x, u(x), Du(x), D^2u(x), \dots, \varepsilon),$$
$$w(x, \varepsilon) := W(x, u(x), Du(x), D^2u(x), \dots, \varepsilon),$$

respectively.

The resulting more general versions of E. Noether's identities and of E. Noether's theorem seem, however, to be rarely used.

Before we turn to examples, we mention that Noether's theorem can easily be carried over to variational integrals

$$\mathcal{F}(u) = \int_\Omega F(x, u, Du, \dots, D^m u)\,dx$$

of m-th order using the formula (14) of Section 3:

$$\delta F(u, \varphi) = L_F(u) \cdot \varphi + \operatorname{div} S_F(u, \varphi),$$

which led to the expression (17) of Section 3 for the general variation of higher order integrals:

$$\delta^* \mathscr{F}(u; \varphi, \mu) = \int_\Omega \{L_F(u) \cdot \varphi + \operatorname{div}[F\mu + S_F(u, \varphi)]\} \, dx.$$

This yields the following analogue of Theorem 2:

Theorem 3. *If there is an expression* $A(u, \omega, \mu) = (A^1, \ldots, A^n)$ *such that*

$$\Phi'(0) + \int_\Omega \operatorname{div} A(u, \omega, \mu) \, dx = 0 \tag{20}$$

holds, ω *and* μ *being the infinitesimal generators defined by* (4), *then we obtain the identity*

$$-L_F(u) \cdot (\omega - Du \cdot \mu) = \operatorname{div}\{A(u, \omega, \mu) + F\mu + S_F(u, \omega - Du \cdot \mu)\}. \tag{21}$$

Thus every extremal u of $\mathscr{F}$ *satisfies*

$$\operatorname{div}\{A(u, \omega, \mu) + F\mu + S_F(u, \omega - Du \cdot \mu)\} = 0. \tag{22}$$

Remark 3. In applications, (1) usually is a one-parameter group leaving the integral $\mathscr{F}(u, \Omega)$ invariant. Clearly, the Theorems 1–3 can easily by carried over to m-parameter groups. If such a group keeps the integral $\mathscr{F}(u, \Omega)$ fixed, we will now obtain m different conservation laws.

One also considers infinite-dimensional transformation groups depending on a number of arbitrary functions. The resulting conservation laws are usually formulated as *Noether's second theorem.* We have refrained from following this custom because Noether's second theorem can be subsumed to the first. Examples [2]–[4] of Section 1 are typical applications of the second Noether theorem.

A detailed account of the relations between variational symmetries and conservation laws can be found in the treatise of Olver [1], pp. 246–377. In particular the one-to-one correspondence between nontrivial conservation laws of the Euler equation and variational symmetries of the associated integral is investigated.

We now consider some examples.

[1] If $n = 1$, $\Omega = (a, b)$, and if the assumptions of the Theorem of Emmy Noether are satisfied, then we infer from (12) that the expression

$$F\mu + F_{p^i} \cdot (\omega^i - u_x^i \mu)$$

is constant on every extremal $z = u(x)$, $a \le x \le b$, and therefore forms a first integral of the Euler equations to $\mathscr{F}$.

Let now $F(z, p)$ be a Lagrangian which does not explicitly depend on the independent variable x. Then

$$\mathscr{F}(u, \Omega) = \int_a^b F(u(x), u'(x)) \, dx$$

does not change its value, if x and u are subject to the family of translations

$$\eta(x, \varepsilon) = x + \varepsilon, \qquad w(x, \varepsilon) = u(x),$$

which has the infinitesimal generators

$$\mu(x) \equiv 1, \qquad \omega(x) \equiv 0 = (0, \ldots, 0) \in \mathbb{R}^N.$$

Then it follows that

$$F - u' \cdot F_p = \text{const}$$

holds for every extremal of $\mathcal{F}$. This is the well known *conservation of energy* that was already stated in 1,*2.2* [7], and in *1* [1].

[2] *The ℓ-body problem and Newton's law of gravitation.* We consider ℓ point masses $m_1, \ldots, m_\ell$ in $\mathbb{R}^3$ whose position at the time t is given by the vectors $X_1(t), \ldots, X_\ell(t)$. We assume that the masses attract each other according to Newton's law of the inverse square. That is, the force F_{ik} exerted from m_k upon m_i is given by

$$F_{ik} = \frac{\kappa m_i m_k}{r_{ik}^3}(X_k - X_i),$$

where $r_{ik} := |X_i - X_k|$.

According to Hamilton's principle, the motion $X(t) := (X_1(t), \ldots, X_\ell(t))$ of the ℓ bodies must be an extremal of the action integral

$$\mathcal{F}(X) = \int_{t_1}^{t_2} F(X, \dot{X})\, dt,$$

the Lagrangian $F(X, Y)$ of which is given by

$$F(X, Y) = T(Y) - V(X),$$

where

$$T(Y) = \tfrac{1}{2} \sum_{j=1}^{\ell} m_j |Y_j|^2, \quad Y = (Y_1, \ldots, Y_\ell),$$

is the expression which, for $Y = \dot{X}(t)$, yields the total kinetic energy of the ℓ points at the time t, and

$$V(X) = -\sum_{i<k} \kappa \frac{m_i m_k}{r_{ik}}$$

is the potential energy of the system, since F_{ik} can be written as

$$F_{ik} = \frac{\partial}{\partial X_i}\left(\kappa \frac{m_i m_k}{r_{ik}}\right)$$

(no summation with respect to i and k), and therefore the total force $F_i = \sum_{k \neq i} F_{ik}$ acting upon m_i is given by

$$F_i = -\operatorname{grad}_{X_i} V;$$

cf. 2,*2* [5] and 2,*3* [3].

Obviously the integral $\mathcal{F}(X)$ does not change its value if X is subject to translations or rotations of $\mathbb{R}^3$ since these transformations neither change the mutual distances r_{ij} of any two points, nor do they change the kinetic energy T.

Denote the old variables by t, x_j, y_j, z_j, and the new ones by t^*, x_j^*, y_j^*, z_j^*, or t, X_j and t^*, X_j^*, respectively.

We begin by considering the 1-*parameter group of translations*

$$t^* = t, \qquad x_j^* = x_j + \varepsilon, \qquad y_j^* = y_j, \qquad z_j^* = z_j \quad (\varepsilon \in \mathbb{R}).$$

Then we have

$$\mu = 0, \qquad \omega = (e_1, e_1, \ldots, e_1),$$

where $e_1 = (1, 0, 0)$, and (11) yields

$$F_{Y_1} \cdot e_1 + F_{Y_2} \cdot e_1 + \cdots + F_{Y_\ell} \cdot e_1 = \text{const}$$

or

$$m_1 \dot{x}_1 + m_2 \dot{x}_2 + \cdots + m_\ell \dot{x}_\ell = \text{const}.$$

An analogous relation holds for the y_j and z_j respectively, and we arrive at

$$\sum_{j=1}^{\ell} m_j \dot{X}_j = A,$$

where A is a constant vector in $\mathbb{R}^3$. This is the law of *conservation of momentum* which can be written as

$$\sum_{j=1}^{\ell} m_j X_j = At + B,$$

with constant vectors $A, B \in \mathbb{R}^3$.

Now we look at the 1-*parameter group of rotations*

$$t^* = t, \quad x_i^* = x_i \cos \varepsilon + y_i \sin \varepsilon,$$
$$y_i^* = -x_i \sin \varepsilon + y_i \cos \varepsilon, \quad z_i^* = z_i,$$

leading to the infinitesimal generators

$$\mu = 0, \qquad \omega = (Z_1, Z_2, \ldots, Z_\ell),$$

where

$$Z_j = (y_j, -x_j, 0).$$

Then (11) implies

$$F_{Y_1} \cdot Z_1 + F_{Y_2} \cdot Z_2 + \cdots + F_{Y_\ell} \cdot Z_\ell = \text{const},$$

or equivalently

$$m_1(y_1 \dot{x}_1 - x_1 \dot{y}_1) + \cdots + m_\ell(y_\ell \dot{x}_\ell - x_\ell \dot{y}_\ell) = \text{const}.$$

This states that the third component of the total angular momentum

$$m_1 X_1 \wedge \dot{X}_1 + \cdots + m_\ell X_\ell \wedge \dot{X}_\ell$$

is preserved. Considering rotations about the y, z-axes or the z, x-axes, we see that also the other two components of the total angular momentum are preserved. Thus we have

$$\sum_{j=1}^{\ell} m_j (X_j \wedge \dot{X}_j) = C,$$

where C is a constant vector in $\mathbb{R}^3$. This is the *law of conservation of the angular momentum.*

The *conservation of energy*, stated in [1], leads to the equation

$$T + V = h,$$

with some constant h. With A, B, C, h we have 10 constants of integration.

The three-body problem requires $18 = 3 \cdot 3 + 3 \cdot 3$ first integrals. Thus 8 first integrals are still missing for the integration of the three-body problem. H. Bruns [1] has shown that, besides of the ten integrals

$$\sum_j m_j X_j, \qquad \sum_j m_j Y_j, \qquad \sum_j m_j X_j \wedge Y_j, \quad \sum_j \frac{m_j}{2} |Y_j|^2 + V(X),$$

there are no other first integrals $\Phi(X, Y)$ of the three-body problem which have algebraic form. For a detailed discussion we refer to the treatise of Whittaker [1], Chapters 13 and 14.

3 *Equilibrium problems in elasticity theory.* Consider an elastic medium such as a *string*, a *membrane* or, in general, any three-dimensional elastic body. One introduces a reference configuration Ω in $\mathbb{R}^n$ ($n = 1, 2, 3$) which can, for instance, be the equilibrium position of the medium in absence of external forces. In case of a taut string we can assume Ω to be a straight interval in $\mathbb{R}$, and for a membrane spanned into a planar curve, Ω can be chosen as a domain in $\mathbb{R}^2$. In general, however, Ω will be a bounded domain in $\mathbb{R}^3$. The deformation, or rather the deviation from the reference configuration, will be described by a mapping $x \to u(x) \in \mathbb{R}^N$, $x \in \Omega$, $N = 1, 2, 3$. In case of the string or membrane, we think of "small" vertical deformations and choose $u(x)$ as height of the distorted medium above or below the straight or planar equilibrium position (i.e., $N = 1$). In general, we have to consider the case $n = N = 3$.

The effect of interior and exterior forces will be accounted for by assuming the existence of an *energy density* $W(x, u, Du)$. The total potential energy $\mathscr{W}(u)$ of the system is then given by

$$\mathscr{W}(u) = \int_\Omega W(x, u, Du)\, dx .$$

For a taut string, the stored energy of the system described by the height function $u(x)$ is assumed to be proportional to the change of length

$$\int_a^b (\sqrt{1 + u'^2} - 1)\, dx ,$$

with respect to the straight equilibrium position.

Since $\sqrt{1 + p^2} - 1 = \frac{1}{2}p^2 +$ higher order terms in p, we obtain in first approximation that the stored energy of a slightly deformed string is described by

$$\frac{\sigma}{2} \int_a^b u'^2\, dx , \quad \sigma = \text{tension factor}.$$

If there acts an external force of density $f(x)$ in vertical direction to the straight reference position, i.e., orthogonally to the x-axis, its potential energy is given by

$$\int_a^b f(x)u(x)\, dx .$$

The total potential energy $\mathscr{W}(u)$ of the string is therefore described by

$$\int_a^b [\tfrac{1}{2}u'^2 + f(x)u]\, dx$$

if we choose $\sigma = 1$ (by suitable scaling).

Similarly the stored energy of a membrane is supposed to be proportional to its change of area

$$\int_\Omega (\sqrt{1 + u_x^2 + u_y^2} - 1)\, dx\, dy$$

in comparison with the planar reference position Ω which is to lie in the x, y-plane. Assuming small deviations and neglecting terms of higher order, we arrive at the expression

$$\frac{\sigma}{2} \int_\Omega (u_x^2 + u_y^2)\, dx\, dy$$

for the stored energy of the membrane, σ being the tension factor. If there again acts a "vertical force" of density $f(x, y)$, the total potential $\mathscr{W}(u)$ of the membrane is given by

$$\int_\Omega [\tfrac{1}{2}(u_x^2 + u_y^2) + f(x, y)u]\, dx\, dy$$

provided that σ is normalized to one.

Sometimes, one has to assume that the energy density W also depends on higher order derivatives: $W = W(x, u, Du, D^2u, \ldots)$. For instance, the potential energy $\mathscr{W}(u)$ of an elastic rod in presence of a vertical force of density $f(x)$ has in first approximation the form

$$\int_a^b [\tfrac{1}{2}(u'')^2 + f(x)u]\, dx,$$

and that of an elastic plate is given by

$$\int_\Omega [(\Delta u)^2 - 2(1-\sigma)(u_{xx}u_{yy} - u_{xy})^2 + f(x, y)u]\, dx\, dy.$$

This is derived by assuming that the density of the stored energy is a quadratic function of the curvature of the rod or of the principal curvatures of the plate, respectively.

If there exist boundary forces, we also must incorporate appropriate boundary terms into the definition of $\mathscr{W}(u)$.

The *equilibrium positions* of a physical system are, by *Johann Bernoulli's principle of virtual work*, characterized as those extremals u of the potential energy $\mathscr{W}(u)$ which, in addition, satisfy prescribed boundary conditions on $\partial\Omega$.

Thus a string in equilibrium satisfies

$$u'' = f(x)$$

and for a membrane in equilibrium, the equation

$$\Delta u = f(x, y)$$

holds. Correspondingly

$$u'''' = -f(x)$$

is fulfilled by a rod in equilibrium, and

$$\Delta\Delta u = -f(x, y)$$

holds for a plate in equilibrium position. (Note that $(u_{xx}u_{yy} - u_{xy}^2) = \det D^2u$ is a null Lagrangian.)

Consider now the stored potential energy

$$\mathscr{W}(u) = \int_\Omega W(x, u(x), Du(x))\, dx$$

of a general elastic body which, in comparison to some reference configuration $\Omega \subset \mathbb{R}^3$, has been deformed into a new position where some point $x \in \Omega$ has been moved into some other position $u(x) \in \mathbb{R}^3$. If there are no body forces, the equilibria are extremals of $\mathscr{W}$.

The structure of $W(x, z, p)$ depends on the nature of the material that forms the body, but in any case it has to be independent of the chosen system of cartesian coordinates in $\mathbb{R}^3$. Therefore we assume W to be invariant with respect to translations $z^* = z + a$ or rotations $z^* = Tz$ of $\mathbb{R}^3$. The first assumption yields

$$W = W(x, p)$$

and therefore the Euler equations become

$$D_\alpha W_{p_\alpha^i}(x, Du) = 0.$$

Applying (11) to $x^* = x$, $z^* = z + \varepsilon a$, we obtain the conservation law $D_\alpha\{W_{p_\alpha^i} a^i\} = 0$, since $\omega = a$ and $\mu = 0$. Thus we merely regain Euler's equations.

If $x^* = x$, $z^* = Tz$, where T is a rotation about the z^3-axis, we obtain[8] $\mu = 0$, $\omega = (u^2, -u^1, 0)$ and (11) yields the conservation law

[8] see 2 of this section.

$$D_\alpha\{W_{p^1_\alpha}u^2 - W_{p^2_\alpha}u^1\} = 0,$$

and similarly two other independent relations which in a uniform way can be written as

$$D_\alpha\{u^i W_{p^k_\alpha} - u^k W_{p^i_\alpha}\} = 0.$$

If the elastic material is thought to be homogeneous, we can assume that W is independent of x, i.e.,

$$W = W(p).$$

Then we have invariance with respect to the translations $x^* = x + \varepsilon a$, $z^* = z$ with $\mu = a$, $\omega = 0$, and (11) implies

$$D_\alpha\{W_{p^i_\alpha}u^i_{x^\beta} - W\delta^\alpha_\beta\} = 0.$$

For isotropic material, the stored energy $\mathscr{W}$ has to be invariant with respect to rotations $x^* = Tx$, $z^* = z$. If, for instance, T is a rotation about the x^3-axis, we find[9] $\mu = (x^2, -x^1, 0)$, and (11) yields

$$D_\alpha\{(x^2\delta^\alpha_1 - x^1\delta^\alpha_2)W + (x^1u^i_{x^2} - x^2u^i_{x^1})W_{p^i_\alpha}\} = 0$$

or, in general,

$$D_\alpha\{(x^\beta\delta^\alpha_\gamma - x^\gamma\delta^\alpha_\beta)W + (x^\gamma u^i_{x^\beta} - x^\beta u^i_{x^\gamma})W_{p^i_\alpha}\} = 0.$$

If, on the other hand, $W(p)$ is assumed to be positively homogeneous of degree s, i.e.,

$$W(tp) = t^s W(p) \quad \text{for } t > 0,$$

then $\mathscr{W}(u)$ turns out to be invariant with respect to the scaling transformations

$$x^* = tx, \quad z^* = t^{1-3/s}z, \qquad t = 1 + \varepsilon,$$

with $\mu(x) = x$, $\omega(x) = \dfrac{s-3}{s}u(x)$. Thus we derive the conservation law

$$D_\alpha\left\{\frac{s-3}{s}u^i W_{p^i_\alpha} + x^\beta(W\delta^\alpha_\beta - W_{p^i_\alpha}u^i_{x^\beta})\right\} = 0.$$

4 *Integration of conservation laws. Hamilton's principle in continuum mechanics.* In physics we often have the situation $n = 4$ and

$$\underline{x} = (x^1, x^2, x^3, x^4) = (x, t) \in \Omega = G \times I,$$

where G is a domain in $\mathbb{R}^3$, $I = (t_1, t_2)$ an interval, $x = (x^1, x^2, x^3)$ the space variables, $t = x^4$ the time. Let $u : \bar{\Omega} \to \mathbb{R}^N$ be an extremal of an integral

$$\mathscr{F}(u) = \int_\Omega F(\underline{x}, u(\underline{x}), Du(\underline{x}))\, d\underline{x},$$

which is invariant with respect to (1), so that we obtain

$$D_\alpha\{F_{p^i_\alpha}\omega^i + (F\delta^\alpha_\beta - F_{p^i_\alpha}u^i_{x^\beta})\mu^\beta\} = 0.$$

Gauss's integration theorem then yields

$$\int_{\partial\Omega} \nu_\alpha\{F_{p^i_\alpha}\omega^i + (F\delta^\alpha_\beta - F_{p^i_\alpha}u^i_{x^\beta})\mu^\beta\}\, d\mathscr{H}^3(x, t) = 0. \tag{23}$$

Suppose now that the part of the boundary integral taken over the cylinder $\partial G \times I$ is zero. Then it follows that the integral

---

[9] see 2 of this section.

$$J(t) := \int_G \{(F_{p_4^i} u_{x^\beta}^i - F\delta_\beta^4)\mu^\beta - F_{p_4^i}\omega^i\}\, dx$$

is independent of t.

Another way to find time-independent integrals is to choose G as a ball $B_R(x_0)$ in $\mathbb{R}^3$, and to let R tend to infinity. Having suitable decay estimates on $u(x, t)$ as $R \to \infty$, the boundary integral over $\partial B_R(x_0) \times I$ will tend to zero. Then it follows that the space integral

$$\int_{\mathbb{R}^3} \{(F_{p_4^i} u_{x^\beta}^i - F\delta_\beta^4)\mu^\beta - F_{p_4^i}\omega^i\}\, dx$$

is independent of t. This situation will occur in *field theories* where fields $u(x, t)$ are spread over all of $\mathbb{R}^3$ and may rapidly decay at infinity.

These results can be considered as an immediate generalization of [1] if we specialize (1) to be the translation group $x^* = x$, $t^* = t + \varepsilon$, $z^* = z$ in t-direction, and assume that the integrand $F(x, u, u_x, u_t)$ does not "explicitly" depend on t. Then (23) is satisfied with $\mu = (0, 0, 0, 1)$ and $\varphi = 0$ and, assuming suitable boundary conditions on ∂G or at infinity, we find that

$$\int_G (F_{p_4^i} u_{x^4}^i - F)\, dx \quad \text{or} \quad \int_{\mathbb{R}^3} (F_{p_4^i} u_{x^4}^i - F)\, dx$$

are time independent.

A particularly important case is that of the integral

$$\mathscr{F}(u) = \int_\Omega \{\tfrac{1}{2}\rho(x)|u_t|^2 - W(x, u, u_x)\}\, dx\, dt, \tag{24}$$

the Lagrangian of which is given by

$$F(x, z, \underline{p}) = \tfrac{1}{2}\rho(x)|p_4|^2 - W(x, z, p),$$

where $\underline{p} = (p_1, p_2, p_3, p_4) = (p, p_4)$, $p = (p_1, p_2, p_3)$. In elasticity theory, $\rho(x) > 0$ denotes the mass density and $W(x, z, p)$ could, for instance, be the stored energy density of an elastic body. Hence, for some domain G in $\mathbb{R}^3$, the integral $T(t) = \int_G \frac{1}{2}\rho|u_t|^2\, dx$ is the total kinetic energy contained in $u(G, t)$ at the time t, and $V(t) = \int_G W(x, u, u_x)\, dx$ is the potential energy of that part of the system which is contained in $u(G, t)$. Then $\mathscr{F}(u)$ is interpreted as *action integral* of a moving body whose motion in time is described by the mapping $u(x, t)$. This motion is ruled by the following general version of

Hamilton's principle. *Among all virtual motions, the actual motion has to satisfy the Euler equations*

$$\rho u_{tt} + L_W(u) = 0$$

of the action integral $\mathscr{F}(u)$.

Here $L_W(u)$ denotes the Lagrange operator

$$L_W(u) = W_z(x, u, u_x) - \frac{\partial}{\partial x^\alpha} W_{p_\alpha}(x, u, u_x)$$

corresponding to W.

Since the action integral (24) clearly is invariant with respect to translations of the time, the integral

$$\int_G \{\tfrac{1}{2}\rho|u_t|^2 + W(x, u, u_x)\}\, dx,$$

expressing the total energy of the system contained in $u(G, t)$ at the time t, will be time independent if the boundary integral

$$\mathscr{R} := \int_{\partial G \times I} \nu_\alpha \{F_{p_\alpha^i}\omega^i + (F\delta_\beta^\alpha - F_{p_\alpha^i}u_{x^\beta}^i)\mu^\beta\}\, d\mathscr{H}^3$$

vanishes, where $\mu = (0, 0, 0, 1)$, $\omega = 0$. Note that $\nu_4 = 0$ on $\partial G \times I$. Therefore $\mathscr{R}$ reduces to

$$\mathscr{R} = -\int_{\partial G \times I} (\nu_1 F_{p_1} + \nu_2 F_{p_2} + \nu_3 F_{p_3}) \cdot u_t \, d\mathscr{H}^3$$

or

$$\mathscr{R} = \int_{\partial G \times I} (\nu_1 W_{p_1} + \nu_2 W_{p_2} + \nu_3 W_{p_3}) \cdot u_t \, d\mathscr{H}^3 .$$

Thus we see that $\mathscr{R}$ vanishes if $u = 0$ on $\partial G \times I$, or if $\nu \cdot W_p = 0$ on $\partial G \times I$, where $\nu = (\nu^1, \nu^2, \nu^3)$ is the exterior normal on ∂G. This expresses the *conservation of total energy* $\int_\Omega \{\frac{1}{2}\rho|u_t|^2 + W\}\, dx\, dt$.

Similarly, we obtain in case of a membrane $u(x^1, x^2, t)$, $(x^1, x^2) \in G$, that the motion is governed by the equation

$$\rho u_{tt} - \Delta u = 0 \quad \text{on } G \times I,$$

$\Delta = D_1^2 + D_2^2$, and the integral

$$\int_G \tfrac{1}{2}\{\rho|u_t|^2 + |u_{x^1}|^2 + |u_{x^2}|^2\}\, dx^1\, dx^2$$

will be time-independent if either

$$u(x^1, x^2, t) = 0 \quad \text{for } (x^1, x^2) \in \partial G,\ t \in I,$$

or if

$$\frac{\partial u}{\partial \nu}(x^1, x^2, t) = 0 \quad \text{on } \partial G \times I,$$

where $\frac{\partial u}{\partial \nu} = \nu_1 u_{x^1} + \nu_2 u_{x^2}$ denotes the normal derivative of $u(x, t)$ in direction of the exterior normal $\nu = (\nu^1, \nu^2)$ to ∂G.

5 *Killing equations.* There is an intimate relation between a Lagrangian $F(x, z, p)$ and the infinitesimal transformations which leave the functional

$$\mathscr{F}(u, \Omega) = \int_\Omega F(x, u(x), Du(x))\, dx$$

infinitesimally invariant. We shall see that the conservation laws implied by Noether's theorem can be used to derive necessary conditions for the "admissible" infinitesimal transformations in form of a system of first order partial differential equations. For this purpose we consider a one-parameter family of transformations

$$\begin{aligned} x^* &= Y(x, z, \varepsilon) = x + \varepsilon\xi(x, z) + o(\varepsilon), \\ z^* &= W(x, z, \varepsilon) = z + \varepsilon\zeta(x, z) + o(\varepsilon) \end{aligned} \tag{25}$$

depending smoothly on ε, $|\varepsilon| < \varepsilon_0$; the infinitesimal transformations of x and z are given by $\xi(x, z)$ and $\zeta(x, z)$ respectively. Now, for an arbitrary C^2-function $z = u(x)$ we introduce

$$\mu(x) := \xi(x, u(x)) \quad \text{and} \quad \omega(x) := \zeta(x, u(x)). \tag{26}$$

Setting $p(x) := Du(x)$, we obtain

$$D_\alpha \mu^\beta = \xi_{x^\alpha}^\beta + \xi_{z^i}^\beta p_\alpha^i, \qquad D_\alpha \omega^i = \zeta_{x^\alpha}^i + \zeta_{z^k}^i p_\alpha^k \tag{27}$$

(where the arguments of $\xi_{x^\alpha}^\beta$, $\xi_{z^i}^\beta$, $\zeta_{x^\alpha}^i$, $\zeta_{z^k}^i$ are to be taken as x, $u(x)$).

Suppose that $\mathscr{F}(\cdot, \Omega)$ is infinitesimally invariant with respect to the family of transformations (25) and that $u(x)$ is an extremal of $\mathscr{F}(\cdot, \Omega)$. Then, by Noether's theorem, we obtain the relation

$$D_\alpha\{F\mu^\alpha - p_\beta^k F_{p_\alpha^k}\mu^\beta + F_{p_\alpha^k}\omega^k\} = 0 \tag{28}$$

along $(x, z, p) = (x, u(x), Du(x))$. By Euler's equations we can replace $D_\alpha F_{p_\alpha^k}$ by F_{z^k}; then we derive from (28) by a straight-forward computation that

$$F_{x^\alpha}\mu^\alpha + FD_\alpha\mu^\alpha + F_{z^k}\omega^k + F_{p_\alpha^k}\{D_\alpha\omega^k - p_\beta^k D_\alpha\mu^\beta\} = 0 \tag{29}$$

holds along the extremal $u(x)$. Employing (27), we arrive at

$$F_{x^\alpha}\xi^\alpha + F\xi^\alpha_{x^\alpha} + F\xi^\alpha_{z^i}p_\alpha^i + F_{z^k}\zeta^k + F_{p_\alpha^k}\{\zeta^k_{x^\alpha} + \zeta^k_{z^i}p_\alpha^i - p_\beta^k[\xi^\beta_{x^\alpha} + \xi^\beta_{z^i}p_\alpha^i]\} = 0 \tag{30}$$

along $(x, z, p) = (x, u(x), Du(x))$.

Now we require for $F(x, z, p)$ a *weak nondegeneracy condition* which is expressed as

Condition (N). *For every point (x_0, z_0, p_0) in $\mathbb{R}^n \times \mathbb{R}^N \times \mathbb{R}^{nN}$ (or in a suitable subdomain thereof) we can find a solution $u(x)$, $|x - x_0| < R \ll 1$, of the Euler equation $L_F(u) = 0$ such that $u(x_0) = z_0$ and $Du(x_0) = p_0$.*

It is easy to see that Condition (N) is satisfied in all "reasonable" geometric or physical applications.

If Condition (N) is satisfied, identity (30) is obviously satisfied for an arbitrary choice of the variables x, z, p. Then, from (30) we can derive first order differential equations for the infinitesimal transformations $\xi(x, z)$ and $\zeta(x, z)$ by differentiating both sides with respect to the variables p_α^i, $p_\beta^k, \ldots$, and then fixing some value of p.

Let us work out this idea in case of the generalized Dirichlet integral

$$\mathscr{F}(u, \Omega) = \tfrac{1}{2}\int_\Omega g_{ik}(u)D_\alpha u^i D_\alpha u^k\,dx \tag{31}$$

which has the Lagrangian

$$F(z, p) = \tfrac{1}{2}g_{ik}(z)p_\alpha^i p_\alpha^k. \tag{32}$$

If the matrix (g_{ik}) is assumed to be symmetric and invertible, the corresponding Euler equations are equivalent to

$$\Delta u^\ell + \Gamma^\ell_{ik}(u)D_\alpha u^i D_\alpha u^k = 0, \tag{33}$$

whence it is easy to see that F satisfies Condition (N). Suppose now that (31) is infinitesimally invariant with respect to a one-parameter family of variations

$$\begin{aligned} x^* &= x, \\ z^* &= z + \varepsilon\zeta(z) + o(\varepsilon), \quad |\varepsilon| < \varepsilon_0. \end{aligned} \tag{34}$$

Then (30) reduces to

$$F_{z^k}\zeta^k + F_{p_\alpha^k}\zeta^k_{z^i}p_\alpha^i = 0. \tag{35}$$

Since (32) yields

$$F_{z^k} = \tfrac{1}{2}\frac{\partial g_{ij}}{\partial z^k}p_\alpha^i p_\alpha^j, \qquad F_{p_\alpha^k} = \tfrac{1}{2}\{g_{kj}p_\alpha^j + g_{ik}p_\alpha^i\},$$

we arrive at

$$\tfrac{1}{2}\left\{\frac{\partial g_{ij}}{\partial z^k}\zeta^k + g_{kj}\zeta^k_{z^i} + g_{ik}\zeta^k_{z^j}\right\}p_\alpha^i p_\alpha^j = 0,$$

whence

$$\frac{\partial g_{ij}}{\partial z^k}\zeta^k + g_{kj}\zeta^k_{z^i} + g_{ik}\zeta^k_{z^j} = 0, \quad 1 \le i, j \le N. \tag{36}$$

These are the *Killing equations* of Riemannian geometry derived by Killing [1] in 1892.

Note that the value of n does not enter. With respect to the one-dimensional Dirichlet integral

$$\mathscr{F}(u, I) = \tfrac{1}{2}\int_a^b g_{ik}(u)\dot{u}^i\dot{u}^k\,dx,$$

$I = (a, b)$, $\dot{u}^k = \dfrac{du^k}{dx}$, our result state that *the infinitesimal generator* $\zeta(z) = (\zeta^1(z), \ldots, \zeta^N(z))$ *of any one-parameter group of isometries*

$$z \to z^* = z + \varepsilon\zeta(z) + o(\varepsilon),$$

with respect to a Riemannian line element

$$ds^2 = g_{ik}(z)\,dz^i\,dz^k$$

satisfies the Killing equations (36).

5. Transformation of the Euler Operator to New Coordinates

In this section we shall investigate how the Euler operator $L_F(u)$ varies with respect to changes of the dependent and the independent variables by diffeomorphisms. As in Sections 3 and 4, the considerations will be purely formal. Therefore we shall not specify where the diffeomorphisms and their inverses are defined. It will be assumed that they are sufficiently often continuously differentiable, that their domains of definition include the sets on which they are supposedly acting, and that they have positive Jacobians.

Actually the discussion of this section is slightly obscured by the fact that we confine our considerations to $\mathbb{R}^N$ instead of working with N-dimensional manifolds. For $\mathbb{R}^N$, it is difficult to see any difference between $\mathbb{R}^N$ and its tangent spaces $T_z\mathbb{R}^N$ since they are isomorphic to each other. Nevertheless the reader should distinguish between points of $\mathbb{R}^N$, tangent vectors to $\mathbb{R}^N$ at some point z of $\mathbb{R}^N$, and tangent and cotangent vector fields. At the present stage of our discussion we can distinguish these quantities only by their different transformation behaviour. For a more thorough discussion we refer the reader to the mathematical literature concerning manifolds and tensor calculus.

We begin by changing the dependent variables, thereby proving that $L_F(u)$ transforms contragrediently to tangent vectors of curves. Thus $L_F(u)$ can be considered as a field of cotangent vectors along the mapping $u: \bar{\Omega} \to \mathbb{R}^N$, $\Omega \subset \mathbb{R}^n$, i.e. as a covector field along u.

Then we vary the independent variables and show that the Euler operator remains essentially invariant.

Finally the transformation character of the Euler operator with respect to changes

$$x^* = Y(x, z), \qquad z^* = W(x, z)$$

of both the dependent and the independent variables are studied.

Consider now a diffeomorphism on $\mathbb{R}^n \times \mathbb{R}^N$, given by

$$x^* = x,$$
$$z^* = W(x, z),$$

with the inverse

$$x = x^*,$$
$$z = Z(x^*, z^*).$$

For a given function $u : \bar{\Omega} \to \mathbb{R}^N$, $\Omega \subset \mathbb{R}^n$, we define the transform $w : \bar{\Omega} \to \mathbb{R}^n$ by

$$w(x) := W(x, u(x)),$$

whence

$$u(x) := Z(x, w(x)).$$

Then

$$\begin{aligned}\mathscr{F}(u) &= \int_\Omega F(x, u, u_x)\, dx \\ &= \int_\Omega F(x, Z(x, w), Z_x(x, w) + Z_{z^*}(x, w)\cdot w_x)\, dx.\end{aligned}$$

Setting

$$G(x, z^*, q) := F(x, Z(x, z^*), Z_x(x, z^*) + Z_{z^*}(x, z^*)\cdot q), \tag{1}$$

we obtain that

$$\mathscr{F}(u) = \int_\Omega F(x, u, u_x)\, dx = \int_\Omega G(x, w, w_x)\, dx = \mathscr{G}(w).$$

Let $\bar{w}(x, \varepsilon) := w(x) + \varepsilon\zeta(x)$ be a variation of $w(x)$ with $\operatorname{supp} \zeta \subset \Omega$. Then $\bar{u}(x, \varepsilon) := Z(x, \bar{w}(x, \varepsilon))$ is a variation of $u(x)$, and $\varphi(x) := \dfrac{\partial \bar{u}}{\partial \varepsilon}(x, 0)$ also has support in Ω. From

$$\mathscr{F}(\bar{u}(\varepsilon)) = \mathscr{G}(\bar{w}(\varepsilon)),$$

it follows that

$$\delta\mathscr{F}(u, \varphi) = \delta\mathscr{F}(w, \zeta),$$

whence

$$\int_\Omega L_F(u)\cdot\varphi\, dx = \int_\Omega L_G(w)\cdot\zeta\, dx.$$

Moreover, the equation $\bar{u}(x, \varepsilon) = Z(x, \bar{w}(x, \varepsilon))$ implies that

$$\varphi(x) = Z_{z*}(x, w(x)) \cdot \zeta(x),$$

and therefore we obtain

$$\int_\Omega \{L_F(u) \cdot Z_{z*}(x, w) - L_G(w)\} \cdot \zeta \, dx = 0$$

for all $\zeta \in C_c^\infty(\Omega, \mathbb{R}^N)$. This implies

(2) $$L_F(u) \cdot Z_{z*}(x, w) = L_G(w).$$

If, on the other hand, $z = c(t)$ is a curve in $\mathbb{R}^N$ and $z^* = W(x, c(t)) =: c^*(t)$ its transform, then we also have $c = Z(x, c^*)$, and therefore the tangent vectors $\dot{c}$ and $\dot{c}^*$ are related to each other by the equation

(3) $$\dot{c} = Z_{z*}(x, c^*) \cdot \dot{c}^*.$$

The transformation rule (2) is contragredient to (3) whence we infer that $L_F(u)$ *can be interpreted as a cotangent vector.*

Let us discuss an example.

[1] We consider the generalized Dirichlet integral

$$\mathscr{D}(u) = \int_\Omega \tfrac{1}{2} g_{ik}(u) D_\alpha u^i D_\alpha u^k \, dx, \quad g_{ik} = g_{ki},$$

that was defined in 2,4 [3]. We had shown that its Euler operator $L(u) = (L_1(u), \ldots, L_N(u))$ is given by

$$-L_k(u) = g_{ik}(u) \Delta u^i + \Gamma_{ikj}(u) D_\alpha u^i D_\alpha u^j.$$

Suppose now that $g_{ik}(z)$ are components of a (0, 2)-tensor field, and that $G(z) = (g_{ik}(z))$ is invertible. Then $G^{-1}(z) = (g^{ik}(z))$ are the components of a (2, 0)-tensorfield, and $L^k(u) := g^{kl}(u) L_l(u)$ are the components of a vector field which can be written as

$$-L^k(u) = \Delta u^k + \Gamma_{ij}^k(u) D_\alpha u^i D_\alpha u^j.$$

Consider the transformation $\bar{z} = W(z)$ or, equivalently, $z = Z(\bar{z})$ which transforms a function $u(x)$ into $w(x) = W(u(x))$. Set

$$\bar{g}_{lm}(\bar{z}) := g_{ik}(Z(\bar{z})) Z^i_{\bar{z}^l}(\bar{z}) Z^k_{\bar{z}^m}(\bar{z}).$$

Then $\bar{g}_{lm}$ are the z-coordinates of the(0, 2)-tensor field, the x-coordinates of which are given by g_{ik}. Since $u(x) = Z(w(x))$, we infer that

$$\tfrac{1}{2} g_{ik}(u) D_\alpha u^i D_\alpha u^k = \tfrac{1}{2} \overline{g_{ik}}(w) D_\alpha w^i D_\alpha w^k$$

or

$$\int_\Omega \tfrac{1}{2} g_{ik}(u) D_\alpha u^i D_\alpha u^k \, dx = \int_\Omega \tfrac{1}{2} \overline{g_{ik}}(w) D_\alpha w^i D_\alpha w^k \, dx.$$

By (2), the Euler operators $L(u)$ and $\bar{L}(w)$ of these two integrals are related to each other by

$$L_k(u) Z^k_{\bar{z}^l}(w) = \bar{L}_l(w),$$

where

$$-\overline{L}_l(w) = \overline{g_{il}}(w)\Delta w^i + \overline{\Gamma}_{ilj}(w)D_\alpha w^i D_\alpha w^j,$$

and $\overline{\Gamma}_{ilj}$ denote the Christoffel symbols for $\overline{g_{ik}}$.

Now we turn to *transformations* $x \mapsto y$ *of the independent variables* x given by

$$y = \eta(x).$$

Denote the inverse mapping by

$$x = \xi(y),$$

and suppose that $\eta(\overline{\Omega}) = \overline{\Omega}^*$, i.e.,

$$\eta : \overline{\Omega} \to \overline{\Omega}^* \quad \text{and} \quad \xi : \overline{\Omega}^* \to \overline{\Omega}.$$

We, in addition, set

$$A(y) := \xi_y^{-1}(y) = \eta_x(\xi(y)), \quad B(y) := \xi_y(y).$$

For any Lagrangian $G(x, z, p)$ on $\overline{\Omega} \times \mathbb{R}^N \times \mathbb{R}^{nN}$ we define a *pullback Lagrangian* $H(y, z, q)$ on $\overline{\Omega}^* \times \mathbb{R}^N \times \mathbb{R}^{nN}$ by

(4) $$H(y, z, q) := G(\xi(y), z, q \cdot A(y)) \det B(y).$$

Then a straight-forward computation shows that

$$\mathscr{G}(w) = \int_\Omega G(x, w, w_x)\, dx = \int_{\Omega^*} H(y, v, v_y)\, dy = \mathscr{H}(v)$$

for every function $z = w(x)$ and its transform $v(y) := w(\xi(y))$.

We want to show that *the Euler operators* $L_G(w)$ *and* $L_H(v)$ *of* $\mathscr{G}(w)$ *and* $\mathscr{H}(v)$ *relate to each other by*

(5) $$(\det \xi_y) L_G(w) \circ \xi = L_H(v).$$

To this end, we consider a variation $\overline{w}(x, \varepsilon) = w(x) + \varepsilon\zeta(x)$ of $w(x)$ with $\operatorname{supp} \zeta \subset \Omega$, and set $\tilde{\zeta} := \zeta \circ \xi$. Then $\overline{v}(y, \varepsilon) = v(y) + \varepsilon\tilde{\zeta}(y)$ is a variation of v with $\operatorname{supp} \tilde{\zeta} \subset \Omega$. From $\mathscr{G}(\overline{w}(\varepsilon)) = \mathscr{H}(\overline{v}(\varepsilon))$ we infer

$$\left.\frac{d}{d\varepsilon}\mathscr{G}(\overline{w}(\varepsilon))\right|_{\varepsilon=0} = \left.\frac{d}{d\varepsilon}\mathscr{H}(\overline{v}(\varepsilon))\right|_{\varepsilon=0},$$

and this implies

$$\delta\mathscr{G}(w, \zeta) = \delta\mathscr{H}(v, \tilde{\zeta}).$$

Consequently, we obtain

$$\begin{aligned}\int_\Omega L_G(w) \cdot \zeta\, dx &= \int_{\Omega^*} L_H(v) \cdot \tilde{\zeta}\, dy \\ &= \int_\Omega \{L_H(v) \circ \eta\} \cdot \zeta \det \eta_x\, dx\end{aligned}$$

since – as always – $\det \eta_x$ is assumed to be positive. On account of the fundamental lemma, we infer that

$$L_G(w) = \{L_H(v) \circ \eta\} \det \eta_x$$

and this is equivalent to (5).

2 Relation (5) can be used to transform differential operators with minimal effort to general curvilinear coordinates if they can be written as Euler operators of suitable Lagrangians. Let us demonstrate this for the Laplace operator $\Delta = \delta^{\alpha\beta} D_\alpha D_\beta$, $D_\alpha = \frac{\partial}{\partial x^\alpha}$. We recall that the Euler operator $L(u)$ of the Dirichlet integral

$$\frac{1}{2}\int_\Omega |Du|^2\, dx$$

is nothing but $L(u) = -\Delta u$. Consider now the transformation

$$y = \eta(x), \qquad x \in \bar{\Omega},$$

and its inverse

$$x = \xi(y), \qquad y \in \bar{\Omega}^*, \quad \text{with } \det \xi_y > 0.$$

For a mapping $u : \bar{\Omega} \to \mathbb{R}$ we consider the transform $v(y) := u(\xi(y))$, $y \in \bar{\Omega}^*$. Then we infer from (4) that the adjoint integrand $H(y, q)$ to $G(p) = \frac{1}{2}|p|^2$ is given by

$$\begin{aligned} H(y, q) &= \tfrac{1}{2}|q \cdot A(y)|^2 \det B(y) \\ &= \tfrac{1}{2} q A(y) A^T(y) q^T \det B(y) \\ &= \tfrac{1}{2} q [B^T(y) B(y)]^{-1} q^T \det B(y). \end{aligned}$$

Let $\gamma_{\alpha\beta}$ be the entries of the matrix $B^T B$ and $\gamma^{\alpha\beta}$ the entries of $(B^T B)^{-1}$, and set $\gamma := \det(\gamma_{\alpha\beta})$. The quantities $\gamma_{\alpha\beta}$, $\gamma^{\alpha\beta}$ and γ are functions of y, and

$$\begin{aligned} \gamma_{\alpha\beta} &= \xi_{y^\alpha} \cdot \xi_{y^\beta} = \delta_{\mu\nu} \xi^\mu_{y^\alpha} \xi^\nu_{y^\beta}, \\ \gamma^{\alpha\beta}(y) &= \delta^{\mu\nu} \eta^\alpha_{x^\mu}(x) \eta^\beta_{x^\nu}(x) \quad \text{with } x = \xi(y), \\ \sqrt{\gamma} &= \det B. \end{aligned}$$

Then we conclude that

$$H(y, q) = \tfrac{1}{2}\gamma^{\alpha\beta}(y) q_\alpha q_\beta \sqrt{\gamma(y)},$$

and therefore

$$\int_\Omega \tfrac{1}{2}|Du|^2\, dx = \int_{\Omega^*} \tfrac{1}{2}\gamma^{\alpha\beta} D_\alpha v D_\beta v \sqrt{\gamma}\, dy, \quad D_\alpha = \frac{\partial}{\partial y^\alpha}.$$

We then infer from (5) that

$$(\Delta u)(x) = \frac{1}{\sqrt{\gamma(y)}} D_\beta\{\sqrt{\gamma(y)}\gamma^{\alpha\beta}(y) D_\alpha v(y)\}, \quad x = \xi(y).$$

Of particular interest are the so-called *orthogonal curvilinear coordinates* y which are defined by the conditions $\gamma_{\alpha\beta} = 0$ for $\alpha \neq \beta$ or, equivalently, by $\gamma^{\alpha\beta} = 0$ for $\alpha \neq \beta$. The latter relations can also be written as

$$\operatorname{grad} \eta^\alpha \cdot \operatorname{grad} \eta^\beta = 0 \quad \text{if } \alpha \neq \beta,$$

which means that the level surfaces

$$\mathscr{S}^\alpha = \{x: \eta^\alpha(x) = \text{const}\}$$

intersect each other orthogonally.

For orthogonal coordinates y the matrices $(\gamma_{\alpha\beta})$ and $(\gamma^{\alpha\beta})$ reduce to

$$(\gamma_{\alpha\beta}) = \begin{bmatrix} g_1 & & & 0 \\ & g_2 & & \\ & & \ddots & \\ 0 & & & g_n \end{bmatrix}, \quad (\gamma^{\alpha\beta}) = \begin{bmatrix} 1/g_1 & & & 0 \\ & 1/g_2 & & \\ & & \ddots & \\ 0 & & & 1/g_n \end{bmatrix}$$

and $\sqrt{\gamma} = \sqrt{g_1 g_2 \cdots g_n}$. Therefore, the transformation formula reduces to

$$\Delta u \circ \xi = \frac{1}{\sqrt{\gamma}}\left\{D_1\left(\frac{\sqrt{\gamma}}{g_1} D_1 v\right) + \cdots + D_n\left(\frac{\sqrt{\gamma}}{g_n} D_n v\right)\right\}.$$

Particular orthogonal coordinates are furnished by the so-called *polar coordinates*. For $n = 2$ we have $x^1 = r\cos\varphi$, $x^2 = r\sin\varphi$, $g_1 = 1$, $g_2 = r^2$, $u(x^1, x^2) = v(r, \varphi)$, and

$$\Delta u = \frac{1}{r}\left\{\frac{\partial}{\partial r}(r v_r) + \frac{\partial}{\partial \varphi}\left(\frac{1}{r} v_\varphi\right)\right\}.$$

For $n = 3$, the polar coordinates r, φ, θ are introduced by $x^1 = r\cos\varphi\sin\theta$, $x^2 = r\sin\varphi\sin\theta$, $x^3 = r\cos\theta$, whence $g_1 = 1$, $g_2 = r^2\sin^2\theta$, $g_3 = r^2$, and we get for $u(x^1, x^2, x^3) = v(r, \varphi, \theta)$ the formula

$$\Delta u = \frac{1}{r^2\sin\theta}\left\{\frac{\partial}{\partial r}(r^2 v_r \sin\theta) + \frac{\partial}{\partial\varphi}\left(\frac{1}{\sin\theta} v_\varphi\right) + \frac{\partial}{\partial\theta}(u_\theta \sin\theta)\right\}.$$

Correspondingly, the Dirichlet integral on a ball of radius R takes the form

$$\frac{1}{2}\int_0^{2\pi}\left(\int_0^R (|v_r|^2 + r^{-2}|v_\varphi|^2) r\, dr\right) d\varphi$$

and

$$\frac{1}{2}\int_0^{2\pi}\left(\int_0^{\pi/2} \sin\theta\left(\int_0^R \left(|v_r|^2 + \frac{1}{r^2\sin\theta}|v_\varphi|^2 + \frac{1}{r^2}|v_\theta|^2\right) r^2\, dr\right) d\theta\, d\varphi\right.$$

if $n = 2$ and 3 respectively.

Other important orthogonal coordinates are the so-called *elliptic coordinates* that were used by Jacobi[10] for determining the geodesics on a general ellipsoid. Elliptic coordinates in the plane were invented by Euler who applied them to treat the motion of a point mass under the influence of two fixed centers of gravity. Legendre employed elliptic coordinates to express the surface area of an ellipsoid in terms of the length of elliptic arcs, just as Archimedes reduced the determination of the area of a sphere to the length of circular arcs. But only Jacobi developed these coordinates to a powerful analytic tool.

[3] *The Laplace–Beltrami operator.*[11] The previous example motivates the definition of the Dirichlet integral

$$\mathscr{D}(u) = \int_\Omega \tfrac{1}{2}\gamma^{\alpha\beta}(x) u_{x^\alpha} u_{x^\beta} \sqrt{\gamma(x)}\, dx$$

---

[10] In his Königsberg lectures [4], pp. 198–211, elliptic coordinates in $\mathbb{R}^n$ are dealt with in full detail. For historical references, cf. Jacobi [3], Vol. 2, pp. 59–63.

[11] For a presentation within the framework of differential geometry we refer the reader to the Supplement.

of a function $u : \overline{\Omega} \to \mathbb{R}$ with respect to a Riemannian line element

$$ds^2 = \gamma_{\alpha\beta}(x)\, dx^\alpha\, dx^\beta$$

and the corresponding volume element

$$\mathrm{dvol} = \sqrt{\gamma(x)}\, dx\,.$$

Here $(\gamma_{\alpha\beta}(x))$ is a positive definite, symmetric matrix, $(\gamma^{\alpha\beta}(x))$ its inverse, and $\gamma(x) = \det(\gamma_{\alpha\beta}(x))$. Then the product $\frac{-1}{\sqrt{\gamma}} L$ of the Euler operator L of the Dirichlet integral $\mathscr{D}(u)$ with the factor $\frac{-1}{\sqrt{\gamma}}$ is called the *Laplace–Beltrami operator* associated[12] with the line element ds^2 and will be denoted by Δ. By the same computation as for [2] we obtain

$$\Delta u = \frac{1}{\sqrt{\gamma}} D_\beta\{\sqrt{\gamma}\gamma^{\alpha\beta} D_\alpha u\}\,.$$

If it should be necessary to distinguish this operator from the standard Laplace operator $\Delta = \delta^{\alpha\beta} D_\alpha D_\beta$, which is the Beltrami operator with respect to the ordinary euclidean line element $ds^2 = \delta_{\alpha\beta}\, dx^\alpha\, dx^\beta$, we will add a subindex to Δ to indicate the corresponding metric.

If the $\gamma_{\alpha\beta}(x)$ are supposed to be the components of a (0, 2)-tensor field, then the differential expression

$$\tfrac{1}{2}\gamma^{\alpha\beta}(x) u_{x^\alpha} u_{x^\beta}$$

as well as $\mathscr{D}(u)$ turn out to be invariant with respect to changes of the independent variables by diffeomorphisms

$$y = \eta(x) \quad \text{or} \quad x = \xi(y)\,.$$

That is, if we set $v(y) = u(\xi(y))$ and

$$\bar{\gamma}_{\alpha\beta}(y) = \gamma_{\mu\nu}(\xi(y))\xi^\mu_{y^\alpha}(y)\xi^\nu_{y^\beta}(y)$$

as well as

$$\bar{\gamma} = \det(\bar{\gamma}_{\alpha\beta}), \qquad (\bar{\gamma}^{\alpha\beta}) = (\bar{\gamma}_{\alpha\beta})^{-1},$$

then we obtain

$$\tfrac{1}{2}\int_\Omega \gamma^{\alpha\beta} u_{x^\alpha} u_{x^\beta} \sqrt{\gamma}\, dx = \tfrac{1}{2}\int_{\Omega^*} \bar{\gamma}^{\alpha\beta} v_{y^\alpha} v_{y^\beta} \sqrt{\bar{\gamma}}\, dy$$

or

$$\mathscr{D}(u, \Omega) = \mathscr{D}(v, \Omega^*),$$

and similarly

$$(\Delta_\Gamma u) \circ \xi = \Delta_{\bar{\Gamma}} v,$$

where

$$\Delta_\Gamma u = \frac{1}{\sqrt{\gamma}} D_\beta\{\sqrt{\gamma}\gamma^{\alpha\beta} D_\alpha u\}, \qquad \Delta_{\bar{\Gamma}} v = \frac{1}{\sqrt{\bar{\gamma}}} D_\beta\{\sqrt{\bar{\gamma}}\bar{\gamma}^{\alpha\beta} D_\alpha v\}\,.$$

This shows that we can define the Dirichlet integral as well as the Laplace–Beltrami operator in a global way on a Riemannian manifold (M, ds^2), i.e., independently from local coordinates on M. The solutions of

[12] The convention about the sign of Δ is not uniform, many authors prefer the opposite sign.

$$\Delta u = 0 \quad \text{on } M$$

are called *harmonic functions on M*.

Let us, finally, consider the special case $n = 2$ of a two-dimensional Riemannian metric

$$ds^2 = e\,dx^2 + 2f\,dx\,dy + g\,dy^2,$$

where e, f, g are functions of $x^1 = x, x^2 = y$. Then the Dirichlet integral of a function $u(x, y), (x, y) \in \Omega$, is given by

$$\mathscr{D}(u) = \int_\Omega F(x, y, u_x, u_y)\sqrt{eg - f^2}\,dx\,dy,$$

with

$$F(x, y, p, q) = \frac{1}{eg - f^2}\{gp^2 - 2fpq + eq^2\},$$

and the invariant differential expression of Laplace–Beltrami is given by

$$\Delta u = \frac{1}{\sqrt{eg - f^2}}\left\{\frac{\partial}{\partial x}\left(\frac{gu_x - fu_y}{\sqrt{eg - f^2}}\right) + \frac{\partial}{\partial y}\left(\frac{-fu_x + eu_y}{\sqrt{eg - f^2}}\right)\right\}.$$

In particular, the Beltrami operator Δ on a two-dimensional surface in $\mathbb{R}^3$ with the parametric representation $z = (z^1, z^2, z^3) = \psi(x, y) \in \mathbb{R}^3$, $(x, y) \in \bar{\Omega}$, $\Omega \subset \mathbb{R}^2$ is to be computed from the line element $ds = |d\psi|^2 = |\psi_x|^2\,dx^2 + 2\psi_x \cdot \psi_y\,dx\,dy + |\psi_y|^2\,dy^2$, i.e.,

$$e = |\psi_x|^2, \qquad f = \psi_x \cdot \psi_y, \qquad g = |\psi_y|^2.$$

4 *Harmonic mappings of Riemannian manifolds.* We want to define a notion of harmonic mappings which comprises all previous examples of harmonic functions and mappings as special cases.

To this end, we choose a Riemannian line element

$$d\sigma^2 = \gamma_{\alpha\beta}(x)\,dx^\alpha\,dx^\beta$$

on $\mathbb{R}^n$, and a second line element

$$ds^2 = g_{ik}(z)\,dz^i\,dz^k$$

on $\mathbb{R}^N$. Define $(\gamma^{\alpha\beta})$ and γ as in 3, and let

$$\Delta = \frac{1}{\sqrt{\gamma}} D_\beta\{\sqrt{\gamma}\gamma^{\alpha\beta} D_\alpha u\}$$

be the Laplace–Beltrami operator corresponding to $d\sigma^2$. Moreover, let

$$\Gamma_{ijk} = \tfrac{1}{2}\{g_{jk,i} - g_{ik,j} + g_{ij,k}\}, \quad \Gamma_{ik}^l = g^{lj}\Gamma_{ijk}$$

be the Christoffel symbols with respect to ds^2. Combining the arguments used in 1 and 3, we infer that both the expression

$$\tfrac{1}{2} g_{ik}(u)\gamma^{\alpha\beta}(x) u_{x^\alpha}^i u_{x^\beta}^k =: e(u)$$

and the generalized Dirichlet integral

$$\mathscr{E}(u) := \int_\Omega e(u)\sqrt{\gamma(x)}\,dx$$

$$= \int_\Omega \frac{1}{2} g_{ik}(u)\gamma^{\alpha\beta}(x) D_\alpha u^i D_\beta u^k \sqrt{\gamma(x)}\,dx,$$

defined for $u \in C^1(\bar{\Omega}, \mathbb{R}^N)$, $\Omega \subset \mathbb{R}^n$, are invariant with respect to transformations of both the dependent and independent variables which are of type

$$y = \eta(x) \quad \text{or} \quad x = \xi(y)$$

and

$$z^* = W(z) \quad \text{or} \quad z = Z(z^*).$$

It is also customary to call $e(u)$ the *energy density of the mapping* $u : \bar{\Omega} \to \mathbb{R}^N$ with respect to the metrics $d\sigma^2$ and ds^2, and $\mathscr{E}(u)$ is sometimes denoted as *energy of the map u*. By the same computations as in 2,4 [3], and in [3] of this section, we compute the Euler operator $L(u) = (L_1(u), \ldots, L_N(u))$ of $\mathscr{E}(u)$ as $-\frac{1}{\sqrt{\gamma}} L_k(u) = g_{ik}(u)\Delta u^i + \Gamma_{ikj}(u)\gamma^{\alpha\beta} D_\alpha u^i D_\beta u^j$. This expression is invariant with respect to changes of the independent variables x and transforms like a covector with respect to changes of the dependent variables z.

Thus we can define $\mathscr{E}(u)$ and $L(u)$ for mappings $u : X \to M$ between two Riemannian manifolds $(X, d\sigma^2)$ and (M, ds^2) in a global way, that is, independently of local coordinates on X and M, and the Euler operator $L(u)$ can be interpreted as a covector field on M along the mapping u, whereas

$$-\frac{1}{\sqrt{\gamma}} L^k(u) := -\frac{1}{\sqrt{\gamma}} g^{ik} L_k(u) = \Delta u^k + \Gamma_{ij}^k(u)\gamma^{\alpha\beta} D_\alpha u^i D_\beta u^j$$

defines a vector field on M along the mapping u.

Note that $L(u)$ is a linear differential operator if $N = 1$, whereas for $N > 1$ the operator $L(u)$ will in general be nonlinear.

The extremals $u : X \to M$ of $\mathscr{E}(u, \Omega)$, $\Omega = \operatorname{int} X$, will be called *harmonic mappings* of X into M. They are the C^2-solutions of the equations

$$\Delta u^k + \Gamma_{ij}^k(u)\gamma^{\alpha\beta} D_\alpha u^i D_\beta u^j = 0,$$

with

$$\Delta u^k = \frac{1}{\sqrt{\gamma}} D_\beta\{\sqrt{\gamma}\gamma^{\alpha\beta} D_\alpha u^k\}.$$

Special cases of harmonic mappings are:

(i) $n = 1$: geodesics in M; cf. 2,4 [3].

(ii) $N = 1$: harmonic functions on X; cf. [3].

(iii) $n = 2$, $N = 3$: conformal representations of minimal surfaces in $\mathbb{R}^3$; cf. Section 2, Remark 2, and in particular formula (21).

At last we shall derive a formula describing the transformation behaviour of the Euler operator with respect to general transformations of the dependent and independent variables.

Theorem. *Consider a transformation* $(x, z) \mapsto (x^*, z^*)$ *of the type*

$$x^* = Y(x, z), \qquad z^* = W(x, z),$$

a function $z = u(x)$, $x \in \bar{\Omega}$, *and a Lagrangian* $F(x, z, p)$.

(i) *Suppose that the mapping* $(x, z) \mapsto (x, z^*)$, *given by*

$$x = x, \qquad z^* = W(x, z),$$

is invertible and has the inverse

$$x = x, \qquad z = Z(x, z^*).$$

Set $w(x) := W(x, u(x))$, $x \in \bar{\Omega}$, *and define the Lagrangian* $G(x, z^*, q)$ *by* (1). *Then we obtain*

$$L_G(w) = L_F(u) \cdot Z_{z^*}(x, w).$$

(*ii*) *Assume that* $y = \eta(x) := Y(x, u(x))$, $x \in \bar{\Omega}$, *defines a diffeomorphism from* $\bar{\Omega}$ *onto* $\bar{\Omega}^*$ *with the inverse* $x = \xi(y)$. *Let* $A = \eta_x \circ \xi$, $B = \xi_y$, $\det B > 0$, $v = w \circ \xi$, *and*

$$H(y, z^*, q) = G(\xi(y), z^*, q \cdot A(y)) \det B(y).$$

Then it follows that

$$L_G(w) = \det \eta_x \cdot \{L_H(v) \circ \eta + [L_H(v) \circ \eta] \cdot w_x \cdot \eta_x^{-1} \cdot T\}, \tag{6}$$

with

$$T(x) := -Y_z(x, u(x)) \cdot W_z^{-1}(x, u(x)).$$

(*iii*) *Combining the results of* (i) *and* (ii), *we infer that*

$$L_F(u) = (\det \eta_x)\{L_H(v) \circ \eta + [L_H(v) \circ \eta] \cdot w_x \cdot \eta_x^{-1} \cdot T\} \cdot W_z(x, u). \tag{7}$$

Proof. (i) The first assertion has been proved at the beginning of this section; cf. formula (2).

(ii) We have also shown that

$$\mathscr{G}(w) = \int_\Omega G(x, w, w_x)\, dx = \int_{\Omega^*} H(y, v, v_y)\, dy = \mathscr{H}(v).$$

Let us now consider a variation $\bar{w}(x, \varepsilon) = w(x) + \varepsilon\zeta(x)$ of $w(x)$ with $\operatorname{supp} \zeta \subset \Omega$. Then $\bar{u}(x, \varepsilon) := Z(x, \bar{w}(x, \varepsilon))$ defines a variation of $u(x)$ with

$$\bar{u}(x, \varepsilon) = u(x) + \varepsilon\varphi(x) + o(\varepsilon),$$

where

$$\varphi(x) = Z_{z^*}(x, w) \cdot \zeta = W_z^{-1}(x, u) \cdot \zeta$$

and $\operatorname{supp} \varphi \subset \Omega$.

Then, for $|\varepsilon| \ll 1$, we obtain by

$$y^* = \bar{\eta}(x, \varepsilon) := Y(x, \bar{u}(x, \varepsilon)),$$

a family of diffeomorphisms of $\bar{\Omega}$ onto itself, the inverses of which be given by

$$x = \bar{\xi}(y^*, \varepsilon), \qquad y^* \in \bar{\Omega}.$$

It, moreover, follows that

$$\bar{\eta}(x, 0) = \eta(x) \quad \text{and} \quad \bar{\xi}(y, 0) = \xi(y)$$

holds as well as

$$\frac{\partial \bar{\eta}}{\partial \varepsilon}(y, 0) = Y_z(x, u(x)) \cdot \varphi(x),$$

whence we infer that

$$y^* = \bar{\eta}(x, \varepsilon) = \eta(x) + \varepsilon Y_z(x, u(x)) \cdot \varphi(x) + o(\varepsilon).$$

Let us introduce the vector field $\lambda(y)$, $y \in \Omega^*$, by

$$\lambda(y) := -Y_z(x, u(x)) \cdot \varphi(x)|_{x=\xi(y)}.$$

Then we can write

$$y^* = y - \varepsilon\lambda(y) + o(\varepsilon) := \sigma(y, \varepsilon).$$

The mapping $\sigma(y, \varepsilon)$ yields a diffeomorphism of Ω onto itself, provided that $|\varepsilon| \ll 1$, and its inverse $\tau(y^*, \varepsilon)$ can be written as

$$y = y^* + \varepsilon\lambda(y^*) + o(\varepsilon) = \tau(y^*, \varepsilon).$$

Together we can write

$$y^* = \bar{\eta}(x, \varepsilon) = \sigma(\eta(x), \varepsilon) = \eta(x) - \varepsilon\lambda(\eta(x)) + o(\varepsilon),$$
$$x = \bar{\xi}(y^*, \varepsilon) = \xi(\tau(y^*, \varepsilon)) = \xi(y^* + \varepsilon\lambda(y^*) + o(\varepsilon)).$$

Finally we introduce

$$\bar{v}(y^*, \varepsilon) := \bar{w}(\bar{\xi}(y^*, \varepsilon), \varepsilon)$$

and the Lagrangian $H^*(y^*, z^*, q^*, \varepsilon)$ which is adjoint to $G(x, z^*, q)$ with respect to the parameter transformation $y^* = \bar{\eta}(x, \varepsilon)$. Since this mapping factors by the two transformations $y^* = \sigma(y, \varepsilon)$ and $y = \eta(x)$, we can also determine $H^*(y^*, z^*, q^*, \varepsilon)$ as adjoint Lagrangian to $H(y, z^*, q)$ with respect to the transformation $y^* = \sigma(y, \varepsilon)$, recalling that H is adjoint to G with respect to the map $y = \eta(x)$. This implies

$$\mathcal{G}(\bar{w}(\varepsilon)) = \int_\Omega G(x, \bar{w}(\varepsilon), \bar{w}_x(\varepsilon))\, dx = \int_{\Omega^*} H^*(y^*, \bar{v}(\varepsilon), \bar{v}_{y*}(\varepsilon), \varepsilon)\, dy^* = \mathscr{H}^*(\bar{v}(\varepsilon), \varepsilon),$$

where

$$H^*(y^*, z^*, q^*, \varepsilon) = H(\tau(y^*, \varepsilon), z^*, q^* A^*(\varepsilon)) \cdot \det B^*(\varepsilon)$$

and

$$A^*(y, \varepsilon) = \sigma_y(y, \varepsilon) = 1 - \varepsilon\lambda_y(y) + o(\varepsilon),$$
$$B^*(y^*, \varepsilon) = \tau_{y*}(y^*, \varepsilon) = 1 + \varepsilon\lambda_{y*}(y^*) + o(\varepsilon),$$

whence also

$$\det B^*(y^*, \varepsilon) = 1 + \varepsilon \operatorname{div}_{y*} \lambda(y^*) + o(\varepsilon).$$

These formulas yield

$$\begin{aligned} &H^*(y^*, z^*, q^*, \varepsilon) \\ &= H(y^* + \varepsilon\lambda(y^*) + \cdots, z^*, q^* \cdot [1 - \varepsilon\lambda_y(y) + \cdots]) \cdot \{1 + \varepsilon \operatorname{div}_{y*} \lambda(y^*) + \cdots\} \\ &= H(\ldots) + \varepsilon\{H_{y*}(\ldots) \cdot \lambda(y^*) + H_q(\ldots) \cdot (-q^* \cdot \lambda_y(y)) + H(\ldots) \operatorname{div}_{y*} \lambda(y^*) + o(\varepsilon)\}, \end{aligned}$$

where $(\ldots)$ stands for (y^*, z^*, q^*). From this we infer

$$\begin{aligned} \frac{d}{d\varepsilon}\mathcal{G}(\bar{w}(\varepsilon))\bigg|_{\varepsilon=0} &= \delta\mathcal{G}(w, \zeta) \\ &= \int_{\Omega^*} H_\varepsilon^*(y, v, v_y, 0)\, dy + \int_{\Omega^*} \left[\frac{d}{d\varepsilon} H^*(y^*, \bar{v}(\varepsilon), \bar{v}_{y*}(\varepsilon), 0)\right]_{\varepsilon=0} dy^* \\ &= \int_{\Omega^*} \{H_y(\ldots) \cdot \lambda(y) + H_q(\ldots) \cdot (-q\lambda_y(y)) + H(\ldots) \operatorname{div} \lambda(y)\}\, dy \\ &\quad + \int_{\Omega^*} \left[\frac{d}{d\varepsilon} H(y^*, \bar{v}(\varepsilon), \bar{v}_{y*}(\varepsilon))\right]_{\varepsilon=0} dy^* = I + II, \end{aligned}$$

where $(\ldots)$ denotes $(y, v(y), v_y(y))$.

On account of (7) and (8) of Section 1, the integral I is just $-\partial\mathscr{H}(v, \lambda)$, whereas II coincides with $\delta^*\mathscr{H}(v; \psi, -\lambda)$, as we see from (7) and (8) of Section 3 as well as from (2) and (4) of Section 3. We have only to note that, in our present case, formula (2) of Section 3 is to be replaced by

$$y^* = \sigma(y, \varepsilon) = y - \varepsilon\lambda(y) + o(\varepsilon),$$

so that μ has to be substituted by $-\lambda$, and the role of $v(x, \varepsilon)$ in 3, (4) is now played by the family of functions $\overline{w}(\xi(y), \varepsilon) = w(\xi(y)) + \varepsilon\zeta(\xi(y))$ whence we infer that $\psi(y) := \zeta(\xi(y))$ has to replace $\varphi(x) = \frac{\partial v}{\partial \varepsilon}(x, 0)$ from 3, (4). Thus we conclude that

$$\delta\mathscr{G}(w, \zeta) = -\partial\mathscr{H}(v, \lambda) + \delta^*\mathscr{H}(v; \psi, -\lambda).$$

By virtue of 3, (11), we see that

$$\partial\mathscr{H}(v, \lambda) = \delta^*\mathscr{H}(v; Dv\cdot\lambda, -\lambda) = -\delta^*\mathscr{H}(v; -Dv\cdot\lambda, \lambda),$$

and we arrive at

$$\delta\mathscr{G}(w, \zeta) = \delta^*\mathscr{H}(v; -Dv\cdot\lambda, \lambda) + \delta^*\mathscr{H}(v; \psi, -\lambda).$$

On account of 3, (9), we then obtain the remarkable formula

$$\delta\mathscr{G}(w, \zeta) = \delta\mathscr{H}(v, \psi - Dv\cdot\lambda). \tag{8}$$

Recall that $\psi = \zeta \circ \xi$, $v = w \circ \xi$, and

$$\lambda(y) = -Y_z(x, u(x))\cdot\varphi(x)|_{x=\xi(y)},$$

$$\varphi = Z_{z*}(x, w)\cdot\zeta = W_z^{-1}(x, u)\cdot\zeta,$$

and note that ψ and λ have compact support in Ω^*, whereas ζ has compact support in Ω. Then we derive from (8) the relation

$$\int_\Omega L_G(w)\cdot\zeta\,dx = \int_{\Omega^*}\{L_H(v)\cdot\zeta(\xi) + L_H(v)\cdot w_x(\xi)\cdot\xi_y\cdot T(\xi)\cdot\zeta(\xi)\}\,dy$$

with

$$T(x) = -Y_z(x, u(x))\cdot W_z^{-1}(x, u(x)).$$

Moreover, we have

$$\int_\Omega L_G(w)\cdot\zeta\,dx = \int_{\Omega^*}[L_G(w)\circ\xi]\cdot\zeta(\xi)\det\xi_y(y)\,dy.$$

Then the fundamental lemma implies

$$(\det\xi_y)L_G(w)\circ\xi = L_H(v) + L_H(v)\cdot w_x(\xi)\cdot\xi_y\cdot T(\xi),$$

whence we obtain (6):

$$L_G(w) = \det\eta_x\cdot\{L_H(v)\circ\eta + [L_H(v)\circ\eta]\cdot w_x\cdot\eta_x^{-1}\cdot T\},$$

where

$$\eta(x) = Y(x, u(x)), \qquad \eta_x(x) = D_x Y(x, u(x)),$$

and the assertion of (ii) is verified.

(iii) Furthermore, the identity

$$z = Z(x, W(x, z))$$

implies that

$$Z_{z*}^{-1}(x, W(x, z)) = W_z(x, z)$$

and therefore

$$Z_{z*}^{-1}(x, w(x)) = W_z(x, u(x))$$

holds. Thus it follows from (i) that

$$L_F(u) = L_G(w) \cdot W_z(x, u)$$

is satisfied. In conjunction with (6), we obtain the desired formula (7). □

6. Scholia

1. The method of "varying the independent variable" was already envisioned by Lagrange (Misc. Taur. *2*, 1760/61), but Euler was the first to give a systematic presentation of this idea (in his *Institutionum calculi integralis* (1770)). However, Euler erred on several occasions, and he sometimes used the δ-procedure rather formalistically, without a sound foundation. As Bolza stated, this purely mechanic application of the δ-algorithm entered Lacroix's textbook (1814) and, from there on, "*it had the most disastrous influence upon the further development of the calculus of variations*" (see [1], p. 13). Occasionally the δ-hobgoblin seems still to be playing evil tricks. On the other hand, some textbook-authors altogether avoid the variation of the independent variables, which is not justified because this technique can be rigorously founded and is furnishing very useful results.

The symbol $\partial\mathscr{F}(u, \lambda)$ is introduced in contrast to $\delta\mathscr{F}(u, \varphi)$, in order to avoid any confusion between the variation of dependent and independent variables. The expression for $\partial\mathscr{F}(u, \lambda)$ implicitely appears in the work of several authors, but it seems to be useful to christen the child. For this reason, we also have introduced the notions of *inner extremals* and *strong inner extremals*, defined by $\partial\mathscr{F}(u, \lambda) = 0$ for suitable vector fields λ. The *Noether equations* are introduced as counterpart to the Euler equations; on account of E. Noether's paper [1], the name might be justified.

Garabedian [1] introduced the notion of *interior variations* of domains in order to give a rigorous mathematical analysis of the so-called *Hadamard variational formula* which represents the first order term in the expansion of Green's function with respect to an infinitesimal displacement of the boundary of the domain. These interior variations are closely related to our inner variations.

2. Equation (16′) in Section *1*,

$$\frac{1}{\rho} = \frac{\partial}{\partial \mathfrak{n}} \log \omega \circ c,$$

was found by Gauss (cf. Bolza [1], pp. 84–85), who formulated it as

$$\frac{\partial \omega}{\partial y} \cos \varphi - \frac{\partial \omega}{\partial x} \sin \varphi = \frac{\omega}{\rho},$$

where $\dfrac{dy}{dx} = \operatorname{tang} \varphi$, i.e., φ is the angle between the tangent and the x-axis. Formula (16′) in *1* was stated by W. Thomson [1].

3. The notion of *mean curvature H* of a surface was introduced by T. Young[13]. Implicitely it is already contained in Meusnier's paper[14], where the Euler equation of the area functional is shown to be equivalent to the equation $H = 0$.

[13] *An essay on the cohesion of fluids*, Phil. Trans. Roy. Soc. London **95**, 65–87 (1805).

[14] *Mémoire sur la courbure des surfaces*, Mém. de Math. Phys. (sav. etrang.) de l'Acad. **10** (1785), 447–550 (lu 1776).

4. Proposition 5 of *1* [4] is essentially taken from Rund [4].

5. The reasoning used in *2* [3] is due to Radó [1] and was elaborated by Courant [2]. Instead of varying the domain, one may as well vary the *conformal structure*. This way, one is lead to a foundation of Teichmüller theory; cf. Lehto [1], Fischer–Tromba [1], and Jost [1], [2]. Applications to minimal surfaces and in particular to Plateau's problem can, for instance, be found in Radó [1], Courant [2], Nitsche [1, 2], Struwe [1], Jost [1], Dierkes–Hildebrandt–Küster–Wohlrab [1].

6. Example [4] in Section 2 was first discussed by Hildebrandt [2]; the important special case $Q(u) = \frac{4}{3}Hu$ was already considered by Heinz [1].

7. The elements of the computation given in Section 3 are due to Euler.[15] Our presentation in essence follows the classical paper of Emmy Noether [1], who was stimulated by ideas of S. Lie and F. Klein. Noether's paper had a number of precursors dealing with particular cases; she herself mentions publications by Hamel, Herglotz, Lorentz, Fokker, Weyl, Klein, and Kneser. However, E. Noether has the merit of mastering the general case and of arranging the computations in a simple, lucid form.

The results of Noether have proved to be important in formulating and exploiting the basic equations of physics. An comprehensive presentation of the development following Noether's paper is given in Olver [1]; there one also finds an extensive bibliography. Olver's monograph contains a description and extension of Noether's theory in the spirit of Lie's application of group theory to partial differential equations. The role of Noether's results is mentioned in Hilbert, Werke [7], Vol. 3, pp. 55–56, and in Noether [1], pp. 240, 256–257. Bessel-Hagen [1] applied Noether's results to electrodynamics; in the same paper, one also finds the treatment of the ℓ-body-problem that we have given in *4* [2].

8. For the history of the theory of elasticity and of continuum mechanics, we refer to Truesdell [1] and [2].

The *energy-momentum tensor*

$$T^\alpha_\beta := W_{p^i_\alpha} u^i_{x^\beta} - W\delta^\alpha_\beta$$

is an expression that plays a fundamental role in physics: cf. for instance Pauli [1], pp. 682–775; Hund [1], pp. 31–32, 100, 243, 294, 319, 331; Dubrovin–Fomenko–Novikov [1], Vol. 1, pp. 202, 305, 381, 389, 425; Olver [1], pp. 282, 288; Landau–Lifschitz [1], Vol. 6, Section 125.

We shall investigate T^α_β together with its role in the theory of Legendre transformations and Haar transformations in Chapter 7. The calculus of these transformations for multiple variational integrals was developed by Carathéodory in the years 1922–1928; cf. the papers XVII–XX in "Gesammelte mathematische Schriften" [16], Vol. 1, pp. 374–426, where one finds the general definition of T^α_β. We also refer to Rund [4], pp. 298–311.

9. The first to investigate *covariance properties* of the differential equations of the calculus of variations apparently was Euler. In Chapter 4 of his *Methodus inveniendi* (E65) from 1744, he discussed the geometrical meaning of the variables x and u appearing in a variational integral

$$\int_a^b F(x, u, u', u'', \ldots)\, dx.$$

These variables have not necessarily to be cartesian coordinates but may be interpreted as general curvilinear coordinates.

---

[15] (E385) *Institutionum calculi integralis*, Vol. 3, Petersburg 1770, appendix; cf. also: A. Kneser, *Variationsrechnung*, Enzykl. math. Wiss., Vol. 2.1, IIA8, pp. 575–576.

Euler's ideas were taken up by Jacobi. In a paper[16] from 1848, he treated the old problem of transforming the Laplace operator from cartesian into curvilinear coordinates by a variational argument. To this end, he first computed the line element $ds^2 = dx^2 + dy^2 + dz^2$ in the new coordinates. From this formula he derived the new expression for $|\nabla u|^2 = u_x^2 + u_y^2 + u_z^2$, and then he obtained the desired formula for Δu with respect to the new variables by the observation that the Dirichlet integral

$$\int_\Omega |\nabla u|^2 \, dx\, dy\, dz,$$

and therefore also its first variation, must have an invariant (covariant) meaning. This exactly is the reasoning that we have applied to establish formula 5, (5); cf. also 5 [2].

Jacobi only used orthogonal coordinates, whereas Beltrami [1], [3] carried the method over to general coordinates. Thus he was led to the definition of the Laplace–Beltrami operator

$$\Delta u = \frac{1}{\sqrt{\gamma}} D_\beta\{\sqrt{\gamma}\gamma^{\alpha\beta} D_\alpha u\}.$$

Other methods to derive differential invariants were developed by Christoffel (1869) and Lipschitz (1870); their ideas led to the definition of *covariant differentiation*, whose importance was particularly stressed by Ricci. This development is described in Vol. 2 of Klein's "*Vorlesungen über die Entwicklung der Mathematik im* 19. *Jahrhundert*" [3].

10. The notion of inner variations plays a role also in establishing so-called *monotonicity formulas*. The importance of such formulas was apparently discovered in geometric measure theory; we particularly mention the work of DeGiorgi and Almgren. However monotonicity formulas for harmonic functions were known since a long time. Such formulas have become of fundamental importance in the regularity theory of elliptic systems, and they have been used to establish the *unique continuation principle* for solutions of nonlinear partial differential equations.

11. We can somewhat modify the reasoning of 2 [3]. Consider the generalized Dirichlet integral

$$\mathscr{F}_\Omega(u) = \int_\Omega \langle D_\alpha u, D_\alpha u\rangle \, dx^1\, dx^2$$

for mappings $u \in C^1(\bar{\Omega}, \mathbb{R}^N)$, where $\langle p_\alpha, p_\beta\rangle$ stands for the inner product

$$\langle p_\alpha, p_\beta\rangle = g_{ik}(u) D_\alpha u^i D_\beta u^k.$$

Set

$$a := \langle D_1 u, D_1 u\rangle - \langle D_2 u, D_2 u\rangle, \qquad b := 2\langle D_1 u, D_2 u\rangle.$$

We have proved that the inner variation $\partial\mathscr{F}_\Omega(u, \lambda)$ of $\mathscr{F}_\Omega$ at u in direction of a vector field $\lambda : \bar{\Omega} \to \mathbb{R}^2$ can be written as

$$(1) \qquad \partial\mathscr{F}_\Omega(u, \lambda) = \int_\Omega \{a[\lambda^1_{x^1} - \lambda^2_{x^2}] + b[\lambda^1_{x^2} + \lambda^2_{x^1}]\} \, dx^1\, dx^2.$$

Let us introduce the complex variable $\zeta = x^1 + ix^2$, the Wirtinger operators $\partial_\zeta = \frac{1}{2}(D_1 - iD_2)$, $\partial_{\bar\zeta} = \frac{1}{2}(D_1 + iD_2)$, and the complex valued functions

---

[16] *Über eine particuläre Lösung der partiellen Differentialgleichung* $\frac{\partial^2 V}{\partial x^2} + \frac{\partial^2 V}{\partial y^2} + \frac{\partial^2 V}{\partial z^2} = 0$, Crelle's Journal f.d. reine u. angew. Math. **36**, 113–134 (1848); cf. also Ges. Werke, Vol. 2, pp. 191–216.

$$f := a - ib, \qquad \lambda := \lambda^1 + i\lambda^2.$$

In other words, $\lambda = \lambda^1 + i\lambda^2$ is the complex notation for the $\mathbb{R}^2$-valued vector field $\lambda = (\lambda^1, \lambda^2)$; both expressions are identified. Similarly we identify the exterior normal $\nu = (\nu_1, \nu_2)$ to $\partial\Omega$ with $\nu = \nu_1 + i\nu_2$. Then we can write the integrand $\{\dots\}$ of (1) as $2\,\mathrm{Re}\{f\,\lambda_{\bar\zeta}\}$, and therefore

$$\partial\mathscr{F}_\Omega(u, \lambda) = \mathrm{Re}\int_\Omega 2f\,\lambda_{\bar\zeta}\,dx^1\,dx^2. \tag{2}$$

If $f \in C^1(\Omega, \mathbb{C})$ and $\operatorname{supp}\lambda \subset \Omega$, an integration by parts yields

$$\tfrac{1}{2}\partial\mathscr{F}_\Omega(u, \lambda) = -\mathrm{Re}\int_\Omega f_{\bar\zeta}\,\lambda\,dx^1\,dx^2. \tag{3}$$

If u is an inner extremal of $\mathscr{F}_\Omega$ we thus obtain

$$\mathrm{Re}\int_\Omega f_{\bar\zeta}\,\lambda\,dx^1\,dx^2 = 0 \quad \text{for all } \lambda \in C_c^1(\Omega, \mathbb{C}). \tag{4}$$

By the fundamental lemma we infer $f_{\bar\zeta} = 0$ which means that f is holomorphic. However, for $u \in C^1(\bar\Omega, \mathbb{R}^2)$ we only get $f \in C^0(\Omega, \mathbb{C})$, and therefore an inner extremal u of $\mathscr{F}_\Omega$ only satisfies

$$\mathrm{Re}\int_\Omega f\,\lambda_{\bar\zeta}\,dx^1\,dx^2 = 0 \quad \text{for all } \lambda \in C_c^1(\Omega, \mathbb{C}). \tag{4'}$$

Using the smoothing device of 2 [3] we can reduce (4′) to (4), and a limiting process yields once again that f is holomorphic.

Suppose now that u is a strong extremal of $\mathscr{F}_\Omega$. Then we have

$$\mathrm{Re}\int_\Omega f\lambda_{\bar\zeta}\,dx^1\,dx^2 = 0 \tag{5}$$

for all $\lambda \in C^1(\bar\Omega, \mathbb{C})$. In fact, for our purposes it suffices to require that *$\lambda|_{\partial\Omega}$ is tangential to $\partial\Omega$*, say,

$$\lambda|_{\partial\Omega} = i\psi\nu, \tag{6}$$

where $\psi : \partial\Omega \to \mathbb{R}$ is of class C^1. Suppose first that $f \in C^1(\bar\Omega, \mathbb{C})$. Then an integration by parts yields

$$\int_\Omega f\,\lambda_{\bar\zeta}\,dx^1\,dx^2 = -\int_\Omega f_{\bar\zeta}\,\lambda\,dx^1\,dx^2 + \int_{\partial\Omega} \tfrac{1}{2} f\lambda\nu\,d\mathscr{H}^1. \tag{7}$$

Since we know already that $f_{\bar\zeta} = 0$ we infer from (5) that

$$\int_{\partial\Omega} \mathrm{Re}(f\lambda\nu)\,d\mathscr{H}^1 = 0 \quad \text{for any } \lambda \text{ of type (6)},$$

i.e.

$$\int_{\partial\Omega} \mathrm{Re}(if\,\nu^2)\psi\,d\mathscr{H}^1 = 0 \quad \text{for all } \psi \in C^1(\partial\Omega). \tag{8}$$

Then the fundamental lemma implies

$$\mathrm{Re}(if|_{\partial\Omega}\nu^2) = 0, \tag{9}$$

and, conversely, (9) implies (5) for all λ of type (6). In particular we have: *The mapping u is a strong inner extremal of $\mathscr{F}_\Omega$ if $(i\nu)^2\lambda$ is real.*

The same holds if we merely know $f \in C^0(\bar\Omega, \mathbb{C})$ (which follows from $u \in C^1(\bar\Omega, \mathbb{R}^N)$). To this end we map Ω conformally to another domain $\tilde\Omega$ such that some part of $\partial\Omega$ is mapped to a straight segment of $\partial\tilde\Omega$, say, to an interval I of the real axis such that, close by, $\tilde\Omega$ lies in the upper halfplane.

The transform of f is for the sake of simplicity again denoted by f. Then we have

$$\int_{B^+} \operatorname{Re}(f\,\lambda_{\bar\zeta})\,dx^1\,dx^2 = 0 \tag{10}$$

for all $\lambda \in C_c^1(B, \mathbb{C})$ with $\lambda|_I$ real, where B is a disk whose intersection with the real axis is I, and $B^+ = B \cap \{\operatorname{Im} \zeta > 0\}$, $B^- = B \cap \{\operatorname{Im} \zeta < 0\}$. We extend f to a function on B by setting $f(\zeta) := f^*(\zeta)$ for $\zeta \in B^-$, where $f^*(\zeta) := \overline{f(\bar\zeta)}$. Clearly f is holomorphic in $B^+ \cup B^-$ and of class $L^1(B, \mathbb{C})$. We are going to prove that the extended function $f: B \to \mathbb{C}$ is weakly holomorphic in B (and therefore also holomorphic). To this end let $\lambda \in C_c^1(B, \mathbb{C})$ and check that $(\lambda^*)_{\bar\zeta} = (\lambda_{\bar\zeta})^*$ for $\lambda^*(\zeta) := \overline{\lambda(\bar\zeta)}$. Then

$$\int_B \operatorname{Re}(f\lambda_{\bar\zeta})\,dx^1\,dx^2 = \int_{B^+} \cdots + \int_{B^-} \cdots$$

$$= \int_{B^+} \operatorname{Re}\{f[\lambda + \lambda^*]\}\,dx^1\,dx^2 = 0,$$

since $\lambda + \lambda^*$ is real on I. Thus we have

$$\int_B \operatorname{Re}(f\,\lambda_{\bar\zeta})\,dx^1\,dx^2 = 0 \quad \text{for all } \lambda \in C_0^1(B, \mathbb{C}).$$

Hence f is holomorphic in B and in particular smooth. If $\partial\Omega$ is smooth, this result in connection with well-known boundary regularity results for conformal mappings implies that the original map $f: \bar\Omega \to \mathbb{C}$ is smooth up to the boundary. Thus our reasoning is justified, and we obtain:

If u is a strong inner extremal of $\mathscr{F}_\Omega$, then f is holomorphic on Ω, smooth up to the boundary $\partial\Omega$, and $v^2 f|_{\partial\Omega}$ is real valued, provided that $\partial\Omega$ is smooth.

If Ω is the unit disk $\{\zeta: |\zeta| < 1\}$ it follows that $g(\zeta) := \zeta^2 f(\zeta)$ is holomorphic in Ω and real on $\partial\Omega$ whence $g(\zeta) \equiv \text{const} = 0$, and therefore $f(\zeta) = 0$. We refer to K. Steffen [1], from whom we have learned the above regularity proof, for generalizations and various applications.

12. **The principle of symmetric criticality (Coleman's principle).** Let $\mathscr{G}$ be a group of mappings $G: \mathscr{C} \to \mathscr{C}$ of a set $\mathscr{C}$ into itself, and let $\mathscr{F}: \mathscr{C} \to \mathbb{R}$ be a $\mathscr{G}$-invariant functional on $\mathscr{C}$, i.e. $\mathscr{F}(Gu) = \mathscr{F}(u)$ for all $u \in \mathscr{C}$ and $G \in \mathscr{G}$. Let $\mathscr{C}_0$ be the fixed-point set of $\mathscr{G}$, and $f := \mathscr{F}|_{\mathscr{C}_0}$ be the restriction of $\mathscr{F}$ to $\mathscr{C}_0$. The elements of $\mathscr{C}_0$ are called **$\mathscr{G}$-symmetric**.

Typically $\mathscr{C}$ is a class of maps $u: \Omega \to M$ from $\Omega \subset \mathbb{R}^n$ into a manifold M, and $\mathscr{G}$ is generated by a group of transformatons $g: M \to M$, setting $(Gu)(x) := g(u(x))$ for $x \in \Omega$ and $u \in \mathscr{C}$.

We say that the **principle of symmetric criticality** holds true for the variational problem "$\mathscr{F} \to$ stat. in $\mathscr{C}$" if the critical points of f in $\mathscr{C}_0$ are necessarily critical points of $\mathscr{F}$ in $\mathscr{C}$.

Sufficient conditions for the validity of this principle were stated by Palais [2] and Kapitanskii-Ladyzhenskaya [1]. The setting of the second authors seems to be more appropriate for applications to the calculus of variations.

Note, however, that the principle is not always true, as one can see from the following simple example pointed out by Kapitanski-Ladyzhenskaya [1]: Let $\mathscr{C}$ be the space of points $x = (x^1, x^2) \in \mathbb{R}^2$, and $\mathscr{G}$ be the group of transformations $G_t: \mathscr{C} \to \mathscr{C}$ defined by $G_t x = (x^1 + tx^2, x^2)$, $t \in \mathbb{R}$. Then $\mathscr{C}_0 = \{x: x^2 = 0\}$, and for any smooth function $\mathscr{F}: \mathscr{C} \to \mathscr{C}$ we have $f(x^1) = \mathscr{F}(x)$ for $x \in \mathscr{C}_0$. Thus the prinicple of symmetric criticality does not hold in this situation.

It would be interesting to find out how the principle is related to E. Noether's theorem.

Part II

The Second Variation and Sufficient Conditions

Chapter 4. Second Variation, Excess Function, Convexity

If $\mathscr{F}(u)$ is a real valued function of a real variable which is of class C^2 on $\mathbb{R}$, then, besides $\mathscr{F}'(u_0) = 0$, also the condition $\mathscr{F}''(u_0) \geq 0$ is necessary for u_0 being a local minimizer of $\mathscr{F}$. Moreover, the conditions

$$\mathscr{F}'(u_0) = 0 \quad \text{and} \quad \mathscr{F}''(u_0) > 0$$

are sufficient for u_0 furnishing a local minimum.

Now let $\mathscr{F}(u)$ be some variational integral of the type

$$\mathscr{F}(u) = \int_\Omega F(x, u, Du)\, dx\,.$$

In the first three chapters we have studied the necessary conditions for a function $u \in C^s(\bar{\Omega}, \mathbb{R}^N)$, $s \geq 1$, to be a minimizer that correspond to the equation $\mathscr{F}'(u) = 0$ in case of a real variable. They are basically expressed by the relations

$$\delta\mathscr{F}(u, \varphi) = 0 \quad \text{and} \quad \partial\mathscr{F}(u, \lambda) = 0\,.$$

We shall now investigate necessary and sufficient conditions for an extremal to furnish a local minimum, which are closely related to the assumptions $\mathscr{F}''(u) \geq 0$ and $\mathscr{F}''(u) > 0$ in case of a function of a real variable.

The formulation of sufficient conditions for an extremal to be a *local* (or *relative*) minimizer clearly depends on the choice of *neighbourhoods* within the set $\mathscr{S}$ where our variational integral $\mathscr{F}(u)$ is considered: One has to define when two admissible functions u and v are to be viewed as "close to each other". In other words, we have to equip the set $\mathscr{S}$ of admissible functions with a topology so that $\mathscr{S}$ becomes a topological space. This remark may seem trivial and not worth mentioning. Yet one should realize that, historically, it took a long time to understand this fact and to see that its neglection may lead to grave errors.[1]

In $\mathbb{R}^n$ the topology is standard; it is generated by the Euclidean norm to which all other norms are equivalent, and more exotic topologies on $\mathbb{R}^n$ are of small importance for differential geometry. Instead we can choose many topologies in the class $\mathscr{S}$ of admissible functions where we look for minimizers

[1] The first to point out this fact were Weierstrass in his lectures on the calculus of variations, and Scheeffer in his very influentual paper [3] from 1886. Cf. also Bolza's treatise [3], Chapter 3, and the historical account in H.H. Goldstine [1], Chapter 5 (in particular 5.5 and 5.11).

of the given functional $\mathscr{F}(u)$, and analysts have learned to play with a combination of different topologies. For our present purpose of deriving "*sufficient conditions*", that is, of establishing conditions guaranteeing that a given extremal is a local minimizer, it is enough to operate with two different kinds of neighbourhoods. Following the classical approach of A. Kneser and Zermelo, we distinguish between C^0- and C^1-neighbourhoods: (i) *Weak neighbourhoods* in the class $\mathscr{S}$ of admissible functions contain functions which are close to each other with respect to the C^1-norm. (ii) *Strong neighbourhoods* in $\mathscr{S}$ consist of comparison functions which are close to each other in the C^0-norm. Correspondingly we shall speak of *weak minimizers* and *strong minimizers*. Unfortunately, the epithets "weak" and "strong" are in the modern literature used in a variety of ways and with a quite different meaning. Therefore we shall also refer to weak minimizers as to functions having the *weak minimum property* $(\mathscr{M}')$, and to strong minimizers as to functions having the *strong minimum property* $(\mathscr{M})$. Clearly strong minimizers are also weak minimizers, but the converse is not true as we can see from *1.1* [1].

If u has the properties $(\mathscr{M})$ or $(\mathscr{M}')$, we infer that, for any function $\varphi \in C_c^\infty(\Omega, \mathbb{R}^N)$, the function

$$\Phi(\varepsilon) := \mathscr{F}(u + \varepsilon\varphi) = \int_\Omega F(x, u + \varepsilon\varphi, Du + \varepsilon D\varphi)\, dx$$

satisfies

$$\Phi(\varepsilon) \geq \Phi(0) \quad \text{for all } \varepsilon \text{ satisfying } |\varepsilon| \ll 1\,.$$

Then Taylor's expansion

$$\Phi(\varepsilon) = \Phi(0) + \varepsilon\Phi'(0) + \tfrac{1}{2}\varepsilon^2\Phi''(0) + o(\varepsilon^2) \quad \text{as } \varepsilon \to 0$$

implies

$$\Phi'(0) = 0 \quad \text{and} \quad \Phi''(0) \geqq 0\,.$$

The first relation is equivalent to the relation

$$\delta\mathscr{F}(u, \varphi) = 0 \quad \text{for all } \varphi \in C_c^\infty(\Omega, \mathbb{R}^N)$$

which we have studied in Chapter 1, and we obtain

$$\mathscr{F}(u + \varepsilon\varphi) = \mathscr{F}(u) + \frac{\varepsilon^2}{2}\delta^2\mathscr{F}(u, \varphi) + o(\varepsilon^2), \tag{1}$$

where $\Phi''(0) = \delta^2\mathscr{F}(u, \varphi)$ is the *second variation of* $\mathscr{F}$ *at* u *in direction of* φ that we had briefly considered already in 1,*1*. The functional $\varphi \to \frac{1}{2}\delta^2\mathscr{F}(u, \varphi)$ is a quadratic integro-differential form with respect to φ which is traditionally called the *accessory integral* related to $\mathscr{F}$, taken at u. It will be denoted by $\mathscr{Q}(\varphi)$, and its Lagrangian $Q(x, z, p)$ is called the *accessory Lagrangian*:

$$\mathscr{Q}(\varphi) = \int_\Omega Q(x, \varphi, D\varphi)\, dx = \tfrac{1}{2}\delta^2\mathscr{F}(u, \varphi)\,. \tag{2}$$

The necessary condition $\delta^2\mathscr{F}(u, \varphi) \geq 0$ then amounts to

(3) $$\mathscr{Q}(\varphi) \geq 0 \quad \text{for all } \varphi \in C_c^\infty(\Omega, \mathbb{R}^N).$$

We shall discuss this condition as well as the accessory integral in *1.1* and *1.2*. It will be shown in *1.3* that, as a consequence of (3), weak minimizers necessarily have to satisfy the so-called *Legendre–Hadamard condition*

(4) $$F_{p_\alpha^i p_\beta^k}(x, u(x), Du(x))\, \xi^i \xi^k \eta_\alpha \eta_\beta \geq 0$$

for all $x \in \Omega$ and all $\xi \in \mathbb{R}^N$, $\eta \in \mathbb{R}^n$. This condition states that F is "rank-one convex" on weak minimizers, or, for $N = 1$, the F is "convex" on weak minimizers.

Secondly we shall prove in *1.4* that strong minimizers u satisfy *Weierstrass's necessary condition*

(5) $$\mathscr{E}_F(x, u(x), Du(x), Du(x) + \pi) \geq 0$$

for all $x \in \Omega$ and for all rank-one matrices π (i.e., for any matrix $\pi = (\pi_\alpha^i)$ whose entries are of the form $\pi_\alpha^i = \xi^i \eta_\alpha$). Here $\mathscr{E}_F$ denotes the *Weierstrass excess function* of the Largrangian F which is defined by

(6) $$\mathscr{E}_F(x, z, p, q) := F(x, z, q) - F(x, z, p) - (q - p) \cdot F_p(x, z, p).$$

We shall also see that Weierstrass's necessary condition implies the Legendre–Hadamard condition. Example [1] in *1.1* shows that the Weierstrass condition actually is more stringent then the Legendre–Hadamard condition.

In contrast to functions of finitely many real variables, the condition of positive definiteness of the second variation, i.e.,

$$\delta^2\mathscr{F}(u, \varphi) > 0 \quad \text{for all } \varphi \in C_c^\infty(\Omega, \mathbb{R}^N) \text{ with } \varphi(x) \not\equiv 0,$$

is no longer sufficient to ensure that u is a strong or even a weak local minimizer (see Chapter 5, and in particular 5,*1.1* [1]). In Section 2 of this chapter as well as in Chapters 5–7 we shall discuss sufficient conditions ensuring that a given extremal actually furnishes a minimum or at least a local minimum on the class of comparison functions.

In Section 2, we begin by deriving global minimum properties of extremals based on *convexity* arguments. The fundamental reasoning of this approach is contained in the following simple observation: *A strictly convex function on a convex domain possesses at most one critical point. Hence, if there exists a critical point in this situation, it will be the unique minimizer of the function.* We shall exploit this observation in various ways. It can be applied very effectively in the form of *calibrators* (see *2.6*), although here convexity enters only in a somewhat hidden way. The role of convexity becomes clear in Chapter 6 where we present the *Weierstrass field theory* for the case of one independent variable ($n = 1$). This theory can be viewed as a first systematic application of the concept of calibrator. Field theory is an appropriate tool to obtain sufficient conditions for an extremal to be a strong minimizer; it efficiently combines geometric and analytic ideas and is closely linked with the *Hamilton–Jacobi-theory* that will be presented in Chapters 7–9. Extensions of the field theory to multiple-dimensional variational problems and for $N \geq 1$ can be found in 6,*3* and in Chapter 7.

Chapter 5 deals with weak minimizers. First we will show that the Euler operator L_Q of the accessory Lagrangian Q to F at u coincides with the linearization $\mathcal{J}_u$ of the Euler operator L_F of the original Lagrangian F taken at u. This linearization $\mathcal{J}_u$ is called the *Jacobi operator* of F along u. Then we shall investigate how the solutions of the Jacobi equation $\mathcal{J}_u v = 0$, the so-called *Jacobi fields* along a given extremal u, are related to the property of u being a *weak minimizer*. To this end we study the *eigenvalue problem*

$$\mathcal{J}_u v = \lambda v \quad \text{in } \Omega, \qquad v = 0 \quad \text{on } \partial\Omega,$$

which is closely related to the Jacobi equation $\mathcal{J}_u v = 0$. There is a sequence of discrete eigenvalues λ_ν, tending to infinity as $\nu \to \infty$. For a weak minimizer u the smallest eigenvalue λ_1 must necessarily be nonnegative, and the assumption $\lambda_1 > 0$ suffices to show that u is a weak minimizer.

For one-dimensional problems these conditions usually are expressed in terms of *conjugate values* and *conjugate points*. In the second part of Chapter 5 we shall develop Jacobi's conjugate point theory for the case $n = N = 1$. The treatment is more or less independent of the rest of Chapter 5 and can be read separately.

1. Necessary Conditions for Relative Minima

Not every extremal u of a variational integral $\mathcal{F}$ is furnishing a relative minimum of $\mathcal{F}$. As we already know, the extremal has necessarily to satisfy the inequality

$$\delta^2 \mathcal{F}(u, \varphi) \geq 0 \quad \text{for all } \varphi \in C_c^\infty(\Omega, \mathbb{R})$$

in order to qualify as a possible minimizer. Requirements of this kind are usually denoted as *necessary conditions*.

We shall derive two types of necessary conditions, the *Legendre–Hadamard condition* and the *Weierstrass condition*. The first must even be satisfied by extremals which only possess the weak minimum property ($\mathcal{M}'$), i.e., by *weak minimizers*, whereas the second is necessary for *strong minimizers*, that is, for extremals with the strong minimum property ($\mathcal{M}$).

We begin by defining in *1.1* weak and strong neighbourhoods as well as weak and strong minimizers. After defining in *1.2* the accessory integral, we shall establish in *1.3* the necessary condition of Legendre and Hadamard of weak minimizers. Finally, in *1.4*, the necessary condition of Weierstrass for strong minimizers will be derived. It will be seen that this condition implies the Legendre–Hadamard condition.

1.1. Weak and Strong Minimizers

In this chapter, we shall always assume that Ω is a bounded domain in $\mathbb{R}^n$ and that $F(x, z, p)$ is a Lagrangian of class C^2 which is defined on some open set $\mathscr{U}$ of $\mathbb{R}^n \times \mathbb{R}^N \times \mathbb{R}^{nN}$.

A function $u \in C^1(\bar{\Omega}, \mathbb{R}^n)$ is said to be *F-admissible* if its 1-graph $\{(x, u(x), Du(x)): x \in \bar{\Omega}\}$ is contained in $\mathscr{U}$.

For F-admissible function u, the integral

$$\mathscr{F}(u) := \int_\Omega F(x, u(x), Du(x))\, dx$$

is well defined.

Consider now some nonempty subset $\mathscr{C}$ of the function space $C^1(\bar{\Omega}, \mathbb{R}^N)$ whose elements u are F-admissible.

If we say that some $u \in \mathscr{C}$ furnishes a *local* or *relative minimum* of $\mathscr{F}$ on $\mathscr{C}$, we mean that

$$\mathscr{F}(u) \le \mathscr{F}(v)$$

holds for all $v \in \mathscr{C}$ which are close to u.

However, in this definition we have left open when two functions are to be considered close to each other. Among the infinitely many possibilities to define vicinities in a function space, two choices are very natural and thrust themselves upon the reader: we can consider proximity with respect to the C^0-norm or the C^1-norm. A minimum in the first sense will be called a *strong minimum*, in the second a *weak minimum*. To make this definition precise, we introduce the notions of weak and strong ε-neighbourhoods $\mathscr{N}_\varepsilon'(u)$ and $\mathscr{N}_\varepsilon(u)$ of a function $z = u(x)$, $x \in \bar{\Omega}$, of class C^1.

Definition. *For $u \in C^1(\bar{\Omega}, \mathbb{R}^N)$ and $\varepsilon > 0$, the set*

$$\mathscr{N}_\varepsilon(u) := \{v \in C^1(\bar{\Omega}, \mathbb{R}^N): \|v - u\|_{C^0(\bar{\Omega})} < \varepsilon\}$$

is called a strong neighbourhood[2] *of u, whereas the ball*

$$\mathscr{N}_\varepsilon'(u) := \{v \in C^1(\bar{\Omega}, \mathbb{R}^N): \|v - u\|_{C^1(\bar{\Omega})} < \varepsilon\}$$

is said to be a weak neighbourhood *of u in $C^1(\bar{\Omega}, \mathbb{R}^N)$.*

Correspondingly u is called a weak minimizer *of $\mathscr{F}$ on $\mathscr{C}$ if there exists a weak neighbourhood $\mathscr{N}_\varepsilon'(u)$ of u such that*

$$\mathscr{F}(u) \le \mathscr{F}(v) \quad \text{for all } v \in \mathscr{C} \cap \mathscr{N}_\varepsilon'(u)$$

holds, and we speak of a strong minimizer *u of $\mathscr{F}$ on $\mathscr{C}$ if there is a strong neighbourhood $\mathscr{N}_\varepsilon(u)$ of u such that*

$$\mathscr{F}(u) \le \mathscr{F}(v) \quad \text{for all } v \in \mathscr{C} \cap \mathscr{N}_\varepsilon(u).$$

---

[2] One might justly object that the C^1-topology is stronger than the C^0-topology. Thus it seems more appropriate to denote weak and strong neighbourhoods as **narrow** and **wide** neighbourhoods respectively.

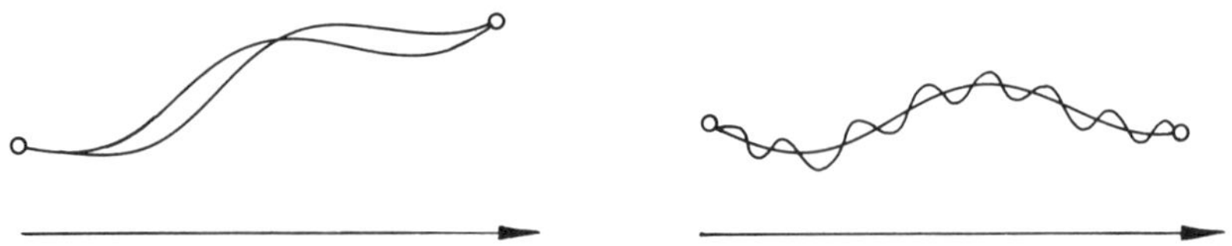

(a) (b)

Fig. 1. The function v lies in **(a)** in a weak neighbourhood of u, or **(b)** in a strong neighbourhood.

It is from now on understood that the epithet "weak" or "strong" in connection with a "minimizer" indicates that we deal with relative (i.e., local) minimizers.

A weak (or strong) minimizer u is said to be *strict* or *isolated* if

$$\mathscr{F}(v) > \mathscr{F}(u) \quad \text{for all } v \in \mathscr{C} \cap \mathscr{N}_\varepsilon'(u) \text{ (or } v \in \mathscr{C} \cap \mathscr{N}_\varepsilon(u)) \text{ with } v \neq u$$

holds.

If the class $\mathscr{C}$ is not particularly specified, it will be assumed to be the set of functions $v \in C^1(\bar{\Omega}, \mathbb{R}^N)$ with $v = u$ on $\partial\Omega$. In other words, *u will be called a weak (or strong) minimizer of $\mathscr{F}$ if it minimizes $\mathscr{F}$ among all functions $v \in C^1(\bar{\Omega}, \mathbb{R}^N)$ which coincide with u on $\partial\Omega$ and are contained in a sufficiently small weak (or strong) neighbourhood of u.*

In order to derive the necessary conditions of Legendre–Hadamard and of Weierstrass, we shall presume the conditions $(\mathscr{M}')$ and $(\mathscr{M})$, respectively, which are slightly differently formulated than the properties of being a weak or a strong minimizer.

Definition of the strong minimum property $(\mathscr{M})$. *We say that $u \in C^1(\bar{\Omega}, \mathbb{R}^N)$ has the strong minimum property $(\mathscr{M})$ if there exists some $\varepsilon > 0$ such that $\mathscr{F}(u) \le \mathscr{F}(u + \varphi)$ holds for every $\varphi \in C_c^\infty(\Omega, \mathbb{R}^N)$ with $\|\varphi\|_{C^0(\bar{\Omega})} < \varepsilon$.*

In the same spirit, we formulate the

Definition of the weak minimum property $(\mathscr{M}')$. *We say that $u \in C^1(\bar{\Omega}, \mathbb{R}^N)$ has the weak minimum property $(\mathscr{M}')$ if there exists some $\varepsilon > 0$ such that $\mathscr{F}(u) \le \mathscr{F}(u + \varphi)$ holds for every $\varphi \in C_c^\infty(\Omega, \mathbb{R}^N)$ with $\|\varphi\|_{C^1(\bar{\Omega})} < \varepsilon$.*

Evidently, any weak minimizer u has the minimum property $(\mathscr{M}')$, and any strong minimizer has the property $(\mathscr{M})$.

If $\partial\Omega$ is sufficiently regular, say, smooth or at least piecewise smooth, also the converse holds true as we can see by an appropriate approximation argument (cf. the remark at the beginning of *2.1*). In this case u is a weak minimizer if and only if it has the property $(\mathscr{M}')$, and $(\mathscr{M})$ is equivalent to u being a strong minimizer.

In the next subsection we shall prove the following two necessary conditions:

Weak minimizers satisfy the Legendre–Hadamard condition

$$F_{p^i_\alpha p^k_\beta}(x, u(x), Du(x))\, \xi^i \xi^k \eta_\alpha \eta_\beta \geq 0,$$

and strong minimizers fulfil the necessary condition of Weierstrass

$$\mathscr{E}_F(x, u(x), Du(x), Du(x) + \pi) \geq 0$$

for every $\pi = (\pi^i_\alpha)$ *such that* $\pi^i_\alpha = \xi^i \eta_\alpha$, *where* $\mathscr{E}_F$ *denotes the excess of the Lagrangian* F *defined by* 4, (6).

Obviously, any strong minimizer also is a weak minimizer, but the converse does in general not hold. For instance, *the function* $u(x) = 0$ *is a weak but not a strong minimizer of the integral*

$$\mathscr{F}(u) = \int_0^1 (u'^2 + u'^3)\, dx,$$

as we shall see in [1] presented below. To treat this example we need a proposition that will also be useful for other purposes as it contains the key to the Jacobi theory of weak minimizers which will be outlined in Chapter 5. In order to formulate this proposition, we recall the definition of the second variation $\delta^2 \mathscr{F}(u, \varphi)$ of $\mathscr{F}$ at u in direction of a function $\varphi \in C^1(\bar{\Omega}, \mathbb{R}^N)$:

$$\delta^2 \mathscr{F}(u, \varphi) := \left.\frac{d^2}{d\varepsilon^2} \mathscr{F}(u + \varepsilon\varphi)\right|_{\varepsilon=0}.$$

The structure of $\delta^2 \mathscr{F}(u, \varphi)$ will be investigated more closely in the next subsection.

Proposition. *Let* F *be a Lagrangian of class* C^2.

(i) *If* u *is a weak minimizer of* $\mathscr{F}$, *then* u *is a weak extremal of* $\mathscr{F}$ *and satisfies*

$$\delta^2 \mathscr{F}(u, \varphi) \geq 0 \quad \textit{for all } \varphi \in C^\infty_c(\Omega, \mathbb{R}^N). \tag{1}$$

(ii) *Conversely, if* u *is a weak extremal of* $\mathscr{F}$ *and if there exists some number* $\lambda > 0$ *such that*

$$\delta^2 \mathscr{F}(u, \varphi) \geq 2\lambda \int_\Omega \{|\varphi|^2 + |D\varphi|^2\}\, dx \quad \textit{for all } \varphi \in C^\infty_c(\Omega, \mathbb{R}^N), \tag{2}$$

then u *is a strict weak minimizer.*

Remark 1. By approximation, inequality (2) implies that

$$\delta^2 \mathscr{F}(u, \varphi) \geq 2\lambda \int_\Omega \{|\varphi|^2 + |D\varphi|^2\}\, dx \quad \text{for all } \varphi \in C^1_0(\bar{\Omega}, \mathbb{R}^N) \tag{2'}$$

holds.

Remark 2. By the reasoning to be given in *1.3* we can also infer that (2) implies the *strict Legendre–Hadamard condition*

$$\tfrac{1}{2}F_{p^i_\alpha p^k_\beta}(x, u(x), Du(x))\xi^i\xi^k\eta_\alpha\eta_\beta \ge \lambda|\xi|^2|\eta|^2. \tag{3}$$

Conversely, (3) does in general not yield the sufficient condition (2), even if $n = N = 1$, as one can easily see.

It is the main goal of the Jacobi theory to supply further conditions which, together with the strict Legendre–Hadamard condition, suffice to demonstrate that a given weak extremal of $\mathscr{F}$ actually is a weak minimizer. The basic features of the Jacobi theory will be presented in Chapter 5.

Proof of the Proposition. For any $u, \varphi \in C^1(\bar{\Omega}, \mathbb{R}^N)$ the function

$$\Phi(\varepsilon) := \mathscr{F}(u + \varepsilon\varphi), \quad |\varepsilon| < \varepsilon_0,$$

is well-defined and of class C^2, $0 < \varepsilon_0 \ll 1$, and, by definition, we have

$$\Phi'(0) = \delta\mathscr{F}(u, \varphi), \qquad \Phi''(0) = \delta^2\mathscr{F}(u, \varphi).$$

(i) If u is a weak minimizer of $\mathscr{F}$, then we have

$$\mathscr{F}(u) \le \mathscr{F}(u + \varepsilon\varphi)$$

for any $\varphi \in C_c^\infty(\bar{\Omega}, \mathbb{R}^N)$ and $|\varepsilon| < \varepsilon_0$, where $0 < \varepsilon_0 \ll 1$. Hence $\Phi(\varepsilon) := \mathscr{F}(u + \varepsilon\varphi)$ has a minimum at $\varepsilon = 0$, and we obtain $\Phi'(0) = 0$ as well as $\Phi''(0) \ge 0$, i.e.,

$$\delta^2\mathscr{F}(u, \varphi) \ge 0.$$

(ii) To prove the converse, we recall that (2) implies (2′). Moreover, by some continuity argument together with the inequality $2ab \le a^2 + b^2$ and the Schwarz inequality, we conclude that there is some number $\varepsilon > 0$ such that

$$|\delta^2F(v, \varphi) - \delta^2F(u, \varphi)| \le \lambda\{|\varphi|^2 + |D\varphi|^2\}$$

holds on $\bar{\Omega}$ for all $v \in \mathscr{N}'_\varepsilon(u)$ and for every $\varphi \in C^1(\bar{\Omega}, \mathbb{R}^N)$. Then it follows from (2′) that

$$\delta^2\mathscr{F}(v, \varphi) \ge \lambda \int_\Omega \{|\varphi|^2 + |D\varphi|^2\}\, dx$$

holds for all $\varphi \in C_0^1(\bar{\Omega}, \mathbb{R}^N)$ and for each $v \in \mathscr{N}'_\varepsilon(u)$.

Let now v be an arbitrary function in $\mathscr{N}'_\varepsilon(u)$ which coincides with u on $\partial\Omega$. Then, for any $t \in [0, 1]$, also the function $\psi(t) := u + t(v - u)$ is contained in $\mathscr{N}'_\varepsilon$ whence we obtain

$$\delta^2\mathscr{F}(\psi(t), v - u) \ge 0 \quad \text{for every } t \in [0, 1].$$

Consider the function $H(t) := \mathscr{F}(\psi(t))$, $0 \le t \le 1$, which is of class C^2. We can write

$$H(1) - H(0) = H'(0) + \int_0^1 (1 - t)H''(t)\, dt.$$

By virtue of

$$H'(0) = \delta\mathcal{F}(u, v - u) = 0,$$

it follows that

$$\mathcal{F}(v) - \mathcal{F}(u) = \int_0^1 (1 - t)\, \delta^2\mathcal{F}(\psi(t), v - u)\, dt \geq 0$$

if we take $H''(t) = \delta^2\mathcal{F}(\psi(t), v - u)$ and the inequality above into account. □

[1] *Scheeffer's example.* Consider now the afore-mentioned variational integral

$$\mathcal{F}(v) = \int_0^1 (v'^2 + v'^3)\, dx,$$

with $n = N = 1$, $\Omega = (0, 1)$, $F(p) = p^2 + p^3$. We shall now see that $\mathcal{F}$ has the strict weak minimizer $u(x) = 0$ which satisfies (2'), but does not provide a strong minimum for $\mathcal{F}$.

In fact,

$$\delta\mathcal{F}(v, \varphi) = \int_0^1 (2v' + 3v'^2)\varphi'\, dx$$

and

$$\delta^2\mathcal{F}(v, \varphi) = \int_0^1 (2 + 6v')\varphi'^2\, dx.$$

Hence $u(x) = 0$, $0 \leq x \leq 1$, is an extremal of $\mathcal{F}$ which satisfies

$$\delta^2\mathcal{F}(u, \varphi) = 2\int_0^1 \varphi'^2\, dx.$$

Moreover, for $\varphi \in C_0^1([0, 1])$, we obtain

$$\varphi(x) = \int_0^x \varphi'(t)\, dt, \quad 0 \leq x \leq 1,$$

whence

$$|\varphi(x)|^2 \leq x \int_0^x |\varphi'(t)|^2\, dt \leq \int_0^1 |\varphi'(t)|^2\, dt$$

and therefore

$$\int_0^1 \varphi^2\, dx \leq \int_0^1 \varphi'^2\, dx.$$

This shows that

$$\delta^2\mathcal{F}(u, \varphi) \geq \int_0^1 (\varphi^2 + \varphi'^2)\, dx \quad \text{for all } \varphi \in C_0^1([0, 1]),$$

that is, $u = 0$ satisfies condition (2'). We then conclude that u is a weak minimizer of $\mathcal{F}$ among all functions $v \in C^1([0, 1])$ with $v(0) = 0$ and $v(1) = 0$.

On the other hand, if we consider the piecewise linear comparison functions

$$v(x) := \begin{cases} -x/h & \text{for } 0 \leq x \leq h^2, \\ h(x - 1)/(1 - h^2) & \text{for } h^2 \leq x \leq 1 \end{cases}$$

for parameter values $h \in (0, 1/2)$, we find that $v'(x) = -1/h$ for $0 \leq x \leq h^2$ and $v'(x) = \dfrac{h}{1 - h^2} < 1$

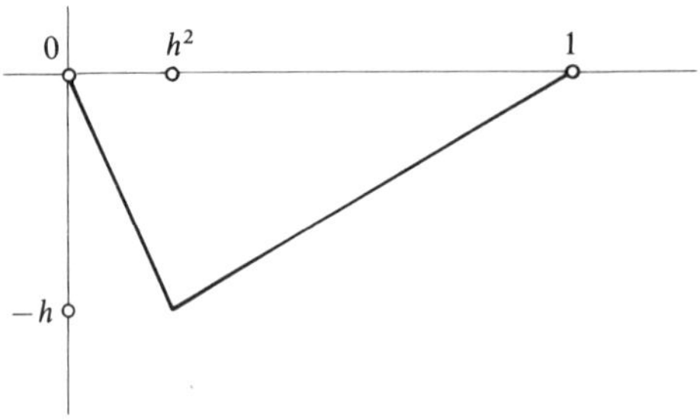

Fig. 2. Scheeffer's example.

for $h^2 \le x \le 1$ whence

$$v'^2(x) + v'^3(x) \le \begin{cases} -1/(2h^3) & \text{for } 0 \le x \le h^2, \\ 2 & \text{for } h^2 \le x \le 1 \end{cases}$$

and, if we write $v_h = v$, it follows that

$$\mathscr{F}(v_h) - \mathscr{F}(u) = \mathscr{F}(v) \le 2 - \frac{1}{2h},$$

and therefore

$$\lim_{h\to 0} [\mathscr{F}(v_h) - \mathscr{F}(u)] = -\infty. \tag{4}$$

We, moreover, have $|v_h(x)| \le h$ for $0 \le x \le 1$ which implies

$$\lim_{h\to 0} \| v_h - u \|_{C^0([0,1])} = 0. \tag{5}$$

By rounding off the corners of the functions $v_h(x)$ at $x = h^2$, we can replace the piecewise linear functions v_h by slightly altered functions v_h of class $C_0^1([0, 1])$ which still satisfy the relations (4) and (5). This finally proves that $u = 0$ *is a weak but not a strong minimizer for* $\mathscr{F}$, *although it satisfies condition* (2′).

We could also have proved that $u = 0$ is not a strong minimizer by verifying that the necessary condition of Weierstrass, to be proved in the next subsection, is violated. In fact, we have

$$\begin{aligned} \mathscr{E}_F(x, z, p, q) &= \frac{(q-p)^2}{2}(2 + 6p) + \frac{(q-p)^3}{6}\cdot 6 \\ &= (q-p)^2(1 + q + 2p), \end{aligned}$$

and therefore

$$\begin{aligned} \mathscr{E}_F(x, u(x), Du(x), Du(x) + \pi) &= \mathscr{E}_F(x, 0, 0, \pi) \\ &= \pi^2(1 + \pi). \end{aligned}$$

This expression is negative for $\pi < -1$.

[2] A similar example can be obtained by considering the functional

$$\mathscr{F}(u) := \int_0^1 (u'^2 - 1)^2 \, dx$$

whose Lagrangian $F(p) = (p^2 - 1)^2$ satisfies $F_p(p) = 4p(p^2 - 1)$, $F_{pp}(p) = 12p^2 - 4$. It can easily be seen that $u(x) := x/\sqrt{2}$ is an extremal satisfying $F_{pp}(u') = 2$ whence

$$\delta^2 \mathscr{F}(u, \varphi) = 2 \int_0^1 |\varphi'(x)|^2 \, dx.$$

We then deduce as in [1] that $u(x) = x/\sqrt{2}$ is a weak minimizer; but

$$\mathscr{E}_F(x, u(x), u'(x), 0) = -1/4 < 0,$$

and therefore $u(x)$ is not a strong minimizer.

1.2. Second Variation: Accessory Integral and Accessory Lagrangian

For the following, we fix some F-admissible function $u \in C^1(\overline{\Omega}, \mathbb{R}^N)$. Then, for each $\varphi \in C^1(\overline{\Omega}, \mathbb{R}^N)$, there exists a number $\varepsilon_0 > 0$ such that the *linear variations*

(1) $$\psi(x, \varepsilon) = u(x) + \varepsilon\varphi(x), \quad |\varepsilon| < \varepsilon_0,$$

are F-admissible. Thus the function

$$\Phi(\varepsilon) := \mathscr{F}(u + \varepsilon\varphi), \quad |\varepsilon| < \varepsilon_0,$$

is well defined and of class C^2, for every $\varphi \in C^1(\overline{\Omega}, \mathbb{R}^N)$. The first and second variations of $\mathscr{F}$ at u are given by

$$\delta\mathscr{F}(u, \varphi) = \Phi'(0), \qquad \delta^2\mathscr{F}(u, \varphi) = \Phi''(0),$$

and we can write

$$\mathscr{F}(u + \varepsilon\varphi) = \mathscr{F}(u) + \varepsilon\delta\mathscr{F}(u, \varphi) + \tfrac{1}{2}\varepsilon^2\delta^2\mathscr{F}(u, \varphi) + o(\varepsilon^2).$$

Hence if u has the weak minimum property $(\mathscr{M}')$, it has necessarily to satisfy the conditions

$$\delta\mathscr{F}(u, \varphi) = 0 \quad \text{and} \quad \delta^2\mathscr{F}(u, \varphi) \geq 0$$

for all $\varphi \in C_c^\infty(\Omega, \mathbb{R}^N)$, as we already saw in *1.1*. The implications of $\delta\mathscr{F}(u, \varphi) = 0$ were investigated in Chapter 1. Now we shall examine which conclusions can be drawn from the nonnegativity of the second variation $\delta^2\mathscr{F}(u, \varphi)$. The *second-variation functional*

(2) $$\delta^2\mathscr{F}(u, \varphi) = \int_\Omega \delta^2 F(u, \varphi) \, dx$$

with the integrand

(3) $$\delta^2 F(u, \varphi)(x) := \frac{d^2}{d\varepsilon^2} F(x, u(x) + \varepsilon\varphi(x), Du(x) + \varepsilon D\varphi(x))\bigg|_{\varepsilon=0}$$

plays a central role for the derivation of both necessary and sufficient conditions for weak extremals having some kind of minimum property. Therefore we shall consider it with some care.

Let us introduce the expression

(4) $$Q(x, z, p) = \frac{1}{2}\frac{d^2}{d\varepsilon^2} F(x, u(x) + \varepsilon z, Du(x) + \varepsilon p)\Big|_{\varepsilon=0},$$

which is a quadratic form in (z, p), as can be seen from the formula

(5) $$Q(x, z, p) = \tfrac{1}{2}[F_{zz}(\ldots)zz + 2F_{zp}(\ldots)zp + F_{pp}(\ldots)pp],$$

where $(\ldots)$ stands for $(x, u(x), Du(x))$. With regard to coordinates $z = (z^i)$, $p = (p^i_\alpha)$, we can also write

(6) $$Q(x, z, p) = \tfrac{1}{2}[F_{z^i z^k}(\ldots)z^i z^k + 2F_{z^i p^k_\beta}(\ldots)z^i p^k_\beta + F_{p^i_\alpha p^k_\beta}(\ldots)p^i_\alpha p^k_\beta].$$

The function $Q(x, z, p)$ is defined for $(x, z, p) \in \bar{\Omega} \times \mathbb{R}^N \times \mathbb{R}^{nN}$; it will be called the *accessory Lagrangian* corresponding to F and u, and the associated variational integral

(7) $$\mathcal{Q}(v) := \int_\Omega Q(x, v(x), Dv(x))\,dx$$

is denoted as *accessory integral* with regard to F and u. It follows from (3) and (4) that

(8) $$\mathcal{Q}(v) = \tfrac{1}{2}\delta^2\mathcal{F}(u, v)$$

holds for all $v \in C^1(\bar{\Omega}, \mathbb{R}^N)$.

Set

(9) $$\begin{aligned} a^{\alpha\beta}_{ik}(x) &:= \tfrac{1}{2}F_{p^i_\alpha p^k_\beta}(x, u(x), Du(x)), \\ b^\beta_{ik}(x) &:= \tfrac{1}{2}F_{z^i p^k_\beta}(x, u(x), Du(x)), \\ c_{ik}(x) &:= \tfrac{1}{2}F_{z^i z^k}(x, u(x), Du(x)). \end{aligned}$$

Then we have the symmetry relations

(10) $$a^{\alpha\beta}_{ik} = a^{\beta\alpha}_{ki}, \qquad c_{ik} = c_{ki},$$

and we can write

(11) $$Q(x, z, p) = a^{\alpha\beta}_{ik}(x)p^i_\alpha p^k_\beta + 2b^\beta_{ik}(x)z^i p^k_\beta + c_{ik}(x)z^i z^k.$$

Thus the first variation $\delta\mathcal{Q}(v, \varphi)$ of the accessory integral $\mathcal{Q}$ at v in direction of φ is given by

(12) $$\begin{aligned} \tfrac{1}{2}\delta\mathcal{Q}(v, \varphi) &= \mathcal{Q}(v, \varphi) \\ &= \int_\Omega \{a^{\alpha\beta}_{ik}(x)D_\alpha v^i D_\beta\varphi^k + b^\beta_{ik}(x)[v^i D_\beta\varphi^k + \varphi^i D_\beta v^k] \\ &\quad + c_{ik}(x)v^i\varphi^k\}\,dx, \end{aligned}$$

where $\mathcal{Q}(v, \varphi) := \frac{1}{2}[\mathcal{Q}(v + \varphi) - \mathcal{Q}(v) - \mathcal{Q}(\varphi)]$ is the symmetric bilinear from associated with the quadratic form $\mathcal{Q}(v)$. If the coefficients $a^{\alpha\beta}_{ik}$ and b^β_{ik} are of class

C^1, we can form the Euler operator L_Q of the accessory Lagrangian Q. This is a linear second order differential operator, the k-th component of which is given by

$$\tfrac{1}{2}(L_Q v)_k = [b^{\alpha}_{ki} D_\alpha v^i + c_{ik} v^i] - D_\beta [a^{\alpha\beta}_{ik} D_\alpha v^i + b^{\beta}_{ik} v^i]. \tag{13}$$

In *1.2* of the next chapter we shall prove that the Euler operator L_Q of the accessory Lagrangian coincides with the *Jacobi operator* $\mathcal{J}_u$ which will be defined as linearization of the Euler operator of F at u.

1.3. The Legendre–Hadamard Condition

In the previous section we have seen that every (F-admissible) function $u \in C^1(\bar{\Omega}, \mathbb{R}^N)$ minimizing $\mathcal{F}$ in the sense of property ($\mathcal{M}'$) satisfies the relation

$$\delta^2 \mathcal{F}(u, \varphi) \geq 0 \quad \text{for all } \varphi \in C^\infty_c(\Omega, \mathbb{R}^N). \tag{1}$$

Equivalently we can say:

If u is a weak minimizer of the variational integral $\mathcal{F}$, then $v = 0$ is a weak minimizer of the accessory integral $\mathcal{Q}$ corresponding to $\mathcal{F}$.

This fact, in turn, implies a *pointwise condition* on u which is stated in the next

Theorem. *If* $u \in C^1(\bar{\Omega}, \mathbb{R}^N)$ *satisfies* (1), *then also the Legendre–Hadamard condition*

$$F_{p^i_\alpha p^k_\beta}(x, u(x), Du(x))\, \xi^i \xi^k \eta_\alpha \eta_\beta \geq 0 \tag{2}$$

holds for all $x \in \Omega$ *and all* $\xi \in \mathbb{R}^N$, $\eta \in \mathbb{R}^n$.

The Legendre–Hadamard condition can equivalently be written as

$$F_{p^i_\alpha p^k_\beta}(x, u(x), Du(x))\, \pi^i_\alpha \pi^k_\beta \geq 0 \tag{2'}$$

for all rank one-matrices $\pi = (\pi^i_\alpha)$, $\pi^i_\alpha = \xi^i \eta_\alpha$.

Matrices π of this type can be written as tensor products $\pi = \xi \otimes \eta$. (Actually we should interpret ξ as vector, η as covector, and π as (1,1)-tensor.)

Proof of the theorem. The Lagrangian $Q(\varphi)$ is quadratic in φ and $D\varphi$. By using highly oscillating functions like $\varphi^i(x) = \sin k \cdot x = \sin(k_1 x^1 + k_2 x^2 + \cdots + k_n x^n)$ (as $k_j \to \infty$) for the components φ^i of φ, which are bounded but whose derivatives are large, we shall see that the F_{pp}-term in $Q(\varphi)$,

$$\tfrac{1}{2} F_{p^i_\alpha p^k_\beta}(x, u(x), Du(x))\, D_\alpha \varphi^i D_\beta \varphi^k,$$

is the dominant term. This is exactly the reason why (1) implies (2). Let us see this in greater detail.

We use the notation introduced in *1.2*. Then condition (1) is equivalent to

(3) $$\mathscr{Q}(\varphi) \geq 0 \quad \text{for all } \varphi \in C_c^\infty(\Omega, \mathbb{R}^N).$$

Now consider an arbitrary function $v \in C_c^\infty(B_1(0), \mathbb{R}^N)$. Then, for arbitrary $x_0 \in \Omega$ and for sufficiently small $\lambda > 0$, the function

$$\varphi(x) = \lambda v\left(\frac{x - x_0}{\lambda}\right)$$

is admissible in (3). If we introduce new cartesian coordinates $y = (y^1, \ldots, y^n)$ by setting $y = \lambda^{-1}(x - x_0)$, and multiply (3) by λ^{-n}, we obtain

$$\int_{B_1(0)} \{a(x_0 + \lambda y)Dv(y)Dv(y) + 2\lambda b(x_0 + \lambda y)v(y)Dv(y)$$
$$+ \lambda^2 c(x_0 + \lambda y)v(y)v(y)\}\, dy \geq 0,$$

where

$$a = (a_{jk}^{\alpha\beta}), \qquad b = (b_{jk}^{\beta}), \qquad c = (c_{jk}).$$

If we let λ tend to zero, it follows that

(4) $$0 \leq \int_{B_1(0)} A_{jk}^{\alpha\beta} D_\alpha v^j D_\beta v^k \, dy$$

for all $v \in C_c^\infty(B_1(0), \mathbb{R}^N)$, where we have set

$$A_{jk}^{\alpha\beta} := a_{jk}^{\alpha\beta}(x_0).$$

We claim that the inequality (4) for real vector functions implies the following condition for complex vector functions:

(5) $$0 \leq \int_{B_1(0)} A_{jk}^{\alpha\beta} D_\alpha v^j \overline{D_\beta v^k} \, dy \quad \text{for all } v \in C_c^\infty(B_1(0), \mathbb{C}^N)$$

holds. In fact, if $v = \varphi + i\psi$, $i = \sqrt{-1}$, $\varphi, \psi \in C_c^\infty(\Omega, \mathbb{R}^N)$, we see that

$$v_{y^\alpha}^j \overline{v_{y^\beta}^k} = \begin{bmatrix} jk \\ \alpha, \beta \end{bmatrix} - i \begin{Bmatrix} jk \\ \alpha, \beta \end{Bmatrix},$$

where the expressions $A_{jk}^{\alpha\beta}$ and

$$\begin{bmatrix} j & k \\ \alpha, & \beta \end{bmatrix} := \varphi_{y^\alpha}^j \varphi_{y^\beta}^k + \psi_{y^\alpha}^j \psi_{y^\beta}^k$$

are symmetric with respect to permutation of the index pairs $\binom{j}{\alpha}$ and $\binom{k}{\beta}$, whereas

$$\begin{Bmatrix} j & k \\ \alpha, & \beta \end{Bmatrix} := \varphi_{y^\alpha}^j \psi_{y^\beta}^k - \psi_{y^\alpha}^j \varphi_{y^\beta}^k$$

is antisymmetric. Hence (5) is an immediate consequence of (4) and $A_{jk}^{\alpha\beta} = A_{kj}^{\beta\alpha}$.

We then insert in (5) the test functions

$$v = (v^1, \ldots, v^N), \qquad v^j(y) = \xi^j e^{it\eta \cdot y} \zeta(y), \quad t > 0,$$

where $\xi = (\xi^1, \dots, \xi^N) \in \mathbb{R}^N$, $\eta = (\eta_1, \dots, \eta_n) \in \mathbb{R}^n$, and $\zeta \in C_c^\infty(B_1(0))$. By multiplying (5) with t^{-2}, we obtain

$$0 \le A_{jk}^{\alpha\beta} \xi^j \xi^k \left\{ \eta_\alpha \eta_\beta \int_{B_1(0)} \zeta^2 \, dy + t^{-2} \int_{B_1(0)} D_\alpha \zeta D_\beta \zeta \, dy \right\}.$$

As $t \to \infty$, it follows that

$$0 \le A_{jk}^{\alpha\beta} \xi^j \xi^k \eta_\alpha \eta_\beta \int_{B_1(0)} \zeta^2 \, dy$$

for each $\zeta \in C_c^\infty(B_1(0))$, which implies

$$0 \le A_{jk}^{\alpha\beta} \xi^j \xi^k \eta_\alpha \eta_\beta,$$

and the assertion is proved. □

Remark. One can avoid the use of complex variables by inserting $v = \xi \sin(t\eta \cdot y)\zeta(y)$ and $v = \xi \cos(t\eta \cdot y)\zeta(y)$, and then adding the results.

In addition to the "necessary" Legendre–Hadamard condition (2), also the strengthened condition

$$\tfrac{1}{2} F_{p_\alpha^i p_\beta^k}(x, u(x), Du(x))\, \xi^i \xi^k \eta_\alpha \eta_\beta \ge \lambda |\xi|^2 |\eta|^2 \tag{6}$$

for some $\lambda > 0$ plays an important role; it is supposed to hold for all $\xi \in \mathbb{R}^N$ and $\eta \in \mathbb{R}^N$. It will be called the *strict Legendre–Hadamard condition*. For reasons to be seen later (cf. Chapter 5), (6) is sometimes called the *sufficient condition of Legendre–Hadamard*. Equivalently we say that the Lagrangian F is *strongly elliptic on the function* $u(x)$, $x \in \Omega$, or that *the Euler operator is strongly elliptic on* u. Note that the strict Legendre–Hadamard condition (6) is not a necessary condition.

A Lagrangian $F(x, z, p)$ will be called (*uniformly*) *strongly elliptic* in general if

$$\tfrac{1}{2} F_{p_\alpha^i p_\beta^k}(x, z, p)\, \xi^i \xi^k \eta_\alpha \eta_\beta \ge \lambda |\xi|^2 |\eta|^2 \tag{7}$$

holds for some $\lambda > 0$ and for arbitrary x, z, p, ξ, η. In contrast, the Lagrangian $F(x, z, p)$ is said to be (*uniformly*) *superelliptic* if

$$\tfrac{1}{2} F_{p_\alpha^i p_\beta^k}(x, z, p)\, \pi_i^\alpha \pi_k^\beta \ge \lambda |\pi|^2 \tag{8}$$

is satisfied for some $\lambda > 0$, and for all x, z, p, and all matrices $\pi = (\pi_i^\alpha)$, not just those of rank one.

Superellipticity is basically equivalent to the convexity of $F(x, z, p)$ with respect to p. In fact, if F is of class C^2 (with respect to p), then the convexity of $F(x, z, p)$ as function of p implies that

$$F_{p_\alpha^i p_\beta^k}(x, z, p)\, \pi_i^\alpha \pi_k^\beta \ge 0$$

holds for all $\pi = (\pi_i^\alpha) \in \mathbb{R}^{nN}$. Conversely, if (8) is fulfilled, then $F(x, z, p)$ is convex in p. (For convexity see e.g. 7,3.)

The condition of strong ellipticity (7) merely yields convexity of $F(x, z, \cdot)$ in rank-one directions.

If $N = 1$ or $n = 1$, the conditions of strong ellipticity and superellipticity (7) and (8) coincide and reduce to the ordinary *ellipticity conditions*

$$\tfrac{1}{2}F_{p_\alpha p_\beta}(x, z, p)\,\eta_\alpha\eta_\beta \geq \lambda|\eta|^2 \quad (\text{for } N = 1)$$

and

$$\tfrac{1}{2}F_{p^i p^k}(x, z, p)\,\xi^i\xi^k \geq \lambda|\xi|^2 \quad (\text{for } n = 1),$$

respectively.

For an arbitrary quadratic functional

$$\mathcal{Q}(u) = \int_\Omega \{a_{ik}^{\alpha\beta}(x)D_\alpha u^i D_\beta u^k + 2b_{ik}^\beta(x)u^i D_\beta u^k + c_{ik}(x)u^i u^k\}\,dx,$$

the strict Legendre–Hadamard condition is just

$$a_{ik}^{\alpha\beta}(x)\,\xi^i\xi^k\eta_\alpha\eta_\beta \geq \lambda|\xi|^2|\eta|^2 \tag{9}$$

for some $\lambda > 0$ and all $x \in \Omega$, $\xi \in \mathbb{R}^N$, $\eta \in \mathbb{R}^n$.

A matrix $(a_{ik}^{\alpha\beta})$ satisfying (9) is said to be *strongly elliptic*. Similarly a differential operator Lu defined by

$$(Lu)_k = -D_\beta\{a_{ik}^{\alpha\beta}(x)D_\alpha u^i\} + \text{lower order terms}$$

is *strongly elliptic* if the coefficients $a_{ik}^{\alpha\beta}$ of the leading term satisfy (9).

If either $n = 1$ or $N = 1$, we can write $a_{ik} = a_{ik}^{11}$ or $a^{\alpha\beta} = a_{11}^{\alpha\beta}$, respectively, and (9) reduces to the definiteness of the matrices (a_{ik}) or $(a^{\alpha\beta})$.

1.4. The Weierstrass Excess Function $\mathcal{E}_F$ and Weierstrass's Necessary Condition

We now will derive a slightly stronger necessary condition than the Legendre–Hadamard condition assuming that u satisfies the minimum property $(\mathcal{M})$. This property is satisfied if u is a strong minimizer, i.e., if $\mathcal{F}(u) \leq \mathcal{F}(v)$ holds for all v in some C^0-neighbourhood of u.

Let us fix the basic assumptions of this subsection. We assume that Ω is a bounded domain in $\mathbb{R}^n$, and that $F(x, z, p)$ is a Lagrangian of class $C^1(\overline{\Omega} \times \mathbb{R}^N \times \mathbb{R}^{nN})$.

Then we introduce the *Weierstrass excess function of* F as the following remainder term:

$$\mathcal{E}_F(x, z, p, q) := F(x, z, q) - F(x, z, p) - (q - p)\cdot F_p(x, z, p). \tag{1}$$

Note that $\mathcal{E}_F$ is a continuous function of the variables (x, z, p, q) on $\overline{\Omega} \times \mathbb{R}^N \times \mathbb{R}^{nN} \times \mathbb{R}^{nN}$.

Theorem. (*Necessary condition of Weierstrass*). *Suppose that* $u \in C^1(\bar{\Omega}, \mathbb{R}^N)$ *has the strong minimum property* ($\mathscr{M}$). *Then for all* $x \in \Omega$ *and for every rank one-matrix* $\pi = (\pi^i_\alpha)$, *the Weierstrass condition*

$$\mathscr{E}_F(x, u(x), Du(x), Du(x) + \pi) \geq 0 \tag{2}$$

is satisfied.

Since each rank one-matrix π *can be written as a tensor product* $\pi = \xi \otimes \eta$ *of some* $\xi \in \mathbb{R}^N$ *and some* $\eta \in \mathbb{R}^N$, *that is,* $\pi^i_\alpha = \xi^i \eta_\alpha$, *we can write the Weierstrass condition in the form*

$$\begin{aligned} F(x, u(x), Du(x) + \xi \otimes \eta) \\ \geq F(x, u(x), Du(x)) + \xi^i \eta_\alpha F_{p^i_\alpha}(x, u(x), Du(x)). \end{aligned} \tag{2'}$$

Before we prove the theorem, we note that *the Weierstrass condition implies the Legendre–Hadamard condition*. In other words, the result of the previous section is a consequence of the theorem just stated, and we therefore have found another proof for it.

In fact, suppose that $F \in C^2$, and that

$$\mathscr{Q}(\varphi) = \tfrac{1}{2}\delta^2\mathscr{F}(u, \varphi) \geq 0 \quad \text{for all } \varphi \in C^\infty_c(\Omega, \mathbb{R}^N)$$

is satisfied. Since $\mathscr{Q}(\varphi)$ vanishes for $\varphi = 0 = (0, \dots, 0)$, it follows that

$$\mathscr{Q}(\varphi) \geq \mathscr{Q}(0) \quad \text{for all } \varphi \in C^\infty_c(\Omega, \mathbb{R}^N).$$

Consequently, the function $v = 0$ is a minimum of the functional $\mathscr{Q}$ in the sense ($\mathscr{M}$) formulated above. Hence condition (2) implies

$$\mathscr{E}_Q(x, 0, 0, \pi) \geq 0,$$

where Q denotes the Lagrangian of the accessory integral $\mathscr{Q}$. This is equivalent to

$$Q(x, 0, \pi) - Q(x, 0, 0) - \pi^i_\alpha Q_{p^i_\alpha}(x, 0, 0) \geq 0,$$

which means

$$a^{\alpha\beta}_{ik}\pi^i_\alpha \pi^k_\beta = Q(x, 0, \pi) \geq 0.$$

Hence we have shown that the Legendre–Hadamard condition follows from Weierstrass's necessary condition applied to the accessory integral $\mathscr{Q}$.

Proof of the theorem. In order to show that the minimum property ($\mathscr{M}$) implies the Weierstrass condition (2), we first note that ($\mathscr{M}$) yields the relation

$$\delta\mathscr{F}(u, \varphi) = 0 \quad \text{for all } \varphi \in C^\infty_c(\Omega, \mathbb{R}^N).$$

Hence ($\mathscr{M}$) is equivalent to the inequality

$$\begin{aligned} \int_\Omega \{F(x, u + \varphi, Du + D\varphi) - F(x, u, Du) \\ - F_z(x, u, Du)\cdot \varphi - F_p(x, u, Du)\cdot D\varphi\}\, dx \geq 0 \\ \text{for all } \varphi \in C^\infty_c(\Omega, \mathbb{R}^N) \text{ with } \|\varphi\|_{C^0(\bar{\Omega})} < \varepsilon. \end{aligned} \tag{3}$$

Now we pick an arbitrary unit vector e and write

$$x = (x \cdot e)e + \bar{x},$$

where $\bar{x}$ is perpendicular to e. Similarly let D_e be the directional derivative in direction of e, i.e., $D_e v = Dv \cdot e$ for any function v. Let $\xi = (\xi^1, \dots, \xi^N)$ be an arbitrary vector in $\mathbb{R}^N$, and x_0 be any point in Ω; without loss of generality we may assume that $x_0 = 0$. Then we choose a number $\lambda_0 \in (0, 1)$ such that the cylinder

$$C_{\lambda_0} = \{x \in \mathbb{R}^n : |x \cdot e| \le \lambda_0, |\bar{x}| \le \lambda_0\}$$

is contained in Ω. Let $\lambda \in (0, \lambda_0)$, $\zeta \in C_c^\infty((-1, 1), \mathbb{R})$, and choose a sequence of functions $\chi_k \in C_c^\infty((-\lambda^2, \lambda), \mathbb{R})$ which converge uniformly to the Lipschitz function ψ_λ, defined by

$$\psi_\lambda(t) := \begin{cases} t + \lambda^2 & \text{for } -\lambda^2 < t \le 0 \\ \lambda(\lambda - t) & \text{for } 0 < t < \lambda \\ 0 & \text{otherwise} \end{cases},$$

such that also the derivatives χ_k' tend uniformly to ψ_λ' on each compact subset of $\{t: -\lambda^2 < t < \lambda, t \ne 0\}$, and that $\sup_{\mathbb{R}} |\chi_k'| \le \sup_{\mathbb{R}} |\psi_\lambda'| = 1$ for all $k = 1, 2, \dots$. Then, for sufficiently small $\lambda > 0$,

$$\varphi_k(x) := \xi \chi_k(x \cdot e)\zeta(|\bar{x}|^2/\lambda^2), \quad k = 1, 2, \dots,$$

is admissible for (3), and, passing to the limit $k \to \infty$, we obtain by a simple reasoning that (3) also holds for

$$\varphi(x) := \xi \psi_\lambda(x \cdot e)\zeta(|\bar{x}|^2/\lambda^2).$$

There is a number K such that $|\zeta| < K$ and $|\zeta'| < K$. We, moreover, have

$$D_e \varphi(x) = \xi \psi_\lambda'(x \cdot e)\zeta(|\bar{x}|^2/\lambda^2)$$

and

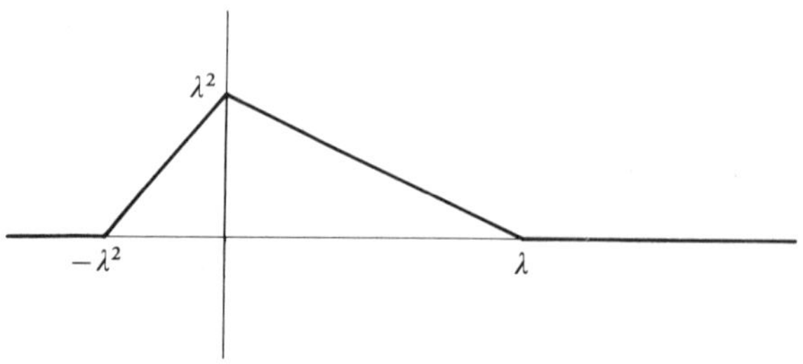

Fig. 3. The graph of $\psi_\lambda(t)$.

$$\bar{D}\varphi(x) = \xi\psi_\lambda(x\cdot e)\zeta'(|\bar{x}|^2/\lambda^2)2\lambda^{-2}\bar{x},$$

where $\bar{D}$ denotes the gradient with respect to $\bar{x}$.

Note that for this choice of φ the integrand of (3) vanishes if $x \notin C_\lambda$. Let C_λ^+ and C_λ^- be the upper half and the lower half of C_λ, that is,

$$C_\lambda^- := C_\lambda \cap \{x\colon x\cdot e \le 0\}, \qquad C_\lambda^+ := C_\lambda \cap \{x\colon x\cdot e > 0\}.$$

Then, for our choice of φ, equation (3) reads

$$\text{(4)} \quad \int_{C_\lambda^-} [F(x, u + \varphi, Du + D\varphi) - F(x, u, Du) - \varphi\cdot F_u(x, u, Du)$$
$$- D\varphi\cdot F_p(x, u, Du)]\,dx + \int_{C_\lambda^+} [\dots] \ge 0 \quad \text{for all } \varphi \in C_c^\infty(\Omega, \mathbb{R}^N).$$

We now introduce new coordinates $y = y^e e + \bar{y}$ by

$$y^e = \frac{x\cdot e}{\lambda^2}, \qquad \bar{y} := \frac{\bar{x}}{\lambda}.$$

Then the sets $C_\lambda^\pm$ correspond to $B_\lambda^\pm$ where

$$B_\lambda^- := \{y\colon |\bar{y}| \le 1, -\lambda^{-1} \le y^e \le 0\},$$
$$B_\lambda^+ := \{y\colon |\bar{y}| \le 1, 0 < y^e \le \lambda^{-1}\},$$

and $dx = \lambda^{n+1}\,dy$. Therefore, multiplying (4) by λ^{-n-1}, we obtain

$$\text{(5)} \qquad \int_{B_1^-} \{\dots\}\,dy + \int_{B_\lambda^- - B_1^-} \{\dots\}\,dy + \int_{B_\lambda^+} \{\dots\}\,dy \ge 0,$$

where x in $\{\dots\}$ has to be replaced by

$$x = x\cdot e + \bar{x} = \lambda^2 y^e + \lambda\bar{y}.$$

Note that

$$D_e\varphi(x) = \xi\,\zeta(|\bar{y}|^2)\begin{cases} 1 & \text{if } -1 < y^e < 0, \\ -\lambda & \text{if } 0 < y^e < \lambda^{-1}, \\ 0 & \text{otherwise } (y^e \ne 0), \end{cases}$$

and that in (5)

$$|\{\dots\}| \le o(\sqrt{|\varphi(x)|^2 + |D\varphi(x)|^2}).$$

Therefore, using the previous estimates on $\varphi(x)$ and $D\varphi(x)$, we obtain

$$|\{\dots\}(y)| \le o(\lambda) \quad \text{as } \lambda \to 0 \text{ for } y \in B_\lambda^+ \cup \{B_\lambda^- - B_1^-\}.$$

Since the measures of B_λ^- and B_λ^+ are bounded by $2^{n-1}/\lambda$, the second and third integral in (5) are of the order $o(1)$ as $\lambda \to 0$. Letting λ tend to zero, we arrive at

$$\int_B \{F(0, u(0), Du(0) + \xi\, \zeta(|\bar{y}|^2)e) - F(0, u(0), Du(0)) - \zeta(|\bar{y}|^2)\xi e \cdot F_p(0, u(0), Du(0))\}\, d\bar{y}^2 \dots d\bar{y}^n \geq 0, \tag{6}$$

where we have set

$$B := \{y\colon y^e = 0, |\bar{y}| \leq 1\},$$

$\bar{y} = (\bar{y}^2, \dots, \bar{y}^n)$, and have observed that

$$D\varphi(x) = D_e\varphi(x)\cdot e + \bar{D}\varphi(x) \to \xi\, \zeta(|\bar{y}|^2)e$$

as $\lambda \to 0$.

We now choose $\zeta = \zeta_\ell$, $\ell = 1, 2, \dots$, where the functions ζ_ℓ satisfy $0 \leq \zeta_\ell \leq 1$ and $\zeta_\ell(t) = 1$ for $|t| < 1 - 1/\ell$. Then, as $\ell \to +\infty$, we infer from (6) that

$$F(0, u(0), Du(0) + \xi e) - F(0, u(0), Du(0)) - \xi e F_p(0, u(0), Du(0)) \geq 0, \tag{7}$$

which is equivalent to (2′). □

2. Sufficient Conditions for Relative Minima Based on Convexity Arguments

Assumptions, which guarantee that a given extremal of a variational integral $\mathcal{F}$ furnishes in some sense a relative or absolute minimum of $\mathcal{F}$, are traditionally called *sufficient conditions*. There are essentially three kinds of sufficient conditions.

Surprisingly, the most elementary type, based on *convexity* arguments, appeared in the history of the calculus of variations at last.

The second and oldest kind of sufficient conditions can be expressed by *eigenvalue criteria for the Jacobi operator*, the Euler operator of the accessory integral.

The third and most effective class of sufficient conditions is based on Weierstrass's idea of using *fields of extremals*, and on *Hilbert's independent integral* which is closely related to the notion of a null Lagrangian. The principal idea of field theory is comprised in the notion of a *calibrator* which will be discussed in *2.6*.

Unfortunately, in the case of several unknown functions of several variables field theory encounters certain technical difficulties which are in the nature of the matter. Therefore we begin in Chapter 6 by giving a detailed discussion of field theory for one dimensional variational problems ($n = 1$, $N > 1$). Field theories for the general case will be presented in Chapter 7.

Jacobi's theory including the eigenvalue criteria will be developed in Chapter 5.

In the present section we begin with convexity arguments. The prototype of this reasoning is the following well-known result: *A strictly convex function* $\mathcal{F} : \mathcal{C} \to \mathbb{R}$ *on a convex domain* $\mathcal{C}$ *of* $\mathbb{R}^n$ *possesses at most one stationary point in* $\mathcal{C}$. *At such a point the function* $\mathcal{F}$ *assumes its absolute minimum on* $\mathcal{C}$.

2.1. A Sufficient Condition Based on Definiteness of the Second Variation

Let $\mathcal{F}(x)$ be a convex function of class C^1 on some convex domain $\mathcal{C}$ of $\mathbb{R}^n$. Then each critical point of $\mathcal{F}$ is an absolute minimizer. If, moreover, $\mathcal{F}$ is strictly convex, then $\mathcal{F}$ has at most one critical point in $\mathcal{C}$, and such a point has to be a strict minimizer. If $\mathcal{F}$ is of class C^2, the strict convexity of $\mathcal{F}$ follows from the positive definiteness of the Hessian matrix $\mathcal{F}_{xx}$. For more information on convex functions cf. for example 7,*3*.

The following result is a straight-forward generalization to variational integrals

$$\mathcal{F}(u) = \int_\Omega F(x, u(x), Du(x))\, dx.$$

Let us for the sake of simplicity assume that the Lagrangian $F(x, z, p)$ is defined on $\bar{\Omega} \times \mathbb{R}^N \times \mathbb{R}^{nN}$, $\Omega \subset \mathbb{R}^n$, as a function of class C^2. Assume also that Ω denotes a bounded domain in $\mathbb{R}^n$, the boundary of which is not too irregular.

To be precise, we need that every function φ of class $C_0^1(\bar{\Omega}, \mathbb{R}^N) := C^1(\bar{\Omega}, \mathbb{R}^N) \cap \{\varphi = 0 \text{ on } \partial\Omega\}$ can be approximated by functions $\varphi_p \in C_c^\infty(\Omega, \mathbb{R}^N)$ in such a way that $\lim_{p\to\infty}\{\|\varphi - \varphi_p\|_{C^0(\bar{\Omega})} + \|\nabla\varphi - \nabla\varphi_p\|_{L^2(\Omega)}\} = 0$ holds. In other words, $C_c^\infty(\Omega, \mathbb{R}^N)$ has to be dense in $C_0^1(\bar{\Omega}, \mathbb{R}^N)$ with respect to the norm $\|\psi\| := \|\psi\|_{C^0(\bar{\Omega})} + \|\nabla\psi\|_{L^2(\Omega)}$. This is certainly the case if $\partial\Omega$ is smooth (i.e., if $\partial\Omega$ is a manifold of class C^1), or at least piecewise smooth; cf. 1,*2.3*.

Proposition. *Let* $\mathcal{C}$ *be a convex set in the function space* $C^1(\bar{\Omega}, \mathbb{R}^N)$ *and suppose that* $u \in \mathcal{C}$ *is a weak extremal of the variational integral* $\mathcal{F}$. *Moreover, assume that*

$$(1) \qquad \delta^2\mathcal{F}(w, \varphi) \ge 0 \quad \textit{for all } w \in \mathcal{C} \textit{ and all } \varphi \in C_c^\infty(\Omega, \mathbb{R}^N).$$

Then we obtain

$$\mathcal{F}(v) \ge \mathcal{F}(u) \quad \textit{for all } v \in \mathcal{C} \textit{ with } v|_{\partial\Omega} = u|_{\partial\Omega}.$$

If instead of (1) *the stronger assumption*

$$(2) \qquad \delta^2\mathcal{F}(w, \varphi) > 0 \quad \textit{for all } w \in \mathcal{C} \textit{ and all } \varphi \in C_0^1(\bar{\Omega}, \mathbb{R}^N) \text{ with } \varphi \ne 0$$

is satisfied, then it follows that

$$\mathcal{F}(v) > \mathcal{F}(u) \quad \textit{for all } v \in \mathcal{C},\ v \ne u, \textit{ with } v|_{\partial\Omega} = u|_{\partial\Omega}.$$

Corollary. (i) *If condition* (1) *is satisfied, then every extremal* $u \in \mathcal{C}$ *of* $\mathcal{F}$ *furnishes an absolute minimum of* $\mathcal{F}$ *on* $\mathcal{C}$.

(ii) *Condition* (2) *implies that an extremal* $u \in \mathscr{C}$ *of* $\mathscr{F}$ *is the unique minimizer of* $\mathscr{F}$ *on* $\mathscr{C}$.

Proof of the proposition. By a suitable approximation procedure we infer that the relations

$$\delta\mathscr{F}(u, \varphi) = 0$$

and

$$\delta^2\mathscr{F}(w, \varphi) \geq 0$$

hold for all $\varphi \in C_0^1(\overline{\Omega}, \mathbb{R}^N)$ and for every $w \in \mathscr{C}$.

Let now v be an arbitrary function in $\mathscr{C}$ which coincides with u on $\partial\Omega$. Then

$$\varphi := v - u$$

is of class $C_0^1(\overline{\Omega}, \mathbb{R}^N)$, and, for every $t \in [0, 1]$, the function

$$\psi(t) := u + t\varphi = tv + (1 - t)u$$

is contained in $\mathscr{C}$. Since $\Phi(t) := \mathscr{F}(\psi(t))$ is of class C^2 on $[0, 1]$, we obtain

$$\Phi(1) - \Phi(0) = \Phi'(0) + \int_0^1 (1 - t)\Phi''(t)\, dt.$$

By virtue of

$$\Phi'(0) = \delta\mathscr{F}(u, \varphi) = 0,$$

it follows that

$$\mathscr{F}(v) - \mathscr{F}(u) = \int_0^1 (1 - t)\delta^2\mathscr{F}(\psi(t), \varphi)\, dt \tag{3}$$

and therefore $\mathscr{F}(v) \geq \mathscr{F}(u)$. If (2) holds, then we even get $\mathscr{F}(v) > \mathscr{F}(u)$, provided that $v \neq u$. □

2.2. Convex Lagrangians

Convexity of the functional $\mathscr{F}$ is certainly guaranteed if, instead of the assumption (1) in *2.1*, we require the much more stringent pointwise condition $\delta^2 F(w, \varphi) \geq 0$ for all $w \in \mathscr{C}$ and all $\varphi \in C^1(\overline{\Omega}, \mathbb{R}^N)$. Therefore the Proposition of *2.1* implies

Proposition 1. *Let* $\mathscr{C}$ *be a convex set in* $C^1(\overline{\Omega}, \mathbb{R}^N)$, *and suppose that* $u \in \mathscr{C}$ *is a weak extremal of* $\mathscr{F}$. *Then we obtain: If* $\delta^2 F(w, \varphi) \geq 0$ *for all* $w \in \mathscr{C}$ *and all* $\varphi \in C_0^1(\overline{\Omega}, \mathbb{R}^N)$, *then* $\mathscr{F}(v) \geq \mathscr{F}(u)$ *for all* $v \in \mathscr{C}$ *with* $v = u$ *on* $\partial\Omega$. *Moreover, if even* $\delta^2 F(w, \varphi)(x) > 0$ *for all* $w \in \mathscr{C}$, $\varphi \in C_0^1(\overline{\Omega}, \mathbb{R}^N)$, *and all* $x \in \Omega$ *with* $(\varphi(x), D\varphi(x)) \neq (0, 0)$, *then* $\mathscr{F}(v) > \mathscr{F}(u)$ *for all* $v \in \mathscr{C}$ *satisfying* $u \neq v$ *and* $v = u$ *on* $\partial\Omega$.

The simplest condition ensuring convexity of $\mathcal{F}$ is convexity of the Lagrangian $F(x, z, p)$ with respect to (z, p), keeping x fixed. Since F is of class C^2, convexity of $F(x, \cdot, \cdot)$ is equivalent to positive semidefiniteness of

$$\begin{bmatrix} F_{zz}, F_{zp} \\ F_{pz}, F_{pp} \end{bmatrix}.$$

Therefore we obtain the following somewhat weaker version of Proposition 1 which is directly applicable to $\mathscr{C} = C^1(\bar{\Omega}, \mathbb{R}^N)$:

Proposition 2. *Suppose that the Lagrangian $F(x, z, p)$ satisfies*

$$(1) \qquad F_{p_\alpha^i p_\beta^k}(x, z, p)\pi_\alpha^i \pi_\beta^k + 2F_{z^i p_\beta^k}(x, z, p)\zeta^i \pi_\beta^k + F_{z^i z^k}(x, z, p)\zeta^i \zeta^k \geq 0$$

for all $x \in \bar{\Omega}$, $z, \zeta \in \mathbb{R}^N$, $p, \pi \in \mathbb{R}^{nN}$. Then it follows that every weak extremal u of $\mathcal{F}$ satisfies $\mathcal{F}(v) \geq \mathcal{F}(u)$ for all $v \in C^1(\bar{\Omega}, \mathbb{R}^N)$ with $v = u$ on $\partial\Omega$. In other words, every weak extremal is an absolute minimizer. If in (1) *strict inequality holds for all $(\zeta, \pi) \neq (0, 0)$, then any weak extremal of $\mathcal{F}$ is the unique minimizer of $\mathcal{F}$ among all C^1-functions with the same boundary values on $\partial\Omega$.*

Unfortunately inequality (1) is difficult to check and, moreover, will in general rarely be satisfied. Yet there is at least one type of Lagrangians including various important examples for which (1) can easily be verified. These are the *Lagrangians F which do not depend on z,*

$$F = F(x, p) \quad \textit{for } (x, p) \in \bar{\Omega} \times \mathbb{R}^{nN},$$

and are convex with respect to p.

In this case (1) reduces to the positive semidefiniteness of the Hessian matrix $F_{pp} = (F_{p_\alpha^i p_\beta^k})$ on $\bar{\Omega} \times \mathbb{R}^{nN}$, i.e., to

$$(2) \qquad F_{p_\alpha^i p_\beta^k}(x, p)\pi_\alpha^i \pi_\beta^k \geq 0$$

for all $x \in \bar{\Omega}$ and all $p, \pi \in \mathbb{R}^{nN}$. The strict inequality

$$(2') \qquad F_{p_\alpha^i p_\beta^k}(x, p)\pi_\alpha^i \pi_\beta^k > 0, \quad \pi \neq 0$$

(implying strict convexity of the function $F(x, \cdot)$, x being fixed) will then imply that every weak extremal of $\mathcal{F}$ is the unique minimizer of $\mathcal{F}$ among all functions with the same boundary values. Since $F \in C^2$, (2′) is just the condition of *superellipticity* which, for $n = 1$ or $N = 1$, reduces to ordinary ellipticity:

$$(3) \qquad F_{p^i p^k}(x, p)\xi^i \xi^k > 0, \quad \xi \neq 0 \ (n = 1)$$

or

$$(4) \qquad F_{p_\alpha p_\beta}(x, p)\eta_\alpha \eta_\beta > 0, \quad \eta \neq 0 \ (N = 1).$$

Actually, we have

Proposition 3. *If the Lagrangian $F(x, p)$ is convex and of class C^1 with respect to p, then every weak extremal is a minimizer.*

Proof. The convexity of F with respect to p yields

$$F(x, q) \geq F(x, p) + F_p(x, p) \cdot (q - p).$$

Thus if u is a weak extremal and if v is a function of class $C^1(\overline{\Omega}, \mathbb{R}^N)$ that agrees with u on $\partial\Omega$, we obtain

$$\int_\Omega F(x, Dv)\, dx \geq \int_\Omega F(x, Du)\, dx + \int_\Omega F_p(x, Du) \cdot (Dv - Du)\, dx$$

$$= \int_\Omega F(x, Du)\, dx. \qquad \square$$

1 Some typical examples in the case $N = 1$ are provided by the *Dirichlet integral*

$$\mathscr{D}(u) = \tfrac{1}{2} \int_\Omega |Du|^2\, dx,$$

with the Lagrangian $F(p) = \frac{1}{2}|p|^2$, by the *generalized Dirichlet integral*

$$\mathscr{D}(u) = \tfrac{1}{2} \int_\Omega \gamma^{\alpha\beta}(x) D_\alpha u D_\beta u \sqrt{\gamma(x)}\, dx$$

with respect to some Riemannian metric $ds^2 = \gamma_{\alpha\beta}(x)\, dx^\alpha\, dx^\beta$, which has the Lagrangian

$$F(x, p) = \sqrt{\gamma(x)} \gamma^{\alpha\beta}(x) p_\alpha p_\beta,$$

and by the area functional

$$\mathscr{A}(u) = \int_\Omega \sqrt{1 + |Du|^2}\, dx,$$

with the Lagrangian

$$F(p) = \sqrt{1 + |p|^2}.$$

For all three examples, condition (4) is not difficult to verify.

2 For $n = 1$, the *length functional*

$$\mathscr{L}(u) = \int_b^a \sqrt{1 + u'^2}\, dx, \quad u = u(x), \ u' = Du,$$

and also the weighted length ($n = N = 1$)

$$\mathscr{L}(u) = \int_a^b \omega(x) \sqrt{1 + u'^2}\, dx, \quad \omega(x) > 0,$$

are typical examples of integrals with Lagrangians $F(x, p)$ satisfying (4), whereas

$$\int_a^b \omega(u) \sqrt{1 + u'^2}\, dx, \quad \omega(z) > 0,$$

does not fall in this category since its Lagrangian is of the kind $F(z, p)$. If, however, the function $z = u(x)$, $a \leq x \leq b$, possesses an inverse $x = \xi(z)$, $\alpha \leq z \leq \beta$, we obtain

$$\int_a^b \omega(u) \sqrt{1 + \left(\frac{du}{dx}\right)^2}\, dx = \int_\alpha^\beta \omega(z) \sqrt{1 + \left(\frac{d\xi}{dz}\right)^2}\, dz,$$

and therefore the problem is in some sense reduced to the previous case.

[3] Condition (1) can also be verified for integrands $F(x, z, p)$ which satisfy the three conditions

$$F_{pp} > 0, \qquad F_{pz} = 0, \qquad F_{zz} \geq 0 \tag{5}$$

on $\bar{\Omega} \times \mathbb{R}^N \times \mathbb{R}^{nN}$, that is, for Lagrangians of the type

$$F(x, p) + G(x, z)$$

with $F_{pp}(x, p) > 0$, and $G(x, z)$ being convex with respect to z.

An example of this type is provided by the variational integral

$$\mathscr{F}(u) = \int_\Omega \{\sqrt{1 + |Du|^2} + \lambda u\}\, dx, \quad N = 1,$$

which we have met at the n-dimensional isoperimetric problem as well as in 1,2.2 [5].

A second example of this kind is furnished by the integral

$$\mathscr{F}(u) = \int_\Omega \{|Du|^2 + G(x, u)\}\, dx, \quad N = 1,$$

with the Lagrangian

$$F(x, z, p) = |p|^2 + G(x, z)$$

satisfying

$$G_{zz}(x, z) \geq 0.$$

This is, for instance, the case if

$$G(z) = \lambda e^z \quad \text{or} \quad G(z) = \lambda z^4, \qquad \lambda \geq 0.$$

The corresponding Euler equation is equivalent to

$$\Delta u = g(x, u) \quad \text{with } g(x, z) := \tfrac{1}{2} G_z(x, z).$$

[4] A fairly obvious generalization allows for certain integrals with $F_{zp} \neq 0$. Suppose, for instance, that $N = 1$, and that there are numbers $\lambda > 0$, $\mu > 0$, $a \geq 0$ such that

$$\begin{gathered} |F_{zp}(x, z, p)| \leq a, \qquad F_{zz}(x, z, p) \geq \mu, \\ F_{p_\alpha p_\beta}(x, z, p)\eta_\alpha \eta_\beta \geq \lambda |\eta|^2 \end{gathered} \tag{6}$$

holds for all $\eta \in \mathbb{R}^n$ and all $(x, z, p) \in \bar{\Omega} \times \mathbb{R}^N \times \mathbb{R}^{nN}$. Then it is easy to see that (1) holds provided that

$$a < \sqrt{\lambda \mu}. \tag{7}$$

Unfortunately, the convexity reasoning already fails for the classical integrals

$$\mathscr{F}(u) = \int_\Omega a(x, u)\sqrt{1 + |Du|^2}\, dx, \quad N = 1,$$

at least, if the reasoning is applied in the straight-forward way described before. Many of these integrals were already treated by Euler[3] in his classical treatise "*Methodus inveniendi*" [2] in 1744.

In fact, suppose that the weight function $a(x, z)$ of the Lagrangian

$$F(x, z, p) = a(x, z)\sqrt{1 + |p|^2}$$

satisfies

$$a(x, z) > 0 \quad \text{and} \quad a_{zz}(x, z) > 0.$$

[3] Cf. Carathéodory [16], Vol. 5, pp. 165–174, and Euler [1], Ser. I, Vol. 24, pp. VIII–LXIII for a list of variational problems studied in Euler [2].

From

$$F_{p_\alpha p_\beta}(x, z, p) = \frac{a(x, z)}{\{1 + |p|^2\}^{3/2}}[(1 + |p|^2)\delta_{\alpha\beta} - p_\alpha p_\beta]$$

we only obtain

$$F_{p_\alpha p_\beta}(x, z, p)\eta_\alpha\eta_\beta \geq \frac{a(x, z)}{\{1 + |p|^2\}^{2/3}}|\eta|^2,$$

and the equality sign is assumed if p and η are collinear. Thus we merely have

$$F_{p_\alpha p_\beta}(x, z, p)\eta_\alpha\eta_\beta > 0 \quad \text{for } \eta \neq 0.$$

This, however, does not suffice to let the preceding method work.

From the preceding discussion we see that the *convexity method* is rather limited, and that we need more refined methods to be able to decide whether a given extremal $u(x)$ furnishes a minimum. This will be achieved by developing eigenvalue criteria for the Jacobi operator $\mathcal{J}_u$, and most effectively by the Weierstrass theory.

Before we turn to the discussion of these more generally applicable methods, we will investigate three different refinements of the convexity method which occasionally work rather well:

(I) the method of coordinate transformation;
(II) application of integral inequalities;
(III) convexity modulo null Lagrangians.

We shall explain the appertaining ideas in the following three subsections. Since in all three cases we find more of an artifice than an elaborated theory, we will restrict ourselves to the consideration of specific examples.

2.3. The Method of Coordinate Transformations

If a mathematical object furnishes a minimum of some functional, then this minimum property will not be lost if we describe the object in terms of new coordinates. Similarly, extremals are mapped into extremals, as we have seen in 3,4, where the general *covariance of the Euler equations* has been proved. This suggests the idea to investigate whether, by introducing new dependent and independent variables, a given functional can be transformed into a convex functional.

The idea of bringing some variational integral by a suitable coordinate transformation into a simple form can already be found in Euler's work. It led him to some covariance principle[4] that, in greater generality, was worked out only much later.

---

[4] Methodus inveniendi [2], Chapter IV, Proposition I. Cf. also Carathéodory [16], Vol. 5, p. 125, and Goldstine [1], pp. 84–92.

Let us restrict ourselves to the case $n = N = 1$, and consider a Lagrangian $F(x, z, p)$. Given any coordinate transformation $z = f(z^*)$, $f'(z^*) > 0$, with the inverse $z^* = g(z)$, we define the *pull-back Lagrangian* $G(x, z^*, q)$ by

$$G(x, z^*, q) := F(x, f(z^*), f'(z^*)q).$$

For a given function $z = u(x)$, we define $z^* = v(x)$ by $v(x) := g(u(x))$, whence $u(x) = f(v(x))$, $u'(x) = f'(v(x))v'(x)$, and therefore

$$F(x, u(x), u'(x)) = G(x, v(x), v'(x)).$$

Consequently, we obtain

$$\mathcal{F}(u) = \mathcal{G}(v)$$

for the corresponding variational integrals $\mathcal{F}$ and $\mathcal{G}$.

Consider, for instance, a Lagrangian $F(z, p)$ defined by

$$F(z, p) = \phi(a(z), b(z)p).$$

We assume that $\phi(\xi, \eta)$ is a convex function on $\mathbb{R}^2$ which is nondecreasing with respect to $\xi \in [0, \infty)$, for every fixed value of η. Secondly we suppose that $a(z)$ and $b(z)$ are smooth, and that $b(z) > 0$ holds (in applications this might only hold on some interval I; then we have to assume that the admissible functions $u(x)$ have values contained in I). We now choose $g(z)$ as primitive of $b(z)$, i.e.,

$$g'(z) = b(z),$$

and let f be the inverse of g. Given $u(x)$, we again introduce $v(x) := g(u(x))$, or $u(x) = f(v(x))$. Then

$$b(u)u' = b(u)f'(v)v' = b(u)\frac{1}{g'(u)}v' = v',$$

and it follows that

$$F(u, u') = \phi(a(f(v)), v') = G(v, v'),$$

where we have set

$$G(z^*, q) := \phi(c(z^*), q), \qquad c(z^*) := a(f(z^*)).$$

The Lagrangian $G(z^*, q)$ is convex in (z^*, q) if the function $c(z^*) := a(f(z^*))$ is convex in z^*.

In particular, consider the variational integrands

$$F(z, p) = a(z)\sqrt{1 + p^2} = \sqrt{a^2(z) + a^2(z)p^2} \quad \text{with } a(z) > 0.$$

Introducing the convex function $\phi(\xi, \eta) = \sqrt{\xi^2 + \eta^2}$, we can write

$$F(z, p) = \phi(a(z), a(z)p).$$

Let $z = f(z^*)$ be the inverse of the function $z^* = g(z)$ defined by

$$g(z) := \int^z a(\underline{z})\, d\underline{z};$$

then $v = g(u)$, $u = f(v)$ satisfy

$$F(u, u') = G(v, v') = \sqrt{c^2(v) + v'^2},$$

with

$$G(z^*, q) := \sqrt{c^2(z^*) + q^2}, \qquad c(z^*) := a(f(z^*)),$$

and $G(z^*, q)$ is a convex function of z^*, q, provided that $c(z^*)$ is a convex function of z^*.

Let us consider some examples where $n = N = 1$.

[1] The Lagrangian $F(z, p) = \sqrt{z^2 + p^2}$ is already convex; and we can write down its extremals. In fact, let us introduce polar coordinates r, φ about the origin in the x, y-plane by

$$x = r\cos\varphi, \qquad y = r\sin\varphi.$$

Then it follows that

$$dx^2 + dy^2 = dr^2 + r^2\,d\varphi^2$$

and therefore

$$\sqrt{1 + \left(\frac{dy}{dx}\right)^2}\,dx = \sqrt{r^2 + \left(\frac{dr}{d\varphi}\right)^2}\,d\varphi.$$

That is, the length functional is in polar coordinates described by

$$\int_{\varphi_1}^{\varphi_2} \sqrt{r^2 + \left(\frac{dr}{d\varphi}\right)^2}\,d\varphi.$$

The extremals of $\int \sqrt{1 + y'^2}\,dx$ are exactly the straight lines that are not parallel to the y-axis; thus they are given by

$$ax + by = c \quad \text{with } b \neq 0.$$

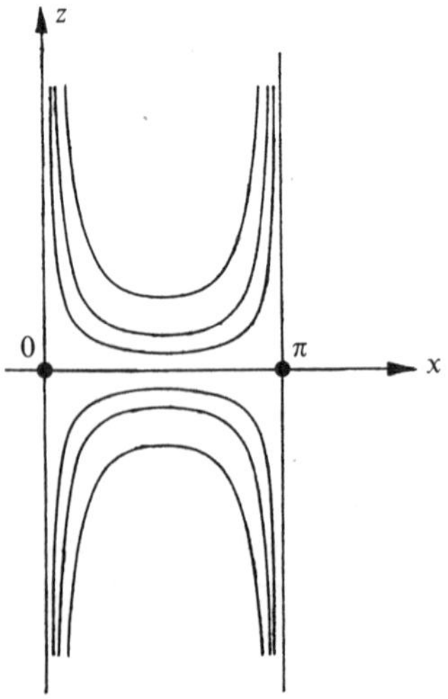

Fig. 4. The family of extremals $u(x) = a/\cos(x - \pi/2)$, $|x| < \pi/2$, of the integral $\int_{x_1}^{x_2} \sqrt{u^2 + u'^2}\,dx$.

Hence the extremals of $\int \sqrt{r^2 + r_\varphi^2}\, d\varphi$ are described by the equation

$$r[a \cos \varphi + b \sin \varphi] = c.$$

By determining some $\varphi_1 \not\equiv \pi/2, 3\pi/2 \bmod 2\pi$ with

$$\cos \varphi_1 = \frac{a}{\sqrt{a^2 + b^2}}, \qquad \sin \varphi_1 = \frac{b}{\sqrt{a^2 + b^2}},$$

and setting

$$d := \frac{c}{\sqrt{a^2 + b^2}},$$

we obtain

$$r \cos(\varphi - \varphi_1) = d$$

as the characterizing equation of geodesics. If $d = 0$, then $\varphi \equiv \varphi_1 + \pi/2 \pmod \pi$. These straight lines cannot be written as graphs $r = r(\varphi)$ and get lost. If $d \neq 0$, we obtain the straight lines

$$r = \frac{d}{\cos(\varphi - \varphi_1)}, \quad |\varphi - \varphi_1| < \pi/2,$$

as extremals.

If we now return to the original Lagrangian $F(z, p) = \sqrt{z^2 + p^2}$, then we obtain the *singular extremal*

$$u(x) = 0,$$

which clearly is the unique minimizer of the integral

$$\mathscr{F}(u) = \int_{x_1}^{x_2} \sqrt{u^2 + u'^2}\, dx$$

among all C^1-curves $z = u(x)$, $x_1 \leq x \leq x_2$, with $u(x_1) = u(x_2) = 0$. In addition, we have

$$u(x) = \frac{a}{\cos(x - x_0)}, \quad |x - x_0| < \pi/2,\ a \neq 0,$$

as the only other extremals.

We conclude that, for given $P_1 = (x_1, z_1)$ and $P_2 = (x_2, z_2)$ with $z_1, z_2 \neq 0$, there is no extremal of $\mathscr{F}(u)$ satisfying $u(x_1) = z_1$ and $u(x_2) = z_2$ if either $|x_1 - x_2| \geq \pi$, or if sign $z_1 \neq$ sign z_2. That is, for arbitrary P_1 and P_2, there is not always a minimizer u of $\mathscr{F}$ that is of class C^2 and connects P_1 with P_2.

Moreover, if there is an extremal $u(x)$, $x_1 \leq x \leq x_2$, with $u(x_1) = z_1$ and $u(x_2) = z_2$, then u is the unique minimizer of the strictly convex functional $\mathscr{F}$ with respect to all C^1-curves satisfying the same boundary conditions.

The gist of this example is that an appropriate geometric interpretation might spare one any computation; here the geometric idea consisted in interpreting the cartesian coordinates x, z as polar coordinates r, φ, thereby linking the given variational problem with another one, the solution of which is known, and then to use the continuance of Euler's equation (i.e., the principle of covariance).

[2] *Carathéodory's example.* $F(z, p) = e^z\sqrt{1 + p^2}$. This example is of the type $F = a(x, z)\sqrt{1 + p^2}$ for which the convexity method fails if it is directly applied. It will turn out that, by a suitable change of the dependent variables, the corresponding integral $\mathscr{F}$ can be transformed into the strictly convex functional appearing in [1], and the extremals of $\mathscr{F}$ will be in correspondence to some of the extremals of [1]. Thus the extremals of $\mathscr{F}(u) = \int_{x_1}^{x_2} e^u\sqrt{1 + u'^2}\, dx$ are recognized as minimizers.

In fact, introducing $v(x)$ by $v = e^u$, we obtain $v^2 + v'^2 = e^{2u}[1 + u'^2]$, and therefore

$$F(u, u') = \sqrt{v^2 + v'^2} := G(v, v')$$

or

$$\mathcal{F}(u) = \mathcal{G}(v) := \int_{x_1}^{x_2} \sqrt{v^2 + v'^2}\, dx.$$

According to [1], the extremals of $\mathcal{G}$ are given by

$$v(x) = \frac{a}{\cos(x - x_0)}, \quad |x - x_0| < \pi/2, \ a \neq 0.$$

Since $v = e^u > 0$, we obtain all extremals of $\mathcal{F}$ by

$$u(x) = z_0 - \log\cos(x - x_0), \quad |x - x_0| < \pi/2,$$

for arbitrary values of x_0 and $z_0 \in \mathbb{R}$. All of them are minimizers of $\mathcal{F}$, for any choice of $x_1, x_2 \in \left(x_0 - \frac{\pi}{2}, x_0 + \frac{\pi}{2}\right)$, among all C^1-curves $z = w(x)$, $x_1 \leq x \leq x_2$, with $w(x_1) = u(x_1)$ and $w(x_2) = u(x_2)$.

There are no extremals joining two given points $P_1 = (x_1, z_2)$ and $P_2 = (x_2, z_2)$ if $|x_1 - x_2| \geq \pi$. Hence there is no minimizer of class C^2 for

$$\mathcal{F}(w) = \int_{x_1}^{x_2} e^w \sqrt{1 + w'^2}\, dx$$

in the class $\mathscr{C}$ of functions $w \in C^1([x_1, x_2])$ with $w(x_1) = z_1$, $w(x_2) = z_2$, if $|x_1 - x_2| \geq \pi$.

Let us consider this minimum problem somewhat closer. Set

$$\alpha := \inf_{\mathscr{C}} \mathcal{F}.$$

We first want to show that $\alpha \leq e^{z_1} + e^{z_2}$ holds, whether $|x_1 - x_2| \geq \pi$ is satisfied or not. To this end, we choose an arbitrary value z_0 with $z_0 < \min\{z_1, z_2\}$, and consider the parametric curve $c(t) = (x(t), w(t))$, $0 \leq t \leq L$, parametrized by the arc length t, which consists of the two vertical pieces $\{(x, z): x = x_i, z_0 \leq z \leq z_i\}$, $i = 1, 2$, and of the horizontal piece $\{(x, z): x_1 \leq x \leq x_2, z = z_0\}$. Then

$$\mathcal{F}(c) := \int_0^L e^w |\dot{c}(t)|\, dt = e^{z_1} + e^{z_2} + e^{z_0}(x_2 - x_1 - 2)$$

and therefore

$$\mathcal{F}(c) \to e^{z_1} + e^{z_2} \quad \text{as } z_0 \to -\infty.$$

Moreover, we can approximate each one of the U-shaped curves c by smooth nonparametric curves $w(x)$, $x_1 \leq x \leq x_2$, of class $\mathscr{C}$, such that the difference

$$\left| \mathcal{F}(c) - \int_{x_1}^{x_2} e^w \sqrt{1 + w'^2}\, dx \right|$$

can be made arbitrarily small. This proves $\alpha \leq e^{z_1} + e^{z_2}$.

Now we want to show that, for $|x_1 - x_2| \geq \pi$, actually $\alpha = e^{z_1} + e^{z_2}$ holds. We recall how the problems of [1] and [2] are connected, and that [1] can be interpreted as the problem of minimal length in $\mathbb{R}^2$, written in polar coordinates. Therefore we write φ, φ_1, φ_2 instead of x, x_1, x_2, respectively, and introduce $r(\varphi) = e^{w(\varphi)}$, $\varphi_1 \leq \varphi \leq \varphi_2$. Then we have

$$\int_{\varphi_1}^{\varphi_2} e^w \sqrt{1 + w'^2}\, d\varphi = \int_{\varphi_1}^{\varphi_2} \sqrt{r^2 + r'^2}\, d\varphi,$$

and the second integral yields the length $\mathscr{L}$ of some curve in the x, y-plane given by $P(\varphi) = (x(\varphi), y(\varphi))$,

$$x(\varphi) := r(\varphi)\cos\varphi, \quad y(\varphi) := r(\varphi)\sin\varphi, \qquad \varphi_1 \leq \varphi \leq \varphi_2.$$

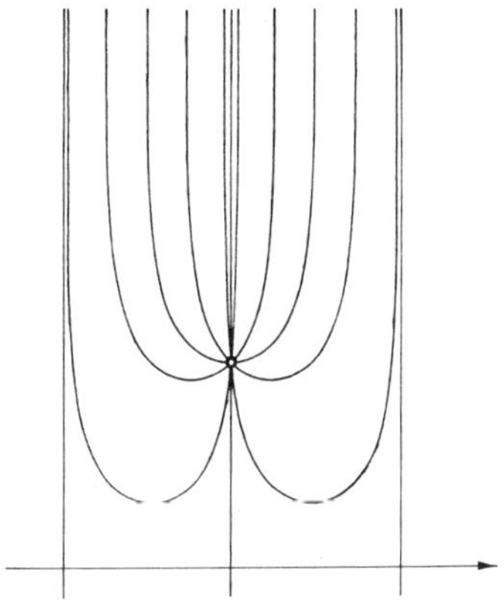

Fig. 5. Carathéodory's example.

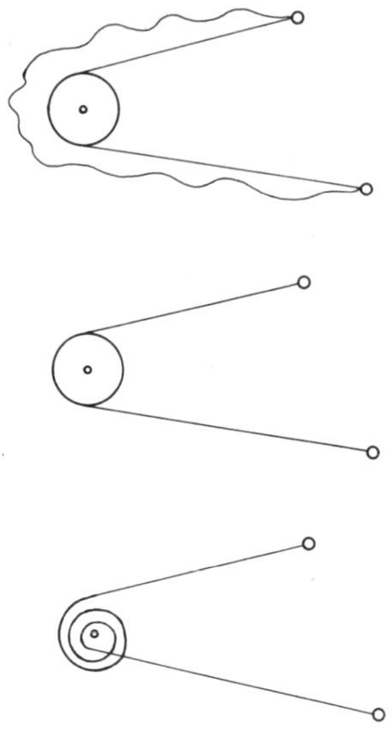

Fig. 6.

For every $w \in \mathscr{C}$, there is some $r_0 > 0$ such that $r(\varphi) \geq r_0$ holds for all $\varphi \in [\varphi_1, \varphi_2]$. Thus the curve $P(\varphi)$ cannot penetrate the disk $B(r_0) := \{(x, y): x^2 + y^2 < r_0^2\}$. Since $|\varphi_1 - \varphi_2| \geq \pi$ is assumed, we then infer that the length $\mathscr{L}$ of the curve $P(\varphi)$, $\varphi_1 \leq \varphi \leq \varphi_2$, must be greater than the sum of the distances of the endpoints $P(\varphi_1)$ and $P(\varphi_2)$ from the origin of the x, y-plane; that is,

$$(*) \qquad \mathscr{L} > r(\varphi_1) + r(\varphi_2) = e^{z_1} + e^{z_2}$$

(see figure 5). This inequality yields another proof of $\alpha \geq e^{z_1} + e^{z_2}$ in case that $|\varphi_1 - \varphi_2| \geq \pi$. At the same time, this geometrical consideration shows how we have to construct a curve $P(\varphi)$ in the x, y-plane that is parametrized by the polar angle $\varphi \in [\varphi_1, \varphi_2]$, $\varphi_2 - \varphi_1 \geq \pi$, which has the endpoints

$$P' := (r_1 \cos \varphi_1, r_1 \sin \varphi_1), \qquad P'' := (r_2 \cos \varphi_2, r_2 \sin \varphi_2),$$

$r_1 := e^{z_1}$, $r_2 := e^{z_2}$, and whose length $\mathscr{L}$ differs as little from $r_1 + r_2$ as one pleases. The principal idea consists in moving from P' to the periphery $C(r_0)$ of some disk $B(r_0)$, $r_0 > 0$ being chosen very small, then walking a certain way on $C(r_0)$ (possibly one has to circulate several times on $C(r_0)$ about the origin if $\varphi_2 - \varphi_1$ is larger than 2π), and finally moving from $C(r_0)$ to P''. If the paths to and fro $C(r_0)$ are basically straight, and if r_0 is sufficiently small, we can achieve that the length $\mathscr{L}$ of the total path differs arbitrarily little from $r_1 + r_2$. However, we have at the same time to secure that the polar angle permanently increases when moving along the curve. Both properties can simultaneously be achieved if and only if $|\varphi_1 - \varphi_2| \geq \pi$ holds (cf. figure 6).

This proves that $\alpha = r_1 + r_2$. Moreover, the previously proved inequality $(*)$ shows that the infimum α cannot be assumed by $\mathscr{F}(w)$ for any $w \in \mathscr{C}$. *Therefore the minimum problem*

$$\text{“}\mathscr{F}(w) \to \min \quad \text{for } w \in \mathscr{C}\text{”}$$

has no solution if $|x_1 - x_2| \geq \pi$ *holds.*

If, on the other hand, $|x_1 - x_2| < \pi$ *and* $x_1 < x_2$*, then there is exactly one extremal joining* P_1 *and* P_2*, and this extremal is the unique minimizer of* $\mathscr{F}$ *in* $\mathscr{C}$.

We leave it to the reader to prove that there is one and only one extremal joining P_1 and P_2, provided that $|x_1 - x_2| < \pi$. Furthermore we note that all extremals $u(x) = z_0 - \log \cos(x - x_0)$

are congruent convex curves. All extremals passing through the same fixed point $P_1 = (x_1, z_2)$ cover both the strip $x_1 - \pi/2 < x < x_1$ and the strip $x_1 < x < x_1 + \pi/2$ in a simple way.

3 *Euler's treatment of the isoperimetric problems.* Let us consider the dual version of the isoperimetric problem: Among all smooth curves, enclosing a given area, to find one of least length.

Besides his multiplier approach, Euler found another way of treating this problem which reminds one of the cutting of the Gordian knot: He removed the subsidiary condition by introducing area as a new independent variable.

Let t, z be cartesian coordinates in the plane, and consider curves which are nonparametrically given as graphs of functions $z = u(t)$, $t_1 \le t \le t_2$. Let us introduce the area

$$\xi(t) = \int_{t_1}^{t} u(\underline{t})\, d\underline{t}$$

under the arc $z = u(\underline{t})$, $t_1 \le \underline{t} \le t$, as new independent variable x, by setting $x = \xi(t)$ and considering the inverse function $t = \tau(x)$. If we define $v(x)$ by

$$v(x) := u(\tau(x)),$$

the curve is now given by $z = v(x)$ as a function of the area x below the graph $z = u(\underline{t})$, $t_1 \le \underline{t} \le t$. The line element ds has the form

$$ds = \sqrt{dt^2 + dz^2} = \sqrt{1 + u'^2}\, dt.$$

On the other hand, we have

$$\xi'(t) = u(t) = v(x) \quad \text{and} \quad \tau'(x) = \frac{1}{\xi'(t)}$$

and

$$u'(t) = v'(x)\xi'(t).$$

Since

$$ds = \sqrt{1 + u'^2}\, dt = \sqrt{1 + v'^2 \xi'^2}\, \tau'\, dx,$$

we arrive at

$$ds = \sqrt{\frac{1}{v^2} + v'^2}\, dx.$$

Thus the length functional $\mathscr{L}(v)$ is given by

$$\mathscr{L}(v) = \int_{x_1}^{x_2} \sqrt{\frac{1}{v^2} + v'^2}\, dx, \quad x_1 = 0,$$

and our problem now requires $\mathscr{L}$ to be minimized. The subsidiary condition "area = constant" miraculously has vanished, and we have an unconstrained problem for $\mathscr{L}$. Since the integrand

$$F(z, p) = \sqrt{\frac{1}{z^2} + p^2}$$

does not depend on the independent variable x, the expression $F - pF_p$ forms a first integral of the extremals $z = v(x)$ of $\mathscr{L}$, whence we obtain that

$$\frac{1}{v^2 \sqrt{\frac{1}{v^2} + v'^2}} = \text{const}$$

holds, and therefore

$$v\sqrt{1+v^2v'^2} = b$$

for some constant b. Inserting $u(t) = v(x)$ and

$$v'(x) = \frac{u'(t)}{u(t)},$$

we obtain for the extremals $z = u(t)$ the equation

$$uu' = \sqrt{b^2 - u^2}.$$

A straight-forward integration yields

$$(t-c)^2 + z^2 = c^2, \quad z = u(t),$$

as defining equation for the extremals which therefore are seen to be circles.

Note that $F(z, p) = \sqrt{\frac{1}{z^2} + p^2}$ is a strictly convex function of z, p; cf. the remarks at the beginning of this subsection. Thus the functional $\mathscr{L}(v) = \int_0^{x_2} \sqrt{v^{-2} + v'^2}\, dx$ is a strictly convex function of v, and its extremals are unique minimizers among all C^1-curves $z = v(x)$ with the same boundary values at $x = 0$ and $x = x_2$. (Actually, the same reasoning yields the minimizing property of the extremals among all AC-curves.)

From this minimum property, one can with some effort derive that, among all simple, closed, regular C^1-curves bounding domains of equal area, exactly the circle has least length.

This yields without difficulty that among all simple, closed, regular C^1-curves of equal length exactly the circle bounds the domain of largest area.

Thus we have proved that the *isoperimetric problem* has a solution, and that the circle is the uniquely determined maximizer. This result is much stronger than the one obtained in 1,*1* [1]. There we had proved that, if the isoperimetric problem possesses a solution, then this must be a circle. It is, however, not a priori clear whether some given minimum or maximum problem will have a solution at all. This fundamental question, which has played an important role in the development of the calculus of variations, will be discussed in Chapter 8 and in another treatise. Here we only note that Euler had the key to a complete proof of the isoperimetric property of the circle in his hands. We, however, do not want to give the impression that Euler really argued in this way. First of all, convexity of functionals was an idea unknown to him. Secondly, and strangely enough, it did not occur to him that a variational problem might have extremals which are neither maxima nor minima. He believed that an extremal necessarily furnishes a minimum if, *for a single comparison surface* (*curve*) satisfying the same boundary conditions, the variational integral has a larger value then for the considered extremal.[5]

[4] *The integrand* $F(z, p) = \frac{1}{\sqrt{z}}\sqrt{1+p^2}$, $z > 0$. Basically this is the Lagrangian appearing in the brachystochrone problem. Since we shall give a fairly complete discussion of this problem in 6,*2.3* [4], we presently only mention that the brachystochrone problem deals with the question to determine the shape of a curve along which a point mass slides in minimal time from some point P_1 to a second point P_2 under the influence of gravity. The extremals are cycloids, and they can be shown to be curves, along which the time of descent is minimized. This is proved by Weierstrass field theory, but an alternate approach is provided by introducing $v(x) := \sqrt{2u(x)}$. This way we obtain

$$\frac{1}{\sqrt{u}}\sqrt{1+u'^2} = \sqrt{\frac{1}{u} + \frac{u'^2}{u}} = \sqrt{2}\sqrt{\frac{1}{v^2} + v'^2}, \quad u(x) > 0,$$

[5] Cf. Carathéodory [16], Vol. 5, p. 118, and Euler, Opera omnia I, Vol. 24.

which basically is the strictly convex integrand from example [3]. Thus we have transformed the variational integral $\int_{x_1}^{x_2} \frac{1}{\sqrt{u}}\sqrt{1+u'^2}\,dx$ into the strictly convex integral $\sqrt{2}\int_{x_1}^{x_2}\sqrt{\frac{1}{v^2}+v'^2}\,dx$, and by our general reasoning it follows that the extremals of $\mathscr{F}(u)$, the cycloids, are indeed minimizers (however, points with $u(x)=0$ require some special care).

2.4. Application of Integral Inequalities

Let $\mathscr{C}$ be a convex set in $C_0^1(\bar{\Omega}, \mathbb{R}^N)$, and let $u \in \mathscr{C}$ be an extremal of some variational integral

$$\mathscr{F}(u) = \int_\Omega F(x, u, Du)\,dx.$$

The Proposition of *2.1* states that u minimizes $\mathscr{F}$ among all functions contained in $\mathscr{C}$, having the same boundary values as u, if $\delta^2\mathscr{F}(w, \varphi) \ge 0$ holds for all $w \in \mathscr{C}$ and all $\varphi \in C_0^1(\bar{\Omega}, \mathbb{R}^N)$, and u even furnishes a strict minimum if

$$\delta^2\mathscr{F}(w, \varphi) > 0 \quad \text{for all } \varphi \in C_0^1(\bar{\Omega}, \mathbb{R}^N) \text{ with } \varphi \ne 0$$

and for all $w \in \mathscr{C}$.

Sometimes this inequality can be obtained by the following approach: One tries to split the Lagrangian F in the form

$$F(x, z, p) = F_1(x, z, p) - F_2(x, z, p),$$

whence it follows that

$$\delta^2\mathscr{F}(w, \varphi) = \delta^2\mathscr{F}_1(w, \varphi) - \delta^2\mathscr{F}_2(w, \varphi),$$

where $\mathscr{F}$, $\mathscr{F}_1$, $\mathscr{F}_2$, denote the variational integrals associated with F, F_1, and F_2. Then the desired inequality is proved if, for every $w \in \mathscr{C}$, one can find some quadratic form $B(\varphi)$ on $C_0^1(\bar{\Omega}, \mathbb{R}^N)$ with $B(\varphi) > 0$ for $\varphi \ne 0$, such that

$$\delta^2\mathscr{F}_1(w, \varphi) \ge c_1 B(\varphi),$$
$$\delta^2\mathscr{F}_2(w, \varphi) \le c_2 B(\varphi)$$

hold for suitable constants c_1 and c_2, $0 < c_2 < c_1$, and for all $\varphi \in C_0^1(\bar{\Omega}, \mathbb{R}^N)$.

Let us consider a very simple example.

[1] Let $N = 1$, $n \ge 3$, and

$$\mathscr{F}(u) = \tfrac{1}{2}\int_\Omega \{|Du|^2 + H(x)u^2\}\,dx.$$

We decompose H into its positive and negative parts H^+ and H^-:

$$H = H^+ - H^-, \quad H^+ \ge 0,\; H^- \ge 0.$$

Then it follows that

$$\begin{aligned}\delta^2\mathscr{F}(w, \varphi) &= \int_\Omega \{|Du|^2 + H(x)\varphi^2\}\,dx \\ &\ge \int_\Omega |D\varphi|^2\,dx - \int_\Omega H^-(x)\varphi^2\,dx\end{aligned}$$

for all $w \in \mathscr{C} := C^1(\overline{\Omega}, \mathbb{R}^N)$. Set $2^* := \frac{2n}{n-2}$; then Hölder's inequality yields

$$\int_\Omega H^- \varphi^2 \, dx \le \left(\int_\Omega |H^-|^{n/2} \, dx\right)^{2/n} \left(\int_\Omega |\varphi|^{2^*} \, dx\right)^{2/2^*}$$

since $1 - \frac{2}{2^*} = \frac{2}{n}$. By Sobolev's inequality, there is a number $S = S(n, \Omega) > 0$ such that

$$\left(\int_\Omega \varphi^{2^*} \, dx\right)^{2/2^*} \le S \int_\Omega |D\varphi|^2 \, dx$$

holds for all $\varphi \in C_0^1(\overline{\Omega}, \mathbb{R}^N)$. Therefore we obtain

$$\int_\Omega H^- \varphi^2 \, dx \le S\left(\int_\Omega |H^-|^{n/2} \, dx\right)^{2/n} \int_\Omega |D\varphi|^2 \, dx.$$

Thus we can choose $B(\varphi) := \int_\Omega |D\varphi|^2 \, dx$, and it follows that *$\mathscr{F}(u)$ is strictly convex on $\mathscr{C} := C_0^1(\overline{\Omega}, \mathbb{R}^N)$ if inequality*

$$S\left(\int_\Omega |H^-|^{n/2} \, dx\right) < 1$$

is satisfied.

2.5. Convexity Modulo Null Lagrangians

We have seen that the convexity condition (1) of *2.2* is rather stringent and hard to verify. On the other hand we know that, by adding an arbitrary null Lagrangian $G(x, z, p)$ to a given Lagrangian $F(x, z, p)$, the Euler equations of F and $F + G$ only formally differ. That is, an extremal of F also is an extremal for $F + G$, and vice versa. Moreover, if

$$\mathscr{F}(v) = \int_\Omega F(x, v, Dv) \, dx, \qquad \mathscr{G}(v) = \int_\Omega G(x, v, Dv) \, dx,$$

then, for a given weak extremal u of $\mathscr{F}$ with $c := \mathscr{G}(u)$, it follows that

$$\mathscr{G}(v) = c \quad \text{for all } v \in C^1(\overline{\Omega}, \mathbb{R}^N) \text{ with } v = u \text{ on } \partial\Omega$$

(cf. 1,*4.1*). Hence u is a minimizer of $\mathscr{F}$ if and only if u is a minimizer of $\mathscr{F} + \mathscr{G}$ (with respect to fixed boundary values on $\partial\Omega$). Therefore we infer from Proposition 2 of *2.2* the following result:

Proposition 1. *Suppose that there is a null Lagrangian $G(x, z, p)$ such that $H := F + G$ satisfies the convexity condition* (1) of *2.2, i.e.,*

(1) $$H_{pp}(x, z, p)\pi\pi + 2H_{zp}(x, z, p)\zeta\pi + H_{zz}(x, z, p)\zeta\zeta \ge 0$$

for all $x \in \overline{\Omega}$, $z, \zeta \in \mathbb{R}^N$, $p, \pi \in \mathbb{R}^{nN}$. Then it follows that every weak extremal u of $\mathscr{F}$ is an absolute minimizer, that is,

$$\mathscr{F}(v) \ge \mathscr{F}(u) \quad \textit{for all } v \in C^1(\overline{\Omega}, \mathbb{R}^N) \textit{ with } v = u \textit{ on } \partial\Omega.$$

In 1,*4.2* we have characterized the null Lagrangians. If $n = 1$ or $N = 1$, there are not many: all of them have the form

$$G(x, z, p) = A(x, z) + B_i^\alpha(x, z)p_\alpha^i,$$

where

$$A_{z^i} = B^\alpha_{i,x^\alpha}, \qquad B^\alpha_{k,z^i} = B^\alpha_{i,z^k}.$$

Hence, for $N = 1$, all null Lagrangians are of the type

$$G(x, z, p) = B^\alpha(x, z)p_\alpha + \int B^\alpha_{x^\alpha}(x, z)\, dz.$$

Thus, by adding a suitable null Lagrangian to F we can get rid of terms $B^\alpha(z)p_\alpha$ with arbitrary functions $B^\alpha(z)$ which is some improvement over the assumptions (6) and (7) of *2.2*.

If $\min\{n, N\} \ge 2$, the class of null Lagrangians is much more extensive but there seems to be no systematic way to decide whether (1) can be achieved.

There is a *second possibility to make use of null Lagrangians*[6] if one wants to verify the minimum property of some extremal by means of the Proposition in *2.1*. Namely, it is conceivable that one can prove condition (1) of *2.1*, although the relation

$$\delta^2 F(w, \varphi) \ge 0 \quad \text{for all } w \in \mathscr{C} \text{ and all } \varphi \in C_0^1(\bar{\Omega}, \mathbb{R}^N)$$

may not be true. In fact, suppose that there is a null Lagrangian G with $\mathscr{G}(\varphi) \le 0$ and

$$\delta^2 F(w, \varphi) + G(x, \varphi, D\varphi) \ge 0 \quad \text{on } \Omega \text{ for } \varphi \in C_0^1(\bar{\Omega}, \mathbb{R}^N).$$

Then, by integrating this inequality, we infer that (1) of *2.1* holds. Since $\mathscr{G}(\varphi) = \mathscr{G}(0)$ for all $\varphi \in C_0^1(\bar{\Omega}, \mathbb{R}^N)$, we obtain

Proposition 2. *Let $\mathscr{C}$ be a convex set in $C^1(\Omega, \mathbb{R}^N)$, and suppose that for every $w \in \mathscr{C}$ there exists a null Lagrangian $G(x, z, p)$ with $\int_\Omega G(x, 0, 0)\, dx \le 0$ such that*

$$\delta^2 F(w, \varphi) + G(x, \varphi, D\varphi) \ge 0$$

holds for all $\varphi \in C_0^1(\bar{\Omega}, \mathbb{R}^N)$, with the strict inequality sign on some subset of positive measure for $\varphi \neq 0$. Then for any weak F-extremal $u \in \mathscr{C}$ it follows that

$$\mathscr{F}(v) > \mathscr{F}(u) \quad \textit{for all } v \in \mathscr{C} \textit{ with } v \neq u \textit{ and } v = u \textit{ on } \partial\Omega.$$

Let us discuss this idea for the following example.

[6] Actually, this time-honoured trick has already been used by Legendre [1] in 1786 to prove the positivity of the second variation. This directly leads to Jacobi's theory for $n = N = 1$; cf. Chapter 5.

[1] Consider the variational integral

$$\mathscr{E}(u) = \int_\Omega \{|Du|^2 + \tfrac{4}{3}Hu\cdot(u_{x^1} \wedge u_{x^2})\}\, dx^1\, dx^2,$$

with the Lagrangian

$$E(z, p) = |p_1|^2 + |p_2|^2 + \tfrac{4}{3}Hz\cdot(p_1 \wedge p_2),$$

where $n = 2$, $N = 3$, $\Omega \subset \mathbb{R}^2$, $u = u(x)$, $x = (x^1, x^2)$, $p_1, p_2 \in \mathbb{R}^3$, $p = (p_1, p_2)$. As we have seen in 3,2 [4], the Euler equations of $\mathscr{E}$ are equivalent to

$$\Delta u = 2Hu_{x^1} \wedge u_{x^2},$$

and every strong inner extremal of $\mathscr{E}$ satisfies the conformality relations

$$|u_{x^1}|^2 = |u_{x^2}|^2, \qquad u_{x^1}\cdot u_{x^2} = 0.$$

Hence every E-extremal u with $\partial\mathscr{E}(u, \lambda) = 0$ for all $\lambda \in C^1(\overline{\Omega}, \mathbb{R}^N)$ is a surface of constant mean curvature provided that $|u_{x^1}| \neq 0$.

For some number $R > 0$, we introduce the set

$$\mathscr{C}_R := C^1(\overline{\Omega}, \mathbb{R}^3) \cap \{v\colon |v(x)| \le R \text{ for } x \in \overline{\Omega}\}.$$

Let $u \in \mathscr{C}_R \cap C^2(\overline{\Omega}, \mathbb{R}^3)$ be an extremal of $\mathscr{E}$, and let v be an arbitrary function in $\mathscr{C}_R$ with $\varphi := v - u \in C_0^1(\overline{\Omega}, \mathbb{R}^3)$. Then, for $t \in [0, 1]$, we have $\psi(t) := u + t\varphi = tv + (1 - t)u \in \mathscr{C}_R$.

For the sake of brevity, we introduce the notations $\psi_\alpha := D_\alpha\psi$, $\varphi_\alpha := D_\alpha\varphi$ $(\alpha = 1, 2)$, and

$$[a, b, c] := \det(a, b, c) = a\cdot(b \wedge c) \quad \text{for } a, b, c \in \mathbb{R}^3.$$

Then we obtain

$$\begin{aligned} &E(\psi + \varepsilon\varphi, D\psi + \varepsilon D\varphi)\\ &\quad = E(\psi, D\psi)\\ &\qquad + 2\varepsilon\left\{D\psi\cdot D\varphi + \frac{2H}{3}([\varphi, \psi_1, \psi_2] + [\psi, \varphi_1, \psi_2] + [\psi, \psi_1, \varphi_2])\right\}\\ &\qquad + \varepsilon^2\{|\varphi_1|^2 + |\varphi_2|^2 + \tfrac{4}{3}H([\psi, \varphi_1, \varphi_2] + [\varphi, \psi_1, \varphi_2] + [\varphi, \varphi_1, \psi_2])\}\\ &\qquad + \varepsilon^3\tfrac{4}{3}H[\varphi, \varphi_1, \varphi_2]. \end{aligned}$$

Suppose presently that also $\varphi \in C^2(\overline{\Omega}, \mathbb{R}^3)$. Then we obtain by a straight forward computation (since the terms involving second derivatives cancel)

$$\begin{aligned} &[\psi, \varphi_1, \psi_2] + [\psi, \psi_1, \varphi_2]\\ &\quad = 2[\varphi, \psi_1, \psi_2] - \{D_1[\varphi, \psi, \psi_2] - D_2[\varphi, \psi, \psi_1]\}, \end{aligned}$$

and, interchanging the roles of φ and ψ, we also obtain

$$\begin{aligned} &[\varphi, \psi_1, \varphi_2] + [\varphi, \varphi_1, \psi_2]\\ &\quad = 2[\psi, \varphi_1, \varphi_2] - \{D_1[\psi, \varphi, \varphi_2] - D_2[\psi, \varphi, \varphi_1]\}. \end{aligned}$$

Therefore,

$$\tfrac{1}{2}\delta E(\psi, \varphi) = D\psi\cdot D\varphi + 2H[\varphi, \psi_1, \psi_2] - \tfrac{2}{3}H\{D_1[\varphi, \psi, \psi_2] - D_2[\varphi, \psi, \psi_1]\}$$

and

$$\delta^2 E(\psi, \varphi) = |D\varphi|^2 + 4H[\psi, \varphi_1, \varphi_2] - \tfrac{4}{3}H\{D_1[\psi, \varphi, \varphi_2] - D_2[\psi, \varphi, \varphi_1]\}. \tag{2}$$

By partial integration, we arrive at

$$\tfrac{1}{2}\delta\mathscr{E}(u, \varphi) = \int_\Omega \varphi\cdot(-\Delta u + 2Hu_{x^1} \wedge u_{x^2})\, dx^1\, dx^2 - \tfrac{2}{3}H\int_{\partial\Omega} \varphi\cdot(u \wedge du)$$

and

$$\delta^2\mathscr{E}(\psi, \varphi) = \int_\Omega \{|D\varphi|^2 + 4H\psi\cdot(\varphi_1 \wedge \varphi_2)\}\, dx^1\, dx^2 - \tfrac{4}{3}H \int_{\partial\Omega} \psi\cdot(\varphi \wedge d\varphi).$$

So far, we have established these two formulas under the additional hypothesis that $\varphi \in C^2(\bar{\Omega}, \mathbb{R}^3)$. By approximating a given C^1-function in the C^1-norm by C^2-functions and passing to the limit, we obtain the two formulas in general. Since φ vanishes on $\partial\Omega$, the two boundary integrals appearing in the two previous formulas vanish, whence

$$-\tfrac{1}{2}\delta\mathscr{E}(u, \varphi) = \int_\Omega (\Delta u - 2Hu_{x^1} \wedge u_{x^2})\cdot \varphi\, dx^1\, dx^2 \tag{3}$$

and

$$\tfrac{1}{2}\delta^2\mathscr{E}(\psi, \varphi) = \int_\Omega \{|D\varphi|^2 + 4H\psi\cdot(\varphi_1 \wedge \varphi_2)\}\, dx^1\, dx^2. \tag{4}$$

We obtain from (3) that the Euler equations are equivalent to

$$\Delta u = 2Hu_{x^1} \wedge u_{x^2}.$$

From $\psi(x, t) = tv(x) + (1 - t)u(x)$, $|u(x)| \le R$, and $|v(x)| \le R$ it follows that $|\psi(x, t)| \le R$ for $x \in \bar{\Omega}$ and $t \in [0, 1]$, and therefore

$$|4H\psi\cdot(\varphi_1 \wedge \varphi_2)| \le 4|H||\psi||\varphi_1||\varphi_2| \le 2R|H||D\varphi|^2.$$

Thus we infer from (4) that

$$\tfrac{1}{2}\delta^2\mathscr{E}(\psi, \varphi) \ge \int_\Omega (1 - 2R|H|)|D\varphi|^2\, dx.$$

Hence we can apply the Proposition of *2.1* and arrive at the following result:

Let $\mathscr{C}_R := \{v \in C^1(\bar{\Omega}, \mathbb{R}^3)\colon |v(x)| \le R\}$, and suppose that $u \in \mathscr{C}_R$ is a weak extremal of $\mathscr{E}$ (or, instead, that $u \in \mathscr{C}_R \cap C^2(\Omega, \mathbb{R}^3)$ and $\Delta u = 2Hu_{x^1} \wedge u_{x^2}$ in Ω). Then, for all $v \in \mathscr{C}_R$ with $v = u$ on $\partial\Omega$, it follows that

$$\mathscr{F}(v) \ge \mathscr{F}(u) \quad \text{if } |H| \le \frac{1}{2R},$$

and that

$$\mathscr{F}(v) > \mathscr{F}(u) \quad \text{if } |H| < \frac{1}{2R} \text{ and } v \ne u.$$

In particular, for every $f \in \mathscr{C}_R$, there exists at most one extremal u of $\mathscr{E}$ such that $u \in \mathscr{C}_R$ and $u = f$ on $\partial\Omega$ hold, provided that $|H| < \dfrac{1}{2R}$ is satisfied.

This result also follows from Proposition 2 above. In fact, formula (2) shows that, for arbitrary $w \in \mathscr{C}_R$, the null Lagrangian $G(x, z, p)$ should be chosen as

$$G(x, z, p_1, p_2) := \tfrac{4}{3}H\{2w(x)\cdot(p_1 \wedge p_2) - z\cdot[w_{x^1}(x) \wedge p_2 + p_1 \wedge w_{x^2}(x)]\}.$$

Notice that

$$G(x, \varphi, D_1\varphi, D_2\varphi) = \tfrac{4}{3}H\{D_1[w, \varphi, D_2\varphi] - D_2[w, \varphi, D_1\varphi]\}.$$

2.6. Calibrators

Now we describe another method based on the use of null Lagrangians which may ascertain that a given extremal is a minimizer. This method rests on a quite simple idea, to prove the existence of a *calibrator*. It will seem surprising that this idea is functioning at all; however, its elaboration in specific cases can be rather involved.

In order to define the notion of a calibrator we fix a Lagrangian $F(x, z, p)$ and a class $\mathscr{C}$ of mappings $u : \Omega \to \mathbb{R}^N$, $\Omega \subset \mathbb{R}^n$, whose domains of definition Ω need not necessarily be the same for all $u \in \mathscr{C}$. We assume that the integral

$$(1) \qquad \mathscr{F}(u) := \int_\Omega F(x, u(x), Du(x))\, dx, \quad \Omega = \text{domain of } u,$$

is defined for all $u \in \mathscr{C}$.

Definition 1. *Let $u_0 : \Omega_0 \to \mathbb{R}^N$ be an arbitrary element of $\mathscr{C}$. Then a* calibrator *for $\{F, u_0, \mathscr{C}\}$ is a Lagrangian $M(x, z, p)$ satisfying the following conditions:*

(i) *The functional*

$$(2) \qquad \mathscr{M}(u) := \int_\Omega M(x, u(x), Du(x))\, dx, \quad \Omega = \textit{domain of } u,$$

is an invariant integral on $\mathscr{C}$, i.e. $\mathscr{M}(u)$ is defined for all $u \in \mathscr{C}$, and we have $\mathscr{M}(u_1) = \mathscr{M}(u_2)$ for any $u_1, u_2 \in \mathscr{C}$.

(ii) $M(x, u_0(x), Du_0(x)) = F(x, u_0(x), Du_0(x))$ *for all* $x \in \Omega_0$.

(iii) $M(x, u(x), Du(x)) \le F(x, u(x), Du(x))$ *for all* $x \in \Omega =$ domain u, $u \in \mathscr{C}$.

Proposition 1. *If there exists a calibrator for the triple $\{F, u_0, \mathscr{C}\}$, $u_0 \in \mathscr{C}$, then u_0 is a minimizer of $\mathscr{F}$ in $\mathscr{C}$.*

Proof. Let M be a calibrator for $\{F, u_0, \mathscr{C}\}$. Then we infer from (ii) that

$$\mathscr{M}(u_0) = \mathscr{F}(u_0).$$

For an arbitrary $u \in \mathscr{C}$ we obtain on account of (i) that

$$\mathscr{M}(u_0) = \mathscr{M}(u),$$

and (iii) implies

$$\mathscr{M}(u) \le \mathscr{F}(u),$$

whence

$$\mathscr{F}(u_0) \le \mathscr{F}(u) \quad \text{for all } u \in \mathscr{C}. \qquad \square$$

Remark 1. Under fairly obvious assumptions on M and $\mathscr{M}$ we infer that a calibrator M has to be a *null Lagrangian*, cf. 1,*4.1*, Proposition 1.

Remark 2. Conversely, in many cases it is easy to prove that a null Lagrangian M satisfying (ii) and (iii) is a calibrator for $\{F, u_0, \mathscr{C}\}$, see 1,*4.1*, Propositions 2 and 3.

Let us illustrate the concept of a calibrator by some examples.

1 The Lagrangian $F(p) = \sqrt{1 + |p|^2}$, $n = 1$, $N \geq 1$, leads to the arc-length functional

$$\mathscr{F}(u) = \int_{x_1}^{x_2} \sqrt{1 + |u'(x)|^2}\, dx, \quad x_1 < x_2. \tag{3}$$

Let $\mathscr{C}$ be the class of functions $u \in C^1([x_1, x_2], \mathbb{R}^N)$ satisfying $u(x_1) = z_1$ and $u(x_2) = z_2$, i.e. $\mathscr{C}$ is the set of nonparametric curves $c(x) = (x, u(x))$, $x_1 \leq x \leq x_2$, connecting the points $P_0 = (x_1, z_1)$ and $P_2 = (x_2, z_2)$. We claim that the linear function

$$u_0(x) = z_1 + \frac{x - x_1}{x_2 - x_1} \cdot (z_2 - z_1), \quad x_1 \leq x \leq x_2,$$

is the minimizer of $\mathscr{F}(u)$ in $\mathscr{C}$. By the invariance of $\mathscr{F}$ with respect to motions of the x, z-space it suffices to prove this assertion for $P_1 = (0, 0)$ and $P_2 = (\xi, 0)$, $\xi > 0$. In this case u_0 is given by

$$u_0(x) = 0, \quad 0 \leq x \leq \xi,$$

and one easily verifies that $M = 1$ is a calibrator for $\{F, u_0, \mathscr{C}\}$. Then the assertion follows from Proposition 1.

2 It is not at all obvious if and how the above reasoning can be generalized to the integrand $F(x, z, p) = \omega(x, z)\sqrt{1 + |p|^2}$, i.e. to the functional

$$\mathscr{F}(u) = \int_{x_1}^{x_2} \omega(x, u(x))\sqrt{1 + |u'(x)|^2}\, dx. \tag{4}$$

Our discussion will be simplified if we replace (4) by the "parametric integral"

$$\mathscr{F}(\xi) = \int_{t_1}^{t_2} \omega(\xi(t))\, |\dot{\xi}(t)|\, dt \tag{5}$$

for curves $\xi : [t_1, t_2] \to \mathbb{R}^{N+1}$. For a nonparametric curve $\xi(t) = (t, u(t))$, $x_1 \leq t \leq x_2$, the integral (5) coincides with (4).

Suppose now that $\omega : \mathbb{R}^{N+1} \to \mathbb{R}$ is a smooth positive function. We consider the functional (5) on C^1-immersions $\xi : [t_1, t_2] \to \mathbb{R}^{N+1}$ whose intervals of definition $[t_1, t_2]$ need not be the same for all ξ. Since (5) is invariant with respect to regular parameter transformations τ, for every such immersion ξ there is a reparametrization $\eta = \xi \circ \tau$ such that $\mathscr{F}(\xi) = \mathscr{F}(\eta)$ and $\omega(\eta)|\dot{\eta}| = 1$. Thus we define $\mathscr{C}$ as class of C^1-representations $\xi : [t_1, t_2] \to \mathbb{R}^{N+1}$ satisfying

$$\omega(\xi(t))\, |\dot{\xi}(t)| \equiv 1 \tag{6}$$

and

$$\xi(t_1) = P_1, \qquad \xi(t_2) = P_2, \tag{7}$$

where P_1 and P_2 are two preassigned points in $\mathbb{R}^{N+1}$, and we also assume that $\xi([t_1, t_2])$ lies in a domain G of $\mathbb{R}^{N+1}$ containing P_1 and P_2 which is still to be chosen. Suppose now that we can choose G in such a way that there is a C^2-solution $S(x)$ of

$$|\nabla S(x)| = \omega(x) \quad \text{on } G \tag{8}$$

satisfying $S(P_1) < S(P_2)$, and let ξ be a curve in $\mathscr{C}$ intersecting the level surfaces $\mathscr{S}_\theta := \{x \in G : S(x) = \theta\}$ perpendicularly, i.e.

$$\dot{\xi}(t) \sim \nabla S(\xi(t)) \quad \text{for } t_1 \leq t \leq t_2. \tag{9}$$

We claim that ξ is a minimizer of $\mathscr{F}$ in $\mathscr{C}$. By virtue of Proposition 1 it is sufficient to show that the Lagrangian $M(x, p)$ defined by

$$M(x, p) := p \cdot \nabla S(x) \tag{10}$$

is a calibrator for the triple $\{F, \xi, \mathscr{C}\}$ where

$$F(x, p) := \omega(x)|p| \tag{11}$$

is the Lagrangian of the variational integral (5). In fact, for any $\eta : [t_1, t_2] \to \mathbb{R}^{N+1}$ of class $\mathscr{C}$ we have

$$\mathscr{M}(\eta) = \int_{t_1}^{t_2} M(\eta, \dot{\eta})\, dt = \int_{t_1}^{t_2} \frac{d}{dt} S(\eta(t))\, dt = S(P_2) - S(P_1),$$

and therefore $\mathscr{M}$ is an invariant integral on $\mathscr{C}$. Moreover, relations (6) and (9) imply that $\xi(t)$ satisfies the differential equation

$$\dot{\xi} = \Psi(\xi), \tag{12}$$

where $\Psi(x)$ denotes the C^1-vector field

$$\Psi(x) := \frac{1}{\omega(x)|\nabla S(x)|} \nabla S(x). \tag{13}$$

We infer from (10), (12) and (13) that $M(\xi, \dot{\xi}) = 1$, and (6) implies that $F(\xi, \dot{\xi}) = 1$. Thus we have

$$M(\xi, \dot{\xi}) = F(\xi, \dot{\xi}).$$

Finally, for any $\eta \in \mathscr{C}$ we have

$$F(\eta, \dot{\eta}) = \omega(\eta)|\dot{\eta}| = 1, \tag{14}$$

and equation (8) yields

$$\frac{1}{\omega(\eta)} |\nabla S(\eta)| = 1. \tag{15}$$

Then, taking (14) and (15) into account, it follows from Schwarz's inequality that

$$M(\eta, \dot{\eta}) = \dot{\eta} \cdot \nabla S(\eta) = \omega(\eta)\dot{\eta} \cdot \frac{1}{\omega(\eta)} \nabla S(\eta) \le 1. \tag{16}$$

From (14) and (16) we now obtain

$$M(\eta, \dot{\eta}) \le F(\eta, \dot{\eta}) \quad \text{for any } \eta \in \mathscr{C}.$$

Thus we have verified that M is a calibrator, and we can state the following result, taking the parameter invariance of the integral (5) into account: *Suppose that $S \in C^2(G)$ is a solution of equation* (8) *in some domain G of $\mathbb{R}^{N+1}$. Then any (regular) orthogonal trajectory $\xi : [t_1, t_2] \to G$ of the one-parameter family of surfaces $\mathscr{S}_\theta$, defined as level surfaces $\mathscr{S}_\theta = \{x \in G: S(x) = \theta\}$ of S, is a minimizer of $\mathscr{F}$ among all regular C^1-curves in G having the same endpoints as ξ.*

Furthermore it is not difficult to verify that the regularity results of 1,*3.1* are applicable in our situation, and so we infer that *the regular orthogonal trajectories of the level surfaces $\mathscr{S}_\theta$ are F-extremals provided that their representations $\xi(t)$ are chosen in such a way that* (6) *holds true.* Clearly every such extremal of (5) has a smooth nonparametric reparametrization $(t, u(t))$, $t_1 \le t \le t_2$, which then is an extremal of the functional

$$\mathscr{F}(u) = \int_{t_1}^{t_2} \omega(t, u(t)) \sqrt{1 + \left|\frac{du}{dt}(t)\right|^2}\, dt,$$

and it minimizes this nonparametric integral among all smooth curves $(t, v(t))$, $t_1 \le t \le t_2$, in G satisfying $v(t_i) = u(t_i)$, $i = 1, 2$. Thus we have found a way to treat, at least in principle, the minimum problem for the functional (4).

Still it is not clear how to find a solution $S(x)$ of (8) such that two given points P_1 and P_2 can be connected by an orthogonal trajectory of the level surfaces of S. It is conceivable and, in fact, true that such an S cannot always be found since it is not always possible to connect any two points by an extremal. Thus it is an interesting problem to find solutions of (8) which lead to a solution of the boundary value problem by extremals (and, in fact, by minimizers) in the way described above. The following idea is often applicable. Suppose that $S(x, a)$, $a = (a_1, \dots, a_N)$, is an N-parameter family of solutions of (8), i.e.

$$|S_x(x, a)| = \omega(x) \quad \text{on } G.$$

Then we determine an N-parameter family of curves $\xi(\cdot, a)$ by solving the initial value problem

$$\dot{\xi}(t, a) = S_x(\xi(t, a), a), \quad \xi(0, a) = P_1 .$$

Finally we try to reach P_2 in a fixed time, say, $t = 1$, by solving the equation

$$\xi(1, a) = P_2 .$$

We shall discuss this approach more closely in Chapter 9.

[3] Let $F(p) = \frac{1}{2}|p|^2$, $n > 1$, $N = 1$, be the Lagrangian of the Dirichlet integral

$$\mathscr{D}(u) = \frac{1}{2}\int_\Omega |Du(x)|^2\,dx, \tag{17}$$

the extremals of which are the harmonic functions in Ω. A simple convexity argument yields that any such function u_0 is a minimizer of $\mathscr{D}(u)$ in the class $\mathscr{C} = \{u \in C^1(\bar{\Omega}): u(x) = u_0(x) \text{ for } x \in \partial\Omega\}$ provided that Ω is bounded and $u_0 \in C^1(\bar{\Omega})$, cf. *2.2*, Proposition 3 and [1]. Let us give another proof by verifying that

$$M(x, p) := \nabla u_0(x)\cdot p - \tfrac{1}{2}|\nabla u_0(x)|^2 \tag{18}$$

is a calibrator for $\{F, u_0, \mathscr{C}\}$. In fact, we know that $M(x, z, p) := A(x, z) + B^\alpha(x, z)p_\alpha$ is a null Lagrangian if $A_z = B^\alpha_{x^\alpha}$. Hence the integrand (18) is a null Lagrangian since div $\nabla u_0(x) = \Delta u_0(x) = 0$, and therefore

$$\mathscr{M}(u) = \int_\Omega \{\nabla u_0(x)\cdot \nabla u(x) - \tfrac{1}{2}|\nabla u_0(x)|^2\}\,dx \tag{19}$$

is an invariant integral on $\mathscr{C}$. Secondly we have

$$F(p) = \tfrac{1}{2}|(p - \nabla u_0) + \nabla u_0|^2 = M(x, p) + \tfrac{1}{2}|p - \nabla u_0(x)|^2, \tag{20}$$

whence we infer that $F(\nabla u_0(x)) = M(x, \nabla u_0(x))$ and

$$F(\nabla u(x)) \le M(x, \nabla u(x)) \quad \text{for all } x \in \Omega .$$

[4] Let Ω be the unit ball $\{x \in \mathbb{R}^n: |x| < 1\}$ in $\mathbb{R}^n$, $N = n \ge 3$, and set

$$\mathscr{C} := \{u \in H^{1,2}(\Omega, \mathbb{R}^n): |u(x)| = 1 \text{ a.e. in } \Omega,\ u(x) = x \text{ on } \partial\Omega\}.$$

Here $H^{1,2}(\Omega, \mathbb{R}^n)$ is the Sobolev space of mappings $u: \Omega \to \mathbb{R}^n$ with generalized first-order derivatives satisfying $\int_\Omega (|u(x)|^2 + |Du(x)|^2)\,dx < \infty$. It is easy to see that $u_0(x) := x/|x|$ for $x \ne 0$ belongs to $\mathscr{C}$ since $n \ge 3$, and that

$$-\Delta u_0 = u_0 |\nabla u_0|^2 \quad \text{on } \Omega - \{0\}, \tag{21}$$

i.e. u_0 furnishes a harmonic mapping of $\Omega - \{0\}$ into the unit sphere S^{n-1}, cf. 2,2 [1]. Furthermore it is not hard to verify that

$$\int_\Omega \{\nabla u_0 \cdot \nabla\varphi - (u_0\cdot\varphi)|\nabla u_0|^2\}\,dx = 0 \tag{22}$$

for all $\varphi \in C_c^\infty(\Omega, \mathbb{R}^N)$, i.e. u_0 is a weak $H^{1,2}$-extremal of the Dirichlet integral $\mathscr{D}(u)$ under the holonomic constraint $|u| = 1$. We want to show that u_0 is a minimizer of $\mathscr{D}(u)$ among all $u \in \mathscr{C}$ by proving that

$$M(p) := \frac{1}{n-2}[(\operatorname{tr} p)^2 - \operatorname{tr}(p\cdot p)] \tag{23}$$

is a calibrator for $\{F, u_0, \mathscr{C}\}$ where

$$F(p) := |p|^2 \tag{24}$$

is twice the Lagrangian of $\mathscr{D}$. Here p denotes the $n \times n$-matrix (p^i_k) with the diagonal elements

$\pi_1 = p_1^1, \pi_2 = p_2^2, \ldots, \pi_n = p_n^n$ and the diagonal $\pi = (\pi_1, \pi_2, \ldots, \pi_n)$. We first verify the inequality

(25) $$(n-2)|p|^2 \geq (\operatorname{tr} p)^2 - \operatorname{tr}(p \cdot p) + (n-3) \sum_{i \neq k} |p_k^i|^2 \quad \text{if } p_n^n = \pi_n = 0.$$

In fact, we infer from $ab \leq a^2/2 + b^2/2$ that

$$\operatorname{tr}(p \cdot p) = p_k^i p_i^k \geq |\pi|^2 - \sum_{i \neq k} |p_k^i|^2,$$

and from $\pi_n = 0$ we conclude that

$$(\operatorname{tr} p)^2 = (\pi_1 + \cdots + \pi_{n-1})^2 \leq (n-1)|\pi|^2,$$

whence

$$(\operatorname{tr} p)^2 - \operatorname{tr}(p \cdot p) + (n-3) \sum_{i \neq k} |p_k^i|^2 \leq (n-2)\left[|\pi|^2 + \sum_{i \neq k} |p_k^i|^2\right] = (n-2)|p|^2.$$

We infer from (25) that

(26) $$M(p) \leq F(p) \quad \text{if } p_n^n = 0.$$

Suppose now that $u \in C^1(\bar{\Omega}, \mathbb{R}^n)$ and $|u(x)|^2 \equiv 1$. Then we obtain $u(x) \cdot D_\alpha u(x) = 0$ for $1 \leq \alpha \leq n$. If $u(x_0) = (0, \ldots, 0, 1) = e_n$ for some $x_0 \in \Omega$ it follows that $u^n_{x^\alpha}(x_0) = 0$ for $1 \leq \alpha \leq n$ and in particular $u^n_{x^n}(x_0) = 0$ whence

(27) $$M(Du(x_0)) \leq F(Du(x_0)) \quad \text{if } u(x_0) = e_n.$$

Since both $F(p)$ and $M(p)$ are invariant under the orthogonal group $O(n)$, we infer from (27) that

(28) $$M(Du(x)) \leq F(Du(x)) \quad \text{for all } x \in \bar{\Omega}.$$

Furthermore an elementary computation yields

(29) $$M(Du_0(x)) = F(Du_0(x)) \quad \text{for all } x \in \bar{\Omega} - \{x_0\}.$$

Finally it follows for $u \in C^2(\bar{\Omega}, \mathbb{R}^n)$ that

$$\operatorname{tr}(Du \cdot Du) - (\operatorname{div} u)^2 = \operatorname{div}[\nabla u \cdot u - (\operatorname{div} u)u]$$

and therefore

(30) $$\int_\Omega [\operatorname{tr}(Du \cdot Du) - (\operatorname{div} u)^2]\, dx = \int_{\partial\Omega} [(\operatorname{div} u)u \cdot \nu - \nu \cdot (\nabla u \cdot u)]\, d\mathscr{H}_{n-1},$$

where $\nu(x)$ is the exterior normal to $\partial\Omega$ at $x \in \partial\Omega$, i.e. $\nu(x) = x$. Thus we have $u \cdot \nu = 1$ on $\partial\Omega$ if u satisfies the boundary condition $u(x) = x$ for $x \in \partial\Omega$. If we also assume that $|u(x)| \equiv 1$ on $\bar{\Omega}$, the computation above yields $(\partial u/\partial\nu) \cdot u = 0$ and $\operatorname{div} u = n-1$ on $\partial\Omega$, whence $\nu \cdot (\nabla u \cdot u) = 0$ and $\operatorname{div} u = n - 1$. Thus the value of the boundary integral in (30) is $(n-1)$ area S^{n-1}, $S^{n-1} = \partial\Omega$. Then we infer from (30) that

(31) $$\mathscr{M}(u) := \int_\Omega M(Du)\, dx = \frac{n-1}{n-2} \cdot \text{area } S^{n-1}$$

for all $u \in C^1(\bar{\Omega}, \mathbb{R}^n)$ satisfying $|u(x)| = 1$ on $\bar{\Omega}$ and $u(x) = x$ on $\partial\Omega$, and (28) implies that

(32) $$\mathscr{M}(u) \leq \mathscr{F}(u)$$

for all such u, where $\mathscr{F}(u) = \int_\Omega F(Du)\, dx = 2\mathscr{D}(u)$. Finally (29) yields

(33) $$\mathscr{M}(u_0) = \mathscr{F}(u_0).$$

Essentially the same reasoning as before shows that (31) and (32) hold for all $u \in \mathscr{C}$, and therefore

$$\mathscr{F}(u_0) = \mathscr{M}(u_0) = \mathscr{M}(u) \leq \mathscr{F}(u) \quad \text{for all } u \in \mathscr{C},$$

i.e. $u_0(x) = \dfrac{x}{|x|}$ is a minimizer of $\mathscr{D}(u)$ in $\mathscr{C}$.

Inspecting the preceding examples we realize that in [1] and [3] we found a calibrator by a rather obvious guesswork whereas the success in [2] and [4] looks like sheer luck. In any case, it seems to be desirable to develop a systematic way for constructing calibrators. This is the objective of various "field theories" which will be discussed in Chapters 6–8. For a given Lagrangian F and an F-extremal u_0 one tries to find a null Lagrangian M and a suitable "neighbourhood" $\mathscr{C}$ of u_0 such that $F^* := F - M$ satisfies

$$F^*(x, u_0(x), Du_0(x)) = 0$$

and

$$F^*(x, u(x), Du(x)) \geq 0 \quad \text{for all } u \in \mathscr{C}.$$

Then M is a calibrator for $\{L, u_0, \mathscr{C}\}$ provided that M leads to an invariant integral on $\mathscr{C}$, which is the case for "reasonable" classes $\mathscr{C}$.

Let us now recall that condition (iii) of Definition 1 leads to the inequality

$$\mathscr{M}(u) \leq \mathscr{F}(u) \quad \text{for all } u \in \mathscr{C}.$$

This motivates

Definition 2. *A calibrator M for $\{F, u_0, \mathscr{C}\}$ is said to be a* strict calibrator *if we know in addition that*

(iv) $$\mathscr{M}(u) < \mathscr{F}(u) \quad \textit{for all } u \in \mathscr{C} - \{u_0\}.$$

The following result is obvious.

Proposition 2. *If there exists a strict calibrator for a triple $\{F, u_0, \mathscr{C}\}$, $u_0 \in \mathscr{C}$, then u_0 is a strict minimizer of $\mathscr{F}$ in $\mathscr{C}$; consequently u_0 is the unique minimizer of $\mathscr{F}$ in $\mathscr{C}$.*

We leave it to the reader to verify the calibrators in [1]–[3] are in fact strict calibrators.

3. Scholia

Section 1

1. In the theory of maxima and minima of ordinary functions, it was rather early discovered that the vanishing of the first derivative is but a necessary condition, and that one has to consider the second (or higher) derivatives, in order to ensure that a real extremum takes place. In the calculus of variations, this insight was only fairly late obtained. Even Euler believed that every extremal must either be a maximizer or a minimizer.[7] After a mistaken attempt by Laplace (1770), the first to successfully discuss this question was Legendre[8] in 1786, who, for this purpose, introduced the *second variation* $\delta^2\mathcal{F}$ of some variational functional $\mathcal{F}$.

Legendre's paper contained a serious gap, and only Jacobi[9] was able to show, when and how this gap can be filled. This investigation led to Jacobi's theory of *conjugate* points which is described in Chapter 5. Jacobi merely gave brief hints how to prove his results; detailed proofs were contributed by Delaunay, Spitzer, Hesse, and others (cf. Todhunter [1], pp. 243–332, and Goldstine [1], pp. 151–189, about the influence of Jacobi's paper on the development of the calculus of variations).

Jacobi still believed that a constant sign of the second variation would imply that the functional really assumes an extremum. This is false, as was noted by Scheeffer [3], p. 197, and also by Weierstrass; cf. 5,*1.1*. Both Scheeffer and Weierstrass discovered that one has to distinguish between weak minimizers (condition ($\mathcal{M}'$)) and strong minimizers (condition ($\mathcal{M}$)). Zermelo [1] introduced the notions of weak and strong neighbourhoods.

The classical calculus of variations was completed by Weierstrass. He not only made it a rigorous field of mathematics, complying to the modern standard of analysis, but he also added many important ideas to the field, thus establishing results of great strength.

Weierstrass lectured about his results at Berlin University over a period of 20 years; yet he never published them. Thus his ideas were mostly distributed by lecture notes of his students, and therefore they but slowly reached the mathematical public. The books of A. Kneser (1900) and Bolza (1904 and 1909) gave the first mathematical expositions of Weierstrass's work, whereas the original notes were only published in 1927,[10] much too late to be of any further influence upon the development of the variational calculus. Carathéodory[11] wrote about Weierstrass:

In his earlier work, prior to the year 1879, he succeeded in removing all the difficulties that were contained in the old investigations of Euler, Lagrange, Legendre, and Jacobi, simply by stating precisely and analyzing carefully the problems involved. In improving upon the work of these men he did several things of paramount importance ...:

(1) he showed the advantages of parametric representation;

(2) he pointed out the necessity of first defining in any treatment of a problem in the Calculus of Variations the class of curves in which the minimizing curve is to be sought, and of subsequently choosing the curves of variation so that they always belong to this class;

(3) he insisted upon the necessity of proving carefully a fact that had hitherto been assumed obvious, i.e., that the first variation does not always vanish unless the differential equation, which is now called the "Euler Equation", is satisfied at all points of the minimizing arc at which the direction of the tangent varies continuously;

(4) he made a very careful study of the second variation and proved for the first time that the condition $\delta^2 I > 0$ is sufficient for the existence of a weak minimum.

... The second part of the work of Weierstrass is directly related to his concept of a strong minimum.

---

[7] Actually, intuition for functions of more than one variable – and especially for variational problems – can be elusive. For instance, there are functions $f \in C^\infty(\mathbb{R}^2)$ with a unique critical point which, in addition, is a local minimizer, but which nevertheless have no global minimizer. An example of this kind is provided by

$$f(x, y) := x^3 - 3x + (e^y - x)^2.$$

[8] *Mémoire sur la manière de distinguer les maxima des minima dans le calcul des variations*, Mém. de l'acad. sci. Paris (1786) 1788, 7–37.

[9] *Zur Theorie der Variations-Rechnung und der Differential-Gleichungen*, Crelle's Journal f. d. reine u. angew. Math. **17**, 68–82 (1837); cf. Werke [3], Vol. 4, pp. 39–55.

[10] *Vorlesungen über Variationsrechnung*, bearbeitet von Rudolf Rothe, Math. Werke [1], Vol. 7, Leipzig, 1927.

[11] Ges. Math. Schriften [16], Vol. 5, pp. 343–347.

... Weierstrass found very early that it is essential to consider the strong minimum as well as the weak, but he became convinced during his research that the classical methods were inadequate for handling it. In 1879 he discovered his E-function and with it was able to establish conditions sufficient for the existence of a strong minimum.

2. The notion of *accessory differential equations* for the Euler equations of the second variation was introduced by von Escherich [1] in 1898. The names *accessory integral* and *accessory Lagrangian* as well as *accessory problem* are coined accordingly.

3. For simple integrals with a Lagrangian $F(x, z, p)$, the "necessary" condition $F_{pp} \geq 0$ and the "sufficient" condition $F_{pp} > 0$ were stated by Legendre in 1786. The generalization to multiple integrals was formulated by Hadamard [1] and [2], p. 252, for $n = N = 2$ and 3, respectively; the general case can, for instance, be found in Boerner [3].

4. The proof of the necessary condition of Weierstrass essentially follows Morrey [1], pp. 10–12. A somewhat different proof can be found in Klötzler [4], pp. 112–116; cf. also McShane [2], and Graves [2]. For $n = 1$, the reasoning can be simplified considerably.

Section 2

1. Amazingly, convex functionals do not play any role in the classical calculus of variations before the turn of the century, although they seem to be predestinated to assume a central place. One has to realize that only Minkowski recognized the notions of *convex set* and *convex function* to be central concepts in mathematics, although the concepts of a convex curve and a convex surface were well-known and often used in ancient times; then convexity meant "locally egg-shaped". Nowadays convex analysis has become one of the important fields in analysis. However, we postpone a more systematic discussion to Chapter 7 and later parts of this treatise.

In Section 2, we have collected a few "differentiable" convexity arguments, that appear scattered in the literature. This presentation is not overly systematic, and very likely the reader will find other instances of convexity reasoning if he glances at the rich literature. An introductory treatment of the calculus of variations on the basis of convexity is given in the textbook by J.L. Troutman [1].

2. A list of the variational problems treated by Euler was compiled by Carathéodory; it comprises 100 different problems, most of which are of the type

$$\int_a^b \omega(x, u)\sqrt{1 + |u'(x)|^2}\, dx,$$

cf. Carathéodory [16], Vol. 5, pp. 165–174, and Euler, Opera omnia [1], Ser. I, Vol. 24, pp. VIII–LXIII.

3. For *2.3* $\boxed{1}$–$\boxed{4}$ we refer the reader also to Carathéodory [10], pp. 305–313, and to Carathéodory [16], Vol. 5, pp. 146–148.

4. The idea described in *2.4* has very often been used. Usually it is applied to $\mathcal{F}$ instead of $\delta^2\mathcal{F}$, in order to prove boundedness of the functional $\mathcal{F}$ from below, or to obtain some kind of coercivity; cf. Bakelman [1], Moser [2], [3]; Emmer [1]; Giaquinta [1]; T. Aubin [1]; Brezis [1] for typical examples.

5. The convexity result for the functional

$$\mathscr{E}(u) = \int_\Omega \{|Du|^2 + \tfrac{4}{3}Hu\cdot(u_x \wedge u_y)\}\,dx\,dy$$

is due to Heinz; cf. Hildebrandt [3], pp. 112–113.

6. The notion of a *calibrator* is derived from the work of Harvey and Lawson [1]; instead of "calibrator" it might be more appealing to use the phrase "minimum verifier". A similar concept was also proposed by Ioffe and Tichomirov [1] ("K-functionals").

7. The calibrator used in *2.6* [4] was found by F.-H. Lin [1].

Chapter 5. Weak Minimizers and Jacobi Theory

If $\mathscr{F}(u)$ is a real-valued function of real variables $u \in \mathbb{R}^n$ which is of class C^2, then the positive definiteness of its Hessian matrix at a stationary point u is sufficient to guarantee that u is a relative minimizer. In other words, the assumption $D^2\mathscr{F}(u) > 0$ for the stationary point u implies that

$$\mathscr{F}(u) \le \mathscr{F}(v)$$

for all v in a sufficiently small neighbourhood of u in $\mathbb{R}^n$.

In contrast, the condition of positive definiteness of the second variation,

$$(1) \qquad \delta^2\mathscr{F}(u, \varphi) > 0 \quad \text{for all } \varphi \in C_0^1(\overline{\Omega}, \mathbb{R}^N) \text{ with } \varphi \neq 0,$$

of the variational integral $\mathscr{F}(u) = \int_\Omega F(x, u, Du)\, dx$ at a stationary point u is no longer sufficient to ensure that u is a strong or even a weak minimizer. An example demonstrating this fact will be provided in *1.1*. However, if an extremal u satisfies the stronger condition

$$(2) \qquad \tfrac{1}{2}\delta^2\mathscr{F}(u, \varphi) \ge \lambda \int_\Omega (|\varphi|^2 + |D\varphi|^2)\, dx \quad \text{for all } \varphi \in C_c^\infty(\Omega, \mathbb{R}^N)$$

and some $\lambda > 0$, then u is a weak extremal as we have seen in the Proposition of 4,*1.1*; however, u need not be a strong minimizer (cf. 4,*1.1* [1]).

Condition (2) will be the starting point for our treatment of Jacobi's theory of the second variation. Following an idea by H.A. Schwarz, we shall formulate Jacobi's theory in terms of an eigenvalue problem for the accessory integral or, equivalently, for the Jacobi operator. The line of reasoning runs as follows: We know already that any extremal u of $\mathscr{F}$ satisfying condition (2) is a weak minimizer of $\mathscr{F}$. Moreover, we infer from 4,*1.3* that (2) implies the strict Legendre–Hadamard condition for u. Conversely, if u satisfies the strict Legendre–Hadamard condition, then the Jacobi operator $\mathscr{J}_u$ with regard to $\mathscr{F}$ is strongly elliptic and, consequently, possesses a discrete spectrum of eigenvalues tending to infinity. The smallest eigenvalue is used to formulate necessary and sufficient conditions for the extremal u to be a weak minimizer.

For one-dimensional problems, the eigenvalue criteria can be expressed in form of the classical *conjugate point theory* which had its origin in the work of Legendre and Jacobi. In Section 2 we shall present an essentially self-contained exposition of this theory in the case $n = N = 1$ which can be read independently of the first section.

In Chapter 6 we shall see that the Jacobi theory plays an essential role for field constructions that lead to sufficient conditions for weak minimizers.

We finally mention that both the conjugate-point theory and the eigenvalue criteria are the historical roots of Morse theory.

1. Jacobi Theory: Necessary and Sufficient Conditions for Weak Minimizers Based on Eigenvalue Criteria for the Jacobi Operator

After an introductory subsection on weak minimizers in *1.2* we shall introduce Jacobi's operator associated with a variational integral $\mathcal{F}$ and a function u. We will show that Jacobi's operator coincides with the Euler operator of the accessory integral to $\mathcal{F}$. Finally, in *1.3* we describe necessary and sufficient eigenvalue criteria for weak minimizers.

We shall throughout assume that $F(x, z, p)$ is a Lagrangian of class C^2 on $\bar{\Omega} \times \mathbb{R}^N \times \mathbb{R}^{nN}$, where Ω is a bounded domain in $\mathbb{R}^n$. There are, of course, many examples where F is only of class C^2 on some subset $\mathcal{U}$ of $\mathbb{R}^n \times \mathbb{R}^N \times \mathbb{R}^{nN}$, or we may only have that F and F_p are of class C^1. We leave it to the reader to formulate corresponding results in those cases.

1.1. Remarks on Weak Minimizers

Recall that $u \in C^1(\bar{\Omega}, \mathbb{R}^N)$ is said to be a weak minimizer of the functional

$$\mathcal{F}(u) = \int_\Omega F(x, u(x), Du(x))\, dx$$

if there exists a weak neighbourhood $\mathcal{N}_\varepsilon'(u)$ of u ($= C^1$-neighbourhood) such that

$$\mathcal{F}(u) \le \mathcal{F}(v) \quad \text{for all } v \in \mathcal{N}_\varepsilon'(u).$$

Any weak minimizer $u \in C^1(\bar{\Omega}, \mathbb{R}^N)$ is necessarily a weak extremal of $\mathcal{F}$, i.e.,

$$(1) \qquad \delta\mathcal{F}(u, \varphi) = \int_\Omega \{F_z(x, u, Du)\cdot\varphi + F_p(x, u, Du)\cdot D\varphi\}\, dx = 0$$

$$\text{for all } \varphi \in C_c^\infty(\Omega, \mathbb{R}^N).$$

If the weak extremal u also is of class $C^2(\Omega, \mathbb{R}^N)$, then u satisfies the Euler equations

$$L_F(u) = 0,$$

i.e., u is an extremal.

In the Proposition of 4,*1.1* we have established the following result:

Let u be a weak extremal of $\mathcal{F}$ satisfying

$$\tfrac{1}{2}\delta^2\mathcal{F}(u,\varphi) \ge \lambda \int_\Omega \{|\varphi|^2 + |D\varphi|^2\}\,dx \quad \textit{for all } \varphi \in C_c^\infty(\Omega, \mathbb{R}^N) \tag{2}$$

and some $\lambda > 0$. Then u is a strict weak minimizer of $\mathcal{F}$.

We infer by approximation that (2) implies

$$\tfrac{1}{2}\delta^2\mathcal{F}(u,\varphi) \ge \lambda \int_\Omega \{|\varphi|^2 + |D\varphi|^2\}\,dx \quad \text{for all } \varphi \in C_0^1(\overline{\Omega}, \mathbb{R}^N). \tag{2'}$$

The following example shows that the assumption

$$\delta^2\mathcal{F}(u,\varphi) > 0 \quad \text{for all } \varphi \in C_0^1(\overline{\Omega}, \mathbb{R}^N), \quad \varphi \ne 0,$$

does not ensure that a weak extremal u is a weak minimizer.

[1] *Scheeffer's example.* Let us consider the variational integral

$$\mathcal{F}(v) = \int_{-1}^{1} [x^2 v'(x)^2 + x v'(x)^3]\,dx.$$

We obtain

$$\delta\mathcal{F}(v,\varphi) = \int_{-1}^{1} [2x^2 v' + 3x(v')^2]\varphi'\,dx$$

and

$$\delta^2\mathcal{F}(v,\varphi) = \int_{-1}^{1} [2x^2 + 6xv'](\varphi')^2\,dx.$$

Thus the function $u(x) \equiv 0$ is an extremal of $\mathcal{F}$, and its second variation satisfies

$$\delta^2\mathcal{F}(u,\varphi) = 2\int_{-1}^{1} x^2(\varphi'(x))^2\,dx > 0 \quad \text{for all } \varphi \in C_0^1(\overline{\Omega}) \text{ with } \varphi(x) \not\equiv 0.$$

We shall now show that $u = 0$ is not a weak minimizer. To this end we consider the family of functions

$$v_{\varepsilon,h}(x) := \begin{cases} \varepsilon(h+x) & \text{for } -h < x \le 0, \\ \varepsilon(h-x) & \text{for } 0 \le x < h, \\ 0 & \text{otherwise}. \end{cases}$$

Then

$$\mathcal{F}(v_{\varepsilon,h}) = \frac{2h^2\varepsilon^2}{3}(h - \tfrac{3}{2}\varepsilon),$$

and, if we choose $h = \frac{3}{4}\varepsilon$ and round off the corners, it is not difficult to see that the modified functions $\tilde{v}_{\varepsilon,(3/4)\varepsilon}$ converge in C^1 to $u(x) \equiv 0$, whereas

$$\mathcal{F}(\tilde{v}_{\varepsilon,(3/4)\varepsilon}) < 0 = \mathcal{F}(0).$$

Hence the example shows that the condtions

$$\delta\mathcal{F}(u,\varphi) = 0 \quad \text{and} \quad \delta^2\mathcal{F}(u,\varphi) > 0$$

for all $\varphi \in C_0^1(\overline{\Omega}, \mathbb{R}^N)$ with $\varphi \ne 0$ are not sufficient to ensure that u is a weak minimizer.

Let us also recall that (2) implies that *strict Legendre–Hadamard* condition

(3) $\frac{1}{2}F_{p_\alpha^i p_\beta^k}(x, u(x), Du(x))\xi^i\xi^k\eta_\alpha\eta_\beta \ge \lambda|\xi|^2|\eta|^2$ for $x \in \bar{\Omega}$, $\xi \in \mathbb{R}^N$, $\eta \in \mathbb{R}^n$,

cf. 4,*1.1* and 4,*1.3*.

1.2. Accessory Integral and Jacobi Operator

Using linear variations $\psi(\varepsilon) = u + \varepsilon\varphi$, $|\varepsilon| < \varepsilon_0$, of a given function u, we had defined the first and second variation of $\mathscr{F}$ at u in direction of φ by

$$\delta\mathscr{F}(u, \varphi) = \frac{d}{d\varepsilon}\mathscr{F}(u + \varepsilon\varphi)\bigg|_{\varepsilon=0}, \qquad \delta^2\mathscr{F}(u, \varphi) = \frac{d^2}{d\varepsilon^2}\mathscr{F}(u + \varepsilon\varphi)\bigg|_{\varepsilon=0} = 2\mathscr{Q}(\varphi).$$

Now we want to see how $\mathscr{F}$ changes along paths described by *nonlinear* (or *general*) *variations* of a given function $u \in C^1(\bar{\Omega}, \mathbb{R}^N)$; see also 1,*2.1*. This is a function $\psi(x, t)$ with values in $\mathbb{R}^N$, defined for $(x, t) \in \bar{\Omega} \times (-t_0, t_0)$, $t_0 > 0$, which satisfies

$$\psi(x, 0) = u(x) \quad \text{for all } x \in \bar{\Omega}.$$

We also assume that, for every $t \in (-t_0, t_0)$, the function $\psi(\cdot, t)$ is F-admissible (cf. 4,*1.1*), and that the derivatives ψ_t and ψ_{tt} exist and are of class $C^1(\bar{\Omega}, \mathbb{R}^N)$. We set

$$\varphi(x) := \psi_t(x, 0) \quad \text{and} \quad \zeta(x) := \psi_{tt}(x, 0). \tag{1}$$

The chain rule implies

$$\frac{\partial}{\partial t}F(x, \psi(x, t), D\psi(x, t)) = \delta F(\psi, \psi_t)(x), \qquad D = D_x,$$

$$\frac{\partial^2}{\partial t^2}F(x, \psi(x, t), D\psi(x, t)) = \delta F(\psi, \psi_{tt})(x) + \delta^2 F(\psi, \psi_t)(x),$$

whence

$$\frac{d}{dt}\mathscr{F}(\psi) = \delta\mathscr{F}(\psi, \psi_t), \tag{2}$$

$$\frac{d^2}{dt^2}\mathscr{F}(\psi) = \delta\mathscr{F}(\psi, \psi_{tt}) + \delta^2\mathscr{F}(\psi, \psi_t) \tag{3}$$

and in particular

$$\frac{d}{dt}\mathscr{F}(\psi)\bigg|_{t=0} = \delta\mathscr{F}(u, \varphi), \tag{2'}$$

$$\frac{d^2}{dt^2}\mathscr{F}(\psi)\bigg|_{t=0} = \delta^2\mathscr{F}(u, \varphi) + \delta\mathscr{F}(u, \zeta) = 2\mathscr{Q}(\varphi) + \delta\mathscr{F}(u, \zeta). \tag{3'}$$

In other words, the first variation of $\mathscr{F}$ remains unchanged if we replace linear by nonlinear variations of u, whereas in the second variation of $\mathscr{F}$ there appears the additional term $\delta\mathscr{F}(u, \zeta)$ which vanishes if u is a weak extremal and $\psi(t)$ is a variation of u with fixed boundary values. Thus we have proved:

Lemma. *If $\psi(x, t)$ is a general variation of u with*

$$\psi(x, 0) = u(x), \qquad \psi_t(x, 0) = \varphi(x), \qquad \psi_{tt}(x, 0) = \zeta(x),$$

such that

$$\delta\mathscr{F}(u, \zeta) = 0$$

holds, then we have

$$\frac{d^2}{dt^2}\mathscr{F}(\psi)\bigg|_{t=0} = \delta^2\mathscr{F}(u, \varphi) = 2\mathscr{Q}(\varphi). \tag{4}$$

Now we turn to the definition of the Jacobi operator. For this purpose we suppose that F is of class C^3. Moreover we consider an arbitrary nonlinear variation $\psi(t)$ of u with

$$\psi(x, 0) = u(x), \quad \psi_t(x, 0) = \varphi(x), \quad \psi_{tt}(x, 0) = \zeta(x), \qquad x \in \bar{\Omega},$$

where, in addition to the previously required assumptions, u, φ, ζ, $\psi(t)$, $\psi_t(t)$, $\psi_{tt}(t)$ are supposed to be functions of $x \in \bar{\Omega}$ which are of class C^2.

Evaluating the Euler operator L_F of F at $\psi = \psi(t)$, we obtain that

$$L_F(\psi) = F_z(x, \psi, D\psi) - D_\alpha F_{p_\alpha}(x, \psi, D\psi),$$

whence

$$\frac{\partial}{\partial t} L_F(\psi) = F_{zz}(\ldots)\cdot\psi_t + F_{zp_\beta}(\ldots)\cdot D_\beta\psi_t$$
$$- D_\alpha\{F_{p_\alpha z}(\ldots)\cdot\psi_t + F_{p_\alpha p_\beta}(\ldots)\cdot D_\beta\psi_t\}$$

and therefore:

$$\frac{\partial}{\partial t} L_F(\psi)|_{t=0} = F_{zz}(x, u, Du)\cdot\varphi + F_{zp_\beta}(x, u, Du)\cdot D_\beta\varphi$$
$$- D_\alpha\{F_{p_\alpha z}(x, u, Du)\cdot\varphi + F_{p_\alpha p_\beta}(\ldots)\cdot D_\beta\varphi\}.$$

This shows that $\frac{\partial}{\partial t} L_F(\psi)\Big|_{t=0}$ depends only on $u(x) = \psi(x, 0)$ and on $\varphi(x) = \frac{\partial\psi}{\partial t}(x, 0)$, and on no other data of $\psi(t)$. Therefore we can, for fixed u, consider this expression as a differential operator

$$\mathscr{J}_u : C^2(\bar{\Omega}, \mathbb{R}^N) \to C^0(\bar{\Omega}, \mathbb{R}^N)$$

defined by

$$(5)\qquad \mathcal{J}_u\varphi := \frac{\partial}{\partial t} L_F(\psi(t))|_{t=0} \quad \text{if } \psi(0) = u,\ \frac{\partial\psi}{\partial t}(0) = \varphi .$$

This linear second order differential operator is called the *Jacobi operator* corresponding to F and u. It is the linearization of the Euler operator L_F at u. Clearly the Jacobi operator is well defined because, for given u and $\varphi \in C^2(\bar{\Omega}, \mathbb{R}^N)$, the linear variation $\psi = u + t\varphi$ is a variation with the required properties $\psi(0) = u$ and $\psi_t(0) = \varphi$. From the previous calculation we infer that

$$(6)\qquad \mathcal{J}_u\varphi = F_{zz}\cdot\varphi + F_{zp}\cdot D\varphi - \operatorname{div}\{F_{pz}\cdot\varphi + F_{pp}\cdot D\varphi\}$$

holds or, in local coordinates,

$$(6')\qquad (\mathcal{J}_u\varphi)_i = F_{z^iz^k}\varphi^k + F_{z^ip^k_\beta}D_\beta\varphi^k - D_\alpha\{F_{p^i_\alpha z^k}\varphi^k + F_{p^i_\alpha p^k_\beta}D_\beta\varphi^k\}$$

for the i-th component of the covector $\mathcal{J}_u\varphi$. In these formulas, we have to choose the arguments of F_{zz}, F_{zp}, F_{pz}, F_{pp} as $(x, u(x), Du(x))$.

Comparing (6′) with the formulas (9)–(13) of 4,*1.2*, we obtain the following result:

Proposition. *The Jacobi operator $\mathcal{J}_u$ coincides with the Euler operator L_Q of the accessory Lagrangian Q of F at u, that is*

$$(7)\qquad \mathcal{J}_u\varphi = L_Q(\varphi) \quad \textit{for all } \varphi \in C^2(\bar{\Omega}, \mathbb{R}^N).$$

Let us give *another proof* which is coordinate-free, assuming that F is of class C^3.

Consider an arbitrary test function $\eta \in C_c^\infty(\Omega, \mathbb{R}^N)$, and let $\psi(t)$ be a general variation of u with the afore-stated assumptions. Then we infer from

$$\delta F(\psi, t\eta) = t\,\delta F(\psi, \eta)$$

that

$$\frac{\partial^2}{\partial t^2}\delta F(\psi, t\eta) = 2\frac{\partial}{\partial t}\delta F(\psi, \eta) + t\frac{\partial^2}{\partial t^2}\delta F(\psi, \eta)$$

holds, whence we conclude that

$$\frac{1}{2}\frac{\partial^2}{\partial t^2}\delta\mathcal{F}(\psi, t\eta)|_{t=0} = \frac{\partial}{\partial t}\delta\mathcal{F}(\psi, \eta)|_{t=0} .$$

Moreover, from

$$\delta\mathcal{F}(\psi, \eta) = \int_\Omega L_F(\psi)\cdot\eta\, dx ,$$

it follows that

$$\frac{\partial}{\partial t}\delta\mathcal{F}(\psi, \eta)|_{t=0} = \int_\Omega \mathcal{J}_u\varphi\cdot\eta\, dx .$$

We also note that the function $\Phi(t, \varepsilon) = \mathcal{F}(\psi(t) + \varepsilon t\eta)$ is of class C^3, and therefore we obtain by taking (3′) into account that

$$\frac{1}{2}\frac{\partial^2}{\partial t^2}\delta\mathscr{F}(\psi, t\eta)\bigg|_{t=0} = \frac{1}{2}\frac{\partial^2}{\partial t^2}\left\{\left[\frac{\partial}{\partial\varepsilon}\mathscr{F}(\psi + t\eta)\right]_{\varepsilon=0}\right\}\bigg|_{t=0} = \frac{1}{2}\frac{\partial^2}{\partial t^2}\Phi_\varepsilon(t, 0)\bigg|_{t=0} = \frac{1}{2}\frac{\partial}{\partial\varepsilon}\Phi_{tt}(0, \varepsilon)\bigg|_{\varepsilon=0}$$

$$= \frac{1}{2}\left\{\frac{\partial}{\partial\varepsilon}\left[\frac{\partial^2}{\partial t^2}\mathscr{F}(\psi + \varepsilon t\eta)\right]_{t=0}\right\}_{\varepsilon=0} = \frac{1}{2}\left\{\frac{\partial}{\partial\varepsilon}[\delta^2\mathscr{F}(u, \varphi + \varepsilon\eta) + \delta\mathscr{F}(u, \zeta)]\right\}_{\varepsilon=0}$$

$$= \frac{\partial}{\partial\varepsilon}\mathscr{Q}(\varphi + \varepsilon\eta)\bigg|_{\varepsilon=0} = \delta\mathscr{Q}(\varphi, \eta) = \int_\Omega L_Q(\varphi)\cdot\eta\, dx.$$

Since we had already shown that

$$\frac{1}{2}\frac{\partial^2}{\partial t^2}\delta\mathscr{F}(\psi, t\eta)\bigg|_{t=0} = \int_\Omega \mathscr{J}_u\varphi\cdot\eta\, dx,$$

we infer that

$$\int_\Omega \{J_u\varphi - L_Q(\varphi)\}\cdot\eta\, dx = 0$$

for all $\eta \in C_0^\infty(\Omega, \mathbb{R}^N)$, and this implies (7).

Corollary. *If $\psi(t)$, $|t| < t_0$, is a one-parameter family of solutions of $L_F(\psi) = 0$ with $\psi(0) = u$, then $\varphi = \frac{\partial\psi}{\partial t}(0)$ is a vector field along the extremal u which satisfies the equation*

(8) $$\mathscr{J}_u\varphi = 0 \quad \text{on } \Omega.$$

The linear second order differential equation (8) is called the *Jacobi equation* corresponding to F and u, and a solution φ is denoted as a *Jacobi field* along u.

Since $\mathscr{J}_u$ is a linear operator, Jacobi fields span a linear subspace of $C^1(\bar{\Omega}, \mathbb{R}^N)$. If $n = 1$, Jacobi fields are solutions of a linear system of second order ordinary differential equations. If the Hessian matrix $F_{pp}(x, u, Du)$ is non-singular, we can prescribe the initial values of φ and φ' in an arbitrary way, and it follows that the space of Jacobi fields has dimension $2N$.

Consider particularly the case $n = N = 1$, and set

$$a(x) := F_{pp}(x, u(x), Du(x)),$$
$$b(x) := F_{zp}(x, u(x), Du(x)),$$
$$c(x) := F_{zz}(x, u(x), Du(x)).$$

Then the accessory integral is given by

(9) $$\mathscr{Q}(\varphi) = \int_\Omega \frac{1}{2}(a\varphi'^2 + 2b\varphi\varphi' + c\varphi^2)\, dx$$

and the Jacobi operator has the form

(10) $$\mathscr{J}_u\varphi = -(a\varphi')' + (c - b')\varphi.$$

Jacobi fields are solutions φ of the equation

$$(11) \qquad -(a\varphi')' + (c - b')\varphi = 0.$$

1.3. Necessary and Sufficient Eigenvalue Criteria for Weak Minima

Let Ω be a bounded domain in $\mathbb{R}^n$ with $\partial\Omega \in C^{2,\alpha}$, $0 < \alpha < 1$. Let $F(x, z, p)$ be a Lagrangian of class $C^{3,\alpha}$ defined on $\overline{\Omega} \times \mathbb{R}^N \times \mathbb{R}^{nN}$, and assume that u is of class $C^{3,\alpha}(\overline{\Omega}, \mathbb{R}^N)$ and represents an extremal of

$$\mathscr{F}(v) = \int_\Omega F(x, v, Dv)\, dx,$$

that is,

$$L_F(u) = 0 \quad \text{on } \Omega.$$

We know that the Legendre–Hadamard condition

$$F_{p_\alpha^i p_\beta^k}(x, u(x), Du(x))\, \xi^i \xi^k \eta_\alpha \eta_\beta \geq 0$$

is a necessary condition for u to be a weak minimizer, and the condition

$$(1) \qquad \delta^2 \mathscr{F}(u, \varphi) \geq 2\lambda \int_\Omega \{|\varphi|^2 + |D\varphi|^2\}\, dx \quad \text{for all } \varphi \in C_0^1(\overline{\Omega}, \mathbb{R}^N)$$

is sufficient for u to be a weak minimizer ($\lambda > 0$). Although (1) implies the strict Legendre–Hadamard condition

$$(2) \qquad \tfrac{1}{2} F_{p_\alpha^i p_\beta^k}(x, u(x), Du(x))\, \xi^i \xi^k \eta_\alpha \eta_\beta \geq \lambda |\xi|^2 |\eta|^2,$$

we cannot conversely infer that (2) yields (1).

The main goal of the following considerations is to find out what hypotheses have to be added to (2) to be able to deduce (1).

In order to analyze the relations between the local condition (2) and the global assumption (1), we shall investigate the *Jacobi eigenvalue problem*

$$(3) \qquad \mathscr{J} v = \lambda v \quad \text{in } \Omega, \qquad v = 0 \quad \text{on } \partial\Omega,$$

for the Jacobi operator $\mathscr{J} := \mathscr{J}_u$. Recall that according to the terminology introduced in 4,*1.2*, the Jacobi operator $\mathscr{J}_u$ is strongly elliptic if the extremal u satisfies the strict Legendre–Hadamard condition (2).

Let us list a few results for this eigenvalue problem which are more or less explicitely contained in the literature. We make the following

General Hypothesis. *Suppose that the Jacobi operator is strongly elliptic on* $\overline{\Omega}$.

This hypothesis is assumed for the rest of this subsection. It will be essential for our considerations.

Theorem 1. (*i*) *There exists a sequence* $\{\lambda_k, v_k\}$ *of real numbers* λ_k *and of functions* $v_k \in C^{2,\alpha}(\bar{\Omega}, \mathbb{R}^N)$ *which satisfy*

$$\mathscr{J} v_k = \lambda_k v_k \quad \text{in } \Omega, \qquad v_k = 0 \quad \text{on } \partial\Omega$$

and

$$\int_\Omega v_k \cdot v_l \, dx = \delta_{kl}$$

(where δ_{kl} *denotes the Kronecker symbol, and no summation is assumed).*

(*ii*) *We, moreover, have* $\lambda_1 \le \lambda_2 \le \lambda_3 \le \ldots$ *and* $\lim_{k\to\infty} \lambda_k = \infty$. *Thus every* $\lambda \in \mathbb{R}$ *can appear at most finitely often in the sequence* $\{\lambda_k\}$.

(*iii*) *If* λ, v *is a solution of* (3) *with* $v \neq 0$, *then the eigenvalue* λ *appears in the sequence* $\{\lambda_k\}$, *and the corresponding eigenfunction* v *is a linear combination of the eigenvectors* v_k *with* $\lambda_k = \lambda$.

(*iv*) *From* (*i*) *we infer that* $\lambda_k = 2\mathscr{Q}(v_k)$. *Moreover, the smallest eigenvalue* λ_1 *is characterized by the minimum property*

$$\lambda_1 = 2\mathscr{Q}(v_1) = \min\left\{2\mathscr{Q}(v) : v \in C_0^1(\bar{\Omega}, \mathbb{R}^N) \quad \text{and} \quad \int_\Omega |v|^2 \, dx = 1\right\}.$$

(*v*) *There exists a number* $\mu > 0$ *with the property that*

$$\mathscr{Q}(v) \ge \mu \int_\Omega \{|v|^2 + |Dv|^2\} \, dx \quad \text{for all } v \in C_0^1(\bar{\Omega}, \mathbb{R}^N)$$

holds if and only if the smallest eigenvalue λ_1 *of the Jacobi operator* $\mathscr{J}$ *is positive.*

These properties of the Jacobi operator imply the following necessary and sufficient conditions, stated as Theorems 2 and 3.

Theorem 2. *If the smallest eigenvalue* λ_1 *of* $\mathscr{J}_u$ *satisfies* $\lambda_1 < 0$, *then the extremal* u *does not furnish a weak minimum for* $\mathscr{F}$.

Proof. By property (iv) of Theorem 1, there is some $\varphi \in C_0^1(\bar{\Omega}, \mathbb{R}^N)$ such that

$$\delta^2\mathscr{F}(u, \varphi) = \mathscr{Q}(\varphi) = 2\lambda_1 < 0.$$

On the other hand, if u were a weak minimizer of $\mathscr{F}$, we would have $\delta^2\mathscr{F}(u, \varphi) \ge 0$ for all $\varphi \in C_0^1(\bar{\Omega}, \mathbb{R}^N)$, a contradiction. □

Theorem 3. *If the smallest eigenvalue* λ_1 *of* $\mathscr{J}_u$ *is positive, then the extremal* u *is a strict weak minimizer of* $\mathscr{F}$.

Proof. The assertion follows from property (v) of Theorem 1 in conjunction with the Proposition of 4,*1.1* that we have restated in *1.1.* □

We now list two more properties of the Jacobi operator $\mathcal{J} = \mathcal{J}_u$.

Consider the Jacobi eigenvalue problem

$$\mathcal{J}v = \lambda v \quad \text{in } \Omega', \qquad v = 0 \quad \text{on } \partial\Omega' \tag{3'}$$

for nonvoid smoothly bounded subdomains Ω' of Ω. Then the corresponding statements to Theorem 1 hold if everywhere Ω is replaced by Ω'. Let $\lambda_k(\Omega')$ be the sequence of eigenvalues of (3') ordered by size:

$$\lambda_1(\Omega') \le \lambda_2(\Omega') \le \lambda_3(\Omega') \le \dots .$$

The dependence of the eigenvalues λ_k on Ω' is in part described by the following results:

Supplement to Theorem 1. (*vi*) *If Ω', Ω'' are subdomains of Ω with $\Omega' \subset \Omega''$, then the inequality*

$$\lambda_k(\Omega'') \le \lambda_k(\Omega')$$

holds for all $k = 1, 2, \dots$. We even have

$$\lambda_k(\Omega'') < \lambda_k(\Omega'),$$

if the interior of $\Omega'' - \Omega'$ is nonvoid and if the Jacobi operator $\mathcal{J}$ fulfils the "unique continuation principle".

(*vii*) *There exists a number $\delta > 0$ such that $\lambda_1(\Omega') > 0$ holds for all subdomains Ω' of Ω with diameter*[1] *less than δ.*

Property (vii) is an immediate consequence of *Gårding*'s and *Poincaré*'s *inequalities*, whereas (vi) follows from the minimum characterization of the eigenvalues in conjunction with the *unique continuation principle.*

On says that a differential operator L on Ω satisfies the *unique continuation principle* if any solution u of

$$L(u) = 0 \quad \text{in } \Omega$$

must vanish identically as soon as it vanishes on some ball contained in Ω, since Ω is connected.

This principle holds for linear elliptic differential operators with real analytic coefficients as well as for nonlinear elliptic operators defined by analytic expressions since, by a fundamental regularity theorem,[2] their solutions are necessarily real analytic.

The unique continuation principle holds for homogeneous second order

---

[1] One can even prove that $\lambda_1(\Omega') > 0$ holds if the measure of Ω' is sufficiently small.

[2] Cf. Morrey [1], Chapter 6.

elliptic operators in the scalar case ($N = 1$), and for homogeneous systems with simple characteristics.[3]

The Jacobi operator $\mathcal{J}_u$ is a linear second order operator on Ω which was supposed to be strongly elliptic. Thus $\mathcal{J}_u$ has the unique continuation property in the scalar case ($N = 1$), and for systems in the case of real analytic Lagrangians F ($N > 1$).

We shall discuss eigenvalue problems for elliptic operators and the unique continuation property in a separate treatise.

For one independent variable ($n = 1$), the unique continuation property of $\mathcal{J}_u$ holds as soon as the extremal u satisfies the *strict Legendre condition*

$$\frac{1}{2}F_{p^i p^k}(x, u(x), Du(x))\, \xi^i \xi^k \geq \lambda |\xi|^2 \tag{4}$$

on Ω for some $\lambda > 0$. This follows from the fact that $\varphi(x) = 0$ is the only solution of the initial value problem

$$\mathcal{J}_u \varphi = 0, \qquad \varphi(x_0) = 0, \qquad \varphi'(x_0) = 0,$$

for any $x_0 \in \Omega \subset \mathbb{R}$.

The next result is a "local version" of Theorem 3 and states that every extremal satisfying the strict Legendre–Hadamard condition (2) is a weak minimizer with respect to perturbations with sufficiently small support. In other words: Adding "small bumps" to u will increase the value of $\mathcal{F}$.

Theorem 4. *There exists a number $\delta > 0$ such that, for every $\Omega' \subset\subset \Omega$ with* diam $\Omega' < \delta$, *the extremal u is a weak minimizer of $\mathcal{F}$ in the class $\mathcal{C}$ defined by*

$$\mathcal{C} := \{v \in C^1(\overline{\Omega}, \mathbb{R}^N): \operatorname{supp}(v - u) \subset \Omega' \subset\subset \Omega\}.$$

The *proof* follows at once from Property (vii) of the "supplement to Theorem 1" in conjunction with Theorem 3.

Finally we prove a necessary condition for a weak minimizer which strengthens Theorem 2.

Theorem 5. *Suppose that, for some nonempty subdomain Ω' of Ω with* $\operatorname{int}(\Omega - \Omega') \neq \emptyset$, *there exists a nontrivial Jacobi field $v \in C^2(\overline{\Omega}, \mathbb{R}^N)$ with* $v|_{\partial\Omega'} = 0$. *Assume also that $\mathcal{J}_u$ satisfies the unique continuation principle. Then u is not a weak minimizer of $\mathcal{F}$.*

Proof. Our assumption implies that zero is an eigenvalue for the Jacobi operator $\mathcal{J}_u$ on Ω' whence, by the supplement to Theorem 1, the smallest eigenvalue of $\mathcal{J}_u$ on Ω is strictly negative. □

[3] Cf. Leis [1], pp. 64–69, 217–219; Hörmander [1], pp. 224–229, and [2], Vol. III, Section 17.2, Vol. IV, Sections 28.1–28.4.

The previous discussion can be summarized and envisioned in the following way.

Consider a family $\{\Omega_t\}_{0<t\le T}$ of expanding domains Ω_t, i.e.,

$$\Omega_t \subsetneqq \Omega_{t'} \quad \text{for } 0 < t < t',$$

which shrinks to some fixed point x_0 as $t \to +0$, in the sense that

$$\lim_{t\to+0} \operatorname{diam} \Omega_t = 0 \quad \text{and} \quad x_0 \in \Omega_t \quad \text{for all } t > 0.$$

Assume also that the Jacobi operator $\mathscr{J}$ has the unique continuation property. Then either

(i) $\lambda_1(\Omega_t) > 0$ for all $t \in (0, T]$,

or

(ii) there is a first positive number $t_0 \le T$ such that $\lambda_1(\Omega_{t_0}) = 0$.

In case (i) the restriction $u|_{\Omega_t}$ is a weak minimizer for all $t \in (0, T)$, whereas in case (ii) the function $u|_{\Omega_t}$ is a weak minimizer for all $t \in (0, t_0)$, but not for any $t > t_0$.

Let us now consider the particular case $n = 1$ where $\mathscr{J}_u$ reduces to a linear ordinary differential operator of second order. Suppose that Ω is the bounded interval (x_1, x_2), and let $\xi \in [x_1, x_2)$. Following Jacobi, a value $\xi^* \in (\xi, x_2]$ is said to be a *conjugate value to* ξ (with respect to the extremal u) if there exists a nontrivial Jacobi field v on $[x_1, x_2]$ which vanishes both at ξ and ξ^*:

$$\mathscr{J}_u v = 0 \quad \text{on } (x_1, x_2), \qquad v(\xi) = 0, \qquad v(\xi^*) = 0.$$

The point $(\xi^*, u(\xi^*))$ on the graph of the extremal u is called a *conjugate point to* $(\xi, u(\xi))$.

Since $\mathscr{J}_u$ satisfies the unique continuation principle[4] if u fulfils the strict Legendre condition (4) which we have throughout supposed, we obtain the following corollaries of the Theorems 5 and 4:

Corollary 1. *If u is a weak minimizer of $\mathscr{F}$ and $\xi \in [x_1, x_2)$, then there exists no conjugate value ξ^* to ξ with $\xi^* \in (\xi, b)$.*

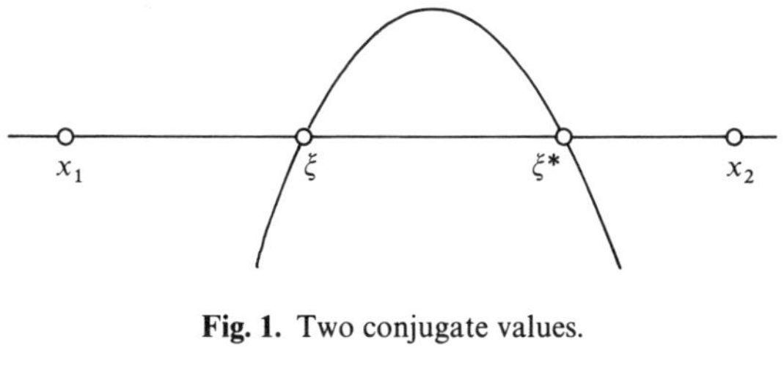

Fig. 1. Two conjugate values.

[4]This is a consequence of the unique solvability of the Cauchy problem for a regular system of ordinary differential equations.

Corollary 2. *For every $\xi \in [x_1, x_2)$, there exists a number ξ' with $\xi < \xi' \le x_2$ such that the interval $(\xi, \xi']$ contains no conjugate value to ξ.*

Consequently, if the interval $(\xi, x_2]$ contains at all a conjugate value to ξ, *there has to be a smallest conjugate value ξ^* with $\xi^* > \xi$. We call ξ^* the next conjugate value to ξ.*

2. Jacobi Theory for One-Dimensional Problems in One Unknown Function

In *1.3*, we have developed the general Jacobi theory relying on the eigenvalue theory for strongly elliptic operators. Although this approach has allowed us to treat the Jacobi theory for all dimensions in a uniform way, a separate inspection of the one-dimensional case might be welcome. Here we shall treat the simplest case $n = N = 1$. This gives us a chance to have a look at the original ideas of Legendre and Jacobi. We shall provide new proofs of all results, independently of any general theory except for a few instances when we would just have to repeat the computations of *1.1–1.3*. The case $n = 1$, $N \ge 1$ will be treated in Chapter 6.

First let us fix the assumptions that are stipulated for the following.

Let G be some domain in $\mathbb{R} \times \mathbb{R}$, and let $F(x, z, p)$ be some Lagrangian of class C^3 which is defined for $(x, z, p) \in G \times \mathbb{R}$, i.e., $n = N = 1$.

Denote by $z = u(x)$, $x \in I = [x_1, x_2]$, a function of class C^3, the graph of which is contained in G, and which is an extremal of

$$\mathscr{F}(u) = \int_{x_1}^{x_2} F(x, u, u')\, dx .$$

It is useful to assume that the extremal can be extended as an extremal of class C^3 to some slightly larger interval $I_0 = (x_1 - \delta_0, x_2 + \delta_0)$, $\delta_0 > 0$. This is no restriction of generality if $F_{pp}(x, u(x), u'(x)) \ne 0$ holds for $x = x_1$ and $x = x_2$ because, in this case, the existence theorem for the Cauchy problem of ordinary differential equations can be used to slightly extend the extremal to both sides of the interval I.

2.1. The Lemmata of Legendre and Jacobi

Recall that the accessory integral $\mathscr{Q}(v)$ corresponding to u and F is given by

$$\mathscr{Q}(v) = \int_{x_1}^{x_2} (av'^2 + 2bvv' + cv^2)\, dx ,$$

where

$$a(x) = \tfrac{1}{2}F_{pp}(x, u(x), u'(x)),$$
$$b(x) = \tfrac{1}{2}F_{zp}(x, u(x), u'(x)),$$
$$c(x) = \tfrac{1}{2}F_{zz}(x, u(x), u'(x)).$$

Its Euler equation is the so-called Jacobi equation

$$(1) \qquad -(av')' + (c - b')v = 0 \quad \text{on } (x_1, x_2),$$

the C^2-solutions v of which are called *Jacobi fields* (along u), and the Jacobi operator $\mathcal{J}_u$ is defined by

$$\mathcal{J}_u v = -(av')' + (c - b')v$$

(cf. formulas (9) and (10) of *1.2*).

Legendre's original idea in treating the accessory integral consists in adding a null Lagrangian

$$G(x, z, p) = 2zpw(x) + z^2w'(x)$$

to the integrand

$$Q(x, z, p) = a(x)p^2 + 2b(x)zp + c(x)z^2$$

of the accessory integral $\mathcal{Q}(v)$ such that the modified Lagrangian becomes a square, except for the factor $a(x)$:

$$Q + G = a\cdot(\ldots)^2.$$

This would give valuable information on $\mathcal{Q}$ as well as on the Jacobi equation since adding a null Lagrangian neither changes the Euler equation of $\mathcal{Q}$ nor the value of $\mathcal{Q}$ among functions with fixed boundary values on ∂I. Clearly, for any $\varphi, w \in C^1(I)$, we obtain

$$\begin{aligned}\mathcal{G}(\varphi) &= \int_{x_1}^{x_2} G(x, \varphi, \varphi')\,dx = \int_{x_1}^{x_2} (2\varphi\varphi' w + \varphi^2 w')\,dx \\ &= \int_{x_1}^{x_2} \frac{d}{dx}(\varphi^2 w)\,dx = [\varphi^2 w]_{x_1}^{x_2},\end{aligned}$$

whence it follows that G is a null Lagrangian, and $\mathcal{G}(\varphi) = 0$ for $\varphi \in C_0^1(I)$. Then we infer that

$$\mathcal{Q}(\varphi) = \int_{x_1}^{x_2} \{a\varphi'^2 + 2(b + w)\varphi\varphi' + (c + w')\varphi^2\}\,dx \quad \text{for all } \varphi \in C_0^1(I).$$

If the discriminant $\alpha\gamma - \beta^2$ of a quadratic form $\alpha\xi^2 + 2\beta\xi\eta + \gamma\eta^2$ vanishes, we can write

$$\alpha\xi^2 + 2\beta\xi + \gamma\eta^2 = \alpha\left(\xi + \frac{\beta}{\alpha}\eta\right)^2$$

provided that $\alpha \neq 0$. Hence, if we can determine $w \in C^1(I)$ in such a way that

$$(b + w)^2 - a(c + w') = 0 \tag{2}$$

holds, the accessory integral can be transformed into

$$\mathcal{Q}(\varphi) = \int_{x_1}^{x_2} a\left(\varphi' + \frac{b + w}{a}\varphi\right)^2 dx \quad \text{for } \varphi \in C_0^1(I),$$

provided that $a(x) \neq 0$. Thus we have derived

Legendre's lemma. *Let $w \in C^1(I)$ be a function which satisfies (2), and let $\varphi \in C_0^1(I)$ be a function such that $a(x) \neq 0$ holds on* supp φ. *Then we obtain*

$$\mathcal{Q}(\varphi) = \int_{x_1}^{x_2} a\left(\varphi' + \frac{b + w}{a}\varphi\right)^2 dx. \tag{3}$$

From this result we infer *Legendre's necessary condition*:

Proposition 1. *If u is a weak minimizer of $\mathcal{F}$, the relation*

$$F_{pp}(x, u(x), u'(x)) \geq 0$$

holds for all $x \in I$.

Proof. We have to show that $a(x) \geq 0$ holds on I. Otherwise there exists some $x_0 \in (x_1, x_2)$ with $a(x_0) < 0$. Then we can find some $\varepsilon > 0$ such that $\mathcal{U}_\varepsilon := (x_0 - \varepsilon, x_0 + \varepsilon)$ is contained in I, and that $a(x) < 0$ holds on $\overline{\mathcal{U}}_\varepsilon$. Consider the initial value problem

$$w' = \frac{(b + w)^2}{a} - c, \qquad w(x_0) = 1.$$

By diminishing $\varepsilon > 0$, we can achieve that there exists a positive solution $w(x)$ in $\overline{\mathcal{U}}_\varepsilon$. Secondly, we define $\varphi \in C_0^1(I)$ by $\varphi(x) := 0$ for $x \notin \mathcal{U}_\varepsilon$, and $\varphi(x) := e^{q(x)}$ if $x \in \mathcal{U}_\varepsilon$, where $q(x) := \dfrac{1}{|x - x_0|^2 - \varepsilon^2}$. Then w and φ are admissible in Legendre's lemma, and it follows that

$$\mathcal{Q}(\varphi) = \int_{x_0-\varepsilon}^{x_0+\varepsilon} a \cdot \left(\varphi' + \frac{b + w}{a}\varphi\right)^2 dx \leq 0.$$

Because u is a weak minimizer of $\mathcal{F}$, we have $\mathcal{Q}(\varphi) \geq 0$ (cf. 4,*1.1* Proposition), whence $\mathcal{Q}(\varphi) = 0$. Taking $a(x) < 0$ on $\overline{\mathcal{U}}_\varepsilon$ into account, we now infer that

$$\varphi' + \frac{b + w}{a}\varphi = 0$$

holds on $\mathcal{U}_\varepsilon$ for all $\varepsilon > 0$ that are sufficiently small. However, this is a contradiction since $\varphi'(x)/\varphi(x)$ depends on ε for $0 < |x - x_0| < \varepsilon$ while $(b + w)/a$ is independent of ε. $\square$

For the sake of completeness, let us recall the following result from 4,*1.1* (cf. Proposition, (ii)):

Proposition 2. *If u is a weak extremal satisfying*

$$\mathcal{Q}(\varphi) \geq \lambda \int_{x_1}^{x_2} (\varphi^2 + \varphi'^2)\, dx \quad \textit{for all } \varphi \in C_c^\infty(I) \tag{4}$$

and some constant $\lambda > 0$, *then u is a weak minimizer of u.*

From (4) we infer that $\varphi = 0$ is the absolute minimizer of $\mathcal{Q}(\varphi) - \lambda \int_{x_1}^{x_2} (\varphi^2 + \varphi'^2)\, dx$. Then Proposition 1 yields the *strict Legendre* condition

$$a(x) = \tfrac{1}{2} F_{pp}(x, u(x), u'(x)) \geq \lambda \quad \text{on } I\,. \tag{5}$$

Now we shall derive a condition which, together with the strict Legendre condition (5), implies that (4) holds for some $\lambda > 0$.

To this end we note that the Legendre equation (2) is a Riccati equation and can be transformed into some homogeneous linear equation. In fact, suppose that $a > 0$ holds on I, and that v is a Jacobi field which is strictly positive on I. Then, as was first observed by Jacobi, it follows that

$$w := -\left(b + a\frac{v'}{v}\right) \tag{6}$$

is a solution of (2), and therefore (3) is satisfied for all $\varphi \in C_0^1(I)$. Moreover, (6) yields

$$\varphi' + \frac{b + w}{a}\varphi = \varphi' - \frac{v'}{v}\varphi = \frac{\varphi' v - v'\varphi}{v} = v\left(\frac{\varphi}{v}\right)'.$$

Thus we have arrived at

Jacobi's lemma. *If* $a > 0$ *holds on I, and if there exists a Jacobi field* $v \in C^2(I)$ *which is strictly positive on I, then the identity*

$$\mathcal{Q}(\varphi) = \int_{x_1}^{x_2} a v^2 \left|\left(\frac{\varphi}{v}\right)'\right|^2 dx \tag{7}$$

holds for all $\varphi \in C_0^1(I)$.

It is not difficult to derive (4) from (7) if we employ

Poincaré's inequality. *For every* $\psi \in C^1(I)$ *with* $\psi(x_1) = 0$, *the inequality*

$$\int_{x_1}^{x_2} \psi^2\, dx \;\leq\; (x_2 - x_1)^2 \int_{x_1}^{x_2} \psi'^2\, dx \tag{8}$$

is true.

Proof. From $\psi(x) = \psi(x) - \psi(x_1) = \int_{x_1}^{x} \psi'(t)\,dt$, $x \in I$, we infer that

$$|\psi(x)|^2 \le \left(\int_{x_1}^{x_2} |\psi'|\,dt\right)^2 \le (x_2 - x_1)\int_{x_1}^{x_2} |\psi'|^2\,dt$$

holds, taking Schwarz's inequality into account. Integrating with respect to x from x_1 to x_2, the desired inequality follows at once. □

Now we conclude:

Theorem. *If $a > 0$ holds on I, and if there exists a Jacobi field $v \in C^2(I)$ with $v > 0$ on I, then u is a strict weak minimizer of $\mathscr{F}$.*

Proof. Let φ be an arbitrary function in $C_0^1(I)$. Then also $\psi := \varphi/v \in C_0^1(I)$, and we infer from Jacobi's lemma and from Poincaré's inequality that

$$\begin{aligned}\mathscr{Q}(\varphi) = \int_{x_1}^{x_2} av^2\psi'^2\,dx &\ge \inf_I \{av^2\}\int_{x_1}^{x_2} \psi'^2\,dx\\ &\ge \inf_I \{av^2\}\frac{1}{|x_2 - x_1|^2}\int_{x_1}^{x_2} \psi^2\,dx\\ &\ge \inf_I \{av^2\}\inf_I\left\{\frac{1}{v^2}\right\}\frac{1}{|x_2 - x_1|^2}\int_{x_1}^{x_2} \varphi^2\,dx.\end{aligned}$$

In other words, there is a constant $\mu > 0$ such that

$$\mathscr{Q}(\varphi) \ge \mu\int_{x_1}^{x_2} \varphi^2\,dx \tag{9}$$

holds for all $\varphi \in C_0^1(I)$.

We now claim that there is some $\lambda > 0$ such that

$$\mathscr{Q}(\varphi) \ge \lambda\int_{x_1}^{x_2} (\varphi'^2 + \varphi^2)\,dx \tag{10}$$

is satisfied for all $\varphi \in C_0^1(I)$; then Proposition 2 yields that u is a weak minimizer.

In order to prove (10), we recall that

$$\mathscr{Q}(\varphi) = \int_{x_1}^{x_2} (a\varphi'^2 + 2b\varphi\varphi' + c\varphi^2)\,dx.$$

Setting

$$\alpha := \inf_I a, \qquad \beta := \sup_I |b|, \qquad \gamma := \sup_I |c|,$$

we obtain that

$$\alpha\int_{x_1}^{x_2} \varphi'^2\,dx \le \mathscr{Q}(\varphi) + 2\beta\int_{x_1}^{x_2} |\varphi||\varphi'|\,dx + \gamma\int_{x_1}^{x_2} \varphi^2\,dx. \tag{11}$$

Now we apply the inequality

$$2|xy| \le \varepsilon x^2 + \varepsilon^{-1} y^2, \quad \varepsilon > 0,$$

to the second integral of (11); it follows that

$$\alpha \int_{x_1}^{x_2} \varphi'^2 \, dx \le \mathcal{Q}(\varphi) + \varepsilon \int_{x_1}^{x_2} \varphi'^2 \, dx + (\varepsilon^{-1}\beta^2 + \gamma) \int_{x_1}^{x_2} \varphi^2 \, dx.$$

Choosing $\varepsilon = \alpha/2$ and taking (9) into account, we arrive at

$$\frac{\alpha}{2} \int_{x_1}^{x_2} \varphi'^2 \, dx \le \mathcal{Q}(\varphi) + (2\alpha^{-1}\beta^2 + \gamma) \int_{x_1}^{x_2} \varphi^2 \, dx \le [1 + \mu^{-1}(2\alpha^{-1}\beta + \gamma)]\mathcal{Q}(\varphi). \tag{12}$$

Then (10) follows from (9) and (12). □

Remark. We shall see below that the converse of Theorem 1 holds true in the sense that the existence of a "nonvanishing" Jacobi field along u is a necessary condition for u to be a weak minimizer. More precisely, we have: *If $a > 0$ on I, and if u is a weak minimizer of $\mathcal{F}$, then for any $x_0 \in [x_1, x_2]$ the solution v of the Cauchy problem*

$$\mathcal{J}_u v = 0, \qquad v(x_0) = 0, \qquad v'(x_0) = 1$$

has no zeros different from x_0 in (x_1, x_2). This fact is often referred to as *Hilbert's necessary condition.*

2.2. Jacobi Fields and Conjugate Values

In order to gain more information about Jacobi fields, we note that Jacobi's equation

$$-(av')' + (c - b')v = 0$$

is equivalent to the equation

$$v'' + pv' + qv = 0, \tag{1}$$

with

$$p := \frac{a'}{a}, \qquad q := \frac{b' - c}{a} \tag{1'}$$

if we assume that $a > 0$ holds on I. For the sake of simplicity, *we will in the following suppose that $a > 0$ is true on $I_0 := (x_1 - \delta_0, x_2 + \delta_0)$, $\delta_0 > 0$.*

We use the fact that, for every $x_0 \in I$ and arbitrary numbers $\alpha, \beta \in \mathbb{R}$, the initial value problem

$$v(x_0) = \alpha, \qquad v'(x_0) = \beta \tag{2}$$

for (1) has exactly one C^2-solution on I which will be denoted by $\omega(x; x_0, \alpha, \beta)$, and we can assume that ω is a C^2-solution of (1) on the larger interval $I_0 = (x_1 - \delta_0, x_2 + \delta_0)$.

Fix some $x_0 \in I$ and set

$$(3') \qquad v_1(x) := \omega(x; x_0, 1, 0), \qquad v_2(x) := \omega(x; x_0, 0, 1).$$

Then $\omega(x; x_0, \alpha, \beta)$ describes all solutions of (1), and we can write

$$(3'') \qquad \omega(x; x_0, \alpha, \beta) = \alpha v_1(x) + \beta v_2(x),$$

taking the uniqueness theorem into account. Hence the set of solutions of (1) forms a two-dimensional linear space X over $\mathbb{R}$. Every pair $\{v_1, v_2\}$ of linearly independent vectors $v_1, v_2 \in X$ is called a *fundamental system* or *base* of X.[5]

Proposition 1. *Let $v \in X$, and suppose that $v(x) \not\equiv 0$. Then the following holds:*

(i) *If $x_0 \in I_0$ and $v(x_0) = 0$, then $v'(x_0) \neq 0$.*

(ii) *The zeros of v are isolated in I_0.*

(iii) *If $x_0 \in I_0$, $v(x_0) = 0$, and $\tilde{v} \in X$, then v, $\tilde{v}$ are linearly dependent if and only if $\tilde{v}(x_0) = 0$.*

Proof. (i) follows from the uniqueness theorem and implies (ii); (iii) is a consequence of the two formulas (3') and (3''). □

Proposition 2. *Suppose that $v_1, v_2 \in X$, and let*

$$W := \begin{vmatrix} v_1 & v_2 \\ v_1' & v_2' \end{vmatrix} = v_1 v_2' - v_2 v_1'$$

be the Wronskian of the pair $\{v_1, v_2\}$. Then we have:

(i) *$\{v_1, v_2\}$ is a base of X exactly if $W(x) \neq 0$.*

(ii) *For every $x_0 \in I_0$, we have*

$$(4) \qquad W(x) = W(x_0) a(x_0) \frac{1}{a(x)} \quad \textit{for all } x \in I_0 .$$

(iii) *If $\{v_1, v_2\}$ is a base of X, then between two neighbouring zeros of v_1 there lies exactly one zero of v_2, and vice versa.*

Proof. (i) follows immediately from the uniqueness theorem.

(ii) A brief computation yields $W' + pW = 0$ and $p = a'/a$, whence $0 = a'W + aW' = (aW)'$; thus $aW = \text{const}$. The last relation implies (4).

(iii) Since $\{v_1, v_2\}$ is a base of X, we have $W(x) \neq 0$, and we may assume that $W(x) > 0$. Let ξ, ξ^* be two neighbouring zeros of v_1, i.e., $v_1(x) \neq 0$ for $\xi < x < \xi^*$.

---

[5] For general results about the theory of ordinary differential equations see e.g. Hartman [1] or Coddington–Levinson [1].

Since $v_1'(\xi) \neq 0$ and $v_1'(\xi^*) \neq 0$, it follows that $v_1'(\xi)$ and $v_1'(\xi^*)$ have opposite signs. From $W(x) > 0$ we infer $-v_1'(\xi)v_2(\xi) > 0$ and $-v_1'(\xi^*)v_2(\xi^*) > 0$, and we arrive at $v_2(\xi) > 0$, $v_2(\xi^*) < 0$ or, vice versa, $v_2(\xi) < 0$, $v_2(\xi^*) > 0$. Thus there exists a zero of v_2 between ξ and ξ^*. If there were a second zero of v_2 in (ξ, ξ^*), then by the same reasoning we could deduce the existence of another zero of v_1 between ξ and ξ^* which is impossible. This proves one implication of (iii), and the other is obvious. □

Property (iii) of the Proposition 2 is known as *Sturm's oscillation theorem*.

For some fixed parameter values $\xi \in I_0$, we now introduce *Jacobi's function*

$$\Delta(x, \xi) := \omega(x; \xi, 0, 1). \tag{5}$$

In other words, $v(x) = \Delta(x, \xi)$ is the uniquely determined Jacobi field on I_0 with $v(\xi) = 0$ and $v'(\xi) = 1$.

Definition. *The (isolated) zeros of $\Delta(\cdot, \xi)$ will be called* conjugate values *to ξ, irrespectively of whether they lie to the left or to the right of ξ.*

Obviously ξ is conjugate to ξ^* exactly if ξ^* is conjugate to ξ. The smallest conjugate value ξ^* with $\xi^* > \xi$ is called the *next conjugate value to ξ* (to the right). If ξ, ξ^* is a pair of conjugate values, then the points $P = (\xi, u(\xi))$ and $P^* = (\xi^*, u(\xi^*))$ will be called a *pair of conjugate points* on the extremal $z = u(x)$.

Proposition 3. *If $(x_1, x_2]$ does not contain any conjugate point to x_1, then there is no pair ξ, ξ^* of conjugate points contained in I.*

Proof. Suppose that $\xi, \xi^* \in I$ are conjugate to each other. By assumption, we have $x_1 < \xi < \xi^* \leq x_2$. Set $v_1(x) := \Delta(x, x_1)$ and $v_2(x) := \Delta(x, \xi)$. From Proposition 1(iii) we infer that v_1, v_2 are linearly independent whence, by Sturm's theorem, v_1 must have a zero between ξ and ξ^*, which, by assumption, is impossible. □

Recall that we are actually interested in the piece $z = u(x)$, $x_1 \leq x \leq x_2$, of the extremal u. We distinguish three possibilities for $I = [x_1, x_2]$:

Case 1: There exists no conjugate value to x_1 in I.

Case 2: There is a conjugate value to x_1 in int $I = (x_1, x_2)$.

Case 3: The value x_2 is the next conjugate value to x_1 in I.

Corresponding to these three cases, we can now state necessary as well as sufficient conditions for u to be a weak minimizer.

Theorem 1. *Suppose that the strict Legendre condition*

$$a(x) = \tfrac{1}{2}F_{pp}(x, u(x), u'(x)) > 0$$

is fulfilled on I.

(i) *If there exists no conjugate value to x_1 in I, then u is a strict weak minimizer.*

(ii) *If there is a conjugate value to* x_1 *in* $(x_1, x_2) = \operatorname{int} I$, *then u is not a weak minimizer.*

(iii) *If the value* x_2 *is the next conjugate value to* x_1 *in I, then it cannot be decided without a more penetrating discussion whether u is a weak minimizer or not.*

Proof. Without loss of generality we can assume that $a(x) > 0$ holds on the larger interval $I_0 = (x_1 - \delta_0, x_2 + \delta_0)$, $0 < \delta_0 \ll 1$.

(i) Suppose now that we are in case 1. Then, by assumption, $\Delta(x, x_1) > 0$ for all $x \in (x_1, x_2]$. On account of Proposition 1(iii), the two Jacobi fields $\Delta(x, x_2)$ and $\Delta(x, x_1)$ have to be independent since $\Delta(x_2, x_2) = 0$.

We claim that there exists some $x_0 \in (x_1 - \delta_0, x_1)$ such that $\Delta(x, x_2) \neq 0$ holds on $[x_0, x_2)$. If not, then there would be some conjugate value to x_2 in $[x_1, x_2)$; let x_2^* be the largest of these. By Sturm's reasoning, $\Delta(x, x_1)$ must have a zero in (x_2^*, x_2) which by assumption is impossible.

With such a value x_0, we form $\Delta(x, x_0)$. Since $\Delta(x_0, x_0) = 0$ but $\Delta(x_0, x_2) \neq 0$, the two Jacobi fields $\Delta(x, x_0)$ and $\Delta(x, x_2)$ are linearly independent. Then it follows that $\Delta(x, x_0) \neq 0$ in $(x_0, x_2]$ (in fact, $\Delta(x, x_0) > 0$). Otherwise Sturm's reasoning would imply that $\Delta(x, x_2)$ possessed a zero in (x_0, x_2), and this is impossible as was shown before.

Hence we have shown that, in case 1, $v(x) := \Delta(x, x_0)$ forms a strictly positive Jacobi field on $I = [x_1, x_2]$. Now we may infer from the theorem of *2.1* that u is a strict weak minimizer of $\mathcal{F}$.

(ii) The assertion for case 2 follows from the next theorem, and in case 3 nothing is to be proved as one can find examples for both situations.[6] □

Theorem 2. *Suppose that the strict Legendre-condition* $a(x) > 0$ *on I holds, and let v be a Jacobi field,* $v(x) \not\equiv 0$, *that vanishes at the endpoints of some proper closed subinterval I′ of I. Then there exists functions* $\varphi \in C_0^1(I)$ *such that* $\mathcal{Q}(\varphi) < 0$, *and consequently u is not a weak minimizer of* $\mathcal{F}$.

Proof. We restrict ourselves to the case where I and I' have the same left endpoint; the general case is handled in the same way.

Let $I' = [x_1, x_1']$, $x_1 < x_1' < x_2$. Since $v \neq 0$ and $v(x_1) = v(x_1') = 0$, the point x_1' is a conjugate value to x_1. We consider the next conjugate value x_1^* to x_1, i.e., $x_1 < x_1^* \le x_1'$, and choose some $\beta \in (x_1^*, x_2)$ such that the interval $(x_1^*, \beta]$ contains no value conjugate to x_1. Then we introduce the two nontrivial Jacobi fields

$$v_1(x) := \Delta(x, x_1), \qquad v_2(x) := -\Delta(x, \beta).$$

Because of $v_1(\beta) \neq 0$ and $v_2(\beta) = 0$, the functions v_1 and v_2 are linearly independent whence, by Sturm's theorem, v_2 possesses exactly one zero α between

[6] Cf. Carathéodory [10], p. 295, and Bolza [3], §47, pp. 357–364.

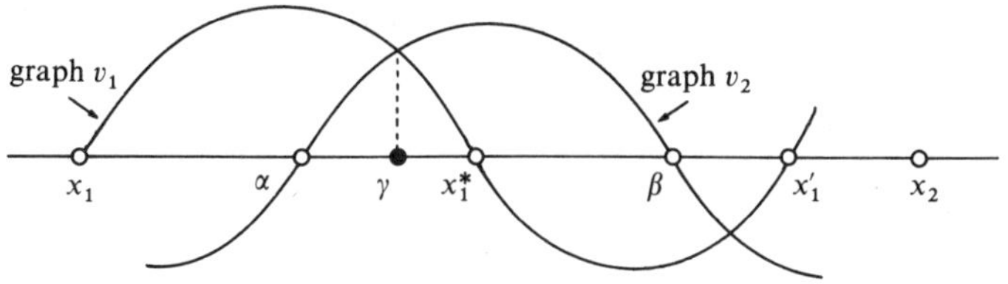

Fig. 2. The functions v_1 and v_2 in Theorem 2.

x_1 and x_1^*, and $v_2(x_1^*) \neq 0$. Moreover, $v_2(x)$ cannot vanish for any $x \in (x_1^*, \beta)$. Otherwise, by Sturm's reasoning, v_1 had to vanish for some value in (x_1^*, β), and this would contradict our choice of β. Hence α is the only zero of v_2 in $[x_1, \beta)$, and $v_2'(\beta) = -1$ implies $v_2(x) > 0$ in (α, β), and therefore $v_2(x) < 0$ in $[x_1, \alpha)$.

Now we consider the Wronskian $W = v_1 v_2' - v_2 v_1'$ which, according to (4), satisfies

$$a(x)W(x) \equiv C \quad \text{on } I,$$

with a constant $C \neq 0$ which can be computed from

$$C = a(x_1)W(x_1).$$

Since $a(x_1) > 0$ and $W(x_1) = -v_2(x_1) > 0$, we infer $C > 0$.

Next we observe that also $v_1, v_2 - v_1$ are linearly independent, and Sturm's theorem yields as well the existence of a zero γ of $v_2 - v_1$ between x_1 and x_1^*, whence $v_1(\gamma) = v_2(\gamma)$ and $\alpha < \gamma < x_1^*$.

Finally we define the test function η by

$$\eta(x) := \begin{cases} v_1(x) & \text{on } [x_1, \gamma], \\ v_2(x) & \text{on } [\gamma, \beta], \\ 0 & \text{on } [\beta, x_2]. \end{cases}$$

Note that η is continuous on I, piecewise of class C^2 and $\eta(x_1) = \eta(x_2) = 0$. Moreover, η' and η'' are only at $x = \gamma$ and $x = \beta$ discontinuous.

Since $Q(x, z, p)$ is a quadratic form with respect to z, p, we have

$$2Q(x, z, p) = zQ_z(x, z, p) + pQ_p(x, z, p)$$

and therefore

$$Q(x, \eta, \eta') = \tfrac{1}{2}\delta Q(\eta, \eta).$$

It follows that

$$\begin{aligned} \mathcal{Q}(\eta) &= \int_{x_1}^{x_2} Q(x, \eta, \eta')\,dx \\ &= \frac{1}{2}\int_{x_1}^{\gamma} \delta Q(v_1, v_1)\,dx + \frac{1}{2}\int_{\gamma}^{\beta} \delta Q(v_2, v_2)\,dx. \end{aligned}$$

Since $v_1(x)$ and $v_2(x)$ are of class C^2 on $[x_1, \gamma]$ and $[\gamma, \beta]$, respectively, we infer

$$\int_{x_1}^{\gamma} \delta Q(v_1, v_1)\, dx = \int_{x_1}^{\gamma} v_1 L_Q(v_1)\, dx + [v_1 Q_p(x, v_1, v_1')]_{x_1}^{\gamma},$$

$$\int_{\gamma}^{\beta} \delta Q(v_2, v_2)\, dx = \int_{\gamma}^{\beta} v_2 L_Q(v_2)\, dx + [v_2 Q_p(x, v_2, v_2')]_{\gamma}^{\beta}.$$

Moreover,

$$Q(x, z, p) = a(x)p^2 + 2b(x)pz + c(x)z^2$$

implies

$$\tfrac{1}{2} Q_p(x, z, p) = a(x)p + b(x)z.$$

Since $v_1(x_1) = 0$, $v_1(\gamma) = v_2(\gamma)$, $v_2(\beta) = 0$, and

$$L_Q(v_1) = \mathcal{J}_u v_1 = 0, \qquad L_Q(v_2) = \mathcal{J}_u v_2 = 0,$$

we obtain

$$\begin{aligned} \mathcal{Q}(\eta) &= a(\gamma)\{v_1(\gamma)v_1'(\gamma) - v_2(\gamma)v_2'(\gamma)\} \\ &= a(\gamma)\{v_2(\gamma)v_1'(\gamma) - v_1(\gamma)v_2'(\gamma)\} \\ &= -a(\gamma)W(\gamma) = -C < 0. \end{aligned}$$

By "smoothing the corners" of η at $x = \gamma$ and $x = \beta$, we can construct a function $\varphi \in C_0^1(I)$ such that also

$$\mathcal{Q}(\varphi) < 0$$

holds; in fact, we can even find some $\varphi \in C_c^\infty(I)$ with $\mathcal{Q}(\varphi) < 0$. Therefore u cannot be a weak minimizer. □

Hilbert's necessary condition formulated in the remark following the theorem in *2.1* is now an immediate consequence of Theorem 1 or Theorem 2.

2.3. Geometric Interpretation of Conjugate Points

Now we turn to a *geometric interpretation of conjugate points* due to Jacobi.

Suppose that the extremal $u(x)$, $x_1 \le x \le x_2$, is embedded in some *one-parameter family of extremals* $\varphi(x, \alpha)$ such that $\varphi(x, \alpha_0) = u(x)$ for $x \in I$. Let φ be of class C^3 on $I \times (\alpha_0 - \delta, \alpha_0 + \delta)$, $\delta > 0$, and assume that $\varphi_{\alpha x}(x, \alpha_0)$ is not identically zero on I. Then $v(x) := \varphi_\alpha(x, \alpha_0)$ is a nontrivial Jacobi field along $u(x)$ as we have stated in the Corollary of *1.2*.

Now we consider the particular case that, for $x = x_1$, all extremals of the family pass through the same point $P_1 = (x_1, z_1)$, i.e.,

$$\varphi(x_1, \alpha) = z_1 \quad \text{for } \alpha_0 - \delta < \alpha < \alpha_0 + \delta.$$

Then we have $v(x_1) = 0$. Suppose that there is a next conjugate point $P_1^*(\alpha_0) = (x_1^*, z_1^*)$, $z_1^* = u(x_1^*)$, $x_1 < x_1^* < x_2$, to P_1 on the extremal $u(x) = \varphi(x, \alpha_0)$. Since $v(x_1) = 0$, it follows that $v(x_1^*) = 0$, and therefore also $v'(x_1) \neq 0$ and $v'(x_1^*) \neq 0$; that is,

$$\varphi_{\alpha x}(x_1, \alpha_0) \neq 0 \quad \text{and} \quad \varphi_{\alpha x}(x_1^*, \alpha_0) \neq 0.$$

Consider now the system of equations

$$\varphi(x, \alpha) - z = 0, \qquad \varphi_\alpha(x, \alpha) = 0 \tag{1}$$

for the unknown variables x, z, α which has the solution x_1^*, z_1^*, α_0. Because of $\varphi_{\alpha x}(x_1^*, \alpha_0) \neq 0$, we can apply the implicit function theorem and obtain a C^2-function $x = \xi(\alpha)$ with $x_1^* = \xi(\alpha_0)$ such that $\varphi_\alpha(\xi(\alpha), \alpha) = 0$ holds. If we set $\zeta(\alpha) := \varphi(\xi(\alpha), \alpha)$, then $(x, z, \alpha) = (\xi(\alpha), \zeta(\alpha), \alpha)$, $\alpha \in \mathscr{U}(\alpha_0)$, is a parameterization of the set of solutions of (1). If we assume that

$$\varphi_{\alpha\alpha}(x_1^*, \alpha_0) \neq 0,$$

then it follows from

$$\varphi_{\alpha\alpha}(\xi(\alpha), \alpha) + \varphi_{\alpha x}(\xi(\alpha), \alpha)\xi_\alpha(\alpha) = 0$$

that $\xi_\alpha(\alpha) \neq 0$ holds in some neighborhood $\mathscr{U}(\alpha_0)$ of α_0, and therefore

$$\mathscr{E} = \{P^*(\alpha)\colon P^*(\alpha) = (\xi(\alpha), \zeta(\alpha)), \alpha \in \mathscr{U}(\alpha_0)\}$$

describes a regular curve of class C^2. It is part of the envelope[7] of the extremals $z = \varphi(x, \alpha)$, $x_1 \le x \le x_2$, emanating from the point $P_1 = (x_1, z_1)$, and the conjugate point $P_1^* = P^*(\alpha_0)$ lies both on $\mathscr{E}$ and on the extremal described by $u(x)$. More general, the condition $\varphi_\alpha(x, \alpha) = 0$ determines points on the extremal $\mathscr{C}(\alpha) = \{(x, z)\colon z = \varphi(x, \alpha), x \in I\}$ which are conjugate to P_1.

If $\varphi_{\alpha\alpha}(x_1^*, \alpha_0) = 0$, the solution set $\mathscr{E}$ of (1) can be rather degenerate, for instance, a curve with a cusp at P_1^*, or even an isolated point (nodal point). In all cases, whether degenerate or not, we shall denote $\mathscr{E}$ as envelope of the family of curves $\mathscr{C}(\alpha)$. Then we can state:

Theorem 1. *Suppose that $\mathscr{C}(\alpha) = \{(x, z)\colon z = \varphi(x, \alpha), x \in I\}$, $\alpha \in (\alpha_0 - \delta, \alpha_0 + \delta)$, is a family of extremals of $\mathscr{F}$ emanating from a fixed point $P_1(x_1, z_1)$, i.e., $\varphi(x_1, \alpha) = z_1$. Assume also that $\varphi(x, \alpha)$ is of class C^3 and that $\varphi_\alpha'(x_1, \alpha) = \varphi_{\alpha x}(x_1, \alpha) \neq 0$ for $|\alpha - \alpha_0| < \delta$. Then the envelope $\mathscr{E}$ of the family $\{\mathscr{C}(\alpha)\}$ consists of all points $P^*(\alpha) \in \mathscr{C}(\alpha)$ conjugate to P_1, α varying in $(\alpha_0 - \delta, \alpha_0 + \delta)$. To be precise, if $P^*(\alpha) \in \mathscr{C}(\alpha) \cap \mathscr{E}$, and if $\mathscr{E}$ is regular at $P^*(\alpha)$, then $P^*(\alpha)$ is a conjugate point to P_1 with respect to the extremal $\mathscr{C}(\alpha)$, and $\mathscr{E}$ is tangent to $\mathscr{C}(\alpha)$ at $P^*(\alpha)$.*

If we imagine the extremals $\mathscr{C}(\alpha)$ as a pencil of light rays, we can interpret the envelope $\mathscr{E}$ as its *caustics*.

[7] We recall that the envelope $\mathscr{E}$ of the family $\varphi(x, \alpha)$ is defined by $\mathscr{E} := \{(x, z)\colon$ there is an α such that $z = \varphi(x, \alpha)$ and $\varphi_\alpha(x, \alpha) = 0\}$.

We may briefly summarize what we so far have found:

Let $u(x)$, $x_1 \le x \le x_2$, be an extremal of the variational integral $\mathcal{F}(u) = \int_{x_1}^{x_2} F(x, u, u')\,dx$ which satisfies the strict Legendre condition (5) *of 2.1 on $I = [x_1, x_2]$. Then the following holds:*

1. *The extremal u is a weak minimizer if I contains no value conjugate to x_1, but u is not a weak minimizer if there exists a conjugate value x_1^* to x_1 in (x_1, x_2). In particular, sufficiently small arcs of an extremal are weak minimizers. Starting at the left endpoint P_1, this minimum property prevails until the first conjugate point P_1^* to P_1 is reached. Beyond P_1^* the extremal arc between P_1 and P_1^* no longer furnishes a weak minimum of $\mathcal{F}$.*
2. *If there exists a nonvanishing Jacobi field on I, then u is a weak minimizer.*
3. *However, u is not a weak minimizer, if there is a nontrivial Jacobi field that vanishes at the endpoints of a proper subinterval of I.*
4. *The conjugate locus of a nontrivial pencil of extremals $\mathscr{C}(\alpha)$ emanating from a fixed point P_1, i.e., the set of points $P^*(\alpha)$ on $\mathscr{C}(\alpha)$ conjugate to P_1 (with respect to $\mathscr{C}(\alpha)$) is the envelope $\mathscr{E}$ of the curves $\mathscr{C}(\alpha)$.*

These are the principal results of the Jacobi theory dealing with the weak-minimizer property of extremals, and thus we could end our discussion at this point. However, we can draw some further information from the conjugate point theory which will be of great value for the Weierstrass theory dealing with sufficient conditions for strong minima. To formulate this result, we need the notion of a *field of extremals.*

Let α be a real parameter varying in some interval $A \subset \mathbb{R}$, and consider a set Γ in $\mathbb{R}^2$ which is of the form

$$\Gamma = \{(x, \alpha): \alpha \in A, x \in I(\alpha)\},$$

where $I(\alpha)$ is an interval in $\mathbb{R}$ with the endpoints $x_1(\alpha)$, $x_2(\alpha)$ which may or may not belong to $I(\alpha)$. We assume that $x_1(\alpha)$ and $x_2(\alpha)$ are continuous functions of $\alpha \in A$. Then int Γ is a simply connected domain in $\mathbb{R}^2$.

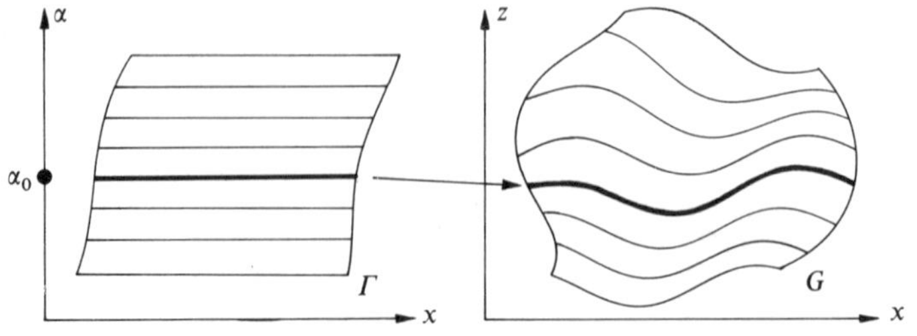

Fig. 3. A field $f: \Gamma \to G$ on G.

Consider now a mapping $f : \Gamma \to \mathbb{R}^2$ given by

$$x = x, \qquad z = \varphi(x, \alpha),$$

i.e., $f(x, \alpha) = (x, \varphi(x, \alpha))$, and suppose that f defines a diffeomorphism of Γ onto $G := f(\Gamma)$ which is of class C^s, $s \geq 1$. This, equivalently, means that $\varphi \in C^s(\Gamma)$, $s \geq 1$, $\varphi_\alpha(x, \alpha) \neq 0$ on Γ, and that for every point $(x, z) \in G$ there is exactly one value $\alpha \in A$ such that $z = \varphi(x, \alpha)$. Denote this α by $a(x, z)$; then we have

$$z = \varphi(x, a(x, z)) \quad \text{and} \quad \alpha = a(x, \varphi(x, \alpha)), \tag{2}$$

for $(x, z) \in G$ and $(x, \alpha) \in \Gamma$, respectively.

For the mapping f, we have the following interpretation: Fixing $\alpha \in A$, the function $\varphi(\cdot, \alpha)$ describes a nonparametric curve

$$\mathscr{C}(\alpha) := \{(x, z) : x \in I(\alpha), z = \varphi(x, \alpha)\}$$

in $\mathbb{R}^2$. Through every point $(x, z) \in G$ passes exactly one of the curves $\mathscr{C}(\alpha)$. Under the previous assumptions we call the family $\{\mathscr{C}(\alpha)\}_{\alpha \in A}$ a *field of curves covering G*.

Consider the derivative $\varphi'(x, \alpha) = \varphi_x(x, \alpha)$ which is the slope of the curve $\mathscr{C}(\alpha)$ at $(x, \varphi(x, \alpha))$ with respect to the x-axis. Introducing

$$\mathscr{P}(x, z) := \varphi'(x, a(x, z)),$$

we obtain that $w(x) = \varphi(x, a)$ is a solution of the differential equation

$$w' = \mathscr{P}(x, w). \tag{3}$$

One calls $\mathscr{P}(x, z)$ the *slope function* of the field $\{\mathscr{C}(\alpha)\}_{\alpha \in A}$. The curves $\mathscr{C}(\alpha)$ "fit" into the direction field $(1, \mathscr{P}(x, z))$ on G. Clearly the slope function of a C^s-field is of class C^{s-1}. Conversely, if G is a convex domain of $\mathbb{R}^2$ and $\mathscr{P}$ an arbitrary function of class $C^{s-1}(G)$, $s \geq 1$, then, by solving (3) with a suitable initial curve, we obtain a field of curves covering G.

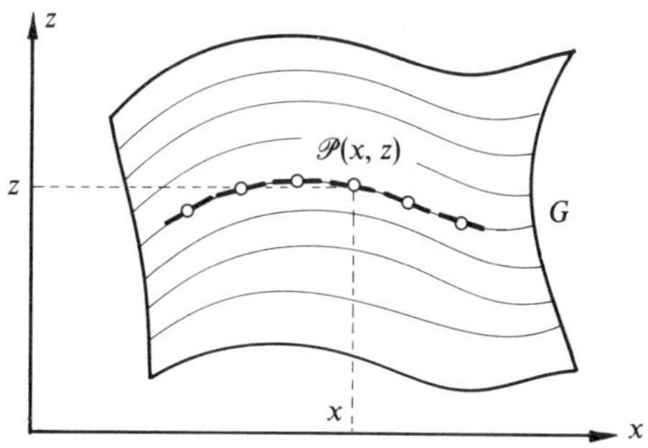

Fig. 4. The slope $P(x, z)$ of a field f.

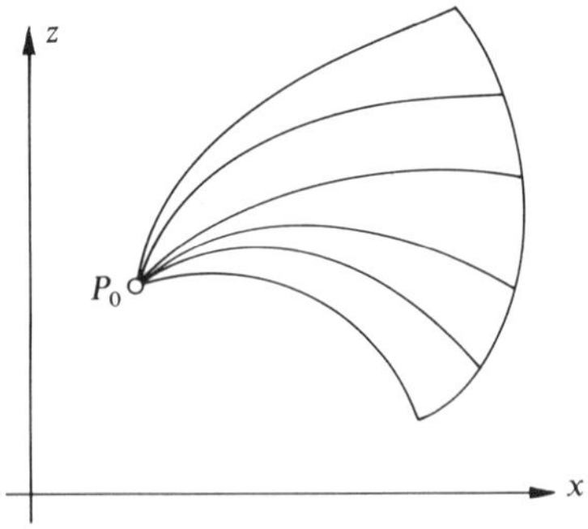

Fig. 5. An improper field: a stigmatic field emanating from P_0.

Sometimes one has also to consider *improper fields.* By this we mean mappings $f : \Gamma \to \mathbb{R}^2$ such that $f|_{\text{int}\,\Gamma}$ is a proper field but f may degenerate at some of the boundary points of $\partial\Gamma$. A typical example of an improper field is provided by a *stigmatic field* (or *central field*) formed by a pencil of curves emanating from some fixed point. If $\varphi_\alpha(x, \alpha) \neq 0$ holds, the curves $\mathscr{C}(\alpha) = \{(x, z): z = \varphi(x, \alpha)\}$ locally form a field but the field property gets lost as soon as we hit the envelope of the $\mathscr{C}(\alpha)$.

We call a curve $\mathscr{C}^*$ *embedded into some field* $\{\mathscr{C}(\alpha)\}_{\alpha \in A}$ if there is some $\alpha_0 \in \text{int}\, A$ such that $\mathscr{C}^*$ is subarc of $\mathscr{C}(\alpha_0)$. If $\mathscr{C}^*$ is given by $z = u(x)$, $x \in I^*$, this means that $I^* \subset I(\alpha_0)$ and $u(x) = \varphi(x, \alpha_0)$ for all $x \in I^*$.

Of particular interest is the question as to whether one can embed a given extremal of $\mathscr{F}$ into some *field of extremals* (compare also Chapter 6). A sufficient condition insuring the possibility of such an embedding is provided by the next result; we stipulate the general requirements stated at the beginning of this section.

Theorem 2. *If the extremal* $u(x)$, $x_1 \leq x \leq x_2$, *satisfies the strict Legendre condition on* $I = [x_1, x_2]$, *and if* I *contains no pair of conjugate values, then* u *can be embedded into a* C^3*-field of extremals given by* $z = \varphi(x, \alpha)$, $\alpha \in A$, $x \in I$.

Proof. In part (i) of the proof of Theorem 1 of 2.2 we have demonstrated that, if I contains no pair of conjugate values of u, then there exists some $x_0 \in (x_1 - \delta_0, x_1)$ such that the Jacobi function $\Delta(x, x_0)$ is positive on $(x_0, x_2]$.

Let us now consider the initial value problem

$$L_F(w) = 0 \quad \text{on } [x_0, x_2], \qquad w(x_0) = u(x_0), \qquad w'(x_0) = \alpha. \tag{4}$$

By virtue of the general theory of ordinary differential equations, there is a number $\rho_0 > 0$ such that, for every $\alpha \in [\alpha_0 - \rho_0, \alpha_0 + \rho_0]$, $\alpha_0 := u'(x_0)$, problem (4) possesses exactly one solution $w(x)$. We shall write

$$w(x) = \varphi(x, \alpha)$$

in order to characterize its dependence on α. Then $\varphi(x, \alpha_0) = u(x)$, $x \in [x_0, x_2]$, and the general theory also yields that $\varphi(x, \alpha)$ is of class C^3 on $[x_0, x_2] \times [\alpha_0 - \rho_0, \alpha_0 + \rho_0]$. Thus $v(x) := \varphi_\alpha(x, \alpha_0)$ is a Jacobi field on $[x_0, x_2]$ which is nontrivial and vanishes at x_0 because of

$$v'(x_0) = \varphi_{x\alpha}(x_0, \alpha_0) = 1, \qquad v(x_0) = \varphi_\alpha(x_0, \alpha_0) = 0.$$

From Proposition 1(iii) of *2.2* we then infer that

$$v(x) = \Delta(x, x_0),$$

whence we obtain

$$v(x) > 0 \quad \text{on } (x_0, x_2].$$

By continuity, there are numbers $\rho \in (0, \rho_0)$ and $\mu > 0$ such that

$$\varphi_\alpha(x, \alpha) \geq \mu \quad \text{for all } (x, \alpha) \in \Gamma,$$

where we have set $\Gamma := I \times A$, $I := [x_1, x_2]$, $A := [\alpha_0 - \rho, \alpha_0 + \rho]$. Thus $f(x, \alpha) = (x, \varphi(x, \alpha))$, $(x, \alpha) \in \Gamma$, describes a field of extremals $\mathscr{C}(\alpha) = \{(x, z): x \in I, z = \varphi(x, \alpha)\}$ into which the given extremal $z = u(x)$, $x \in I$, is embedded. □

Remark. Recapitulating the proof, we see that the given extremal was embedded into some stigmatic field emanating from a point $P_0 = (x_0, u(x_0))$ which lies on the "prolongation of the given extremal to the left". Choosing x_0 sufficiently close to x_1, $x_0 < x_1$, the assumption on u guarantees that some sufficiently thin pencil of these extremals forms a field containing the given extremal in its interior. Clearly this field construction is not unique; for instance, the nodal point P_0 can arbitrarily be chosen in some neighborhood $(x_1 - \delta, x_1)$ to the left of x_1. Many other field constructions can be carried out as well.

Finally, let us see how conjugate points are related to any two-parameter family of extremals.

Consider some two-parameter family of solutions $\psi(x, \alpha, \beta)$ of the Euler equation $L_F(w) = 0$ on I, $w = \psi(\cdot, \alpha, \beta)$, where (α, β) varies in some neighbourhood of the point (α_0, β_0) with the property that $u(x) = \psi(x, \alpha_0, \beta_0)$ for $x \in I$. We assume that $\psi(x, \alpha, \beta)$ is of class C^3, and that

$$\frac{\partial(\psi, \psi')}{\partial(\alpha, \beta)} \neq 0 \tag{5}$$

holds, $\psi' := \psi_x$. Setting

$$v_1(x) := \psi_\alpha(x, \alpha_0, \beta_0), \qquad v_2(x) := \psi_\beta(x, \alpha_0, \beta_0), \tag{6}$$

we know that v_1, v_2 are Jacobi fields along the extremal $z = u(x)$, $x_1 \leq x \leq x_2$. By virtue of Proposition 1(iii) of *2.2* we infer from (5) that v_1 and v_2 are linearly independent, and that not both $v_1(x_1)$ and $v_2(x_1)$ are zero. Thus

$$v(x) := v_1(x)v_2(x_1) - v_2(x)v_1(x_1)$$

is a nontrivial solution of the Jacobi equation satifying $v(x_1) = 0$, and therefore v and $\Delta(\cdot, x_1)$ must be linearly dependent. Hence there is a constant $c \neq 0$ such that

$$\Delta(x, x_1) = cv(x) \quad \text{on } I$$

holds, and we have found:

Proposition 1. *If $\psi(x, \alpha, \beta)$ is a two-parameter family of solutions of the Euler equation $L_F(w) = 0$, $w = \psi(\cdot, \alpha, \beta)$, containing the extremal $u(x)$ in its interior, i.e., $u(x) = \psi(x, \alpha_0, \beta_0)$ for some pair α_0, β_0, and if* (5) *holds, then the conjugate values to x_1 are the roots x of the equation*

$$\psi_\alpha(x, \alpha_0, \beta_0)\psi_\beta(x_1, \alpha_0, \beta_0) - \psi_\beta(x, \alpha_0, \beta_0)\psi_\alpha(x_1, \alpha_0, \beta_0) = 0. \tag{7}$$

The gist of this result is the observation that, having found a *complete solution* $\psi(\cdot, a, b)$ of the original equation $L_F(\psi) = 0$, we need not derive the Jacobi equation and to integrate it. It suffices to form the partial derivatives ψ_α and ψ_β and to determine the roots x of equation (7), if we want to find the conjugate values to x_1.

2.4. Examples

In this final subsection we shall discuss a few significant examples.

1 The quadratic integral

$$\mathcal{F}(u) = \frac{1}{2}\int_0^a (u'^2 - Ku^2)\, dx, \quad K \in \mathbb{R},$$

has the extremal $u(x) \equiv 0$, $0 \le x \le a$. The accessory integral $\mathcal{Q}(v)$, corresponding to $\mathcal{F}$ and u, is nothing but $\mathcal{F}(v)$, and therefore the Jacobi equation is given by

$$u'' + Ku = 0.$$

We obtain the Jacobi function $\Delta(x, 0)$ as

$$\Delta(x, 0) = \begin{cases} \dfrac{1}{\sqrt{-K}} \sinh \sqrt{-K}\, x & \text{if } K < 0, \\ x & \text{if } K = 0, \\ \dfrac{1}{\sqrt{K}} \sin \sqrt{K}\, x & \text{if } K > 0. \end{cases}$$

Hence there exist no conjugate values to $x = 0$ if $K \le 0$. If $K > 0$, the first conjugate value to $x = 0$ is $\xi := \pi/\sqrt{K}$. Thus the extremal $u(x) \equiv 0$ on $[0, a]$ has no conjugate value to $x = 0$ in $[0, a]$ if $a < \xi$, whereas for $a \ge \xi$ there is the conjugate value ξ.

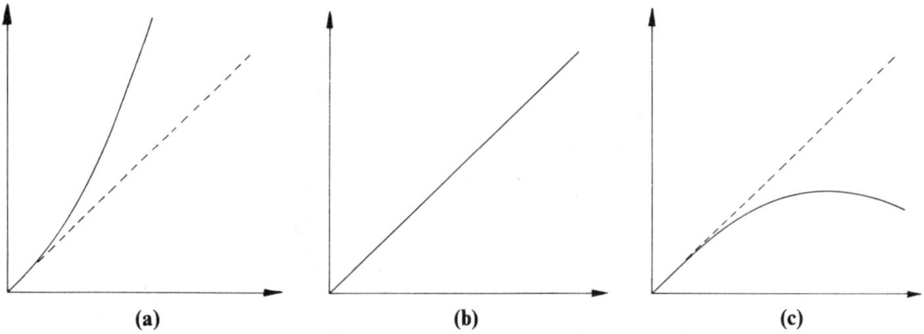

Fig. 6. The Jacobi function $\Delta(x, 0)$ for **(a)** $K < 0$, **(b)** $K = 0$, **(c)** $K > 0$.

2 **Sturm's comparison theorem.** *Suppose that $v(x)$ and $w(x)$ are nontrivial solutions of*

$$v'' + a(x)v = 0$$

and

$$w'' + b(x)w = 0,$$

respectively, in $\alpha \le x \le \beta$, and assume that $a(x) \le b(x)$. Then between two consecutive zeros of $w(x)$ there can be at most one zero of $v(x)$.

Proof. Let $\xi_1 < \xi_2$ be two consecutive zeros of w in $\alpha \le x \le \beta$, and suppose that v has two consecutive zeros η_1 and η_2 such that $\xi_1 \le \eta_1 < \eta_2 \le \xi_2$. We can assume that $v(x) > 0$ and $w(x) > 0$ in $\eta_1 \le x \le \eta_2$. Then, multiplying the equation $w'' + b(x)w = 0$ by v and the first equation $v'' + a(x)v = 0$ by w, and subtracting the second result from the first, we obtain by an integration that

$$\begin{aligned} 0 &= \int_{\eta_1}^{\eta_2} [v(w'' + bw) - (v'' + av)w]\,dx \\ &= \int_{\eta_1}^{\eta_2} (b - a)vw\,dx + v'(\eta_1)w(\eta_1) - v'(\eta_2)w(\eta_2). \end{aligned}$$

However we infer from $v(x) > 0$ in $\eta_1 < x < \eta_2$, $v(\eta_1) = 0$, $v(\eta_2) = 0$, and from $v'' + a(x)v = 0$ that $v'(\eta_1) > 0$ and $v'(\eta_2) < 0$. Hence the second integral above is positive, and we have a contradiction. □

Sturm's theorem implies that the zeros of nontrivial solutions v of

$$v'' + a(x)v = 0$$

draw together if $a(x)$ increases. If, for instance,

$$K_1^2 \le a(x) \le K_2^2, \quad K_1, K_2 > 0,$$

we infer from Example 1 that, for two consecutive zeros x_1, x_2 of v, the estimates

$$\pi/K_2 \le |x_2 - x_1| \le \pi/K_1$$

hold.

3 *Geodesics with respect to a two-dimensional Riemannian line element.* Consider some line element

$$ds^2 = e(x, y)\,dx^2 + 2f(x, y)\,dx\,dy + g(x, y)\,dy^2$$

defined on $\mathbb{R}^2$. The geodesics $c(t) = (x(t), y(t))$ with respect to this metric are the extremals of the generalized Dirichlet integral

$$\mathscr{D}(c) = \frac{1}{2}\int_{t_1}^{t_2} [e(x, y)\dot{x}^2 + 2f(x, y)\dot{x}\dot{y} + g(x, y)\dot{y}^2]\, dt,$$

the Euler equations of which were derived in 2,4 [3]. The geodesics on a surface $\mathscr{S}$ in $\mathbb{R}^3$ are a special case of this problem.[8] In fact, let $z = (z^1, z^2, z^3)$ be Cartesian coordinates on $\mathbb{R}^3$, and suppose that $\mathscr{S}$ is given as embedding of a planar parameter domain $\Omega \subset \mathbb{R}^2$ into $\mathbb{R}^3$ by some regular smooth parameter representation $z = \psi(x, y)$, $(x, y) \in \Omega$. Then the line element ds^2 on $\mathscr{S}$ is given by

$$e = |\psi_x|^2, \qquad f = \psi_x \cdot \psi_y, \qquad g = |\psi_y|^2.$$

With a slight abuse of language, points $(x, y) \in \Omega$ will be identified with their image points $\psi(x, y)$ on $\mathscr{S}$. Correspondingly, a geodesic can either mean a curve $c(t) = (x(t), y(t))$ in Ω, or the lifted curve $\psi(c, t))$, and in the same way we can lift the notion of conjugate points from Ω to $\mathscr{S}$.

Let us now treat the special case of a line element

$$ds^2 = e(x, y)\, dx^2 + dy^2,$$

where $e > 0$, $e(x, 0) = 1$, $e_y(x, 0) = 0$. Here the corresponding geodesics $c(t) = (x(t), y(t))$ are extremals of the integral

$$\frac{1}{2}\int_{t_1}^{t_2} [e(x, y)\dot{x}^2 + \dot{y}^2]\, dt,$$

which has the Euler equations

$$\frac{d}{dt}[e(x, y)\dot{x}] - \frac{1}{2}e_x(x, y)\dot{x}^2 = 0,$$

$$\ddot{y} - \frac{1}{2}e_y(x, y)\dot{x}^2 = 0.$$

These equations obviously have the solution $c(t) = (x(t), 0)$ where $x(t)$ is to be determined from the equation

$$\ddot{x} - \tfrac{1}{2}e_x(x, 0)\dot{x}^2 = 0.$$

This might look very special but it is not. In fact, if $c(t)$ is some geodesic with respect to some general line element ds^2, then we introduce orthogonal curvilinear coordinates x, y on some part of $\mathbb{R}^2$ containing c such that c corresponds to the line $y = 0$, whereas the lines $x = \text{const}$ are chosen as geodesics orthogonal to c (with respect to the general metric ds^2). With this choice of x and y one reduces the general situation to the special one which was considered before. Here the extremal c can be considered as piece of the x-axis, say, $x_1 \le x \le x_2$, and it therefore makes sense to compare it with nonparametric curves $(x, u(x))$, $x_1 \le x \le x_2$. Since we now have fixed the parameter representation, we are no longer allowed to work with the Dirichlet integral (the expression $1 + u'^2$ will not be constant) but have to investigate the length functional

$$\mathscr{L}(u) = \int_{x_1}^{x_2} \sqrt{e(x, u) + u'^2}\, dx$$

instead of the Dirichlet integral. The Euler equation of $\mathscr{L}$ has the form

$$\left\{\frac{u'}{\sqrt{e(x, u) + u'^2}}\right\}' - \frac{1}{2}\frac{e_y(x, u)}{\sqrt{e(x, u) + u'^2}} = 0,$$

and we read off that $u(x) \equiv 0$ is a solution as was to be expected. The accessory integral $\mathscr{Q}(\varphi)$

---

[8] Since our present notation is inadequate, the reader may get the impression that the following discussion only applies to suitably small pieces of $\mathscr{S}$. This, however, is not the case.

corresponding to $\mathscr{L}$ and the extremal $u = 0$ turns out to be

$$\mathscr{Q}(\varphi) = \frac{1}{2}\int_{x_1}^{x_2} [\varphi'^2 + \tfrac{1}{2}e_{yy}(x, 0)\varphi^2]\, dx.$$

On the other hand, by Gauss's *theorema egregium*, the Gaussian curvature $K(x)$ of the metric ds^2 at the point $(x, 0)$ of the geodesic is given by

$$K(x) = -\tfrac{1}{2}e_{yy}(x, 0).$$

Thus we arrive at the formula

$$\mathscr{Q}(\varphi) = \frac{1}{2}\int_{x_1}^{x_2} [\varphi'^2 - K(x)\varphi^2]\, dx$$

for the second variation.

Applying the results of [1] and [2], we see that *the extremal* $u(x) \equiv 0$ *contains no pair of conjugate points if* $K(x) \le 0$. If $K(x) \ge \kappa > 0$, the distance of two consecutive conjugate values ξ_1, ξ_2 is bounded from above by

$$|\xi_1 - \xi_2| \le \pi/\sqrt{\kappa}.$$

Since $e(x, 0) = 1$, the Euclidean distance $|\xi_1 - \xi_2|$ of the two conjugate values ξ_1 and ξ_2 coincides with the geodesic distance $\int_{\xi_1}^{\xi_2} \sqrt{e(x, u(x)) + u'(x)^2}\, dx$ of the two points $P_1 = (\xi_1, 0)$ and $P_2 = (\xi_2, 0)$ measured along the geodesic $u(x) \equiv 0$. Therefore we have found that *the geodesic distance of two consecutive conjugate points on the geodesic represented by* $u(x) \equiv 0$ *is at most* $\pi/\sqrt{\kappa}$ *if* $K(x)$ *satisfies the inequality* $K(x) \ge \kappa > 0$.

By the remarks made at the beginning, these results hold for any general two-dimensional metric ds^2 and for every geodesic because the general situation can always be transformed to the special one.

For a sphere $\mathscr{S}$ of radius R, the next conjugate point P^* to some given point P comes exactly at a distance of πR, i.e., P^* is the antipodal point of P on $\mathscr{S}$. Hence the *first conjugate locus* of all geodesics on $\mathscr{S}$ emanating from a fixed point $P \in \mathscr{S}$ consists of exactly one point, the antipodal point P^* of P. This is a very special case because Carathéodory has shown that, for a compact (closed) convex surface $\mathscr{S}$, the first conjugate locus of some point P contains at least four cusps, except if it degenerates to a single point (cf. Blaschke [5], pp. 231–232), and for an ellipsoid, the first conjugate locus in general has exactly four cusps (see von Mangoldt [1], and von Braunmühl [2], [3]).

[4] *Parabolic orbits and Galileo's law.* Let us consider the variational integral

$$\mathscr{F}(u) := \int_{x_1}^{x_2} \sqrt{H - u}\,\sqrt{1 + u'^2}\, dx,$$

with the Lagrangian

$$F(x, z, p) = \omega(x, z)\sqrt{1 + p^2}, \qquad \omega(x, z) = \sqrt{H - z},$$

where H denotes a positive real number. Its extremals are closely related to the motions $c(t) = (x(t), z(t))$ of a point mass $m = 1$ subjected to Galileo's law

$$\ddot{x} = 0, \qquad \ddot{z} = -g, \quad g > 0.$$

The constant field of forces $(0, -g)$ possesses the potential energy $V(x, z) = gz$. The solutions $(x(t), z(t))$ of this system of differential equations describe the motion of a (very small) body in the gravity field of the earth which we assume to act in the direction of the negative z-axis; the ground be described by the x-axis $\{z = 0\}$.

Suppose that some projectile (point mass) is shot from the point $(0, 0)$ with the speed $v_0 > 0$ at an angle α. That is, $x(t)$ and $z(t)$ are to satisfy the initial conditions

$$x(0) = 0, \quad z(0) = 0, \quad \dot{x}(0) = v_0 \cos\alpha, \quad \dot{z}(0) = v_0 \sin\alpha$$

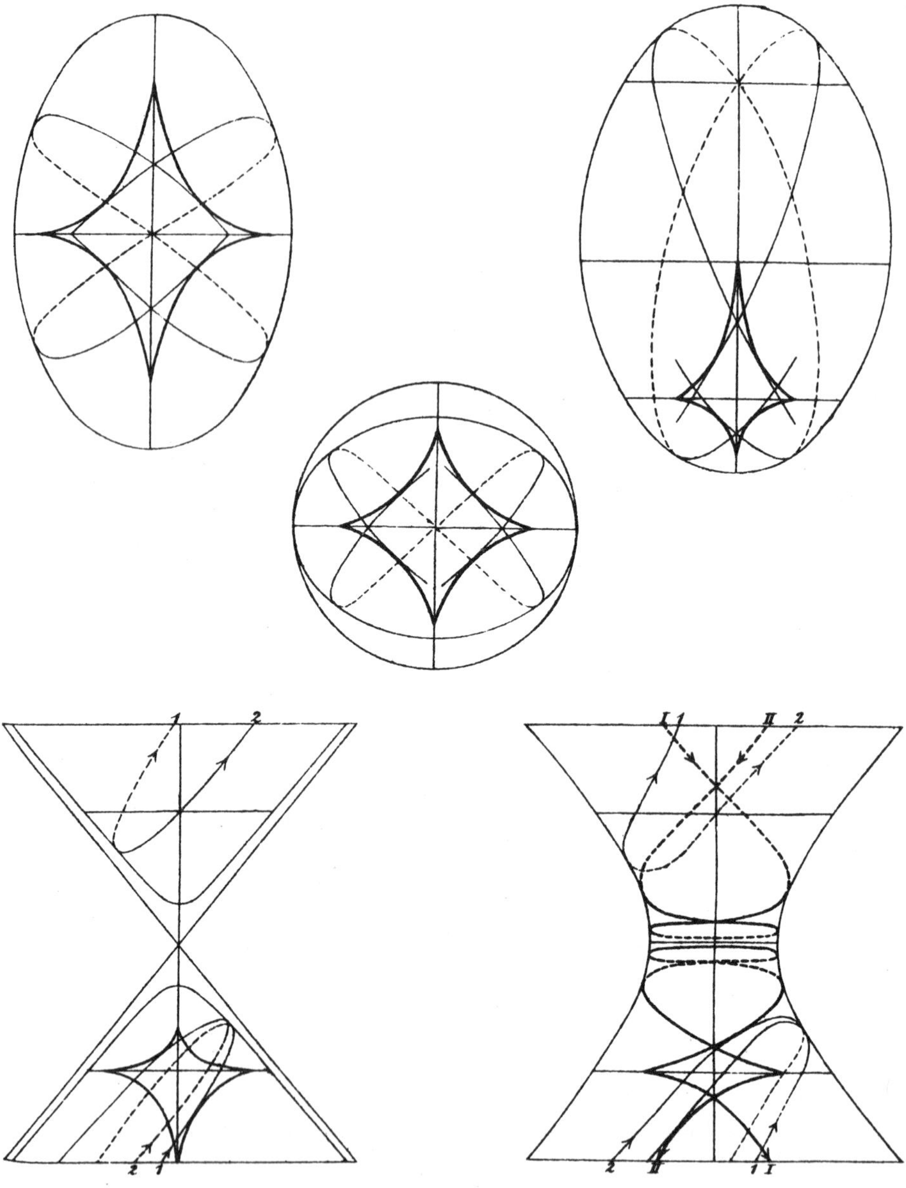

Fig. 7. Envelopes of geodesic flows (after v. Braunmühl [1]).

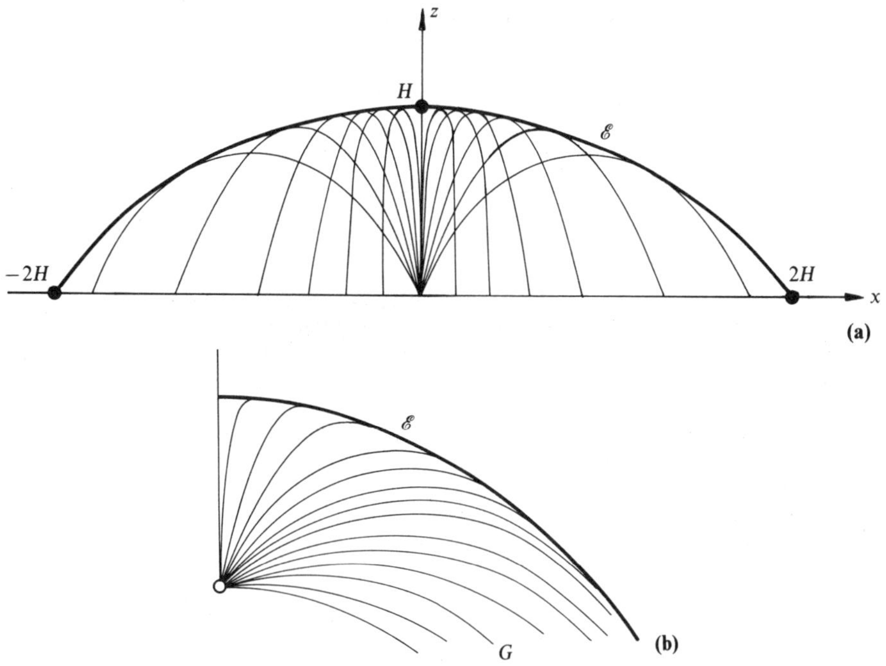

Fig. 8. (a) The envelope $\mathscr{E}$ of Galilei parabolae with a fixed value of v_0. **(b)** A stigmatic field of Galilei parabolae.

at some fixed time $t = 0$. Integration yields

$$x(t) = tv_0 \cos\alpha, \qquad z(t) = -\tfrac{1}{2}gt^2 + tv_0 \sin\alpha .$$

If $\alpha \neq \pm\pi/2$, the orbits of these motions are given by $z = u(x)$, $x \in \mathbb{R}$, where

$$u(x) := -\frac{gx^2}{2v_0^2\cos^2\alpha} + x \operatorname{tang}\alpha .$$

If we set

$$a := \operatorname{tang}\alpha, \qquad H := \frac{v_0^2}{2g},$$

we obtain

$$u(x) = \varphi(x, a) := -\frac{1+a^2}{4H}x^2 + ax, \quad a \in \mathbb{R}.$$

This describes a parabola $\mathscr{C}(a)$ with vertex at $x = \dfrac{2aH}{1+a^2}$. The envelope $\mathscr{E}$ of this one-parameter family of parabolas $\mathscr{C}(a)$ is defined by

$$z = \varphi(x, a), \qquad \varphi_a(x, a) = 0,$$

and a brief computation shows that $\mathscr{E}$ is given by

$$z = -\frac{x^2}{4H} + H .$$

Thus $\mathscr{E}$ is a parabola with the vertex $(0, H)$ which intersects the x-axis at $x = \pm 2H$. The parabola $\mathscr{C}(a)$ touches $\mathscr{E}$ at exactly one point $P^*(a) = (\xi(a), \zeta(a))$ given by

$$\xi(a) = \frac{2H}{a}, \quad \zeta(a) = \frac{a^2 - 1}{a^2} H, \qquad \text{if } a \neq 0.$$

Set $v(t) := \sqrt{\dot{x}^2(t) + \dot{y}^2(t)}$. Then conservation of total energy yields

$$\tfrac{1}{2}v^2 + gz = \text{const} = \tfrac{1}{2}v_0^2 = gH$$

or

$$v = \sqrt{2g}\sqrt{H - z}.$$

On account of 3,*1* [3], we infer that the parabolic orbits $z = \varphi(x, a)$, $x_1 \le x \le x_2$, are extremals of the action functional $\mathscr{F}$ defined at the beginning. The point $P^*(a)$ is the only point on $\mathscr{C}(a)$ which is conjugate to $P_0 = (0, 0)$. If $a > 0$, the only conjugate value $\xi(a)$ to 0 lies to the right of $x = 0$, and it lies to the left if $a < 0$. In the exceptional case $a = 0$ there is no value conjugate to $x = 0$ (with respect to $\mathscr{C}(0)$). Hence

$$u(x) = -\frac{1 + a^2}{4H} x^2 + ax, \quad x_1 \le x \le x_2,$$

is a weak minimizer of $\mathscr{F}$ as long as $x_2 < \xi(a)$ holds. This minimizing property gets lost as soon as $x_2 > \xi(a)$, that is, after the parabola $\mathscr{C}(a)$ has touched the envelope $\mathscr{E}$.

We finally note that $f : \Gamma \to \mathbb{R}^2$ defined by $\Gamma := \{(x, a) \colon a \in \mathbb{R}, 0 < x < \xi(a)\}$ and $f(x, a) := (x, \varphi(x, a))$ defines a field of extremals $z = \varphi(x, a)$, $0 < x < \xi(a)$, covering the domain $G := \left\{(x, z)\colon 0 < x < \infty, z < H - \frac{x^2}{4H}\right\}$.

By a simple coordinate transformation we can reduce the integral

$$\int_{x_1}^{x_2} \sqrt{bu + c}\sqrt{1 + u'^2}\, dx, \quad b, c \in \mathbb{R},\ b \neq 0,$$

to the integral treated before; in particular, we can handle

$$\int_{x_1}^{x_2} \sqrt{u}\sqrt{1 + u'^2}\, dx.$$

[5] *Minimal surfaces of revolution.* In this example we want to discuss the family of catenaries $\mathscr{C}(\alpha)$, given by $z = \varphi(x, \alpha)$ with

$$\varphi(x, \alpha) := \frac{z_1}{\cosh(\alpha)} \cosh\left(\alpha + \frac{x - x_1}{z_1}\cosh\alpha\right),$$

which emanate from a fixed point $P_1 = (x_1, z_1)$ in the upper halfplane of the x, z-space. In particular we shall describe the envelope $\mathscr{E}$ of the catenaries $\mathscr{C}(\alpha)$; it is the locus of the points conjugate to P_1. Furthermore, we shall determine all those points $P_2 = (x_2, z_2)$ with $z_2 > 0$ which can be connected with P_1 by a catenary $\mathscr{C}(\alpha)$. We note that our discussion could be simplified if we would only consider points P_1 and P_2 of equal height above the x-axis (i.e., $z_1 = z_2$).

The reader, who wants to see a collection of the essential results known for the catenaries $\mathscr{C}(\alpha)$ and the associated minimal surfaces of revolution, should consult the list at the end of this example.

Minimal surfaces of revolution are the extremals of the functional

$$\mathscr{A}(u) = 2\pi \int_{x_1}^{x_2} u\sqrt{1 + u'^2}\, dx.$$

As we have seen in 1,*2.2* [7], the extremals are catenaries of the type

$$u(x) = a \cosh\frac{x - b}{a}, \quad x \in \mathbb{R},$$

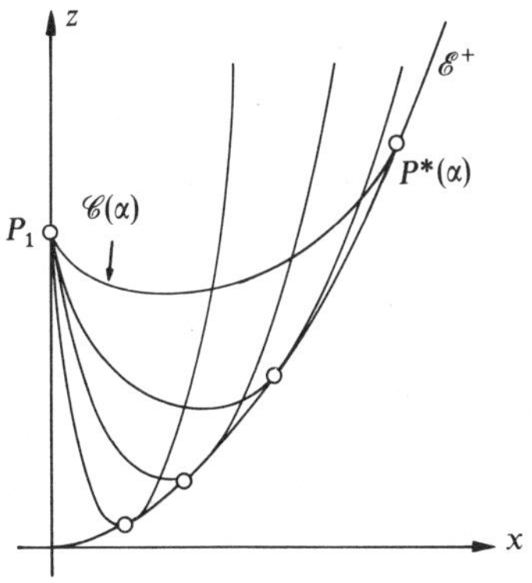

Fig. 9. Catenoids emanating from P_1 and their envelope $\mathscr{E}^+$.

where a, b are arbitrary constants, $a > 0$. For the sake of brevity we write

$$c(x) := \cosh x, \qquad s(x) := \sinh x.$$

Let us fix some point $P_1 = (x_1, z_1)$ in the upper halfplane, i.e., $z_1 > 0$. By means of a new parameter $\alpha \in \mathbb{R}$ we can express the catenaries passing through P_1 by

$$a = \frac{z_1}{c(\alpha)}, \qquad b = x_1 - z_1 \frac{\alpha}{c(\alpha)}.$$

In other words, the catenaries through P_1 are written as one-parameter family $z = \varphi(x, \alpha)$, $x \in \mathbb{R}$, where

$$\varphi(x, \alpha) := \frac{z_1}{c(\alpha)} c\left(\alpha + \frac{x - x_1}{z_1} c(\alpha)\right).$$

Denoting the differentiation with respect to x by $'$, it follows that

$$\varphi'(x, \alpha) = s\left(\alpha + \frac{x - x_1}{z_1} c(\alpha)\right).$$

In particular,

$$\varphi(x_1, \alpha) = z_1, \qquad \varphi'(x_1, \alpha) = s(\alpha).$$

The vertex $P_0(\alpha) = (x_0(\alpha), z_0(\alpha))$ of the catenary $\mathscr{C}(\alpha)$ is given by

$$x_0(\alpha) = x_1 - \frac{z_1 \alpha}{c(\alpha)}, \qquad z_0(\alpha) = \frac{z_1}{c(\alpha)}.$$

On every catenary $\mathscr{C}(\alpha)$, for which P_1 does not fall into the vertex $P_0(\alpha)$, there is exactly one conjugate point $P^*(\alpha)$ to P_1. It is obtained by a construction due to L. Lindelöf, described in the following

Proposition 1. *If P_1 coincides with the vertex $P_0(\alpha)$ of $\mathscr{C}(\alpha)$, i.e., $x_1 = x_0(\alpha)$, then there is no conjugate point to P_1 on $\mathscr{C}(\alpha)$. If $P_1 \neq P_0(\alpha)$, then there exists exactly one conjugate point $P^*(\alpha) = (\xi(\alpha), \zeta(\alpha))$ to P_1 on $\mathscr{C}(\alpha)$, and we have*

$$\xi(\alpha) > x_0(\alpha) \quad \textit{if } x_1 < x_0(\alpha),$$

$$\xi(\alpha) < x_0(\alpha) \quad \textit{if } x_1 > x_0(\alpha),$$

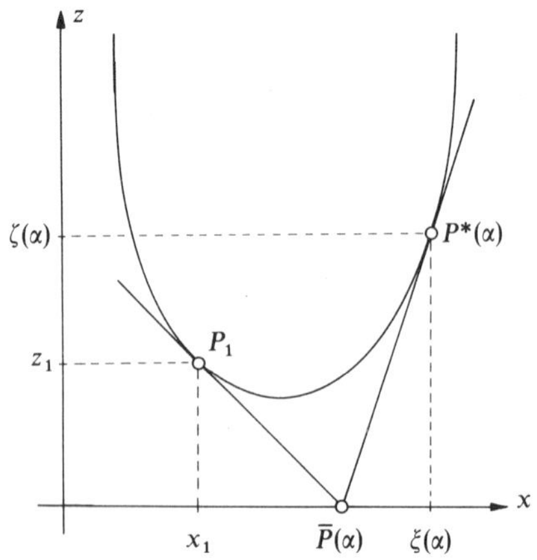

Fig. 10. Lindelöf's construction of the conjugate point $P^*(\alpha)$ to P_1 on $C(\alpha)$.

that is, P_1 and its conjugate point $P^*(\alpha)$ lie on opposite branches of the catenary with respect to the vertex $P_0(\alpha)$.

In order to obtain $P^*(\alpha)$, we draw the tangent $\mathscr{T}_1(\alpha)$ to $\mathscr{C}(\alpha)$ at P_1 and determine its intersection point $\bar{P}(\alpha)$ with the x-axis. Then we draw the tangent $\mathscr{T}(\alpha)$ from $\bar{P}(\alpha)$ to the opposite branch of $\mathscr{C}(\alpha)$. It touches $\mathscr{C}(\alpha)$ exactly at the conjugate point $P^*(\alpha)$.

Proof. Differentiating $\varphi(x, \alpha)$ with respect to α, we obtain

$$\varphi_\alpha(x, \alpha) = -\frac{z_1}{c^2(\alpha)} c'(\alpha) c\left(\alpha + \frac{x - x_1}{z_1} c(\alpha)\right)$$
$$+ \frac{z_1}{c(\alpha)} c'\left(\alpha + \frac{x - x_1}{z_1} c(\alpha)\right)\left[1 + \frac{x - x_1}{z_1} c'(\alpha)\right],$$

which can be rewritten as

$$\varphi_\alpha(x, \alpha) = \frac{1}{c(\alpha)}\{-\varphi(x, \alpha)s(\alpha) + z_1\varphi'(x, \alpha) + x\varphi'(x, \alpha)s(\alpha) - x_1\varphi'(x, \alpha)s(\alpha)\}.$$

If $P_1 = P_0(\alpha)$, we have $\varphi'(x_1, \alpha) = s(\alpha) = 0$, and therefore

$$\varphi_\alpha(x, \alpha) = \frac{z_1\varphi'(x, \alpha)}{c(\alpha)}.$$

But for $x \neq x_1 = x_0(\alpha)$ it follows that $\varphi'(x, \alpha) \neq 0$, and consequently $\varphi_\alpha(x, \alpha) \neq 0$. That is, P_1 possesses no conjugate point on $\mathscr{C}(\alpha)$ if it coincides with the vertex $P_0(\alpha)$ of $\mathscr{C}(\alpha)$.

Suppose now that $P_1 \neq P_0(\alpha)$, i.e., $s(\alpha) \neq 0$, and consider some point $P = (x, z)$ on $\mathscr{C}(\alpha)$ different from P_1. Let $\mathscr{T}_1(\alpha)$ and $\mathscr{T}(\alpha)$ be the tangent lines to $\mathscr{C}(\alpha)$ at P_1 and P, respectively. Since $\mathscr{C}(\alpha)$ is strictly convex, $\mathscr{T}_1(\alpha)$ and $\mathscr{T}(\alpha)$ are not parallel and intersect at some point $\bar{P}(\alpha) = (\bar{\xi}(\alpha), \bar{\zeta}(\alpha))$.

If $\varphi'(x, \alpha) = 0$, then $\mathscr{T}(\alpha)$ is the parallel to the x-axis through $P_0(\alpha)$, and we obtain

$$\bar{\zeta}(\alpha) = z_0(\alpha).$$

If, however, $\varphi'(x, \alpha) \neq 0$, the coordinates of the intersection point $\bar{P}(\alpha)$ have to satisfy

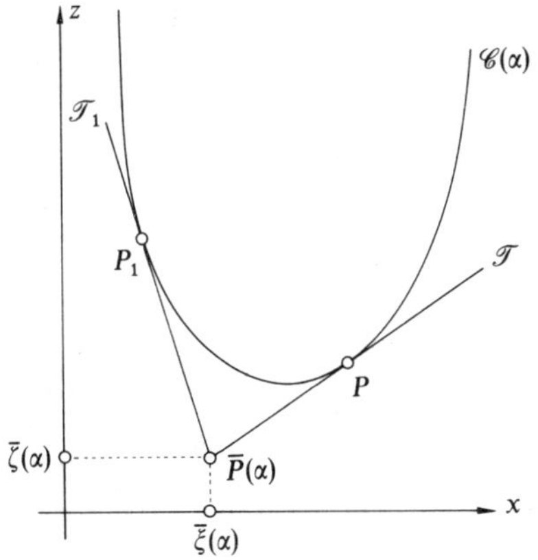

Fig. 11. The tangent $\mathscr{T}$ and $\mathscr{T}_1$ in the proof of Proposition 1.

$$\bar{\xi}(\alpha) = x_1 + \frac{\bar{\zeta}(\alpha) - z_1}{s(\alpha)} \quad \text{and} \quad \bar{\xi}(\alpha) = x + \frac{\bar{\zeta}(\alpha) - \varphi(x, \alpha)}{\varphi'(x, \alpha)},$$

whence we obtain

$$\bar{\zeta}(\alpha) = \frac{\varphi'(x, \alpha)s(\alpha)}{\varphi'(x, \alpha) - s(\alpha)} \left\{ x - x_1 - \frac{\varphi(x, \alpha)}{\varphi'(x, \alpha)} + \frac{z_1}{s(\alpha)} \right\}.$$

This together with the previous formula for $\varphi_\alpha(x, \alpha)$ implies

$$\bar{\zeta}(\alpha) = \frac{c(\alpha)\varphi_\alpha(x, \alpha)}{\varphi'(x, \alpha) - s(\alpha)}.$$

These two formulas for $\bar{\zeta}(\alpha)$ imply the general result

$$\varphi_\alpha(x, \alpha) = \frac{\varphi'(x, \alpha) - s(\alpha)}{c(\alpha)} \bar{\zeta}(\alpha),$$

from which the other claim of the proposition follows at once if we also take the strict convexity of $\mathscr{C}(\alpha)$ into account. □

The complete description of the situation is contained in the following

Proposition 2. (*i*) *The branch $\mathscr{E}^+ := \mathscr{E} \cap \{x > x_1\}$ of the envelope $\mathscr{E}$ that is contained in the quarter plane $\{x > x_1, z > 0\}$ can be described as graph of a strictly convex, real analytic function $h(x)$, $x_1 < x < \infty$, which satisfies $h'(x) > 0$, $h''(x) > 0$, and*

$$\lim_{x \to x_1 + 0} h(x) = 0, \qquad \lim_{x \to x_1 + 0} h'(x) = 0,$$
$$\lim_{x \to \infty} h(x) = \infty, \qquad \lim_{x \to \infty} h'(x) = \infty.$$

(*ii*) *Let P be some point in the quarter plane $\{x > x_1, z > 0\}$, and denote by G the set above $\mathscr{E}^+$, or, more precisely,*

$$G = P_1 \cup \{(x, z) \colon x > x_1, h(x) \le z < \infty\}.$$

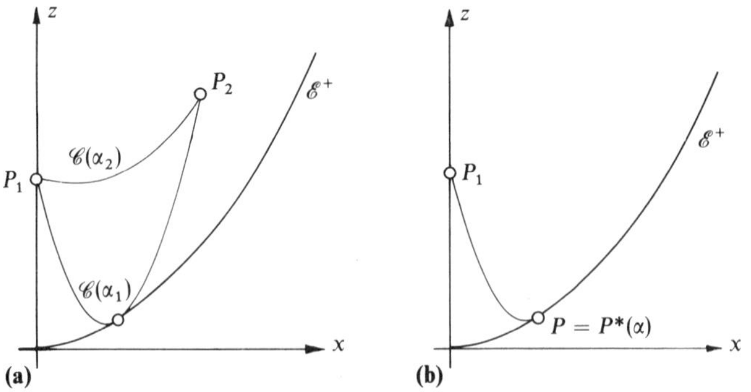

Fig. 12. Catenary arcs joining P_1 and P_2 **(a)** $P_2 \in \operatorname{int} G$; **(b)** $P_2 \in \mathscr{E}^+$.

If $P \notin G$, it cannot be connected with P_1 by any of the catenaries $\mathscr{C}(\alpha)$. Secondly, if $P \in \mathscr{E}^+$, it can be connected with P_1 by exactly one catenary $\mathscr{C}(\alpha)$, and P is conjugate to P_1 on $\mathscr{C}(\alpha)$, that is, $P = P^*(\alpha)$. Thirdly, every point P in $\operatorname{int} G$ can be connected with P_1 by exactly two catenaries $\mathscr{C}(\alpha_1)$ and $\mathscr{C}(\alpha_2)$, $\alpha_1 < \alpha_2$. On $\mathscr{C}(\alpha_1)$, the conjugate point $P^*(\alpha_1)$ lies between P_1 and P, whereas on $\mathscr{C}(\alpha_2)$ the conjugate point $P^*(\alpha_2)$ comes before P_1 or past P. That is, $\mathscr{C}(\alpha_1)$ touches $\mathscr{E}^+$ between P_1 and P and $\mathscr{C}(\alpha_2)$ after P or not at all. Therefore, between x_1 and x, $\varphi(\cdot, \alpha_2)$ is a weak minimizer while $\varphi(\cdot, \alpha_1)$ is not weakly minimizing.

(iii) *The curves $z = \varphi(x, \alpha)$, $x_1 < x < \xi(\alpha)$, $\alpha \in \mathbb{R}$, form a field of extremals on G, and their extension to $x_1 \le x \le \xi(\alpha)$ provides an improper field (in fact, a stigmatic field) covering G.*

The proof of this result is not difficult but tedious; so we break it up in several steps.

Lemma 1. *Let $x > x_1$ be a conjugate value to x_1 with respect to $\mathscr{C}(\alpha)$, i.e., $\varphi_\alpha(x, \alpha) = 0$. Then it follows that $\alpha < 0$, $\varphi'(x, \alpha) > 0$, and*

$$\varphi''(x, \alpha) > 0, \qquad \varphi'_\alpha(x, \alpha) < 0, \qquad \varphi_{\alpha\alpha}(x, \alpha) > 0.$$

Proof. The assertions $\alpha < 0$ and $\varphi'(x, \alpha) > 0$ are a consequence of Proposition 1, and $\varphi''(x, \alpha) > 0$ follows from

$$\varphi''(x, \alpha) = \left|\frac{c(\alpha)}{z_1}\right|^2 \varphi(x, \alpha).$$

Secondly,

$$\begin{aligned}\varphi'_\alpha(x, \alpha) &= c\left(\alpha + \frac{x - x_1}{z_1} c(\alpha)\right)\left[1 + \frac{x - x_1}{z_1} s(\alpha)\right] \\ &= \varphi(x, \alpha) c(\alpha) s(\alpha) \frac{1}{z_1^2}\left[\frac{z_1}{s(\alpha)} + x - x_1\right].\end{aligned}$$

On the other hand, we had proved that $\varphi_\alpha(x, \alpha) = 0$ implies the relation

$$(*) \qquad x - x_1 - \frac{\varphi(x, \alpha)}{\varphi'(x, \alpha)} + \frac{z_1}{s(\alpha)} = 0.$$

Therefore,

$$\varphi'_\alpha(x, \alpha) = \frac{s(\alpha) c(\alpha) \varphi^2(x, \alpha)}{z_1^2 \varphi'(x, \alpha)} < 0.$$

Finally, using $\varphi_\alpha(x, \alpha) = 0$, we obtain

$$\varphi_{\alpha\alpha}(x, \alpha) = \frac{\varphi'(x, \alpha)s^2(\alpha)}{z_1^2}\left\{\frac{\varphi(x, \alpha)\varphi'_\alpha(x, \alpha)z_1^2}{s(\alpha)c(\alpha)\varphi'(x, \alpha)^2} - \frac{z_1^3}{s(\alpha)^3}\right\}.$$

Inserting the expression for $\varphi'_\alpha(x, \alpha)$ derived before, we arrive at

$$\varphi_{\alpha\alpha}(x, \alpha) = \frac{\varphi'(x, \alpha)s^2(\alpha)}{z_1^2}\left\{\left[\frac{\varphi(x, \alpha)}{\varphi'(x, \alpha)}\right]^3 - \left[\frac{z_1}{s(\alpha)}\right]^3\right\}.$$

Moreover, relation $(*)$ yields

$$\frac{\varphi(x, \alpha)}{\varphi'(x, \alpha)} = x - x_1 + \frac{z_1}{s(\alpha)} > \frac{z_1}{s(\alpha)},$$

whence it follows that

$$\varphi_{\alpha\alpha}(x, \alpha) > 0. \qquad \square$$

Lemma 2. *For any fixed value x with $x > x_1$, there is a value $\sigma(x) < 0$ such that $\varphi(x, \alpha)$ is decreasing for $-\infty < \alpha < \sigma(x)$ and increasing for $\sigma(x) < \alpha < \infty$, and $\lim_{\alpha \to \pm\infty} \varphi(x, \alpha) = \infty$. Moreover, we have $\varphi_\alpha(x, \alpha) < 0$ for $-\infty < \alpha < \sigma(x)$ and $\varphi_\alpha(x, \alpha) > 0$ for $\sigma(x) < \alpha < \infty$. The value x is conjugate to x_1 with respect to the catenary $\mathscr{C}(\sigma(x))$.*

Proof. The relations $\lim_{\alpha \to \pm\infty} \varphi(x, \alpha) = \infty$ are easily verified, and they imply the existence of some $\alpha_0 \in \mathbb{R}$ such that $\varphi(x, \alpha_0) = \inf_{\alpha \in \mathbb{R}} \varphi(x, \alpha)$ holds, whence we infer $\varphi_\alpha(x, \alpha_0) = 0$. Therefore x is conjugate to x_1 with respect to $\mathscr{C}(\alpha_0)$. On account of Proposition 1, it follows that $s(\alpha_0) = \varphi'(x_1, \alpha_0) < 0$, and therefore $\alpha_0 < 0$.

By Lemma 1, we have $\varphi_{\alpha\alpha}(x, \alpha) > 0$ for every α with $\varphi_\alpha(x, \alpha) = 0$. That is, $\varphi(x, \cdot)$ is strictly convex in a neighbourhood of any stationary point α. We claim that α_0 is the only critical point of $\varphi(x, \cdot)$. In fact, if we had two stationary points α_1 and α_2 with $\alpha_1 < \alpha_2$, both were strict minima. Then there had to be at least one relative maximizer α_3 in (α_1, α_2) which, by $\varphi_{\alpha\alpha}(x, \alpha_3) > 0$, had simultaneously to be a strict minimizer. Since this is impossible, α_0 is the only critical point of $\varphi(x, \cdot)$. Then we infer from $\lim_{\alpha \to \pm\infty} \varphi(x, \alpha) = \infty$ that $\varphi(x, \alpha)$ is decreasing and satisfies $\varphi_\alpha(x, \alpha) < 0$ for $-\infty < \alpha < \alpha_0$, and that $\varphi(x, \alpha)$ is increasing and fulfils $\varphi_\alpha(x, \alpha) > 0$ for $\alpha_0 < \alpha < \infty$. Setting $\sigma(x) := \alpha_0$, the assertion is proved. $\square$

Lemma 3. *Let $x > x_1$, and set $h(x) := \varphi(x, \sigma(x)) = \inf_{\alpha \in \mathbb{R}} \varphi(x, \alpha)$. Then, for every $z \in (h(x), \infty)$, there exist uniquely determined values $\alpha_1 \in (-\infty, \sigma(x))$ and $\alpha_2 \in (\sigma(x), \infty)$ such that $z = \varphi(x, \alpha_1)$ and $z = \varphi(x, \alpha_2)$ holds. Moreover, we have $x_1 < \xi(\alpha_1) < x$ and also $x < \xi(\alpha_2)$ provided that $\alpha_2 < 0$. Finally, the relation $h(x) = \varphi(x, \alpha)$ implies $\alpha = \sigma(x)$.*

The *proof* follows directly from Proposition 1 and Lemma 2.

Lemma 4. *The graph of the function $h(x)$, $x > x_1$, describes the branch $\mathscr{E}^+$ of the envelope $\mathscr{E}$. The function $h(x)$ is real analytic and satisfies $h'(x) > 0$, $h''(x) > 0$, $\lim_{x \to x_1 + 0} h(x) = \lim_{x \to x_1 + 0} h'(x) = 0$, and $\lim_{x \to \infty} h(x) = \lim_{x \to \infty} h'(x) = \infty$.*

Proof. The assertion $\mathscr{E}^+ = \mathrm{graph}(h)$ follows from Lemma 2. From the relation

$$h(x) = \varphi(x, \sigma(x)),$$

we obtain that $h(x)$ is real analytic as soon as we have proved that $\sigma(x)$ is real analytic, and this, by virtue of the implicit function theorem, follows from the fact that $\alpha = \sigma(x)$ is the uniquely determined solution of the equation $\varphi_\alpha(x, \alpha) = 0$ which, according to Lemma 1, satisfies $\varphi_{\alpha\alpha}(x, \sigma(x)) > 0$. Differentiating $h(x) = \varphi(x, \sigma(x))$, we obtain

$$h'(x) = \varphi'(x, \sigma(x)) + \varphi_\alpha(x, \sigma(x))\sigma'(x) = \varphi'(x, \sigma(x)) > 0$$

and

$$h''(x) = \varphi''(x, \sigma(x)) + \varphi'_\alpha(x, \sigma(x))\sigma'(x).$$

On the other hand, $\varphi_\alpha(x, \sigma(x)) = 0$ yields

$$\varphi'_\alpha(x, \sigma(x)) + \varphi_{\alpha\alpha}(x, \sigma(x))\sigma'(x) = 0,$$

and we arrive at

$$h''(x) = \varphi''(x, \sigma(x)) - \frac{|\varphi'_\alpha(x, \sigma(x))|^2}{\varphi_{\alpha\alpha}(x, \sigma(x))}.$$

Let us temporarily set $\alpha = \sigma(x)$. In the proof of Lemma 1, we have derived formulas for $\varphi'_\alpha(x, \alpha)$ and $\varphi_{\alpha\alpha}(x, \alpha)$ which imply

$$\frac{|\varphi'_\alpha(x, \alpha)|^2}{\varphi_{\alpha\alpha}(x, \alpha)} = \frac{c^2(\alpha)}{z_1^2} \frac{\varphi^4(x, \alpha)s^3(\alpha)}{\varphi^3(x, \alpha)s^3(\alpha) - z_1^3\varphi'(x, \alpha)^3} > 0.$$

Since $\alpha < 0$, it follows that $s(\alpha) < 0$ and therefore

$$z_1^3\varphi'(x, \alpha)^3 - \varphi^3(x, \alpha)s^3(\alpha) > 0.$$

Combining these relations, we infer

$$h''(x) = \frac{\varphi(x, \alpha)z_1\varphi'(x, \alpha)^3c^2(\alpha)}{z_1^3\varphi'(x, \alpha)^3 - \varphi^3(x, \alpha)s^3(\alpha)} > 0.$$

It remains to investigate the limit behaviour of $h(x)$ and $h'(x)$ as $x \to +0$ and $x \to \infty$, respectively.

Consider the vertex

$$P_0(\alpha) = \left(x_1 - \frac{z_1\alpha}{c(\alpha)}, \frac{z_1}{c(\alpha)}\right)$$

of the catenary $\mathscr{C}(\alpha)$ for $\alpha < 0$. Since $P_0(\alpha)$ lies above $\mathscr{E}^+$, we infer from $\lim_{\alpha \to -\infty} P_0(\alpha) = (x_1, 0)$ that $\lim_{x \to x_1 + 0} h(x) = 0$, and secondly we get

$$h(x_0) \le \frac{z_1}{c(\alpha)} \quad \text{for } x_0 = x_1 - \frac{\alpha z_1}{c(\alpha)}.$$

Since we can extend $h(x)$ continuously to $x_1 \le x < \infty$ by setting $h(x_1) = 0$, we can apply the mean value theorem to $h(x_0) - h(x_1)$. Hence there is some $\theta \in (0, 1)$ such that

$$h(x_0) = h(x_0) - h(x_1) = h'(x_1 + \theta(x_0 - x_1))(x_0 - x_1),$$

whence

$$h'(x_1 + \theta(x_0 - x_1)) = \frac{h(x_0)}{x_0 - x_1} < -\frac{1}{\alpha}.$$

Letting $\alpha \to -\infty$, we obtain $h'(x_1) = 0$.

The relation $\lim_{x \to \infty} h(x) = \infty$ is an immediate consequence of the inequalities $h'(x) > 0$ and $h''(x) > 0$. Finally, from

$$\sigma'(x) = -\frac{\varphi'_\alpha(x, \sigma(x))}{\varphi_{\alpha\alpha}(x, \sigma(x))}$$

we infer that $\sigma'(x) > 0$, taking Lemma 1 into account. Fixing some $\underline{x} > x_1$ and setting $\underline{\alpha} := \sigma(\underline{x})$, we then have $\sigma(x) > \underline{\alpha}$ for $x > \underline{x}$, and therefore

$$h'(x) = \varphi'(x, \sigma(x)) = s\left(\sigma(x) + \frac{x - x_1}{z_1}c(\sigma(x))\right) > s\left(\underline{\alpha} + \frac{x - x_1}{z_1}c(\underline{\alpha})\right),$$

since s is increasing. From this equation, we infer

$$\lim_{x\to\infty} h'(x) = \infty,$$

and the lemma is proved. □

The *proof of Proposition 2* can now be given without difficulty by combining the Lemmata 1–4.

We shall carry on the discussion of the catenaries and their minimizing properties in 6,*2.3*, using Weierstrass field theory, and with more details in Chapter 8.

For the reader, who would like to see a *survey of the properties of the catenaries* $\mathscr{C}(\alpha)$, we present the following brief

List of results for the catenaries $\mathscr{C}(\alpha)$:

(i) *The branch $\mathscr{E}^+ := \mathscr{E} \cap \{x > x_1\}$ of the envelope $\mathscr{E}$ of the catenaries $\mathscr{C}(\alpha)$ emanating from the point $P_1 = (x_1, z_1)$, $z_1 > 0$, is the graph of a strictly convex, real analytic function $h(x)$, $x_1 < x < \infty$ with $\lim_{x\to x_1+0} h(x) = 0$ and $\lim_{x\to\infty} h(x) = \infty$.*

(ii) *$\mathscr{E}^+$ is the locus of the conjugate points $P^*(\alpha)$ to the right of P_1.*

(iii) *On every catenary $\mathscr{C}(\alpha)$, there is exactly one conjugate point $P^*(\alpha)$ with respect to P_1, except if P_1 coincides with the vertex of $\mathscr{C}(\alpha)$. The point $P^*(\alpha)$ lies to the right of P_1 if P_1 is contained in the descending branch of $\mathscr{C}(\alpha)$; otherwise $P^*(\alpha)$ lies to the left of P_1.*

(iv) *If P_2 is a point on $\mathscr{E}^+$, then there is exactly one catenary $\mathscr{C}(\alpha)$ connecting P_1 and P_2, and P_2 is the conjugate point $P(\alpha)$ on $\mathscr{C}(\alpha)$.*

(v) *If $P_2 = (x_2, z_2)$, $x_1 < x_2$, $z_2 > 0$, then P_1 and P_2 can be connected by two catenaries $\mathscr{C}(\alpha_1)$ and $\mathscr{C}(\alpha_2)$ provided that P_2 lies in the domain between $\mathscr{E}^+$ and the axis $\{x = x_1\}$. The "lower" catenary*

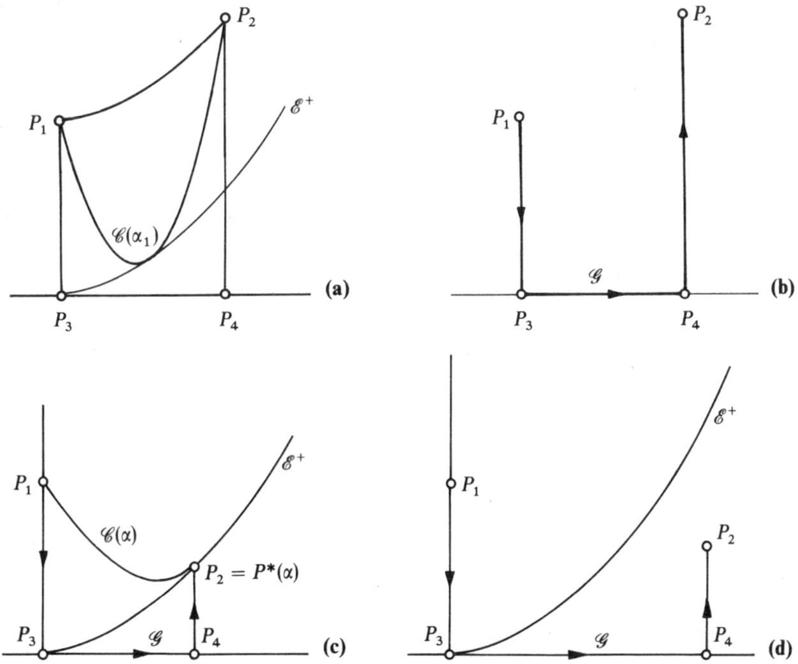

Fig. 13. **(a)** P_1 and P_2 are connected by two catenary arcs $C(\alpha_1)$, $C(\alpha_2)$ and the Goldschmidt curve $\mathscr{G}$. **(b)** Goldschmidt curve $\mathscr{G}$. **(c)** P_1 and P_2 are connected by one catenary and $\mathscr{G}$. **(d)** P_1 and P_2 are only connected by $\mathscr{G}$; there is no catenary linking P_1 with P_2.

$\mathscr{C}(\alpha_1)$ touches $\mathscr{E}^+$ before P_2 and thus contains the conjugate point $P^*(\alpha)$ between P_1 and P_2. Consequently the arc of $\mathscr{C}(\alpha_1)$ between P_1 and P_2 is not minimizing, not even in the weak sense. The "upper" catenary $\mathscr{C}(\alpha_2)$ lies above $\mathscr{C}(\alpha_1)$ and contains no conjugate point between P_1 and P_2. The arc of $\mathscr{C}(\alpha_2)$ between P_1 and P_2 is a strict strong minimizer.

On the other hand, P_1 and P_2 cannot be connected by any catenary $\mathscr{C}(\alpha)$ if P_2 lies to the right of $\mathscr{E}^+$. In this case, the Goldschmidt curve $\mathscr{G}$ connecting P_1 and P_2 (cf. Fig. 13) is the unique absolute minimizer; moreover, there is no other relative minimizer.

(*vi*) *In any case, the Goldschmidt curve $\mathscr{G}$ connecting P_1 and P_2 is a strict strong minimizer. If P_2 is "sufficiently close" to P_1, then the upper catenary $\mathscr{C}(\alpha_2)$ yields the absolute minimizer; otherwise $\mathscr{G}$ furnishes a smaller value for $\mathscr{F}$ than $\mathscr{C}(\alpha_2)$. The switch from $\mathscr{C}(\alpha_2)$ to $\mathscr{G}$ being the absolute minimizer occurs on a parabola-like curve $\mathscr{M}$ between $\mathscr{E}^+$ and the axis $\{x = x_1\}$.*

3. Scholia

1. The eigenvalue criteria for the second variation have a long history with many ramifications to other fields, in particular to eigenvalue problems of differential operators. We shall not try to describe this fascinating development which has led to some of the most important discoveries in mathematics, but we will only mention some of the chief steps connected with the investigation of the second variation.

The history begins with H.A. Schwarz's celebrated paper "*Über ein die Flächen kleinsten Inhalts betreffendes Problem der Variationsrechnung*" (Acta soc. sci. Fenn. *15*, 315–362 (1885), cf. Ges. Math. Abh. [2], Vol. 1, pp. 223–269). There, for instance, one finds for the first time the minimum characterization of the smallest eigenvalue of an elliptic operator. Schwarz's discovery greatly influenced the further development of the eigenvalue theory by E. Schmidt, H. Weyl, and R. Courant. In particular, it led to the maximum-minimum characterization of eigenvalues by Fischer–Weyl–Courant; cf. Courant-Hilbert [1, 2], Vol. 1 and the last chapter of Vol. 2.

Combining the ideas of Schwarz with Hilbert's *analysis of infinitely many variables*, Lichtenstein developed a powerful method, by which he successfully treated the second variation theory of single and double integrals. E. Hölder used Lichtenstein's method to complete the one-dimensional theory in several respects; in particular, he treated the Lagrange problem. Other important contributions are due to Boerner, Radon, G.D. Birkhoff, M. Morse and the Chicago school (for instance, Bliss, Reid, Hestenes). Morse discovered the connection between the eigenvalue problem for the second variation of geodesics and the topology of Riemannian manifolds, thus founding global analysis. A *selection of pertinent references* is Lichtenstein [1]–[10], in particular [6]; E. Hölder [1], [3], [4], [6]–[10]; Boerner [1]; Radon [3], [4]; Bliss [1]–[5]; Bliss–Hestenes [1]; Morse [1]–[4]; Reid [1, 3, 4, 5]; McShane [2], [3]; Hestenes [1, 2, 4, 5] cf. also *Contributions to the Calculus of Variations* 1938–1945 (University of Chicago).

The development of functional analysis and of the theory of strongly elliptic systems of partial differential equations permitted to simplify Lichtenstein's method considerably, and to apply it to multiple integrals. We particularly refer to van Hove [4], Hestenes [4], E. Hölder [8], [9], Klötzler [2], [4], and Hildebrandt [1].

2. In *3.5*, the Jacobi theory for one-dimensional problems is developed independently of the eigenvalue method of Schwarz–Lichtenstein, and very close to the original ideas of Legendre and Jacobi. Except for a few minor simplifications, we have essentially followed the presentation of Bolza [3], Chapters 2 and 3.

3. Sturm's celebrated *oscillation theorem* as well as his *comparison theorem* have been proved[9]

[9] Sturm, *Mémoire sur les équations différentielles du second ordre*, Journal de Liouville *1*, 106–186 (1836).

in 1836; the first rigorous proofs are due to M. Bôcher (1898).[10] Today, many extensions of Sturm's theory have been found. Of particular importance are Sturm's comparison theorems in differential geometry where they are used in a form discovered by Rauch; they are known to differential geometers as *Rauch's comparison theorems*. A presentation of these results can be found in Gromoll–Klingenberg–Meyer [1] and Cheeger–Ebin [1].

4. The notion of a *field of curves* (*extremals*) is due to Weierstrass. It has been suggested to leave the term *field* to the algebrists and to replace it by some more modern expression, say, *foliation* (cf. Hadamard [4], L.C. Young [1]); but, in the whole, analysts seem to stick to the old name.

5. In the discussion of minimal surfaces of revolution (*2.4* [5]) we have essentially followed Bliss [2]. We refer the reader also to the treatments in Carathéodory [10], pp. 281–283, and in Bolza [3].

6. Lindelöf's construction of conjugate points. In *2.4* [5] we have seen how pairs of conjugate points of extremals for the integral $\int u\sqrt{1+u'^2}\,dx$ can be found by a geometric construction, which is due to E.L. Lindelöf [1]. According to this construction two points P and P^* on an extremal $\mathscr{C}$ are conjugate to each other if the tangents to $\mathscr{C}$ at P and P^* respectively intersect the x-axis in the same point. This construction remains valid for a considerably larger class of functionals. In fact, let

$$\mathscr{F}(u) := \int_{x_1}^{x_2} F(u(x), u'(x))\,dx \quad (n = N = 1), \tag{1}$$

be a functional with the property that for any extremal u of $\mathscr{F}$ and for all real values of α and β, $\alpha > 0$, the function

$$\varphi(x, \alpha, \beta) := \frac{1}{\alpha} u(\alpha(x + \beta)), \quad x_1 \le x \le x_2 \tag{2}$$

defines a two-parameter family of extremals of $\mathscr{F}$. This, for instance, holds true if $\mathscr{F}$ admits the 2-parameter group of similarity transformations of $\mathbb{R}^2$ mapping the x-axis into itself, because the transformations $(x, z) \mapsto (x^*, z^*)$ are of the form

$$x^* = ax + b, \qquad z^* = az.$$

Since the integrand of $\mathscr{F}$ is of the form $F(z, p)$, the functional $\mathscr{F}$ is certainly invariant under the

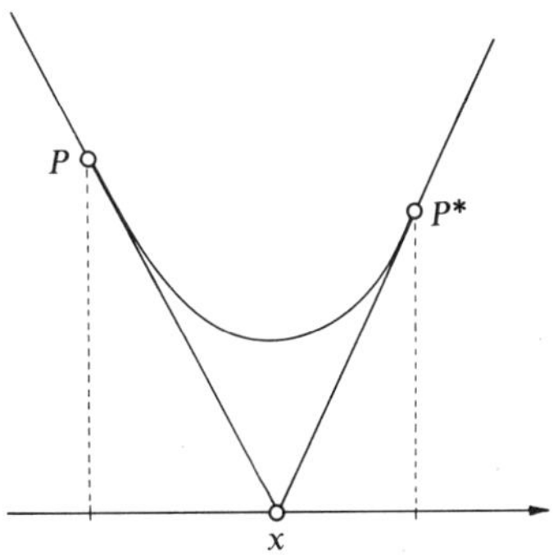

Fig. 14. Lindelöf's construction of the conjugate point P^* to P.

[10] Bull. American Math. Soc. *4*, 295–313 & 365–376 (1898); cf. also Kamke [3], pp. 125–128, and 260–261.

translations $x^* = x + b$, $z^* = z$. Therefore we have only to assume that $\mathcal{F}$ is invariant with respect to the group of similarity transformations $x^* = ax$, $z^* = az$ or, what might be a weaker assumption, that for any extremal $u(x)$ of $\mathcal{F}$ also $\psi(x, \alpha) := (1/\alpha)u(\alpha x)$, $\alpha > 0$, is an extremal of $\mathcal{F}$.

By Jacobi's theorem (cf. the Corollary in *1.2*) the functions

$$v_1(x) := \varphi_\alpha(x, 1, 0) = xu'(x) - u(x)$$

and

$$v_2(x) := \varphi_\beta(x, 1, 0) = u'(x)$$

are Jacobi fields for the extremal $u(x)$. If $uu'' \neq 0$ then v_1, v_2 are linearly independent, and therefore every Jacobi field v along the extremal u can be written in the form $v = \lambda\varphi_\alpha + \mu\varphi_\beta$, $\lambda, \mu \in \mathbb{R}$, that is,

$$v(x) = \mu u'(x) + \lambda[xu'(x) - u(x)]. \tag{3}$$

If $\xi, \xi^* \in [x_1, x_2]$ is a pair of conjugate values for u, then there is a Jacobi field (3) vanishing for $x = \xi$ and $x = \xi^*$. If $u'(\xi) \neq 0$ and $u'(\xi^*) \neq 0$, this implies

$$-\mu/\lambda = \xi - u(\xi)/u'(\xi) = \xi^* - u(\xi^*)/u'(\xi^*) \tag{4}$$

which means that the tangent lines $\ell(x) := u(\xi) + u'(\xi)(x - \xi)$ and $\ell^*(x) := u(\xi^*) + u'(\xi^*)(x - \xi^*)$ of graph u at the points $P = (\xi, u(\xi))$ and $P^* = (\xi^*, u(\xi^*))$ respectively intersect the abscissa in the same point $x = -\mu/\lambda$. Hence P and P^* are related to each other by Lindelöf's construction. On the other hand, if e.g. $u'(\xi) = 0$ and $v(\xi) = 0$ then $\lambda u(\xi) = 0$, and $u(\xi) \neq 0$ implies $\lambda = 0$, i.e. $v(x) = \mu u'(x)$. Therefore we have $v(\xi^*) \neq 0$ for all $\xi^* \neq \xi$ if we assume that the extremal is strictly convex in the sense that $u'' > 0$, and so this case cannot occur if P and P^* are a pair of conjugate points. Thus we have proved the following result: *Let $u(x)$, $x_1 \leq x \leq x_2$, be an extremal of* (1) *such that $u > 0$ and $u'' > 0$, and suppose that $\psi(x, \alpha) := (1/\alpha)u(\alpha x)$, $x_1 \leq x \leq x_2$, is an F-extremal for all $\alpha > 0$. Then any two conjugate points P and P* are related to each other by Lindelöf's construction.*

One easily checks that the latter property is satisfied if $F(z, p)$ is of the form

$$F(z, p) = f(z)g(p) \quad \textit{where } f(z) \textit{ is positively homogeneous}. \tag{5}$$

Hence we in particular obtain: *All strictly convex extremals $u > 0$ of a variational integral of the kind*

$$\int_{x_1}^{x_2} u^s\sqrt{1 + u'^2}\, dx, \tag{6}$$

$s \in \mathbb{R}$, *have the property that any two of its conjugate points are related by Lindelöf's construction.*

Furthermore, the derivative of the Jacobi field (3) is given by

$$v'(x) = (\mu + \lambda x)u''(x);$$

thus $v(x)$ is nontrivial if $u'' > 0$ and $\sqrt{\lambda^2 + \mu^2} > 0$. Therefore *Lindelöf's construction is in our case not only a necessary, but even a sufficient condition for pairs of conjugate points.*

7. A similar reasoning shows that for variational integrals of the kind considered in 6 one can construct the focal points from the knowledge of the two-parameter family of solutions (2), provided that "free transversality" is the same as "orthogonality", which is the case for variational integrals of the kind

$$\int_{x_1}^{x_2} f(u)\sqrt{1 + u'^2}\, dx,$$

see 2,*4*, Remark 2. This construction of focal points is sketched in Funk [1], Kap. IV, 2, Section 9. For the definition of focal points cf. 6,*2.4*.

8. Consider a general variational integral $\mathcal{F}(u) = \int F(x, u, u')\, dx$, $n = N = 1$, and a one-parameter family of extremal curves $\mathcal{C}(\alpha)$ given by

$$z = \varphi(x, \alpha), \quad x \in I,\ \alpha_1 < \alpha < \alpha_2.$$

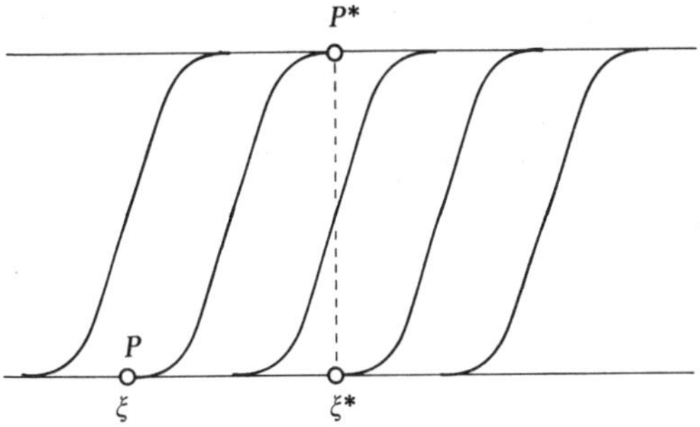

Fig. 15.

For a fixed value α_0 of α, the function $v(x) := \varphi_\alpha(x, \alpha_0)$ is a Jacobi field along $u(x) = \varphi(x, \alpha_0)$, and we assume that $v(x)$ is nontrivial. Then the computation in the beginning of *2.3* shows that $x = \xi$ is a zero of $v(x)$ if and only if the envelope ξ of the family $\mathscr{C}(\alpha)$ touches the extremal curve $\mathscr{C}(\alpha_0)$ at $P = (\xi, u(\xi))$.

Suppose now that $\mathfrak{G} = \{\mathscr{T}_\alpha\}$ is a one-parameter group of transformations $\mathscr{T}_\alpha : \mathbb{R}^2 \to \mathbb{R}^2$ such that $\mathscr{C}(\alpha)$ is the image of a fixed extremal curve $\mathscr{C}$ under $\mathscr{T}_\alpha$, i.e. $\mathscr{T}_\alpha \mathscr{C} = \mathscr{C}(\alpha)$, and assume that $\mathscr{C}$ is not (part of) an orbit of the group $\mathfrak{G}$. Then each contact point P of $\mathscr{C}$ with an orbit of $\mathfrak{G}$ lies on the envelope $\mathscr{E}$ of the family $\{\mathscr{C}(\alpha)\}$. Thus we obtain the following

Theorem. *Suppose that $\mathfrak{G}$ is a one-parameter group of transformations $\mathbb{R}^2 \to \mathbb{R}^2$ mapping extremals into extremals. Then two consecutive contact points of an extremal $\mathscr{C}$ with orbits of $\mathfrak{G}$ are consecutive conjugate points on $\mathscr{C}$.*

For example, the integral $\int F(u, u')\,dx$ admits the group $\mathfrak{G}$ of translations $x^* = x + \alpha$, $z^* = z$. Let $u(x)$ be an extremal, and assume that ξ, ξ^* are two consecutive zeros of $u'(x)$. Then $\varphi(x, \alpha) := u(x + \alpha)$ is a one-parameter family of extremals, and the two parallels of the abscissa $\mathscr{L} := \{(x, z) \colon z = u(\xi)\}$ and $\mathscr{L}^* := \{(x, z) \colon z = u(\xi^*)\}$ belong to the envelope $\mathscr{E}$ of the family of extremal curves $\mathscr{C}(\alpha) := \{(x, z) \colon z = \varphi(x, \alpha)\}$. Moreover, $\mathscr{L}$ and $\mathscr{L}^*$ are orbits of $\mathfrak{G}$ which touch $\mathscr{C} := \{(x, z) \colon z = u(x)\}$ at the points $P = (\xi, u(\xi))$ and $P^* = (\xi^*, u(\xi^*))$. According to the above theorem, P and P^* are (consecutive) conjugate points of $\mathscr{C}$. Of course, this statement is an immediate consequence of the fact that $x = \xi$ and $x = \xi^*$ are (consecutive) zeros of the Jacobi field $v(x) := \varphi_\alpha(x, 0) = u'(x)$ of the extremal $u(x)$.

9. Scheeffer's example presented in *1.1* was published in 1885 in his paper [2]. Scheeffer's papers [1]–[4] were quite influential for the development of analysis, see Goldstine [1], pp. 237–245.

10. Traditionally the *necessary conditions* are listed as follows:

(I) Euler equations for weak local minimizers of class C^2;
(II) Weierstrass's necessary condition for strong minimizers;
(III) Legendre's condition for weak minimizers;
(IV) Jacobi's conjugate point condition.

References:
for (II): 4,*1.2* and *2.1*, Proposition 1;
for (III): 4,*1.3*;
for (IV): *2.2*, Theorem 2.

Chapter 6. Weierstrass Field Theory for One-Dimensional Integrals and Strong Minimizers

The main goal of this chapter is the derivation of *sufficient conditions* for one-dimensional variational problems. That is, we want to establish criteria ensuring that a given extremal u of a variational integral

$$\mathscr{F}(u) = \int_a^b F(x, u(x), u'(x))\, dx$$

is, in fact, a *strong minimizer*. This will be achieved by a method the elements of which were developed by Weierstrass. One of its basic ideas is to consider a whole bundle of extremals instead of a single one, just as one investigates in optics *ray bundles* instead of isolated single rays. This poses the problem to embed a given extremal in an entire pencil of extremals. Among bundles of extremals, those free of singularities are particularly important; they are called *extremal fields*. The curves of such a field, the field lines, cover some domain G of the configuration space simply, i.e., through every point of G there passes exactly one field line. It will turn out that a special kind of extremal fields satisfying certain integrability conditions will be particularly useful for the calculus of variations; these are the so-called *Mayer fields*. A particular feature of every such field is that it defines a scalar function S whose level surfaces form a one-parameter family of hypersurfaces which are transversal to all field curves. This function S is, up to an additive constant, uniquely determined by the field; we call it the *eikonal* (or *optical distance function*) of the field.

Analytically the relations between a Mayer field and its eikonal S are expressed by the *Carathéodory equations*

(1) $$S_x = \overline{F} - \mathscr{P} \cdot \overline{F}_p, \qquad S_{z^i} = \overline{F}_{p^i},$$

where the bar over F and F_p indicates that the arguments (x, z, p) of these functions are to be taken as $(x, z, \mathscr{P}(x, z))$, i.e. $\overline{F}(x, z) := F(x, z, \mathscr{P}(x, z))$, etc., and $\mathscr{P}: G \to \mathbb{R}^N$ denotes the *slope* of the field. That means, the field lines $(x, u(x))$ are defined by the ordinary differential equation $u' = \mathscr{P}(x, u)$.

The use of Mayer fields for the calculus of variations becomes apparent from *Weierstrass's representation formula* which states the following: Let G be a domain in the configuration space covered by some Mayer field. Consider two points P_1 and P_2 which are connected by a field line $(x, u(x))$, $a \le x \le b$, and let $(x, v(x))$, $a \le x \le b$, be an arbitrary nonparametric C^1-curve in G connecting P_1 and P_2. Then we have

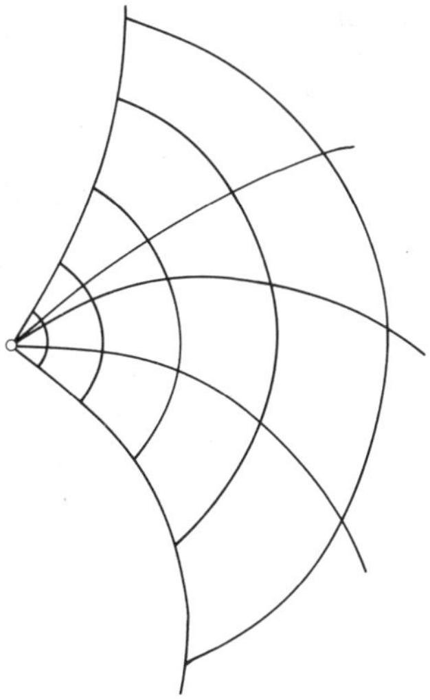

Fig. 1. Light rays and wave fronts.

(2)
$$\mathcal{F}(v) = \mathcal{F}(u) + \int_a^b \mathscr{E}_F(x, v(x), \mathscr{P}(x, v(x)), v'(x))\, dx,$$

where $\mathscr{E}_F$ denotes the excess function of the Lagrangian F.

Consequently, if the excess function is nonnegative, the subarcs of any field line turn out be strong minimizers of $\mathcal{F}$. *Hence, a given extremal is a strong minimizer if it can be embedded into some Mayer field, and if* $\mathscr{E}_F \geq 0$. Thus the problem of establishing sufficient conditions is essentially reduced to the following interesting geometric question: *When is it possible to embed a given extremal into a Mayer field?*

The in some sense optimal answer to this question links the embedding problem with the *theory of conjugate points* discussed in Chapter 5. The link is provided by a special class of Mayer fields, the so-called *stigmatic ray bundles.* Those are bundles of extremals emanating from a single point, just as light rays emanate from a point source.

Using such bundles in an appropriate way, we shall see that *every extremal arc can be embedded into a Mayer field if it does not contain a pair of consecutive conjugate points.* This result then completes our reasoning, and we have found a very satisfactory sufficient condition. By using other Mayer fields to be chosen in a suitable way we can similarly obtain sufficient conditions which guarantee that a given extremal is a minimizer with respect to free or partially free boundary conditions.

Central tools in our discussion will be the notions of *invariant integral* and of *transversality.* The invariant integral links the concepts of Mayer field and null Lagrangian, whereas transversality relates Mayer fields to free boundary conditions.

A different method for deriving sufficient conditions is based on *matrix Riccati equations.*[1] This elegant and very flexible analytic technique is also of classical origin. We shall prefer the geometric approach since it is more intuitive and reflects the close connection between the calculus of variations and geometrical optics. This is important since geometrical optics is the origin of the *Hamilton–Jacobi theory*.

Field theory for multiple integrals is much more complicated and less satisfactorily developed. One of the reasons for this is that there exists a multitude of null Lagrangians if both $n > 1$ and $N > 1$ while there is essentially only one type of null Lagrangians if either $n = 1$ or $N = 1$. Therefore we shall defer the treatment of field theories for multiple integrals to Chapter 7, except for the case of codimension $N = 1$ which is treated in Section 3 of the present chapter.

1. The Geometry of One-Dimensional Fields

In this section we shall present the basic concepts of the one-dimensional field theory developed by Weierstrass, A. Kneser, Zermelo, Hilbert, Mayer and Carathéodory.

We begin in *1.1* by introducing the notions of a *Mayer field* and a *Mayer bundle*, and we state the Carathéodory equations

$$S_x = \bar{F} - \mathscr{P} \cdot \bar{F}_p, \qquad S_z = \bar{F}_p \tag{1}$$

linking the *slope* $\mathscr{P}$ and the *eikonal* S of a Mayer field. The computations are simplified by the use of differential forms. Particularly important is the 1-form

$$\gamma = (F - p \cdot F_p)\, dx + F_{p^i}\, dz^i, \tag{2}$$

which will be denoted as *Beltrami form*. We can use it to write (1) as

$$dS = \not{p}^* \gamma, \tag{3}$$

where $\not{p}(x, z) = (x, z, \mathscr{P}(x, z))$ denotes the *slope field* characterizing the Mayer field, and $\not{p}^*\gamma$ is the pull-back of γ with respect to $\not{p}$.

A field will turn out to be a Mayer field if and only if its *Lagrange brackets* vanish on all of its rays. This way we can easily construct examples of Mayer fields, for instance, the *stigmatic ray bundles* consisting of rays which emanate from a fixed nodal point.

In *1.2* we shall explain why Mayer fields are useful for the calculus of variations, using an idea of Carathéodory. This approach has been called "Carathéodory's royal road to the calculus of variations".

[1] See I.M. Gelfand and S.V. Fomin [1], Chapter 6, and also F.H. Clarke and V. Zeidan [1]. A comprehensive treatment of matrix Riccati equation can also be found in the monograph of W.T. Reid [6].

Then, in *1.3*, we shall present the formalism of field theory using *Hilbert's invariant integral*. The main result is *Weierstrass's representation formula* from which one can read off the minimizing property of all field lines of a Mayer field if the excess function $\mathscr{E}_F$ is nonnegative. This leads to the concept of an *optimal field*. A further consequence of the representation formula is A. Kneser's *transversality theorem* linking the field lines of a Mayer field with the level surfaces of its eikonal. In particular we shall discuss these ideas for *improper fields* such as stigmatic fields.

The problem of embedding an extremal arc into a Mayer field and related issues and examples will be discussed in Section 2.

1.1. Formal Preparations: Fields, Extremal Fields, Mayer Fields and Mayer Bundles, Stigmatic Ray Bundles

The main feature of Weierstrass's field theory to be developed in the sequel is that, instead of a single curve, one always considers curve bundles which simply cover a fixed domain G of the configuration space $\mathbb{R} \times \mathbb{R}^N \triangleq \mathbb{R}^{N+1}$, thought to be the x, z-space. That means that, through every point of G, there passes exactly one curve of the bundle. In the calculus of variations, such a curve bundle is traditionally called a *field on G* (or a *line field*). Since we presently investigate nonparametric variational integrals, *all field lines are thought to be nonparametric curves.* As we saw earlier in the case $N = 1$, extremal fields (i.e., fields of extremal curves) can be used to establish sufficient conditions for weak minima. In the sequel we shall realize that fields can even be employed to derive sufficient conditions for strong minima.

To make the theory of sufficient conditions more transparent and to free it as much as possible from formal computations, it seems advisable to investigate first some of the main features of extremal fields. In particular we wish to characterize such fields in terms of their slopes with respect to the x-axis. This will

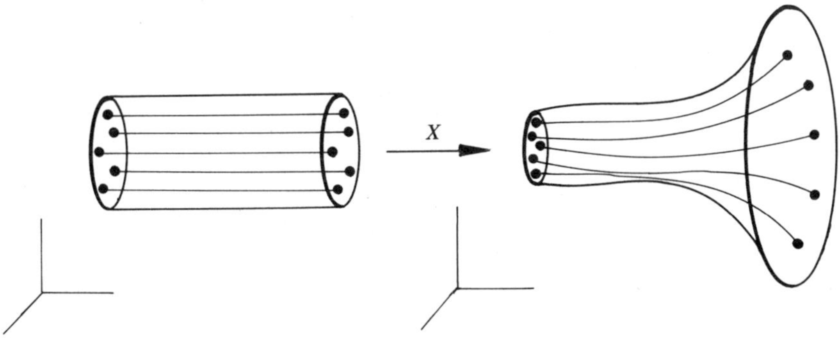

Fig. 2. A field.

be achieved by deriving the *modified Euler equations*. The attempt to write these new equations in a particularly simple form will lead us in a natural way to the notion of a *Mayer field*, to the *Carathéodory equations* and to the *eikonal*. In this way we arrive at the basic notions of the Weierstrass field theory as conceived by Hilbert. Using Carathéodory's approach, we shall see in the next subsection why these notions are basic for the calculus of variations.

To give a precise definition of fields, we adjust the definition of 5,2.3 to the present general case, except that we lessen the regularity requirements to allow for suitable generality. Note that, by definition, fields will only be considered on simply connected sets G of the configuration space $\mathbb{R} \times \mathbb{R}^N$.

Definition 1. *A* field *on a simply connected subset G of the configuration space is a C^1-diffeomorphism $f : \Gamma \to G$ of a set Γ in $\mathbb{R} \times \mathbb{R}^N$ onto G with the following properties:*

(*i*) *The mapping f is of the form*

$$f(x, c) = (x, \varphi(x, c)),$$

and its set of definition Γ is a subset of $\mathbb{R} \times \mathbb{R}^N$ which can be written in the form

$$\Gamma = \{(x, c): c \in I_0, x \in I(c)\},$$

where I_0 is a (nonempty) parameter set in $\mathbb{R}^N$, and $I(c)$ denotes intervals on the real axis with endpoints $x_1(c)$ and $x_2(c)$ which may or may not belong to $I(c)$.

(*ii*) *The partial derivative $f'(x, c)$ $(= f_x(x, c))$ is of class C^1.*

In many situations it will be sufficient to assume that $I(c)$ does not depend on c; in such case $\Gamma = I \times I_0$ is a cylinder above I_0.

The *slope* (or: the *slope function*) $\mathscr{P}(x, z)$ of a field $f : \Gamma \to G$ with $f(x, c) = (x, \varphi(x, c))$ is the uniquely determined C^1-map $\mathscr{P} : G \to \mathbb{R}^N$ such that

$$\varphi' = \mathscr{P}(f), \tag{1}$$

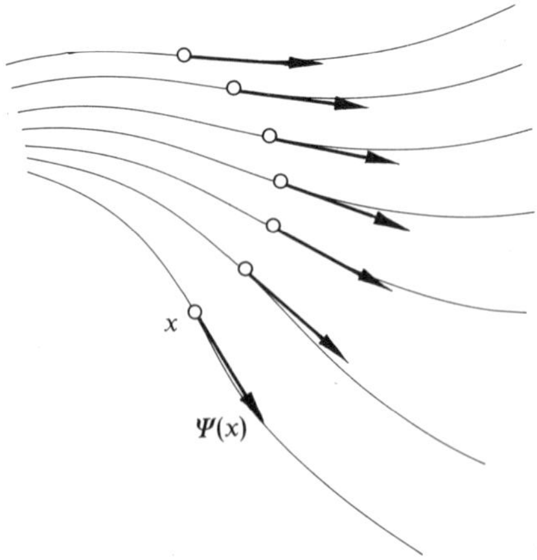

Fig. 3. A direction field of a field.

or equivalently that

$$(1') \qquad f' = (1, \mathscr{P}(f)),$$

holds true. It is easy to express $\mathscr{P}$ in terms of f and of its inverse g. Clearly we can write $g := f^{-1}$ in the form

$$g(x, z) = (x, a(x, z)).$$

Then we have

$$(2) \qquad \varphi(x, a(x, z)) = z, \qquad a(x, \varphi(x, c)) = c,$$

and φ, φ' as well as a, a' are of class C^1. Moreover, (1) is equivalent to

$$(3) \qquad \varphi'(x, c) = \mathscr{P}(x, \varphi(x, c)).$$

Replacing c by $a(x, z)$ and taking (2) into account, we obtain

$$(4) \qquad \mathscr{P}(x, z) = \varphi'(x, a(x, z)).$$

This computation firstly shows that $\mathscr{P}$ exists and is uniquely determined by f, and we secondly see that $\mathscr{P}$ is of class $C^1(G, \mathbb{R}^N)$.

The mapping $f : \Gamma \to G$ and its slope $\mathscr{P} : G \to \mathbb{R}^N$ are interpreted in the following way: If we fix $c = (c^1, \ldots, c^N) \in I_0$, the function $\varphi(\cdot, c)$ describes a nonparametric curve

$$\mathscr{C}(c) = \text{graph } \varphi(\cdot, c) = \{(x, z): x \in I(c), z = \varphi(x, z)\}.$$

Through every point (x, z) of G passes exactly one of the curves $\mathscr{C}(c)$, and $(1, \mathscr{P}(x, z))$ is its direction at (x, z). In other words, (1) expresses the fact that the vector field $\Pi : G \to \mathbb{R}^{N+1}$ defined by $\Pi := (1, \mathscr{P})$ is the *direction field* of the bundle of curves $\mathscr{C}(c)$ which fiberize G, and

$$\mathscr{P}(x, z) = (\mathscr{P}^1(x, z), \mathscr{P}^2(x, z), \ldots, \mathscr{P}^N(x, z))$$

yields the slopes of the corresponding tangent vectors with respect to the z^i-axis.

The curves $\mathscr{C}(c)$ are called the *field lines* of f.

Note that (1) can be written as

$$(5) \qquad f' = \Pi(f),$$

and the equivalent equation (3) means that $u := \varphi(\cdot, c)$ is a solution of the differential equation

$$(6) \qquad u' = \mathscr{P}(x, u).$$

This is expressed by saying that the curves $\mathscr{C}(c)$ "fit" into the direction field $\Pi(x, z) = (1, \mathscr{P}(x, z))$ on G. More generally we say that a curve $\mathscr{C}$, given by

$$\mathscr{C} = \{(x, z): z = u(x), \alpha \le x \le \beta\},$$

fits into the slope field $\mathscr{P}$ (or into Π) if it is contained in G and satisfies (6). In this case we also say that *$\mathscr{C}$ is embedded into the field f with the field lines $\mathscr{C}(c)$*.

We can recover a field f from its slope function $\mathscr{P}$ by solving suitable initial value problems. Hence, slope functions $\mathscr{P}: G \to \mathbb{R}^N$ and curve fields $f: \Gamma \to G$ are essentially equivalent objects since one can obtain one from the other. The special connection (5) or (6) respectively between a field f and its slope function $\mathscr{P}$ is expressed by saying that the curves $\mathscr{C}(c)$ of the field f "fit" into the direction field $\Pi = (1, \mathscr{P})$ on G.

With any slope function $\mathscr{P}$ on G we associate a linear partial differential operator $D_{\mathscr{P}}$ defined by

$$D_{\mathscr{P}} := \frac{\partial}{\partial x} + \mathscr{P}^i(x, z)\frac{\partial}{\partial z^i}\,. \tag{7}$$

We can use this operator to state the Euler equations in a different form called *modified Euler equations.* To this end we introduce the notion of an *extremal field* which will just be a "field of extremals". That is, among arbitrary fields, extremal fields are characterized by the Euler equation $L_F(\varphi) = 0$. Our aim is to characterize extremal fields by a condition on their slope function, and this condition will be expressed by the "modified Euler equation". So let us first precisely define extremal fields.

Definition 2. *A field $f: \Gamma \to G$, given by $f(x, c) = (x, \varphi(x, c))$, is said to be an* extremal field *(with respect to the Lagrangian $F(x, z, p)$) if each of the functions $\varphi(\cdot, c)$ is an extremal with respect to F.*

With any field $f(x, c) = (x, \varphi(x, c))$ we associate its prolongation into the phase space $\mathbb{R} \times \mathbb{R}^N \times \mathbb{R}^N \triangleq \mathbb{R}^{2N+1}$ defined by

$$e(x, c) := (x, \varphi(x, c), \pi(x, c)) \tag{8}$$

where we have set

$$\pi(x, c) := \varphi'(x, c). \tag{9}$$

By definition, a field f is an extremal field if its prolongation e satisfies the Euler equations

$$F_{z^i}(e) - \frac{\partial}{\partial x}F_{p^i}(e) = 0, \quad 1 \le i \le N, \tag{10}$$

in short, if

$$L_F(\varphi) = 0. \tag{10'}$$

For any triple $e(x, c) = (x, \varphi(x, c), \pi(x, c))$, we have the identities

$$F_{z^i}(e) - \frac{\partial}{\partial x}F_{p^i}(e) = F_{z^i}(e) - F_{p^i x}(e) - F_{p^i z^k}(e)\varphi^k_x - F_{p^i p^k}(e)\varphi^k_{xx}\,. \tag{11}$$

If $f(x, c) = (x, \varphi(x, c))$ is a field with the slope $\mathscr{P}(x, z)$, then

$$\varphi' = \mathscr{P}(x, \varphi), \tag{12}$$

and therefore

$$\varphi'' = \mathscr{P}_x(\cdot, \varphi) + \mathscr{P}_{z^\ell}(\cdot, \varphi)\mathscr{P}^\ell(\cdot, \varphi). \tag{13}$$

By inserting (12) and (13) into (11), we arrive at

$$\begin{aligned} F_{z^i}(e) - \frac{\partial}{\partial x} F_{p^i}(e) &= F_{z^i}(e) - F_{p^i x}(e) - F_{p^i z^\ell}(e)\mathscr{P}^\ell(\cdot, \varphi) \\ &\quad - F_{p^i p^k}(e)\mathscr{P}^k_x(\cdot, \varphi) - F_{p^i p^k}(e)\mathscr{P}^k_{z^\ell}(\cdot, \varphi)\mathscr{P}^\ell(\cdot, \varphi). \end{aligned} \tag{14}$$

Because of (9) and (12), the right-hand side of this identity is just the function

$$F_{z^i}(x, z, \mathscr{P}(x, z)) - D_{\mathscr{P}} F_{p^i}(x, z, \mathscr{P}(x, z))$$

composed with the mapping $(x, z) = (x, \varphi(x, c)) = f(x, c)$. On the other hand, the left-hand side of (14) is the i-th component of the Euler operator L_F applied to φ. Hence we can write (14) as

$$L_F(\varphi) = [F_z(\not{p}) - D_{\mathscr{P}} F_p(\not{p})] \circ f, \tag{15}$$

where

$$\not{p}(x, z) = (x, z, \mathscr{P}(x, z)) \tag{16}$$

denotes the "full slope field" consisting of the base points (x, z) and the corresponding slopes $\mathscr{P}(x, z)$.

Agreement. *Given a fixed slope field $\not{p} : G \to \mathbb{R}^{2N+1}$, and an arbitrary function $\phi(x, z, p)$ on $\mathbb{R}^{2N+1}$, we shall write*

$$\bar{\phi} := \phi(\not{p}) = \phi \circ \not{p}, \tag{17}$$

that is,

$$\bar{\phi}(x, z) = \phi(x, z, \mathscr{P}(x, z)) \quad \text{for any} \quad (x, z) \in G. \tag{17'}$$

Then the identity (15) takes the form

$$L_F(\varphi) = (\bar{F}_z - D_{\mathscr{P}} \bar{F}_p) \circ f, \tag{18}$$

and we arrive at the following result:

Proposition 1. *A field $f(x, c) = (x, \varphi(x, c))$ is an extremal field with respect to F (i.e., the relation $L_F(\varphi) = 0$ holds true) if and only if its slope field $\not{p}(x, z) = (x, z, \mathscr{P}(x, z))$ satisfies*

$$\bar{F}_z - D_{\mathscr{P}} \bar{F}_p = 0, \tag{19}$$

i.e., *if and only if*

(19′) $$F_{z^i}(x, z, \mathscr{P}(x, z)) - D_{\mathscr{P}} F_{p^i}(x, z, \mathscr{P}(x, z)) = 0, \quad 1 \le i \le N.$$

Equations (19) or (19′) are the *modified Euler equations* that we were looking for. These new equations characterize extremal fields in terms of their slope functions.

For later use we shall state another and equivalent version of the modified Euler equations (19′) which, in explicit form, read as

$$\bar{F}_{z^i} - \{\bar{F}_{p^i x} + \bar{F}_{p^i p^k}\mathscr{P}^k_x\} - [\bar{F}_{p^i z^\ell} + \bar{F}_{p^i p^k}\mathscr{P}^k_{z^\ell}]\mathscr{P}^\ell = 0.$$

Adding to the left-hand side $\bar{F}_{p^k}\mathscr{P}^k_{z^i} - \bar{F}_{p^\ell}\mathscr{P}^\ell_{z^i} = 0$, we arrive at

$$(\bar{F}_{z^i} + \bar{F}_{p^k}\mathscr{P}^k_{z^i}) - (\bar{F}_{p^i x} + \bar{F}_{p^i p^k}\mathscr{P}^k_x) - [\mathscr{P}^\ell_{z^i}\bar{F}_{p^\ell} + \mathscr{P}^\ell(\bar{F}_{p^i z^\ell} + \bar{F}_{p^i p^k}\mathscr{P}^k_{z^\ell})] = 0,$$

which is equivalent to

(20) $$\frac{\partial}{\partial z^i}\bar{F} - [\mathscr{P}^\ell_{z^i}\bar{F}_{p^\ell} + \mathscr{P}^\ell \frac{\partial}{\partial z^\ell}\bar{F}_{p^i}] = \frac{\partial}{\partial x}\bar{F}_{p^i}.$$

If the conditions

(21) $$\frac{\partial}{\partial z^i}\bar{F}_{p^k} = \frac{\partial}{\partial z^k}\bar{F}_{p^i}$$

are satisfied, the bracket [...] can be written as $\frac{\partial}{\partial z^i}(\mathscr{P}^\ell \bar{F}_{p^\ell})$, and thus equations (20) can be stated as

(22) $$\frac{\partial}{\partial z^i}(\bar{F} - \mathscr{P}^\ell \bar{F}_{p^\ell}) = \frac{\partial}{\partial x}\bar{F}_{p^i}.$$

Let us summarize these results.

Proposition 2. *A field $f : \Gamma \to G$ is an extremal field with respect to the Lagrangian F if and only if its slope function $\mathscr{P} : G \to \mathbb{R}^N$ satisfies equations* (20). *These equations are equivalent to* (22) *if* (21) *holds true.*

This leads us to a particular subclass of extremal fields which will be called *Mayer fields.*

Definition 3. *A field $f : \Gamma \to G$ is said to be a* Mayer field (*with respect to the Lagrangian F*) *if and only if its slope function $\mathscr{P} : G \to \mathbb{R}^N$ satisfies the first order system of partial differential equations*

(23) $$\frac{\partial}{\partial x}\bar{F}_{p^i} = \frac{\partial}{\partial z^i}(\bar{F} - \mathscr{P} \cdot \bar{F}_p), \qquad \frac{\partial}{\partial z^k}\bar{F}_{p^i} = \frac{\partial}{\partial z^i}\bar{F}_{p^k}.$$

Remark 1. If $N = 1$, the system (21) reduces to a single equation which is satisfied by any field. Therefore *every extremal field is automatically a Mayer field* if $N = 1$ while, for $N > 1$, the class of extremal fields turns out to be much larger than the class of Mayer fields.

As we shall see in the next subsections, Mayer fields play a fundamental role if one wants to formulate sufficient conditions for strong minimizers. At the present stage, Mayer fields are not much more than a formal curiosity, as for this special class of extremal fields the modified Euler equations can be written in the particularly elegant form (22). Their relevance for the calculus of variations, which temporarily lies in the dark, will immediately become clear by means of *Carathéodory's approach* which Boerner [5] has called a royal road to the calculus of variations. This approach will be explained in the next subsection.

Presently we content ourselves by giving an interpretation of the defining equations (23) of Mayer fields which will lead to the so-called *Carathéodory equations*, and these will turn out to be the fundamental equations of field theory. The solution of these equations will allow us to compute the integral

$$\int_\alpha^\beta F(x, u(x), u'(x))\, dx$$

along all field curves $\mathscr{C}(c)$ of a Mayer field $f : \Gamma \to G$ in terms of a single scalar function S associated with the field f.

In fact, equations (23) just state that

$$\omega := (\bar{F} - \mathscr{P} \cdot \bar{F}_p)\, dx + \bar{F}_{p^i}\, dz^i \tag{24}$$

is a closed differential form of degree one on G, that is, (23) is equivalent to the relation

$$d\omega = 0. \tag{25}$$

Since G is simply connected, this means that ω is an exact 1-form, i.e., that there is a function $S : G \to \mathbb{R}$ such that

$$\omega = dS. \tag{26}$$

(To avoid any difficulties we think all Lagrangians F to be at least of class C^2.) Equation (26) is equivalent to the first order system of partial differential equations

$$S_x = \bar{F} - \mathscr{P} \cdot \bar{F}_p, \qquad S_{z^i} = \bar{F}_{p^i}, \tag{27}$$

which will be denoted as *Carathéodory equations*. By means of the Schwarzian relations

$$S_{xz^i} = S_{z^i x}, \qquad S_{z^i z^k} = S_{z^k z^i},$$

we can interpret (23) as integrability conditions for the sytem (27); this just rephrases the relation $d\omega = 0$. Since G is simply connected, the integrability conditions (23) are both necessary and sufficient for solving (27). Thus, to any Mayer field f on G with slope $\mathscr{P} : G \to \mathbb{R}^N$, there exists a solution $S : G \to \mathbb{R}$ of the corresponding Carathéodory equations (27). This solution is uniquely determined up to an additive constant and of class $C^2(G)$.

Consider an arbitrary function $u \in C^2([\alpha, \beta], \mathbb{R}^N)$ such that graph $u = \{(x, z): z = u(x), x \in [\alpha, \beta]\}$ "fits" into a Mayer field f on G with the slope $\mathscr{P}$ and with the solution S of (27). By this we mean that

$$\text{graph } u \subset G \tag{28}$$

and that

$$u'(x) = \mathscr{P}(x, u(x)), \quad \alpha \le x \le \beta. \tag{29}$$

In other words, graph u is supposed to be a subarc of one of the field lines $\mathscr{C}(c)$ of the Mayer field f. Let us introduce the mapping $w \in C^2([\alpha, \beta], \mathbb{R}^{N+1})$ by

$$w(x) := (x, u(x)), \quad \alpha \le x \le \beta. \tag{30}$$

Then (28) means that $w([\alpha, \beta]) \subset G$, and (29) can equivalently be written as

$$(\not{p} \circ w)(x) = \not{p}(w(x)) = (x, u(x), u'(x)),$$

whence

$$(\bar{F} \circ w)(x) = F(x, u(x), u'(x)).$$

Moreover, we have

$$0 = du^i - u^{i'}(x)\, dx = du^i - \mathscr{P}^i(x, u(x))\, dx,$$

that is, the pull-back $w^*\theta^i$ of the 1-forms

$$\theta^i := dz^i - \mathscr{P}^i(x, z)\, dx \tag{31}$$

on G vanishes,

$$w^*\theta^i = 0. \tag{32}$$

Hence we obtain for the pull-back $w^*\omega$ of the 1-form

$$\omega = \bar{F}\, dx + \bar{F}_{p^i}\theta^i,$$

the formula

$$w^*\omega = F(x, u(x), u'(x))\, dx. \tag{33}$$

On the other hand, (26) yields

$$w^*\omega = w^*\, dS = d(S \circ w) = dS(x, u(x)), \tag{34}$$

whence

$$F(x, u(x), u'(x))\, dx = dS(x, u(x)), \tag{35}$$

and therefore

$$\int_\alpha^\beta F(x, u(x), u'(x))\, dx = S(\beta, u(\beta)) - S(\alpha, u(\alpha)). \tag{36}$$

Let us write

$$P_1 := (\alpha, u(\alpha)) = w(\alpha), \qquad P_2 := (\beta, u(\beta)) = w(\beta)$$

for the endpoints of $\mathscr{C} :=$ graph u. Then (36) can be brought into the form

$$\mathscr{F}(u) = S(P_2) - S(P_1) \tag{37}$$

if, as usual, the left-hand side of (36) is denoted by $\mathscr{F}(u)$. This means that the integral $\int_{\mathscr{C}} F\,dx$ along any arc $\mathscr{C}$ fitting into the Mayer field f can be expressed as difference of the value of S at the endpoints of $\mathscr{C}$. Interpreting $ds = F(x, u(x), u'(x))\,dx$ as a "Finsler metric", the integral $\int_{\mathscr{C}} F\,dx$ can be viewed as a "distance" between the two points P_1 and P_2 along the "ray" $\mathscr{C}$, and the formula

$$\int_{\mathscr{C}} F\,dx = S(P_2) - S(P_1)$$

tells us that this distance can be computed from the scalar function S. The function S has therefore the meaning of a *distance function* on the field f. In geometrical optics one interprets Mayer fields as *ray bundles* emanating from a light source, and the corresponding functions S are called *optical distance functions* or *phase functions* since the level surfaces $\{S = \text{const}\}$ are viewed as *wave fronts* of light. This is the first glimpse at the particle – wave dualism in optics that in the sequel will be explored time and again. For the sake of brevity we shall denote a solution S of the Carathéodory equations (27) corresponding to the slope $\mathscr{P}$ of a Mayer field f as *eikonal* of the field. This notation, coined by the astronomer Bruns, has also its origin in geometrical optics; its meaning will be discussed in the context of the Hamilton–Jacobi theory.

Remark 2. As a precaution we have to mention that many of the classical authors and also Carathéodory ([10], p. 209) use the notation "extremal field" for the object that we have called "Mayer field". This usage seems somewhat unfortunate to us. On other occasions, Carathéodory denotes Mayer fields as "geodesic fields", and this name has become quite customary. We prefer the notation "Mayer field", used in the American literature, which appears to go back to Bolza (see [3], p. 643). Moreover, we note that the original eikonal of Bruns is not S itself but a function closely related to S.

Sometimes it will be useful to consider *improper fields*. By this we mean mappings $f : \Gamma \to \mathbb{R}^{N+1}$ into the configuration space such that $f|_{\text{int}\,\Gamma}$ is a proper field in the sense of Definition 1.

An important example of an improper field is furnished by *stigmatic bundles of extremals* which are defined as follows (we are not aiming for the most general situation):

Definition 4. *A* stigmatic bundle of extremals (*or*: *a* stigmatic ray bundle) *is a mapping $f : \Gamma \to \mathbb{R}^{N+1}$ of the form*

$$f(x, c) = (x, \varphi(x, c)), \quad (x, c) \in \Gamma,$$

which is defined on a cylinder $\Gamma = I \times I_0$ above some simply connected parameter

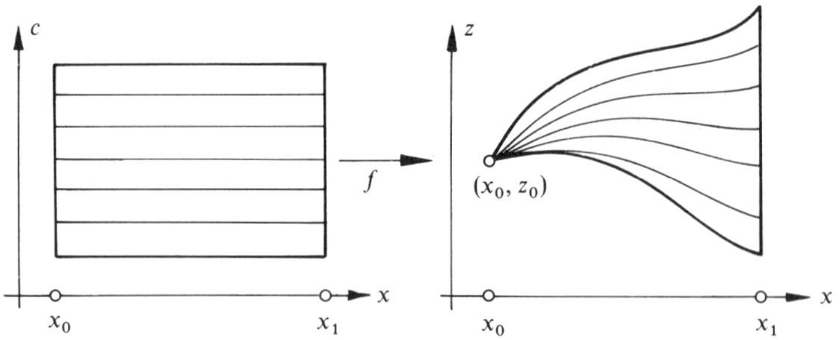

Fig. 4. A stigmatic field.

domain $I_0 \subset \mathbb{R}^N$ and satisfies the following conditions:

(i) f and f' are of class C^1.

(ii) We have $L_F(\varphi) = 0$ for some C^2-Lagrangian F defined on an open set containing $f(\Gamma)$, i.e., the mapping $\varphi(\cdot, c): I \to \mathbb{R}^N$ is an F-extremal for every $c \in I_0$.

(iii) There is some $(x_0, z_0) \in f(\Gamma)$, i.e., $x_0 \in I$ and $z_0 \in \varphi(x_0, I_0)$, such that

$$\varphi(x_0, c) = z_0 \quad \textit{and} \quad \det \varphi_c'(x_0, c) \neq 0$$

for all $c \in I_0$.

The point $P_0 = (x_0, z_0)$ is called a *nodal point* (or *stigma*) of any field f as in Definition 4 since all curves $\mathscr{C}(c) = \text{graph } f(\cdot, c)$ intersect at P_0.

Note that a stigmatic ray bundle can easily be generated by solving the family of initial value problems

$$L_F(\varphi, (\cdot, c)) = 0, \qquad \varphi(x_0, c) = z_0, \qquad \varphi'(x_0, c) = c.$$

We only need to assume some nondegeneracy condition such as $\det F_{pp} \neq 0$ in order to transform the Euler equation $L_F(u) = 0$ into the normal form

$$u'' = \phi(x, u, u'),$$

and $F \in C^3$, so that $\phi \in C^1$.

Let us now consider a stigmatic ray bundle $f : \Gamma \to G$ in the neighbourhood of one of its nodal points (x_0, z_0). The Taylor expansion of $\varphi(x, c)$ with respect to the variable x at the point $x = x_0$ yields

$$\varphi(x, c) = z_0 + \varphi'(x_0, c)(x - x_0) + \dots,$$

where ... stands for higher-order terms in $x - x_0$. Differentiation of $\varphi(x, c)$ with respect to c yields

$$\varphi_c(x, c) = \varphi_c'(x_0, c)(x - x_0) + \dots,$$

whence

(38) $$\det \varphi_c(x, c) = (x - x_0)^N \det \varphi_c'(x_0, c) + \dots .$$

By passing from I_0 to a sufficiently small neighbourhood of some point $c_0 \in I_0$ which will again be denoted by I_0 we can assume that

$$|\det \varphi_c'(x_0, c)| \geq \delta$$

holds true for some $\delta > 0$ and for all $c \in I_0$. Moreover, the Jacobian of f is given by

$$\det Df = \begin{vmatrix} 1 & 0 \\ \varphi' & \varphi_c \end{vmatrix} = \det \varphi_c .$$

Hence we infer that, for any pair $\varepsilon, \varepsilon'$ with $0 < \varepsilon' < \varepsilon \ll 1$, there is a neighbourhood I_0' of c_0 such that f restricted to $[x_0 + \varepsilon', x_0 + \varepsilon] \times I_0'$ or to $[x_0 - \varepsilon, x_0 - \varepsilon'] \times I_0'$ is invertible. Thus we have found:

Proposition 3. *If $f : \Gamma \to \mathbb{R}^{N+1}$ is a stigmatic ray bundle with the nodal point (x_0, z_0), then, to the left and to the right of x_0, the bundle f locally forms a field provided that we stay sufficiently close to the plane $\{x = x_0\}$.*

It is a more subtle question to decide how far to the left and to the right of a nodal point a stigmatic ray bundle locally remains a field. This will be answered later by means of the theory of conjugate points.

Now we want to show that *whenever a part of a stigmatic ray bundle is a field, it is not only an extremal field but even a Mayer field.*

Let us begin by considering an arbitrary n-parameter bundle of extremals $f : \Gamma \to \mathbb{R}^{N+1}$, given by

$$f(x, c) = (x, \varphi(x, c)), \quad (x, c) \in \Gamma,$$

where Γ is a "normal set" as in Definition 1. We assume that $f, f' \in C^1(\Gamma)$, and that $F \in C^2(f(\Gamma) \times \mathbb{R}^N)$; some computations need even $f \in C^2(\Gamma)$. The extremal property of f means that the Euler equation $L_F(\varphi) = 0$ is satisfied.

Denote by $e : \Gamma \to \mathbb{R}^{2N+1}$ the prolongation of f into the phase space which is defined by

(39) $$e(x, c) := (x, \varphi(x, c), \pi(x, c)), \quad \pi(x, c) := \varphi'(x, c).$$

The key result of our computations will be that the so-called *Lagrange brackets*

(40) $$[c^\alpha, c^\beta] := \frac{\partial \tilde{F}_{p^i}}{\partial c^\alpha} \frac{\partial \varphi^i}{\partial c^\beta} - \frac{\partial \tilde{F}_{p^i}}{\partial c^\beta} \frac{\partial \varphi^i}{\partial c^\alpha}$$

of the extremal bundle f are independent of x, i.e., the brackets $[c^\alpha, c^\beta]$ are invariant on every extremal orbit $\mathscr{C}(c) = \operatorname{graph} \varphi(\cdot, c)$. Here the superscript $\sim$ means the composition of F_{p^i} with the mapping e:

(41) $$\tilde{F}_{p^i} := F_{p^i} \circ e = F_{p^i}(e).$$

The invariance of the Lagrange brackets becomes fairly obvious if we use the calculus of differential forms. Our starting point is *Beltrami's form* γ defined on the phase space (or, rather, on an open neighbourhood $\mathscr{U}$ of $f(\Gamma) \times \mathbb{R}^N$) by

(42) $$\gamma := (F - p^i F_{p^i})\, dx + F_{p^i}\, dz^i .$$

The usefulness of this 1-form is suggested by the formulas (24)–(27); note that the 1-form ω defined by (24) can be expressed as $\omega = \not{p}^*\gamma$, i.e., as pull-back of the Beltrami form γ under the slope-field mapping $\not{p}$. Next we introduce on $\mathbb{R}^{2N+1}$ (or on $\mathscr{U}$) the 1-forms

(43) $$\theta^i := dz^i - p^i\, dx$$

and

(44) $$\eta_i := F_{z^i}\, dx - dF_{p^i},$$

$1 \le i \le N$. Then γ can be written as

(45) $$\gamma = F\, dx + F_{p^i}\theta^i,$$

and its exterior differential $d\gamma$ becomes

(46) $$d\gamma = \theta^i \wedge \eta_i .$$

In fact, we have

$$d\gamma = dF \wedge dx + dF_{p^i} \wedge \theta^i + F_{p^i}\, d\theta^i .$$

Since

$$dF \wedge dx = F_{z^i}\, dz^i \wedge dx + F_{p^i}\, dp^i \wedge dx$$

and

$$d\theta^i = -dp^i \wedge dx,$$

we obtain that

$$\begin{aligned} d\gamma &= F_{z^i}\, dz^i \wedge dx + dF_{p^i} \wedge \theta^i \\ &= (dz^i - p^i\, dx) \wedge F_{z^i}\, dx + dF_{p^i} \wedge \theta^i, \end{aligned}$$

and this is just formula (46).

Moreover, the 1-forms η_i can be used to express the Euler equation $L_F(\varphi) = 0$. In fact, if we freeze c and set $\underline{e} := e(\cdot, c)$, the Euler equation is equivalent to the system of equations

$$\underline{e}^*\eta_i = 0, \quad i = 1, 2, \ldots, N .$$

If we allow c to vary, we instead obtain

$$e^*\eta_i = -\frac{\partial \tilde{F}_{p^i}}{\partial c^\alpha}\, dc^\alpha \quad \text{and} \quad e^*\theta^i = \frac{\partial \varphi^i}{\partial c^\beta}\, dc^\beta .$$

Therefore, it follows that

$$
(47)\qquad
\begin{aligned}
e^*(\theta^i \wedge \eta_i) = e^*\theta^i \wedge e^*\eta_i &= -\left(\frac{\partial\varphi^i}{\partial c^\beta}\, dc^\beta\right) \wedge \left(\frac{\partial\tilde{F}_{p^i}}{\partial c^\alpha}\, dc^\alpha\right)\\
&= \sum_{(\alpha,\beta)} [c^\alpha, c^\beta]\, dc^\alpha \wedge dc^\beta,
\end{aligned}
$$

where the summation $\sum_{(\alpha,\beta)}$ is to be extended over all pairs (α, β) with $\alpha < \beta$.

Since $e^*(d\gamma) = d(e^*\gamma)$, we infer from (46) and (47) that

$$
(48)\qquad d(e^*\gamma) = \sum_{(\alpha,\beta)} [c^\alpha, c^\beta]\, dc^\alpha \wedge dc^\beta.
$$

If F and e are of class C^3, we can once again apply the exterior derivative d to both sides of (48) and, on account of the formula $d^2 = dd = 0$, we obtain

$$
\begin{aligned}
0 = &\sum_{(\alpha,\beta)} \frac{\partial}{\partial x}[c^\alpha, c^\beta]\, dx \wedge dc^\alpha \wedge dc^\beta\\
&+ \sum_{(\alpha,\beta)} \frac{\partial}{\partial c^\sigma}[c^\alpha, c^\beta]\, dc^\sigma \wedge dc^\alpha \wedge dc^\beta.
\end{aligned}
$$

This, in particular, implies the desired relations

$$
(49)\qquad \frac{\partial}{\partial x}[c^\alpha, c^\beta] = 0,
$$

which express the fact that the Lagrange brackets are not depending on x.

Since we have only assumed that $F \in C^2$, $f, f' \in C^1$ (instead of $F \in C^3$ and $\varphi \in C^3$), we are not allowed to apply d to (48), and therefore these computations are not quite justified. We could try to make them rigorous by using a suitable approximation device, but it is easier to proceed by a direct computation. First we infer from the Euler equations

$$
(50)\qquad \frac{\partial}{\partial x}\tilde{F}_{p^i} = \tilde{F}_{z^i}
$$

that the functions $\frac{\partial}{\partial x}\tilde{F}_{p^i}$ are of class C^1. Hence Schwarz's theorem implies that $\frac{\partial}{\partial x}\frac{\partial}{\partial c^\alpha}\tilde{F}_{p^i}$ exists and that

$$
\frac{\partial}{\partial x}\frac{\partial}{\partial c^\alpha}\tilde{F}_{p^i} = \frac{\partial}{\partial c^\alpha}\frac{\partial}{\partial x}\tilde{F}_{p^i} = \frac{\partial}{\partial c^\alpha}\tilde{F}_{z^i},
$$

taking (50) into account. Hence the Lagrange brackets $[c^\alpha, c^\beta]$ are differentiable with respect to x, and we obtain

$$\frac{\partial}{\partial x}[c^\alpha, c^\beta] = \frac{\partial}{\partial c^\alpha}\frac{\partial}{\partial x}\tilde{F}_{p^i}\cdot\frac{\partial \varphi^i}{\partial c^\beta} + \frac{\partial}{\partial c^\alpha}\tilde{F}_{p^i}\cdot\frac{\partial \pi^i}{\partial c^\beta}$$

$$- \frac{\partial}{\partial c^\beta}\frac{\partial}{\partial x}\tilde{F}_{p^i}\cdot\frac{\partial \varphi^i}{\partial c^\alpha} - \frac{\partial}{\partial c^\beta}\tilde{F}_{p^i}\cdot\frac{\partial \pi^i}{\partial c^\alpha}$$

$$= \left\{\frac{\partial \tilde{F}_{z^i}}{\partial c^\alpha}\frac{\partial \varphi^i}{\partial c^\beta} + \frac{\partial \tilde{F}_{p^i}}{\partial c^\alpha}\frac{\partial \pi^i}{\partial c^\beta} - \frac{\partial \tilde{F}_{z^i}}{\partial c^\beta}\frac{\partial \varphi^i}{\partial c^\alpha} - \frac{\partial \tilde{F}_{p^i}}{\partial c^\beta}\frac{\partial \pi^i}{\partial c^\alpha}\right\}.$$

A straight-forward computation shows that $\{\ldots\} = 0$, and (49) is finally established.

If f is a stigmatic ray bundle with the nodal point (x_0, z_0), we have $\varphi(x_0, c) = z_0$ whence

$$\varphi_{c^\alpha}(x_0, c) = 0 \quad \text{for all } c \in I_0\,, \ \alpha = 1, \ldots, N\,.$$

This implies

$$[c^\alpha, c^\beta]|_{x=x_0} = 0\,,$$

and since $[c^\alpha, c^\beta]$ is independent of x we arrive at

(51) $$[c^\alpha, c^\beta] = 0 \quad \text{for } 1 \le \alpha, \beta \le N\,.$$

Let us summarize these results.

Proposition 4. *The Lagrange brackets $[c^\alpha, c^\beta]$ of an arbitrary field of extremals f are independent of x, and they vanish identically if f is a stigmatic ray bundle.*

Combining this result with formula (48), we obtain

Proposition 5. *If f is a stigmatic ray bundle and e its prolongation into the phase space, defined by* (39), *then we have*

(52) $$d(e^*\gamma) = 0\,,$$

i.e., the pull-back of the Beltrami form γ with respect to e is a closed 1*-form. Since Γ is simply connected, the form $e^*\gamma$ is even exact, i.e., there is a function $\Sigma(x, c)$ of class $C^2(\Gamma)$ such that*

(53) $$e^*\gamma = d\Sigma\,.$$

We shall see later that stigmatic ray bundles are not the only extremal bundles which satisfy (52) or, equivalently, (53). This leads us to the following

Definition 5. *A* Mayer bundle *is an extremal bundle whose prolongation e into the phase space satisfies $d(e^*\gamma) = 0$.*

Proposition 6. *If $f: \Gamma \to \mathbb{R}^{N+1}$ is a Mayer bundle with the property that, for some subset Γ' of Γ, the restriction $f|_{\Gamma'}$ is a field, then $f|_{\Gamma'}$ is a Mayer field. In*

particular, $f|_{\Gamma'}$ is a Mayer field if $f: \Gamma \to \mathbb{R}^{N+1}$ is a stigmatic ray bundle and if $f|_{\Gamma'}$ is a field, $\Gamma' \subset \Gamma$.

Proof. Suppose that $f: \Gamma \to \mathbb{R}^{N+1}$ is a Mayer bundle and that $f|_{\Gamma'}$ is a field for some set Γ' in Γ. Let $\mathscr{P}(x, z)$ be the slope function of $f|_{\Gamma'}$ on $G' = f(\Gamma')$, and let $\not{p}(x, z) = (x, z, \mathscr{P}(x, z))$ be the corresponding slope field. To shorten notation, we agree to write f for the restriction $f|_{\Gamma'}$, and g be its inverse $f^{-1}: G' \to \Gamma'$. Then we have

$$\varphi' = \pi = \mathscr{P}(\cdot, \varphi)$$

or, equivalently

$$e = f^* \not{p}.$$

By (52), it follows that

$$0 = d((f^*\not{p})^*\gamma) = d(f^*(\not{p}^*\gamma)) = f^*\, d(\not{p}^*\gamma).$$

Pulling both sides back to G' by means of g, we arrive at

$$d(\not{p}^*\gamma) = 0.$$

Since $\omega := \not{p}^*\gamma$ is just the 1-form

$$\omega = (\bar{F} - \mathscr{P}^i \bar{F}_{p^i})\, dx + \bar{F}_{p^i}\, dz^i$$

introduced in (24), we have the relation

$$d\omega = 0,$$

which is equivalent to the Carathéodory equations (23). Hence, f (or, rather, $f|_{\Gamma'}$) is a Mayer field. This proves the first assertion of Proposition 6. In conjunction with Propositions 4 and 5 we then obtain also the second assertion.

□

1.2. Carathéodory's Royal Road to Field Theory

In principle we are now prepared to set up the Weierstrass field theory and to derive "sufficient conditions" for minima of one-dimensional variational problems. Using the basic notion of a Mayer field we could proceed rather quickly by a few formal computations, thereby obtaining all basic results of field theory. However, this approach might leave the reader somewhat unsatisfied as it lacks motivation and geometric insight. Therefore we dispose it to *1.3*. Instead we presently want to investigate why the notions"Mayer field", "eikonal", "Carathéodory equations" are basic tools of the calculus of variations, and we want to motivate the approach to sufficient conditions to be followed in *1.3*. For this purpose we want to describe a beautiful scheme known as Carathéodory's "royal road to field theory".

In order to explain Carathéodory's approach, we consider a Lagrangian $F(x, z, p)$ which is defined for $(x, z) \in \mathscr{U}$ and $p \in \mathbb{R}^N$ where $\mathscr{U}$ is an open set in $\mathbb{R} \times \mathbb{R}^N$, and we suppose that F is of class C^2. Given an arbitrary F-extremal $u : [a, b] \to \mathbb{R}^N$ with graph $u \subset G$, we want to find out if u is a minimizer of the functional

$$\mathscr{F}(v) := \int_a^b F(x, v(x), v'(x))\, dx. \tag{1}$$

Then Carathéodory reasons as follows: Suppose that u is embedded in a field f on $G \subset \mathscr{U}$ with the slope $\mathscr{P}(x, z)$ which has the nice property that, for any point $(x, z) \in G$, the Lagrangian $F(x, z, p)$ vanishes if p coincides with the slope $\mathscr{P}(x, z)$ and is positive otherwise. That is, we assume that

$$F(x, z, \mathscr{P}(x, z)) = 0, \tag{2}$$

$$F(x, z, p) > 0 \quad \text{if } p \neq \mathscr{P}(x, z) \tag{3}$$

for all $(x, z) \in G$. Then it is perfectly obvious that u is a strong minimizer of $\mathscr{F}$. In fact, if $v : [a, b] \to \mathbb{R}^N$ is an arbitrary C^1-map with graph $v \subset G$, we infer from (2) and (3) that

$$\mathscr{F}(u) = 0, \qquad \mathscr{F}(v) \geq 0,$$

since $u'(x) = \mathscr{P}(x, u(x))$ for all $x \in [a, b]$, i.e.,

$$\mathscr{F}(u) \leq \mathscr{F}(v),$$

and a closer inspection even yields $\mathscr{F}(v) > 0$ and, therefore,

$$\mathscr{F}(u) < \mathscr{F}(v)$$

if $v \neq u$. In other words, u is a strict minmizer of $\mathscr{F}$ among all C^1-mappings $v : [a, b] \to \mathbb{R}^N$ with their graph contained in G which have the same boundary values as u. In particular, *u is a strong minimizer if G is open, and therefore it is also an extremal.*

However, the situation just described is certainly too nice to be true in general: the assumptions (2) and (3) are not very realistic. So we might like try the second best, to subtract from F a null Lagrangian M and to hope that instead of F the modified Lagrangian $F^* := F - M$ satisfies (2) and (3). Then the previous reasoning would imply that $\mathscr{F}^*(u) < \mathscr{F}^*(v)$ holds true for every $v \in C^1([a, b], \mathbb{R}^N)$ with graph $v \subset G$ which has the same boundary values as u and satisfies $v \neq u$. Since the functional $\mathscr{M}$ corresponding to the null Lagrangian M would fulfil $\mathscr{M}(u) = \mathscr{M}(v)$ we would also have the desired inequality $\mathscr{F}(u) < \mathscr{F}(v)$.

Amazingly enough this idea works perfectly well if one does not insist it to function for all fields. In fact, this cannot be expected because the previous reasoning shows that any field, for which the procedure is working, has to consist solely of extremals. Thus we can hope to find a suitable null Lagrangian only if the given field f is an extremal field, that is, a field whose field lines are each and all extremals. However, the assumption that f be extremal is sufficient only if $N = 1$ while, for $N \geq 1$, we shall see that the above construction can be carried out if and only if f is a Mayer field. Moreover, the null Lagrangian M of the construction cannot be chosen independently of f but will turn out to be the expression

$$M(x, z, p) = S_x(x, z) + S_{z^i}(x, z)p^i,$$

where S is the eikonal of f. The gist of the following discussion will be that, given some F-extremal $u : [a, b] \to \mathbb{R}^N$, we have to find a solution $\{S, \mathscr{P}\}$ of the Carathéodory equations such that u fits into the Mayer field defined by the slope $\mathscr{P}$.

Let us now consider Carathéodory's approach in some detail. For a given field $f : \Gamma \to G$ with the slope function $\mathscr{P}(x, z)$, and for a fixed Lagrangian F on $G \times \mathbb{R}^N = f(\Gamma) \times \mathbb{R}^N$ we try to determine some null Lagrangian $M(x, z, p)$ such that the modified Lagrangian

$$F^*(x, z, p) := F(x, z, p) - M(x, z, p) \tag{4}$$

satisfies for all $(x, z) \in G$ the equations

$$F^*(x, z, \mathscr{P}(x, z)) = 0, \tag{5}$$

$$F^*(x, z, p) > 0 \quad \text{if } p \neq \mathscr{P}(x, z), \tag{6}$$

which are the analogues of (2) and (3). By Proposition 1 of 1,4.2, any null Lagrangian $M(x, z, p)$ is of the form

$$M(x, z, p) = A(x, z) + B_i(x, z)p^i,$$

where the coefficients $A(x, z)$, $B_1(x, z), \ldots, B_N(x, z)$ satisfy

$$A_{z^i} = B_{i,x}, \qquad B_{i,z^k} = B_{k,z^i}.$$

(The coefficients A and B_i are assumed to be of class $C^1(G)$ whence the Euler operator corresponding to M is well defined.) Then there is a function $S(x, z)$ on G such that

$$A = S_x, \qquad B_1 = S_{z^1}, \ldots, B_N = S_{z^N}.$$

Clearly, S is of class $C^2(G)$, and we can write the null Lagrangian M as

$$M(x, z, p) = S_x(x, z) + S_z(x, z) \cdot p. \tag{7}$$

Then the modified Lagrangian F^* takes the form

$$F^*(x, z, p) = F(x, z, p) - S_x(x, z) - p \cdot S_z(x, z). \tag{8}$$

From (5) and (6) we see that the function $F^*(x, z, \cdot)$ assumes its strict absolute minimum exactly at $p = \mathscr{P}(x, z)$. This implies the two relations

$$F_p^*(x, z, \mathscr{P}(x, z)) = 0 \tag{9}$$

and

(10) $$F^*_{pp}(x, z, \mathscr{P}(x, z)) \geq 0 .$$

Equations (5) and (9) are equivalent to the *Carathéodory equations*

(11) $$\begin{aligned} S_x(x, z) &= F(x, z, \mathscr{P}(x, z)) - \mathscr{P}(x, z) \cdot F_p(x, z, \mathscr{P}(x, z)), \\ S_{z^i}(x, z) &= F_{p^i}(x, z, \mathscr{P}(x, z)), \end{aligned}$$

which we can write in the abbreviated form

(11′) $$S_x = \bar{F} - \mathscr{P} \cdot \bar{F}_p , \qquad S_z = \bar{F}_p ,$$

using the notation $\bar{F}(x, z) := F(x, z, \mathscr{P}(x, z))$, etc., introduced in *1.1*. This shows that Carathéodory's method can be carried out if and only if $\{S, \mathscr{P}\}$ is a solution of the Carathéodory equations (11), that is, if and only if f is a Mayer field. (Recall that the notions "extremal field" and "Mayer field" coincide exactly if $N = 1$.)

Furthermore, the necessary condition (10) is equivalent to the "necessary Legendre condition"

(12) $$\bar{F}_{pp} \geq 0 .$$

If the Carathéodory equations (11′) hold true, the sharp inequality (6) is just

$$F(x, z, p) - F(x, z, \mathscr{P}(x, z)) - [p - \mathscr{P}(x, z)] \cdot F_p(x, z, \mathscr{P}(x, z)) > 0$$

provided that $p \neq \mathscr{P}(x, z)$. By means of the Weierstrass excess function $\mathscr{E}_F(x, z, q, p)$ for the Lagrangian F which (by 4, (6)) is defined as

$$\mathscr{E}_F(x, z, q, p) = F(x, z, p) - F(x, z, q) - (p - q) \cdot F_p(x, z, q),$$

the last inequality can be written as

(13) $$\mathscr{E}_F(x, z, \mathscr{P}(x, z), p) > 0 \quad \text{for } (x, z) \in G \text{ and } p \neq \mathscr{P}(x, z).$$

This is the so-called *sufficient* (or *strict*) *Weierstrass condition.*

Thus we have proved:

Proposition 1. *Let $S(x, z)$ be a scalar function of class $C^2(G)$, and set*

$$F^*(x, z, p) := F(x, z, p) - S_x(x, z) - p \cdot S_z(x, z) .$$

Then the Carathéodory equations (11) *or* (11′) *in connection with the strict Weierstrass condition* (13) *are equivalent to the basic relations* (5) *and* (6) *of Carathéodory's approach. Secondly, equations* (5) *and* (9) *are equivalent to the Carathéodory equations* (11).

According to the reasoning given at the beginning of this subsection we derive the following result from Proposition 1:

Proposition 2. *If an F-extremal $u : [a, b] \to \mathbb{R}^N$ can be embedded into a Mayer field $f : \Gamma \to G$ with the slope $\mathscr{P}$ such that the strict Weierstrass condition* (13) *is satisfied, then we have*

$$\mathscr{F}(u) \le \mathscr{F}(v)$$

for all $v \in C^1([a, b], \mathbb{R}^N)$ *such that* graph $v \subset G$ *and* $v(a) = u(a)$, $v(b) = u(b)$, *and we even obtain*

$$\mathscr{F}(u) < \mathscr{F}(v)$$

if also $v \neq u$ *holds true.*

In other words, an extremal $u : [a, b] \to \mathbb{R}^N$ is shown to be a strict strong minimizer of the functional

$$\int_a^b F(x, v(x), v'(x))\, dx$$

among all C^1-curves $v : [a, b] \to \mathbb{R}^N$ within a domain $G \subset \mathbb{R} \times \mathbb{R}^N$ having the same boundary values as u if one can embed u into a Mayer field on G satisfying the strict Weierstrass condition. This embedding problem is essentially solved if we combine the construction of stigmatic ray bundles considered in *1.1* (see Propositions 4–6 of *1.1*) with the theory of conjugate points investigated in Chapter 4. A detailed discussion of this procedure will be given in *2.1*.

We can summarize the results of this subsection by stating that the notions "Mayer field" and "Carathéodory equations" are the key to sufficient conditions for minimizers, and that this becomes evident via Carathéodory's approach.

We finally mention that the eikonal of a Mayer field satisfies a partial differential equation of first order. In fact, assuming

$$\det F_{pp}(x, z, \mathscr{P}(x, z)) \neq 0,$$

we can eliminate $\mathscr{P}(x, z)$ from the equation

$$F_p(x, z, \mathscr{P}(x, z)) = S_z(x, z)$$

and express it in terms of x, z, and $S_z(x, z)$. Inserting this expression into the equation

$$F(x, z, \mathscr{P}(x, z)) - \mathscr{P}(x, z) \cdot F_p(x, z, \mathscr{P}(x, z)) = S_x(x, z),$$

we arrive at a scalar equation of the form

$$\Phi(x, z, \operatorname{grad} S) = 0. \tag{14}$$

This elimination process will be carried out in Chapter 7 as it is greatly simplified by using a suitable Legendre transformation. The resulting equation will be the so-called *Hamilton–Jacobi equation.*

Note that the equation for S is not uniquely determined since (14) implies that

$$\Psi(\Phi(x, z, \operatorname{grad} S)) = 0$$

holds true for every function $\Psi(s)$, $s \in \mathbb{R}$, satisfying $\Psi(0) = 0$.

1.3. Hilbert's Invariant Integral and the Weierstrass Formula. Optimal Fields. Kneser's Transversality Theorem

Now we resume the discussion where we left it at the end of *1.1*. In principle we do not need the results of *1.2* since they follow from the Weierstrass formula to be stated in the sequel. However, the ideas presented in *1.2* will motivate the coming considerations. In particular, they provide a motivation for the following

Definition 1. *Let $S \in C^2(G)$ be the eikonal of some Mayer field f on G, and set*

$$M(x, z, p) := S_x(x, z) + p \cdot S_z(x, z) \tag{1}$$

and

$$\mathscr{M}(v) := \int_a^b M(x, v(x), v'(x))\, dx. \tag{2}$$

Then $\mathscr{M}(v)$ is called Hilbert's invariant integral *for the Mayer field f.*

Clearly, $\mathscr{M}(v)$ is defined for every C^1-mapping $v: [a, b] \to \mathbb{R}^N$ such that graph $v \subset G$. Since S is uniquely determined by f modulo some additive constant, Hilbert's invariant integral is uniquely determined in terms of f. Moreover, $\mathscr{M}(v)$ is even defined for functions $v \in D^1([a, b], \mathbb{R}^N)$ with graph $v \subset G$. Here $D^1([a, b], \mathbb{R}^N)$ denotes the class of continuous mappings $u: [a, b] \to \mathbb{R}^N$ which are piecewise continuously differentiable in $[a, b]$. By this we mean that there is a decomposition

$$a = \xi_0 < \xi_1 < \xi_2 < \cdots < \xi_k = b$$

of $[a, b]$ into subintervals $I_j := [\xi_{j-1}, \xi_j]$ such that $u|_{I_j} \in C^1(I_j)$. That is, the one-sided derivatives $u'_+(\xi_j)$ and $u'_-(\xi_j)$ exist but, possibly, $u'_+(\xi_j) \neq u'_-(\xi_j)$ for $1 \le j \le k - 1$.

Note that

$$M(x, v(x), v'(x)) = \frac{d}{dx} S(x, v(x)) \tag{3}$$

for any $v \in C^1([a, b], \mathbb{R}^N)$ with graph $v \subset G$ whence

$$\mathscr{M}(v) = S(b, v(b)) - S(a, v(a)). \tag{4}$$

The same formula holds for any $v \in D^1([a, b], \mathbb{R}^N)$ with graph $v \subset G$. We merely have to apply (4) to the intervals I_j where v' is continuous and then to add the resulting formulas whence the desired result follows at once:

$$\begin{aligned}\mathscr{M}(v) &= \int_a^b M(x, v, v')\, dx = \sum_{j=1}^k \int_{\xi_{j-1}}^{\xi_j} M(x, v, v')\, dx \\ &= \sum_{j=1}^k [S(x, v(x))]_{x=\xi_{j-1}}^{x=\xi_j} = S(b, v(b)) - S(a, v(a)).\end{aligned}$$

Now we derive a formula linking the notions of Lagrangian, excess function, Mayer field and eikonal. This formula comprises the ideas and results of *1.2* and will be the key to all of the following results. To avoid unnecessary repetitions, we agree upon some notations:

Let G be a fixed simply connected set in the x, z-space $\mathbb{R} \times \mathbb{R}^N$, the configuration space. This set will later be thought to carry a Mayer field. Secondly let $F(x, z, p)$ be a fixed Lagrangian of class C^2 defined on $\mathcal{U} \times \mathbb{R}^N$ where $\mathcal{U}$ is an open neighbourhood of G, and $\mathcal{E}_F(x, z, q, p)$ be its excess function. All notions like extremal field, Mayer field, eikonal etc. are to be understood with respect to this fixed Lagrangian.

Theorem 1. (Weierstrass representation formula). *Let f be a Mayer field on G which has the slope $\mathcal{P}$ and the eikonal S. Then the functional*

$$\mathcal{F}(u) := \int_a^b F(x, u(x), u'(x))\, dx \tag{5}$$

is defined for every $u \in D^1([a, b], \mathbb{R}^N)$ with graph $u \subset G$, *and it can be expressed by the formula*

$$\mathcal{F}(u) = S(b, u(b)) - S(a, u(a)) + \int_a^b \mathcal{E}_F(x, u(x), \mathcal{P}(x, u(x)), u'(x))\, dx. \tag{6}$$

Proof. Consider Hilbert's invariant integral $\mathcal{M}(u)$ associated with the Mayer field f. By virtue of the Carathéodory equations

$$S_x = \bar{F} - \mathcal{P} \cdot \bar{F}_p, \qquad S_z = \bar{F}_p,$$

we can express the Lagrangian $M(x, z, p)$ of $\mathcal{M}(u)$ in the form

$$M = S_x + p \cdot S_z = (\bar{F} - \mathcal{P} \cdot \bar{F}_p) + p \cdot \bar{F}_p = \bar{F} + (p - \mathcal{P}) \cdot \bar{F}_p,$$

i.e.,

$$M(x, z, p) = F(x, z, \mathcal{P}(x, z)) + [p - \mathcal{P}(x, z)] \cdot F_p(x, z, \mathcal{P}(x, z)). \tag{7}$$

Then the modified Lagrangian

$$F^*(x, z, p) := F(x, z, p) - M(x, z, p) \tag{8}$$

can be written as

$$F^*(x, z, p) = F(x, z, p) - F(x, z, \mathcal{P}(x, z)) - [p - \mathcal{P}(x, z)] \cdot F_p(x, z, \mathcal{P}(x, z)),$$

which states that

$$F^*(x, z, p) = \mathcal{E}_F(x, z, \mathcal{P}(x, z), p). \tag{9}$$

Thus we obtain the crucial formula

$$F(x, z, p) = M(x, z, p) + \mathcal{E}_F(x, z, \mathcal{P}(x, z), p) \tag{10}$$

for all $(x, z, p) \in G \times \mathbb{R}^N$. Inserting $z = u(x)$ and $p = u'(x)$ for some

$u \in C^1([a, b], \mathbb{R}^N)$, we arrive at

$$F(x, u(x), u'(x)) = \frac{d}{dx} S(x, u(x)) + \mathscr{E}_F(x, u(x), \mathscr{P}(x, u(x)), u'(x)), \tag{11}$$

taking (3) into account. Integrating both sides of (11) with respect to x from a to b, we obtain the desired representation formula (6). By decomposing $[a, b]$ into appropriate subintervals, the reasoning leading to (4) can be used to establish (6) even for functions $u : [a, b] \to \mathbb{R}^N$ of class D^1. □

Since $\mathscr{E}_F(x, z, p, q) = 0$ if $p = q$, we obtain as an immediate application of Theorem 1 the following result:

Theorem 2. *Let f be a Mayer field on G with the slope $\mathscr{P}$ and the eikonal S. Then, for any extremal $u : [a, b] \to \mathbb{R}^N$ fitting into the field f (i.e., $u'(x) = \mathscr{P}(x, u(x))$ for all $x \in [a, b]$), we have*

$$\mathscr{F}(u) := \int_a^b F(x, u(x), u'(x))\, dx = S(b, u(b)) - S(a, u(a)). \tag{12}$$

Therefore, the relation

$$\mathscr{F}(v) = \mathscr{F}(u) + \int_a^b \mathscr{E}_F(x, v(x), \mathscr{P}(x, v(x)), v'(x))\, dx \tag{13}$$

holds true for every $v \in D^1([a, b], \mathbb{R}^N)$ such that graph $v \subset G$ *and $v(a) = u(a)$, $v(b) = u(b)$. We conclude that*

$$\mathscr{F}(v) \geq \mathscr{F}(u) = S(b, u(b)) - S(a, u(a)) \tag{14}$$

if $\mathscr{E}_F(x, v(x), \mathscr{P}(x, v(x)), v'(x)) \geq 0$ on $[a, b]$, and even

$$\mathscr{F}(v) > \mathscr{F}(u) \tag{15}$$

if, in addition, $\mathscr{E}_F(x, v(x), \mathscr{P}(x, v(x)), v'(x))$ does not vanish identically.

By means of Taylor's formula we obtain

$$\begin{aligned} &\mathscr{E}_F(x, v(x), \mathscr{P}(x, v(x)), v'(x)) \\ &\quad = [v'(x) - \mathscr{P}(x, v(x))] \cdot A(x) [v'(x) - \mathscr{P}(x, v(x))], \end{aligned} \tag{16}$$

where

$$A(x) := \tfrac{1}{2} F_{pp}(x, v(x), \mathscr{P}(x, v(x)) + \theta(x)\{v'(x) - \mathscr{P}(x, v(x))\}), \quad 0 < \theta < 1.$$

Thus we can derive from Theorem 2 the following

Corollary 1. *Let $u : [a, b] \to \mathbb{R}^N$ be an extremal fitting into a Mayer field on an (open) domain G. Then u is a weak minimizer of the functional $\mathscr{F}$ if it satisfies the strict Legendre condition*

$$F_{pp}(x, u(x), u'(x)) > 0 \quad \text{for all } x \in [a, b], \tag{17}$$

and u is even a strong minimizer of $\mathscr{F}$ if it satisfies the strong Legendre condition

$$F_{pp}(x, u(x), p) > 0 \quad \text{for all } x \in [a, b] \text{ and all } p \in \mathbb{R}^N. \tag{18}$$

In the next subsection we shall prove that any extremal $u: [a, b] \to \mathbb{R}^N$ is a strong minimizer of $\mathscr{F}(u) = \int_a^b F(x, u(x), u'(x))\,dx$ if it satisfies the strong Legendre condition (18) and if, in addition, the interval $[a, b]$ contains no pair of conjugate values of u. This will be the principal "sufficient condition". Presently we can verify this result only for $N = 1$. Recall that in 5,*2.3* we have developed a criterion which ensures that a given extremal can be embedded into a field of extremals which then is a Mayer field since $N = 1$. In fact, the field construction leading to Theorem 2 of 5,*2.3* was based on the choice of a suitable stigmatic field which, in a neighbourhood of the given extremal, formed a proper field. In conjunction with Theorem 2 we obtain the following preliminary result for C^3-Lagrangians F:

Theorem 3 ($N = 1$). *An F-extremal $u: [a, b] \to \mathbb{R}$ is a (strict) strong minimizer of $\mathscr{F}(u) = \int_a^b F(x, u(x), u'(x))\,dx$ if (i) the interval $[a, b]$ contains no pair of conjugate values of u, and if (ii) there exists an open neighbourhood $\mathscr{U}$ of* graph *u in $\mathbb{R}^2$ such that $F_{pp}(x, z, p) > 0$ holds true for all $(x, z, p) \in \mathscr{U} \times \mathbb{R}$.*

This a remarkable strengthening of Theorem 1 of 5,*2.2*. Note that assumption (ii) implies

$$\mathscr{E}_F(x, z, q, p) > 0 \quad \text{for } (x, z) \in \mathscr{U} \text{ and } p \neq q, \tag{19}$$

which in turn yields the strict Weierstrass condition

$$\mathscr{E}_F(x, z, \mathscr{P}(x, z), p) > 0 \quad \text{for } (x, z) \in \mathscr{U} \text{ and } p \neq \mathscr{P}(x, z). \tag{20}$$

In view of Theorem 2 we give the following

Definition 2. *A Mayer field f on G with the eikonal S is called* optimal field *if the functional $\mathscr{F}(v) := \int_a^b F(x, v(x), v'(x))\,dx$ satisfies*

$$\mathscr{F}(v) \geq S(b, v(b)) - S(a, v(a)) \tag{21}$$

for every interval $[a, b]$ and every $v \in D^1([a, b], \mathbb{R}^N)$ such that graph $v \subset G$. *Moreover, a Mayer field f on G is called* Weierstrass field *if its slope $\mathscr{P}$ satisfies the strong Weierstrass condition*

$$\mathscr{E}_F(x, z, \mathscr{P}(x, z), p) > 0 \tag{22}$$

for all $(x, z) \in G$ and all $p \in \mathbb{R}^N$ such that $p \neq \mathscr{P}(x, z)$.

As a consequence of Theorem 2, we obtain

Corollary 2. *Every Weierstrass field is an optimal field.*

Before we go on with our discussion, we want to formulate part of our previous results in a different way. Recall the definition of the Beltrami form γ given in *1.1*, (42):

$$\gamma = (F - p \cdot F_p)\, dx + F_{p^i}\, dz^i = F\, dx + F_{p^i}\theta^i, \tag{23}$$

where

$$\theta^i = dz^i - p^i\, dx.$$

The pull-back $\mathscr{p}^*\gamma$ of γ with respect to the slope field

$$\mathscr{p}(x, z) = (x, z, \mathscr{P}(x, z))$$

of a slope function $\mathscr{P}(x, z)$ is just the 1-form

$$\omega = (\bar{F} - \mathscr{P} \cdot \bar{F}_p)\, dx + \bar{F}_{p_i}\, dz^i = \bar{F}\, dx + \bar{F}_{p^i}\mathscr{p}^*\theta^i \tag{24}$$

defined in (24) of *1.1*. If $\mathscr{P}$ is the slope of a Mayer field f on G, then we have

$$\omega = \mathscr{p}^*\gamma = dS, \tag{25}$$

where S is the eikonal of f (see (26) of *1.1*). Hence, for any C^1-curve $v: [a, b] \to \mathbb{R}^N$ with graph $v \subset G$ we have

$$\mathscr{M}(v) = \int_{\text{graph } v} \omega. \tag{26}$$

That is, the line integral $\int_{\text{graph } v} \omega = \int_a^b w^*\omega$, $w(x) := (x, v(x))$, is just Hilbert's invariant integral. Let $u: [a, b] \to \mathbb{R}^N$ be an extremal fitting into the Mayer field f. Then we have

$$\int_{\text{graph } u} \omega = \int_{\text{graph } v} \omega \tag{27}$$

provided that $u(a) = v(a)$, $u(b) = v(b)$, and the proof of Theorem 2 is essentially contained in the following sequence of equations:

$$\begin{aligned} \mathscr{F}(v) - \mathscr{F}(u) &= \mathscr{F}(v) - \int_{\text{graph } u} \omega = \mathscr{F}(v) - \int_{\text{graph } v} \omega \\ &= \int_a^b \mathscr{E}_F(x, v(x), \mathscr{P}(x, v(x)), v'(x))\, dx. \end{aligned} \tag{28}$$

Now we shall discuss some relations between the field curves $f(\cdot, c)$ of a Mayer field $f: \Gamma \to G$ and the level surfaces

$$\mathscr{S}_\theta := \{(x, z) \in G: S(x, z) = \theta\}$$

of its eikonal S. The main result will be *A. Kneser's transversality theorem.*

Recall that the slope $\mathscr{P}: G \to \mathbb{R}^N$ of the Mayer field f on G and its eikonal $S: G \to \mathbb{R}$ are linked by the Carathéodory equations

$$\operatorname{grad} S = (\bar{F} - \mathscr{P} \cdot \bar{F}_p, \bar{F}_p), \tag{29}$$

where $\operatorname{grad} S = (S_x, S_z)$ is the gradient of S, and $\bar{F}$, $\bar{F}_p$ are as usual defined by

$$\bar{F} := F \circ \mathscr{p}, \qquad \bar{F}_p := F_p \circ \mathscr{p},$$

where

$$\mathscr{p}(x, z) = (x, z, \mathscr{P}(x, z))$$

is the full slope field of f.

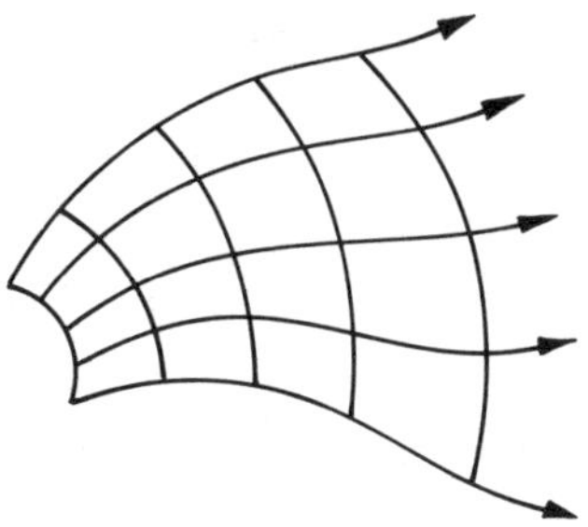

Fig. 5. The complete figure of a field.

Since grad S is perpendicular to the level surfaces $\mathscr{S}_\theta$, we infer that the covector field

$$(30) \qquad \nu := (\bar{F} - \mathscr{P} \cdot \bar{F}_p, \bar{F}_p)$$

is perpendicular to the surfaces $\mathscr{S}_\theta$. This means that $\not{p}: G \to \mathbb{R}^{2N+1}$ *is a field of line elements which is transversal to the level surface* $\mathscr{S}_\theta$. (Here and in the sequel, *transversality* means *free transversality* in the sense of 2,4.). Therefore, the surfaces $\mathscr{S}_\theta$ are called *transversal surfaces* of the Mayer field f. Carathéodory has denoted the system of field lines $f(\cdot, c)$ of a Mayer field f together with its 1-parameter family of transveral surfaces $\mathscr{S}_\theta$ as the *complete figure* of the field f.

Moreover, we infer from (29) that grad $S(x, z)$ vanishes if and only if both $\bar{F}(x, z)$ and $\bar{F}_p(x, z)$ are zero, and we see by the relation

$$(31) \qquad \operatorname{grad} S(x, z) \cdot (1, \mathscr{P}(x, z)) = \bar{F}(x, z)$$

that the vector $(1, \mathscr{P}(x, z))$ is nontangential to the level surface $\mathscr{S}_\theta$ passing through (x, z) if $\bar{F}(x, z) \neq 0$. Thus we obtain

Proposition 1. *A (nonempty) level surface $\mathscr{S}_\theta$ of the eikonal of a Mayer field is a regular surface (i.e., an N-dimensional submanifold of G) if*

$$(32) \qquad (F(x, z, \mathscr{P}(x, z)), F_p(x, z, \mathscr{P}(x, z))) \neq 0$$

holds true for all $(x, z) \in \mathscr{S}_\theta$. Hence the family of transveral surfaces $\mathscr{S}_\theta$ provides a foliation of G by regular leaves if (32) is satisfied on G. Moreover, if

$$(33) \qquad F(x, z, \mathscr{P}(x, z)) \neq 0 \quad \text{on } \mathscr{S}_\theta,$$

none of the field lines $f(\cdot, c)$ is tangent to the transveral surface $\mathscr{S}_\theta$.

Remark 1. Condition (33) is clearly satisfied if we assume that

$$(34) \qquad F(x, z, p) > 0 \quad \text{for all } (x, z, p) \in G \times \mathbb{R}^N.$$

In this case, every Mayer field on G consists of curves which meet the corresponding transveral surfaces of any complete figure on G are transversal to each other in the sense of algebraic geometry. This may be not the case if we only require that

$$F(x, z, p) \geq 0 \quad \text{for all } (x, z, p) \in G \times \mathbb{R}^N,$$

or if we even allow F to change its sign. In fact, if there is a field curve $f(x, c_0) = (x, u(x))$, $a \leq x \leq b$, such that $F(x, u(x), u'(x)) \equiv 0$, then graph u is completely contained in some transversal surface $\mathcal{S}_\theta$ which may or may not be regular.

Let us look at some simple examples.

1 Let $N \geq 1$ and

$$F(p) = |p|^2 - 1.$$

Then the stigmatic bundle

$$f(x, c) = (x, cx), \quad x \geq 0, \; c \in \mathbb{R}^N,$$

of extremals $\varphi(x, c) = cx$ with the nodal point $(x, z) = (0, 0)$ forms a Mayer field on $G = \{(x, z) \in \mathbb{R} \times \mathbb{R}^N : x > 0\}$ with the slope $\mathcal{P}(x, z) = \frac{z}{x}$. Since

$$F_p(p) = 2p, \qquad F(p) - p \cdot F_p(p) = -|p|^2 - 1,$$

the Carathéodory equations (29) for the eikonal $S(x, z)$ of the field are

$$S_x = -\frac{|z|^2}{x^2} - 1, \qquad S_z = \frac{2z}{x}.$$

The solution, normalized by the condition $\lim_{x \to 0} S(f(x, c)) = 0$ for all $c \in \mathbb{R}^N$, is given by

$$S(x, z) = \frac{|z|^2}{x} - x.$$

Note that

$$\int_0^x F(\varphi'(t, c))\, dt = S(f(x, c)).$$

The transversal surfaces $\mathcal{S}_\theta = \{(x, z) \in G : S(x, z) = \theta\}$ are the quadrics

$$|z|^2 - x^2 - \theta x = 0, \quad x > 0,$$

which sweep G if θ runs through all real numbers. Since $F^2 + |F_p|^2 \geq 1$, all transversal surfaces are regular (note that 0, the vertex of the half cone $\mathcal{S}_0$, is not contained in G). However, we have

$$F(\mathcal{P}(x, z)) = 0 \text{ if and only if } (x, z) \in \mathcal{S}_0.$$

Hence all of the transversal surfaces $\mathcal{S}_\theta$ with $\theta \neq 0$ are nontangentially intersected by the field lines $f(\cdot, c)$ whereas the half-cone $\mathcal{S}_0$ is tangent to the field. In fact, $\mathcal{S}_0$ consists precisely of the rays $f(x, c) = (x, cx)$, $x > 0$, with $|c| = 1$. Note that all $\mathcal{S}_\theta$ for $\theta < 0$ and all $\mathcal{S}_\theta \cup \{0\}$ for $\theta > 0$ are sheets of rotationally symmetric hyperboloids.

2 For $N \geq 1$ and

$$F(p) = |p|^2,$$

again the stigmatic ray bundle

$$f(x, c) = (x, cx), \quad x \geq 0, \; c \in \mathbb{R}^N,$$

considered in [1] forms a Mayer field on the right halfspace $G = \{(x, z): x > 0\}$, and its slope is given by $\mathscr{P}(x, z) = z/x$. The Carathéodory equations for the corresponding eikonal are then

$$S_x = -\frac{|z|^2}{x^2}, \qquad S_z = \frac{2z}{x},$$

and

$$S(x, z) = \frac{|z|^2}{x}$$

is the unique solution if we require $\lim_{x \to 0} S(f(x, c)) = 0$. Note that

$$\int_0^x F(\varphi'(t, c))\, dt = S(f(x, c)).$$

The transversal surfaces $\mathscr{S}_\theta$ are described by the equations

$$|z|^2 = \theta x, \quad x > 0, \ \theta \geq 0.$$

These surfaces are regular if $\theta > 0$; in fact, they are paraboloids touching the hyperplane $\{x = 0\}$ while $\mathscr{S}_0$ is degenerate: $\mathscr{S}_0$ is just the positive x-axis which is also a ray of the bundle f. For $\theta > 0$, all $\mathscr{S}_\theta$ are intersected by the rays of f under an angle different form zero.

[3] Let $N = 1$ and

$$F(z, p) = p^2 - Kz^2.$$

The corresponding Euler equations are equivalent to

$$u'' = -Ku,$$

and we want to consider the stigmatic bundle $f(x, c) = (x, \varphi(x, c))$, $x \geq 0$, of extremals emanating from the origin. That is, we require

$$\varphi''(x, c) = -K\varphi(x, c), \qquad \varphi(0, c) = 0, \qquad \varphi'(0, c) = c.$$

Then we have

$$\varphi(x, c) = \begin{cases} c \sin(\sqrt{K}x) & \text{if } K > 0, \\ cx & \text{if } K = 0, \\ c \sinh(\sqrt{-K}x) & \text{if } K < 0. \end{cases}$$

If $K \leq 0$, then f furnishes a Mayer field on the right halfplane $\{(x, z): x > 0\}$ whereas f forms a Mayer field on the strip $\{(x, z): 0 < x < \pi/\sqrt{K}\}$ if $K > 0$. In the latter case we have two nodal points $P_0 = (0, 0)$ and $P_1 = (x_1, 0)$, $x_1 = \pi/\sqrt{K}$.

The corresponding slope fields are given by

$$\mathscr{P}(x, z) = \begin{cases} z\sqrt{K} \cot(\sqrt{K}x) & \text{if } K > 0, \\ z/x & \text{if } K = 0, \\ z\sqrt{-K} \coth(\sqrt{-K}x) & \text{if } K < 0. \end{cases}$$

Moreover, we have

$$F - pF_p = -p^2 - Kz^2, \quad F_p = 2p,$$

and the corresponding Carathéodory equations lead to

$$S(x, z) = z^2\sqrt{K} \cot(\sqrt{K}x) \quad \text{if } K > 0, \qquad S(x, z) = \frac{z^2}{x} \quad \text{if } K = 0,$$

$$S(x, z) = z^2\sqrt{-K} \coth(\sqrt{-K}x) \quad \text{if } K < 0$$

if we require $\lim_{x\to 0} S(f(x,c)) = 0$. One computes that

$$\int_0^x F(\varphi(t,c), \varphi'(t,c))\,dt = S(\varphi(x,c)).$$

As in [2], only $\mathscr{S}_0$ is degenerate. This set is just the positive x-axis, i.e., $\mathscr{S}_0$ coincides with the extremal curve $f(x,0) = (x,0)$ if $K \le 0$; if $K > 0$, then $\mathscr{S}_0$ is the interval $(0, \pi/\sqrt{K})$ on the x-axis. The surfaces $\mathscr{S}_\theta$, $\theta > 0$, are regular and nontangential to the field curves. Close to the origin, the complete figure looks approximatively the same in all three cases $K > 0$, $K = 0$, $K < 0$ as one verifies by a simple limit consideration.

[4] The situation described in [3] is more or less typical for quadratic Lagrangian of the kind

$$Q(x,z,p) = a_{ik}(x)p^i p^k + 2b_{ik}(x)z^i p^k + c_{ik}z^i z^k,$$

where $A = (a_{ik})$ is positive definite. For instance, the accessory Lagrangian $Q(x,z,p)$ of some Lagrangian F and some F-extremal u is of this form if $A(x) = F_{pp}(x, u(x), u'(x)) > 0$.

Let $v_1, v_2, \ldots, v_N$ be N linearly independent Q-extremals, i.e., solutions of the Euler equations

$$\frac{d}{dx} Q_p(\cdot, v, v') - Q_z(\cdot, v, v') = 0.$$

Since these equations are linear, it follows that every linear combination $v = c^1 v_1 + \cdots + c^N v_N$, $c^i \in \mathbb{R}$, is a Q-extremal. Set $w_i := Q_p(\cdot, v_i, v_i')$. Then the family of extremals

$$\zeta(x,c) := c^1 v_1(x) + \cdots + c^N v_N(x)$$

defines a bundle

$$\phi(x,c) = (x, \zeta(x,c)),$$

which is a Mayer bundle if and only if all of the brackets

$$[i,k] := w_i \cdot v_k - w_k \cdot v_i$$

vanish.

A system $\{v_1, v_2, \ldots, v_N\}$ of N linearly independent Q-extremals with $[i,k] = 0$ is called a *conjugate base of Q-extremals*.

Since $\zeta(x,0) \equiv 0$, the trivial Q-extremal $v_0(x) \equiv 0$ belongs to the family $\zeta(x,c)$.

Set

$$\Delta(x) := \det(v_1(x), \ldots, v_N(x)).$$

Since we have

$$\Delta = \det \zeta_c = \det D\phi,$$

the determinant $\Delta(x)$ is just the Jacobian of the mapping ϕ. Hence, between two consecutive zeros x_1 and x_2 of $\Delta(x)$, the map ϕ defines a Q-Mayer field on the strip $G = \{(x,z)\colon x_1 < x < x_2, z \in \mathbb{R}^N\}$. We note that [2] and [3] are special cases of the present example where $x_1 = 0$ and $x_2 = \infty$ or $\pi/\sqrt{K}$ respectively.

The zeros of $\Delta(x)$ are called *focal points* of the bundle ϕ. It is one of the principal results of the linear theory that all focal points are isolated (see *2.4*).

[5] The preceding examples show that "transversality" does in general not mean "orthogonality". In fact, we have

Proposition 2. *A positive Lagrangian F of class $C^1(G \times \mathbb{R}^N)$ has the property that F-transversality is equivalent to orthogonality if and only if it is of the form*

$$F(x,z,p) = \omega(x,z)\sqrt{1 + |p|^2}, \tag{35}$$

where $\omega(x,z)$ denotes some positive function of class $C^1(G)$.

Proof. We have pointed out in Remark 2 of 2,4 that (35) with $\omega > 0$ is certainly sufficient for the equivalence of orthogonality and F-transversality. Thus it remains to show that the condition is also necessary.

In fact, if (free) F-transversality is the same as orthogonality, we have for all $(x, z, p) \in G \times \mathbb{R}^N$ that

$$(F(x, z, p) - p \cdot F_p(x, z, p), F_p(x, z, p)) \sim (1, p).$$

Omitting the arguments x, z, p, we can write this relation in the form

$$F - p \cdot F_p = \lambda, \quad F_p = \lambda p$$

for some $\lambda \in \mathbb{R}$, whence

$$pF - p(p \cdot F_p) = F_p,$$

and therefore

$$|p|^2 F - |p|^2 p \cdot F_p = p \cdot F_p.$$

Since $F > 0$, we infer that

$$\frac{|p|^2}{1 + |p|^2} = \frac{p \cdot F_p}{F},$$

which is equivalent to

$$p^i \frac{\partial}{\partial p^i} \log \frac{F}{\sqrt{1 + |p|^2}} = 0.$$

Fix $(x, z) \in G$ and consider $\psi(p) := \log \dfrac{F(x, z, p)}{\sqrt{1 + |p|^2}}$ as a function of p alone. Then we have

$$p^i \frac{\partial \psi}{\partial p^i}(p) = 0$$

for all $p \in \mathbb{R}^N$. By Euler's theorem we infer that

$$\psi(\lambda p) = \psi(p) \quad \text{for all } \lambda > 0 \text{ and all } p \in \mathbb{R}^N.$$

Since ψ is a continuous function on $\mathbb{R}^N$, this relation is even satisfied for $\lambda = 0$, whence

$$\psi(p) = \psi(0) \quad \text{for all } p \in \mathbb{R}^N.$$

This implies (35) where $\omega(x, z) := F(x, z, 0)$. □

Now we state A. Kneser's transversality theorem which is a direct consequence of the Weierstrass representation formula written in the form (12) or (13). In differential geometry, when dealing with geodesics, the content of this theorem is known as Gauss's lemma.

Theorem 4 (Kneser's transversality theorem). *Let $f : \Gamma \to G$ be a Mayer field on G whose slope $\mathscr{P}$ satisfies $F(x, z, \mathscr{P}(x, z)) > 0$ on G, and denote by S the eikonal of f. Consider any two level surfaces $\mathscr{S}_{\theta_1}$ and $\mathscr{S}_{\theta_2}$ of S with $\theta_1 < \theta_2$, and let $I_{1,2}$ be the set of $c \in I_0$ such that the field curves $f(\cdot, c) : I(c) \to G$ are intersected both by $\mathscr{S}_{\theta_1}$ and $\mathscr{S}_{\theta_2}$.*

(i) *For $c \in I_{1,2}$, the transversal surfaces $\mathscr{S}_{\theta_1}$ and $\mathscr{S}_{\theta_2}$ excise subarcs $\mathscr{C}^*(c)$ from the field curves $f(\cdot, c)$ such that the variational integral $\int F\,dx$ has a constant value*

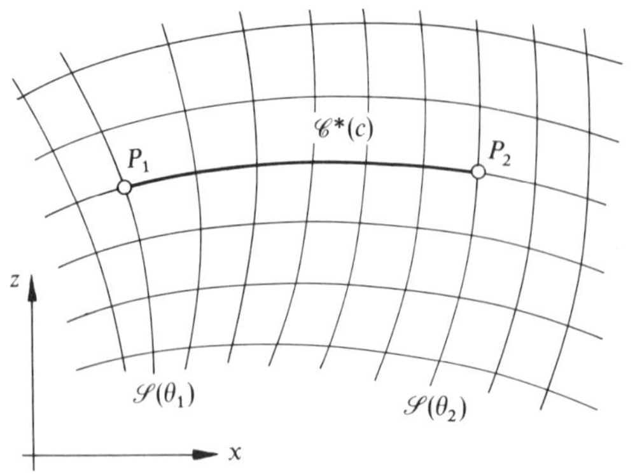

Fig. 6. Kneser's transversality theorem.

along all excised arcs $\mathscr{C}^*(c)$. In fact, we have

$$\int_{\mathscr{C}^*(c)} F\,dx = \theta_2 - \theta_1 .$$

(ii) Let m be the infimum of the values $\int_{x_1}^{x_2} F(x, u(x), u'(x))\,dx$ among all D^1-curves in G with endpoints on $\mathscr{S}_{\theta_1}$ and $\mathscr{S}_{\theta_2}$ respectively. If $I_{1,2}$ is nonempty and if f is an optimal field, this infimum is $\theta_2 - \theta_1$, and it will be assumed on all excised subarcs $\mathscr{C}^*(c)$, $c \in I_{1,2}$, and only on these arcs if f is even a Weierstrass field.

6 *Fermat's principle.* In geometrical optics, one is concerned with the investigation of *light rays* or, rather, of *bundles of light rays*. One imagines that light penetrates optical media. Such a medium is thought to be a set G in the configuration space (= x, z-space) which is equipped with an *index of refraction* $\omega(x, z, p) > 0$. This is to say, the refraction of a ray passing the medium in general depends not only on the locus (x, z) but also on the direction $(1, p)$ of the ray at (x, z). Such a general medium is called *anisotropic*. If ω is merely a function of the locus, the medium is said to be *isotropic*, and it is called *homogeneous* if ω is independent of the locus. According to *Fermat's principle*, every light ray penetrating the medium $\{G, \omega\}$ is thought to be an extremal of the integral

$$\mathscr{F}(u) = \int_{x_1}^{x_2} \omega(x, u(x), u'(x))\sqrt{1 + |u'(x)|^2}\,dx,$$

whose Lagrangian is given by

$$F(x, z, p) = \omega(x, z, p)\sqrt{1 + |p|^2}. \tag{36}$$

If we interpret $ds = \sqrt{1 + |u'|^2}\,dx = \sqrt{dx^2 + |du|^2}$ as the line element along the curve $(x, u(x))$ and $v = 1/\omega(x, u(x), u'(x))$ as the speed of a fictive light particle moving along the curve, we have $v = \dfrac{ds}{dt}$, and therefore $\omega\,ds = \dfrac{1}{v}\,ds = dt$. That is, $\mathscr{F}(u)$ is viewed as the time T needed by the light particle to move along $\mathscr{C} = \{(x, u(x)) \colon x_1 \le x \le x_2\}$ from the point $P_1 = (x_1, u(x_1))$ to the point $P_2 = (x_2, u(x_2))$. Fermat's principle distinguishes the true light rays from all virtual paths by the requirement that they are to afford a stationary value to the total time T needed by the light to travel from one point to another. (This model tacitly assumes that the light is monochromatic, and it does not take phenomena like interference into account which are connected with the fact that light has a finite wave length.)

Mayer bundles of the Lagrangian (36) are interpreted as bundles of light rays in the medium $\{G, \omega\}$, and every light-ray bundle is thought to be a Mayer bundle. If a ray bundle is free of singularities, it forms a field on G. Let S be the eikonal of such a Mayer field of rays, and let $\mathcal{S}_\theta = \{(x, z) \in G \colon S(x, z) = \theta\}$ be the corresponding transversal surfaces. By Kneser's theorem, the value $\theta_2 - \theta_1$ is the time needed by a light particle to pass from a surface $\mathcal{S}_{\theta_1}$ to another surface $\mathcal{S}_{\theta_2}$ along a ray, $\theta_1 < \theta_2$. Hence, we can interpret the transversal surfaces $\mathcal{S}_\theta$ as the *wave fronts* of light. For instance, if in every point of a given surface $\mathcal{S}_0$ at a fixed time $\theta = 0$ there starts a light particle, then, for a sufficiently small time interval $0 \le \theta \le T$, the particles move on rays forming a Mayer bundle and, after a time $\theta \le T$, they have arrived at the transversal surface $\mathcal{S}_\theta$. This explains why we can view the eikonal S of an optimal field of light rays as an *optical distance function.*

Any complete figure of Carathéodory consists of a field of light rays and the transversal family of wave fronts whose optical distance is just the time needed by light to travel from one front to another. If the optical medium is isotropic, then the corresponding Lagrangian is of the kind (35). Exactly in this case transversality means orthogonality (cf. Proposition 2). Hence isotropic media are precisely those optical materials where light rays meet the wave fronts perpendicularly.

For an isotropic homogeneous medium, the index ω is a positive constant. In such a medium, all extremals are linear functions, and the associated extremal curves are straight lines. Hence, in an isotropic homogeneous medium, the ray bundles are Mayer bundles of straight lines, and the corresponding transversal wave fronts form a one-parameter family of equidistant surfaces (in the sense of Euclidean geometry) which are perpendicularly intersected by the rays.

Remark 2. It is interesting to note that also the converse of Kneser's transversality theorem is true. Precisely speaking we have the following result: *Let $F \neq 0$, and suppose that $f: \Gamma \to G$ is an extremal field on G and $\{\mathcal{S}_t\}$ a foliation of G by regular surfaces $\mathcal{S}_t = \{(x, z) \in G \colon \Sigma(x, z) = t\}$, $\Sigma \in C^2(G)$, $\nabla\Sigma(x, z) \neq 0$ on G such that the field curves f intersect the leaves $\mathcal{S}_t$ transversally. Then f is a Mayer field and the surfaces $\mathcal{S}_t$ are just the level surfaces of the eikonal S corresponding to f.*

Actually, a much stronger result holds true which we are now going to derive:

Theorem 5. *A field $f: \Gamma \to G$ of extremals is automatically a Mayer field if there is a (possibly degenerate) hypersurface $\mathcal{S}$ in G which is transversally intersected by each of the field lines $f(\cdot, c)$.*

This result immediately follows from the next proposition if we take *1.1*, Proposition 6 into account.

Proposition 3. *An extremal bundle $f: \Gamma \to G$ is a Mayer bundle if there is a hypersurface $\mathcal{S}$ in G which is transversally intersected by each of the field curves $f(\cdot, c)$.*

Proof. Let $e(x, c) = (x, \varphi(x, c), \pi(x, c))$ with $\pi(x, c) = \varphi'(x, c)$ be the prolongation of the extremal bundle $f(x, c) = (x, \varphi(x, c))$ into the phase space where

$(x, c) \in \Gamma$ and

$$\Gamma = \{(x, c): c \in I_0, x \in I(c)\}. \tag{37}$$

To state the transversality assumption in a precise way, we assume that $\mathcal{S}$ is a hypersurface in G given by the parametric representation $x = \xi(c)$, $z = \zeta(c) = \varphi(\xi(c), c)$, $c \in I_0$, which is of class C^2. Set $\lambda(c) := \pi(\xi(c), c)$, $\sigma(c) := (\xi(c), \zeta(c))$, $\varepsilon(c) := (\xi(c), \zeta(c), \lambda(c))$. Then $\mathcal{S} = \sigma(I_0)$, and $\varepsilon : I_0 \to \mathbb{R}^{2N+1}$ is a field of line elements along $\mathcal{S}$ which is tangent to the extremal field f. By assumption, the line element $\varepsilon(c)$ is transversal to the corresponding tangent space of $\mathcal{S}$ spanned by the tangent vectors $\dfrac{\partial \sigma(c)}{\partial c^\alpha}$, $\alpha = 1, \ldots, N$.

This means that the vectors $\dfrac{\partial \sigma}{\partial c^\alpha}(c)$ are perpendicular to the vector $v(c)$ defined by

$$v(c) := (F(\varepsilon(c)) - \lambda(c) \cdot F_p(\varepsilon(c)), F_p(\varepsilon(c))). \tag{38}$$

In terms of the Beltrami form $\gamma = (F - p \cdot F_p)\, dx + F_{p^i}\, dz^i$, this orthogonality relation can be expressed by the relation

$$\varepsilon^* \gamma = 0 \tag{39}$$

which implies

$$d(\varepsilon^* \gamma) = 0. \tag{40}$$

Let us introduce the initial value map $a : I_0 \to \Gamma$ defined by

$$a(c) := (\xi(c), c), \quad c \in I_0.$$

Then we have $\varepsilon = e \circ a = a^* e$, and we infer from (40) the relation

$$0 = d(\varepsilon^* \gamma) = d(a^*(e^* \gamma)) = a^*(d(e^* \gamma)). \tag{41}$$

On the other hand, we have shown in *1.1*, (48) that

$$d(e^* \gamma) = \sum_{(\alpha, \beta)} [c^\alpha, c^\beta]\, dc^\alpha \wedge dc^\beta. \tag{42}$$

By (41) we conclude that $[c^\alpha, c^\beta] \circ a = 0$, and the invariance of the Lagrange brackets along extremals (cf. *1.1*, Proposition 4) then yields $[c^\alpha, c^\beta] = 0$ on Γ, whence

$$d(e^* \gamma) = 0. \tag{43}$$

□

For the convenience of the reader we include a proof of Theorem 5 which does not directly use Proposition 6 of *1.6*; however, the underlying ideas are very similar.

Proof of Theorem 5. Since Γ is simply connected, we infer from (43) that there is a function $\Sigma \in C^2(\Gamma)$ such that

(44) $$e^*\gamma = d\Sigma.$$

Moreover, if $\not{p} : G \to \mathbb{R}^{2N+1}$ is the slope field of f, we have

(45) $$e = f^*\not{p},$$

whence

(46) $$f^*(\not{p}^*\gamma) = d\Sigma.$$

Let $g : G \to \Gamma$ be the inverse of the mapping $f : \Gamma \to G$. If we pull both sides of (46) back by g, we obtain

(47) $$\not{p}^*\gamma = g^*(d\Sigma) = d(g^*\Sigma) = d(\Sigma \circ g).$$

Introducing the function

(48) $$S := \Sigma \circ g \in C^2(G),$$

we arrive at

(49) $$\not{p}^*\gamma = dS,$$

which states that f is a Mayer field with the eikonal S. □

Remark 3. Note that, in Proposition 3 and Theorem 5, the representation $\sigma : I_0 \to G$ was not assumed to be an immersion, i.e., the "transversal surface" $\mathcal{S} = \sigma(I_0)$ can be highly degenerated and may even be a point. Of course, $f : \Gamma \to G$ cannot be a field if σ is not an imbedding; in this case we have to pass to some suitable part $f|_{\Gamma'}$ of f such as in Proposition 6 of *1.1*. If $\mathcal{S}$ is degenerated to a point, Theorem 5 is essentially the statement of *1.1*, Proposition 6.

Remark 4. The transversal hypersurface in Theorem 5 turns in hindsight out to be one of the level surface (= wave fronts) of the eikonal S of the field $f : \Gamma \to G$. Namely, $S \circ \sigma = \sigma^*S = a^*\Sigma$ whence $d(S \circ \sigma) = d(a^*\Sigma) = a^*(d\Sigma) = a^*(e^*\gamma) = \varepsilon^*\gamma = 0$ by virtue of (44) and (39). This shows that S is constant on $\mathcal{S}$.

Remark 5. We can use Theorem 5 to construct Mayer fields which emanate from an arbitrarily chosen regular (embedded) hypersurface $\mathcal{S}$ in $\mathbb{R}^{N+1}$ in such a way that $\mathcal{S}$ becomes a level surface of the corresponding eikonal. For this purpose we choose a parametric representation $\sigma : I_0 \to \mathbb{R}^{N+1}$ of $\mathcal{S}$ which we write as $x = \xi(c)$, $z = \zeta(c)$, i.e., $\sigma = (\xi, \zeta)$. Then we determine a field of line elements $\varepsilon(c) = (\xi(c), \zeta(c), \lambda(c))$ such that $\varepsilon^*\gamma = 0$, i.e., the line elements $\varepsilon(c)$ are chosen transversally to $\mathcal{S}$. Finally we construct an N-parameter family $\varphi(\cdot, c)$ of solutions of the Euler equation $L_F\varphi(\cdot, c) = 0$ satisfying the initial value conditions $\varphi(\xi(c), c) = \zeta(c)$, $\varphi'(\xi(c), c) = \lambda(c)$. Sufficiently close to $\mathcal{S}$, the bundle $f(x, c) = (x, \varphi(x, c))$ then forms a Mayer field which has $\mathcal{S}$ as one of the level surfaces $\mathcal{S}_\theta$ of its eikonal. The level surfaces $\mathcal{S}_\theta$ are equidistant surfaces with respect to the "metric" provided by the "distance function" $\int F\,dx$ (in the sense described before). One can use (θ, c) as *geodesic coordinates* on the domain G

where f is defined as a field. In differential geometry it has become customary to denote such coordinates as *Fermi coordinates* around $\mathcal{S}$.

Of this construction, only the determination of the $\lambda(c)$ has remained. We dispose this question to *2.4* and Chapter 10. (In full generality this problem can be settled by means of Legendre transformations and Hölder transformations.)

Let us briefly return to the notion of an optimal field given in Definition 2. Per definitionem, any optimal field is a Mayer field. Interestingly, we can characterize optimal fields by means of the inequality (21) alone if we add the assumption that the equality sign is assumed if graph v fits into the field. This fact can be seen as a global analogue of Carathéodory's approach in *1.2* which is more of a local nature. Precisely, speaking, we have the following result:

Theorem 6. *A field f on a domain $G \subset \mathbb{R}^{N+1}$ is optimal if and only if it has the following two properties:*

(i) There is a function $S \in C^2(G)$ such that the inequality

$$\int_a^b F(x, u(x), u'(x))\, dx \geq S(b, u(b)) - S(a, u(a)) \tag{50}$$

holds for all $u \in C^1([a, b], \mathbb{R}^N)$, $a < b$, with graph $u \subset G$.

(ii) The equality sign in (50) is assumed if u fits in the field, i.e., if $u'(x) = \mathcal{P}(x, u(x))$ holds for $a \leq x \leq b$ where $\mathcal{P}$ denotes the slope function of f.

Moreover, if (i) and (ii) hold true, the function S is just the eikonal of the Mayer field f.

Proof. The necessity of (i) and (ii) follows directly form the definition of an optimal field. So we have only to show that (i) and (ii) imply that f is a Mayer field with the eikonal S.

For this purpose we fix some $u \in C^1([x - \delta, x + \delta], \mathbb{R}^N)$, $\delta > 0$, with graph $u \subset G$. If $\delta \ll 1$, the function

$$\Phi(t) := \int_x^t F(\xi, u(\xi), u'(\xi))\, d\xi - [S(t, u(t)) - S(x, u(x))]$$

is well defined for $|t - x| < \delta$ and of class C^1, and

$$\Phi'(t) = F(t, u(t), u'(t)) - S_x(t, u(t)) - u'(t) \cdot S_z(t, u(t)). \tag{51}$$

Since (50) implies $\Phi(t) \geq 0 = \Phi(x)$, it follows that $\Phi'(x) \geq 0$, and (51) yields

$$F(x, u(x), u'(x)) - S_x(x, u(x)) - u'(x) \cdot S_z(x, u(x)) \geq 0. \tag{52}$$

Because of (ii), the equality sign holds true if $u'(t) = \mathcal{P}(t, u(t))$ for $|t - x| < \delta$.

Given $(x, z) \in G$ and $p \in \mathbb{R}^N$, we can always find some $\delta > 0$ and some $u \in C^1([x - \delta, x + \delta], \mathbb{R}^N)$ with graph $u \subset G$ and $z = u(x)$, $p = u'(x)$. Hence we infer from (50) that

$$F(x, z, p) - S_x(x, z) - p \cdot S_z(x, z) \geq 0$$

and

$$F(x, z, \mathcal{P}(x, z)) - S_x(x, z) - \mathcal{P}(x, z) \cdot S_z(x, z) = 0$$

for all $(x, z) \in G$. These are the relations (5) and (6) of *1.2* (except that we do not have the strict inequality in (6) which, however, is inessential for the following reasoning). By the straightforward arguments of *1.2*, we arrive at the Carathéodory equations for $\{\mathcal{P}, S\}$, and therefore f is a Mayer field with the eikonal S. □

We close this subsection with some results on *stigmatic fields*. These are *improper fields* on a set $\hat{G}$ of $\mathbb{R} \times \mathbb{R}^N$ which become true fields if one removes their nodal point from G. To have a clearcut situation, we assume the following:

Let $f : \hat{\Gamma} \to \mathbb{R} \times \mathbb{R}^N$ be a stigmatic ray bundle which is defined on a parameter domain

$$\hat{\Gamma} = \{(x, c) \in \mathbb{R} \times \mathbb{R}^N : x_0 \le x \le b(c), c \in I_0\}$$

and has $P_0 = (x_0, z_0)$ as its nodal point, i.e. $f(x_0, c) = P_0$ for all $c \in I_0$. The parameter set I_0 is supposed to be the closure of a bounded domain. (Note that we have slightly altered the form of $\hat{\Gamma}$ in comparison with Definition 4 of *1.1*. This, however, is of no consequence for the validity of Propositions 3–6 in *1.1*.) Moreover, we set

$$\Gamma := \{(x, c) \in \mathbb{R} \times \mathbb{R}^N : x_0 < x \le b(c), c \in I_0\},$$

and we assume that $f|_\Gamma$ is a field. By Proposition 6 in *1.1*, the field $f|_\Gamma$ is even a Mayer field on $G := f(\Gamma)$. Set $\hat{G} := G \cup \{P_0\} = f(\hat{\Gamma})$.

Definition 3. *A mapping $f : \hat{\Gamma} \to \hat{G}$ with the afore-stated properties is called a* stigmatic field *on $\hat{G}$ with the* nodal point P_0.

We shall see that the eikonal S of a stigmatic field $f : \hat{\Gamma} \to \hat{G}$ can be computed in a simple way via the so-called *value function* $\Sigma(x, c)$ defined in formula (54).

Theorem 7. *Let $f : \hat{\Gamma} \to \hat{G} = G \cup \{P_0\}$ be a stigmatic field with the nodal point P_0 in the sense of the preceding definition, $f(x, c) = (x, \varphi(x, c))$, and let $g(x, z) = (x, a(x, z))$ be the inverse $g : G \to \Gamma$ of $f|_\Gamma$. Then the eikonal $S(x, z)$ of $f|_\Gamma$ is obtained by*

$$S(x, z) = \Sigma(x, a(x, z)), \quad (x, z) \in G, \tag{53}$$

where $\Sigma(x, c)$ is defined by

$$\Sigma(x, c) := \int_{x_0}^{x} F(\xi, \varphi(\xi, c), \varphi'(\xi, c))\, d\xi \tag{54}$$

for all $(x, c) \in \hat{\Gamma}$, and we have $\lim_{P \to P_0} S(P) = 0$ (for $P \in G$).

Proof. First we verify that $\Sigma \in C^1(\hat{\Gamma})$, $S \in C^1(G)$, and that

$$S_z(x, z) = \Sigma_c(x, a(x, z)) \cdot a_z(x, z). \tag{55}$$

Moreover, the identity $z = \varphi(x, a(x, z))$ implies

$$a_z(x, z) = \varphi_c^{-1}(x, a(x, z)),$$

whence

$$S_z(x, z) = \Sigma_c(x, a(x, z)) \cdot \varphi_c^{-1}(x, a(x, z)). \tag{56}$$

On the other hand, (54) yields

$$\Sigma_c(x, c) = \int_{x_0}^{x} L_F(\varphi(\cdot, c)) \cdot \varphi_c(\cdot, c)\, dx + [F_p(x, \varphi(x, c), \varphi'(x, c)) \cdot \varphi_c(x, c)]_{x_0}^{x}$$

$$= F_p(x, \varphi(x, c), \varphi'(x, c)) \cdot \varphi_c(x, c),$$

and, by virtue of (56), it follows that

$$S_z(x, z) = F_p(x, z, \mathscr{P}(x, z)) \tag{57}$$

if we also take the differential equation

$$\varphi'(x, c) = \mathscr{P}(x, \varphi(x, c)) \quad \text{for } (x, c) \in \hat{\Gamma}$$

into account.

Secondly, from

$$\Sigma(x, c) = S(x, \varphi(x, c))$$

we infer the equation

$$\Sigma_x(x, c) = S_x(x, \varphi(x, c)) + S_z(x, \varphi(x, c)) \cdot \varphi'(x, c),$$

which is equivalent to

$$\Sigma_x(x, a(x, z)) = S_x(x, z) + S_z(x, z) \cdot \mathscr{P}(x, z). \tag{58}$$

On the other hand, from (54) we derive the equation

$$\Sigma_x(x, c) = F(x, \varphi(x, c), \varphi'(x, c))$$

or, equivalently,

$$\Sigma_x(x, a(x, c)) = F(x, z, \mathscr{P}(x, z)). \tag{59}$$

Combining equations (57)–(59), it follows that

$$S_x(x, z) = F(x, z, \mathscr{P}(x, z)) - \mathscr{P}(x, z) \cdot F_p(x, z, \mathscr{P}(x, z)). \tag{60}$$

However, equations (57) and (60) are just the Carathéodory equations

$$S_x = \bar{F} - \mathscr{P} \cdot \bar{F}_p, \qquad S_z = \bar{F}_p$$

for the pair $\{\mathscr{P}, S\}$. This proves that S is the eikonal of the Mayer field $f|_\Gamma$, in particular that $S \in C^2$. Moreover, we have

$$\Sigma(x, c) = S(z, \varphi(x, c)), \tag{61}$$

and definition (54) implies that

$$\lim_{x \to x_0 + 0} \Sigma(x, c) = 0$$

uniformly in $c \in I_0$. Thus we finally obtain

$$\lim_{P \to P_0} S(P) = 0$$

for $P \in G$. □

Note that in the proof of the relation "$S(P) \to 0$ as $P \to P_0$" we have used the fact that I_0 is a compact set. If we want to allow for more general stigmatic fields, only a less strong result may be true, but we can always show that $S(P) \to 0$ as P moves to P_0 along a field line.

Combining Theorem 7 with the preceding reasoning, we obtain the following generalization of Theorem 2 for stigmatic fields:

Theorem 8. *Let $f: \hat{\Gamma} \to \hat{G} = G \cup \{P_0\}$ be a stigmatic field with the nodal point P_0 (in the sense of Definition 3), and suppose that the strict Weierstrass condition*

$$\mathscr{E}_F(x, z, \mathscr{P}(x, z), p) > 0 \quad \textit{for } (x, z) \in G \textit{ and } p \neq \mathscr{P}(x, z)$$

holds true. Finally, let $P_1 = (x_1, z_1)$ be an arbitrary point in G, and denote by $f(\cdot, c_1)$ the uniquely determined field curve passing through P_1 which satisfies $f(x_1, c_1) = P_1$. Then $u(x) := \varphi(x, c_1)$ is the uniquely determined minimizer of the functional

$$\mathscr{F}(v) := \int_{x_0}^{x_1} F(x, v(x), v'(x))\, dx$$

among all functions $v: [x_0, x_1] \to \mathbb{R}^N$ with graph $v \subset \hat{G}$ *and $v(x_0) = z_0$, $v(x_1) = z_1$, which are of class C^1 or D^1.*

Proof. By assumption, $f|_\Gamma$ is a Weierstrass field and, therefore, also an optimal field on G. Let us now choose an arbitrary function v as described in the assertion of the theorem, and some ξ with $x_0 < \xi < x_1$. Then the Weierstrass representation formula yields

$$\int_{\xi}^{x_1} F(x, v(x), v'(x))\, dx = S(P_1) - S(P) + \int_{\xi}^{x_1} \mathscr{E}_F(x, v(x), \mathscr{P}(x, v(x)), v'(x))\, dx \tag{62}$$

where $P := (\xi, v(\xi))$, and S is the eikonal of $f|_\Gamma$. If $\xi \to x_0 + 0$, then $S(P) \to 0$ (by virtue of Theorem 7) and

$$\lim_{\xi \to x_0 + 0} \int_{\xi}^{x_1} F(x, v(x), v'(x))\, dx = \mathscr{F}(v).$$

Therefore also the integral on the right-hand side of (62) tends to a limit, and we obtain

$$\mathscr{F}(v) = \mathscr{F}(u) + \int_{x_0}^{x_1} \mathscr{E}_F(x, v(x), \mathscr{P}(x, v(x)), v'(x))\, dx, \tag{63}$$

since we easily see that $\mathscr{F}(u) = S(P_1)$. If $v(x) \not\equiv u(x)$, then the strict Weierstrass condition implies

$$\mathscr{F}(v) > \mathscr{F}(u).$$

□

The presuppositions of this theorem can be weakened in various ways which is seen best by looking at specific examples.

The construction of the distance function which led to Theorem 7 can be carried over to certain other improper fields. This can be seen by strengthening the reasoning used for the proof of Theorem 5. Another proof can be obtained by using the arguments employed in the proof of Theorem 7. To this end we consider an improper field $f: \hat{\Gamma} \to G$ which is defined in the same way as a stigmatic field except that Γ and $\hat{\Gamma}$ are replaced by

$$\Gamma = \{(x, c): \xi(c) < x \le b(c), c \in I_0\},$$

$$\hat{\Gamma} = \{(x, c): \xi(c) \le x \le b(c), c \in I_0\},$$

and the stigmatic condition is substituted by a condition

$$\varphi(\xi(c), c) = \zeta(c).$$

Moreover, we assume that

$$\mathscr{S} := \{(\xi(c), \zeta(c)): c \in I_0\}$$

is transversally intersected by f, and we have to set

$$G := f(\Gamma), \qquad \hat{G} := f(\hat{\Gamma}) = G \cup \mathscr{S}.$$

If we now replace (54) by

$$\Sigma(x, c) := \int_{\xi(c)}^{x} F(t, \varphi(t, c), \varphi'(t, c))\, dt,$$

then it follows that

$$\Sigma_c(x, c) = -F(t, \varphi(t, c), \varphi'(t, c))|_{t=\xi(c)} \cdot \xi_c(c) + [F_p(t, \varphi(t, c), \varphi'(t, c)) \cdot \varphi_c(t, c)]_{t=\xi(c)}^{t=x}$$

and

$$\varphi_c(\xi(c), c) = -\varphi'(\xi(c), c)\xi_c(c) + \zeta_c(c).$$

This implies

$$\Sigma_c(x, c) = F_p(x, \varphi(x, c), \varphi'(x, c)) \cdot \varphi_c(x, c) - \{\tilde{F}_p \cdot \zeta_c(c) + [\tilde{F} - \varphi'(\xi(c), c) \cdot \tilde{F}_p]\xi_c(c)\},$$

where the superscript $\sim$ indicates that the argument has to be taken as $\xi(c)$, $\varphi(\xi(c), c)$, $\varphi'(\xi(c), c)$.

Since $\mathscr{S}$ is transversally intersected by the improper field, we have $\{\dots\} = 0$. Thus we arrive at

$$\Sigma_c(x, c) = F_p(x, \varphi(x, c), \varphi'(x, c)) \cdot \varphi_c(x, c).$$

From here on we can repeat the computations of the proof of Theorem 7, and we see that

$$S(x, z) := \Sigma(x, a(x, z))$$

is the eikonal of the field $f|_\Gamma$ with the inverse $g(x, z) = (x, a(x, z))$. We leave it to the reader to complete the discussion and to formulate complete results analogous to Theorems 7 and 8.

2. Embedding of Extremals

We shall now discuss the problem of embedding a given extremal into a Mayer field.

In *2.1* we shall see that an extremal arc can be embedded into a Mayer field if it does not contain a pair of consecutive conjugate points. This embedding is only local, i.e., it will be carried out on a sufficiently small C^0-neighbourhood of the given extremal arc. In *2.2* we shall see that this construction even leads to a global embedding provided that $n = N = 1$. As an interesting by-product we obtain *Jacobi's envelope theorem* which generalizes classical results on bundles of straight lines and on the evolute and the involute of a given curve. In *2.3* we shall consider certain integrals of the kind $\int_a^b \omega(x, u)\sqrt{1 + u'^2}\, dx$ for $n = N = 1$, and we discuss the *brachystochrone* in some detail since its weight function $\omega(x, z) = 1/\sqrt{H - z}$ is singular at the left end point $P = (a, H)$.

Finally, in *2.4* we consider certain generalizations of stigmatic bundles which are called *field-like Mayer bundles*, and their singular points which are denoted as *focal points*. The focal points form the *focal set* (or *caustic*) of the field-like bundle. Between two consecutive sheets of its caustic, a field-like bundle is, in fact, a Mayer field, and it will turn out that focal points play the same role for field-like Mayer bundles as conjugate points for stigmatic bundles.

2.1. Embedding of Regular Extremals into Mayer Fields

Fix some Lagrangian $F(x, z, p)$ which is defined on $\Omega \times \mathbb{R}^N$ where Ω is an open set in the x, z-space (= configuration space), and let $u : [a, b] \to \mathbb{R}^N$ with graph $u \subset \Omega$ be an F-extremal. Since we want to work with the Jacobi equations

$$\mathcal{J}_u v = 0 \tag{1}$$

corresponding to the extremal u, we assume for the following discussion that F is of class C^3. Moreover, we shall suppose that all line elements $\ell(x) = (x, u(x), u'(x))$ of the extremal u are *regular*, i.e., the matrix

$$A(x) := (F_{p^i p^k}(x, u(x), u'(x))) \tag{2}$$

is supposed to be invertible for all $x \in [a, b]$. Then we can write (1) in the form

$$v'' = H(x)v + K(x)v', \tag{3}$$

with suitable matrix functions $H(x) = (h_{ij}(x))$ and $K(x) = (k_{ij}(x))$ which can be computed in terms of the matrices $A := F_{pp} \circ \ell$, $B := F_{pz} \circ \ell$, $C := F_{zz} \circ \ell$ and their first derivates. The matrices $H(x)$ and $K(x)$ depend continuously on x.

Then we can extend u as an F-extremal to some slightly larger interval $[a_0, b_0]$ such that $a_0 < a < b < b_0$, and that $A(x)$ is also invertible on $[a_0, b_0]$. Correspondingly we consider the Jacobi equations (1) or (3) on the larger interval $[a_0, b_0]$.

Let us recall that the solutions $v : [a_0, b_0] \to \mathbb{R}^N$ of (1) or (3) are denoted as *Jacobi fields* (with respect to the extremal u). The set of all Jacobi fields forms a real vector space which will be denoted by $\mathcal{V}$. For any $\xi \in [a_0, b_0]$, we introduce the linear subspace

(4) $$\mathscr{V}_0(\xi) := \{v \in \mathscr{V} : v(\xi) = 0\}$$

of Jacobi fields v that vanish at $x = \xi$.

Let $e_1 = (1, 0, \dots, 0), \dots, e_N = (0, \dots, 0, 1)$ be the canonical base of $\mathbb{R}^N$. For any $\xi \in [a_0, b_0]$ we consider the Jacobi fields $v_1, \dots, v_N, w_1, \dots, w_N$ satisfying

(5) $$v_j(\xi) = e_j, \qquad w_j(\xi) = 0,$$
$$v_j'(\xi) = 0, \qquad w_j'(\xi) = e_j.$$

For prescribed values of $v(\xi)$ and $v'(\xi)$, equation (3) possesses exactly one solution $v : [a_0, b_0] \to \mathbb{R}^N$. This implies that

(6) $$\mathscr{V} = \operatorname{span}\{v_1, \dots, v_N, w_1, \dots, w_N\}$$

and

(7) $$\mathscr{V}_0(\xi) = \operatorname{span}\{w_1, \dots, w_N\}.$$

It is now an immediate consequence of (5) that the $2N$ Jacobi fields $v_1, \dots, v_N$, $w_1, \dots, w_N$ are linearly independent elements of $\mathscr{V}$. Hence they form a base of $\mathscr{V}$, and $w_1, \dots, w_N$ provides a base of $\mathscr{V}_0(\xi)$. In particular, we have

(8) $$\dim \mathscr{V} = 2N, \qquad \dim \mathscr{V}_0(\xi) = N.$$

Now we want to characterize pairs of conjugate values $\xi, \xi^* \in [a_0, b_0]$ with $\xi \neq \xi^*$ for the extremal $u : [a_0, b_0] \to \mathbb{R}^N$, extending this way the results of 5,2.2 to the case $N > 1$. Recall that ξ, ξ^* are said to be a *pair of conjugate values in* $[a_0, b_0]$ if $\xi \neq \xi^*$ and if there is some $v \in \mathscr{V}$ which is nontrivial (i.e. $v \neq 0$) and satisfies $v(\xi) = 0$, $v(\xi^*) = 0$.

Lemma 1. *Let $w_1, \dots, w_N$ be a base of $\mathscr{V}_0(\xi)$ for some $\xi \in [a_0, b_0]$, and set*

(9) $$\Delta(x) := \det(w_1(x), \dots, w_N(x)) \quad \textit{for } x \in [a_0, b_0].$$

Then, for $\xi^* \in [a_0, b_0] - \{\xi\}$, the couple ξ, ξ^* is a pair of conjugate values if and only if $\Delta(\xi^*) = 0$.

Proof. (i) Suppose that ξ, ξ^* is a conjugate pair of values in $[a_0, b_0]$. Then there is some $v \in \mathscr{V}_0(\xi)$ such that $v \neq 0$ and $v(\xi^*) = 0$. Since $v \neq 0$, there exists a nontrivial linear combination $\lambda^j w_j$ of $w_1, \dots, w_N$ such that $v = \lambda^j w_j$. By $v(\xi^*) = 0$ we obtain $\lambda^1 w_1(\xi^*) + \dots + \lambda^N w_N(\xi^*) = 0$, and since $\sum_{j=1}^N |\lambda^j|^2 \neq 0$ it follows that $\Delta(\xi^*) = 0$.

(ii) Conversely, if $\Delta(\xi^*) = 0$, then there is a nontrivial N-tuple $(\lambda^1, \dots, \lambda^N) \in \mathbb{R}^N$ such that $\lambda^j w_j(\xi^*) = 0$. Set $v := \lambda^j w_j$. Then $v \neq 0$, $v \in \mathscr{V}$, and $v(\xi) = 0$, $v(\xi^*) = 0$. □

Another characterization of pairs of conjugate values is provided by

Lemma 2. *Let $v_1, v_2, \dots, v_{2N}$ be an arbitrary base of $\mathscr{V}$, and set*

$$D(x, \xi) := \det \begin{pmatrix} v_1(x), \ldots, v_{2N}(x) \\ v_1(\xi), \ldots, v_{2N}(\xi) \end{pmatrix} \quad \textit{for } x \in [a_0, b_0], \tag{10}$$

where ξ is a fixed value in $[a_0, b_0]$. Then, for $\xi^* \in [a_0, b_0] - \{\xi\}$, the couple ξ, ξ^* is a pair of conjugate values if and only if $D(\xi^*, \xi) = 0$.

Proof. (i) If (ξ, ξ^*) is a conjugate pair, there is some $v \in \mathscr{V}$, $v \neq 0$, such that $v(\xi) = 0$ and $v(\xi^*) = 0$. Hence there is some $2N$-tuple $c = (c^1, \ldots, c^{2N}) \neq 0$ such that $v = c^1 v_1 + \cdots + c^{2N} v_{2N} = c^\alpha v_\alpha$. We infer from $c^\alpha v_\alpha(\xi) = 0$ and $c^\alpha v_\alpha(\xi^*) = 0$ that $D(\xi^*, \xi) \neq 0$ since $c \neq 0$.

(ii) Conversely, the relation $D(\xi^*, \xi) = 0$ implies the existence of some $c = (c^1, \ldots, c^{2N}) \in \mathbb{R}^{2N}$, $c \neq 0$, such that the system

$$c^\alpha v_\alpha(\xi) = 0, \qquad c^\alpha v_\alpha(\xi^*) = 0$$

is satisfied. Then $v := c^\alpha v_\alpha$ is a nontrivial Jacobi field vanishing at $x = \xi$ and at $x = \xi^*$. □

We can use Lemma 2 to show that nonexistence of pairs of conjugate values in the interval $[a, b]$ is a stable property with respect to small changes of the endpoints a, b.

Proposition 1. *If there is no pair of conjugate values ξ, ξ^* (with $\xi \neq \xi^*$) in $[a, b]$, then, for $0 < \delta \ll 1$, also the interval $[a - \delta, b + \delta]$ does not contain pairs of conjugate values.*

Proof. Choose $\delta > 0$ so small that $a_0 \le a - \delta \le b + \delta \le b_0$. Then, for $\xi \in [a - \delta, b + \delta]$, we consider a base $v_j(\cdot, \xi)$, $1 \le j \le 2N$, of $\mathscr{V}$ whose elements satisfy the initial conditions

$$v_j(\xi, \xi) = e_j, \qquad v_j'(\xi, \xi) = 0, \qquad v_{N+j}(\xi, \xi) = 0, \qquad v_{N+j}'(\xi, \xi) = e_j$$

for $1 \le j \le N$. Let $D(x, \xi)$ be the corresponding determinant defined by (10); for the sake of brevity, we write $v_j(x)$ instead of $v_j(x, \xi)$. Then our assumption implies that

$$D(x, \xi) \neq 0 \quad \text{for } x \in [a, b],\ x \neq \xi,\ (\xi \in [a, b]), \tag{11}$$

if we take Lemma 2 into account.

Moreover, we can write

$$v_j(x) - v_j(\xi) = (x - \xi)\omega_j(x, \xi),$$

where

$$\omega_j(x, \xi) := \int_0^1 v_j'(\xi + t(x - \xi))\, dt.$$

We infer that

$$D(x,\xi) = \det\begin{pmatrix} v_1(x) - v_1(\xi), \ldots, v_{2N}(x) - v_{2N}(\xi) \\ v_1(\xi), \ldots, v_{2N}(\xi) \end{pmatrix}$$

$$= (x-\xi)^n \det\begin{pmatrix} \omega_1(x,\xi), \ldots, \omega_{2N}(x,\xi) \\ v_1(\xi), \ldots, v_{2N}(\xi) \end{pmatrix}.$$

Let us denote the last determinant by $\tilde{D}(x, \xi)$. Then we have

(12) $$D(x,\xi) = (x-\xi)^n \tilde{D}(x,\xi) \quad \text{for } x, \xi \in [a-\delta, b+\delta],$$

and $\omega_j(\xi, \xi) = v_j'(\xi)$ implies that

(13) $$\tilde{D}(\xi,\xi) = \pm 1 \quad \text{for all } \xi \in [a-\delta, b+\delta].$$

It follows from (11)–(13) that

(14) $$\tilde{D}(x,\xi) \neq 0 \quad \text{for all } x, \xi \in [a, b].$$

Well-known results on the dependence of solutions of initial value problems on their initial data imply that $\tilde{D}(x, \xi)$ is a continuous function of (x, ξ) on the rectangle $[a-\delta, b+\delta] \times [a-\delta, b+\delta]$. Then we infer from (14) that

$$\tilde{D}(x,\xi) \neq 0 \quad \text{for all } x, \xi \in [a-\delta, b+\delta]$$

if $0 < \delta \ll 1$. Hence, by (12), we obtain that

$$D(x,\xi) \neq 0 \quad \text{for all } x, \xi \in [a-\delta, b+\delta] \text{ with } x \neq \xi.$$

By virtue of Lemma 2, we conclude that $[a-\delta, b+\delta]$ does not contain a pair of conjugate values ξ, ξ^* with $\xi \neq \xi^*$ provided that $0 < \delta \ll 1$. □

Let us now return to the extremal $u: [a, b] \to \mathbb{R}^N$ that we considered in the very beginning. It was thought to be extended to a larger interval $[a_0, b_0]$, $a_0 < a < b < b_0$. Set

(15) $$z_0 := u(a_0), \qquad c_0 := u'(a_0),$$

and consider the initial value problem

(16) $$L_F(\varphi(\cdot, c)) = 0, \quad \varphi(a_0, c) = z_0, \quad \varphi'(a_0, c) = c$$

for the function $z = \varphi(x, c)$. If we restrict c to a sufficiently small neighbourhood

$$I_0(\varepsilon) := \{c \in \mathbb{R}^N : |c - c_0| \leq \varepsilon\}, \quad \varepsilon > 0,$$

of c_0, the uniquely determined maximal solution $\varphi(\cdot, c)$ of the initial value problem (16) is defined on $\hat{I} := [a_0, b_0]$. Thus we can view φ as a mapping of $\hat{\Gamma} := \hat{I} \times I_0(\varepsilon)$ into $\mathbb{R}^N$, and $f(x, c) := (x, \varphi(x, c))$, $(x, c) \in \hat{\Gamma}$, defines a *stigmatic ray bundle* in the sense of Definition 4 of *1.1*, since the initial conditions of (16) imply

$$\varphi(a_0, c) = z_0, \qquad \det \varphi_c'(a_0, c) = 1.$$

The nodal point of the bundle $f: \hat{\Gamma} \to \mathbb{R}^{N+1}$ is the point $P_0 = (a_0, z_0)$. By Proposition 3 of *1.3* it follows that *f is a Mayer bundle.*

Moreover, by the results of Chapter 5, the functions

$$w_\alpha(x) := \frac{\partial \varphi}{\partial c^\alpha}(x, c_0), \quad 1 \le \alpha \le N, \tag{17}$$

are Jacobi fields along the extremal u, i.e., $w_1, w_2, \ldots, w_N \in \mathscr{V}$, and we infer from the initial conditions of (16) that

$$w_\alpha(a_0) = 0, \quad w'_\alpha(a_0) = e_\alpha, \qquad 1 \le \alpha \le N. \tag{18}$$

Hence $w_1, w_2, \ldots, w_N$ is a base of $\mathscr{V}_0$.

Suppose now that $[a, b]$ does not contain a pair of conjugate values ξ, ξ^* with $\xi \neq \xi^*$. Then the same holds true for $[a_0, b_0]$ if $a - a_0$ and $b_0 - b$ are sufficiently small (see Proposition 1), and, by Lemma 1, (17) and (18), we obtain that

$$\det(\varphi_{c^1}(x, c_0), \ldots, \varphi_{c^N}(x, c_0)) \neq 0 \quad \text{for } a_0 < x \le b_0 .$$

Hence, for any $x_0 \in (a_0, a]$, we can find a sufficiently small value of $\varepsilon > 0$ such that

$$\det \varphi_c(x, c) \neq 0 \quad \text{if } x_0 \le x \le b_0 \text{ and } |c - c_0| \le \varepsilon .$$

Because of

$$Df = \begin{pmatrix} 1, & \varphi' \\ 0, & \varphi_c \end{pmatrix},$$

the Jacobian $\mathscr{J}_f$ of the mapping $f : \hat{\Gamma} \to \mathbb{R}^{N+1}$ is given by

$$\mathscr{J}_f = \det \varphi_c .$$

Then the implicit function theorem in conjunction with a simple covering argument yields that the restriction of f to $[x_0, b_0] \times I_0(\varepsilon)$ is a field provided that $0 < \varepsilon \ll 1$. Combining this result with Proposition 3 of *1.1*, we infer that $f|_\Gamma$ is a field on $G = f(\Gamma)$ if we define Γ by

$$\Gamma := I \times I_0(\varepsilon), \quad I := (a_0, b_0], \qquad 0 < \varepsilon \ll 1,$$

and Proposition 6 of *1.1* implies that *$f|_\Gamma : \Gamma \to G$ is a Mayer field on G*. According to Definition 3 of *1.3*, *the mapping $f : \hat{\Gamma} \to \hat{G} := G \cup \{P_0\} = f(\hat{\Gamma})$ is a stigmatic field on $\hat{G}$ with the nodal point $P_0 = (a_0, z_0)$*.

Let us summarize the results just obtained:

Theorem 1. *Let $F(x, z, p)$ be a C^3-Lagrangian on $\Omega \times \mathbb{R}^N$, and $u : [a, b] \to \mathbb{R}^N$ be an F-extremal in Ω with regular line elements, i.e., the matrix $F_{pp}(x, u(x), u'(x))$ be invertible for all $x \in [a, b]$. Suppose also that the interval $[a, b]$ contains no pair of conjugate values of u. Then we can embed u into a stigmatic field $f : \hat{\Gamma} \to \hat{G} = G \cup \{P_0\}$ with the nodal point $P_0 = (a_0, z_0)$, where $\hat{\Gamma} = \hat{I} \times I_0(\varepsilon)$, $\hat{I} = [a_0, b_0]$, $a_0 < a < b < b_0$, $I_0(\varepsilon) = \{c : |c - c_0| \le \varepsilon\}$, $\varepsilon > 0$, and $f(x, c_0) = (x, u(x))$ for $a \le x \le b$. The restriction of f to $\Gamma = I \times I_0(\varepsilon)$, $I = (a_0, b_0]$, is a proper Mayer field on $G = f(\Gamma)$.*

By the results of *1.3*, we then obtain the following generalization of Theorem 3 in *1.3*:

Theorem 2. *An F-extremal* $u: [a, b] \to \mathbb{R}^N$ *is a (strict) strong minimizer of* $\mathcal{F}(u) = \int_a^b F(x, u(x), u'(x))\,dx$ *if the following conditions are satisfied:*

(i) $F_{pp}(x, u(x), u'(x))$ *is invertible for all* $x \in [a, b]$;

(ii) $\mathcal{E}_F(x, u(x), u'(x), p) > 0$ *for all* $x \in [a, b]$ *and all* $p \in \mathbb{R}^N - \{u'(x)\}$;

(iii) *the interval* $[a, b]$ *contains no pair of (different) conjugate values of u.*

Remark 1. We can replace (i) and (ii) by the more stringent assumption that $F_{pp}(x, z, p) > 0$ holds true for all $(x, z, p) \in \mathcal{U} \times \mathbb{R}^N$ where $\mathcal{U}$ denotes a sufficiently small neighbourhood of graph u in $\mathbb{R} \times \mathbb{R}^N$.

Remark 2. The condition (ii) of the preceding theorem implies that $F_{pp}(x, u(x), u'(x)) \geq 0$. Hence, (i) is equivalent to

$$F_{pp}(x, u(x), u'(x)) > 0 \quad \text{for all } x \in [a, b]$$

if also (ii) is required.

2.2. Jacobi's Envelope Theorem

We now want to consider singular Mayer fields in the case $N = 1$. Such a field is just a one-parameter family of extremals loosing the field property at its boundary.

Let us begin by looking at a stigmatic bundle f of class C^2,

$$f(x, \alpha) = (x, \varphi(x, \alpha)), \quad x_0 \leq x < b(\alpha), \ \alpha \in A, \tag{1}$$

with the nodal point $P_0 = (x_0, z_0) \in \mathbb{R}^2$. We assume that $\varphi(\cdot, \alpha)$ is a one-parameter family of extremals with respect to a C^3-Lagrangian F where α is a real parameter varying in an interval A, i.e.,

$$L_F(\varphi(\cdot, \alpha)) = 0. \tag{2}$$

We assume that F is elliptic on the line elements of f, that is,

$$F_{pp}(x, \varphi(x, \alpha), \varphi'(x, \alpha)) > 0, \tag{3}$$

and we suppose that $[x_0, b(\alpha))$ is the maximal interval of existence to the right of the solution $\varphi(\cdot, \alpha)$ of (2). Moreover, we have

$$\varphi(x_0, \alpha) = z_0, \quad \text{i.e., } f(x_0, \alpha) = P_0 \tag{4}$$

and

$$\varphi'_\alpha(x_0, \alpha) \neq 0 \quad \text{for all } \alpha \in A, \tag{5}$$

since f is assumed to be a stigmatic bundle with the nodal point P_0.

Then, for every $\alpha \in A$, the function $v(x) := \varphi_\alpha(x, \alpha)$ defines a nontrivial Jacobi field v along the extremal $\varphi(\cdot, \alpha)$, and $v(x_0) = 0$. It follows that $v'(x) \neq 0$ whenever $v(x)$ vanishes. That is, we have $\varphi_{\alpha x}(x, \alpha) \neq 0$ at all values x where $\varphi_\alpha(x, \alpha) = 0$, and therefore we have found:

For all $x \in (x_0, b(\alpha))$ conjugate to x_0 we have $\varphi'_\alpha(x, \alpha) \neq 0$.

The envelope $\mathscr{E}$ of the bundle f is just the *conjugate locus of f*, i.e., the locus of the conjugate points of the extremal curves $f(\cdot, \alpha)$. It is described

(6) $$\mathscr{E} = \{(x, \varphi(x, \alpha)) \colon x_0 < x < b(\alpha), \alpha \in A, \varphi_\alpha(x, \alpha) = 0\}.$$

(Precisely speaking, it is the envelope of the "open" bundle $f(x, \alpha)$, $x_0 < x < b(\alpha)$, $\alpha \in A$.)

Denote by $\mathscr{E}_1$ the *first conjugate locus* of f consisting of all conjugate points $f(\xi, \alpha)$ of the extremal curves

(7) $$\mathscr{C}(\alpha) := f(\cdot, \alpha),$$

such that ξ is the first conjugate value of the extremal $\varphi(\cdot, \alpha)$ to the right of $x = x_0$. Of course, there might not exist a conjugate point of $\mathscr{C}(\alpha)$ and, in particular, $\mathscr{E}_1$ could be empty.

Let us presently assume that, for every $\alpha \in A$, there is a first conjugate value $\xi(\alpha) \in (x_0, b(\alpha))$ to $x = x_0$ for the extremal curve $\mathscr{C}(\alpha)$. Then

(8) $$P(\alpha) = (\xi(\alpha), \zeta(\alpha)), \quad \zeta(\alpha) := \varphi(\xi(\alpha), \alpha)$$

is the first conjugate point of $\mathscr{C}(\alpha)$ to the right of P_0. However, we are not allowed to conclude that $P(\alpha)$ is the only intersection point of $\mathscr{E}_1$ with the curve $\mathscr{C}(\alpha)$.

We infer from (6) that

(9) $$\varphi_\alpha(\xi(\alpha), \alpha) = 0 \quad \text{for all } \alpha \in A,$$

and by the afore-stated remark we have

(10) $$\varphi_{\alpha x}(\xi(\alpha), \alpha) \neq 0 \quad \text{for all } \alpha \in A.$$

Hence we can apply the implicit function theorem and obtain that $\xi(\alpha)$ is a C^1-function of $\alpha \in A$. By differentiating (9) with respect to α, we obtain

$$\varphi_{\alpha x}(\xi(\alpha), \alpha)\, \xi_\alpha(\alpha) + \varphi_{\alpha\alpha}(\xi(\alpha), \alpha) = 0,$$

whence

(11) $$\frac{d\xi}{d\alpha}(\alpha) = -\frac{\varphi_{\alpha\alpha}(\xi(\alpha), \alpha)}{\varphi_{\alpha x}(\xi(\alpha), \alpha)}.$$

Moreover, we infer from (8) and (9) that

(12) $$\frac{d\zeta}{d\alpha}(\alpha) = \varphi_x(\xi(\alpha), \alpha)\frac{d\xi}{d\alpha}(\alpha).$$

Therefore, the first conjugate locus $\mathscr{E}_1 = \{P(\alpha) \colon \alpha \in A\}$ is a regular C^1-curve if we assume that

$$\varphi_{\alpha\alpha}(\xi(\alpha), \alpha) \neq 0 \quad \text{for all } \alpha \in A. \tag{13}$$

In fact, the relations (11) and (13) imply that $\frac{d\xi}{d\alpha}(\alpha) \neq 0$ for all $\alpha \in A$. Hence the function $x = \xi(\alpha)$ possesses a C^1-inverse $\alpha = a(x)$ defined on some interval I on the x-axis, and $\mathscr{E}_1$ turns out to be the graph of the function $h(x) := \zeta(a(x))$, $x \in I$, i.e.,

$$\mathscr{E}_1 = \{(x, h(x)) \colon x \in I\}. \tag{14}$$

Now we introduce the sets $\hat{\Gamma}$ and Γ by

$$\begin{aligned} \hat{\Gamma} &:= \{(x, \alpha) \colon x_0 \le x \le \xi(\alpha), \alpha \in A\}, \\ \Gamma &:= \{(x, \alpha) \colon x_0 < x < \xi(\alpha), \alpha \in A\}. \end{aligned} \tag{15}$$

We claim that $f|_\Gamma$ is a field, and that each "restricted" curve

$$\mathscr{C}^*(\alpha) := f(\cdot, \alpha)|_{(x_0, \xi(\alpha))} \tag{16}$$

meets the first conjugate locus $\mathscr{E}_1$ exactly at the point $P(\alpha)$.

It suffices to show this assertion under the additional assumptions

$$\varphi_\alpha(x, \alpha) > 0 \quad \text{for } x_0 < x < \xi(\alpha),$$

$$\varphi_{\alpha\alpha}(\xi(\alpha), \alpha) < 0 \quad \text{for } \alpha \in A,$$

and

$$\varphi_{\alpha x}(\xi(\alpha), \alpha) > 0 \quad \text{for } \alpha \in A.$$

Then $\xi(\alpha)$ is a strictly increasing function of $\alpha \in A$, $a(x)$ is a strictly increasing function of $x \in I$, and $\varphi(x, \alpha)$ is strictly increasing for $\alpha \in A(x) := A \cap \{\alpha \ge a(x)\}$. Therefore, f maps the segment $\{x\} \times A(x)$ bijectively onto the segment $\{x\} \times \varphi(x, A(x))$. This proves the assertion.

Furthermore, one infers from (12) that $h(x)$ is strictly monotonic if

$$\varphi_x(\xi(\alpha), \alpha) \neq 0 \quad \text{for all } \alpha \in A. \tag{17}$$

Let us summarize our results as

Theorem 1. *Let $f(x, \alpha) = (x, \varphi(x, \alpha))$, $\alpha \in A$ and $x_0 \le x < b(\alpha)$, be a stigmatic bundle of extremals ($N = 1$) with the nodal point $P_0 = (x_0, z_0) = f(x_0, \alpha)$, and suppose that, for every $\alpha \in A$, there is a first conjugate value $\xi(\alpha) \in (x_0, b_0(\alpha))$ to x_0 with respect to the extremal $\varphi(\cdot, \alpha)$. Suppose also that* (3) *holds and that*

$$\varphi_{\alpha\alpha}(\xi(\alpha), \alpha) \neq 0 \quad \textit{for all } \alpha \in A.$$

Then f is injective on $\{(x, \alpha) \colon \alpha \in A, x_0 < x \le \xi(\alpha)\}$ and the restriction $f|_\Gamma$ to $\Gamma := \{(x, \alpha) \colon \alpha \in A, x_0 < x < \xi(\alpha)\}$ is a Mayer field. The first conjugate locus $\mathscr{E}_1 = \{P \colon P = f(\xi(\alpha), \alpha), \alpha \in A\}$ is the graph of a C^1-function $z = h(x)$, $x \in I$ (= interval in $\mathbb{R}$) which is strictly monotonic if (17) *holds true. Each field*

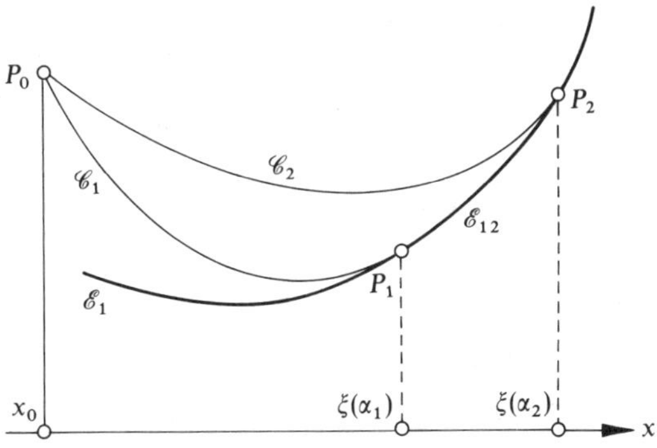

Fig. 7. Jacobi's envelope theorem for extremals emanating from a fixed point P_0.

curve $\mathscr{C}^*(\alpha) = f(\cdot, \alpha)|_{(x_0, \xi(\alpha))}$ *meets* $\mathscr{E}_1$ *exactly at its first conjugate point* $P(\alpha) := f(\xi(\alpha), \alpha)$. *In particular, the restriction* $f|_{\hat{\Gamma}}$ *to* $\hat{\Gamma} := \{(x, \alpha): \alpha \in A, x_0 \le x \le \xi(\alpha)\}$ *is a stigmatic field. The first conjugate locus is part of the envelope* $\mathscr{E}$ *of the bundle* f, *and this envelope is defined by the equations*

$$z = \varphi(x, \alpha) \quad \text{and} \quad \varphi_\alpha(x, \alpha) = 0.$$

Note that this result can be viewed as a "global" embedding theorem of a given extremal up to the first conjugate point to its left end point.

An obvious variant of the preceding theorem is that a stigmatic bundle furnishes a stigmatic field if its envelope is empty, or, equivalently, if its (first) conjugate locus is empty, provided that $N = 1$.

We remark that here as well as in Theorem 1 the assumption $N = 1$ is crucial.

Theorem 2. (Jacobi's envelope theorem). *Suppose that* $N = 1$, *and that we have the situation of Theorem* 1. *Then, for two arbitrary values* $\alpha_1, \alpha_2 \in A$ *with* $\xi(\alpha_1) < \xi(\alpha_2)$, *it follows that*

$$\int_{x_0}^{\xi(\alpha_1)} F(x, \varphi(x, \alpha_1), \varphi'(x, \alpha_1))\, dx + \int_{\xi(\alpha_1)}^{\xi(\alpha_2)} F(x, h(x), h'(x))\, dx = \int_{x_0}^{\xi(\alpha_2)} F(x, \varphi(x, \alpha_2), \varphi'(x, \alpha_2))\, dx, \tag{18}$$

or, in a sloppy but more suggestive notation

$$\mathscr{F}(\mathscr{C}_1) + \mathscr{F}(\mathscr{E}_{12}) = \mathscr{F}(\mathscr{C}_2). \tag{18'}$$

Here $\mathscr{C}_1$ *and* $\mathscr{C}_2$ *denote the extremal curves* $f(x, \alpha_1)$, $x_0 \le x \le \xi(\alpha_1)$, *and* $f(x, \alpha_2)$, $x_0 \le x \le \xi(\alpha_2)$, *respectively, and* $\mathscr{E}_{12}$ *stands for the arc* $z = h(x)$, $\xi(\alpha_1) \le x \le \xi(\alpha_2)$, *of the first conjugate locus between the endpoints of* $\mathscr{C}_1$ *and* $\mathscr{C}_2$.

Proof. Let S be the eikonal of the field $f: \Gamma \to G$ where $G := f(\Gamma)$. We can extend S from G to $G \cup \{P_0\}$ such that $S(P) \to S(P_0) = 0$ if P tends to P_0 along an extremal emanating from P_0 (see Theorem 7 of *1.3*). Secondly we can extend the slope function $\mathscr{P}(x, z)$ to a continuous function on $G \cup \mathscr{E}_1$ by setting

$$\mathscr{P}(x, h(x)) := h'(x) = \varphi'(x, a(x)),$$

where $\alpha = a(x)$ is the inverse of $x = \xi(\alpha)$. Then, by employing the Carathéodory equations

$$S_x = \bar{F} - \mathscr{P} \cdot \bar{F}_p, \qquad S_z = \bar{F}_p,$$

we can extend $S(x, z)$ to a continuous function on $G \cup \mathscr{E}_1$. It follows that

$$\frac{d}{dx} S(x, h(x)) = S_x(x, h(x)) + S_z(x, h(x))h'(x) = F(x, h(x), h'(x)),$$

whence we obtain

$$\int_{\xi(\alpha_1)}^{\xi(\alpha_2)} F(x, h(x), h'(x))\, dx = S(P(\alpha_2)) - S(P(\alpha_1)).$$

On the other hand, we have

$$\int_{x_0}^{\xi(\alpha_i)} F(x, \varphi(x, \alpha_i), \varphi'(x, \alpha_i))\, dx = S(P(\alpha_i))$$

for $i = 1, 2$, and the assertion is proved. □

Remark. We already know that an extremal cannot furnish a minimum and not even a weak minimum if its interval of definition contains a pair of conjugate values. That is, an extremal arc $f(\cdot, \alpha)$ emanating from $P_0 = (x_0, z_0)$ to the right loses its minimizing property beyond the first conjugate point $P(\alpha)$ to P_0. *In the situation assumed in the envelope theorem it is clear that none of the extremals*

$$\mathscr{C}^*(\alpha) := \{(x, \varphi(x, \alpha)) : x_0 \le x \le \xi(\alpha)\}$$

between P_0 and $P(\alpha) = (\xi(\alpha), \zeta(\alpha))$ is a minimizer or even a weak minimizer.

In fact, if $\alpha_2 \in A$ is not the left endpoint of A, then there is an $\varepsilon_0 > 0$ such that $\alpha \in A$ if $\alpha_2 - \varepsilon_0 < \alpha < \alpha_2$. We now set

$$v(x) := \begin{cases} \varphi(x, \alpha_1) & \text{if } x_0 \le x \le \xi(\alpha_1), \\ h(x) & \text{if } \xi(\alpha_1) \le x \le \xi(\alpha_2), \end{cases}$$

where α_1 is an arbitrary number in $(\alpha_2 - \varepsilon_0, \alpha_2)$ and $\xi(\alpha)$ is assumed to be increasing. Then v is of class C^1, has the same boundary values as $\varphi(\cdot, \alpha_2)$ and satisfies

$$\mathscr{F}(v) = \mathscr{F}(\varphi(\cdot, \alpha_2))$$

by virtue of (18). It is easy to see that, therefore, the function v is a weak minimizer of $\mathscr{F}$ if α_1 is sufficiently close to α_2, provided that $\varphi(\cdot, \alpha_2)$ is a weak minimizer. In particular, the restriction $h|_I$ of h to the interval $I := [\xi(\alpha_1), \xi(\alpha_2)]$ has to be a weak minimizer. Because of (3), the Legendre condtion

$$(19) \qquad F_{pp}(x, h(x), h'(x)) > 0$$

holds true, and therefore $h|_I$ is an F-extremal. Moreover, the Cauchy problem

$$L_F(u) = 0, \quad u(x) = h(x), \quad u'(x) = h'(x) \qquad \text{for } x = \xi(\alpha_2)$$

has only one solution whence $h(x) \equiv \varphi(x, \alpha_2)$ for all $x \in I$ since $h(x) = \varphi(x, \alpha_2)$ and $h'(x) = \varphi'(x, \alpha_2)$ at $x = \xi(\alpha_2)$. Thus we have arrived at a contradiction to Theorem 1.

The loss of minimum property exactly at the conjugate point with regard to the left endpoint occurs in most cases, but there are exceptional situations. It was discovered by A. Kneser[2] that one has such an exceptional case if the envelope has a cusp at the point $P(\alpha)$ conjugate to P_0 which points towards P_0.

As we are in the case $N = 1$, no new ideas are needed to carry over the preceding results to an arbitrary bundle of extremals

$$f(x, \alpha) = (x, \varphi(x, \alpha)), \quad x_0(\alpha) \le x < b(\alpha), \ \alpha \in A,$$

satisfying

$$\varphi_\alpha(x_0(\alpha), \alpha) \neq 0 \quad \text{for all } \alpha \in A.$$

We know that f is a Mayer bundle. Suppose that $x_0 \in C^2$ and $x_{0\alpha} \neq 0$. Then the left endpoints $P_0(\alpha) = (x_0(\alpha), z_0(\alpha))$ with $z_0(\alpha) := \varphi(x_0(\alpha), \alpha)$ of the rays $f(\cdot, \alpha)$ form a regular curve

$$\mathscr{S}_0 := \{P_0(\alpha) \colon \alpha \in A\}.$$

Let $x = \xi(\alpha)$ be the first zero of $\varphi_\alpha(x, \alpha)$ to the right of $x = x_0$, and set $\zeta(\alpha) := \varphi(\xi(\alpha), \alpha)$, $P(\alpha) := (\xi(\alpha), \zeta(\alpha))$. Then $P(\alpha)$ is called the *first focal point* on the extremal curve $\mathscr{C}(\alpha) := f(\cdot, \alpha)$, and

$$\mathscr{E}_1 := \{P(\alpha) \colon \alpha \in A\}$$

is said to be the (first) *focal curve* of the bundle f. *Assume that on every $\mathscr{C}(\alpha)$ there is a first focal point and that*

$$\varphi_{\alpha\alpha}(\xi(\alpha), \alpha) \neq 0.$$

Then the focal curve is a regular C^1-curve which has a nonparametric representation as graph of a C^1-function $z = h(x)$, and the rays $\mathscr{C}^*(\alpha)$ of f between $\mathscr{S}_0$ and $\mathscr{E}_1$ form an (improper) field. This is proved in exactly the same way as Theorem 1.

Suppose now that $\mathscr{S}_0$ is *transversal* to the rays of the improper field f. Then $\mathscr{S}_0$ is a level line of the eikonal S of f. Let us consider two rays $\mathscr{C}_1$ and $\mathscr{C}_2$ of the field whose left endpoints on the tranversal line are P_0' and P_0'', and which meet the focal line $\mathscr{E}_1$ at $P_1 = (\xi_1, \zeta_1)$ and $P_2 = (\xi_2, \zeta_2)$ respectively. Let $\xi_1 < \xi_2$ and denote by $\mathscr{E}_{12}$ the subarc of $\mathscr{E}_1$ connecting P_1 and P_2. Then *Jacobi's envelope theorem* takes once again the form

$$(20) \qquad \mathscr{F}(\mathscr{C}_1) + \mathscr{F}(\mathscr{E}_{12}) = \mathscr{F}(\mathscr{C}_2).$$

We can view (20) as a generalization of a classical result on *evolutes* (or *involutes*). The evolute $\mathscr{E}$ of

[2] Cf. Carathéodory [10], p. 295, and Bolza [3], pp. 363–364.

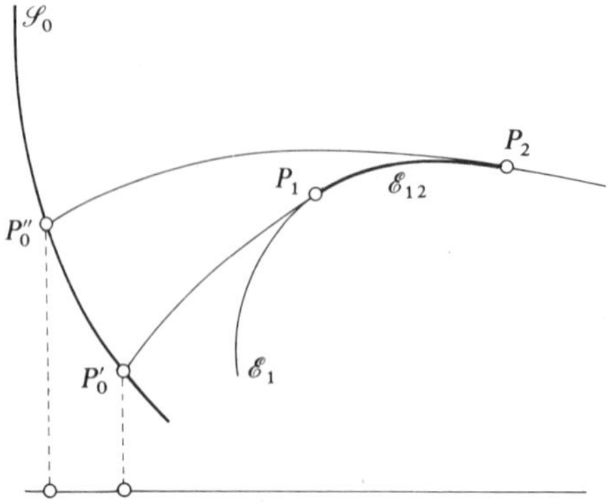

Fig. 8. Jacobi's envelope theorem for extremals intersecting a curve S_0 tranversally.

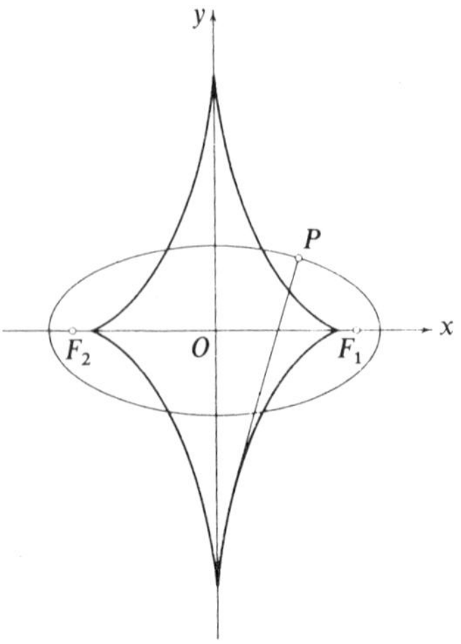

Fig. 9. Evolute of an ellipse.

a given curve $\mathscr{C}$ is by definition the envelope of the normals to $\mathscr{C}$. Consider two normals $\mathscr{N}_1$ and $\mathscr{N}_2$ to $\mathscr{C}$ emanating from P_0' and $P_0'' \in \mathscr{C}$ and touching $\mathscr{E}$ at P_1 and P_2 respectively. Then $|P_2 - P_0''| - |P_1 - P_0'|$ is equal to the arc $\mathscr{E}_{12}$ on the evolute $\mathscr{E}$ between P_1 and P_2, provided that $\mathscr{E}$ is regular. Conversely, if $\mathscr{E}$ is a given curve, one obtains the involute $\mathscr{C}$ corresponding to $\mathscr{E}$ by the following geometric construction: We imagine a flexible but inextensible thread which is attached to some subarc of the curve $\mathscr{E}$ and stretched so that the nonattaching part of it extends tangentially away from the curve. Then, as we unwind the thread from $\mathscr{E}$, its free endpoint P_0 will describe a curve $\mathscr{C}$ which is called the involute of $\mathscr{E}$. Clearly, the original curve $\mathscr{E}$ is the evolute of its involute $\mathscr{C}$.

For instance, the involute of a *cycloid* is itself a cycloid which can be obtained from it by a rigid motion. This fact was discovered by Huygens and used for the construction of a cycloidal pendulum. It is well known that such a pendulum can be used to build a "perfect" pendulum clock as every cycloid is a tautochrone.

The involute of the *circle* $x = \cos \varphi$, $y = -\sin \varphi$ is seen to be the spiral-like curve

$$x = \cos \varphi + \varphi \sin \varphi, \qquad y = -\sin \varphi + \varphi \cos \varphi,$$

and the involute of an *astroid* is an *ellipse* or, vice versa, the evolute of an ellipsoid is an astroid. This provides us with an example of a transversal surface $\mathscr{S}_0$ for which the envelope (= focal line) of the rays orthogonal to $\mathscr{S}_0$ has *cusps*.

2.3. Catenary and Brachystochrone

Many examples lead to variational integrals of the kind

$$\mathscr{F}(u) = \int_{x_1}^{x_2} \omega(x, u)\sqrt{1 + u'^2}\, dx,$$

with the Lagrangian

(1) $$F(x, z, p) = \omega(x, z)\sqrt{1 + p^2},$$

where $\omega(x, z)$ is defined on $G_0 \subset \mathbb{R}^2$, and $p \in \mathbb{R}$.

Suppose that ω is sufficiently smooth on G_0 and satisfies $\omega(x, y) > 0$ on G_0 (we can, for instance assume $\omega \in C^3$ to fulfil the general assumptions of this section but the reader will easily verify that usually less will do). From

$$\begin{aligned}\mathscr{E}_F(x, z, p, q) &= F(x, z, q) - F(x, z, p) - (q - p)F_p(x, z, p)\\ &= \tfrac{1}{2}(q - p)^2 F_{pp}(x, z, p + \delta(q - p)), \quad \delta \in (0, 1)\end{aligned}$$

and

$$F_{pp}(x, z, p) = \frac{\omega(x, z)}{\{1 + p^2\}^{3/2}},$$

we infer that

$$\mathscr{E}_F(x, z, p, q) = \tfrac{1}{2}(q - p)^2 \omega(x, z)\{1 + \tilde{p}^2\}^{-3/2},$$

where $\tilde{p} = p + \delta(q - p)$, $0 < \delta < 1$, and therefore

$$\mathscr{E}_F(x, z, p, q) > 0 \quad \text{if } p \neq q.$$

Thus we have proved: *Every extremal field of* (1) *is an optimal field.*

By virtue of the results in *1.3*, *2.1* and *2.2* we obtain:

Let f be an extremal field covering $G \subset G_0$, and assume that $u(x)$, $x_1 \le x \le x_2$, is an extremal which fits into f. Then u is the unique minimizer of the functional (1) *among all D^1-curves $v(x)$, $x_1 \le x \le x_2$, with graph$(v) \subset G$ which have the same boundary values as u.*

Similarly we obtain:

Let $u : [x_1, x_2] \to \mathbb{R}$ be an extremal with the left endpoint $P_1 = (x_1, z_1)$, and suppose that u fits into a stigmatic field of extremals with the nodal point P_1. Then u is a strict minimizer of $\mathscr{F}$ among all D^1-curves in $G \cup \{P_1\}$.

Recall that for Lagrangians of type (1), free *transversality means orthogonality* provided that $\omega > 0$. Moreover, for every $\omega > 0$, any corresponding stigmatic field with the nodal point P_1 looks like a pencil of straight lines emanating from P_1, and the associated transversal lines are nearly circular arcs centered at P_1 (cf. Figs. 10a and b). For singular densities $\omega(x, z)$ vanishing at P_1, the situation can be more complicated (see Fig. 10c). In such a case, a careful analysis of the behaviour of the eikonal S in the neighbourhood of P_1 is needed. An example of this type is furnished by the brachystochrone problem to be discussed below.

Let us mention that the expression

$$ds = \omega(x, z)\sqrt{dx^2 + dz^2}$$

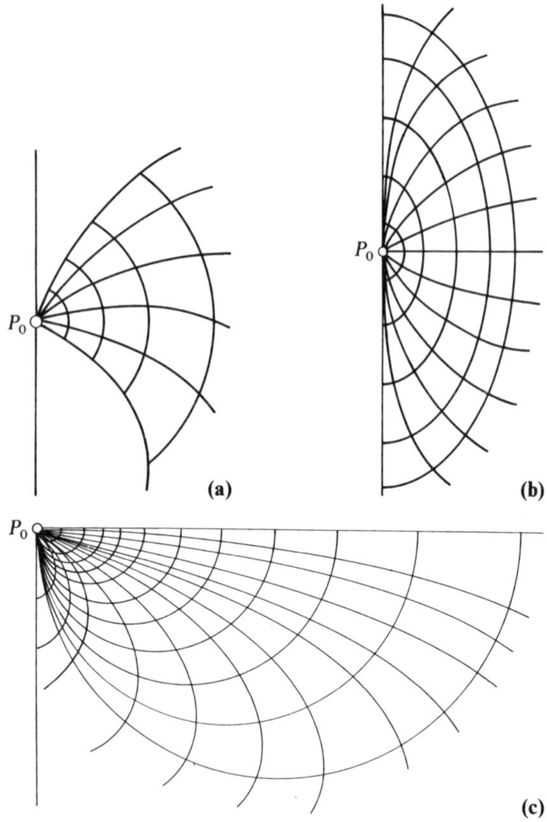

Fig. 10. (a–c).

defines a non-Euclidean line element which is conformally equivalent to the Euclidean line element $\sqrt{dx^2 + dz^2}$ provided that we assume $\omega(x, z) > 0$ on G_0. Then G_0 equipped with the metric ds becomes a two-dimensional Riemannian manifold. Its Gauss curvature $K(x, z)$ at the point (x, z) can be computed by the formula

$$K = -\frac{1}{\omega^2}\Delta \log \omega,$$

where $\Delta = \frac{\partial^2}{\partial x^2} + \frac{\partial^2}{\partial z^2}$ is the ordinary Laplacian (see Supplement).

If $G_0 = \{(x, z) \in \mathbb{R}^2 : z > 0\}$ is the upper half-plane and $\omega(x, z) = 1/z$, we obtain $K(x, z) \equiv -1$, i.e., (G_0, ds) is a two-dimensional model space of constant negative curvature. We shall look at this example in $\boxed{3}$.

The integral

$$\mathcal{L}(c) = \int_a^b \omega(c)|\dot{c}|\, dt$$

is the length of a curve $c : [a, b] \to G_0$. In particular, $\mathcal{F}(u) = \int_{x_1}^{x_2} \omega(x, u(x))\sqrt{1 + u'(x)^2}\, dx$ is the length of a curve in G_0 which is represented as a graph of some function u, i.e., $c(x) = (x, u(x))$.

A good illustration of the Weierstrass field theory is provided by 5,2.4 [4] and [5] where we had constructed stigmatic fields together with their envelopes $\mathscr{E}$.

[1] Consider the integral

$$\mathscr{F}(u) = \int_0^{x_1} \sqrt{H-u}\sqrt{1+u'^2}\,dx,$$

with $H > 0$, $x_1 > 0$. In 5,2.4 [4] we had seen that the Galilei parabolas

$$z = \varphi(x, a) := -\frac{1+a^2}{4H}x^2 + ax, \quad 0 \le x < \xi^*(a), \; a \in \mathbb{R}$$

(with $\xi^*(a) = \infty$ if $a \le 0$, $\xi^*(a) = 2H/a$ if $a > 0$) form a stigmatic field of extremals covering the domain

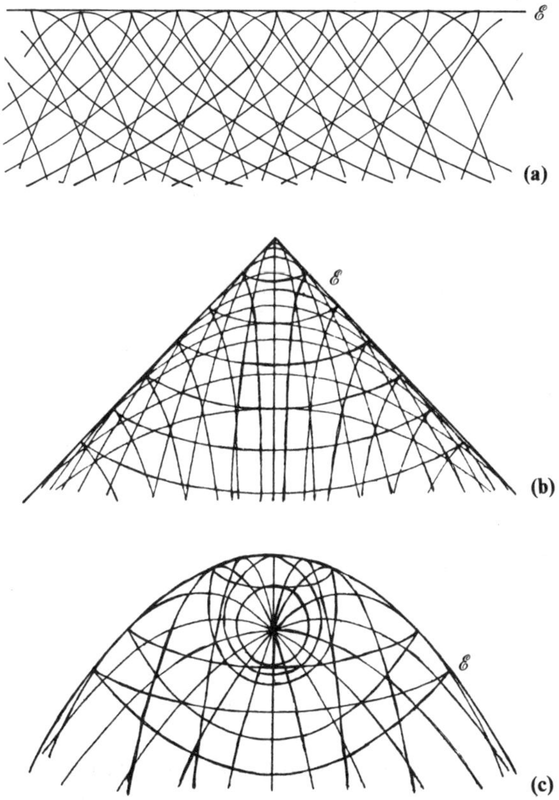

Fig. 11. (a) A family of congruent parabolas with their vertex on $\mathscr{E}$ and their axes perpendicular to $\mathscr{E}$, together with their orthogonal trajectories (= lines of constant action = wave fronts), which are Neil parabolas. **(b)** Parabolic orbits of particles in the (vertical) gravitational field of the earth. The particles are projected horizontally in a vertical plane from points on the same vertical line with such velocities that their total energies are equal to a fixed constant h, together with their orthogonal trajectories (= lines of constant action). **(c)** Parabolas emanating from a fixed point P_0 describing orbits of a particle within the gravitation field of the earth having fixed initial speed v; their envelope $\mathscr{E}$ and their orthogonal trajectories (= lines of constant action = wave fronts).

$$G = \left\{(x, z):\quad 0 < x < \infty, z < H - \frac{x^2}{4H}\right\},$$

which lies below the envelope $\mathscr{E}$ and to the right of the z-axis. Hence $P_0 = (0, 0)$ *and any point* $P_1 = (x_1, z_1)$ *in* G *can be connected by an arc* $z = \varphi(x, a)$, $0 \le x \le x_1 < \xi^*(a)$, *of one of the field curves, and* $u(x) = \varphi(x, a)$ *minimizes* $\mathscr{F}(u)$ *among all* $v \in C^1([0, x_1])$ *with* $\operatorname{graph}(v) \subset G$ *and* $v(x_1) = z_1$, $v(0) = 0$.

Note that $P_0, P_1 \in G$ can be connected by exactly two parabolic arcs $z = \varphi(x, a)$, $0 \le x \le x_1$, but for one of them $0 < \xi(a) < x_1$ holds true. This arc is not contained in G, and it is not minimizing since there is a point $P^*(a)$ conjugate to P_0 on the arc between P_0 and P_1.

[2] Secondly we consider the area functional

$$\mathscr{A}(u) = 2\pi \int_{x_1}^{x_2} u\sqrt{1 + u'^2}\, dx$$

of surfaces of revolution generated by meridian curves $z = u(x)$, $x_1 \le x \le x_2$, with $u(x) > 0$. Its extremals are the catenaries $u(x) = a \cosh \dfrac{x - b}{a}$, $x \in \mathbb{R}$, with $a > 0$, $b \in \mathbb{R}$. The one-parameter family of catenaries

$$\varphi(x, \alpha) = \frac{z_1}{\cosh \alpha} \cosh\left(\alpha + \frac{x - x_1}{z_1} \cosh \alpha\right), \quad \alpha \in \mathbb{R},$$

describes all extremals passing through the point $P_1 = (x_1, z_1)$. They cover the set

$$G = \{P_1\} \cup \{(x, z) \colon x > x_1, h(x) \le z < \infty\}$$

between the positive z-axis and the branch $\mathscr{E}^+$ of their envelope which is given by the parabola-like function $z = h(x)$, $x > x_1$. Every point $P_2 \in \operatorname{int} G$ can be connected with P_1 by exactly two extremals $\varphi(x, \alpha_1)$ and $\varphi(x, \alpha_2)$. If $\alpha_1 < \alpha_2$, then $\varphi(x, \alpha_1) < \varphi(x, \alpha_2)$, and the conjugate point $P^*(\alpha_1)$ to P_1 lies between P_1 and P_2 whereas the conjugate point $P^*(\alpha_2)$ comes behind P_2 (if there exists a conjugate point to P_1 at all to the right of P_1).

The catenaries $z = \varphi(x, \alpha)$, $x_1 \le x \le \xi(\alpha)$ for $\alpha < 0$ and $x_1 \le x < \infty$ for $\alpha \ge 0$ form an improper stigmatic field covering G, which is proper on $\operatorname{int} G$.

Therefore every point $P_2 = (x_2, z_2)$ in $\operatorname{int} G$ can be connected with P_1 by exactly one minimizing catenary. *This catenary is the unique minimizer of the functional* $\mathscr{A}$ *among all* $u \in C^1([x_1, x_2])$ *with* $u(x_1) = z_1$, $u(x_2) = z_2$ *and* $\operatorname{graph}(u) \subset P_1 \cup \operatorname{int} G$.

However, if we also admit parametric curves $c(t) = (x(t), z(t))$ contained in the half plane $\{z \ge 0\}$, this minimizing curve will not always be the absolute minimizer.

Consider the so-called *Goldschmidt curve* $\mathscr{G}$ formed by the vertical connections P_1P_3 and P_2P_4 of P_1 and P_2 with the x-axis, and by the horizontal piece P_3P_4 on the x-axis, connecting the foots P_3 and P_4 of P_1 and P_2, respectively. It can happen that $\mathscr{G}$ is the absolute minimizer of the surface area

$$2\pi \int_{P_1}^{P_2} z\sqrt{dx^2 + dz^2},$$

and in any case $\mathscr{G}$ is a relative minimizer of surface area if we admit all rectifiable curves $c(t)$ in $\{z \ge 0\}$ for competition. Here we have an example of a so-called[3] *discontinuous solution* of a variational problem. For instance, if P_1 and P_2 are "far apart", there will be no catenary joining these points. Nevertheless, there is always a Goldschmidt curve connecting P_1 and P_2 which then will furnish the absolute minimum of $2\pi \int_{t_1}^{t_2} z|\dot{c}|\, dt$ among D^1-functions $c(t)$ such that $|\dot{c}(t)| \ne 0$.

---

[3] Clearly $c(t)$ is not discontinuous in the present-day terminology but only nonsmooth, i.e., $\dot{c}(t)$ is discontinuous. The name derives from the old notation of the Leibnizian age where "continuous" roughly speaking meant: representable by analytic expressions.

A detailed account on minimal surfaces of revolution and on the role played by the Goldschmidt curve will be given in Chapter 8.

3 Now we consider the Lagrangian $F(x, z, p) = \omega(x, z)\sqrt{1 + p^2}$ with

$$\omega(x, z) = \frac{1}{z}$$

corresponding to the line element

$$ds^2 = \frac{1}{z^2}(dx^2 + dz^2).$$

As we had remarked earlier, the upper halfplane $G_0 = \{(x, z) \in \mathbb{R}^2 : z > 0\}$ equipped with the metric ds becomes a two-dimensional Riemannian manifold with the Gauss curvature $K = -1$. Every (nonparametric) extremal $u(x)$ of

$$\mathscr{F}(u) = \int_{x_1}^{x_2} \frac{1}{u}\sqrt{1 + u'^2}\, dx$$

describes a circle, the center of which lies on the x-axis. Hence any two points $P_1 = (x_1, z_1)$ and $P_2 = (x_2, z_2)$ in G_0 with $x_1 < x_2$ can be connected by exactly one extremal arc. Let $Q_1 = (\xi_1, 0)$ and $Q_2 = (\xi_2, 0)$ be the intersection points of the full circle determined by this arc with the x-axis, and let ζ_1 and ζ_2 be the two complex numbers defined by $\zeta_1 := x_1 + iz_1$, $\zeta_2 := x_2 + iz_2$. Then the length $\mathscr{F}(u)$ of the extremal arc $(x, u(x))$, $x_1 \le x \le x_2$, connecting P_1 and P_2 is found to be

$$\mathscr{F}(u) = \log D(\xi_1, \zeta_2, \zeta_1, \xi_2),$$

where

$$D(\xi_1, \zeta_2, \zeta_1, \xi_2) = \frac{\xi_1 - \zeta_2}{\xi_1 - \zeta_1} : \frac{\xi_2 - \zeta_2}{\xi_2 - \zeta_1}$$

is the cross-ratio of the four complex numbers $\xi_1, \zeta_2, \zeta_1, \xi_2$.

The two-dimensional space (G_0, ds) is known as *Poincare's model* of the hyperbolic plane. It is a model of a two-dimensional non-Euclidean geometry where the extremal circles are the non-Euclidean straight lines. In order to complete the model we have to add the lines $\{x = \text{const}\}$ which are extremals of the parametric integral

$$\int_a^b \frac{1}{z}\sqrt{\dot{x}^2 + \dot{z}^2}\, dt.$$

It can be seen that all extremals u of $\mathscr{F}(u)$ in the upper half plane are global minimizers.

The next example will also show the necessity of suitable enlargement of the class of admissible functions beyond C^1.

4 *The brachystochrone problem*, posed by Johann Bernoulli in 1696, is the problem of quickest descent:

For two given points P_1 and P_2 in a vertical plane, find a line connecting them, on which a movable point M descends from P_1 to P_2 under the influence of gravitation in the quickest possible way.

We assume that the point slides frictionless in the x, z-plane from $P_1 = (x_1, z_1)$ to $P_2 = (x_2, z_2)$, $x_1 < x_2$, $z_1 > z_2$, where gravity is acting in direction of the negative z-axis. Assume that only nonparametric paths $z = u(x)$, $x_1 \le x \le x_2$, are admitted for competition. The actual motion of the point will be denoted by

$$c(t) = (x(t), z(t)), \quad t_1 \le t \le t_2,$$

where $c(t_1) = P_1$ and $c(t_2) = P_2$. Since we consider a *constrained motion* along an arbitrarily chosen path $z = u(x)$, there will be reaction forces $\lambda(r)\mathfrak{n}(t)$ perpendicular to the path direction forcing the point mass to stay on the path. Thus Newton's equations of motion take the form

$$\ddot{c} = -ge_2 + \lambda\mathfrak{n},$$

where g is the gravity acceleration, $e_2 = (0, 1)$ the unit vector in direction of the z-axis, and $\mathfrak{n}$ the unit normal of the curve. As pointed out in 3,*1* [2] we have

$$\ddot{c} = \dot{v}\mathfrak{t} + \frac{v^2}{\rho}\mathfrak{n},$$

where $\mathfrak{t} = \dfrac{\dot{c}}{|\dot{c}|}$ is the tangent vector, $v = |\dot{c}|$, and ρ is the curvature radius of the path. For the tangential component we obtain the equation

$$\dot{v} = -g\frac{\dot{z}}{v},$$

i.e.

$$(\tfrac{1}{2}v^2)^{\cdot} = v\dot{v} = -g\dot{z},$$

and therefore

$$\tfrac{1}{2}v^2 + gz = \text{const} := h.$$

If $c(t_1) = P_1 = (x_1, z_1)$ and $|\dot{c}(t_1)| = v_1 \geq 0$, it follows that

$$h = \tfrac{1}{2}v_1^2 + gz_1.$$

In the classical brachystochrone problem it is assumed that, at $t = t_1$, the point starts to fall with the initial velocity zero, i.e., $v_1 = 0$. As this will lead to a singular integral, we first assume $v_1 > 0$, and then $v_1 = 0$ is treated by a limit consideration.

Introduce the arc length $s = s(t)$ by $\dot{s} = v$. Then it follows that

$$\frac{ds}{dt} = \sqrt{2(h - gz)},$$

and therefore the total fall time $T = \int_{t_1}^{t_2} dt$ from P_1 to P_2 is given by

$$T = \int_{s_1}^{s_2} \frac{ds}{\sqrt{2(h - gz)}} = \frac{1}{\sqrt{2g}} \int_{s_1}^{s_2} \frac{ds}{\sqrt{H - z}},$$

where

$$H := z_1 + k, \quad k := \frac{v_1^2}{2g}.$$

The required path $z = u(x)$, $x_1 \leq x \leq x_2$, has to minimize T among all virtual paths connecting P_1 and P_2. Since the factor $\dfrac{1}{\sqrt{2g}}$ is irrelevant and $ds = \sqrt{1 + u'^2}\, dx$, we have to minimize the integral

$$\mathscr{F}_H(u) := \int_{x_1}^{x_2} \frac{1}{\sqrt{H - u}} \sqrt{1 + u'^2}\, dx \tag{2}$$

among all C^1-curves $z = u(x)$, $x_1 \leq x \leq x_2$, with $u(x_1) = z_1$, $u(x_2) = z_2$, and $u(x) \leq H$. If we cut off all bumps of $u(x)$ which lie above the linear function $l(x) = ax + b$ with $l(x_1) = z_1$, $l(x_2) = z_2$ and replace u by the function $w(x)$ defined by

$$w(x) := \begin{cases} l(x) & \text{if } u(x) \geq l(x), \\ u(x) & \text{if } u(x) < l(x), \end{cases}$$

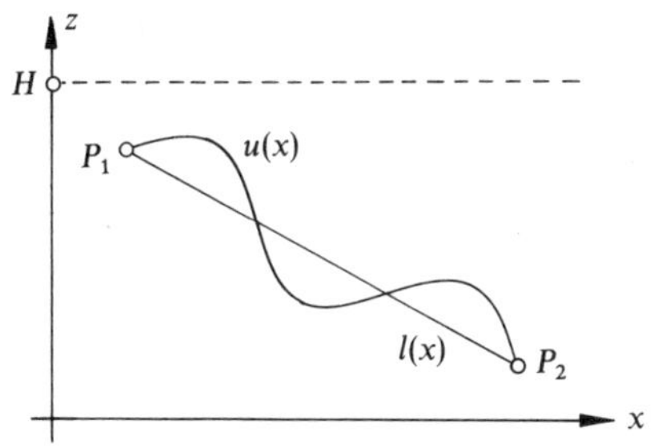

Fig. 12.

then $\mathscr{F}_H(w) < \mathscr{F}_H(u)$ if there is some $x \in [x_1, x_2]$ with $u(x) > l(x)$. This suggests that, for determining a minimizer of $\mathscr{F}_H$, we can restrict our investigation to functions u with $u(x) \le l(x)$. Unfortunately, the function $w(x)$ constructed above will not be of class C^1 and, in general, not even of class D^1 but only of class AC. Although this would not pose a serious problem, we want to avoid the use of absolutely continuous functions at the present level. However, a slight modification of the previous reasoning by rounding-off corners shows that every admissible u can be replaced by another admissible C^1-function w satisfying $w(x) < z_1$ for $x \in (x_1, x_2]$ and $\mathscr{F}_H(w) < \mathscr{F}_H(u)$. Thus we may require that admissible functions $u(x)$ are satisfying $u(x) < z_1$ if $x_1 < x \le x_2$.

Now we are going to determine an extremal $z = u(x)$ connecting P_1 and P_2. In fact, we shall embed this extremal into some extremal field covering the quarter plane $G = \{(x, z): x > x_1, z < z_1\}$ provided that $k > 0$ (i.e. $v_1 > 0$).

By 1,*2.2* 7, we know that the expression $F - pF_p$ is a first integral of the extremals $u(x)$ of $\mathscr{F}_H$. Thus a brief computation yields

$$\sqrt{1 + u'^2}\sqrt{H - u} = \text{const}$$

or

$$(1 + u'^2)(H - u) = 2r \tag{3}$$

for a suitable constant $r > 0$. This first order differential equation can explicitly be integrated if we introduce for z the new variable τ, setting

$$H - z = r(1 - \cos\tau) = 2r\sin^2\tau/2 .$$

Then $z = u(x)$ is transformed to some function $\tau(x)$, and from (3) we derive the equation

$$4r^2\tau'^2 \sin^4 \tau/2 = 1 ,$$

and therefore

$$r\tau'(1 - \cos\tau) = \pm 1 .$$

We choose the branch with $+1$ on the right-hand side and integrate with respect to x. Then we obtain

$$x = x_0 + r(\tau - \sin\tau),$$

where x_0 and $r > 0$ are arbitrary integration constants. Together with the defining relation for τ we arrive at the equations

$$\begin{aligned} x &= x_0 + r(\tau - \sin\tau), \\ z &= H - r(1 - \cos\tau), \end{aligned} \tag{4}$$

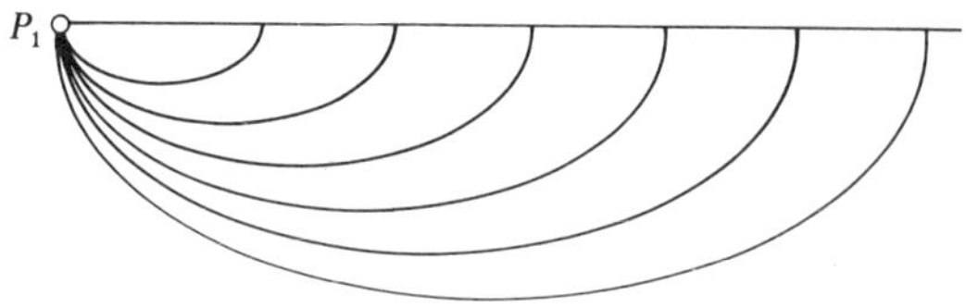

Fig. 13.

which describe a two-parameter family of cycloids. For fixed r, every such cycloid is generated as orbit of some point on the periphery of a circular wheel of radius r that rolls on the lower side of the straight line $z = H$ towards the right. The cusps of these cycloids lie on the line $z = H$ and point in direction of the z-axis.

Consider the function $x(\tau) = x_0 + r(\tau - \sin\tau)$ for $0 \le \tau \le 2\pi$. Since $\frac{dx}{d\tau}(\tau) = r(1 - \cos\tau) > 0$ for $0 < \tau < 2\pi$, the mean value theorem implies that $x(\tau)$ is strictly increasing for $0 \le \tau \le 2\pi$. Hence there is an inverse function $\tau = \tau(x)$ which is continuous on the closed interval $[x_0, x_0 + 2\pi r]$ and real analytic in its interior. Thus we can write (4) nonparametrically as $z = z(\tau(x)) = u(x)$, or, if we mark the parameters, as

$$z = \psi(x; H, x_0, r), \quad x_0 \le x \le x_0 + 2\pi r.$$

For $x_0 < x < x_0 + 2\pi r$ and arbitrary H, $x_0 \in \mathbb{R}$, $r > 0$, ψ is a real analytic function of its four variables, and the expansion

$$\psi(x; H, x_0, r) = H - \tfrac{1}{2} r^{1/3} 6^{2/3} (x - x_0)^{2/3} + \ldots \quad \text{for } x \to x_0 + 0 \tag{5}$$

shows the asymptotic behaviour as $x \to x_0 + 0$.

Fix x_0 and H. Then (4), restricted to $0 \le \tau < 2\pi$, describes a family of cycloidal arcs $\mathscr{C}(r)$ which are similar to each other with respect to the center of similarity $P_0(H) := (x_0, H)$. Consider for instance the arc $\mathscr{C}(1)$. It is not only convex, but every straight line emanating from $P_0(H)$ and not parallel to the x- and z-axis intersects $\mathscr{C}(1)$ in exactly one point different from $P_0(H)$. Hence, for fixed x_0 and H, the family

$$z = \psi(x; H, x_0, r), \quad x_0 \le x < x_0 + 2\pi r,$$

describes a "stigmatic extremal field" for the functional $\mathscr{F}_H$ covering $G := \{(x, z): x > x_0, z < H\}$ with the nodal point $P_0(H) = (x_0, H)$. In particular, for every point $P_2 = (x_2, z_2) \in G$, there is exactly one value of r such that $P_0(H)$ and P_2 are connected by $\mathscr{C}(r)$, i.e., $P_2 = \psi(x_2; H, x_0, r)$. (Note that the family $\{\mathscr{C}(r)\}$ merely describes a weak stigmatic field since the assumptions of Definition 4 in *1.1* are not satisfied. The following discussion will show how to remedy this weakness.) Since $P_0(H) = P_1 = (x_1, z_1)$ if $k = 0$, the *boundary value problem* to connect P_1 and P_2 with an extremal of $\mathscr{F}_H$ is solved, provided that $H = z_1$ (or $k = 0$), and simultaneously we have solved the *embedding problem*: to embed this extremal into some "stigmatic field" with center P_1.

Now we want to do the same if $H > z_1$ (or $k > 0$). For $2r > H - z_1$ we have $0 < \frac{H - z_1}{2r} < 1$, and thus we can find some $\tau_1 > 0$ such that

$$\tau_1 = 2 \arcsin \sqrt{\frac{H - z_1}{2r}}$$

(where arc sin denotes the principal branch of arcus sinus). For fixed H and z_1, the quantity τ_1 is to be considered as function of r: $\tau_1 = \tau_1(r)$. Then we define the function $x_0(r)$ by

$$x_0(r) := x_1 - r(\tau_1(r) - \sin\tau_1(r)).$$

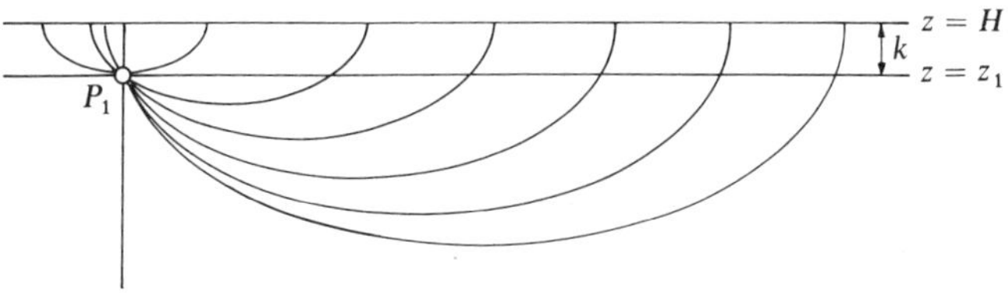

Fig. 14.

This way we select from (4) the one-parameter family of cycloidal arcs

$$x = x_0(r) + r(\tau - \sin\tau), \qquad (\tau_1(r) < \tau < 2\pi),$$
$$z = H - r(1 - \cos\tau),$$

which, on account of

$$H - z_1 = 2r\sin^2\tau_1/2 = r(1 - \cos\tau_1),$$

all pass through $P_1 = (x_1, z_1)$. This family has the nonparametric representation

$$z = \varphi(x, r), \quad x_1 \le x < x^*(r), \ r > \tfrac{1}{2}(H - z_1),$$

where we have set

$$\varphi(x, r) := \psi(x; H, x_0(r), r),$$

and $x^*(r) \in (x_1, x_0(r) + 2\pi r)$ is to be determined from $\varphi(x^*(r), r) = z_1$.

From the discussion of the case $H = z_1$, it is not difficult to derive that $f(x, r) = (x, \varphi(x, r))$ describes a stigmatic field with the nodal point P_1 which covers

$$G := \{(x, z): x > x_1, z < z_1\}.$$

Thus we have solved the boundary value problem and the corresponding embedding problem for $h > z_1$:

The points P_1 and P_2 can be connected by an extremal of $\mathcal{F}_H$, and this extremal can be embedded into a stigmatic field $f(x, r)$ covering the quadrant G and having P_1 as its center.

For $H > z_1$, the function $\varphi(x, r)$ is real analytic for $x_1 \le x \le x^*(r)$, $r > \frac{1}{2}(H - z_1)$. Moreover, for integrands of the type

$$F(x, z, p) = \omega(x, z)\sqrt{1 + p^2},$$

transversality means orthogonality. Thus the level lines $\mathcal{S}_\theta = \{(x, z): S(x, z) = \theta\}$ of the eikonal S of the field f intersect the field curves orthogonally. Inspecting the field, it then becomes evident that $\lim_{P \to P_1} S(P) = 0$ for $P \in G$. By virtue of *1.3*, Theorem 8, there exists a unique minimizer of $\mathcal{F}_H$ among all nonparametric curves $u \in C^1([x_1, x_2])$ with $u(x_1) = z_1$, $u(x_2) = z_2$ and $u(x) < z_1$ for $x \in (x_1, x_2]$, given by a cycloidal arc. By the remarks made at the beginning, this arc even minimizes among all curves with $u(x) \le H$, and we have proved:

If the initial speed v_1 is positive, then the brachystochrone problem has exactly one solution for any two boundary points $P_1 = (x_1, z_1)$ and $P_2 = (x_2, z_2)$ with $x_1 < x_2$ and $z_1 > z_2$. This solution is given by a cycloidal arc $z = \varphi(x, r)$, $x_1 \le x \le x_2$, without any cusp.

For $v_1 = 0$, basically the same is true except that the arc which is expected to be the solution of the brachystochrone problem, now has a singularity at $x = x_1$ since it is vertical, and thus the solution will not be in $C^1([x_1, x_2])$. In other words, the singularity of the integrand $\sqrt{1 + p^2}/\sqrt{z_1 - z}$ at $z = z_1$ forces us to enlarge the class $\mathcal{A}$ of admissible functions in order to find a minimizer within the class. We define $\mathscr{A}$ as

$$\mathscr{A} := \{w : w \in C^0 \text{ on } [x_1, x_2], w \in C^1 \text{ on } (x_1, x_2], w(x) < z_1 \text{ for } x > x_1, w(x_1) = z_1, w(x_2) = z_2\}.$$

Moreover, let u_H be the cycloidal arc joining P_1 and P_2 which is an extremal for $\mathscr{F}_H$. By (5), it follows that

$$\mathscr{F}_{z_1}(u_{z_1}) < \infty,$$

and Lebesgue's theorem on dominated convergence yields

$$\lim_{H \to z_1 + 0} \mathscr{F}_H(u_H) = \mathscr{F}_{z_1}(u_{z_1}).$$

Let now w be an arbitrary function in $\mathscr{A}$. If $\mathscr{F}_{z_1}(w) = \infty$, then trivially $\mathscr{F}_{z_1}(w) > \mathscr{F}_{z_1}(u_{z_1})$ holds. If $\mathscr{F}_{z_1}(w) < \infty$, then also $\mathscr{F}_H(w) < \infty$ for every $H > z_1$. We now can proceed as in the proof of Theorem 2 of *1.3* and obtain

$$\mathscr{F}_H(w) = \mathscr{F}_H(u_H) + \int_{x_1}^{x_2} \mathscr{E}^H(x, w, \mathscr{P}_H(x, w), w')\, dx, \tag{6}$$

where $\mathscr{E}^H$ denotes the excess function of $\mathscr{F}_H$, and $\mathscr{P}_H$ is the slope function corresponding to the field for the parameter value H (cf. *1.3*, formula (6)). Since $\mathscr{F}_{z_1}(w) < \infty$, Lebesgue's theorem on dominated convergence implies

$$\lim_{H \to z_1 + 0} \mathscr{F}_H(w) = \mathscr{F}_{z_1}(w).$$

By the discussion in the beginning we have $\mathscr{E}^H \ge 0$, and therefore we can apply Fatou's lemma to $\int_{x_1}^{x_2} \mathscr{E}^H(\ldots)\, dx$. Then we infer from (6) that

$$\mathscr{F}_{z_1}(w) \ge \mathscr{F}_{z_1}(u_{z_1}) + \int_{x_1}^{x_2} \mathscr{E}^{z_1}(x, w, \mathscr{P}_{z_1}(x, w), w')\, dx$$

holds for every $w \in \mathscr{A}$, whence $\mathscr{F}_{z_1}(w) > \mathscr{F}_{z_1}(u_{z_1})$ if $w \ne u_{z_1}$. Consequently also for $H = z_1$, the cycloidal arc $z = u_{z_1}(x)$, $x_1 \le x \le x_2$, joining P_1 and P_2 is the unique minimizer of $\mathscr{F}_{z_1}$ within the class $\mathscr{A}$. Together with the remark made at the beginning we conclude that u_{z_1} *is the unique solution of the brachystochrone problem for* $v_1 = 0$ *among all nonparametric paths from* P_1 *to* P_2 *below* $z = z_1$ *which are continuous on* $[x_1, x_2]$ *and of class* C^1 *on* $(x_1, x_2]$.

We note that the minimizing property of the cycloids can also be proved by the convexity method, either by reversing the roles of x and z (that is, x becomes the dependent variable and z the independent one), or by transforming the variational integral $\mathscr{F}$ into a strictly convex functional; cf. 4,*2.2* and *2.3* [4]. The "method of reversing the roles of x ad z" only functions if the prospective minimizers have an inverse. In other words, at most the first half of a cycloidal "festoon" can this way shown to be a minimizer.

2.4. Field-like Mayer Bundles, Focal Points and Caustics

In *2.1* we had solved the embedding problem of extremals by means of stigmatic bundles. Now we want to carry over some of these results to general Mayer bundles. This will lead us to the concepts of *focal point* and *caustic* of a regular Mayer bundle, and we shall be able to derive sufficient conditions for an extremal to minimize the integral

$$\mathscr{F}(u) = \int_a^b F(x, u(x), u'(x))\, dx,$$

with respect to free boundary conditions.

Let us recall some facts on stigmatic fields. Such a field is a mapping f whose field lines $x \to f(x, c) = (x, \varphi(x, c))$ are extremal curves, all of which pass through a single point $P_0 = (x_0, z_0)$ which is called nodal point of the field f. The singular points of the field curves of f different from P_0 form the conjugate locus with respect to P_0, and this conjugate locus (together with P_0) is just the envelope of the field curves of f.

Now we want to replace stigmatic fields by regular Mayer bundles. Following Carathéodory, we shall substitute the epithet "regular" by the notation "field-like". Then the analoga to conjugate points of a stigmatic field will be the focal points of a field-like Mayer bundle whose union is the caustic of the bundle, and the caustic corresponds to the conjugate locus in the former case.

Definition 1. *A Mayer bundle* $f(x, c) = (x, \varphi(x, c))$, $(x, c) \in \Gamma \subset \mathbb{R} \times \mathbb{R}^N$, *is called* field-like *or* regular *if the matrix function*

$$\begin{pmatrix} \varphi_c \\ \varphi_c' \end{pmatrix} = \begin{pmatrix} \varphi^i_{c^\alpha} \\ \varphi^{i\prime}_{c^\alpha} \end{pmatrix}_{1 \le i, \alpha \le N}$$

has rank N everywhere on Γ.

In order to interpret this definition appropriately, we need a few simple facts on Jacobi fields which are more or less obvious. Consider some F-extremal $u : [a, b] \to \mathbb{R}^N$ which is to be *elliptic*, i.e., we assume that

$$(1) \qquad F_{pp}(x, u(x), u'(x)) > 0 \quad \text{for all } x \in [a, b].$$

Let $\mathscr{V}$ be the $2N$-dimensional space of Jacobi fields v along u, that is, of the solutions v of the Jacobi equation $\mathscr{J}_u v = 0$. Because of (1), we can write the Jacobi equation in the equivalent form

$$(2) \qquad v'' = H(x)v + K(x)v',$$

where $H(x)$ and $K(x)$ are continuous $N \times N$-matrics. We infer from (2) that the assumptions

$$v \in \mathscr{V} \quad \text{and} \quad v(\xi) = 0, v'(\xi) = 0 \quad \text{for some } \xi \in [a, b]$$

imply that $v(x) \equiv 0$, i.e., $v = 0$. This observation immediately yields

Lemma 1. *Let $v_1, v_2, \ldots, v_k \in \mathscr{V}$. Then the rank of the matrix*

$$\begin{pmatrix} v_1(x), v_2(x), \ldots, v_k(x) \\ v_1'(x), v_2'(x), \ldots, v_k'(x) \end{pmatrix}$$

is constant on $[a, b]$.

Nearly as obvious is the following result.

Lemma 2. *Let $v_1, v_2, \ldots, v_k$, and set* $\mathscr{T} := \begin{pmatrix} v_1, v_2, \ldots, v_k \\ v_1', v_2', \ldots, v_k' \end{pmatrix}$. *Then we have*

$$(3) \qquad \dim \operatorname{span}\{v_1, v_2, \ldots, v_k\} = \operatorname{rank} \mathscr{T}.$$

Proof. Note that $\operatorname{span}\{v_1, \ldots, v_k\}$ is a linear subspace of $\mathscr{V}$. Thus the left-hand side of (3) is well defined. The right-hand side is to be interpreted as rank $\mathscr{T}(x)$, and this function is constant because of Lemma 1. Hence equation (3) makes sense. Clearly, it suffices to prove that

$$\operatorname{rank} \mathscr{T} = k \text{ if and only if } \dim \operatorname{span}\{v_1, \ldots, v_k\} = k.$$

(i) Suppose that rank $\mathscr{T}(x_0) < k$ for some $x_0 \in [a, b]$. Then there exists some $\lambda = (\lambda^1, \ldots, \lambda^k) \in \mathbb{R}^k$ such that $\lambda \neq 0$ and

$$\lambda^1 \begin{pmatrix} v_1(x_0) \\ v_1'(x_0) \end{pmatrix} + \cdots + \lambda^k \begin{pmatrix} v_k(x_0) \\ v_k'(x_0) \end{pmatrix} = 0.$$

Set $v := \lambda^1 v_1 + \cdots + \lambda^k v_k$. Then $v \in \mathscr{V}$, and we have $v(x_0) = v'(x_0) = 0$ whence $v = 0$, i.e., the Jacobi fields $v_1, \ldots, v_k$ are linearly dependent. Consequently we have $\dim \operatorname{span}\{v_1, \ldots, v_k\} < k$.

(ii) Conversely, if $\dim \operatorname{span}\{v_1, \ldots, v_k\} < k$, then there is a nontrivial k-tuple $(\lambda^1, \ldots, \lambda^k) \in \mathbb{R}^k$ such that $\lambda^1 v_1 + \cdots + \lambda^k v_k = 0$ whence $\lambda^1 v_1' + \cdots + \lambda^k v_k' = 0$, and consequently

$$\lambda^1 \begin{pmatrix} v_1(x) \\ v_1'(x) \end{pmatrix} + \cdots + \lambda^k \begin{pmatrix} v_k(x) \\ v_k'(x) \end{pmatrix} = 0 \quad \text{for all } x \in [a, b].$$

This implies rank $\mathscr{T}(x) < k$ for all $x \in [a, b]$. □

If $f(x, c) = (x, \varphi(x, c))$ is an N-parameter family of F-extremals, then, for any admissible $c = (c^1, \ldots, c^N)$, the vectors

$$v_1 := \varphi_{c^1}(\cdot, c), \ldots, v_N := \varphi_{c^N}(\cdot, c) \tag{4}$$

are Jacobi fields along the extremal $u := \varphi(\cdot, c)$. The extremal bundle f is a field-like Mayer bundle if the Jacobi fields $v_1, \ldots, v_N$ satisfy

$$\dim \operatorname{span}\{v_1, v_2, \ldots, v_N\} = N \tag{5}$$

and

$$w_\alpha \cdot v_\beta - w_\beta \cdot v_\alpha = 0 \quad \text{for } 1 \leq \alpha, \beta \leq N, \tag{6}$$

where

$$w_\alpha := \frac{\partial}{\partial c^\alpha} F_p(\cdot, \varphi(\cdot, c), \varphi'(\cdot, c)). \tag{7}$$

If we introduce the accessory Lagrangian $Q(x, z, p)$ by

$$Q(x, z, p) := \tfrac{1}{2}\{p \cdot A(x)p + 2z \cdot B(x)p + z \cdot C(x)z\}, \tag{8}$$

where

$$\begin{aligned} A(x) &:= F_{pp}(x, u(x), u'(x)), \\ B(x) &:= F_{zp}(x, u(x), u'(x)), \\ C(x) &:= F_{zz}(x, u(x), u'(x)), \end{aligned} \tag{9}$$

we obtain from (7) that

$$w_\alpha = Av'_\alpha + B^T v_\alpha = Q_p(\cdot, v_\alpha, v'_\alpha). \tag{10}$$

These formulas motivate

Definition 2. *Let $u : [a, b] \to \mathbb{R}^N$ be an F-extremal with the accessory Lagrangian Q defined by* (8) *and* (9). *Then, a system $\{v_1, v_2, \dots, v_N\}$ of Jacobi fields v_α along u is called a* conjugate base *of Jacobi fields if the following two conditions are satisfied*:

(*i*) dim span$\{v_1, v_2, \dots, v_N\} = N$.

(*ii*) *Let $w_\alpha := Q_p(\cdot, v_\alpha, v'_\alpha)$, $1 \le \alpha \le N$. Then we have*

$$w_\alpha \cdot v_\beta - w_\beta \cdot v_\alpha = 0 \quad \textit{for } 1 \le \alpha, \beta \le N. \tag{11}$$

This notation, due to von Escherich [2], is somewhat misleading since a conjugate base $\{v_1, \dots, v_N\}$ of Jacobi fields spans an N-dimensional subspace of the $2N$-dimensional space $\mathscr{V}$ of Jacobi fields along u, and it does not form a true base of $\mathscr{V}$.

We can write (10) in the form

$$\begin{pmatrix} v_\alpha \\ w_\alpha \end{pmatrix} = \begin{pmatrix} E & 0 \\ B^T & A \end{pmatrix} \begin{pmatrix} v_\alpha \\ v'_\alpha \end{pmatrix}, \quad 1 \le \alpha \le N, \tag{12}$$

where E is the N-dimensional unit matrix. By (1), we have $\det A > 0$, and therefore

$$\det \begin{pmatrix} E & 0 \\ B^T & A \end{pmatrix} > 0. \tag{13}$$

We infer from (12), (13) and Lemma 2 the following result:

Lemma 3. *Let $\{v_1, \dots, v_N\}$ be a conjugate base of Jacobi fields, and let $w_\alpha = Q_p(\cdot, v_\alpha, v'_\alpha)$ be the "momenta" of the vectors v_α. Then both the matrices*

$$\mathscr{T}(x) := \begin{pmatrix} v_1(x), \dots, v_N(x) \\ v'_1(x), \dots, v'_N(x) \end{pmatrix} \quad \textit{and} \quad \mathscr{U}(x) := \begin{pmatrix} v_1(x), \dots, v_N(x) \\ w_1(x), \dots, w_N(x) \end{pmatrix} \tag{14}$$

have constant rank N.

Furthermore, we have

Lemma 4. *Let $\{v_1, \dots, v_N\}$ be a conjugate base of Jacobi fields, and suppose that $\bar v_1, \dots, \bar v_N$ are N linearly independent vectors in* span$\{v_1, \dots, v_N\}$. *Then the system $\{\bar v_1, \dots, \bar v_N\}$ is also a conjugate base.*

Proof. If $v = \lambda^\alpha v_\alpha$ and $w_\alpha = Q_p(\cdot, v_\alpha, v'_\alpha)$, we infer from (10) that $w = Q_p(\cdot, v, v')$ is given by $w = \lambda^\alpha w_\alpha$. Hence, if $\bar w_\alpha = Q_p(\cdot, \bar v_\alpha, \bar v'_\alpha)$ are the momenta of N independent vectors $\bar v_1, \dots, \bar v_N$ in span$\{v_1, \dots, v_N\}$, it follows from (11) that

$$\bar w_\alpha \cdot \bar v_\beta - \bar w_\beta \cdot \bar v_\alpha = 0. \qquad \square$$

Let us now introduce the *Mayer determinant* $\Delta(x)$ of a conjugate base of Jacobi fields $v_1, \dots, v_N$ by

$$\Delta(x) := \det(v_1(x), v_2(x), \dots, v_N(x)). \tag{15}$$

Although the matrix $\mathcal{T}(x)$ defined by (14) satisfies rank $\mathcal{T}(x) \equiv N$, the determinant $\Delta(x)$ may very well vanish. However, we shall see that its zeros are isolated.

Proposition 1. *The zeros of the Mayer determinant $\Delta(x)$ of a conjugate base of Jacobi fields $v_1, \dots, v_N$ are isolated.*

Proof. Set $\mathcal{V}_N^* := \operatorname{span}\{v_1, \dots, v_N\}$, and let $\bar{v}_1, \dots, \bar{v}_N$ be a system of N independent vectors in $\mathcal{V}_N^*$. Then $\{\bar{v}_1, \dots, \bar{v}_N\}$ is another conjugate base spanning $\mathcal{V}_N^*$ whose Mayer determinant be denoted by $\bar{\Delta}$. It is easy to see that there is some number $c \neq 0$ such that $\Delta(x) \equiv c\,\bar{\Delta}(x)$.

Consider any zero x_0 of $\Delta(x)$; we can assume that $x_0 = 0$, i.e., $\Delta(0) = 0$.Let

$$k := \dim \operatorname{span}\{v_1(0), \dots, v_N(0)\},$$

$0 \le k \le N$. If $k = 0$, then we have $v_1(0) = \cdots = v_N(0) = 0$. Since rank $\mathcal{T}(0) = N$, it follows that

$$d(0) := \det(v_1'(0), \dots, v_N'(0)) \neq 0.$$

On the other hand, we have

$$v_\alpha(x) = x v_\alpha'(0) + O(x^2)$$

and therefore

$$\Delta(x) = x^N d(0) + O(x^{N+1}).$$

Consequently, $x = 0$ is an isolated zero of $\Delta(x)$ if $k = 0$. Thus we may assume that $1 \le k \le N$. Let $\{e_1, e_2, \dots, e_N\}$ be an orthonormal base of $\mathbb{R}^N$ such that

$$\operatorname{span}\{v_1(0), \dots, v_N(0)\} = \operatorname{span}\{e_1, \dots, e_k\}$$

and

$$(\operatorname{span}\{v_1(0), \dots, v_N(0)\})^\perp = \operatorname{span}\{e_{k+1}, \dots, e_N\}.$$

We can choose a base $\{\bar{v}_1, \dots, \bar{v}_N\}$ of $\mathcal{V}_N^*$ such that

$$\bar{v}_j(0) = e_j \quad \text{if } 1 \le j \le k, \qquad \bar{v}_\ell(0) = 0 \quad \text{if } k+1 \le \ell \le N. \tag{16}$$

Let $\bar{w}_\alpha = Q_p(\cdot, \bar{v}_\alpha, \bar{v}_\alpha')$ be the momenta of the Jacobi fields $\bar{v}_\alpha$. Since

$$\bar{w}_\alpha \cdot \bar{v}_\beta - \bar{w}_\beta \cdot \bar{v}_\alpha = 0,$$

it follows that

$$e_j \cdot \bar{w}_\ell(0) = 0 \quad \text{for } 1 \le j \le k, \quad k+1 \le \ell \le N. \tag{17}$$

Thus we have

$$\bar{\mathcal{U}}(0) = \begin{pmatrix} e_1, \dots, e_k, & 0 & , \dots, & 0 \\ *, \dots, *, & \bar{w}_{k+1}(0), & \dots, & \bar{w}_N(0) \end{pmatrix},$$

where $\overline{\mathscr{U}}$ and $\overline{\mathscr{T}}$ be defined in terms of $\bar{v}_1, \dots, \bar{v}_N$ as $\mathscr{U}$ and $\mathscr{T}$ are defined in terms of $v_1, \dots, v_N$ (see (14)). On account of Lemmata 3 and 4 we have

$$\operatorname{rank} \overline{\mathscr{U}}(0) = N.$$

Consequently, the vectors $\bar{w}_{k+1}(0), \dots, \bar{w}_N(0)$ are linearly independent. Therefore we can choose $\bar{v}_1, \dots, \bar{v}_N$ even in such a way that

$$\bar{w}_\ell(0) e_\ell \quad \text{if } k+1 \le \ell \le N, \tag{18}$$

if we also take (17) into account. Hence we have

$$\overline{\mathscr{U}}(0) = \begin{pmatrix} e_1, \dots, e_k, & 0, \dots, 0 \\ *, \dots, *, & e_{k+1}, \dots, e_N \end{pmatrix}. \tag{19}$$

Suppose in addition that

$$A(0) = E := \text{identity}. \tag{20}$$

Then equations (10) and (19) imply that

$$\bar{v}_j(0) = e_j \quad \text{if } 1 \le j \le k, \qquad \bar{v}'_\ell(0) = e_\ell \quad \text{if } k+1 \le \ell \le N,$$

and we obtain the Taylor expansions

$$\bar{v}_j(x) = e_j + O(x) \quad \text{if } 1 \le j \le k,$$

$$\bar{v}_\ell(x) = x e_\ell + O(x^2) \quad \text{if } k+1 \le \ell \le N.$$

Therefore, the Mayer determinant $\bar{\Delta}(x)$ of $\{\bar{v}_1, \dots, \bar{v}_N\}$ can be writen as

$$\bar{\Delta}(x) = \pm x^{N-k} + O(x^{N-k+1}),$$

since $\det(e_1, \dots, e_N) = \pm 1$, and it follows that $x = 0$ is an isolated zero of $\bar{\Delta}(x)$. Because of the identity

$$\Delta(x) \equiv c\, \bar{\Delta}(x)$$

for some $c \ne 0$, we now infer that $x = 0$ is also an isolated singularity of $\Delta(x)$.

Finally we note that the assumption (20) is no restriction of generality as we can reduce the general case to the special one by a suitable affine transformation of the dependent variables, taking $A(0) > 0$ into account. □

As a consequence of Proposition 1 and of the formulas (4)–(10) we arrive at the following result.

Proposition 2. *Let $f(x, c) = (x, \varphi(x, c))$ be a field-like Mayer bundle. Then, for every admissible c, the Jacobi fields*

$$v_1 := \varphi_{c^1}(\cdot, c), \qquad v_2 := \varphi_{c^2}(\cdot, c), \quad \dots, \quad v_N := \varphi_{c^N}(\cdot, c)$$

along the extremal $u := \varphi(\cdot, c)$ form a conjugate base. Therefore, for every admissible c, the zeros x of the Mayer determinant

$$\Delta(x, c) := \det \varphi_c(x, c) \tag{21}$$

are isolated.

Definition 3. *The zeros* $x = x_0$ *of the Mayer determinant* $\Delta(x, c)$ *of a field-like Mayer bundle* $f(x, c) = (x, \varphi(x, c))$ *are called* focal values *of the extremal* $u = \varphi(\cdot, c)$, *and the corresponding points* $P_0 = (x_0, z_0) = f(x_0, c)$ *are called* focal points *of the bundle* f *corresponding to the extremal* $f(\cdot, c)$. *The set of focal points* P_0 *of* f *is denoted as* focal surface *or* caustic *of the field-like bundle.*

For a fixed $c_0 = (c_0^1, \ldots, c_0^N)$, the zeros $x = x_0$ of $\Delta(x, c_0)$ are isolated. Hence the focal values x_0 and the focal points $P_0 = (x_0, z_0)$ of a field-like Mayer bundle f corresponding to the extremal $f(\cdot, c_0)$ are isolated. Nevertheless there might exist other focal points of the bundle on the curve $f(\cdot, c_0)$ corresponding to other values of the parameter c. In fact, it is not difficult to find examples of field-like bundles where the focal points fill a whole subarc of a fixed field curve $f(\cdot, c_0)$; cf. Fig. 15 and 7,*2.2* [1].

Clearly, any stigmatic field $f(x, c) = (x, \varphi(x, c))$ is a field-like Mayer bundle. In fact, if $P_0 = (x_0, z_0)$ is the nodal point of f, then we have

$$\begin{pmatrix} \varphi_c(x_0, c) \\ \varphi_c'(x_0, c) \end{pmatrix} = \begin{pmatrix} 0 \\ \varphi_c'(x_0, c) \end{pmatrix}$$

and

$$\operatorname{rank} \varphi_c'(x_0, c) = N .$$

Because of Lemma 1, it follows that

$$\operatorname{rank} \begin{pmatrix} \varphi_c \\ \varphi_c' \end{pmatrix} = N .$$

Furthermore, we know already that every stigmatic field is a Mayer bundle whence the assertion is proved. Comparing our preceding results with those of *2.1*, we see that the focal points of a stigmatic field are just the conjugate points of this field with respect to its nodal point P_0, and the caustic of a stigmatic field is the conjugate locus corresponding to P_0.

Now we want to formulate sufficient conditions for partially free boundary problems analogous to the results of *2.1*.

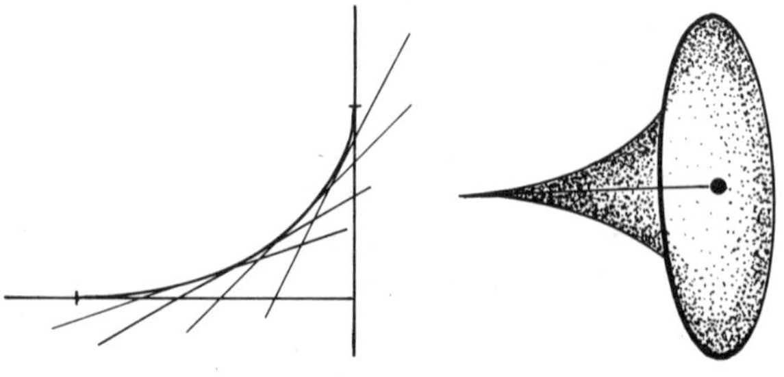

Fig. 15.

Assumption (T). *Let $\mathcal{S}$ be a regular hypersurface of class C^2 in the configuration space $\mathbb{R} \times \mathbb{R}^N$ carrying a normal vector field of the form*

$$\mathcal{N}(x, z) = (-1, \nu(x, z)), \tag{22}$$

where $\nu : \mathcal{S} \to \mathbb{R}^N$ is of class C^1, and let P be some point outside of $\mathcal{S}$.

Moreover, we assume that $u : [a, b] \to \mathbb{R}^N$ is an F-extremal such that the left endpoint $P_0 = (a, u(a))$ of the corresponding extremal curve

$$\mathcal{C}_0 = \{(x, z): z = u(x), a \le x \le b\}$$

lies on $\mathcal{S}$, whereas its right endpoint $(b, u(b))$ coincides with P. Suppose that $\mathcal{C}_0$ minimizes the functional

$$\mathcal{F}(\mathcal{C}) := \int_{x_1}^{b} F(x, v(x), v'(x))\, dx \tag{23}$$

among all C^1-curves

$$\mathcal{C} = \{(x, z): z = v(x), x_1 \le x \le b\}$$

connecting $\mathcal{S}$ with P. In other words, the left endpoint P_1 of the admissible comparison curves $\mathcal{C}$ may freely vary on $\mathcal{S}$ while their right endpoint is kept fixed and is to be the preassigned point P. We know already that the minimum property of $\mathcal{C}_0$ implies that $\mathcal{C}_0$ meets the support surface $\mathcal{S}$ transversally at P_0, i.e., the line element $\ell_0 := (a, u(a), u'(a))$ tangent to $\mathcal{C}_0$ at $P_0 = (a, u(a))$ is transversal to the surface element $\sigma_0 := (P_0, \nu(P_0))$ tangent to $\mathcal{S}$ at P_0. Set $p_0 := u'(a)$. Then the transversality is expressed by

$$(-1, \nu(P_0)) \sim (F(\ell_0) - p_0 \cdot F_p(\ell_0), F_p(\ell_0))$$

or, equivalently, by

$$\nu(P_0) = -\frac{F_p(\ell_0)}{F(\ell_0) - p_0 \cdot F_p(\ell_0)}, \tag{24}$$

if we assume that

$$F(\ell_0) - p_0 \cdot F_p(\ell_0) \ne 0. \tag{25}$$

Now we turn to the converse question:

Given some extremal curve $\mathcal{C}_0$ = graph *u with the endpoints P_0 and P which meets $\mathcal{S}$ transversally at P_0, is it a minimizer of the functional $\mathcal{F}(\mathcal{C})$ among all curves $\mathcal{C}$ =* graph *v sufficiently close to u, which are of class C^1 (or D^1) and connect $\mathcal{S}$ with the preassigned point P?*

To obtain a positive answer to this question, we shall try to embed the given extremal u in a Mayer field whose field lines intersect the surface $\mathcal{S}$ transversally. That is, the field is to be constructed in such a way that $\mathcal{S}$ becomes a level surface of the corresponding eikonal. The first step of this construction is to secure that, close to P_0, the surface $\mathcal{S}$ carries a uniquely determined direction field $\ell(x, z) = (x, z, \pi(x, z))$ which is transversal to the field of surface elements $\sigma(x, z) = (x, z, v(x, z))$ of $\mathcal{S}$ and satisfies $\sigma(P_0) = \ell_0$ where $P_0 = (a, u(a))$, $\ell_0 = (P_0, p_0)$, $p_0 = u'(a)$.

To perform this first step we make the following assumption.

Assumption (T1). *The Lagrangian F satisfies*

$$F(x, z, p) \neq 0, \quad F(x, z, p) - p \cdot F_p(x, z, p) \neq 0, \quad \det F_{pp}(x, z, p) \neq 0$$

for all $(x, z, p) \in \mathcal{S} \times \mathbb{R}^N$.

It will turn out that (T1) implies

Assumption (T2). *For* $0 < \varepsilon \ll 1$, *there is a uniquely determined mapping* $\pi : \mathcal{S} \cap B_\varepsilon(P_0) \to \mathbb{R}^N$ *of class* C^1 *such that* $\pi(P_0) = p_0$ *and*

$$v(x, z) = -\frac{F_p(x, z, \pi(x, z))}{F(x, z, \pi(x, z)) - \pi(x, z) \cdot F_p(x, z, \pi(x, z))} \tag{26}$$

for all $(x, z) \in \mathcal{S} \cap B_\varepsilon(P_0)$.

Remark 1. In order to show that (T1) implies (T2), one conveniently uses both Legendre's transformation and E. Hölder's transformation. This will be carried out in Chapter 10. Presently we just note that it is easy to verify (T1) and (T2) for Lagrangians of the form

$$F(x, z, p) = \omega(x, z)\sqrt{1 + |p|^2}, \tag{27}$$

where $\omega(x, z) > 0$. In fact, we have

$$F(x, z, p) - p \cdot F_p(x, z, p) = \frac{\omega(x, z)}{\sqrt{1 + |p|^2}}, \tag{28}$$

$$\frac{F_p}{F - p \cdot F_p} = p, \tag{29}$$

and from the equation

$$F_{p^i p^k}(x, z, p) = \frac{\omega(x, z)}{(1 + |p|^2)^{3/2}} [(1 + |p|^2)\delta^{ik} - p^i p^k] \tag{30}$$

it follows easily that $F_{pp} > 0$.

We remark that formula (26) expresses the fact that the field of line elements $\ell(x, z) = (x, z, \pi(x, z))$ on $\mathcal{S}$ is transversal to the field of surface elements $\sigma(x, z) = (x, z, v(x, z))$ tangential to $\mathcal{S}$, and (29) expresses the fact that, for

Lagrangians of the form (27), transversality is the same as orthogonality, and that we have $\pi(x, z) = \nu(x, z)$.

Let us now choose a parametric representation

(31) $$x = \xi(c), \quad z = \zeta(c), \qquad c \in I_0 \subset \mathbb{R}^N,$$

for $\mathcal{S} \cap B_\varepsilon(P_0)$ which is of class C^2 and satisfies $\det \zeta_c \neq 0$ and $P_0 = (\xi(c_0), \zeta(c_0))$. Then we define an N-parameter family of extremals

$$z = \varphi(x, c), \quad x \in I(c), \qquad c \in I_0,$$

which satisfy the initial value conditions

(32) $$\varphi(\xi(c), c) = \zeta(c), \qquad \varphi'(\xi(c), c) = \pi(\xi(c), \zeta(c)).$$

Then, by Proposition 3 of *1.3*, the extremal bundle $f(x, c) := (x, \varphi(x, c))$ is a Mayer bundle since its rays $f(\cdot, c)$ are transversally intersected by the surface $\mathcal{S}$. Moreover, we have $f(x, c_0) = (x, u(x))$ for $a \le x \le b$ because of $\varphi(a, c_0) = u(a)$ and $\varphi'(a, c_0) = u'(a)$. By a standard reasoning, we conclude that there is some $\delta > 0$ such that $\overline{B}_\delta(c_0) \subset I_0$ and $[\xi(c), b] \subset I(c)$, i.e., the extremals $\varphi(x, c)$ are at least defined on $[\xi(c), b]$ if $|c - c_0| \le \delta$.

We claim that

(33) $$\det \varphi_c(\xi(c), c) \neq 0 \quad \text{for } |c - c_0| \le \delta.$$

Otherwise, we could find some $\lambda = (\lambda^1, \ldots, \lambda^N) \in \mathbb{R}^N$, $\lambda \neq 0$, such that

$$\lambda^\alpha \varphi_{c^\alpha}(\xi(c), c) = 0.$$

Because of

$$\zeta_{c^\alpha}(c) = \varphi'(\xi(c), c)\xi_{c^\alpha}(c) + \varphi_{c^\alpha}(\xi(c), c),$$

it follows that

$$\lambda^\alpha \xi_{c^\alpha}(c) = \varphi'(\xi(c), c)\tau,$$

where we have set

$$\tau := \lambda^\alpha \xi_{c^\alpha}(c).$$

Since $\det \zeta_c \neq 0$, we have $\lambda^\alpha \zeta_{c^\alpha} \neq 0$, and therefore $\tau \neq 0$; hence we may assume that $\tau = 1$, i.e.,

$$\varphi'(\xi(c), c) = \lambda^\alpha \zeta_{c^\alpha}(c).$$

A geometric interpretation of this relation can be obtained as follows. Let $\gamma : [0, 1] \to \mathbb{R}^N$ be a curve in the parameter domain I_0 such that $\gamma(0) = c$ and $\dot{\gamma}(0) = \lambda$. Then

$$P(t) := (\xi(\gamma(t)), \zeta(\gamma(t))), \quad 0 \le t \le 1,$$

defines a curve on $\mathcal{S}$ with the initial point

$$P(0) = (\xi(c), \xi(c))$$

and the initial velocity

$$V(0) := \dot{P}(0) = (1, \varphi'(\xi(c), c)).$$

Consequently, the tangent vector $f'(\xi(c), c) = (1, \varphi'(\xi(c), c))$ to the ray $f(\cdot, c)$ at the point $f(\xi(c), c)$ would be tangent to $\mathcal{S}$. By virtue of Proposition 1 in *1.3*, this is impossible since $F > 0$ and $\mathcal{S}$ is a transversal surface. Thus we have verified (33).

Let us now restrict $f(x, c)$ to the set

$$\Gamma := \{(x, c) \colon \xi(c) \le x \le b, |c - c_0| \le \delta\}.$$

Then the preceding results show that $f : \Gamma \to \mathbb{R} \times \mathbb{R}^N$ *is a field-like Mayer bundle, and* $f(\cdot, c_0)$ *is just the given extremal curve* $\mathscr{C}_0 = \text{graph } u$.

Suppose now that there is no focal value of the extremal $\varphi(\cdot, c_0) = u$ in $[a, b]$, i.e., there is no focal point of the bundle f on the extremal arc $\mathscr{C}_0$ corresponding to this arc. Then we have

$$\det \varphi_c(x, c_0) \ne 0 \quad \text{for } a \le x \le b$$

and, by the reasoning of *2.1*, we infer that f furnishes a diffeomorphism of Γ onto $G := f(\Gamma)$ provided that $0 < \delta \ll 1$. (Cf. *2.1*, Theorem 1.)

Thus we have embedded u in a Mayer field of extremals provided that $\mathscr{C}_0 = \text{graph } u$ *contains no focal point of f.*

Remark 2. Since the construction of the Mayer bundle f only depends on the data of $\mathcal{S}$ and u, the absence of focal points on $\mathscr{C}_0 = \text{graph } u$ is merely an assumption on the extremal u and the transversal surface $\mathcal{S}$ in a small neighbourhood $B_\varepsilon(P_0)$ of the left endpoint P_0 of $\mathscr{C}_0$. Thus, any focal point P^* of f on $\mathscr{C}_0$ (corresponding to P_0) could be called *a focal point of* $\mathcal{S}$ *on* $\mathscr{C}_0$.

By the results of *1.3* we then obtain (see also *2.1*, Theorem 2):

Theorem 1. *Let* $\mathscr{C}_0 = \text{graph } u$ *be an extremal curve with a preassigned right endpoint P which meets a support surface* $\mathcal{S}$ *transversally at its left endpoints* P_0. *Suppose also that the following assumptions are satisfied*:

(*i*) (*T*), (*T1*) *and* (*T2*) *hold true*;
(*ii*) $F_{pp}(x, u(x), p) > 0$ *for all* $x \in [a, b]$;
(*iii*) $\mathscr{C}_0$ *contains no focal point of* $\mathcal{S}_\varepsilon := \mathcal{S} \cap B_\varepsilon(P_0)$, $0 < \varepsilon \ll 1$.

Then there is an open neighbourhood G_0 *of* $\mathscr{C}_0$ *in the configuration space* $\mathbb{R} \times \mathbb{R}^N$ *such that*

$$\mathcal{F}(\mathscr{C}_0) < \mathcal{F}(\mathscr{C})$$

for all D^1*-curves* $\mathscr{C} = \text{graph } v$ *contained in* G_0 *which connect* $\mathcal{S}$ *with the point P.*

We recall that (T1) implies (T2), and that both assumptions can easily be verified for Lagrangians of the kind $F(x, z, p) = \omega(x, z)\sqrt{1 + |p|^2}$, $\omega > 0$; cf. Remark 1. Thus we obtain the following corollary of Theorem 1:

Theorem 2. *Let F be a Lagrangian of the form*

$$F(x, z, p) = \omega(x, z)\sqrt{1 + |p|^2}, \quad \omega(x, z) > 0,$$

and let $\langle \mathscr{S}, P \rangle$ be a boundary configuration consisting of a surface $\mathscr{S}$ and a point P which satisfy (T). Moreover, suppose that $\mathscr{C}_0 = \mathrm{graph}\, u$ is an extremal curve which transversally meets $\mathscr{S}$ at its left endpoint and has P as right endpoint. Finally, assume that $\mathscr{C}_0$ contains no focal point of $\mathscr{S}_\varepsilon := \mathscr{S} \cap B_\varepsilon(P_0)$, $0 < \varepsilon \ll 1$. Then there is an open neighbourhood G_0 of $\mathscr{C}_0$ in $\mathbb{R} \times \mathbb{R}^N$ such that $\mathscr{F}(\mathscr{C}_0) < \mathscr{F}(\mathscr{C})$ holds true for all D^1-curves $\mathscr{C} = \mathrm{graph}\, v$ contained in G_0 which connect $\mathscr{S}$ with P.

If $N = 1$, we can easily formulate global embedding results by deriving an analogue to Theorem 1 of *2.2*. Such a result yields global minimality of an extremal curve $\mathscr{C}_0$ in a boundary configuration $\langle \mathscr{S}, P \rangle$ provided that $\mathscr{C}_0$ meets $\mathscr{S}$ perpendiculary at its left endpoint. However, if $N > 1$, then the question of global minimality is much more subtle.

More generally, we can consider boundary configurations $\langle \mathscr{S}, P \rangle$ where $\mathscr{S}$ is an m-dimensional submanifold of $\mathbb{R} \times \mathbb{R}^N$, $0 \le m \le N$, and P is some point outside of $\mathscr{S}$. If $m = N$, we have the case that was just treated, and if $m = 0$, then $\mathscr{S}$ reduces to a single point, and we have the ordinary two-point boundary problem studied in *1.3* and in *2.1–2.3*. Hence it remains to consider the cases $m = 1, 2, \dots, N - 1$. In principle, one can proceed as for $m = N$. First one determines all line elements $\ell(x, z) = (x, z, \pi(x, z))$ which are transversal to $\mathscr{S}$. By this we mean that $\ell(x, z)$ is transversal to any surface element $(x, z, v(x, z))$ which is tangent to $\mathscr{S}$ at the point $(x, z) \in \mathscr{S}$. Then we define an extremal bundle $f(x, c) = (x, \varphi(x, c))$ emanating from $\mathscr{S}$ in the transversal directions that were just determined. This bundle is seen to be a field-like Mayer bundle if we take the necessary precautions similar to assumptions (T), (T1) and (T2). If we exclude focal points by choosing sufficiently small parts of f, we obtain a Mayer field to which the reasoning of *1.3* can be applied.

Finally, we can treat the minimum problem for boundary configurations $\langle \mathscr{S}_1, \mathscr{S}_2 \rangle$ consisting of two hypersurfaces $\mathscr{S}_1$ and $\mathscr{S}_2$. Let $\mathscr{C}_0$ be an extremal curve which transversally meets $\mathscr{S}_1$ and $\mathscr{S}_2$ at its endpoints P_1 and P_2 respectively. Then we first construct the field-like Mayer bundle f whose

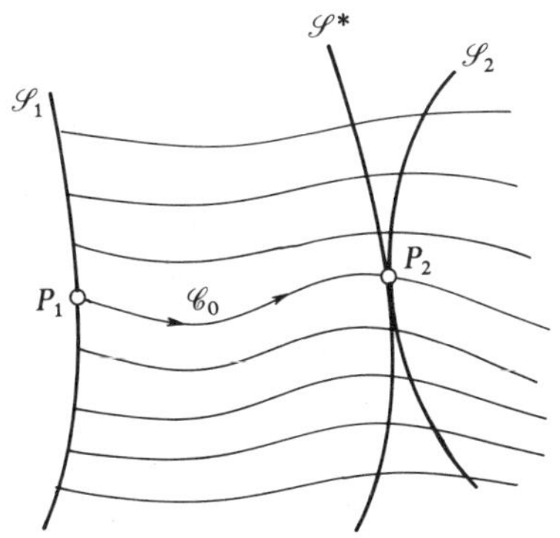

Fig. 16.

rays are transversal to $\mathscr{S}_1$. If no focal point of f lies on $\mathscr{C}_0$, we can assume that f is a Mayer field, by passing to a sufficiently thin pencil of rays close to $\mathscr{C}_0$. Let S be the eikonal of this field, and be $\mathscr{S}^*$ that level surface of S which passes through P_2. Then, obviously, $\mathscr{C}_0$ minimizes $\mathscr{F}(\mathscr{C})$ among all curves $\mathscr{C} = \text{graph } v$ which lie in the domain covered by f and whose endpoints lie on $\mathscr{S}_1$ and $\mathscr{S}_2$ respectively, provided that $\mathscr{S}^*$ is tangent to $\mathscr{S}_2$ at P_2, and that $\mathscr{S}_2$ is curved away from $\mathscr{S}^*$ in the opposite direction with respect to $\mathscr{C}_0$ (see Fig. 16). This turns out to be the case if the first focal point P' of $\mathscr{S}_1$ on $\mathscr{C}_0$ (to the right of $\mathscr{S}_1$) lies behind the first focal point P'' of $\mathscr{S}_2$ on $\mathscr{C}_0$ (*sufficient condition of Bliss*). Cf. Bolza [3], pp. 327–331.

3. Field Theory for Multiple Integrals in the Scalar Case: Lichtenstein's Theorem

In this section we want to extend the results of Sections 1 and 2 to the case $n > 1$, $N = 1$, i.e. to scalar valued extremals of multiple integrals. The presentation will not be self-contained as we shall rely on results for elliptic operators, in particular, on regularity theory.

H.A. Schwarz [1] was the first to derive *sufficient conditions* for extremals of the two-dimensional area functional. Lichtenstein [1] extended Schwarz's result to scalar extremals of general two-dimensional variational integrals, and Morrey [1] carried Lichtenstein's method over to n-dimensional integrals.

The main task in Lichtenstein's approach is to embed a given extremal in a field of extremals, i.e. in a foliation by extremal surfaces of codimension one. This embedding problem is of greater technical difficulty than the corresponding problem in Section 2 since we have now to solve a boundary value problem for a partial differential equation instead of an ordinary differential equation. On the other hand, the task is conceptually simpler than the previous one since we have not to distinguish between extremal fields and Mayer fields: no integrability condition enters.

We begin by defining the notion of a *field of* (*nonparametric*) *hypersurfaces in* $\mathbb{R}^{n+1} = \mathbb{R}^n \times \mathbb{R}$. Let σ be a real parameter varying in some compact interval $\Sigma \subset \mathbb{R}$, and consider a set Γ in $\mathbb{R}^{n+1}$ which is of the form

$$\Gamma = \{(x, \sigma) \colon \sigma \in \Sigma, x \in \overline{\Omega}(\sigma)\},$$

where $\Omega(\sigma)$ are bounded domains in $\mathbb{R}^n$. We also assume that $\operatorname{int} \Gamma$ is a domain in $\mathbb{R}^{n+1}$, given by

$$\operatorname{int} \Gamma = \{(x, \sigma) \colon \sigma \in \operatorname{int} \Sigma, x \in \Omega(\sigma)\}.$$

Consider now a mapping $f \colon \Gamma \to \mathbb{R}^{n+1}$ defined by

$$x = x, \qquad z = \varphi(x, \sigma),$$

that is, $f(x, \sigma) = (x, \varphi(x, \sigma))$, and suppose that f defines a diffeomorphism of Γ onto $G := f(\Gamma)$ which is of class C^m, $m \geq 1$. This means that $\varphi \in C^m(\Gamma)$ and $\varphi_\sigma(x, \sigma) \neq 0$ on Γ, and that, for every point $(x, z) \in G$, there is exactly one value $\sigma \in \Sigma$ such that $z = \varphi(x, \sigma)$. Denote this σ by $s(x, z)$. Then we have

(1) $$z = \varphi(x, s(x, z)) \quad \text{and} \quad \sigma = s(x, \varphi(x, \sigma))$$

for $(x, z) \in G$ and $(x, \sigma) \in \Gamma$, respectively.

We interpret the mapping f as family of nonparametric surfaces

$$\mathscr{S}(\sigma) := \{(x, z) : x \in \overline{\Omega}(\sigma),\ z = \varphi(x, \sigma)\}$$

in $\mathbb{R}^{n+1}$. The hypersurfaces $\mathscr{S}(\sigma)$ form the leaves of a foliation of G: through every point (x, z) of G there passes exactly one leaf. As customary in the calculus of variations, we call the family $\{\mathscr{S}(\sigma)\}_{\sigma \in \Sigma}$ a *field of hypersurfaces covering G*; equivalently we denote the mapping $f : \Gamma \to G$ a *field on G*. If the domains $\Omega(\sigma)$ do not depend on σ, i.e., if $\Omega(\sigma) \equiv \Omega$, then $f : \Gamma \to G$ is called a *field over* $\overline{\Omega}$ *covering G*.

The *slope* $\mathscr{P}(x, z) = (\mathscr{P}_1(x, z), \ldots, \mathscr{P}_n(x, z))$ *of the field* $f : \Gamma \to \overline{G}$ is defined by

(2) $$\mathscr{P}_\alpha(x, z) := \varphi_{x^\alpha}(x, s(x, z)),$$

whence we infer that

(3) $$\begin{aligned} \varphi_{x^\alpha}(x, \sigma) &= \mathscr{P}_\alpha(x, \varphi(x, \sigma)), \\ \varphi_{x^\alpha x^\beta}(x, \sigma) &= \mathscr{P}_{\alpha, x^\beta}(x, \varphi(x, \sigma)) + \mathscr{P}_{\alpha, z}(x, \varphi(x, \sigma))\mathscr{P}_\beta(x, \varphi(x, \sigma)). \end{aligned}$$

Clearly, $\mathscr{P}$ is of class C^{m-1} if f is of class C^m.

Let $\mathscr{S}_0$ be some nonparametric hypersurface in $\mathbb{R}^{n+1}$ given by $z = u_0(x)$, $x \in \overline{\Omega}_0$. *We say that* $\mathscr{S}_0$ *is embedded into some field* $\{\mathscr{S}(\sigma)\}_{\sigma \in \Sigma}$ *if there is some* $\sigma_0 \in \operatorname{int} \Sigma$ *such that* $\mathscr{S}_0$ *is part of the leaf* $\mathscr{S}(\sigma_0)$, *i.e. if* $\Omega_0 \subset \Omega(\sigma_0)$ *and* $u_0(x) = \varphi(x, \sigma_0)$ *for all* $x \in \overline{\Omega}_0$.

Assume now that $F(x, z, p)$ is some Lagrangian of class $C^3(\mathbb{R}^{2n+1})$. A field $f : \Gamma \to G$ is said to be an *extremal field* with respect to F if every leaf $z = \varphi(x, \sigma)$, $x \in \overline{\Omega}(\sigma)$, is an F-extremal.

In the following discussion we assume that $f : \Gamma \to G$ is a field of class C^2 with a slope function $\mathscr{P} = (\mathscr{P}_1, \ldots, \mathscr{P}_n)$ of class C^1. As in *1.1* we introduce the following *notation*:

For any function $A = A(x, z, p)$ *on* $\overline{G} \times \mathbb{R}^n$ *we define the function* $\overline{A} = \overline{A}(x, z)$ *on* $\overline{G}$ *by*

$$\overline{A}(x, z) := A(x, z, \mathscr{P}(x, z)).$$

Then we obtain the following result:

Proposition 1. *The field f is an extremal field if and only if the equation*

(4) $$\frac{\partial}{\partial z}[\overline{F} - \mathscr{P}_\alpha \overline{F}_{p_\alpha}] = \frac{\partial}{\partial x^\alpha} \overline{F}_{p_\alpha}$$

holds true in G.

Proof. Suppose that $f(x, \sigma) = (x, \varphi(x, \sigma))$ is an extremal field, i.e.,

$$F_z(x, \varphi, D\varphi) - D_\alpha F_{p_\alpha}(x, \varphi, D\varphi) = 0.$$

This means that

$$F_z - F_{p_\alpha x^\alpha} - F_{p_\alpha z}\varphi_{x^\alpha} - F_{p_\alpha p_\beta}\varphi_{x^\alpha x^\beta} = 0,$$

where $(x, \varphi(x, \sigma), D\varphi(x, \sigma))$ are the arguments of F_z, $F_{p_\alpha x^\alpha}$, $F_{p_\alpha z}$, and $F_{p_\alpha p_\beta}$. On account of (3) we can write this equation in the form

$$\bar{F}_z - \{\bar{F}_{p_\alpha x^\alpha} + \bar{F}_{p_\alpha p_\beta}\mathscr{P}_{\alpha, x^\beta}\} - [\bar{F}_{p_\alpha z} + \bar{F}_{p_\alpha p_\beta}\mathscr{P}_{\beta, z}]\mathscr{P}_\alpha = 0.$$

Adding the relation

$$\bar{F}_{p_\alpha}\mathscr{P}_{\alpha, z} - \bar{F}_{p_\alpha}\mathscr{P}_{\alpha, z} = 0,$$

we arrive at

$$(\bar{F}_z + \bar{F}_{p_\alpha}\mathscr{P}_{\alpha, z}) - \{\bar{F}_{p_\alpha}\mathscr{P}_{\alpha, z} + [\bar{F}_{p_\alpha z} + \bar{F}_{p_\alpha p_\beta}\mathscr{P}_{\beta, z}]\mathscr{P}_\alpha\} = \bar{F}_{p_\alpha x^\alpha} + \bar{F}_{p_\alpha p_\beta}\mathscr{P}_{\alpha, x^\beta},$$

which is equivalent to (4).

Since we can reverse this computation, also the converse is true. □

Proposition 2. *A C^3-field f over $\bar{\Omega}$ with the slope $\mathscr{P}$ is an extremal field if and only if there is a function $S(x, z) = (S^1(x, z), \ldots, S^n(x, z))$ of class $C^1(G, \mathbb{R}^n)$ such that the equations*

$$S^\alpha_{x^\alpha} = \bar{F} - \mathscr{P}_\alpha \bar{F}_{p_\alpha}, \qquad S^\alpha_z = \bar{F}_{p_\alpha} \tag{5}$$

are satisfied for $1 \le \alpha \le n$.

Proof. Set

$$\begin{aligned} A &:= \bar{F} - \mathscr{P}_\alpha \bar{F}_{p_\alpha}, \qquad B^\alpha := \bar{F}_{p_\alpha}, \\ M(x, z, p) &:= A(x, z) + B^\alpha(x, z)p_\alpha. \end{aligned} \tag{6}$$

Then (4) is equivalent to the equation $A_z = B^\alpha_{x^\alpha}$. By virtue of Proposition 2 of 1,4.2, there exists some function $S \in C^1(G, \mathbb{R}^n)$ such that

$$M(x, z, p) = S^\alpha_{x^\alpha}(x, z) + S^\alpha_z(x, z)p_\alpha \tag{7}$$

holds for all $p \in \mathbb{R}^n$ and all $(x, z) \in G$. From (6) and (7) we infer that

$$A = S^\alpha_{x^\alpha}, \qquad B^\alpha = S^\alpha_z$$

is satisfied, and thus (5) is verified.

Conversely, if for a given field f with slope $\mathscr{P}$ there is a function $S \in C^1(G, \mathbb{R}^n)$ satisfying (5). Thus we obtain $S^\alpha_z \in C^1(G)$ whence $S^\alpha_{zx^\beta} = S^\alpha_{x^\beta z}$. Then we have $S^\alpha_{x^\alpha z} = S^\alpha_{zx^\alpha}$, which implies (4). By virtue of Proposition 1 we now conclude that f is an extremal field. □

Using the Addendum to Proposition 2 in 1,*4.2*, we obtain an analogous result for extremal fields f of class C^2.

Equations (5) are the n-dimensional analogue of the Carathéodory equations for the case $n = 1$ stated in *1.1*. Hence we shall call (5) the *Carathéodory equations* for $N = 1, n > 1$. A C^2-field f on G with a slope $\mathscr{P}$ is called a "*Mayer field*" if there is a function $S = (S^1, \ldots, S^n)$ such that S and f are linked by the Carathéodory equations (5).

Then we can rephrase Proposition 2 as follows: *For $N = 1, n > 1$ (or $n \geq 1$), a field f is a Mayer field if and only if it is an extremal field.*

Let f be an extremal field on G with the slope $\mathscr{P}$, and let S be a solution of the Carathéodory equations (5). By Propositions 1 and 2 of 1,*4.2*, the function M in (7),

$$M(x, z, p) = S^\alpha_{x^\alpha}(x, z) + S^\alpha_z(x, z)p_\alpha,$$

is a null Lagrangian. Hence, for any function $u \in C^1(\bar{\Omega})$ with graph$(u) \subset G$, the integral

$$\mathscr{M}(u) := \int_\Omega M(x, u(x), Du(x))\, dx \tag{8}$$

is an *invariant integral* (cf. Proposition 2 of 1,*3.2*), that is,

$$\mathscr{M}(u) = \mathscr{M}(v)$$

for any two functions $u, v \in C^1(\bar{\Omega})$ with graph$(u) \subset G$, graph$(v) \subset G$ and $u|_{\partial\Omega} = v|_{\partial\Omega}$. In fact, if $\partial\Omega$ is piecewise smooth, we have

$$\mathscr{M}(u) = \int_{\partial\Omega} \nu_\alpha(x) \cdot S^\alpha(x, u(x))\, d\mathscr{H}^{n-1}(x), \tag{9}$$

where $\nu = (\nu_1, \ldots, \nu_n)$ is the exterior normal on $\partial\Omega$, or simply

$$\mathscr{M}(u) = \int_{\partial\Omega} \nu \cdot S(x, u)\, d\mathscr{H}^{n-1}. \tag{9'}$$

In other words, $\mathscr{M}(u)$ is the flow of the vector field $S(x, u(x))$ through the boundary of Ω. One denotes $\mathscr{M}(u)$ *as Hilbert's invariant integral* associated with the extremal field f. By means of the Carathéodory equations (5) the integrand $M(x, z, p)$ of Hilbert's integral can be written as

$$M(x, z, p) = F(x, z, \mathscr{P}(x, z)) + [p_\alpha - \mathscr{P}_\alpha(x, z)] \cdot F_{p_\alpha}(x, z, \mathscr{P}(x, z)) \tag{10}$$

Using Weierstrass's excess function $\mathscr{E}_F$ we obtain

$$F(x, z, p) - M(x, z, p) = \mathscr{E}_F(x, z, \mathscr{P}(x, z), p). \tag{11}$$

Then, for any $v \in C^1(\bar{\Omega})$ with graph$(v) \subset G$, we obtain

$$\int_\Omega F(x, v, Dv)\, dx = \int_\Omega M(x, v, Dv)\, dx + \int_\Omega \mathscr{E}_F(x, v, \mathscr{P}(x, v), Dv)\, dx, \tag{12}$$

that is

$$(12') \qquad \mathscr{F}(v, \Omega) = \mathscr{M}(v, \Omega) + \int_\Omega \mathscr{E}_F(x, v, \mathscr{P}(x, v), Dv)\, dx .$$

This is the analogue of the *Weierstrass representation formula* of *1.3* for the case $n > 1$.

Proposition 3. *Suppose that the extremal $u \in C^2(\bar{\Omega})$ is embedded in an extremal field f on G with the slope $\mathscr{P}$. Moreover let $v \in C^1(\bar{G})$ be an arbitrary function with* graph$(v) \subset G$ *and $v = u$ on $\partial\Omega$. Then it follows that*

$$(13) \qquad \mathscr{F}(v, \Omega) = \mathscr{F}(u, \Omega) + \int_\Omega \mathscr{E}_F(x, v, \mathscr{P}(x, v), Dv)\, dx .$$

Proof. Since $\mathscr{M}$ is an invariant integral, we infer from the assumptions on u and v that $\mathscr{M}(v, \Omega) = \mathscr{M}(u, \Omega)$ is satisfied. Furthermore, since u fits into the field f, we have $Du(x) = \mathscr{P}(x, u(x))$, and it follows from (10) that $M(x, u(x), Du(x)) = F(x, u(x), Du(x))$ for all $x \in \bar{\Omega}$. Thus we obtain $\mathscr{M}(u, \Omega) = \mathscr{F}(u, \Omega)$, and therefore also $\mathscr{M}(v, \Omega) = \mathscr{F}(u, \Omega)$. Then formula (13) follows at once from (12'). □

Corollary 1. *If also the strict Legendre condition*

$$(14) \qquad F_{p_\alpha p_\beta}(x, u(x), Du(x))\, \xi_\alpha \xi_\beta \ge \lambda |\xi|^2, \quad \lambda > 0,$$

is satisfied for all $x \in \bar{\Omega}$ and all $\xi \in \mathbb{R}^n$, then u is a weak minimizer of $\mathscr{F}(\cdot, \Omega)$.

Corollary 2. *If in addition to the assumptions of Proposition 3 the extremal field f satisfies the "sufficient Weierstrass condition"*

$$(15) \qquad \mathscr{E}_F(x, z, \mathscr{P}(x, z), p) > 0 \quad \text{for } (x, z) \in G \text{ and } p \ne \mathscr{P}(x, z),$$

then it follows that $\mathscr{F}(v, \Omega) > \mathscr{F}(u, \Omega)$ for all $v \in C^1(\bar{\Omega})$ with graph$(v) \subset G$, $v = u$ *on $\partial\Omega$, and $v(x) \ne u(x)$, i.e., u is the uniquely determined minimizer of $\mathscr{F}(\cdot, \Omega)$ among all functions $v \in C^1(\bar{\Omega})$ with* graph$(v) \subset G$ *and $v = u$ on $\partial\Omega$.*

Corollary 1 is an obvious consequence of (13). Thus we turn to the

Proof of Corollary 2. Again, the inequality $\mathscr{F}(v, \Omega) \ge \mathscr{F}(u, \Omega)$ is an immediate consequence of (13), and by assumption (15) we even have $\mathscr{F}(v, \Omega) > \mathscr{F}(u, \Omega)$, if there is at least one point $x \in \Omega$ such that $Dv(x) \ne \mathscr{P}(x, v(x))$ holds true. Hence it suffices to show that the relation

$$(16) \qquad Dv(x) = \mathscr{P}(x, v(x)) \quad \text{for all } x \in \Omega$$

implies $u(x) \equiv v(x)$.

To this end we consider the function $s(x, z)$ corresponding to $f(x, \sigma) = (x, \varphi(x, \sigma))$ via the relation

$$z = \varphi(x, s(x, z)).$$

Differentiating this equation with respect to x^α and z respectively we infer that

$$0 = \varphi_{x^\alpha}(x, s(x, z)) + \varphi_\sigma(x, s(x, z))s_{x^\alpha}(x, z),$$

and

$$1 = \varphi_\sigma(x, s(x, z))s_z(x, z),$$

and we note that $\varphi_\sigma \neq 0$. Furthermore, consider the function $\omega(x) := s(x, v(x))$. From

$$\omega_{x^\alpha} = s_{x^\alpha}(x, v) + s_z(x, v)v_{x^\alpha}$$

and the previous formulas we infer that

$$\omega_{x^\alpha} = \frac{1}{\varphi_\sigma(x, s(x, v))}[v_{x^\alpha} - \varphi_{x^\alpha}(x, s(x, v))].$$

Since

$$\varphi_{x^\alpha}(x, \sigma) = \mathscr{P}_\alpha(x, \varphi(x, \sigma)) \quad \text{and} \quad v(x) = \varphi(x, s(x, v(x))),$$

it follows that

$$\varphi_{x^\alpha}(x, s(x, v)) = \mathscr{P}_\alpha(x, v).$$

Therefore relation (16) yields $D\omega = 0$, i.e. $\omega = \text{const}$ on every connected component of $\bar{\Omega}$. On the other hand the relations $u|_{\partial\Omega} = v|_{\partial\Omega}$ and $u(x) = \varphi(x, \sigma_0)$ for some σ_0 and all $x \in \bar{\Omega}$ imply

$$\omega(x) = s(x, u(x)) = s(x, \varphi(x, \sigma_0)) = \sigma_0 \quad \text{for all } x \in \partial\Omega,$$

whence $\omega(x) \equiv \sigma_0$ on $\bar{\Omega}$, that is,

$$s(x, v(x)) = \sigma_0 \quad \text{for all } x \in \bar{\Omega}.$$

This relation is equivalent to

$$v(x) = \varphi(x, \sigma_0) = u(x) \quad \text{for all } x \in \bar{\Omega}. \qquad \square$$

It remains to investigate under which assumptions a given extremal can be embedded into a field of extremals. A positive answer to this question leads to a *sufficient condition ensuring that in this case the given extremal actually furnishes a strong minimum.*

Now we are going to state *Lichtenstein's theorem* which is the analogue of Theorem 2 of *2.1*, only that the "conjugate point condition" is now replaced by an eigenvalue criterion.

For the following we require of F, Ω and u that they satisfy

Assumption (A). *Let $u \in C^{2,\mu}(\bar{\Omega})$ be an F-extremal defined on a bounded domain Ω of $\mathbb{R}^n$ with $\partial\Omega \in C^{2,\mu}$, $0 < \mu < 1$, satisfying the strict Legendre condition*

$$\tfrac{1}{2}F_{p_\alpha p_\beta}(x, u(x), Du(x))\,\eta_\alpha\eta_\beta \geq c|\eta|^2 \tag{17}$$

for some $c > 0$ *and for all* $\eta \in \mathbb{R}^n$ *and* $x \in \bar{\Omega}$. *Furthermore, the Lagrangian* $F(x, z, p)$ *be of class* C^5 *on* $\bar{\Omega} \times \mathbb{R} \times \mathbb{R}^n$.

Let $\mathcal{J} = \mathcal{J}_u$ be the Jacobian operator of F with respect to u, and let λ_1 be the smallest eigenvalue of $\mathcal{J}$ on Ω, i.e. the smallest number $\lambda \in \mathbb{R}$ such that there is a nontrivial solution v of

$$\mathcal{J}v = \lambda v \text{ in } \Omega, \quad v = 0 \text{ on } \partial\Omega .$$

Recall also that $\mathcal{J}$ is the Euler operator of the accessory integral

$$\mathcal{Q}(v) := \tfrac{1}{2}\delta^2\mathcal{F}(u, v).$$

According to Theorem 1,(iv) of 5,*1.3*, we have

$$\lambda_1 = 2 \inf\{\mathcal{Q}(v) \colon v \in C_0^1(\bar{\Omega}, \mathbb{R}^N) \text{ and } \int_\Omega v^2\,dx = 1\}. \tag{18}$$

We are now going to prove

Proposition 4. *Suppose that the extremal* $u \in C^2(\bar{\Omega})$ *satisfies assumption (A) and that, in addition, the smallest eigenvalue* λ_1 *of the Jacobi operator* $\mathcal{J} = \mathcal{J}_u$ *is positive. Then u can be embedded into a* C^2*-extremal field over* $\bar{\Omega}$.

Combining Corollary 2 of Proposition 3 with Proposition 4, we arrive at

Lichtenstein's theorem. *Suppose that the F-extremal* $u \in C^2(\bar{\Omega})$ *satisfies assumption (A) and that the smallest eigenvalue* λ_1 *of the Jacobi operator* $\mathcal{J}_u$ *on* Ω *is positive. Assume finally that* $\mathcal{E}_F(x, z, p, q) > 0$ *holds true for all* x, z, p, q *with* $x \in \bar{\Omega}$, $|z - u(x)| < \varepsilon_0$, $p, q \in \mathbb{R}^n$, *and* $p \neq q$, *where* ε_0 *is a fixed positive number independent of* $x \in \bar{\Omega}$. *Then u is a strict strong minimizer of the variational integral*

$$\mathcal{F}(v) = \int_\Omega F(x, v, Dv)\,dx .$$

Thus it remains to give a *Proof of Proposition 4*, which will be carried out in two steps.

First we construct a family $\varphi(x, \sigma)$, $x \in \bar{\Omega}$, $|\sigma| \le \sigma_0$, of solutions of

$$L_F(\varphi(\cdot, \sigma)) = 0 \quad \text{in } \Omega \tag{19}$$

and

$$\varphi(x, \sigma) = u(x) + \sigma \quad \text{on } \partial\Omega , \tag{20}$$

which is a sufficiently smooth function of $(x, \sigma) \in \bar{\Omega} \times [-\sigma_0, \sigma_0]$ and satisfies in particular

$$u(x) = \varphi(x, 0) \quad \text{on } \bar{\Omega}. \tag{21}$$

Secondly we shall verify that $\varphi_\sigma(x, \sigma) > 0$ holds on $\bar{\Omega} \times [-\sigma_0, \sigma_0]$, after replacing σ_0 by an appropriately small number $\tilde{\sigma}_0 \in (0, \sigma_0]$ which will again be denoted by σ_0.

Step 1. We try to construct $\varphi(x, \sigma)$ in the form

$$\varphi(x, \sigma) = u(x) + \sigma\zeta(x, \sigma), \tag{22}$$

where $\zeta(x, \sigma)$ is to satisfy

$$\zeta(x, \sigma) = 1 \quad \text{on } \partial\Omega. \tag{23}$$

Equations (22) and (23) clearly imply relations (20) and (21).

We now introduce the function

$$\psi(x, t) = u(x) + t\sigma\zeta(x, \sigma), \quad x \in \bar{\Omega},\ t \in [0, 1],$$

for some fixed $\sigma \in \mathbb{R}$. It follows that

$$\psi(x, 0) = u(x), \qquad \psi(x, 1) = \varphi(x, \sigma), \qquad \psi_t(x, t) = \sigma\zeta(x, \sigma).$$

Since both u and $\varphi(\cdot, \sigma)$ are to be extremals, we obtain

$$L_F(\psi(\cdot, 1)) - L_F(\psi(\cdot, 0)) = 0.$$

This can be written as

$$\int_0^1 \frac{d}{dt} L_F(\psi(\cdot, t))\, dt = 0.$$

On account of formula (5) of 5,*1.2* we obtain

$$\frac{d}{dt} L_F(\psi(\cdot, t)) = \sigma \mathscr{J}_{\psi(t)}\zeta(\cdot, \sigma),$$

where $\mathscr{J}_{\psi(t)}$ denotes the Jacobi operator cooresponding to $\psi(t)$. Set

$$L(t) := \mathscr{J}_{\psi(t)}, \qquad \Lambda := L(0) = \mathscr{J}_u, \quad \zeta = \zeta(\cdot, \sigma),$$

keeping σ fixed. Then we arrive at

$$\sigma\Lambda\zeta = \sigma \int_0^1 \{L(0) - L(t)\}\zeta\, dt.$$

This equation suggests that we should determine $\zeta = \zeta(\cdot, \sigma)$ as a solution of the boundary value problem

$$\begin{aligned} \Lambda\zeta &= \left(\int_0^1 \{L(0) - L(t)\}\, dt\right)\zeta \quad \text{in } \Omega, \\ \zeta &= 1 \quad \text{on } \partial\Omega. \end{aligned} \tag{24}$$

The operator-valued function $L(0) - L(t)$ can be written in the form

$$L(0) - L(t) = -\int_0^t \left\{\frac{d}{ds} L(s)\right\} ds.$$

Taking formula (6) of 5,*1.2* into account, it follows that

$$\int_0^1 \{L(0) - L(t)\}\zeta\, dt = -\int_0^1 \int_0^t \left\{\frac{d}{ds} L(s)\right\}\zeta\, ds\, dt$$
$$= \sigma \cdot g(x, \zeta, D\zeta, D^2\zeta, \sigma),$$

where g depends smoothly on σ. Thus we can write (24) as

$$\text{(25)} \qquad \begin{aligned} \Lambda\zeta &= \sigma g(x, \zeta, D\zeta, D^2\zeta, \sigma) \quad \text{in } \Omega, \\ \zeta &= 1 \quad \text{on } \partial\Omega. \end{aligned}$$

The assumption $\lambda_1 > 0$ implies that the accessory integral $\mathcal{Q}(v)$ to u and F satisfies

$$\text{(26)} \qquad \mathcal{Q}(v) \geq \mu \| v \|^2_{H^{1,2}(\Omega)} \quad \text{for all } v \in H_0^{1,2}(\Omega)$$

and some constant $\mu > 0$, cf. 5,*1.3*, Theorem 1, (v).

Here $\| v \|_{H^{1,2}(\Omega)}$ denotes the Sobolev space norm defined by

$$\| v \|^2_{H^{1,2}(\Omega)} := \int_\Omega \{|v|^2 + |Dv|^2\}\, dx,$$

and $H_0^{1,2}(\Omega)$ is the subspace of $H^{1,2}(\Omega)$ consisting of functions with generalized boundary values zero. Note that $H_0^{1,2}(\Omega)$ is the completion of $C_c^\infty(\Omega)$ with respect to the norm of $H^{1,2}(\Omega)$.

Inequality (26) and Schauder theory imply that there is exactly one solution $\zeta_1 \in C^{2,\mu}(\overline{\Omega})$ of the boundary value problem

$$\text{(27)} \qquad \begin{aligned} \Lambda\zeta_1 &= 0 \quad \text{in } \Omega, \\ \zeta_1 &= 1 \quad \text{on } \partial\Omega. \end{aligned}$$

Set

$$\omega := \zeta - \zeta_1$$

and

$$h(x, \omega, D\omega, D^2\omega, \sigma) := g(x, \zeta_1 + \omega, D\zeta_1 + D\omega, D^2\zeta_1 + D^2\omega, \sigma).$$

By virtue of (27) we can write (25) in the equivalent form

$$\text{(28)} \qquad \begin{aligned} \Lambda\omega &= \sigma h(x, \omega, D\omega, D^2\omega, \sigma) \quad \text{in } \Omega, \\ \omega &= 0 \quad \text{on } \partial\Omega. \end{aligned}$$

The theory of elliptic boundary value problems in conjunction with (26) yields that

$$\Lambda : C_0^{2,\mu}(\overline{\Omega}) \to C^{0,\mu}(\overline{\Omega})$$

is a bounded linear bijective mapping of $C_0^{2,\mu}(\overline{\Omega})$ onto $C^{0,\mu}(\overline{\Omega})$. The inverse G of this mapping is called *Green's operator*. Then G furnishes a bounded linear mapping of $C^{0,\mu}(\overline{\Omega})$ onto $C_0^{2,\mu}(\overline{\Omega})$; cf. for instance Gilbarg–Trudinger [1], Chapter 6. The construction of G enables us to transform (28) into an equivalent integral equation

$$\omega = \sigma G h(\cdot, \omega, D\omega, D^2\omega, \sigma), \tag{29}$$

which, for small absolute values of σ, can be solved by Picard iterations, i.e., by means of Banach's fixed point theorem.

By a standard procedure we may now conclude that $\zeta(x, \sigma) = \zeta_1(x) + \omega(x, \sigma)$ and therefore also $\varphi(x, \sigma) = u(x) + \sigma\zeta(x, \sigma)$ is of class $C^2(\Gamma)$ on $\Gamma := \{(x, \sigma): |\sigma| \le \sigma_0, x \in \overline{\Omega}\}$ for some $\sigma_0 > 0$.

Step 2. In order to show that $f : \Gamma \to \mathbb{R}^{n+1}$, defined by $f(x, \sigma) := (x, \varphi(x, \sigma))$, is a field, it remains to show that $\varphi_\sigma(x, \sigma) > 0$ holds on Γ. Clearly it suffices to prove that $\varphi_\sigma(x, 0) = \zeta(x, 0) > 0$ on $\overline{\Omega}$ because we are allowed to replace σ_0 by some smaller positive number. Set

$$\zeta(x) := \zeta(x, 0) = \varphi_\sigma(x, 0).$$

We know that ζ is a Jacobi field for the given extremal u, that is,

$$\Lambda\zeta = 0 \quad \text{in } \Omega, \tag{30}$$

and in addition the boundary condition

$$\zeta = 1 \quad \text{on } \partial\Omega \tag{31}$$

is satisfied.

Because of (26), ζ is minimizing the accessory integral $\mathcal{Q}$ among all functions in $H^{1,2}(\Omega)$ having the boundary values 1 on $\partial\Omega$ (in the weak sense).

Write

$$\zeta = \zeta^+ - \zeta^- \quad \text{with } \zeta^+(x) := \max\{\zeta(x), 0\}, \quad \zeta^-(x) := \max\{-\zeta(x), 0\},$$

and set

$$\Omega^+ := \{x \in \Omega: \zeta(x) > 0\}, \quad \Omega^0 := \{x \in \Omega: \zeta(x) = 0\},$$
$$\Omega^- := \{x \in \Omega: \zeta(x) < 0\}.$$

Since $\zeta \in C^0(\overline{\Omega})$ and $\zeta|_{\partial\Omega} = 1$, we have $\Omega^- \cup \Omega^0 \subset\subset \Omega$. Moreover, Ω^0 is compact, and Ω^-, Ω^+ are open. One can prove that $\zeta^+, \zeta^- \in H^{1,2}(\Omega)$, and that $\zeta^+ = 1$ on $\partial\Omega$, $\zeta^- \in H_0^{1,2}(\Omega^-)$. We also obtain

$$D\zeta(x) = D\zeta^+(x) \quad \text{on } \Omega^+, \qquad D\zeta(x) = -D\zeta^-(x) \quad \text{on } \Omega^-,$$
$$D\zeta(x) = D\zeta^+(x) = D\zeta^-(x) = 0 \text{ a.e. on } \Omega^0.$$

The minimum property of ζ in conjunction with $\zeta - \zeta^+ \in H_0^{1,2}(\Omega)$ implies that

$$\mathcal{Q}(\zeta) \le \mathcal{Q}(\zeta^+),$$

whence

$$\mathcal{Q}_{\Omega^-}(\zeta) \le \mathcal{Q}_{\Omega^-}(\zeta^+) = 0.$$

On the other hand the minimum characterization of the smallest eigenvalue λ_1

of the Jacobi operator $\Lambda = \mathcal{J}_u$ on Ω yields

$$\lambda_1 \int_\Omega |\zeta^-|^2\,dx \le 2\mathcal{Q}(\zeta^-).$$

This inequality does not change if ζ^- is replaced by $-\zeta^-$, and thus we arrive at

$$\lambda_1 \int_{\Omega-} |\zeta|^2\,dx \le 2\mathcal{Q}_{\Omega^-}(\zeta) \le 0,$$

whence it follows that $\zeta(x) \equiv 0$ on Ω^-, i.e., Ω^- is void. Therefore we have $\zeta(x) \ge 0$ on Ω and $\Omega = \Omega^+ \cup \Omega^0$.

Now we want to show that Ω^0 is empty. Otherwise there exists some point x_0 in $\partial\Omega^0$. Since Ω^0 is closed and $\Omega^0 \subset\subset \Omega$, we have $\zeta(x_0) = 0$, and for every $R > 0$ there is some $x \in \Omega \cap B_R(x_0)$ such that $\zeta(x) > 0$ holds. This leads to a contradition, because we shall prove that for some sufficiently small $R > 0$ the function $\zeta(x)$ has to vanish on the ball $B_R(x_0)$.

For this purpose we write the Jacobi operator Λ in the nondivergence form

$$-\Lambda = A^{\alpha\beta}(x)D_\alpha D_\beta + B^\alpha(x)D_\alpha + C(x), \quad A^{\alpha\beta} = A^{\beta\alpha},$$

where the coefficients $A^{\alpha\beta}$, B^α, C are continuous on $\bar{\Omega}$ and satisfy the ellipticity condition

$$A^{\alpha\beta}(x)\eta_\alpha\eta_\beta \ge \lambda|\eta|^2$$

for some $\lambda > 0$ and all $x \in \bar{\Omega}$ and $\eta \in \mathbb{R}^n$.

Choose some $R_0 > 0$ such that $B_{R_0}(x_0) \subset \Omega$, and some $K > 0$ such that

$$|B^1| \le K, \qquad |C| \le K$$

holds on $B_{R_0}(x_0)$. Then we claim that, for a suitable choice of $s > 0$ and $R \in (0, R_0)$, the function

$$p(x) := 1 - Ke^{s(x^1 - x_0^1 + R)}$$

satisfies the inequalities

$$-\Lambda p \le 0 \text{ and } p \ge \tfrac{1}{2}$$

on $B_R(x_0)$.

In fact, if we choose $s > 0$ in such a way that

$$s^2\lambda - (s+1)K \ge 1,$$

then we obtain

$$\begin{aligned}-\Lambda p(x) &= -Ke^{s(x^1 - x_0^1 + R)}[s^2A^{11}(x) + sB^1(x) + C(x)] + C(x)\\ &\le -K[s^2\lambda - (s+1)K] + K \le 0\end{aligned}$$

on $B_R(x_0)$, since $x^1 - x_0^1 + R \ge 0$, and $p(x) \ge 1 - Ke^{2sR} \ge \frac{1}{2}$ if $R < \dfrac{1}{2s}\log\dfrac{1}{2K}$.

For $x \in \overline{B}_R(x_0)$ we introduce the function $w := \zeta/p$, and the differential operator $L := -\frac{1}{p}\Lambda(p\cdot)$. By inserting $\zeta = p\cdot w$ into the equation $\Lambda\zeta = 0$, we infer that

$$Lw = 0 \quad \text{in } B_R(x_0)$$

and

$$L = A^{\alpha\beta}(x)D_\alpha D_\beta + \left\{\frac{2}{p(x)}A^{\alpha\beta}(x)p_{x^\beta}(x) + B^\alpha(x)\right\}D_\alpha + \tilde{C}(x),$$

$$\tilde{C}(x) := -\frac{1}{p(x)}\cdot \Lambda p(x) \le 0 \quad \text{on } B_R(x_0).$$

Moreover, we have $w(x_0) = 0$ and $w(x) \ge 0$ for $x \in \overline{B}_R(x_0)$. That is, the function $w(x)$, $x \in \overline{B}_R(x_0)$, has a nonnegative interior minimum at $x = x_0$. Then Hopf's maximum principle implies that $w(x) \equiv w(x_0) = 0$, and we have arrived at a contradiction (see Gilbarg–Trudinger [1], Section 3.2).

That we finally conclude that Ω^0 is empty, and therefore $\zeta(x) > 0$ holds on $\overline{\Omega}$.

The experienced reader will have no difficulties to fill in the gaps left in the proof of step 1, using Schauder estimates and the difference-quotient technique. The statement about ζ^+ and ζ^- in step 2 can easily be derived from well-known results about Sobolev functions.

We finally mention that similar results can be derived for geometric variational problems whose extremals are surfaces of codimension 1. The situation is much more complicated if both $n > 1$ and $N > 1$, and it seems to be impossible to proceed in the same way. Nevertheless there exist certain field theories for this case leading to *local sufficiency conditions*. For this we refer to Chapter 7.

4. Scholia

1. As Carathéodory noted in his thesis [1], the first satisfactory and completely rigorous solution of a variational problem was given in a paper[4] by Johann Bernoulli. There we find the first glimpse at field theory. In the special case of geodesics Gauss had a clear picture of field theory. His *Disquisitiones* [2] contain geodesic polar coordinates as well as parallel coordinates on a surface, and the concepts of stigmatic geodesic fields and of transversal surfaces were completely clear to Gauss. A. Kneser [3] extended these ideas to general one-dimensional variational problems, stimulated by work of Darboux[5] that became exemplary for differential geometers.

---

[4] Johann Bernoulli, *Remarques sur qu'on a donné jusqu'ici de solutions des problèmes sur les isoperimetres*, Mémoires de l'Acad. Roy. Sci. Paris 1718 (in Latin: Acta Eruditorum Lips. 1718). Cf. Opera Omnia, Bousquet, Lausanne and Geneve 1742, Vol. 2, Nr. CIII, pp. 235–269 and Fig. 6 on p. 270, in particular pp. 266–269. Cf. also: *Die Streitschriften von Jacob und Johann Bernoulli* [1].

[5] G. Darboux [1], Vol. 2, nr. 521.

The integrability conditions

$$\frac{\partial}{\partial x}\bar{F}_p = \frac{\partial}{\partial z}(\bar{F} - \mathscr{P}\cdot\bar{F}_p)$$

for the eikonal S were derived by Beltrami [1] who then concluded the existence of a function S satisfying the Carathéodory equations

$$S_x = \bar{F} - \mathscr{P}\cdot\bar{F}_p, \qquad S_z = \bar{F}_p.$$

Moreover he stated that S has to satisfy an equation of the form

$$\Phi(x, z, S_x, S_z) = 0,$$

and he expressed S in form of Hilbert's independent integral. Also the 1-form

$$\gamma = (F - p^i F_{p^i})\,dx + F_{p^i}\,dz^i$$

can be found in Beltrami's work.[6] The results of Beltrami were rediscovered by Hilbert in 1900.[7] Nevertheless the name *Carathéodory equations* seems to be justified because Carathéodory, who called them *fundamental equations of the calculus of variations* ("Fundamentalgleichungen der Variationsrechnung"; cf. [10], p. 201), realized that they can be used as a key to the calculus of variations, in particular to sufficient conditions. Therefore Boerner [5], [6] has called Carathéodory's approach a *royal road to the calculus of variations.*

In Carathéodory's treatise [10] this road is somewhat hidden because Carathéodory had discovered it while reading the galley proofs of his book, and only in the last minute he managed to include it into [10].

Undoubtedly Carathéodory's royal road is nowadays the quickest and most elegant approach to *sufficient conditions.* In addition, it can easily be carried over to multiple integrals. Thus it may be suprising to learn that Carathéodory was led to his approach by Johann Bernoulli's paper from 1718, the essence of which he had already described in the appendix to his thesis (1904); cf. Carathéodory [16], Vol. 1, pp. 69–78. As we noted before, Bernoulli's paper was the first where an extremal was shown to furnish a minimum. In Vol. 2 of [16], pp. 97–98, Carathéodory wrote:

"*In the ancient oriental courts there was often besides the official history written by an appointed historian, a secret history that was not less thrilling and interesting than the former. Something of this kind can be traced also in the history of the Calculus of Variations.*

It is a known fact that the whole of the work on Calculus of Variations during the eighteenth century deals only with necessary conditions for the existence of a minimum and that most of the methods employed during that time do not allow even to separate the cases in which the solution yields a maximum from those in which a minimum is attained. According to general belief Gauss[8] *in* 1823 *was the first to give a method of calculation for the problem of geodesics which was equivalent to the sufficient conditions emphasized fifty years later by Weierstrass for more general cases.*

It is therefore important to know that the very first solution which John Bernoulli found for the problem of the quickest descent contains a demonstration of the fact that the minimum is really attained for the cycloid and it is more important still to learn from a letter which Bernoulli addressed to Basnage, in 1697, *that he himself was thoroughly aware of the advantages of his method*[9] (...). *But just as in the case of the problem of geodesics Bernoulli did not publish this most interesting result until* 1718 *and he did this on the very last pages of a rather tedious tract.*[10]

---

[6] For a detailed survey of some historical roots of field theory we refer to Bolza [1], in particular pp. 52–70.

[7] See Hilbert [1], Problem 23, and [5]. Cf. also Bolza [3], p. 107, and [1], p. 62.

[8] *Disquisitiones generales circa superficies curvas*, Section 15; Werke [1], Vol. 4, pp. 239–241.

[9] Joh. Bernoulli, Opera omnia, tom. I, no. XXXVIII, p. 194.

[10] Loc. cit., tom.. II, no. CIII, pp. 235–269; see particularly p. 267.

Thus this method of Bernoulli, in which something of the field theory of Weierstrass appears for the first time (...) did not attract the attention even of his contemporaries and remained completely ignored for nearly two hundred years.

These two pages of Bernoulli, which I discovered by chance more than thirty years ago [Thesis, 1904], *have had a very decisive influence on the work which I myself did in the Calculus of Variations. I succeeded gradually in simplifying the exposition of this theory and came finally to a point where I found to my astonishment that the method to which I had been directed through long and hard work was contained, at least in principle, in the "Traité de la lumière" of Christian Huygens.*"

2. The importance of "*Mayer bundles*" and "*Mayer fields*" for the calculus of variations was discovered by A. Mayer [10] in 1905. The name was essentially coined by Bolza who introduced the notation *Mayersche Extremalenschar*; cf. Bolza [3], p. 643. In fact, Mayer invented this tool to treat Lagrange problems for one-dimensional variational integrals using Hilbert's *invariant integral.* This integral was already known to Beltrami (see nr. 1). Hilbert rediscovered it about 1900 and emphasized its importance for the calculus of variations by posing Problem 23 of his celebrated list of problems (cf. Hilbert [1, 3]). This problem is less specific than the other 22 questions and, roughly speaking, asks for the further development of the calculus of variations by his *independency theorem* ("*Unabhängigkeitssatz*"). Mayer immediately realized the importance of Hilbert's discovery, and in two profound papers he developed Hilbert's program for Lagrange problems (see Mayer [9, 10]). Shortly thereafter, D. Egorov [1] extended Hilbert's independency theorem to Mayer problems. Hilbert himself returned to the problem once again in his paper [5].

Carathéodory's approach can be viewed as the final and highly elegant version of the *theory of sufficient conditions* created by Weierstrass, A. Kneser, Hilbert and Mayer. It took Carathéodory about thirty years until he had reached the form presented in the textbook [10] and in the survey [11] on geometrical optics. Boerner popularized Carathéodory's ideas in his two survey papers [5] and [6].

Let us note that Marston Morse's monograph [3] presents a masterly exposition of field theory.

A critical appreciation of Mayer's contributions to the calculus of variations was given by Klötzler [8]. A rich source of historical remarks on field theory is Bolza's treatise [3].

The first presentation of Weierstrass' field theory was given in A. Kneser's textbook [3] which appeared in 1900. According to Carathéodory[11] this treatise has played a role which is as important as rare: "*One can say without exaggeration that almost all researchers who since* 1900 *have worked on the calculus of variations were more or less influenced by Kneser since all textbooks which appeared in the mean time were dependent on this first presentation of the modern calculus of variations.*"

In the next decade Bolza and Hadamard published their extensive monographs. Hadamard emphasized the functional analytic point of view in the calculus of variations, and he coined the notion of a *functional.* When in 1927 Weierstrass' *Vorlesungen über Variationsrechnung* were eventually published as Vol. 7 of his collected papers, they offered nothing new to the mathematical public, contrary to previous rumours, and therefore Carathéodory ([16], Vol. 5, p. 343) wrote:

"*By the present publication of Weierstrass's Calculus of Variations a fairy tale has ceased to exist which almost nobody had anymore believed in the last years but which had been wide-spread in former days, particularly abroad: According to that, mysterious discoveries of Weierstrass in the field of the calculus of variations – I don't know which – were hidden in his papers and withheld from the mathematical public. So the publication of the seventh volume of Weierstrass's works comes too late to still have a perciptible influence on the calculus of variations.*"[12]

---

[11] See Carathéodory [16], Vol. 5, p. 337.

[12] "*Durch die jetzige Veröffentlichung der Weierstraßschen Variationsrechnung hat also ein Märchen ausgelebt, an welches in den letzten Jahren fast niemand mehr geglaubt hat, das aber früher, besonders im Auslande, ziemlich verbreitet war: danach sollten – ich weiß nicht welche – geheimnisvolle Entdekkungen von Weierstraß auf dem Gebiete der Variationsrechnung in seinen Papieren noch verborgen und*

In 1935 Carathéodory published his own treatise [10] which not only gave a masterly presentation of the classical one-dimensional calculus of variations but also strongly influenced the future development of optimization and control theory.

We finally mention that the state of the art about 1900 is briefly described in the two surveys of A. Kneser and of E. Zermelo and H. Hahn, published in the *Encyklopädie der mathematischen Wissenschaften*, Vol. 2.1 (IIA8 and IIA8a).

An account of the work of Weierstrass, Kneser and Hilbert can be found in Chapters 5 and 7 of Goldstine's historical study [1].

A rich source of historical information are Vols. 1 and 2 of Carathéodory's Schriften [16] and in particular Vol. 5, containing his reviews of the books of Bolza, Hancock, Hadamard, A. Kneser, Forsyth and Weierstrass on the calculus of variations.

3. The usefulness of differential forms for the calculus of variations became clear by the work of E. Cartan and of the Belgian mathematicians De Donder and Lepage. Later on, Boerner [4] showed how the one-dimensional variational calculus can be formulated in terms of differential forms, and in [3] he developed the corresponding formulas for multiple integrals. The key to Boerner's approach is the one-form

$$\gamma_F = (F - p \cdot F_p)\,dx + F_p \cdot dz = F\,dx + F_{p^i}\theta^i$$

that we have denoted as *Beltrami form.* We shall see later that its Legendre transform is the *Cartan form*

$$\kappa_H = -H\,dx + y_i\,dz^i,$$

where $y_i = F_{p^i}$ are the canonical momenta and H is the Hamiltonian corresponding to F. The Cartan form is a central object of the Hamilton–Jacobi theory to be discussed in Chapters 7 and 9. A brief modern discussion of Carathéodory's ideas for multiple integrals can be found in Chapter 36 of the second edition of Hermann [1]. A more recent exposition of the calculus of variations by Griffith [1] is exclusively based on the use of differential form. This way a formally very elegant treatment of Lagrangian subsidiary conditions is achieved.

4. In the special case of geodesics the transversality theorem was already discovered by Gauss,[13] and modern differential geometers call it *Gauss's lemma.* This lemma states that geodesics emanating from a fixed point P of a Riemannian manifold are perpendicularly intersecting any geodesic sphere centered at P, i.e. to any hypersurface consisting of points at a fixed distance from P.

The contributions of Gauss to the calculus of variations are described in a survey by Bolza [1] which is published in Vol. 10 of Gauss's Werke, cf. in particular Part II, pp. 52–70. There Bolza points out the contributions of Malus, Hamilton, Gauss, Minding, Beltrami, Darboux, Thomson and Tait, Kneser and Hilbert to the theory of sufficient conditions and to the transversality theorem.

5. The so-called *Jacobi envelope theorem* was first formulated by Darboux for geodesics on a surface (see [1], Vol. 2, p. 88) and in general by Zermelo in his thesis [1]. Generalizations of this theorem were found by A. Kneser [3], p. 93; (second edition p. 116); [2], p. 27. We refer also to the account given by Goldstine [1], Chapter 7.

6. The notion of an *optimal field* was apparently introduced by Hestenes.

7. The terminology of geometrical optics is well suited for the calculus of variations. Later we shall justify the usage of the notions of *ray*, *wave front*, *eikonal* (or *wave function*). The notion of an

---

dem mathematischen Publikum vorenthalten sein. Die Publikation des siebenten Bandes der Werke von Weierstraß kommt also zu spät, um einen erkennbaren Einfluß auf die Variationsrechnung noch zu veranlassen."

[13] Disquisitiones generales circa superficies curvas, Werke [1], Vol. 4, Section 15, pp. 239–241.

eikonal was introduced by the astronomer Bruns [2] in 1895 as a new tool in the ray optics. Felix Klein immediately pointed out the close connection of this tool with the Hamilton–Jacobi theory, in particular with Hamilton's ray optics; cf. Klein [1], Vol. 2, pp. 601–602, 603–606, 607–613.

Hamilton's collected papers have appeared in 1931; cf. [1]. We also refer to Prange [1], and to F. Klein [3], Vol. 1, pp. 194–202.

Actually, Brun's eikonal is somewhat different from our optical distance function S, see Carathéodory [11], Sections 43–55, and also Chapter 9 of our treatise. However, since a short name for the function S is missing in the literature on the calculus of variations, the term "eikonal" might be acceptable. A synonym would be *distance function*, *wave function* or *wave front function*. In mechanics the function S has the meaning of an *action*. Therefore, in the context of mechanics where the variable x is to be interpreted as a *time parameter*, the reader might prefer "action" instead of "eikonal" as terminology. In optimization theory the eikonal is often called *value function*.

The close connection between mechanics, optics and the calculus of variations was discovered very early in the history of mathematics. This can be seen in the work of Fermat, Johann Bernoulli, Newton, Maupertuis, Euler, Lagrange and many others. Finally Hamilton discovered that point mechanics, geometrical optics and the one-dimensional calculus of variations furnish three different pictures of essentially the same mathematical topic. A very readable brief survey on the history of geometrical optics can be found in the introduction to Carathéodory's notes [11].

8. The generalization of Jacobi's theory of conjugate points to the case $N > 1$ is due to Clebsch, von Escherich and, in particular, to Mayer. We refer to Chapter 6 of Goldstine's historical study [1] for a detailed account.

9. Euler did not pay much attention to the question whether or not one can connect two given points by one or more extremals. As Carathéodory pointed out there are several examples among the problems treated by Euler where one cannot draw any extremal from one point to another if these two points are clumsily chosen. Nevertheless Euler repeatedly claimed that there always exists exactly one solution of a given n-th order differential equation fitting n arbitrarily prescribed points (cf. Carathéodory [16], Vol. 5, p. 119).

The question of solving the boundary problem for nonlinear differential equations is by no means trivial. For Euler equations the boundary problem is usually treated by the so-called *direct methods*; other methods use topological concepts.

10. The results of Section 3 were proved by Lichtenstein [1] in 1917. He treated the case of a real analytic, elliptic Lagrangian F for $n = 2$, generalizing Schwarz's celebrated field construction for minimial surfaces mentioned in 5,*3*, Nr. 1. By means of elliptic regularity theory Lichtenstein's result can be carried over to smooth elliptic Lagrangians and to $n \geq 2$; this has been sketched in Morrey's treatise [1], pp. 12–15. Section 3 is an elaborated version of Morrey's presentation. More generally Lichtenstein's method can be extended to n-dimensional surfaces embedded into $\mathbb{R}^{n+1}$ which are not necessarily graphs; cf. N. Smale [1]. Further applications of Lichtenstein's method were given by J.C.C. Nitsche (cf. [1], pp. 99, 702) and X. Li-Jost [1], [2].

Supplement. Some Facts from Differential Geometry and Analysis

This supplement to Vol. 1 is not meant to be a systematic introduction to analysis, linear algebra, and differential geometry, but its only intention is to fix some terminology and notation used in this book and to remind the reader of some basic facts from analysis and geometry.

In Section 1 we recall the meaning of vectors and covectors and their notations by means of upper and lower indices. Also, we fix some terminology on sets in Euclidean space.

In Section 2 we define the basic classes of differentiable mappings of sets in $\mathbb{R}^n$ as well as norms on these classes.

Section 3 is closely related to the first and recalls the notions of vector fields and covector fields on domain G of $\mathbb{R}^N$ as well as of the tangent bundle TG and the cotangent bundle T^*G. Basically this section is meant to fix our terminology and to link the notation of old-fashioned tensor analysis with the modern one.

In Section 4 we outline the calculus of differential forms on domains of $\mathbb{R}^N$, in particular the pull-back operation and the exterior differential, which are indispensable if one wants to keep some computations of the calculus of variations within reasonable limits. We have avoided the exterior calculus in the first five chapters to keep this introductory part as elementary as possible; however, the section on null Lagrangians in Chapter 1 would certainly have profited if we had already there used the exterior calculus. From Chapter 6 on we frequently use the calculus of forms, since it not only simplifies computations, but also puts many things in the right perspective.

In the last two sections we recollect some terminology, notations, and facts on curves and surfaces in $\mathbb{R}^N$; in particular those concerned with curvature properties.

1. Euclidean Spaces

By $\mathbb{R}$ and $\mathbb{C}$ we denote the fields of real and complex numbers respectively.

Let E be an N-dimensional vector space over $\mathbb{R}$ and $\{e_1, \ldots, e_N\}$ a basis of E. Then, for every $x \in E$, there are uniquely determined numbers $x^1, \ldots. x^N \in \mathbb{R}$, the coordinates (or components) of x with respect to $\{e_1, \ldots, e_N\}$, such that

(1) $$x = x^1 e_1 + \cdots + x^N e_N .$$

Using *Einstein's summation convention*[1] we can write (1) as

(1′) $$x = x^i e_i .$$

If $\{f_1, \ldots, f_N\}$ is another base of E we can analogously represent x as

(1″) $$x = y^i f_i .$$

In particular there are numbers $a_i^k \in \mathbb{R}$ such that

(2) $$f_i = a_i^k e_k .$$

Then we obtain from (1′) and (1″) that

$$x^k e_k = a_i^k y^i e_k ,$$

whence

(3) $$x^k = a_i^k y^i , \quad 1 \le k \le N .$$

Let us introduce the $N \times N$-matrix $A = (a_i^k)$ where i is the row index and k the column index of A. Correspondingly we define the row matrices X and Y by

(4) $$X := (x^1, \ldots, x^N), \qquad Y := (y^1, \ldots, y^N).$$

Then (3) can be written as

(5) $$X = YA$$

or, correspondingly, as

(5′) $$X^T = A^T Y^T ,$$

where A^T is the transpose of A, and similarly

(6) $$X^T = \begin{bmatrix} x^1 \\ \vdots \\ x^N \end{bmatrix} , \qquad Y^T = \begin{bmatrix} y^1 \\ \vdots \\ y^N \end{bmatrix}$$

are the transposed matrices of X and Y respectively. Applying (5) successively to $x = e_1, e_2, \ldots, e_N$ we obtain

$$E = CA$$

for a suitable $N \times N$-matrix C and the unit matrix $E = (\delta_i^k)$ where $\delta_i^k = 0$ or 1 if $i \ne k$ or $i = k$ respectively; hence $1 = \det C \cdot \det A$, and therefore A is invertible. Thus we can write (5′) as

(7) $$Y^T = A^{T-1} X^T .$$

Let E^* be the dual space of the N-dimensional linear space E. By defin-

[1] All indices $i, k, \ldots$ that appear twice in an expression are to be summed from 1 to N.

tion E^* consists of all linear forms $\omega : E \to \mathbb{R}^N$; the space E^* is again an N-dimensional linear space. If $\{e_1, \ldots, e_N\}$ is a base of E, then the set of linear forms $\{e^1, \ldots, e^N\}$ defined by

$$e^i(e_k) = \delta^i_k \tag{8}$$

is a base of E^*, called the dual base of E^* (i.e. $\{e_1, \ldots, e_N\}$ is dual to $\{e^1, \ldots, e^N\}$). Any $\omega \in E^*$ can uniquely be represented as

$$\omega = \omega_i e^i, \tag{9}$$

with $\omega_1, \ldots, \omega_N \in \mathbb{R}^N$. Let $\{f_1, \ldots, f_N\}$ be another base of E connected with $\{e_1, \ldots, e_N\}$ by (2), and let $\{f^1, \ldots, f^N\}$ be the dual base of E^* with respect to $\{f_1, \ldots, f_N\}$, i.e.

$$f^i(f_k) = \delta^i_k. \tag{10}$$

Then we obtain

$$e^i(f_k) = e^i(a^l_k e_l) = a^l_k e^i(e_l) = a^l_k \delta^i_l = a^i_k,$$

that is,

$$e^i(f_k) = a^i_k. \tag{11}$$

On the other hand we can express ω as

$$\omega = \varphi_i f^i, \tag{12}$$

with $\varphi_1, \ldots, \varphi_N \in \mathbb{R}^N$. Evaluating ω at f_k we obtain

$$\omega(f_k) = \varphi_i f^i(f_k) = \omega_i e^i(f_k).$$

By (10) and (11) it follows that

$$\varphi_k = a^i_k \omega_i. \tag{13}$$

Forming the columns

$$\Omega := \begin{bmatrix} \omega_1 \\ \vdots \\ \omega_N \end{bmatrix}, \qquad \Phi := \begin{bmatrix} \varphi_1 \\ \vdots \\ \varphi_N \end{bmatrix}, \tag{14}$$

we can write

$$\Phi = A\Omega. \tag{15}$$

Let us compare the formulae (7) and (15) with (2) which formally could be expressed as

$$f = Ae, \tag{16}$$

where

$$e := \begin{bmatrix} e_1 \\ \vdots \\ e_N \end{bmatrix}, \qquad f := \begin{bmatrix} f_1 \\ \vdots \\ f_N \end{bmatrix}.$$

We see that the columns Ω and Φ are related in the same way as e and f, whereas X^T and Y^T are related in an inverse manner. Accordingly one says that the coordinate vectors Ω and Φ of a linear form Ω with respect to the old base $\{e^1, \dots, e^N\}$ and the new base $\{f^1, \dots, f^N\}$ transform *covariantly* (with respect to the base vectors e_i and f_k), while the coordinate vectors X^T and Y^T of a vector x with respect to the old base $\{e^1, \dots, e^N\}$ and the new $\{f^1, \dots, f^N\}$ transform *contragrediently*. Therefore one denotes the elements of E^* as *covectors*.

The above considerations easily show that all linear spaces of the same dimension N are linearly isomorphic, and therefore they can be identified with the model space of $\mathbb{R}^N$ consisting of the ordered N-tuples $x = (x^1, x^2, \dots, x^N)$ which can be written as $x = x^i e_i$ with respect to the canonical base

$$(17) \qquad e_1 = (1, 0, \dots, 0), \quad e_2 = (0, 1, 0, \dots, 0), \quad \dots, \quad e_N = (0, \dots, 0, 1),$$

the dual base $\{e^1, \dots, e^N\}$ of which is given by the relations

$$(18) \qquad e^1(x) = x^1, e^2(x) = x^2, \dots, e^N(x) = x^N \quad \text{for any } x \in \mathbb{R}^N.$$

Any N-tuple $(\omega_1, \dots, \omega_N)$ of real numbers ω_i defines an N-form ω by

$$(19) \qquad \omega(x) = \omega_1 x^1 + \cdots + \omega_N x^N = \omega_i x^i$$

for any $x = x^i e_i = (x^1, \dots, x^N) \in \mathbb{R}^N$. Given an arbitrary N-tuple of real numbers we may interpret it either as the set of components of a vector $x \in \mathbb{R}^N$ with respect to the canonical base (17) or as the set of components of a covector $\omega \in \mathbb{R}^{N*}$ with respect to the dual canonical base (18). In this sense we can identify vectors with covectors and $\mathbb{R}^N$ with $\mathbb{R}^{N*}$ if the base of E and E^* remain fixed. However, if we change bases it is essential to know whether an N-tuple of real numbers are the components of a vector or a covector as those transform contragrediently to each other. The components x^i of a vector $x \in \mathbb{R}^N$ transform according to (3) or (7), while the components ω_i of a covector $x \in \mathbb{R}^{N*}$ transform by (13) or (15). In the old literature an N-tuple of numbers x^i transforming by the rule (7) is called a *contravariant vector* while the N-tuples of numbers ω_i transforming by (15) are said to be *covariant vectors*; but we shall instead speak of vectors and covectors. The components x^i of vectors x are characterized by *raised indices i*, while the components ω_i of covectors ω carry *lowered indicies i*. It is quite useful for computational purposes to keep this convention in mind.

From the rigorous point of view it would be quite preferable to work in "abstract" N-dimensional spaces E and their duals E^* instead of $\mathbb{R}^N$ and its dual $\mathbb{R}^{N*}$ which is even identified with $\mathbb{R}^N$. On the other hand the identifications $E = \mathbb{R}^N = \mathbb{R}^{N*}$ are quite convenient and not dangerous as long as one takes suitable precautions. However we also use some sloppy notation which may occasionally irritate the reader. For example we usually identify the elements x with the rows X and the columns X^T, while in matrix computations X and X^T are different objects. A quadratic form $a_{ik}x^i x^k$ is written as xAx or Axx while the matrix representation might look like $x^T \cdot Ax$ if we interpret x as a column. Here the reader usually has to guess what is meant, but mostly we write at least once the formulas in coordinates as, for instance, $a_{ik}x^i x^k$ in the present case, from where one can derive the meaning of the other expressions appearing in the same context. This notational ambivalence simplifies formulas and usually helps to convey their meaning. However, sometimes more precision is needed; then we either use coordinate notation or matrix calculus (as for instance in many parts of Chapter 9).

Let us now endow $\mathbb{R}^N$ with the *Euclidean scalar product* (*inner product*) $\langle \cdot, \cdot \rangle : \mathbb{R}^N \times \mathbb{R}^N \to \mathbb{R}$, which for $x = (x^1, \ldots, x^N)$, $y = (y^1, \ldots, y^N) \in \mathbb{R}^N$ is defined by

$$\langle x, y \rangle := \delta_{ij} x^i y^j = x^1 y^1 + \cdots + x^N y^N, \tag{20}$$

where δ_{ij} is the usual Kronecker symbol,

$$\delta_{ij} = \begin{cases} 0 & \text{for } i \neq j, \\ 1 & \text{for } i = j. \end{cases}$$

Instead of (20) we mostly write $x \cdot y$,

$$x \cdot y = \delta_{ij} x^i y^j, \tag{20'}$$

which in matrix notation would read $x^T \cdot y$ if x and y are to be interpreted as columns. We have $x \cdot x \geq 0$, $x \cdot x = 0$ if and only if $x = 0$, $x \cdot y = y \cdot x$, $(\lambda x + \mu y) \cdot z = \lambda x \cdot z + \mu y \cdot z$ for x, y, $z, \in \mathbb{R}^N$ and λ, $\mu \in \mathbb{R}$, and $x \cdot y$ satisfies Schwarz's inequality

$$|x \cdot y|^2 \leq (x \cdot x)(y \cdot y).$$

Then the expression

$$|x| := \sqrt{x \cdot x} = \sqrt{\delta_{ik} x^i y^j} \tag{21}$$

is a norm, i.e. $|x| \geq 0$, $|x| = 0$ if and only if $x = 0$, $|\lambda x| = |\lambda| \, |x|$ for any $\lambda \in \mathbb{R}$ and $x \in \mathbb{R}^N$, and we have the triangle inequality

$$|x + y| \leq |x| + |y| \quad \text{for any } x, y \in \mathbb{R}^N.$$

We call $|\cdot| : \mathbb{R}^N \to \mathbb{R}$ the *Euclidean norm* on $\mathbb{R}^N$, and $\mathbb{R}^N$ endowed with the distance function $d : \mathbb{R}^N \times \mathbb{R}^N \to \mathbb{R}$ defined by

$$d(x, y) := |x - y| = \{|x^1 - y^1|^2 + \cdots + |x^N - y^N|^2\}^{1/2} \tag{22}$$

is said to be the *N-dimensional Euclidean space*, and the elements of $\mathbb{R}^N$ are interpreted either as *vectors* of the linear space $\mathbb{R}^N$ or as *points* of the matrix space $\{\mathbb{R}^N, d\}$, which is clearly *complete*.

Open sets in $\mathbb{R}^N$ are usually denoted by the symbol Ω. Arc-wise connected open sets of $\mathbb{R}^N$ are said to be *domains in* $\mathbb{R}^N$; they are usually denoted by G. Open balls in $\mathbb{R}^N$ of radius $R > 0$ centered at x_0 are denoted by $B_R(x_0)$ or $B(x_0, R)$, i.e.

$$B_R(x_0) = \{x \in \mathbb{R}^N : |x - x_0| < R\},$$

and $\mathscr{K}_R(x_0)$ often denotes the closed ball $\overline{B_R(x_0)}$ where $\overline{M}$ is the closure of a set M in $\mathbb{R}^N$, $\mathring{M} = \operatorname{int} M$ its interior, and ∂M its boundary. An open set Ω of $\mathbb{R}^N$ is said to be *nonsingular* if $\Omega = \operatorname{int} \overline{M}$.

Let $A, A' \subset \mathbb{R}^N$. Then A' is said to be compactly contained in A if $\bar{A}' \subset A$ and $\bar{A}'$ is compact. In particular $\Omega' \subset\subset \Omega$ implies that $\operatorname{dist}(\Omega', \partial\Omega) > 0$.

We shall not recall the definition of p, q-tensors

(23) $$t = (t^{i_1 \dots i_p}_{k_1 \dots k_q})$$

and the properties of the tensor algebra; for this we refer to any reasonable treatise of linear algebra. We just remind the reader that the components of such a tensor transform like

(24) $$s^{j_1 \dots j_p}_{l_1 \dots l_q} = t^{i_1 \dots i_p}_{k_1 \dots k_q} a^{k_1}_{l_1} \dots a^{k_q}_{l_q} b^{j_1}_{i_1} \dots b^{j_p}_{i_p},$$

where $A = (a^k_l)$, $A^{-1} = B = (b^j_i)$, that is,

(25) $$a^k_i b^l_k = \delta^l_i \quad \text{and} \quad b^k_i a^l_k = \delta^l_i .$$

2. Some Function Classes

Let A be a nonempty subset of $\mathbb{R}^N$ and $u: A \to \mathbb{R}^N$ a mapping from A into $\mathbb{R}^N$. Then

(1) $$u(A) := \{u(x) \colon x \in A\}$$

is called the *range of* u, and A is said to be the *set of definition* of u, or the *domain* (*of definition*) of u, even if A is not a "domain".

The set of mapping $u: A \to \mathbb{R}^N$ having certain properties $\mathscr{P}$ is denoted by $P(A, \mathbb{R}^N)$. If $N = 1$ we simply write $P(A)$, i.e. $P(A) = P(A, \mathbb{R})$.

A particularly important class is the class of continuous mapping $u: A \to \mathbb{R}^N$ which is denoted by $C^0(A, \mathbb{R}^N)$, and $C^0(A)$ is the class of continuous functions $u: A \to \mathbb{R}$, i.e. $C^0(A) = C^0(A, \mathbb{R})$.

If k is an integer with $k \ge 1$ and Ω a nonempty open set in $\mathbb{R}^N$, then $C^k(\Omega, \mathbb{R}^N)$ denotes the class of functions $u: \Omega \to \mathbb{R}^N$ which are continuously differentiable in Ω up to the order k, and $C^\infty(\Omega, \mathbb{R}^N) := \bigcap_{k \ge 1} C^k(\Omega, \mathbb{R}^N)$ is the class of mappings $u: \Omega \to \mathbb{R}^N$ which are infinitely often continuous differentiable. Furthermore, suppose that $\Omega \subset A \subset \bar{\Omega}$. We say that $u: A \to \mathbb{R}^N$ *is of class* C^k, or: $u \in C^k(A, \mathbb{R}^N)$, $1 \le k \le \infty$, if $v := u|_\Omega \in C^k(\Omega, \mathbb{R}^N)$ and if $v, Dv, D^2v, \dots, D^s v, \dots, 0 \le s \le k$, can be extended from Ω to continuous mappings of A. Here $D^s v = (D^\alpha v)$ is the s-th gradient of v consisting of all partial derivatives $D^\alpha v$ of order s, i.e. of all derivatives

(2) $$D^\alpha v = D_1^{\alpha_1} D_2^{\alpha_2} \dots D_n^{\alpha_n}, \qquad D_\beta = \frac{\partial}{\partial x^\beta}, \quad \beta = 1, \dots, n,$$

$\alpha = (\alpha_1, \alpha_2, \dots, \alpha_n)$, $|\alpha| := \alpha_1 + \alpha_2 + \dots + \alpha_n = s$, $D^0_\beta = 1$. Another and often more convenient definition is: The map $u: A \to \mathbb{R}^N$ is of class C^k if there is an open set Ω_0 containing A and a mapping $w \in C^k(A, \mathbb{R}^N)$ such that $u = w|_A$. In many cases these two definitions coincide, for instance, if $A = \bar{\Omega}$ and $\partial\Omega$ is sufficiently smooth (i.e. a sufficiently smooth manifold of dimension $n-1$). This condition in particular implies that Ω is nonsingular, which is a reasonable assumption if one wants to extend $u:\bar{\Omega}\to\mathbb{R}^N$ to a smooth function $w:\Omega_0\to\mathbb{R}^N$ on an open set Ω_0 containing $\bar{\Omega}$.

The support of a continuous mapping $u:\mathbb{R}^n\to\mathbb{R}^N$, denoted by supp u, is the closure of the set $\{x\in\mathbb{R}^n: u(x)\neq 0\}$. Then $C_c^k(\Omega,\mathbb{R}^N)$ denotes the set of functions $u:\Omega\to\mathbb{R}^N$ which can be extended to functions $w\in C^k(\mathbb{R}^n,\mathbb{R}^N)$ such that supp $w\subset\subset\Omega$. Equivalently we can say that $u\in C_c^k(\Omega,\mathbb{R}^N)$ if $u\in C^k(\Omega,\mathbb{R}^N)$ and supp $u\subset\subset\Omega$ where the support of u is now defined as the closure in Ω of the set $\{x\in\Omega: u(x)\neq 0)\}$. In particular $C_c^\infty(\Omega)$ is the class of infinitely often continuous differentiable functions $u:\Omega\to\mathbb{R}$ with compact support in Ω.

It is well known and easy to verify that all these classes are linear vector spaces over $\mathbb{R}$ and even algebras if $N=1$, where Leibniz's rule holds true: If $u, v\in C^k(\Omega)$ and $\alpha=(\alpha_1,\ldots,\alpha_n)$ is a multi-index with $|\alpha|\le k$, then

(3) $$D^\alpha(uv)=\sum_{\substack{\beta,\gamma\\ \beta+\gamma=\alpha}}\frac{\alpha!}{\beta!\gamma!}D^\beta u D^\gamma v,$$

where $\alpha!=\alpha_1!\,\alpha_2!\ldots\alpha_n!$, and β, γ are multi-indices with $\alpha=\beta+\gamma$, i.e. $\alpha_1=\beta_1+\gamma_1,\ldots,\alpha_n=\beta_n+\gamma_n$.

Let K be a compact subset of $\mathbb{R}^N$. Then

(4) $$\|u\|_{C^0(K)}:=\max_K|u(x)|\quad\text{for } u\in C^0(K,\mathbb{R}^N)$$

defines a norm on $C^0(K,\mathbb{R}^N)$, and $C^0(K,\mathbb{R}^N)$ equipped with this norm forms a Banach space. For the sake of brevity we often write

(4′) $$\|u\|_{0,K}:=\max_K|u(x)|\quad\text{for } u\in C^0(K,\mathbb{R}^N).$$

Similarly the expression

(5) $$\|u\|_{C^k(\bar{\Omega})}:=\sum_{|\alpha|\le k}\sup_\Omega|D^\alpha u(x)|$$

defines a norm on $C^k(\bar{\Omega},\mathbb{R}^N)$ if Ω is a bounded open subset of $\mathbb{R}^N$, and then $C^k(\bar{\Omega},\mathbb{R}^N)$ equipped with this norm is a Banach space.

Occasionally we also use the class $C_0^1(\bar{\Omega},\mathbb{R}^N)$ which is defined by

(6) $$C_0^1(\bar{\Omega},\mathbb{R}^N):=\{u\in C^1(\bar{\Omega},\mathbb{R}^N): u|_{\partial\Omega}=0\}.$$

Clearly $C_c^1(\Omega,\mathbb{R}^N)$ is a proper subset of $C_0^1(\bar{\Omega},\mathbb{R}^N)$. Note that, differing from our notation, some authors denote the space $C_c^\infty(\Omega,\mathbb{R}^N)$ by $C_0^\infty(\Omega,\mathbb{R}^N)$.

Let K be a compact subset of $\mathbb{R}^N$, and suppose $u:K\to\mathbb{R}^N$ is a mapping with the following property: There is a number α with $0<\alpha\le 1$ and a constant $c>0$ such that

(7) $$|u(x)-u(y)|\le c|x-y|^\alpha\quad\text{for all } x, y\in K.$$

Obviously such a function u is continuous on K, and any function $u:K\to\mathbb{R}^N$ of this kind is said to be *Hölder continuous* or *Lipschitz continuous* if $0<\alpha<1$

or $\alpha = 1$ respectively, and condition (7) is called a *Hölder condition* on K or a *Lipschitz condition* if $0 < \alpha < 1$ or $\alpha = 1$ respectively. By $Lip(K, \mathbb{R}^N)$ we denote the class of *Lipschitz (continuous) functions* $u : K \to \mathbb{R}^N$ whose *Lipschitz constant* is defined by

(8) $$Lip_K(u) := \sup\left\{\frac{|u(x) - u(y)|}{|x - y|} : x, y \in K, x \neq y\right\}.$$

Note that $Lip(K, \mathbb{R}^N)$ is a Banach space with respect to the norm

(9) $$\|u\|_{Lip(K)} := \|u\|_{C^0(K)} + Lip_K(u).$$

Furthermore, $C^{0,\alpha}(K, \mathbb{R}^N)$ denotes the class of Hölder continuous functions $u : K \to \mathbb{R}^N$ with the same *Hölder exponent* $\alpha \in (0, 1)$ in (7). On this function class we can define the *Hölder seminorm* $[u]_{\alpha,K}$ by

(10) $$[u]_{\alpha,K} := \sup\left\{\frac{|u(x) - u(y)|}{|x - y|^\alpha} : x, y \in K, x \neq y\right\}$$

and the Hölder norm

(11) $$\|u\|_{C^{0,\alpha}(K)} := \|u\|_{C^0(K)} + [u]_{\alpha,K}.$$

Again, $C^{0,\alpha}(K, \mathbb{R}^N)$ equipped with the norm (11) is a Banach space.

Furthermore let Ω be a bounded open set in $\mathbb{R}^N$, and denote by $C^{k,\alpha}(\overline{\Omega}, \mathbb{R}^N)$, $k = 1, 2, 3, \ldots, 0 < \alpha < 1$ the class of functions $u \in C^k(\overline{\Omega}, \mathbb{R}^N)$ whose derivatives of order k are Hölder continuous on $\overline{\Omega}$ with the exponent α. Then

(12) $$\|u\|_{C^{k,\alpha}(\overline{\Omega})} := \|u\|_{C^k(\overline{\Omega})} + \sum_{|\beta|=k} [D^\beta u]_{\alpha,\overline{\Omega}}$$

defines a norm on $C^{k,\alpha}(\overline{\Omega}, \mathbb{R}^N)$, and this class equipped with the norm (12) is a Banach space. Finally $C^{k,\alpha}(\Omega, \mathbb{R}^N)$ denotes the class of functions $u : \Omega \to \mathbb{R}^N$ such that $u|_{\Omega'} \in C^{k,\alpha}(\overline{\Omega}', \mathbb{R}^N)$ for any $\Omega' \subset\subset \Omega$. However, we define $Lip(\Omega, \mathbb{R}^N)$ as the class of functions $u : \Omega \to \mathbb{R}^N$ such that there is a constant $c = c(u)$ such that

$$|u(x) - u(y)| \le c|x - y| \quad \text{for all } x, y \in \Omega.$$

It is easy to see that each $u \in Lip(\Omega, \mathbb{R}^N)$ can be extended in a uniquely determined way to some Lipschitz map $\overline{\Omega} \to \mathbb{R}^N$ without increasing its Lipschitz constant, and thus we can identify $Lip(\Omega, \mathbb{R}^N)$ with $Lip(\overline{\Omega}, \mathbb{R}^N)$ if Ω is bounded.

Let $I = [a, b]$ be a closed interval in $\mathbb{R}$. A map $u : I \to \mathbb{R}^N$ is said to be *piecewise continuous* or *of class D^0* if there is a decompostion

(13) $$a = \xi_0 < \xi_1 < \cdots < \xi_{\nu+1} = b$$

of I such that the restriction $u|_{(\xi_{j-1}, \xi_j)}$ can be extended to a continuous function $[\xi_{j-1}, \xi_j] \to \mathbb{R}^N$ for every $j = 1, \ldots, \nu + 1$. By $D^0(I, \mathbb{R}^N)$ we denote the class of piecewise continuous functions $u : I \to \mathbb{R}^N$.

Moreover we say that $u : I \to \mathbb{R}^N$ is of class D^1 or $u \in D^1(I, \mathbb{R}^N)$, if u is continuous on I and if there exists a decomposition (13) of I such that the restrictions $u|_{[\xi_{j-1}, \xi_j]}$ are of class C^1.

Similarly one can define the classes $D^0(\bar{G}, \mathbb{R}^N)$ and $D^1(\bar{G}, \mathbb{R}^N)$ for a bounded domain G in $\mathbb{R}^N$; but here it is not quite clear how a (finite) decomposition of $\bar{G}$ is to be defined as there are many possibilities. We leave this question undecided at present since these classes are rarely used. We refrain from defining the Lebesgue classes $L^p(\Omega, \mathbb{R}^N)$ and the Sobolev classes $H^{k,p}(\Omega, \mathbb{R}^N)$ of maps $u : \Omega \to \mathbb{R}^N$ as they are rarely used in our book; we here mostly want to stay within the framework of differentiable maps.

3. Vector and Covector Fields. Transformation Rules

Let us recall some elementary facts on vector and covector fields defined on domains G of $\mathbb{R}^N$. Here we use the naive old-fashioned concept of vector fields as in Carathéodory [10], Sections 68–81. For a modern presentation within the framework of manifolds we refer e.g. to Abraham-Marsden [1]; a brief survey can be found in Vol. 2 of our treatise (see 9,*3.7*).

Let G be a domain in $\mathbb{R}^N$. Then we attach to each point $x \in G$ a copy of $\mathbb{R}^N$ and call $T_xG := \{x\} \times \mathbb{R}^N$ the *tangent space* of G at x. Its elements (x, v) are said to be the *tangent vectors* of G at x. Each tangent vector (x, v) can be interpreted as first order initial values of a curve $c : [0, 1] \to G$ in the sense that

$$c(0) = x, \qquad \dot{c}(0) = v. \tag{1}$$

The set of all tangent spaces is the tangent bundle TG of G,

$$TG := \{T_xG\}_{x \in G} = G \times \mathbb{R}^N. \tag{2}$$

In other words, each tangent vector (x, v) consists of a footpoint x to which some direction v is attached. Somewhat sloppily but conveniently one denotes as well v as tangent vector at x, and identifies T_xG with $\mathbb{R}^N$.

Similarly the *cotangent bundle* T^*G is defined by

$$T^*G := G \times \mathbb{R}^{N^*}. \tag{3}$$

Its elements (x, π) are said to be *cotangent vectors* (at x), and $T_x^*G := \{x\} \times \mathbb{R}^{N^*}$ is called the *cotangent space* of G at x. Again, one often identifies T_x^*G with $\mathbb{R}^{N^*}$ and calls π a cotangent vector attached to x.

A (smooth) *vector field f on G* is a (smooth) mapping $f : G \to TG$ of the form $f(x) = (x, p(x))$, $x \in G$, where

$$p(x) = (p^1(x), \ldots, p^N(x)) = p^i(x)e_i$$

and $\{e_1, \ldots, e_N\}$ is the canonical base of $\mathbb{R}^N$. Usually there is no harm in calling $p : G \to \mathbb{R}^N$ a vector field on G, thereby identifying f and p, but sometimes it is helpful to keep f and p apart. Then we shall denote $p : G \to \mathbb{R}^N$ as *direction field* of the vector field f.

Similarly a (smooth) *covector field* σ on G is a (smooth) mapping $\sigma : G \to T^*G$ of the form $\sigma(x) = (x, \omega(x))$, $x \in G$, where

$$\omega(x) = (\omega_1(x), \dots, \omega_N(x)) = \omega_i(x)e^i$$

and $\{e^1, \dots, e^N\}$ is the canonical base of $\mathbb{R}^{N*}$, i.e. $e^i(e_k) = \delta^i_k$. The special covector fields $x \mapsto (x, e^i)$ are denoted by dx^i. Defining the addition and multiplication of tangent and cotangent vectors fiberwise, i.e.

$$(4) \qquad \begin{aligned} (x, v) + (x, w) &:= (x, v + w), \qquad & \lambda(x, v) &:= (x, \lambda v), \\ (x, \pi) + (x, \omega) &:= (x, \pi + \omega), \qquad & \lambda(x, \pi) &:= (x, \lambda\pi), \end{aligned}$$

we then can write

$$(5) \qquad \sigma(x) = (x, \omega(x)) = \omega_i(x)\, dx^i.$$

In this way of writing one denotes covector fields σ as 1-*forms* on G, and not much harm is done if we "identify" σ with ω and write

$$(6) \qquad \omega = \omega_i\, dx^i,$$

or even more sloppily

$$(6') \qquad \omega = \omega_i(x)\, dx^i,$$

just to indicate the independent variables x.

Consider now a diffeomorphism $\varphi : G \to G^*$, i.e. a mapping $x \mapsto y$ given by

$$(7) \qquad y = \varphi(x), \quad x \in G,$$

which maps G smoothly and bijectively (i.e. 1–1) onto $G^* \subset \mathbb{R}^N$. If $\varphi(x) = (\varphi^1(x), \dots, \varphi^N(x))$ we write (7) as

$$(7') \qquad y^k = \varphi^k(x^1, x^2, \dots, x^N), \quad 1 \le k \le N.$$

A smooth curve $c : [0, 1] \to G$ in G is mapped into a smooth curve $\gamma : [0, 1] \to G^*$ in G which is given by

$$(8) \qquad \gamma = \varphi \circ c, \quad \text{i.e. } \gamma(t) = \varphi(c(t)), \quad 0 \le t \le 1.$$

Then the chain rule yields

$$(9) \qquad \dot\gamma = ((D\varphi) \circ c)\, \dot c, \quad \text{i.e. } \dot\gamma(t) = D\varphi(c(t))\, \dot c(t),$$

which in coordinates means

$$(9') \qquad \dot\gamma^k(t) = \varphi^k_{x^i}(c(t))\dot c^i(t).$$

Hence the diffeomorphism $\varphi : G \to G^*$ maps the 1-initial values x, v of a curve $c : [0, 1] \to G$ onto the 1-initial values y, w of the image curve $\gamma = \varphi \circ c$ by means of the transformation rules

$$(10) \qquad y = \varphi(x), \qquad w = D\varphi(x)v.$$

In coordinates this reads as

$$(10') \qquad y^i = \varphi^i(x), \qquad w^i = \varphi^i_{x^k}(x)v^k.$$

Introducing $B = (b^i_k)$ by

(11) $$b^i_k := \varphi^i_{x^k}(x),$$

we can write the second formula of (10′) as

(12) $$w^i = b^i_k v^k.$$

This leads us to the following result:

Any C^s-diffeomorphism $\varphi : G \to G^*$ can be extended to a C^{s-1}-diffeomorphism $\Phi : TG \to TG^*$ of the corresponding tangent bundles by setting

(13) $$\Phi(x, v) = (y, w), \quad y = \varphi(x), \ w = D\varphi(x)v$$

for any $(x, v) \in TG$.

If we want that tangent vectors and vector fields are notions which remain invariant under arbitrary coordinate transformations $\varphi : x \mapsto y$ defined by diffeomorphisms φ *we simply require that they have to transform according to rule* (13). Corresponding, any vector field which in x-coordinates has the form

$$f(x) = (x, p(x)), \quad x \in G,$$

takes in y-coordinates the form

(14a) $$g(y) = (y, q(y)), \quad y \in G^*,$$

where

(14b) $$y = \varphi(x), \quad q(y) = D\varphi(x)p(x), \qquad G^* = \varphi(G).$$

Now we consider the inverse $\psi : G^* \to G$ of a diffeomorphism $\varphi : G \to G^*$ of G onto G^*, which is a diffeomorphism of the same class as φ. If x and y are related by $y = \varphi(x)$, then $x = \psi(y)$, and vice versa. Let $S : G \to \mathbb{R}$ be a smooth function on G. Then its push-forward to G^* under the diffeomorphism $\varphi : G \to G^*$ is defined as the pull-back

(15) $$\Sigma := S \circ \psi$$

of S under the inverse map $\psi = \varphi^{-1} : G^* \to G$, i.e.

(15′) $$\Sigma(y) = S(\psi(y)), \quad y \in G.$$

We shall often denote $S \circ \psi$ as $\psi^* S$, that is,

(15″) $$(\psi^* S)(y) = (S \circ \psi)(y) = S(\psi(y)), \quad y \in G.$$

By the chain rule we have

(16) $$\Sigma_{y^i}(y) = S_{x^k}(x)\psi^k_{y^i}(y) \quad \text{where } x = \psi(y).$$

Now $S_{x^1}, \ldots, S_{x^N}$ are the components of the covector field $dS : G \to T^*G$ defined by

(17) $$dS(x) := (x, S_{x^k}(x)e^k).$$

This is the total differential of S at x, and we can also write

$$dS = S_{x^k}\, dx^k \tag{17'}$$

on account of our definition of the 1-forms dx^k. Similarly we have

$$d\Sigma(y) = (y, \Sigma_{y^i}(y) f^i)$$

or

$$d\Sigma = \Sigma_{y^i}\, dy^i .$$

(Here $\{f_1, \ldots, f_N\}$ and $\{f^1, \ldots, f^N\}$ are the canonical bases in the y-coordinates.) On account of (16) we obtain that the diffeomorphism $\varphi : G \to G^*$ leads to a mapping of the covector (x, ω) with $\omega = S_x(x)$ to the covector (y, π) with

$$y = \varphi(x), \qquad \pi = \psi_y(y)\omega ; \tag{18}$$

in coordinates:

$$\pi_i = \psi^k_{y^i}(y)\omega_k . \tag{18'}$$

Introducing $A = (a^k_i)$ by

$$a^k_i := \psi^k_{y^i}(y), \tag{19}$$

we can write this formula as

$$\pi_i = a^k_i \omega_k . \tag{20}$$

Now each $\omega \in \mathbb{R}^{N*}$ can be written as $S_x(x)$ at some fixed $x \in G$ for a suitable function S. Thus it is convincing to use (18) as definition of the transformation rule for covector (x, ω): *A covector* $(x, \omega) \in T^*G$ *is transformed by a coordinate transformation* $\varphi : x \mapsto y$ *to the covector* $(y, \pi) \in T^*G^*$ by

$$y = \varphi(x), \qquad \pi = \psi_y(y)\omega, \qquad G^* = \varphi(G). \tag{21}$$

Furthermore we infer from $\psi(\varphi(x)) = x$ that

$$\psi^i_{y^k}(y)\varphi^k_{x^l}(x) = \delta^i_l \quad \text{if } y = \varphi(x), \tag{22}$$

that is,

$$a^i_k b^k_l = d^i_l \tag{23}$$

or $B = A^{-1}$.

Comparing formulae (12), (20), (23) with relations (3), (13), and (25) of Section 1 we see that tangent vectors $(x, v) \in TG$ transform like *contravariant vectors*, whereas cotangent vectors $(x, \pi) \in T^*G$ transform like *covariant vectors*. In other words, *tangent vectors and cotangent vectors transform contragrediently to each other, and the same holds for vector fields and covector fields.*

Finally we write grad S for dS, that is, we introduce the *gradient (field) of S* by

$$\operatorname{grad} S(x) = (x, \omega(x)) \quad \text{with } \omega(x) = (S_{x^1}(x), \ldots, S_{x^N}(x)) \tag{24}$$

or, more sloppily,

$$\text{grad}\, S(x) = (S_{x^1}(x), \ldots, S_{x^N}(x)). \tag{24'}$$

Thus grad S *is a covector field on* G. (Note, however, that in geometry the gradient is usually thought to be a vector field ∇S whose i-th component is given by

$$(\nabla S)^i = g^{ik} S_{x^k},$$

where (g^{ik}) is the inverse of (g_{ik}), the components of the fundamental tensor defining the Riemannian line element $ds = \{g_{ik}(x)\, dx^i\, dx^k\}^{1/2}$ on G.)

4. Differential Forms

This section provides a brief collection of some definitions and facts on differential forms, essentially without any proofs. This survey is just meant as a reminder and has the additional purpose of fixing notation. For a coherent treatment and for proofs we refer the reader to the literature, e.g. Abraham–Marsden [1], Boothby [1], Choquet–Bruhat [1], Dubrovin–Fomenko–Novikov [1], Vol. 1, Flanders [1], Guillemin–Pollack [1], Hermann [1], Spivak [1], or Choquet–Bruhat/DeWitt–Morette/Dillard–Bleick [1].

Let $V = \mathbb{R}^n$ and $V^* = \mathbb{R}^{n^*}$ = dual space of V. Let $\{e_1, \ldots, e_n\}$ be the canonical base of V and $\{\eta^1, \ldots, \eta^n\}$ the corresponding dual base of V^* defined by $\eta^i(e_k) = \delta^i_k$. (In Sections 1 and 3 we have used e^i instead of η^i.) If $v = v^i e_i \in V$ and $\omega = \omega_i \eta^i \in V^*$ we have

$$\omega(v) = \omega_i \eta^i.$$

Instead of $\omega(v)$ we also write $\omega \cdot v$ or $v \lrcorner \omega$.

An *exterior form of degree* p (or, simply, a p-form) is a multilinear form $\omega: V^p \to \mathbb{R}$ on the p-fold Cartesian product $V^p := V \times V \times \cdots \times V$ of V which is alternating, that is,

$$\omega(\lambda a + \mu b, v_2, \ldots, v_p) = \lambda\omega(a, v_2, \ldots, v_p) + \mu\omega(b, v_2, \ldots, v_p) \tag{i}$$

for any $a, b, v_2, \ldots, v_p \in V$ and $\lambda, \mu \in \mathbb{R}$;

$$\omega(v_1, v_2, \ldots, v_p) = (\text{sign}\, \pi)\omega(v_{i_1}, v_{i_2}, \ldots, v_{i_p}) \tag{ii}$$

if $\pi = (i_1, i_2, \ldots, i_p)$ is a permutation of the numbers $1, 2, \ldots, p$; in particular, we have

$$\omega(v_1, v_2, \ldots) = -\omega(v_2, v_1, \ldots), \quad \text{etc.}$$

The set of all p-forms, $p \geq 2$, is denoted by $\Lambda_p V$. In addition we introduce $\Lambda_0 V := \mathbb{R}$ and $\Lambda_1 V := V^*$. Each set $\Lambda_p V$ is in a natural way a real vector space.

For any $v \in V$ and any $\omega \in \Lambda_p V$, $p \geq 1$, we define the *inner product* $i_v \omega$ as a $(p-1)$-form given by

$$i_v\omega(v_2, \ldots, v_p) := \omega(v, v_2, \ldots, v_p)$$

for any $v_2, \ldots, v_p \in V$. The inner product $i_v\omega$ is also denoted as the *contraction* $v \lrcorner \omega$ of ω by v, i.e.

$$(v \lrcorner \omega)(v_2, \ldots, v_p) = \omega(v, v_2, \ldots, v_p).$$

For $\alpha \in \Lambda_p V$ and $\beta \in \Lambda_q V$, $p, q \geq 0$, one defines the *exterior product* (or *wedge product*) $\alpha \wedge \beta$ of α and β as a $(p+q)$-form given by

$$(\alpha \wedge \beta)(v_1, \ldots, v_p, v_{p+1}, \ldots, v_{p+q}) := \sum_\pi (\text{sign } \pi)\alpha(v_{j_1}, \ldots, v_{j_p})\beta(v_{j_{p+1}}, \ldots, v_{j_{p+q}})$$

for any $v_1, \ldots, v_{p+q} \in V$ where the summation is to be extended over all permutations

$$\pi = (j_1, j_2, \ldots, j_{p+q})$$

of the numbers $1, 2, \ldots, p+q$ such that

$$j_1 < j_2 < \cdots < j_p \quad \text{and} \quad j_{p+1} < j_{p+2} < \cdots < j_{p+q}.$$

If $p = 0$ or $q = 0$, then the wedge product $\alpha \wedge \beta$ just reduces to the scalar multiplication $\alpha\beta$.

If, for instance, $p = q = 1$, then

(1) $$(\alpha \wedge \beta)(v_1, v_2) = \alpha(v_1)\beta(v_2) - \alpha(v_2)\beta(v_1),$$

and for $p = 2$, $q = 1$ we have

$$(\alpha \wedge \beta)(v_1, v_2, v_3) = \alpha(v_1, v_2)\beta(v_3) + \alpha(v_2, v_3)\beta(v_1) + \alpha(v_3, v_1)\beta(v_2).$$

The wedge product $\alpha \wedge \beta$ is bilinear, associative, and skew symmetric in the sense that

$$\alpha \wedge \beta = (-1)^{pq}\beta \wedge \alpha.$$

Thus a multiple wedge product $\alpha^1 \wedge \alpha^2 \wedge \cdots \wedge \alpha^r$ can be written without parentheses, and for $\omega^1, \omega^2, \ldots, \omega^n \in \Lambda_1 V$, $V = \mathbb{R}^n$, we obtain

(2) $$(\omega^1 \wedge \omega^2 \wedge \cdots \wedge \omega^n)(v_1, v_2, \ldots, v_n) = \det(\omega^i(v_k)).$$

We note that the p-forms $\eta^{i_1} \wedge \eta^{i_2} \wedge \cdots \wedge \eta^{i_p}$ with $i_1 < i_2 < \cdots < i_p$ form a base of $\Lambda_p V$. Hence

$$\dim \Lambda_p V = \binom{n}{p} \quad \text{if } 0 \leq p \leq n, \qquad \Lambda_p V = \{0\} \quad \text{if } p > n.$$

With respect to this base, every p-form ω, $p \geq 1$, possesses the unique representation

(3) $$\omega = \sum_{(i_1, \ldots, i_p)} \omega_{i_1 \ldots i_p} \eta^{i_1} \wedge \cdots \wedge \eta^{i_p},$$

with

(4) $$\omega_{i_1 \ldots i_p} := \omega(e_{i_1}, \ldots, e_{i_p}),$$

where the summation is to be extended over all ordered p-tupels $(i_1, \ldots, i_p)$ of numbers $i_\nu \in \{1, 2, \ldots, n\}$, i.e., over all p-tupels satisfying $i_1 < i_2 < \cdots < i_p$. This follows from

$$(5) \qquad (\eta^{i_1} \wedge \eta^{i_2} \wedge \cdots \wedge \eta^{i_p})(e_{j_1}, e_{j_2}, \ldots, e_{j_p}) = \delta^{i_1}_{j_1} \delta^{i_2}_{j_2} \ldots \delta^{i_p}_{j_p}$$

for $i_1 < \cdots < i_p$ and $j_1 < \cdots < j_p$.

Now we consider *differential forms of degree p* on some domain G of $\mathbb{R}^n$. Such a differential from ω is defined as a mapping $\omega: G \to G \times \Lambda_p \mathbb{R}^n := \Lambda_p TG$ which maps any point $x \in G$ to $(x, \omega(x)) \in \Lambda_p TG$ where $\omega(x)$ is a p-form on $\mathbb{R}^n$ (with a slight abuse of notation we identify $(x, \omega(x))$ with $\omega(x)$). Such a differential form is also denoted as a *p-form on G*. Similarly the 1-forms η^i are identified with $dx^i = (x, \eta^i)$. On account of (3) we can write every p-form ω on G as

$$(6) \qquad \omega(x) = \sum_{(i_1, \ldots, i_p)} \omega_{i_1 \ldots i_p}(x)\, dx^{i_1} \wedge \cdots \wedge dx^{i_p}.$$

This sum over p-tuples $i_1, \ldots, i_p$ ordered by $i_1 < \cdots < i_p$ will be writen in the abbreviated form

$$(6') \qquad \omega = \omega_{(i_1 \ldots i_p)}\, dx^{i_1} \wedge \cdots \wedge dx^{i_p}.$$

We can write ω in the form

$$(6'') \qquad \omega = \frac{1}{p!} \omega_{i_1 \ldots i_p}\, dx^{i_1} \wedge \cdots \wedge dx^{i_p}$$

if we assume that $\omega_{i_1 \ldots i_p}$ is skew symmetric in $i_1 \ldots i_p$. We call a p-form ω of class C^s if its coefficients $\omega_{i_1 \ldots i_p}(x)$ are functions of class C^s. In the sequel we assume all p-forms to be of class C^s with s sufficiently large.

For any differentiable function $f: G \to \mathbb{R}$ the differential df can be viewed as a 1-form on G, which for any vector field $v(x) = v^i(x) e_i$ is defined by

$$(7) \qquad df(x)(v) = f_{x^i}(x) v^i(x).$$

If in particular $f(x) = x^i$, then

$$df(x)(v) = v^i(x) = \eta^i(v) = dx^i(v),$$

i.e. $dx^i = df$, which motivates the notation dx^i. To any p-form ω on G we define a $(p + 1)$-form on G denoted by $d\omega$ which is called the *exterior derivative of* ω. If ω is a 0-form, i.e. function f on G, then df is defined as the classical differential df given by (7). If ω is a p-form, $p \geq 1$, given by (6), then $d\omega$ is defined by

$$(8) \qquad d\omega := \sum_{(i_1, \ldots, i_p)} (d\omega_{i_1 \ldots i_p}) \wedge dx^{i_1} \wedge \cdots \wedge dx^{i_p}.$$

The *exterior differential* d is to be viewed as a map $d: \Lambda_p TG \to \Lambda_{p+1} TG$. On account of the computational rules

$$(9) \qquad \begin{aligned} &dx^i \wedge dx^k = -dx^k \wedge dx^i, \\ &dx^i \wedge dx^i = 0, \\ &f(dx^i) = (dx^i) f \quad \text{for any } f: G \to \mathbb{R}, \end{aligned}$$

we obtain the following fundamental properties of d:

(i) For any smooth $f : G \to \mathbb{R}$, df is the ordinary differential of f.

(ii) Linearity: $d(\alpha + \beta) = d\alpha + d\beta$.

(iii) Multiplication rule:

$$d(\alpha \wedge \beta) = (d\alpha) \wedge \beta + (-1)^p \alpha \wedge d\beta \tag{10}$$

if α is a p-form.

(iv) Cocycle condition: $d(d\omega) = 0$.

It can be shown that the operator d defined by (8) is the only operator satifying (i)–(iv). Thus we could use (i)–(iv) to give a coordinate-free definition of the exterior derivative $d\omega$.

If $\omega = \omega_i \, dx^i$ is a 1-form, then $d\omega$ is given by

$$d\omega = \sum_{(i,k)} \left(\frac{\partial \omega_k}{\partial x^i} - \frac{\partial \omega_i}{\partial x^k} \right) dx^i \wedge dx^k, \tag{11}$$

where the summation is to be extended over all pairs (i, k) satisfying $i < k$ ($=$ ordered pairs). If on the other hand ω is an $(n-1)$-form on $G \subset \mathbb{R}^n$, then we can write ω in the form

$$\omega = \sum_{i=1}^{n} \omega_i (dx)_i \, . \tag{12}$$

Here dx denotes the n-form

$$dx := dx^1 \wedge dx^2 \wedge \cdots \wedge dx^n \tag{13}$$

spanning $\Lambda_n TG$, and

$$(dx)_i := e_i \lrcorner \, dx, \qquad (dx)_{ik} := e_k \lrcorner (e_i \lrcorner \, dx), \quad \text{etc.} \tag{14}$$

Then we have for instance

$$(dx)_i = (-1)^{i-1} \, dx^1 \wedge \cdots \wedge \widehat{dx^i} \wedge \cdots \wedge dx^n, \tag{15}$$

where the symbol $\wedge$ above dx^i means that the factor dx^i is to be deleted. It follows that

$$d\omega = \left(\frac{\partial \omega_1}{\partial x^1} + \frac{\partial \omega_2}{\partial x^2} + \cdots + \frac{\partial \omega_n}{\partial x^n} \right) dx. \tag{16}$$

A differential p-form ω on G is said to be *closed* if $d\omega = 0$, and it is called *exact* if there is a $(p-1)$-form α on G such that $\omega = d\alpha$.

Because of $d(d\alpha) = 0$ every exact form is closed while the converse is in general not true. For instance the 1-form $\omega = (x/r)\,dy - (y/r)\,dx$, $r = \sqrt{x^2 + y^2}$, on $G = \{(x, y) \in \mathbb{R}^2 : r \neq 0\}$ is closed but not exact as $\int_\gamma \omega = 2\pi$ for each path winding once around the origin 0 (in the positive sense). However, for special domains it can be proved that a closed form is necessarily exact. The simplest result of this kind is the

Lemma of Poincaré. *If ω is a closed differential form of degree $p \geq 1$ on a star-shaped domain U of $\mathbb{R}^n$, then ω is exact.*

Proof (*Sketch*). Let x_0 be the star-point of U. Then we define a $(p-1)$-form α on U by

$$\alpha(x)(v_1, \ldots, v_{p-1}) := \int_0^1 t^{p-1}\omega(x_0 + t(x - x_0))(x - x_0, v_1, \ldots, v_{p-1})\, dt$$

for any $(p-1)$-tuple of vectors $v_1, \ldots, v_{p-1}$. One computes that $d\alpha = \omega$. □

In order to generalize this result, we define the *pull-back $u^*\omega$ of a p-form ω with respect to a differentiable mapping u*. Let Ω be a domain of $\mathbb{R}^m$ and G a domain of $\mathbb{R}^n$, and consider some differentiable map $u: \Omega \to \mathbb{R}^n$ with $u(\Omega) \subset G$. Then, for any $y \in \Omega$, the differential $du(y)$ of u at y defines a linear mapping $\mathbb{R}^m \to \mathbb{R}^n$ which maps any vector $v = (v^1, v^2, \ldots, v^m)$ of $\mathbb{R}^m$ onto some vector $du(y)(v) = (u^1_{y^k}(y)v^k, u^2_{y^k}(y)v^k, \ldots, u^n_{y^k}(y)v^k)$ of $\mathbb{R}^n$. For any p-form ω on G we define the *pull-back* $u^*\omega$ of ω as a p-form on Ω by setting

(i) $u^*\omega := \omega \circ u$ if $p = 0$;

(ii) $(u^*\omega)(y)(v_1, \ldots, v_p) := \omega(u(y))(du(y)(v_1), \ldots, du(y)(v_p))$ for any $y \in \Omega$ and $v_1, \ldots, v_p \in \mathbb{R}^m$.

Note that we allow m and n to be different. Thus the pull-back $u^*\omega$ of a differential ω is defined with respect to any differentiable mapping u. We also remark that, for any p-form ω on $\mathbb{R}^n_y$,

$$\omega = \omega_{i_1 \ldots i_p}\, dy^{i_1} \wedge \cdots \wedge dy^{i_p}$$

and for any mapping $u: \mathbb{R}^m_x \to \mathbb{R}^n_y$, the pull-back $u^*\omega$ of ω is the p-form

$$u^*\omega = \omega_{i_1 \ldots i_p}(u(x))\, du^{i_1} \wedge \cdots \wedge dy^{i_p}. \tag{17}$$

The pull-back mapping for differential forms is compatible both with respect to exterior multiplication and exterior differentiation, that is, we have the formulas

$$u^*(\alpha \wedge \beta) = (u^*\alpha) \wedge (u^*\beta), \tag{18}$$

$$u^*(d\omega) = d(u^*\omega). \tag{19}$$

Moreover, for any two p-forms α and β on G we have

$$u^*(\alpha + \beta) = (u^*\alpha) + (u^*\beta). \tag{20}$$

If ω is a p-form on $G \subset \mathbb{R}^n$, $\Omega_1 \subset \mathbb{R}^m$, $\Omega_2 \subset \mathbb{R}^l$, and if $u_1 : \Omega_1 \to \mathbb{R}^n$, $u_2 : \Omega_2 \to \mathbb{R}^m$ are differentiable mappings such that $u_1(\Omega_1) \subset G$, $u_2(\Omega_2) \subset \Omega_1$, then we have

$$(u_1 \circ u_2)^*\omega = u_2^*(u_1^*\omega). \tag{21}$$

Now we define the *inner product* $i_a\omega$ of some vector field $a: G \to \mathbb{R}^n$ and some p-form $\omega: G \to \Lambda_p\mathbb{R}^n$ (or: the *contraction* $a \lrcorner \omega$ of the form ω by the vector field a). We set

$$i_a\omega := 0 \quad \text{if } p = 0 \tag{22}$$

and

$$(i_a\omega)(v_1, \ldots, v_{p-1}) := \omega(a, v_1, \ldots, v_{p-1}) \quad \text{if } p \geq 1. \tag{23}$$

Thus $i_a\omega$ is a $(p-1)$-form on G, and obviously $i_a(i_a\omega) = 0$. Besides obvious linearity properties of the operator i_a we have the following computational rules:

$$i_a(\alpha \wedge \beta) = (i_a\alpha) \wedge \beta + (-1)^p \alpha \wedge (i_a\beta) \quad \text{if } \alpha \text{ is a } p\text{-form}, \tag{24}$$

$$i_a df = df(a) =: L_a f \quad \text{and} \quad i_{f_a}\omega = f\, i_a\omega \quad \text{for any function } f: G \to \mathbb{R}. \tag{25}$$

Next we define the *Lie derivative* $L_a\omega$ of some p-form ω; it will again be a p-form on G. To this end we consider the local phase flow ϕ^t generated by the vector field $a: G \to \mathbb{R}^n$. We define

$$L_a\omega := \frac{d}{dt}(\phi^{t*}\omega)|_{t=0}. \tag{26}$$

If $\omega = \omega_i\, dx^i$ is a 1-form (= cotangent vector field) on G, then its Lie derivative $L_a\omega$ with respect to $a = (a^1, \ldots, a^n)$ is given by

$$L_a = \left(a^k \frac{\partial \omega_i}{\partial x^k} + \frac{\partial a^k}{\partial x^i}\omega_k\right) dx^i. \tag{27}$$

This formula can equivalently be expressed by

$$(L_a\omega)(b) = L_a[\omega(b)] - \omega(L_a b), \tag{28}$$

where a, b are arbitrary vector fields on G; note that $L_a b = [a, b] = (a^k b^i_{x^k} - b^k a^i_{x^k})e_i$. If ω is a p-form, $p \geq 1$, we have

$$\begin{aligned}(L_a\omega)&(b_1, b_2, \ldots, b_p) \\ &= L_a[\omega(b_1, b_2, \ldots, b_p)] - \omega(L_a b_1, b_2, \ldots, b_p) - \cdots - \omega(b_1, \ldots, b_{p-1}, L_a b_p).\end{aligned} \tag{29}$$

Besides obvious linearity properties of the operator L_a we have the following properties:

$$L_a(\alpha \wedge \beta) = (L_a\alpha) \wedge \beta + \alpha \wedge (L_a\beta), \tag{30}$$

$$L_a(df) = d(L_a f) \quad \text{for } f: G \to \mathbb{R}, \tag{31}$$

$$L_a(i_b\omega) - i_b(L_a\omega) = i_{[a,b]}\omega, \tag{32}$$

$$L_{fa}\omega = f L_a\omega + (df) \wedge (i_a\omega) \quad \text{for } f: G \to \mathbb{R}, \tag{33}$$

$$L_a\omega = i_a(d\omega) + d(i_a\omega). \tag{34}$$

The last formula is of particular importance as it relates the exterior differential d and the Lie derivative L_a.

Applying the pull-back operation we immediately obtain the following consequence of Poincaré's lemma:

If the domain G in $\mathbb{R}^n$ is diffeomorphic to an open ball B then each closed p-form on G, $p \geq 1$, is necessarily exact.

In fact, let $u : B \to G$ be a diffeomorphism of B onto G. Then the pull-back $u^*\omega$ of ω under u is a closed form on B since $d\omega = 0$ implies

$$d(u^*\omega) = u^*(d\omega) = 0 .$$

As B is star-shaped, there is a $(p-1)$-form β on B such that $u^*\omega = d\beta$. Let $v := u^{-1}$ and $\alpha := v^*\beta$. Then we obtain

$$\omega = (u \circ v)^*\omega = v^*(u^*\omega) = v^* d\beta = d(v^*\beta) = d\alpha ,$$

and we see that ω is exact.

Somewhat more subtle results are obtained by de Rham cohomology. Let us for simplicity confine our discussion to p-forms of class C^∞ on a domain G in $\mathbb{R}^n$. In the real vector space $C^\infty(G, \Lambda_p TG)$ both the closed and the exact p-forms define linear subspaces, and their quotient space

$$H^p(G) := \frac{\{\text{closed } p\text{-forms on } G\}}{\{\text{exact } p\text{-forms on } G\}}$$

defines the p-th *cohomology group* of the domain G. The dimension $b_p(G)$ of the space $H^p(G)$ is called the p-th *Betti number of the domain G. Hence* $b_p(G) = 0$ *means that every closed p-form on G is exact.*

Let us call two closed p-forms ω and ω' on G *cohomologous* (symbol: $\omega \sim \omega'$) if their difference is exact, i.e., $\omega - \omega' = d\alpha$; the relation $\sim$ is a equivalence relation, and $H^p(G)$ is the vector space of all cohomology classes of closed p-forms on G.

If $\Omega \subset \mathbb{R}^M$, $G \subset \mathbb{R}^n$ and $u : \Omega \to G$ is a C^∞-map of Ω into G, then the pull-back $u^*\omega$ of any closed (exact) p-form ω on G is a closed (exact) p-form on Ω, and $\omega \sim \omega'$ on G implies $u^*\omega \sim u^*\omega'$ on Ω. Thus u^* pulls cohomology classes on G back to cohomology classes on Ω, i.e., u^* defines a mapping $u^\# : H^p(G) \to H^p(\Omega)$ which turns out to be linear. If $u_2 : \Omega_2 \to \Omega_1$ and $u_1 : \Omega_1 \to G$ are differentiable mappings, then

$$(u_1 \circ u_2)^\# = u_2^\# \circ u_1^\# .$$

Hence, if G and Ω are diffeomorphic, then the vector spaces $H^p(G)$ and $H^p(\Omega)$ are linearly isomorphic; in particular, $b_p(G) = b_p(\Omega)$. This result implies $b_p(G) = 0$ for $p \geq 1$ if G is diffeomorphic to a ball, and we obtain the earlier proved result that every closed p-form on G is exact if $p \geq 1$ and G is diffeomorphic to some ball. Working with cohomology groups we can generalize this result in serveral directions, either by applying homotopies to G, or by replacing domains in $\mathbb{R}^n$ by n-dimensional manifolds. Let us only pursue the first idea. Consider C^∞ maps $u_1 : \Omega \to G$ and $u_2 : \Omega \to G$ which are C^∞-homotopic. Then one can prove a generalization of the Poincaré lemma which implies that $u_1^\# = u_2^\#$. Moreover, one easily sees that $H^p(G) = 0$ if $p > n = \dim G$. This motivates the definition $H^p(\{x_0\}) = 0$ if $p > 0$ for the cohomology groups of a single point, and the homotopy reasoning implies the following result:

If G is contractible (i.e., *if the identity map of G is homotopic to some constant map $G \to \{x_0\}$ where x_0 is a point of G*), *then all* Betti *numbers $b_p(G)$ vanish for $p > 0$. Hence every closed p-form, $p \geq 1$, on a contractible domain G of $\mathbb{R}^n$ is exact.*

Once one has defined $H^p(M)$ for arbitrary manifolds M, the same result can be established for arbitrary contractible manifolds.

Let $\langle \cdot, \cdot \rangle : TG \times TG \to \mathbb{R}$ be a symmetric, bilinear, positive definite form on TG which associates with any pair (x, v), (x, w) of tangent vectors in T_xG a real number $\langle v, w \rangle_x$ such that $\langle v, w \rangle_x$ is linear both in v and w, $\langle v, w \rangle_x = \langle w, v \rangle_x$, $\langle v, v \rangle_x \geq 0$, $\langle v, v \rangle_x = 0$ if and only if $v = 0$. Let

(35) $$g_{ik}(x) := \langle e_i, e_k \rangle_x .$$

Then $(g_{ik}(x))$ is a symmetric, positive definite matrix function whose entries $g_{ik}(x)$ can be interpreted as components of a 0, 2-tensor field on G. We say that $\langle \cdot, \cdot \rangle$ defines a *Riemannian metric* on G with the line element ds on T_xG defined by

(36) $$ds^2 = g_{ik}(x)\, dx^i\, dx^k,$$

since we can associate with any C^1-curve $c : [a, b] \to G$ an *arc length* $\mathscr{L}(c)$ by

(37) $$\mathscr{L}(c) = \int_a^b \langle \dot{c}, \dot{c} \rangle^{1/2}\, dt = \int_a^b \sqrt{g_{ik}(c(t))\dot{c}^i(t)\dot{c}^k(t)}\, dt .$$

With respect to a fixed Riemann metric we define for each $p = 0, 1, \ldots, n$ a map $* : \Lambda_p TG \to \Lambda_{n-p} TG$ which associates with any p-form

(38) $$\omega = \frac{1}{p!} \omega_{i_1 \ldots i_p}\, dx^{i_1} \wedge \cdots \wedge dx^{i_p},$$

with skew-symmetric coefficients $\omega_{i_1 \ldots i_p}$ an $(n - p)$-form

(39) $$*\omega = \frac{1}{(n-p)!} \omega^*_{k_1 \ldots k_{n-p}}\, dx^{k_1} \wedge \cdots \wedge dx^{k_{n-p}},$$

with skew-symmetric indices $\omega^*_{k_1 \ldots k_{n-p}}$ which will be defined below. Using summation over ordered p-tuples $(i_1 \ldots i_p)$ and ordered $(n - p)$-tuples $(k_1 \ldots k_{n-p})$ respectively we can write (38) and (39) as

(38′) $$\omega = \omega_{(i_1 \ldots i_p)}\, dx^{i_1} \wedge \cdots \wedge dx^{i_p},$$

(39′) $$*\omega = \omega^*_{(k_1 \ldots k_{n-p})}\, dx^{k_1} \wedge \cdots \wedge dx^{k_{n-p}}.$$

We call $*\omega$ the *dual form* to ω.

Let (g_{ik}) be the *fundamental tensor* described by (35) and set

(40) $$g := \det(g_{ik}), \qquad (g^{ik}) := (g_{ik})^{-1},$$

(41) $$G^{(i_1 \ldots i_p)(k_1 \ldots k_p)} := \det(g^{i_r k_s})_{r,s=1,\ldots,p} .$$

Then we first define $\omega^{(i_1 \ldots i_p)}$ by

(42) $$\omega^{(i_1 \ldots i_p)} = \omega_{k_1 \ldots k_p} g^{i_1 k_1} g^{i_p k_p}$$

and

(43) $$\varepsilon_{i_1 \ldots i_p j_1 \ldots j_{n-p}} := \operatorname{sign} \pi$$

for $\pi = (i_1 \ldots i_p j_1 \ldots j_{n-p})$. Here sign π is 1 or -1 if π is an even or odd permutation of the numbers 1, 2, ..., n which sign $\pi = 0$ if π is not a permutation of 1, ..., n. Now we define

(44) $$\omega^*_{j_1 \ldots j_{n-p}} := \sqrt{g}\,\varepsilon_{(i_1 \ldots i_p) j_1 \ldots j_{n-p}} \omega^{(i_1 \ldots i_p)}.$$

The operator $*$ has the following properties

(i) $*(\omega + \sigma) = *\omega + *\sigma$, $*(f\omega) = f(*\omega)$ if ω, σ are p-forms and f is function on G;

(ii) $*(*\omega) = (-1)^{pn+p}\omega$ for any p-form ω;

(iii) $\omega \wedge *\sigma = \sigma \wedge *\omega$ for any two p-forms ω, σ.

Note that $\omega \wedge *\sigma$ is an n-form, hence a multiple of the basic n-form

(45) $$dx = dx^1 \wedge \cdots \wedge dx^n = *1,$$

where $*1$ is the dual form of the 0-form $f(x) \equiv 1$. In fact, if

$$\omega = \omega_{(i_1 \ldots i_p)}\, dx^{i_1} \wedge \cdots \wedge dx^{i_p}, \qquad \sigma = \sigma_{(i_1 \ldots i_p)}\, dx^{i_1} \wedge \cdots \wedge dx^{i_p},$$

then

(46) $$\omega \wedge *\sigma = \omega_{(i_1 \ldots i_p)} \sigma^{(i_1 \ldots i_p)} * 1;$$

is particular we have $\omega \wedge *\omega = 0$ if and only if $\omega = 0$. It follows that

(47) $$(\omega, \sigma) := \int_G \omega \wedge *\sigma$$

is a scalar product on the class of p-forms ω with $\int_M \omega \wedge *\omega < \infty$, the completion of which with respect to the norm

(48) $$\|\omega\| := (\omega, \omega)^{1/2}$$

yields the Hilbert space $L^2(G, \Lambda_p TG)$ of square integrable p-forms on G.

We then define the *codifferential* δ by

(49) $$\delta := (-1)^{np+n+1} * d*$$

and the *Laplace operator* Δ on p-form by

(50) $$\Delta := d\delta + \delta d.$$

We have

(i) $\delta(\omega + \sigma) = \delta\omega + \delta\sigma$;

(ii) $\delta\delta\omega = 0$;

(iii) $*\delta\omega = (-1)^p d * \omega$, $*d\omega = (-1)^{p+1}\delta * \omega$.

Suppose now that ω is a p-form, and that φ is a $(p+1)$-form with compact support in G. Then $\omega \wedge *\varphi$ is an $(n-1)$-form. Hence $d(\omega \wedge *\varphi)$ is of the form $f(*1)$ where f is the divergence of a vector field with compact support in G (cf. (16)), and then integration by parts yields

$$0 = \int_G d(\omega \wedge *\varphi) = \int_G d\omega \wedge *\varphi + (-1)^p \int_G \omega \wedge d*\varphi$$

on account of (10). This can be written as

(51) $$(d\omega, \varphi) = (\omega, \delta\varphi) \quad \text{if } \operatorname{supp} \varphi \subset\subset G.$$

Similarly if ω and φ are of degree p and $p-1$ respectively then

(52) $$(\omega, d\varphi) = (\delta\omega, \varphi) \quad \text{if } \operatorname{supp} \varphi \subset\subset G.$$

Let us introduce the Dirichlet integral $\mathscr{D}(\omega)$ of a p-form ω on G by

(53) $$\mathscr{D}(\omega) := (d\omega, d\omega) + (\delta\omega, \delta\omega),$$

and suppose that $\mathscr{D}(\omega) < \infty$. Let φ be a p-form on G with $\operatorname{supp} \varphi \subset\subset G$. Then

$$\begin{aligned} \delta\mathscr{D}(\omega, \varphi) &= \frac{d}{d\varepsilon} \mathscr{D}(\omega + \varepsilon\varphi)\bigg|_{\varepsilon=0} = 2(d\omega, d\varphi) + 2(\delta\omega, \delta\varphi) \\ &= 2(\Delta\omega, \varphi). \end{aligned}$$

Hence we have $\delta\mathscr{D}(\omega, \varphi) = 0$ for all φ with $\operatorname{supp} \varphi \subset\subset G$ if and only if $\Delta\omega = 0$, i.e. the Euler equation of (53) is

(54) $$\Delta\omega = 0.$$

This equation is satisfied if

(55) $$d\omega = 0 \quad \text{and} \quad \delta\omega = 0,$$

but the converse is in general not true. *Solutions ω of* (54) *or* (55) *are said to be harmonic in the sense of Kodaira or of Hodge respectively.*

5. Curves in $\mathbb{R}^N$

A smooth mapping $c: I \to \mathbb{R}^N$ given by

(1) $$c(t) = (c^1(t), \ldots, c^N(t)), \quad t \in I = [a, b]$$

is said to be a *smooth curve in* $\mathbb{R}^N$, more precisely, a *parametric representation* of a curve $\mathscr{C}$ in $\mathbb{R}^N$. If we interpret t as a time parameter, (1) describes the *motion of some point* $P \in \mathbb{R}^N$, and $c(t)$ is the position of P at the time t. Furthermore $\dot{c}(t)$ is the *velocity* (or *tangent*) *vector* and $\ddot{c}(t)$ the *acceleration vector* of this motion at the time t. From the geometric point of view two motions $c: I \to \mathbb{R}^N$ and $\tilde{c}: \tilde{I} \to \mathbb{R}^N$ are said to be *equivalent* if there is a diffeomorphism $\tau: \tilde{I} \to I$ from $\tilde{I}$

into I with $\dot{\tau} > 0$ such that $\tilde{c} = c \circ \tau$; we write $c \sim \tilde{c}$. Clearly, $\sim$ is an equivalence relation (the differentiability class of τ is assumed to be the same as that of c and $\tilde{c}$), and the actual geometric object $\mathscr{C}$ described by a motion c is the equivalence class $[c]$ to which c belongs.

Let $c: I \to \mathbb{R}^N$ be a smooth curve; then points $t \in I$ where $\dot{c}(t) = 0$ are called *singular points* of c, while t is said to be regular if $\dot{c}(t) \neq 0$. At a singular point (time) $t \in I$ the motion $c(t)$ is in rest. The curve representation is said to be *regular* if all its points are regular. Regular curves are called *immersions*. We say that $c: I \to \mathbb{R}^N$ is an *embedding* if $t \neq t'$ implies $c(t) \neq c(t')$ for any $t, t' \in I$. Suppose that $c: I \to \mathbb{R}^N$, $I = [a, b]$ is an immersion of class C^k. Then we can introduce the arc-length function $\sigma: I \to \mathbb{R}$ by $\dot{\sigma} = |\dot{c}|$, $\sigma(a) = 0$, i.e.

$$\sigma(t) = \int_a^t |\dot{c}(t)|\, dt. \tag{2}$$

The transformations $s = \sigma(t)$ yields a diffeomorphism of I onto $[0, l]$ where $l = \int_a^b |\dot{c}|\, dt$ is the total length of c, and σ is of class C^k. Let $\tau = \sigma^{-1}$ be the inverse of σ, and set $r := c \circ \tau$. Then the two curve representations are equivalent, $r \sim c$, in particular $r \in C^k([0, l], \mathbb{R}^N)$, and r has the special property that $|\dot{r}(t)| \equiv 1$. In other words, an immersion $c: I \to \mathbb{R}^N$ represents a curve $\mathscr{C}$ which can be representated by a distinguished motion $r(s)$ of speed 1. The *curvature* $\kappa(s)$ of $r(s)$ at $s \in [0, l]$ is defined by

$$\kappa(s) := |\ddot{r}(s)|. \tag{3}$$

Differentiating the relation $|\dot{r}(s)|^2 = 1$ we obtain

$$\dot{r}(s) \cdot \ddot{r}(s) = 0, \tag{4}$$

i.e. $\ddot{r}(s)$ is perpendicular to $\dot{r}(s)$. Suppose that $\kappa(s) \neq 0$, and let $\mathfrak{n}(s)$ be the unit vector in direction of $\ddot{r}(s)$. We call $\mathfrak{t}(s) := \dot{r}(s)$ the *tangent vector* and $\mathfrak{n}(s) := \ddot{r}(s)/|\ddot{r}(s)|$ the *principal normal vector* of $r(s)$. Differentiating $|\mathfrak{n}(s)|^2 = 1$ we obtain

$$\mathfrak{n}(s) \cdot \dot{\mathfrak{n}}(s) = 0. \tag{5}$$

Let us consider two important special cases.

(i) *Plane curves*: $N = 2$. We have

$$|\mathfrak{t}| = 1, \qquad |\mathfrak{n}| = 1, \qquad \mathfrak{t} \cdot \mathfrak{n} = 0, \qquad \dot{\mathfrak{t}} = \kappa \mathfrak{n}, \tag{6}$$

i.e. we view $\{\mathfrak{t}(s), \mathfrak{n}(s)\}$ as a *moving orthonormal frame* whose origin is thought to be attached to the curve point $r(s)$. From (5) and (6) we infer that

$$\dot{\mathfrak{n}} = \alpha \mathfrak{t}$$

for some smooth function $\alpha(s)$ whence

$$0 = \frac{d}{ds}(\mathfrak{t} \cdot \mathfrak{n}) = \dot{\mathfrak{t}} \cdot \mathfrak{n} + \mathfrak{t} \cdot \dot{\mathfrak{n}} = \kappa + \alpha,$$

whence $\dot{\mathfrak{n}} = -\kappa \mathfrak{t}$. Thus we obtain the *two-dimensional Frenet formulae*

$$\dot{\mathscr{t}} = \kappa \mathscr{n}, \qquad \dot{\mathscr{n}} = -\kappa \mathscr{t}, \tag{7}$$

provided that $\kappa(s) \neq 0$.

By definition the curvature is nonnegative. For plane curves we can define a new curvature $\kappa(s)$ which carries a sign. To this end we define a new moving frame $\{\mathscr{t}(s), \mathscr{n}(s)\}$ where $\mathscr{t}(s)$ is as before defined by $\mathscr{t} = \dot{c}$ while $\mathscr{n}(s)$ is a unit vector pependicular to $\mathscr{t}(s)$ such that the base $\{\mathscr{t}(s), \mathscr{n}(s)\}$ is in the same way oriented as the canonical base $\{e_1, e_2\}$ *of* $\mathbb{R}^2$. Then the curvature with sign, κ, is defined by the condition

$$\dot{\mathscr{t}} = \kappa \mathscr{n}. \tag{8}$$

Let us now return to the original representation c which is connected with the distinguished representation r by

$$c = r \circ \sigma, \qquad \dot{\sigma} = |\dot{c}|. \tag{9}$$

Let $k := \kappa \circ \sigma$ be the pull-back of the curvature function κ to c, and $\{\mathfrak{t}, \mathfrak{n}\}$ be the pull-back of the moving frame $\{\mathscr{t}, \mathscr{n}\}$ to c, i.e.

$$\mathfrak{t} = \mathscr{t} \circ \sigma, \qquad \mathfrak{n} = \mathscr{n} \circ \sigma. \tag{10}$$

Then we have $\dot{c} = (\dot{r} \circ \sigma)\dot{\sigma} = (\mathscr{t} \circ \sigma)|\dot{c}|$, and therefore the velocity vector $v = \dot{c}$ is given by

$$v = |v|\mathfrak{t}, \tag{11}$$

and the acceleration vector $a = \dot{v} = \ddot{c}$ can be expressed as

$$a = \frac{d}{dt}(|v|\mathfrak{t}) = |v|^{\cdot}\mathfrak{t} + |v|\frac{d}{dt}(\mathscr{t} \circ \sigma)$$

$$= \frac{v \cdot \dot{v}}{|v|}\mathfrak{t} + k|v|^2\mathfrak{n},$$

whence

$$a = (a \cdot \mathfrak{t})\mathfrak{t} + k|v|^2\mathfrak{n}. \tag{12}$$

Thus

$$k = \frac{1}{|v|^2}|a - (a \cdot \mathfrak{t})\mathfrak{t}|, \tag{13}$$

that is,

$$k = \frac{1}{|\dot{c}|^2}\left|\ddot{c} - \frac{\dot{c} \cdot \ddot{c}}{|\dot{c}|^2}\dot{c}\right|. \tag{14}$$

If $c(t)$ is given by $c(t) = (x(t), y(t))$, we have

$$k^2 = \frac{(\ddot{x}\dot{y} - \ddot{y}\dot{x})^2}{(\dot{x}^2 + \dot{y}^2)^3}, \tag{15}$$

and for a "nonparametric" curve $c(t) = (t, u(t))$ we obtain

$$k = \frac{|\ddot{u}|}{(1 + \dot{u}^2)^{3/2}} = \left| \frac{d}{dt}\left(\frac{\dot{u}}{\sqrt{1 + \dot{u}^2}} \right) \right|, \tag{16}$$

whereas the signed curvature becomes

$$k = \frac{d}{dt}\left(\frac{\dot{u}}{\sqrt{1 + \dot{u}^2}} \right). \tag{17}$$

(ii) *Space curves*: $N = 3$. Here we use the moving orthonormal frame $\{\mathscr{t}(s), \mathscr{n}(s), \mathscr{b}(s)\}$ attached to $r(s)$ as origin, where $\mathscr{b}(s)$ is the *binormal vector* defined as

$$\mathscr{b}(s) := \mathscr{t}(s) \wedge \mathscr{n}(s). \tag{18}$$

We know already that

$$\dot{\mathscr{t}} = \kappa \mathscr{n}. \tag{19}$$

From (5) it follows that

$$\dot{\mathscr{n}} = \alpha \mathscr{t} + \tau \mathscr{b}.$$

As in (i) it follows that $\alpha = -\kappa$, and thus we have

$$\dot{\mathscr{n}} = -\kappa \mathscr{t} + \tau \mathscr{b}. \tag{20}$$

The function

$$\tau(s) := \dot{n}(s) \cdot \mathscr{b}(s) \tag{21}$$

is called the *torsion* of r (note that some authors use $\tau(s) := -\dot{\mathscr{n}}(s) \cdot \mathscr{b}(s)$). Furthermore $\mathscr{b} \cdot \mathscr{n} = 0$ yields

$$\dot{\mathscr{b}} \cdot \mathscr{n} = -\mathscr{b} \cdot \dot{\mathscr{n}} = -\tau,$$

and $|\mathscr{b}|^2 = 1$ gives

$$\mathscr{b} \cdot \dot{\mathscr{b}} = 0,$$

whereas $\mathscr{b} \cdot \mathscr{t} = 0$ leads to

$$\dot{\mathscr{b}} \cdot \mathscr{t} = -\dot{\mathscr{t}} \cdot \mathscr{b} = 0$$

on account of (19). Therefore we have obtained the *three-dimensional Frenet formulae*

$$\dot{\mathscr{t}} = \kappa \mathscr{n}, \qquad \dot{\mathscr{n}} = -\kappa \mathscr{t} + \tau \mathscr{b}, \qquad \dot{\mathscr{b}} = -\tau \mathscr{n}, \quad (\kappa > 0). \tag{22}$$

One easily computes that

$$\tau = -\mathscr{n} \cdot \dot{\mathscr{b}} = [\dot{r}, \ddot{r}, \dddot{r}]\,|\ddot{r}|^{-2}, \tag{23}$$

where $[e, f, g] := e \cdot (f \wedge g) = \det(e, f, g)$ for $e, f, g \in \mathbb{R}^3$. If we introduce the pull-backs $k := \kappa \circ \sigma$, $\mathscr{T} := \tau \circ \sigma$, it follows that

(24)
$$k = \frac{\sqrt{|\dot{c}|^2|\ddot{c}|^2 - (\dot{c}\cdot\ddot{c})^2}}{|\dot{c}|^3} = \frac{|\dot{c}\times\ddot{c}|}{|\dot{c}|^3}$$

and

(25)
$$\mathscr{T} = \frac{[\dot{c}, \ddot{c}, \dddot{c}]}{|\dot{c}|^2|\ddot{c}|^2 - (\dot{c}\cdot\ddot{c})^2} = \frac{[\dot{c}, \ddot{c}, \dddot{c}]}{|\dot{c}\times\ddot{c}|^2}\,.$$

For planar curves we have $\tau(s) \equiv 0$. By integrating (22) we see that the functions $\kappa(s)$ and $\tau(s)$ determine $c(s)$ uniquely up to a motion (i.e. an isometry) of $\mathbb{R}^3$. One calls $\kappa = \kappa(s)$, $\tau = \tau(s)$ the *natural equations* of a space curve.

6. Mean Curvature and Gauss Curvature

Let $N = n + \nu$. A set M in $\mathbb{R}^N$ is said to be an *n-dimensional submanifold of* $\mathbb{R}^N$ *of class* C^s, $s \geqq 1$, if for every point x_0 in M there is an open neighbourhood Ω of x_0 in $\mathbb{R}^N$ and a function $f \in C^s(\Omega, \mathbb{R}^\nu)$ with rank $Df = \nu$ such that $M \cap \Omega = \{x \in \Omega : f(x) = 0\}$.

By means of the implicit function theorem one can prove that $M \subset \mathbb{R}^N$ is an n-dimensional submanifold of $\mathbb{R}^N$ of class C^s, $s \geqq 1$, if and only if the following holds true: *For each* $x_0 \in M$ *there is an open neighbourhood* Ω *of* x_0 *in* $\mathbb{R}^N$ *and a regular embedding* $X \in C^s(B, \mathbb{R}^N)$ *of a domain* $G \subset \mathbb{R}^n$ *into* $\mathbb{R}^N$ *such that* $M \cap \Omega = X(B)$. Here X is said to be *regular* if it is an *immersion*, i.e. rank $DX = n$, and X is called an *embedding* if it is injective, i.e. one-to-one. The mapping $X : B \to \mathbb{R}^N$ is said to be a *parameter representation of* $M \cap \Omega$.

A pair $(x, v) \in M \times \mathbb{R}^N$ is called a *tangent vector* of M at x if there is a C^1-curve $c : [0, 1]\ \mathbb{R}^N$ with $c([0, 1]) \subset M$ such that $c(0) = x$ and $\dot{c}(0) = v$. Mostly we omit x and call v a tangent vector at x.

The set T_xM consisting of all tangent vectors of M at x is the *tangent space* of M at x; for an n-dimensional submanifold M of $\mathbb{R}^N$ it is an n-dimensional subspace of $\mathbb{R}^N$. Its orthogonal complement $T_x^\perp M$ is called the *normal space* to M at x; it is a ν-dimensional subspace of $\mathbb{R}^N$. Suppose that M is locally described by the equations

(1)
$$f^1(x) = 0, f^2(x) = 0, \ldots, f^\nu(x) = 0, \quad x \in \Omega \subset \mathbb{R}^N,$$

or by the representation

(2)
$$x = X(u), \quad u \in B \subset \mathbb{R}^N,$$

where rank $f_x = \nu$ and rank $X_u = n$. Then $T_x^\perp M$ is spanned by $f_x^1(x), \ldots, f_x^\nu$, whereas T_xM is spanned by $X_{u^1}(u), \ldots, X_{u^n}(u)$ if $x = X(u)$ and $u = (u^1, \ldots, u^n)$. The *tangent bundle* TM of M is defined by

$$TM := \bigcup_{x\in M} T_xM,$$

and similarly one defines the *cotangent bundle* T^*M and the *normal bundle* $T^\perp M$ by

$$T^*M := \bigcup_{x \in M} (T_x M)^*, \qquad T^\perp M := \bigcup_{x \in M} T_x^\perp M.$$

In the sequel we consider parameter representations (2) of class C^s which are still immersions, i.e. rank $X_u = n$ but not necessarily embeddings; then $S : B \to \mathbb{R}^N$ is said to be an *immersed surface* or a *regular surface* in $\mathbb{R}^N$ of dimension n and codimension ν, $N = n + \nu$. Let

(3) $$G := X_u^T \cdot X_u, \qquad g := \det G,$$

and suppose that $X \in C^1(\bar{B}, \mathbb{R}^N)$, $B \subset \mathbb{R}^n$, and rank $X_u = n$. Then $G(u)$ is positive definite and $g(u) > 0$ on $\bar{B}$. Suppose first that X is an embedding and set $S := X(\bar{B})$. Then we define for $f \in C^0(S, \mathbb{R}^k)$:

(4) $$\int_S f\, dA := \int_B f \circ X \sqrt{g}\, du, \quad g = \det(X_u^T \cdot X_u).$$

The integral $\int_S f\, dA$ is independent of the representation X of S since we obtain the same value if we replace X by an "equivalent" representation $\tilde{X} := X \circ \tau$, by virtue of the transformation theorem (here τ is a change of the parameters by means of a diffeomorphism). Sometimes we write the area element $dA = \sqrt{g}\, du$ as $d\mathscr{H}^n$ where $\mathscr{H}^n$ means the n-dimensional Hausdorff measure in $\mathbb{R}^N$. In particular

(5) $$A(S) := \int_B \sqrt{g}\, du$$

denotes the area of S. If X is merely an immersion or, even less, if only $X \in C^1(\bar{B}, \mathbb{R}^N)$, we still can define an *area of* X by

(6) $$A(X) := \int_B \sqrt{g}\, du, \quad g := \det(X_u^T \cdot X_u).$$

In particular if $n = 2$ and $N = 3$ we write $u^1 = u$, $u^2 = v$, i.e. (2) becomes $x = X(u, v)$, and then (6) can be written as

(7) $$A(X) = \int_B |X_u \wedge X_v|\, du\, dv.$$

If X is regular, we can introduce the *surface normal* $\mathscr{N}$ by

(8) $$\mathscr{N} := \frac{X_u \wedge X_v}{|X_u \wedge X_v|}.$$

Let us return to the general case $n < N = n + \nu$ and suppose that $X \in C^1(\bar{B}, \mathbb{R}^N)$ is a regular surface representation. Then the vectors $X_{u^1}(u), \ldots, X_{u^n}(u)$ are linearly independent, and so

(9) $$T_u X := \operatorname{span}\{X_{u_1}(u), \ldots, X_{u^n}(u)\}$$

is an n-dimensional subspace of $\mathbb{R}^N$, called the *tangent space of X at $u \in \bar{B}$*. If X is even an embedding we can identify $T_u X$ with the tangent space $T_x S$ of $S := X(\bar{B})$ at $x = X(u)$.

Suppose now that $c: I \to \mathbb{R}^N$, $I = [a, b]$, is a "curve on X" of the form $c = X \circ \gamma$ where $\gamma: I \to \mathbb{R}^n$ is a C^1-curve in the parameter domain B. Then we have

$$\dot{c}(t) = X_u(\gamma(t))\,\dot{\gamma}(t) = X_{u^\alpha}(\gamma(t))\,\dot{\gamma}^\alpha(t) \tag{10}$$

(summation with respect to repeated Greek indices from 1 to n) whence

$$|\dot{c}|^2 = \langle X_u(\gamma)\dot{\gamma}, X_u(\gamma)\dot{\gamma}\rangle = \langle G(\gamma)\dot{\gamma}, \dot{\gamma}\rangle, \tag{11}$$

$$\begin{aligned} G(u) &= X_u^T(u)\cdot X_u(u) = (g_{\alpha\beta}(u)),\\ g_{\alpha\beta}(u) &:= X_{u^\alpha}(u)\cdot X_{u^\beta}(u). \end{aligned} \tag{12}$$

Thus the length $\mathscr{L}(c) = \int_a^b |\dot{c}|\,dt$ of a curve $c = X \circ \gamma$ on X is given by

$$\mathscr{L}(c) = \int_a^b \sqrt{g_{\alpha\beta}(\gamma)\dot{\gamma}^\alpha\dot{\gamma}^\beta}\,dt. \tag{13}$$

If we introduce the bilinear form $I(V, W)$ for tangent vectors $V = v^\alpha X_{u^\alpha}(u)$, $W = w^\alpha X_{u^\alpha}(u) \in T_u X$ by

$$I(V, W) := V\cdot W = g_{\alpha\beta}(u)v^\alpha w^\beta, \tag{14}$$

with the associated quadratic form

$$I(V) := I(V, V) = g_{\alpha\beta}(u)v^\alpha v^\beta, \tag{14'}$$

we have

$$\mathscr{L}(c) = \int_a^b \sqrt{I(\dot{c})}\,dt,$$

since $\dot{c} = \dot{\gamma}^\alpha X_{u^\alpha}(\gamma)$. We call $I(V)$ the *first fundamental form* of X.

Next we want to define the second fundamental form of X on $T_u X$; we restrict ourselves to the case $N = n + 1$, i.e. to immersed hypersurfaces X in $\mathbb{R}^{n+1}$. Generalizing (8) we define the *surface normal* $\mathscr{N}(u)$ as that uniquely defined unit vector in $\mathbb{R}^N$ which is perpendicular to $T_u X$, i.e.

$$\mathscr{N}(u)\cdot X_{u^\alpha}(u) = 0, \quad |\mathscr{N}(u)| = 1, \tag{15}$$

such that $\det(X_{u^1}, \dots, X_{u^n}, \mathscr{N}) > 0$. From $|\mathscr{N}(u)|^2 = 1$ we infer that $\mathscr{N}(u)\cdot \mathscr{N}_{u^\alpha}(u) = 0$; therefore $\mathscr{N}_{u^\alpha}(u) \in T_u X$. Hence we can define a mapping $S(u): T_u X \to T_u X$ as follows. For any tangent vector $V = v^\alpha X_{u^\alpha}(u)$ we define $S(u)V$ by

$$S(u)V := -v^\alpha \mathscr{N}_{u^\alpha}(u). \tag{16}$$

The mapping $S(u)$ is obviously linear, and we shall see that it is selfadjoint with respect to the scalar product $I(V, W) = V\cdot W$ on $T_u X$. In fact, from (15_1) we infer

$$-\mathcal{N}_{u^\alpha} \cdot X_{u^\beta} = \mathcal{N} \cdot X_{u^\alpha u^\beta} = -\mathcal{N}_{u^\beta} \cdot X_{u^\alpha}, \tag{17}$$

whence for $V = v^\alpha X_{u^\alpha}$, $W = w^\alpha X_{u^\alpha} \in T_u X$:

$$\langle SV, W \rangle = -\mathcal{N}_{u^\alpha} \cdot X_{u^\beta} v^\alpha w^\beta = -X_{u^\alpha} \cdot \mathcal{N}_{u^\beta} v^\alpha w^\beta = \langle V, SW \rangle.$$

Set

$$II(V, W) := \langle SV, W \rangle, \qquad II(V) := II(V, V). \tag{18}$$

One calls $II(V)$ the *second fundamental form* of X. By introducing

$$b_{\alpha\beta}(u) := \mathcal{N}(u) \cdot X_{u^\alpha u^\beta} = -\mathcal{N}_{u^\alpha} \cdot X_{u^\beta}(u), \tag{19}$$

we can write $II(V)$ for $V = v^\alpha X_{u^\alpha}(u)$ as

$$II(V) = b_{\alpha\beta}(u) v^\alpha v^\beta, \tag{20}$$

where the matrix $(b_{\alpha\beta})$ is symmetric. Consider the eigenvalue problem

$$S(u)V = \kappa V \tag{21}$$

for the mapping $S(u): T_u X \to T_u X$. Since $S(u)$ is symmetric there are n real eigenvalues $\kappa_1, \ldots, \kappa_n$ with $\kappa_1 \leqq \kappa_2 \leqq \cdots \leqq \kappa_n$ and corresponding eigenvectors $V_1, \ldots, V_n \in T_u X$ such that $V_i \cdot V_k = I(V_i, V_k) = \delta_{ik}$. These eigenvectors are the stationary points of the Rayleigh quotient $II(V)/I(V)$, and κ_i are the corresponding stationary values $II(V_i)/I(V_i)$. One calls $\kappa_i = \kappa_i(u)$ the *principal curvatures* of the surface X at u, and the corresponding eigenvectors $V_i(u)$ are the *directions of principal curvature*.

To understand the geometric meaning of the second fundamental form and the curvatures $\kappa_1, \ldots, \kappa_n$ we consider a smooth curve $\gamma: [0, 1] \to B$ such that $\gamma(0) = u$. Then $c(t) := X(\gamma(t))$, $t \in I$, defines a smooth curve on X satisfying $c(0) = X(u) =: x$, $\dot{c}(0) = X_{u^\alpha}(u)\dot{\gamma}^\alpha(0) := V \in T_u X$. Let us presently suppose that t is the parameter of the arc length s of c. Then $\mathfrak{t}(s) = \dot{c}(s)$ is the unit tangent vector of the curves $c(s)$, $\kappa(s) = |\dot{\mathfrak{t}}(s)|$ is its curvature, and for $\kappa(s) \neq 0$ the principal normal $\mathfrak{n}(s)$ of $c(s)$ is uniquely defined by the relation

$$\dot{\mathfrak{t}}(s) = \kappa(s)\mathfrak{n}(s).$$

Differentiating

$$\mathfrak{t}(s) = \dot{c}(s) = X_{u^\alpha}(\gamma(s))\,\dot{\gamma}^\alpha(s),$$

we obtain

$$\dot{\mathfrak{t}}(s) = \ddot{c}(s) = X_{u^\alpha u^\beta}(\gamma(s))\,\dot{\gamma}^\alpha(s)\dot{\gamma}^\beta(s) + X_{u^\alpha}(\gamma(s))\,\ddot{\gamma}(s),$$

whence

$$\begin{aligned}\mathcal{N}(u) \cdot \dot{\mathfrak{t}}(0) &= \mathcal{N}(u) \cdot X_{u^\alpha u^\beta}(u)\,\dot{\gamma}^\alpha(0)\dot{\gamma}^\beta(0)\\ &= b_{\alpha\beta}(u)\,\dot{\gamma}^\alpha(0)\dot{\gamma}^\beta(0) = \dot{c}(0) \cdot S(u)\dot{c}(0)\end{aligned}$$

and therefore

$$\mathcal{N}(u) \cdot \dot{\mathfrak{t}}(0) = II(\dot{c}(0)) \tag{22}$$

and

$$\kappa(0)\mathcal{N}(u)\cdot\mathbb{n}(0) = II(\dot{c}(0)). \tag{23}$$

Since $\mathbb{t}\cdot\dot{\mathbb{t}} = 0$ we can write $\dot{\mathbb{t}}$ as

$$\dot{\mathbb{t}}(s) = \kappa_g(s)\mathbb{s}(s) + \kappa_n(s)\mathfrak{N}(s), \tag{24}$$

where $\mathfrak{N} := \mathcal{N}\circ\gamma$ is the surface normal pulled back to γ and $\mathbb{s}(s)\in T_{\gamma(s)}X$ is a unit vector perpendicular to $\mathbb{t}(s)$ and $\mathfrak{N}(s)$. One calls $\kappa_g(s)$ the *geodesic curvature* and $\kappa_n(s)$ the *normal curvature* of $c(s)$ at s.

For $N = 3$ the vector field $\mathbb{s}(s)$ along $c(s)$ is uniquely defined by the requirement that the orthonormal moving frame $\{\mathbb{t}(s), \mathbb{s}(s), \mathfrak{N}(s)\}$ satisfies $\det(\mathbb{t}, \mathbb{s}, \mathfrak{N}) = 1$; one sometimes calls $\mathbb{s}(s)$ the *side normal* of $c(s)$.

From (23) and (24) we obtain

$$\kappa_n(0) = II(\dot{c}(0)) \quad \text{where } |\dot{c}(0)| = 1. \tag{25}$$

If we drop the condition $|\dot{c}(s)| = 1$ and return to an arbitrary parametrization $c(t)$, we have to replace $\dot{c}(0)$ in (25) by $\dot{c}(0)/\sqrt{I(\dot{c}(0))}$, and we obtain in general that

$$\kappa_n = \frac{II(V)}{I(V)} \tag{25'}$$

for the normal curvature of a curve $c:[0,1]\to\mathbb{R}^N$ at $c(0) = x = X(u)$ if $\dot{c}(0) = V$. If, in particular, c is a *normal section*, i.e. the curve obtained by intersecting X with a 2-dimensional plane span $\{V, \mathcal{N}(u)\}$, $V\in T_uX$, then $\mathcal{N}(u)$ and the principal normal of c at $x = X(u)$ are collinear, i.e. the curvature κ of c at x is given by $\pm\kappa_n$; therefore

$$\kappa = \pm\frac{II(V)}{I(V)}. \tag{26}$$

Thus we see that the Rayleigh quotient II/I measures the curvatures κ of all possible normal sections of the surface X at the parameter point u.

Of particular importance are the elementary symmetric functions of the principal curvatures $\kappa_1, \ldots, \kappa_n$, in particular $\kappa_1 + \cdots + \kappa_n$ and $\kappa_1 \ldots \kappa_n$. One calls the function $H(u)$ defined by

$$H := \frac{1}{n}(\kappa_1 + \cdots + \kappa_n), \tag{27}$$

the *mean curvature* of the hypersurface $X: B\to\mathbb{R}^{n+1}$, $B\subset\mathbb{R}^n$. If $X: B\to\mathbb{R}^3$, $B\subset\mathbb{R}^2$, is a surface in $\mathbb{R}^3$, then the mean curvature H of X is given by

$$H = \tfrac{1}{2}(\kappa_1 + \kappa_2), \tag{27'}$$

while the *Gauss curvature* K is defined by

$$K = \kappa_1\kappa_2. \tag{28}$$

Note that the principal curvatures $\kappa_1, \ldots, \kappa_n$ are the roots of the polynomial

$$p(\lambda) := \det(\kappa g_{\alpha\beta} - b_{\alpha\beta});$$

therefore

$$\begin{aligned} p(\lambda) &= \det(g_{\alpha\beta})(\kappa - \kappa_1)(\kappa - \kappa_2)\ldots(\kappa - \kappa_n) \\ &= \det(g_{\alpha\beta})\det(\kappa\delta_\beta^\alpha - b_\beta^\alpha), \end{aligned}$$

where we have set $b_\beta^\alpha := g^{\alpha\sigma}b_{\sigma\beta}$, $(g^{\alpha\beta}) := (g_{\alpha\beta})^{-1}$. Comparing coefficients we obtain

$$\kappa_1\kappa_2\ldots\kappa_n = \det(b_\beta^\alpha), \qquad \kappa_1 + \kappa_2 + \cdots + \kappa_n = \operatorname{trace}(b_\beta^\alpha),$$

whence

$$\kappa_1\kappa_2\ldots\kappa_n = \det(b_{\alpha\beta})/\det(g_{\alpha\beta}), \qquad \kappa_1 + \kappa_2 + \cdots + \kappa_n = b_{\alpha\beta}g^{\alpha\beta}. \tag{29}$$

Hence the mean curvature function is given by

$$H = \frac{1}{n} b_{\alpha\beta}g^{\alpha\beta}. \tag{30}$$

In particular, if $n = 2$, $N = 3$, we have

$$H = \tfrac{1}{2}b_{\alpha\beta}g^{\alpha\beta}, \qquad K = \frac{\det(b_{\alpha\beta})}{\det(g_{\alpha\beta})}. \tag{31}$$

In this case one often uses the Gauss notation

$$\begin{aligned} &E = g_{11}, \quad F = g_{12} = g_{21}, \quad G = g_{22}, \quad W = \sqrt{g} = \sqrt{EF - G^2}, \\ &L = b_{11}, \quad M = b_{12} = b_{21}, \quad N = b_{22}, \end{aligned} \tag{32}$$

which of course cannot be applied here as some of these letters have a different meaning in our context. Then

$$H = \frac{LG + NE - 2MF}{2(EG - F^2)}, \qquad K = \frac{LN - M^2}{EG - F^2}. \tag{33}$$

If in particular $X(u, v)$ for $n = 2$ is given in the nonparametric form $u = x$, $v = y$, $X(u, v) = (x, y, z(x, y))$, then

$$\begin{aligned} \sqrt{g(u, v)} &= \sqrt{1 + z_x^2 + z_y^2}, \\ \mathcal{N}(u, v) &= \frac{1}{\sqrt{1 + z_x^2 + z_y^2}}(-z_x, -z_y, 1), \end{aligned} \tag{34}$$

and a straight-forward computation yields

$$H = \frac{(1 + z_y^2)z_{xx} - 2z_xz_yz_{xy} + (1 + z_x^2)z_{yy}}{2(1 + z_x^2 + z_y^2)^{3/2}}, \tag{35}$$

$$K = \frac{z_{xx}z_{yy} - z_{xy}^2}{(1 + z_x^2 + z_y^2)^2}. \tag{36}$$

For an n-dimensional nonparametric surface $x \mapsto (x, z(x))$, $x \in B \subset \mathbb{R}^n$, we obtain

$$\sqrt{g} = \sqrt{1 + |Dz|^2}, \tag{37}$$

$$\mathcal{N} = \frac{1}{\sqrt{1 + |Dz|^2}}(-Dz, 1). \tag{38}$$

$$nH = \operatorname{div} \frac{Dz}{\sqrt{1 + |Dz|^2}} = D_\alpha \frac{D_\alpha z}{\sqrt{1 + |Dz|^2}}. \tag{39}$$

A List of Examples

Under this headline we have collected a list of facts, ideas and principles illustrating the general theory in specific relevant situations. So our "examples" are not always examples in the narrow sense of the word; rather they often are the starting point of further and more penetrating investigations.

The reader might find this collection useful for a quick orientation, as our examples are spread out over the entire text and need some effort to be located.

Length and Geodesics

The arc-length integral: 1,*2.2* [5]; 4,*2.6* [1]; 8,*1.1* [1] [2] [3] [4]

Arcs of constant curvature: 1,*2.2* [5]

Minimal surfaces of revolution: 1,*2.2* [7]; 5,*2.4* [5]; 6,*2.3* [2]; 8,*4.3*

Catenaries or chain lines: 1,*2.2* [7]; 2,*1* [5]; 2,*3* [2]; 6,*2.3*

Shortest connections: 2,*2* [2]; 2,*4* [1]

Obstacle problem: 1,*3.2* [8]

Geodesics: 2,*2* [2] [3] [4] and 2,5 nrs. 14, 15; 3,*1* [2]; 5,*2.4* [3]; 8,*4.4*; 9,*1.7* [3]

Weighted-length functional: 1,*2.2* [6] [7]; 2,*1* [5]; 2,*4* [1]; 3,*1* [2]; 4,*2.2* [2]; 4,*2.3* [1] [2] [3] [4]; 4,*2.6* [2]; 5,*2.4* [4] [5]; 6,*1.3* [5]; 6,*2.3*; 6,*2.4*; 8,*1.1* [1] [2] [3] [4] [5] [6] [7]; 8,*2.3* [1]; 9,*3.3* [2]; 10,*3.2* [4]

Brachystochrone and cycloids: 6,*2.3* [4]; 9,*3.3* [2]

Isoperimetric problem: 2,*1* [1]; 4,*2.3* [3]

Parameter invariant integrals: 3,*1* [2]

Conjugate points: 5,*2.4* [1] [5]

Goldschmidt curve: 8,*4.3*

Poincaré's model of the hyperbolic plane: 6,*2.3* [3]

Area, Minimial Surfaces, *H*-Surfaces

Area functional: 1,*2.2* [5]; 1,*2.4* [2]; 1,6 nr. 5 [1]; 3,*1* [3] [4]; 4,*2.2* [1]

Minimal surfaces of revolution: 1,*2.2* [7]; 5,*2.4* [5]; 6,*2.3* [2]; 8,*4.3*

Dirichlet Integral and Harmonic Maps

Curvature Functionals

Null Lagrangians

The divergence: 1,*4*

The Jacobian determinant: 1,*4*

The Hessian determinant: 1,*5* $\boxed{3}$

Cauchy's integral theorem: 1,*4.1* $\boxed{1}$

Rotation number of a closed curve: 1,*5* $\boxed{6}$

Gauss-Bonnet theorem: 1,*5* $\boxed{4}$

Calibrators: 4,*2.6* $\boxed{1}$ $\boxed{2}$ $\boxed{3}$ $\boxed{4}$

Counterexamples

Nonsmooth extremals: 1,*3.1* $\boxed{1}$ $\boxed{2}$ $\boxed{3}$

Euler's paradox: 1,*3.1* $\boxed{4}$

Weierstrass's example: 1,*3.2* $\boxed{1}$

Non-existence of minimizers: 1,*3.2* $\boxed{2}$ $\boxed{3}$ $\boxed{4}$

Extremals and inner extremals: 3,*1*

Scheeffer's examples: 4,*1.1* $\boxed{1}$; 5,*1.1* $\boxed{1}$

The Lagrangian $\sqrt{u^2 + p^2}$: 4,*2.3* $\boxed{1}$

Carathéodory's example: 4,*2.3* $\boxed{2}$

Mechanics

Newton's variational problem: 1,*6* Section 2 nr. 13; 8,*1.1* $\boxed{5}$

Hamilton's principle of least action: 2,*2* $\boxed{5}$; 2,*3* $\boxed{3}$; 2,*5* $\boxed{8}$; 3,*1* $\boxed{2}$

Lagrange's version of the least action principle: 2,*3* $\boxed{3}$

Maupertuis's principle of least action: 2,*3* $\boxed{3}$

Elastic line: Chapter 2 Scholia nr. 16

Jacobi's geometric version of the least action principle: 3,*1* $\boxed{2}$; 8,*1.1* $\boxed{8}$; 8,*2.2*; 9,*3.5*

Hamilton's principle: 3,*4* $\boxed{4}$

Conservation of energy and conservation laws: 1,*2.2* $\boxed{7}$; 2,*2* $\boxed{7}$; 3,*1* $\boxed{1}$; 3,*2* $\boxed{1}$; 3,*4* $\boxed{1}$ $\boxed{2}$ $\boxed{3}$

The *n*-body problem: 2,*2* $\boxed{5}$; 2,*2* $\boxed{3}$; 3,*4* $\boxed{2}$

Pendulum equation: 2,*2* [6]

Harmonic oscillator: 9,*3.1* [1]; 9,*3.3* [1]

Equilibrium of a heavy thread: 2,*3* [2]

Galileo's law: 5,*2.4* [4]; 6,*2.3* [1]

The brachystochrone: 6,*2.3* [4]; 9,*3.3* [2]

Vibrating string: 2,*1* [2]; 5,*2.4* [1] [2]

Vibrating membrane: 2,*1* [3]

Thin plates: 1,*5* [1]

Fluid flows: 3,*3* [1] [2]

Solenoidal vector fields: 2,*3* [4]

Elasticity: 3,*4* [3]

Motion in a central field: 9,*1.6* [1]

Kepler's problem: 9,*1.6* [2]

The two-body problem: 9,*1.6* [2]

Toda lattices: 9,*1.7* [1] [2]

The motion in a field of two fixed centers: 9,*3.5* [1]

The regularization of the 3-body problem: 9,*3.5* [2]

Optics

Fermat's principle: 6,*1.3* [6]; 7,*2.2* [1]; 8,*1.3* [2]

Law of refraction: 8,*1.3* [2]

Huygens's principle: 8,*3.4*; 10,*2.6*

Canonical and Contact Transformations

Elementary canonical transformations: 9,*3.2* [4]

Poincaré transformation: 9,*3.2* [5]

Levi-Civita transformation: 9,*3.2* [6]

Homogeneous transformations: 9,*3.2* [7]

Legendre's transformation: 10,*2.1* [3]

Euler's contact transformation: 10,*2.1* [4]

Ampère's contact transformation: 10,*2.1* [4]

The 1-parameter group of dilatations: 10,*2.1* [5]

Prolonged point transformation: 10,*2.1* [6]

The pedal transformation: 10,*2.4* [4]

Apsidal transformation: 10,*2.4* [10]

Lie's G–K transformation: 10,*2.4* [11] [12]

Bonnet's transformation: 10,*2.4* [12]

Bibliography

Abbatt, R.
1. A treatise on the calculus of variations. London, 1837

Abraham, R. and Marsden, J.
1. Foundation of mechanics. Benjamin/Cummings, Reading, Mass. 1978, 2nd edition

Akhiezer, N.I.
1. Lectures on the calculus of variations. Gostekhizdat, Moscow, 1955 (in Russian). (Engl. transl.: The calculus of variations. Blaisdell Publ., New York 1962)

Alekseevskij, D.V., Vinogradov, A.M. and Lychagin, V.L.
1. Basic ideas and concepts of differential geometry. Encyclopaedia of Mathematical Sciences, vol. 28: Geometry I. Springer, Berlin Heidelberg New York 1991

Alexandroff, P. and Hopf, H.
1. Topologie. Springer, Berlin 1935. (Reprint: Chelsea Publ. Co., New York 1965)

Allendorfer, C.B. and Weil, A.
1. The Gauss-Bonnet theorem for Riemann polyhedra. Trans. Am. Math. Soc. **53** 101–129 (1943)

Almquist, E.
1. De Principiis calculi variationis. Upsala 1837

Appell, P.
1. Traité de Mécanique Rationelle. 5 vols. 2nd edn. Gauthier-Villars, Paris 1902–1937

Arnold, V.I.
1. Small divisor problems in classical and celestial mechanics. Usp. Mat. Nauk **18** (114) 91–192 (1963)
2. Mathematical methods of classical mechanics. Springer, New York Heidelberg Berlin 1978
3. Ordinary differential equations. MIT-Press, Cambridge, Mass. 1978
4. Geometrical methods in the theory of ordinary differential equations. Grundlehren der mathematischen Wissenschaften, Bd. 250. Springer, Berlin Heidelberg New York 1988. 2nd edn.

Arnold, V.I. and Avez, A.
1. Ergodic problems of classical mechanics. Benjamin, New York 1968

Arnold, V.I. and Givental, A.B.
1. Symplectic geometry. Encyclopaedia of Mathematical Sciences, vol. 4. Springer, Berlin Heidelberg New York 1990, pp. 1–136

Arnold, V.I., Gusein-Zade, S.M. and Varchenko, A.N.
1. Singularities of differentiable maps I. Birkhäuser, Boston Basel Stuttgart 1985

Arnold, V.I. and Il'yashenko, Y.S.
1. Ordinary differential equations. Encyclopaedia of Mathematical Sciences, vol. 1. Dynamical systems I, pp. 1–148. Springer, Berlin Heidelberg New York 1988

Arnold, V.I., Kozlov, V.V. and Neishtadt, A.I.
1. Mathematical aspects of classical and celestial mechanics. Encyclopaedia of Mathematical Sciences, vol. 3: Dynamical Systems III. Springer, Berlin Heidelberg New York 1988

Arthurs, A.
1. Calculus of variations. Routledge and Kegan Paul, London 1975

Asanov, G.
1. Finsler geometry, relativity and gauge theories. Reidel Publ., Dordrecht 1985

Aubin, J.-P.
1. Mathematical methods in game theory. North-Holland, Amsterdam 1979

Aubin, J.P. and Cellina, A.
1. Differential inclusions. Set-valued maps and viability theory. Grundlehren der mathematischen Wissenschaften, Bd. 264. Springer, Berlin Heidelberg New York 1984

Aubin, J.-P. and Ekeland, I.
1. Applied nonlinear analysis. Wiley, New York 1984

Aubin, T.
1. Nonlinear analysis on manifolds. Monge-Ampère equations. Springer, New York Heidelberg Berlin 1982

Bagnera, G.
1. Lezioni sul calcolo delle variazioni. Palermo, 1914

Bakelman, I.Y.
1. Mean curvature and quasilinear elliptic equations. Sib. Mat. Zh. **9** 1014–1040 (1968)

Baule, B.
1. Variationsrechnung. Hirzel, Leipzig 1945

Beckenbach, E.F. and Bellman, R.
1. Inequalities. Springer, Berlin Heidelberg New York 1965. 2nd revised printing.

Beem, J.K. and Ehrlich, P.E.
1. Global Lorentzian geometry. Dekker, New York 1981

Bejancu, A.
1. Finsler geometry and applications. Ellis Horwood Ltd., Chichester 1990

Bellman, R.
1. Dynamic Programming. Princeton Univ. Press, Princeton 1957
2. Dynamic programming and a new formalism in the calculus of variations. Proc. Natl. Acad. Sci. USA, **40** 231–235 (1954)
3. The theory of dynamic programming. Bull. Am. Math. Soc. **60** 503–516 (1954)

Beltrami, E.
1. Ricerche di Analisi applicata alla Geometria. Giornale di Matematiche **2** 267–282, 297–306, 331–339, 355–375 (1864)
2. Ricerche di Analisi applicata alla Geometria. Giornale di Matematiche **3** 15–22, 33–41, 82–91, 228–240, 311–314 (1865). (Opere Matematiche, vol. I, nota IX, pp. 107–198)
3. Sulla teoria delle linee geodetiche. Rend. R. Ist. Lombardo, A **(2)** **1** 708–718 (1868). (Opere Matematiche, vol. I., nota XXIII, pp. 366–373).
4. Sulla teoria generale dei parametri differentiali. Mem. Accad. Sci. Ist. Bologna, ser. II, **8** 551–590 (1868). (Opere Matematiche, vol II, nota XXX, pp. 74–118)

Benton, S.
1. The Hamilton-Jacobi equation. A global approach. Academic Press, New York San Francisco London 1977

Berge, C.
1. Espaces topologiques. Fonctions multivoques. Dunod, Paris 1966

Bernoulli, Jacob
1. Jacob Bernoulli, Basileensis, Opera, 2 vols. Cramer et Philibert, Geneva 1744

Bernoulli, Johann
1. Johannis Bernoulli, Opera Omnia, 4 vols. Bousquet, Lausanne and Geneva 1742

Bernoulli, Jacob and Johann
1. Die Streitschriften von Jacob und Johann Bernoulli. Bearbeitet u. Komment. von H.H. Goldstine. Hrg. von D. Speiser. Birkhäuser, Basel 1991

Bessel-Hagen, E.
1. Über die Erhaltungssätze der Elektrodynamik. Math. Ann. **84** 258–276 (1921)

Birkhoff, G.D.
1. Dynamical Systems, vol. IX of Am. Math. Soc. Am. Math. Soc. Coll. Publ., New York 1927

Bittner, L.
1. New conditions for the validity of the Lagrange multiplier rule. Math. Nachr. **48** 353–370 (1971)

Blanchard, P. and Brüning, E.
1. Direkte Methoden der Variationsrechnung. Springer, Wien 1982

Blaschke, W.
1. Über die Figuratrix in der Variationsrechnung. Arch. Math. Phys. **20** 28–44 (1913)
2. Kreis und Kugel. W. de Gruyter, Berlin 1916
3. Räumliche Variationsprobleme mit symmetrischer Transversalitätsbedingung. Ber. kgl. Sächs. Ges. Wiss., Math. Phys. Kl. **68** 50–55 (1916)
4. Geometrische Untersuchungen zur Variationsrechnung I. Über Symmetralen. Math. Z. **6** 281–285 (1920)
5. Vorlesungen über Differentialgeometrie, vols. 1–3. Springer, Berlin 1923–30. Vol. 1: Elementare Differentialgeometrie (3rd edition 1930). Vol. 2: Affine Differentialgeometrie, prepared by K. Reidemeister (1923). Vol. 3: Differentialgeometrie der Kreise und Kugeln, prepared by G. Thomson (1929)
6. Integralgeometrie, XI. Zur Variationsrechnung. Abh. Math. Semin. Univ. Hamb. **11** 359–366 (1936)
7. Zur Variationsrechnung. Rev. Fac. Sci. Univ. Istanbul, Sér. A. **19** 106–107 (1954)

Bliss, G.A.
1. Jacobi's condition for problems of the calculus of variations in parametric form. Trans. Am. Math. Soc. **17** 195–206 (1916)
2. Calculus of variations. M.A.A., La Salle, Ill. 1925. Carus Math. Monographs.
3. A boundary value problem in the calculus of variations. Publ. Am. Math. Soc. **32** 317–331 (1926)
4. The problem of Bolza in the calculus of variations. Ann of Math. **33** 261–274 (1932)
5. Lectures on the calculus of variations. The University of Chicago Press, Chicago 1946

Bliss, G.A. and Hestenes, M.R.
1. Sufficient conditions for a problem of Mayer in the calculus of variations. Trans. Am. Math. Soc. **35** 305–326 (1933)

Bliss, G.A. and Schoenberg, I.J.
1. On separation, comparison and oscillation theorems for self-adjoint systems of linear second order differential equations. Am. J. Math., **53** 781–800, 1931

Bochner, S.
1. Harmonic surfaces in Riemannian metric. Trans. Am. Math. Soc., **47** 146–154, 1940

Boerner, H.
1. Über einige Eigenwertprobleme und ihre Anwendungen in der Variationsrechnung. Math. Z. **34** 293–310 (1931) and Math. Z. **35** 161–189 (1932)
2. Über die Extremalen und geodätischen Felder in der Variationsrechnung der mehrfachen Integrale. Math. Ann. **112** 187–220 (1936)
3. Über die Legendresche Bedingung und die Feldtheorien in der Variationsrechnung der mehrfachen Integrale. Math. Z. **46** 720–742 (1940)
4. Variationsrechnung aus dem Stokesschen Satz. Math. Z. **46** 709–719 (1940)
5. Carathéodory's Eingang zur Variationsrechnung. Jahresber. Deutsche Math.-Ver. **56** 31–58 (1953)

6. Variationsrechnung à la Carathéodory und das Zermelo'sche Navigationsproblem. Selecta Mathematica V, Heidelberger Taschenbücher Nr. 201. Springer, Berlin Heidelberg New York 1979, pp. 23–67

Boltyanskii, V.G., Gamkrelidze, R.V. and Pontryagin, L.S.
1. On the theory of optimal processes. Dokl Akad. Nauk SSSR **110** 7–10 (1956)

Boltzmann, L.
1. Vorlesungen über die Prinzipe der Mechanik, vol. 1 and 2. Johann Ambrosius Barth, Leipzig 1897 and 1904

Bolza, O.
1. Gauss und die Variationsrechnung. In Vol. 10 of Gauss, Werke.
2. Lectures on the calculus of variations. University of Chicago Press, Chicago 1904
3. Vorlesungen über Variationsrechnung. B.G. Teubner, Leipzig 1909. (Reprints 1933 and 1949)
4. Über den Hilbertschen Unabhängigkeitssatz beim Lagrangeschen Variationsproblem. Rend. Circ. Mat. Palermo **31** 257–272 (1911); (zweite Mitteilung) **32** 111–117 (1911)

Bonnesen, T. and Fenchel, W.
1. Theorie der konvexen Körper. Ergebnisse der Mathematik und ihrer Grenzgebiete, vol. 3, Heft I. Springer, Berlin 1934

Boothby, W.M.
1. An introduction to differentiable manifolds. Academic Press, 1986

Bordoni, A
1. Lezioni di calcolo sublime, vol. 2. Giusti Tip., Milano 1831

Born, M.
1. Untersuchung über die Stabilität der elastischen Linie in Ebene und Raum. Thesis, Göttingen 1909

Born, M. and Jordan, P.
1. Elementare Quantenmechanik. Springer, Berlin 1930

Bottazini, U.
1. The higher calculus. A history of real and complex analysis from Euler to Weierstrass. Springer, Berlin (1986). (Ital. ed. 1981)

Braunmühl, A.V.
1. Über die Enveloppen geodätischer Linien. Math. Ann. **14** 557–566, (1879)
2. Geodätische Linien auf dreiachsigen Flächen zweiten Grades. Math. Ann. **20** 557–586 (1882)
3. Notiz über geodätische Linien auf den dreiachsigen Flächen zweiten Grades, welche sich durch elliptische Funktionen darstellen lassen. Math. Ann. **26** 151–153 (1885)

Brechtken-Manderscheid, U.
1. Einführung in die Variationsrechnung. Wiss. Buchgesellschaft, Darmstadt 1983

Brezis, H.
1. Some variational problems with lack of compactness. Proc. Symp. Pure Math. **45** Part 1, 165–201 (1986)

Brown, A.B.
1. Functional dependence. Trans. Am. Math. Soc. **38** 379–394 (1935)

Brunacci, V.
1. Corso di matematica sublime, vol. 4. Pietro Allegrini, Firenze 1808

Brunet, P.
1. Maupertuis: Etude biographique. Blanchard, Paris 1929
2. Maupertuis: L'Oeuvre et sa place dans le pensée scientifique et philosophique du XVIIIe siècle. Blanchard, Paris 1929

Bruns, H.
1. Über die Integrale des Vielkörperproblems. Acta Math. **11** 25–96 (1887–1888); cf. also: Berichte der königl. Sächs. Ges. Wiss. (1887)

2. Das Eikonal. Abh. Sächs. Akad. Wiss. Leipzig, Math.-Naturwiss. Kl., **21** 323–436 (1895) also: Abh. der königl. Sächs. Ges. Wiss. **21** (1895)

Bruun, H.
1. A manual of the calculus of variations. Odessa 1848 (in Russian)

Bryant, R.L.
1. A duality theorem for Willmore surfaces. J. Differ. Geom. **20** 23–53 (1984)

Bryant, R.L., and Griffiths, P.
1. Reduction of order for the constrained variational problem and $\frac{1}{2}\int k^2\,ds$. Am. J. Math. **108**, 525–570 (1986)

Bulirsch, R. and Pesch, H.J.
1. The maximum principle, Bellmann's equation, and Carathéodory's work. Technical Report No. 396, Technische Universität, München, 1992. Schwerpunktprogramm der DFG: Anwendungsbezogene Optimierung und Steuerung

Buquoy, G. von
1. Zwei Aufsätze. Eine eigene Darstellung der Grundlehren der Variationsrechnung. Breitkopf und Härtel, Leipzig 1812 pp. 57–70

Busemann, H.
1. The geometry of geodesics. Acad. Press, New York 1955

Buslayev, W.
1. Calculus of variations. Izdatelstvo Leningradskovo Universiteta, Leningrad, 1980 (in Russian)

Buttazzo, G., Ferone, V. and Kawohl, B.
1. Minimum problems over sets of concave functions and related questions. Math. Nachr. **173** 71–89 (1995)

Buttazzo, G., Kawohl, B.
1. On Newton's problem of minimal resistance. Math. Intelligencer **15**, No. 4, 7–12 (1993)

Carathéodory, C.
1. Über die diskontinuierlichen Lösungen in der Variationsrechnung. Thesis, Göttingen 1904. Schriften I, pp. 3–79
2. Über die starken Maxima und Minima bei einfachen Integralen. Math. Ann. **62** 449–503 (1906). Schriften I, pp. 80–142
3. Über den Variabilitätsbereich der Fourierschen Konstanten von positiven harmonischen Funktionen. Rend. Circ. Mat. Palermo, **32** 193–217 (1911). Schriften III, pp. 78–110
4. Die Methode der geodätischen Äquidistanten und das Problem von Lagrange. Acta Math. **47** 199–236 (1926). Schriften I, pp. 212–248
5. Über die Variationsrechnung bei mehrfachen Integralen. Acta Math. Szeged **4** (1929). Schriften I, pp. 401–426
6. Untersuchungen über das Delaunaysche Problem der Variationsrechnung. Abh. Math. Semin. Univ. Hamb., **8** 32–55 (1930). Schriften I, pp. 12–39
7. Bemerkung über die Eulerschen Differentialgleichungen der Variationsrechnung. Göttinger Nachr., pp. 40–42 (1931). Schriften I, pp. 249–252
8. Über die Existenz der absoluten Minima bei regulären Variationsprobleme auf der Kugel. Ann. Sc. Norm. Super Pisa Cl. Sec., IV. Ser. (2), **1** 79–87 (1932)
9. Die Kurven mit beschränkten Biegungen. Sitzungsber. Preuss. Akad. Wiss., pp. 102–125 (1933). Schriften I, pp. 65–92
10. Variationsrechnung und partielle Differentialgleichungen erster Ordnung. B.G. Teubner, Berlin 1935. Second German Edition: Vol. 1, Teubner 1956, annotated by E. Hölder, Vol. 2, Teubner 1993, with comments and supplements by R. Klötzler. (Engl. transl.: Chelsea Publ. Co., 1982)
11. Geometrische Optik, vol. 4 of Ergebnisse der Mathematik und ihrer Grenzgebiete. Springer, Berlin 1937
12. The beginning of research in calculus of variations. Osiris III, Part I, 224–240 (1937). Schriften II, pp. 93–107

13. E. Hölder. Die infinitesimalen Berührungstransformationen der Variationsrechnung. Report in: Zentralbl. Math. **21** 414 (1939). Schriften V, pp. 360–361
14. Basel und der Beginn der Variationsrechnung. Festschrift zum 60. Geburtstag von Prof. A. Speiser, Zürich, pp. 1–18 (1945). Schriften II, pp. 108–128
15. Einführung in Eulers Arbeiten über Variationsrechnung. Leonhardi Euleri Opera Omnia I 24, Bern, pp. VIII–LXII (1952). Schriften V, pp. 107–174
16. Gesammelte mathematische Schriften, vols. I–V. C.H. Beck, München 1954–1957

Carll, L.B.
1. A treatise on the calculus of variations. Macmillan New York and London 1885

Cartan, E
1. Leçons sur les invariants integraux. Hermann, Paris 1922
2. Les espaces métriques fondés sur la notion d'aire. Actualités scientifiques n. 72, Paris 1933
3. Les espaces de Finsler. Actualités scientifiques n. 79, Paris 1934
4. Les systémes differentiels extérieurs et leurs applications géometriques. Actualités scientifiques n.994, Paris 1945
5. Géométrie des espaces de Riemann. Gauthier-Villars, Paris 1952
6. Oeuvres complètes, 3 vols. in 6 parts. Gauthier-Villars, Paris 1952–55

Castaing, C. and Valadier, M.
1. Convex analysis and measurable multifunctions. Lecture Notes Math., vol. 580. Springer, Berlin Heidelberg New York 1977

Cauchy, A.
1. Exercises d'analyse et de physique mathematique. Bachelier, Paris. tome 1 (1840), tome 2 (1841), tome 3 (1844)
2. Note sur l'intégration des équations aux differences partielles du premier ordre à un nombre quelconque de variables. Bull. Soc. philomathique de France, pp. 10–21 (1819)

Cayley, A.
1. Collected Mathematical Papers. Cambridge Univ. Press, Cambridge 1890

Cesari, L.
1. Optimization theory and applications. Applications of Mathematics, vol. 17. Springer, New York 1983

Charlier, C.L.
1. Die Mechanik des Himmels. Veit & Co. Leipzig. 2 vols, 1902, 1907

Chasles, M.
1. Aperçu historique sur l'origine et développement des méthodes en géométrie. First ed. 1837. Third ed. Gauthier–Villars 1889

Cheeger, J. and Ebin, D.G.
1. Comparison Theorems in Riemannian Geometry. North-Holland and American Elsevier, Amsterdam-Oxford and New York 1975

Chern, S.S.
1. A simple intrinsic proof of the Gauss-Bonnet formula for closed Riemannian manifolds. Ann. Math. **45** 747–752 (1944)

Choquet-Bruhat, Y.
1. Géométrie différentielle et systèmes extérieurs. Dunod, Paris 1968

Choquet-Bruhat, Y., DeWitt-Morette, C. and Dillard-Bleick, M.
1. Analysis, manifolds, and physics. North-Holland, Amsterdam New York Oxford 1982. Revised edition

Clarke, F. and Zeidan, V.
1. Sufficiency and the Jacobi condition in the calculus of variations. Can. J. Math. **38** 1199–1209 (1986)

Clarke, F.H.
1. Optimization and nonsmooth analysis. Wiley, New York 1983

Clegg, J.
1. Calculus of Variations. Oliver & Boyd, Edinburgh 1968

Coddington, E.A. and Levinson, N.
1. Theory of ordinary differential equations. McGraw-Hill, New York Toronto London 1955

Courant, R.
1. Calculus of variations. Courant Inst. of Math. Sciences, New York 1946. Revised and amended by J. Moser in 1962, with supplementary notes by M. Kruskal and H. Rubin
2. Dirichlet's principle, conformal mapping, and minimal surfaces. Interscience, New York London 1950

Courant, R. and Hilbert, D.
1. Methoden der mathematischen Physik, vol. 1. Springer, Berlin 1924. 2nd edition 1930
2. Methoden der mathematischen Physik, vol. 2. Springer, Berlin 1937
3. Methods of Mathematical Physics, vol. 1. Wiley-Interscience, New York 1953
4. Methods of Mathematical Physics, vol. 2. Wiley Interscience Publ., New York 1962

Courant, R. and John, F
1. Introduction to Calculus and Analysis, vols. 1 and 2. Wiley-Interscience, New York 1974

Crandall, M.G., Ishii, H., and Lions, P.L.
1. User's guide to viscosity solutions of second order partial differential equations. Bull. Am. Math. Soc. **27** 1–67 (1992)

Dadok, J. and Harvey, R.
1. Calibrations and spinors. Acta Math. **170** 83–120 (1993)

Damköhler, W.
1. Über indefinite Variationsprobleme. Math. Ann. **110** 220–283 (1934)
2. Über die Äquivalenz indefiniter mit definiten isoperimetrischen Variationsproblemen. Math. Ann. **120** 297–306 (1948)

Damköhler, W. and Hopf, E.
1. Über einige Eigenschaften von Kurvenintegralen und über die Äquivalenz von indefiniten mit definiten Variationsproblemen. Math. Ann. **120** 12–20 (1947)

Darboux, G.
1. Leçons sur la théorie generale des surfaces, vols. 1–4. Gauthier-Villars, Paris 1887–1896

Debever, R.
1. Les champs de Mayer dans le calcul des variations des intégrales multiples. Bull. Acad. Roy. Belg., Cl. Sci. **23** 809–815 (1937)

Dedecker, P.
1. Sur les integrales multiples du calcul des variations. C.R. du IIIe Congrès Nat. Sci., Bruxelles **2** 29–35 (1950)
2. Calcul des variations, formes differentielles et champs géodésiques. In Géometrie Differentielle, Strasbourg 1953, pp. 17–34, Paris, 1953. Coll. Internat. CNRS nr. 52
3. Calcul des variations et topologie algebrique. Mém. Soc. Roy. Sci. Liège **19** (4e sér.), Fasc. I, (1957)
4. A property of differential forms in the calculus of variations. Pac. J. Math. **7** 1545–1549 (1957)
5. On the generalization of symplectic geometry to multiple integrals in the calculus of variations. In: K. Bleuler and A. Reetz (eds.) Diff. Geom. Methods in Math. Phys. Lecture Notes in Mathematics, vol. 570. Springer, Berlin Heidelberg New York 1977, pp. 395–456

De Donder, T.
1. Sur les equations canoniques de Hamilton-Volterra. Acad. Roy. Belg., Cl. Sci. Mém., **3**, p. 4 (1911)
2. Sur le théorème d'independence de Hilbert. C.R. Acad. Sci. Paris, **156** 868–870 (1913)
3. Théorie invariantive de calcul des variations. Hyez, Bruxelles 1935 Nouv. ed.: Gauthier–Villars, Paris 1935

Dienger, J.
1. Grundriss der Variationsrechnung. Vieweg, Braunschweig, 1867

Dierkes, U.
1. A Hamilton–Jacobi theory for singular Riemannian metrics. Arch. Math. **61**, 260–271 (1993)

Dierkes, U., Hildebrandt, S., Küster, A. and Wohlrab, O.
1. Minimal surfaces I (Boundary value problems), II (Boundary regularity). Grundlehren der mathematischen Wissenschaften, vols. 295–296. Springer, Berlin Heidelberg New York 1992

Dirac, P.A.M.
1. Homogeneous variables in classical mechanics. Proc. Cambridge Phil. Soc., math. phys. sci. **29** 389–400 (1933)
2. The principles of quantum mechanics. Oxford University Press, Oxford 1944. 3rd edition

Dirichlet, G.L.
1. Werke, vols. 1 and 2. G. Reimer, Berlin 1889–1897

Dirksen, E.
1. Analytische Darstellung der Variationsrechnung. Schlesinger, Berlin 1823

Doetsch, G.
1. Die Funktionaldeterminante als Deformationsmass einer Abbildung und als Kriterium der Abhängigkeit von Funktionen. Math. Ann. **99** 590–601 (1928)

Dombrowski, P.
1. Differentialgeometrie. Ein Jahrhundert Mathematik, Festschrift zum Jubiläum der DMV. Vieweg, Braunschweig-Wiesbaden 1990

Dörrie, H.
1. Einführung in die Funktionentheorie. Oldenburg, München 1951

Douglas, J.
1. Extremals and transversality of the general calculus of variations problems of first order in space. Trans. Am. Math. Soc. **29** 401–420 (1927)
2. Solutions of the inverse problem of the calculus of variations. Trans. Am. Math. Soc. **50** 71–128 (1941)

Du Bois-Reymond, P.
1. Erläuterungen zu den Anfangsgründen der Variationsrechnung. Math. Ann. **15** 283–314 (1879)
2. Fortsetzung der Erläuterungen zu den Anfangsgründen der Variationsrechnung. Math. Ann. **15** 564–578 (1879)

Dubrovin, B.A., Fomenko, A.T. and Novikov, S.P.
1. Modern geometry – methods and applications, vols. 1, 2, 3. Springer, New York Berlin Heidelberg 1984–1991. Vol. 1: The geometry of surfaces, transformation groups, and fields (1984). Vol. 2: The geometry and topology of manifolds (1985). Vol. 3: Introduction to homology theory (1991)

Duvaut, G. and Lions, J.L.
1. Inequalities in Mechanics and Physics. Grundlehren der mathematischen Wissenschaften, vol. 219. Springer, Berlin Heidelberg New York 1976

Eells, J. and Lemaire, L.
1. A report on harmonic maps. Bull. Lond. Math. Soc. **10** 1–68 (1978)
2. Selected topics in harmonic maps. C.B.M.S. Regional Conf. Series 50. Amer. Math. Soc. 1983
3. Another report on harmonic maps. Bull. Lond. Math. Soc. **20** 385–524 (1988)

Eggleston, H.G.
1. Convexity. Cambridge Univ. Press, London New York 1958

Egorov, D.
1. Die hinreichenden Bedingungen des Extremums in der Theorie des Mayerschen Problems. Math. Ann. **62** 371–380 (1906)

Eisenhart, L.P.
1. Continuous groups of transformations. Dover Publ., 1961 (First printing 1933, Princeton University Press).
2. Riemannian geometry. Princeton University Press, Princeton, 1964 Fifth printing. (First printing 1925)

Ekeland, I.
1. Periodic solutions of Hamilton's equations and a theorem of P. Rabinowitz. J. Differ. Equations, **34** 523–534 (1979)
2. Une théorie de Morse pour les systèmes Hamiltoniens convexes. Ann. Inst. Henri Poincaré, Anal. Non Linéaire, **1** 19–78 (1984)

Ekeland, I. and Hofer, H.
1. Symplectic topology and Hamiltonian dynamics I, II. Math. Z. **200** 335–378 (1989); **203** 553–567 (1990)

Ekeland, I. and Lasry, J.M.
1. On the number of closed trajectories for a Hamiltonian flow on a convex energy surface. Ann. Math. **112** 283–319 (1980)

Ekeland, I. and Temam, R.
1. Analyse convexe et problèmes variationnels. Dunod/Gauthiers-Villars, Paris-Bruxelles-Montréal 1974

Eliashberg, Y. and Hofer, H.
1. An energy-capacity inequality for the symplectic holonomy of hypersurfaces flat at infinity. Proceedings of a Workshop on Symplectic Geometry, Warwick, 1990

Elsgolts, L.
1. Calculus of variations. Addison-Wesley Publ. Co., Reading 1962. Translated from the Russian (Nauka, Moscow 1965)
2. Differential equations and the calculus of variations. Mir Publ., Moscow 1970

Emmer, M.
1. Esistenza, unicità e regolarità nelle superfici di equilibrio nei capillari. Ann. Univ. Ferrara Nuova Ser., Sez. VII **18** 79–94 (1973)

Engel, F. and Faber, K.
1. Die Liesche Theorie der partiellen Differentialgleichungen erster Ordnung. Teubner, Leipzig Berlin 1932

Engel, F. and Liebmann, H.
1. Die Berührungstransformationen. Geschichte und Invariantentheorie. Zwei Referate. Jahresber. Dtsch. Math.-Ver. 5. Ergänzungsband, 1–79 (1914)

Epheser, H.
1. Vorlesung über Variationsrechnung. Vandenhoeck & Ruprecht, Göttingen 1973

Erdmann, G.
1. Über unstetige Lösungen in der Variationsrechnung. J. Reine Angew. Math. **82** 21–33 (1877)

Escherich, G. von
1. Die zweite Variation der einfachen Integrale. Wiener Ber., Abt. IIa **17** 1191–1250, 1267–1326, 1383–1430 (1898)
2. Die zweite Variation der einfachen Integrale. Wiener Ber., Abt. IIa **18** 1269–1340 (1899)

Euler, L.
1. Opera Omnia I–IV. Birkhäuser, Basel. Series I (29 vols.): Opera mathematica. Series II (31 vols.): Opera mechanica et astronomica. Series III (12 vols.): Opera physica, Miscellanea. Series IV (8 + 7 vols.): Manuscripta. Edited by the Euler Committee of the Swiss Academy of Sciences, Birkhäuser, Basel; formerly: Teubner, Leipzig, and Orell Füssli, Turici
2. Methodus inveniendi lineas curvas maximi minimive proprietate gaudentes, sive solutio problematis isoperimetrici lattisimo sensu accepti. Bousquet, Lausannae et Genevae 1744. E65A. O.O. Ser. I, vol. 24

3. Analytica explicatio methodi maximorum et minimorum. Novi comment. acad. sci. Petrop. **10** 94–134 (1766). O.O. Ser. I, vol. 25, 177–207
4. Elementa calculi variationum. Novi comment. acad. sci. Petrop. **10** 51–93 (1766) O.O. Ser. I, vol. 25, 141–176
5. Institutionum calculi integralis volumen tertium, cum appendice de calculo variationum. Acad. Imp. Scient., Petropoli 1770. O.O. Ser. I, vols. 11–13 (appeared as: Institutiones Calculi Integralis)
6. Methodus nova et facilis calculum variationum tractandi. Novi comment. acad. sci. Petrop. **16** 3–34 (1772). O.O. Ser. I. vol. 25, 208–235
7. De insigni paradoxo, quod in analysi maximorum et minimorum occurit. Mém. acad. sci. St. Pétersbourg **3** 16–25 (1811). O.O. Ser. I, vol. 25, 286–292

Ewing, G.
1. Calculus of variations with applications. Norton, New York 1969

Fenchel, W.
1. On conjugate convex functions. Can. J. Math. **1** 73–77 (1949)
2. Convex Cones, Sets and Functions. Princeton Univ. Press, Princeton 1953. Mimeographed lecture notes

Fierz, M.
1. Vorlesungen zur Entwicklungsgeschichte der Mechanik. Lecture Notes in Physics, Nr. 15. Springer, Berlin Heidelberg New York 1972

Finn, R.
1. Equilibrium capillary surfaces. Springer, New York Berlin Heidelberg 1986

Finsler, P.
1. Kurven und Flächen in allgemeinen Räumen. Thesis, Göttingen 1918. Reprint: Birkhäuser, Basel 1951

Flanders, H.
1. Differential forms with applications to the physical sciences. Academic Press, New York London 1963

Flaschka, H.
1. The Toda lattice I. Phys. Rev. **9** 1924–1925 (1974)

Fleckenstein, O.
1. Über das Wirkungsprinzip. Preface of the editor J.O. Fleckenstein to: L. Euler, Commentationes mechanicae. Principia mechanica. O.O. Ser. II, vol. 5, pp. VII–LI.

Fleming, W.H.
1. Functions of several variables. Addison-Wesley, Reading, Mass. 1965

Fleming, W.H. and Rishel, R.W.
1. Deterministic and stochastic optimal control. Springer, Berlin Heidelberg New York 1975

Floer, A. and Hofer, H.
1. Symplectic Homology I. Open Sets in $\mathbb{C}^n$. Math. Z. **215** 37–88 (1994)

Forsyth, A.
1. Calculus of variations. University Press, Cambridge 1927

Fox, C.
1. An introduction to calculus of variations. Oxford University Press, New York 1950

Friedrichs, K.O.
1. Ein Verfahren der Variationsrechnung, das Maximum eines Integrals als Maximum eines anderen Ausdrucks darzustellen. Göttinger Nachr., pp. 13–20 (1929)
2. On the identity of weak and strong extensions of differential operators. Trans. Am. Math. Soc. **55** 132–151 (1944)
3. On the differentiability of the solutions of linear elliptic equations. Commun. Pure Appl. Math. **6** 299–326 (1953)
4. On differential forms on Riemannian manifolds. Commun. Pure Appl. Math. **8** 551–558 (1955)

Fuller, F.B.
1. Harmonic mappings. Proc. Natl. Acad. Sci. **40** 987–991 (1954)

Funk, P.
1. Variationsrechnung und ihre Anwendung in Physik und Technik. Grundlehren der mathematischen Wissenschaften, Bd. 94. Springer, Berlin Heidelberg New York; 1962 first edition, 1970 second edition

Fučik, S., Nečas, J. and Souček, V.
1. Einführung in die Variationsrechnung. Teubner-Texte zur Mathematik. Teubner, Leipzig 1977

Gähler, S. and Gähler, W.
1. Über die Existenz von Kurven kleinster Länge. Math. Nachr. **22** 175–203 (1960)

Garabedian, P.
1. Partial differential equations. Wiley, New York 1964

Garber, W., Ruijsenaars, S., Seiler, E. and Burns, D.
1. On finite action solutions of the nonlinear σ-model. Ann. Phys., **119** 305–325 (1979)

Gauss, C.F.
1. Werke, vols. 1–12. B.G. Teubner, Leipzig 1863–1929
2. Disquisitiones generales circa superficies curvas. Göttinger Nachr. **6** 99–146 (1828). Cf. also Werke, vol. 4, pp. 217–258. (German transl.: Allgemeine Flächentheorie, herausg. v. A. Wangerin, Ostwald's Klassiker, Engelmann, Leipzig 1905. English transl.: General investigations of curved surfaces. Raven Press, New York 1965)
3. Principia generalia theoriae figurae fluidorum in statu aequilibrii. Göttingen 1830, and also Göttinger Abh. **7** 39–88 (1832), cf. Werke 5, 29–77

Gelfand, I.M. and Fomin, S.V.
1. Calculus of variations. Prentice-Hall, Inc., Englewood Cliffs 1963. Russian ed. Fizmatgiz, 1961

Gericke, H.
1. Zur Geschichte des isoperimetrischen Problems. Mathem. Semesterber., **29** 160–187 (1982)

Giaquinta, M.
1. On the Dirichlet problem for surfaces of prescribed mean curvature. Manuscr. Math. **12** 73–86 (1974)

Gilbarg, D. and Trudinger, N.S.
1. Elliptic partial differential equations. Springer, Berlin Heidelberg New York 1977 first edition, 1983 second edition

Goldschmidt, B.
1. Determinatio superficiei minimae rotatione curvae data duo puncta jungentis circa datum axem ortae. Thesis, Göttingen 1831

Goldschmidt, H. and Sternberg, S.
1. The Hamilton-Cartan formalism in the calculus of variations. Ann. Inst. Fourier (Grenoble) **23** 203–267 (1973)

Goldstein, H.
1. Classical mechanics. Addison-Wesley, Reading, Mass. and London 1950

Goldstine, H.H.
1. A history of the calculus of variations from the 17th through the 19th century. Springer, New York Heidelberg Berlin 1980

Goursat, E.
1. Leçons sur l'intégration des équations aux dérivées partielles du premier ordre. Paris 1921, 2nd edition
2. Leçons sur le problème de Pfaff. Hermann, Paris 1922

Graves, L.M.
1. Discontinuous solutions in space problems of the calculus of variations. Am. J. Math. **52** 1–28 (1930)

2. The Weierstrass condition for multiple integral variation problems. Duke Math. J. **5** 656–658 (1939)

Griffiths, P.
1. Exterior differential systems and the calculus of variations. Birkhäuser, Boston 1983

Gromoll, D., Klingenberg, W. and Meyer, W.
1. Riemannsche Geometrie im Großen. Lecture Notes in Mathematics, vol. 55. Springer, Berlin Heidelberg New York 1968

Gromov, M.
1. Pseudoholomorphic curves in symplectic manifolds. Invent. Math. **82** 307–347 (1985)

Grüss, G.
1. Variationsrechnung. Quelle & Meyer, Leipzig 1938. 2nd edition, Heidelberg 1955

Grüter, M.
1. Über die Regularität schwacher Lösungen des Systems $\Delta x = 2H(x)x_u \wedge x_v$. Thesis, Düsseldorf 1979
2. Regularity of weak H-surfaces. J. Reine Angew. Math. **329** 1–15 (1981)

Guillemin, V. and Pollack, A.
1. Differential topology. Prentice Hall, Englewood Cliffs, N. J. 1974

Guillemin, V. and Sternberg, S.
1. Geometric asymptotics. Am. Math. Soc. 1977. Survey vol. 14

Günther, C.
1. The polysymplectic Hamiltonian formalism in the field theory and calculus of variations. I: The local case. J. Differ. Geom. **25** 23–53 (1987)

Günther, N.
1. A course of the calculus of variations. Gostekhizdat, 1941 (in Russian)

Haar, A.
1. Zur Charakteristikentheorie. Acta Sci. Math. **4** 103–114 (1928)
2. Sur l'unicité des solutions des équations aux derivées partielles. C.R. **187** 23–25 (1928)
3. Über adjungierte Variationsprobleme und adjungierte Extremalflächen. Math. Ann., **100** 481–502 (1928)
4. Über die Eindeutigkeit und Analytizität der Lösungen partieller Differentialgleichungen. Atti del Congr. Int. Mat., Bologna 3–10 Sett. 1928, pp. 5–10 (1930)

Hadamard, J.
1. Sur quelques questions du Calcul des Variations. Bull. Soc. Math. Fr., **30** 153–156 (1902)
2. Leçons sur la propagation des ondes et les équations de l'hydrodynamique. Paris 1903
3. Sur le principe de Dirichlet. Bull. Soc. Math. Fr., **24** 135–138 (1906), cf. also Oeuvres, t. III, pp. 1245–1248
4. Leçons sur le calcul des variations. Hermann, Paris 1910
5. Le calcul fonctionelles. L'Enseign. Math., pp. 1–18 (1912), cf. Oeuvres IV, pp. 2253–2266
6. Le développement et le rôle scientifique du calcul fonctionelle. Int. Math. Congr., Bologna 1928
7. Œuvres, volume I–IV. Edition du CNRS, Paris 1968

Hagihara, Y.
1. Celestial mechanics, volume 1–V. M.I.T. Press, Cambridge, MA 1970

Hamel, G.
1. Über die Geometrien, in denen die Geraden die kürzesten sind. Thesis, Göttingen 1901
2. Über die Geometrien, in denen die Geraden die kürzesten Linien sind. Math. Ann. **57** 231–264 (1903)

Hamilton, W.R.
1. Mathematical papers. Cambridge University Press. Vol. 1: Geometrical Optics (1931), ed. by Conway and Synge; Vol. 2: Dynamics (1940), ed. by Conway and McConnel; Vol. 3: Algebra (1967), ed. by Alberstam and Ingram

Hancock, H.
1. Lectures on the calculus of variations. Univ. of Cincinnati Bull. of Mathematics, Cincinnati 1904

Hardy, G.H. and Littlewood, J.E. and Pólya, G.
1. Inequalities. Cambridge Univ. Press, Cambridge 1934

Hartman, P.
1. Ordinary differential equations. Birkhäuser, Boston Basel Stuttgart 1982. 2nd edition

Harvey, R.
1. Calibrated geometries. Proc. Int. Congr. Math., Warsaw, pp. 727–808 (1983)
2. Spinors and calibrations. Perspectives in Math. 9. Acad. Press, New York, 1990

Harvey, R. and Lawson, B.
1. Calibrated geometries. Acta Math. **148** 47–157 (1982)
2. Calibrated foliations (foliations and mass-minimizing currents). Am. J. Math. **104** 607–633 (1982)

Haupt, O. and Aumann, G.
1. Differential- und Integralrechnung, vols. I–III. Berlin 1938

Hawking, S.W. and Ellis, G.F.R.
1. The large scale structure of space-time. Cambridge University Press, London New York 1973

Heinz, E.
1. Über die Existenz einer Fläche konstanter mittlerer Krümmung bei vorgegebener Berandung. Math. Ann. **127** 258–287 (1954)
2. An elementary analytic theory of the degree of mapping in n-dimensional space. J. Math. Mech. **8** 231–247 (1959)
3. On the nonexistence of a surface of constant mean curvature with finite area and prescribed rectifiable boundary. Arch. Ration. Mech. Anal. **35** 249–252 (1969)
4. Über das Randverhalten quasilinearer ellipischer Systeme mit isothermen Parametern. Math. Z. **113** 99–105 (1970)

Henriques, P.G.
1. Calculus of variations in the context of exterior differential systems. Differ. Geom. Appl. **3** 331–372 (1993)
2. Well-posed variational problem with mixed endpoint conditions. Differ. Geom. Appl. **3** 373–392 (1993)
3. The Noether theorem and the reduction procedure for the variational calculus in the context of differential systems. C.R. Acad. Sci. Paris **317** (Ser. I), 987–992 (1993)

Herglotz, G.
1. Vorlesungen über die Theorie der Berührungstransformationen. Göttingen, Sommer, 1930. (Lecture Notes kept in the Library of the Dept. of Mathematics in Göttingen)
2. Vorlesungen über die Mechanik der Kontinua. Teubner-Archiv zur Mathematik, Teubner, Leipzig 1985. (Edited by R.B. Guenther and H. Schwerdtfeger; based on lectures by Herglotz held in Göttingen in 1926 and 1931)
3. Gesammelte Schriften. Edited by H. Schwerdtfeger. Van den Hoek & Ruprecht, Göttingen 1979

Hermann, R.
1. Differential geometry and the calculus of variations. Academic Press, 1968. Second enlarged edition by Math. Sci. Press, 1977

Herzig, A. and Szabó, I.
1. Die Kettenlinie, das Pendel und die "Brachistochrone" bei Galilei. Verh. Schweiz. Naturforsch. Ges. Basel **91** 51–78 (1981)

Hestenes, M.R.
1. Sufficient conditions for the problem of Bolza in the calculus of variations. Trans. Am. Math. Soc. **36** 793–818 (1934)
2. A sufficiency proof for isoperimetric problems in the calculus of variations. Bull. Am. Math. Soc. **44** 662–667 (1938)

3. A general problem in the calculus of variations with applications to paths of least time. Technical Report ASTIA Document No. AD 112382, RAND Corporation RM-100, Santa Monica, California 1950
4. Applications of the theory of quadratic forms in Hilbert space to the calculus of variations. Pac. J. Math. **1** 525–581 (1951)
5. Calculus of variations and optimal control theory. Wiley, New York London Sydney 1966

Hilbert, D.
1. Mathematische Probleme. Göttinger Nachrichten, pp. 253–297 (1900). Vortrag, gehalten auf dem internationalen Mathematikerkongreß zu Paris 1900
2. Über das Dirichletsche Prinzip. Jahresber. Dtsch. Math.-Ver., **8** 184–188, 1990. (Reprint in: Journ. reine angew. Math. **129** 63–67 (1905)
3. Mathematische Probleme. Arch. Math. Phys., (3) **1** 44–63 and 213–137 (1901), cf. also Ges. Abh., vol. 3, 290–329. (English transl.: Mathematical problems. Bull Amer. Math. Soc. **8** 437–479 (1902). French transl.: Sur les problèmes futurs des Mathématiques. Compt. rend. du deux. congr. internat. des math., Paris 1902, pp. 58–114)
4. Über das Dirichletsche Prinzip. Math. Ann. **59** 161–186 (1904). Festschrift zur Feier des 150-jährigen Bestehens der Königl. Gesell. d. Wiss. Göttingen 1901; cf. also Ges. Abhandl., vol. 3, pp. 15–37
5. Zur Variationsrechnung. Math. Ann. **62** 351–370 (1906). Also in: Göttinger Nachr. (1905) 159–180, and in: Ges. Abh., vol. 3, 38–55
6. Grundzüge einer allgemeinen Theorie der linearen Integralgleichungen. B.G. Teubner, Leipzig Berlin 1912
7. Gesammelte Abhandlungen, vols. 1–3. Springer, Berlin 1932–35

Hildebrandt, S.
1. Rand- und Eigenwertaufgaben bei stark elliptischen Systemen linearer Differentialgleichungen. Math. Ann. **148** 411–429 (1962)
2. Randwertprobleme für Flächen vorgeschriebener mittlerer Krümmung und Anwendungen auf die Kapillaritätstheorie, I: Fest vorgegebener Rand. Math. Z. **112** 205–213 (1969)
3. Über Flächen konstanter mittlerer Krümmung. Math. Z. **112** 107–144 (1969)
4. Contact transformations. Huygens's principle, and Calculus of Variations. Calc. Var. **2** 249–281 (1994)
5. On Hölder's transformation. J. Math. Sci. Univ. Tokyo. **1**, 1–21 (1994)

Hildebrandt, S. and Tromba, A.
1. Mathematics and optimal form. Scientific American Library, W.H. Freeman and Co., New York 1984 (German transl.: Panoptimum, Spektrum der Wiss., Heidelberg 1987. French translation: Pour la Science, Diff. Belin, Paris 1986. Dutch edition. Wet. Bibl., Natuur Technik, Maastricht 1989. Spanish edition: Prensa Cientifica, Viladomat, Barcelona 1990)

Hölder, E.
1. Die Lichtensteinsche Methode für die Entwicklung der zweiten Variation, angewandt auf das Problem von Lagrange. Prace mat.-fiz. **43** 307–346 (1935)
2. Die infinitesimalen Berührungstransformationen der Variationsrechnung. Jahresber. Dtsch. Math.-Ver. **49** 162–178 (1939)
3. Entwicklungssätze aus der Theorie der zweiten Variation. Allgemeine Randbedingungen. Acta Math. **70** 193–242 (1939)
4. Reihenentwicklungen aus der Theorie der zweiten Variation. Abh. Math. Semin. Univ. Hamburg **13** 273–283 (1939)
5. Stabknickung als funktionale Verzweigung und Stabilitätsproblem. Jahrb. dtsch. Luftfahrtforschung, pp. 1799–1819 (1940)
6. Einordnung besonderer Eigenwertprobleme in die Eigenwerttheorie kanonischer Differentialgleichungssysteme. Math. Ann. **119** 22–66 (1943)
7. Das Eigenwertkriterium der Variationsrechnung zweifacher Extremalintegrale. VEB Deutscher Verlag der Wissenschaften, pp. 291–302 (1953). (Ber. Math.-Tagung Berlin 1953)

8. Über die partiellen Differentialgleichungssysteme der mehrdimensionalen Variationsrechnung. Jahresber. Dtsch. Math.-Ver. **62** 34–52 (1959)
9. Beweise einiger Ergebnisse aus der Theorie der 2. Variation mehrfacher Extremalintegrale. Math. Ann. **148** 214–225 (1962)
10. Entwicklungslinien der Variationsrechnung seit Weierstraß (with appendices by R. Klötzler, S. Gähler, S. Hildebrandt). Arbeitsgemeinschaft für Forschung des Landes Nordrhein-Westfalen, **33** 183–240 (1966). Westdeutscher Verlag, Köln Opladen

Hölder, O.
1. Über die Prinzipien von Hamilton und Maupertuis. Göttinger Nachr., pp. 1–36 (1896)
2. Über einen Mittelwertsatz. Nachr. Ges. Wiss. Göttingen pp. 38–47 (1889)

Hofer, H.
1. On the topological properties of symplectic maps. Proc. R. Soc. Edinburg **115A** 25–83 (1990)
2. Symplectic invariants. Proceedings Internat. Congress of Math., Kyoto, 1990. Springer, Tokyo 1991.
3. Symplectic capacities. Lond. Math. Soc. Lect. Note Ser. **152** 1992

Hofer, H. and Zehnder, E.
1. A new capacity for symplectic manifolds. Analysis et cetera, Acad. Press, 1990, edited by P. Rabinowitz and E. Zehnder, pp. 405–428
2. Symplectic invariants and Hamiltonian dynamics. Birkhäuser, Basel 1994

Hopf, E.
1. Generalized solutions of non-linear equations of first order. J. Math. Mech. **14** 951–974 (1965)

Hopf, H.
1. Über die Curvatura integra geschlossener Hyperflächen. Math. Ann. **95** 340–367 (1925)

Hopf, H. and Rinow, W.
1. Über den Begriff der vollständigen differentialgeometrischen Fläche. Comment. Math. Helv. **3** 209–225 (1931)

Hörmander, L.
1. Linear Partial Differential Operators. Springer, Berlin Göttingen Heidelberg 1963
2. The analysis of linear partial differential operators, volume I–IV. Springer, Berlin Heidelberg New York 1983–85

Hove, L. van
1. Sur la construction des champs de De Donder-Weyl par la méthode des charactéristiques. Bull. Acad. Roy. Belg., Cl. Sci. V **31** 278–285 (1945)
2. Sur les champs de Carathéodory et leur construction par la méthode des characteristiques. Bull. Acad. Roy. Belg., Cl. Sci. V **31** 625–638 (1945)
3. Sur l'extension de la conditions de Legendre du calcul des variations aux intégrales multiples à plusieurs fonctions inconnues. Nederl. Akad. Wetensch. Proc. Ser. A, **50** 18–23 (1947). (Indag. Math. 9, 3–8)
4. Sur le signe de la variation seconde des intégrales multiples à plusieurs fonctions inconnues. Acad. Roy. Belg. Cl. Sci. Mém. Coll. (2) **24** 65 pp. (1949)

Huke, A.
1. An historical and critical study of the fundamental Lemma of the calculus of variations. Contributions to the calculus of variations 1930. The University of Chicago, Chicago 1931. Reprint: Johnson, New York 1965

Hund, F.
1. Materie als Feld. Springer, Berlin Göttingen Heidelberg 1954

Huygens, C.
1. Horologium oscillatorium sive de motu pendulorum ad horologia aptato demonstrationes geometricae. Muguet, Paris 1673
2. Traité de la Lumière. Avec un discours de la cause de la pesanteur. Vander Aa, Leiden 1690
3. Oeuvres complètes, 22 vols. M. Nijhoff, Den Haag 1888–1950

Ioffe, A. and Tichomirov, V.
1. Theory of extremal problems. Nauka, Moscow 1974 (In Russian). (Engl. transl.: North-Holland, New York 1978)

Irrgang, R.
1. Ein singuläres bewegungsinvariantes Variationsproblem. Math. Z. **37** 381–401 (1933)

Isaacs, R.
1. Games of pursuit. Technical Report Paper-No. P-257, RAND Corporation, Santa Monica, California 1951
2. Differential games. Wiley, New York 1965. 3rd printing: Krieger, New York 1975
3. Some fundamentals in differential games. In: A. Blaquière (ed.) Topics in Differential Games. North-Holland, Amsterdam 1973

Jacobi, C.G.J.
1. Zur Theorie der Variations-Rechnung und der Theorie der Differential-Gleichungen. Crelle's J. Reine Angew. Math. **17** 68–82 (1837). (See Werke, vol. 4, pp. 39–55)
2. Variationsrechnung. 1837/38. (Lectures Königsberg, Handwritten Notes by Rosenhain).
3. Gesammelte Werke, vols. 1–7. G. Reimer, Berlin 1881–1891
4. Vorlesungen über Dynamik, Supplementband der Ges. Werke. G. Reimer, Berlin 1884. (Lectures held at Königsberg University, Wintersemester 1842–43; Lecture notes by C.W. Borchardt; first edition by A. Clebsch, 1866; revised edition from 1884 by E. Lottner)

Jellett, J.H.
1. An elementary treatise on the calculus of variations. Dublin 1850. (German transl.: Die Grundlehren der Variationsrechnung, frei bearbeitet von C.H. Schnuse. E. Leibrock, Braunschweig 1860)

Jensen, J.L.W.V.
1. Om konvexe Funtioner og Uligheder mellem Middelvaerdier. Nyt Tidsskr. Math. **16B** 49–69 (1905)
2. Sur les fonctions convexes et les inegalités entre les valeurs moyennes. Acta Math. **30** 175–193 (1906)

John, F.
1. Partial differential equations. Springer, New York Heidelberg Berlin 1981. Fourth edition

Jost, J.
1. Two-dimensional geometric variational problems. Wiley-Interscience, Chichester New York 1991
2. Riemannsche Flächen. Springer, Berlin 1994

Kähler, E.
1. Einführung in die Theorie der Systeme von Differentialgleichungen. Hamburger Math. Einzelschriften Nr. 16. Teubner, Leipzig Berlin 1934

Kamke, E.
1. Abhängigkeit von Funktionen und Rang der Funktionalmatrix. Math. Z. **39** 672–676 (1935)
2. Differentialgleichungen reeller Funktionen. Akad. Verlagsgesellschaft, Leipzig 1950
3. Differentialgleichungen. Lösungsmethoden und Lösungen, vol. 1: Gewöhnliche Differentialgleichungen, 5th edition; vol. 2: Partielle Differentialgleichungen erster Ordnung für eine gesuchte Funktion, 3rd edition. Akad. Verlagsgesellschaft, Leipzig 1956

Kapitanskii, L.V. Ladyzhanskaya, D.A.
1. Coleman's principle for the determination of the stationary points of invariant functions. J. Soviet Math. **27** 2606–2616 (1984). Russian Orig.: Zap. Nauch. Sem. Leningradskovo Otdel. Mat. Inst. Steklova **127**, 84–102 (1982)

Kastrup, H.A.
1. Canonical theories of Lagrangian dynamical systems in physics. Physics Reports (Review Section of Physics Letters) **101** 1–167 (1983)

Kaul, H.
1. Variationsrechnung und Hamiltonsche Mechanik. Lecture Notes, Tübingen 1979/80

Kijowski, J., Tulczyjew, W.M.
1. A symplectic framework for field theories. Lecture Notes Math. **107**. Springer, Berlin Heidelberg New York 1979
Killing, W.
1. Über die Grundlagen der Geometrie. J. Reine Angew. Math., **109** 121–186 (1892)
Kimball, W.
1. Calculus of variations by parallel displacement. Butterworths Scientific Publ., London 1952
Klein, F.
1. Gesammelte mathematische Abhandlungen, vols. 1–3. Springer, Berlin 1921–1923
2. Vorlesungen über höhere Geometrie. Springer, Berlin 1926. (Edited by Blaschke, with Supplements by Blaschke, Radon, Artin, and Schreier)
3. Vorlesungen über die Entwicklung der Mathematik im 19. Jahrhundert, vols. 1 and 2. Springer, Berlin 1926/1927
4. Vorlesungen über nicht-euklidische Geometrie. Grundlehren der mathematischen Wissenschaften, vol. 26. Springer, Berlin 1928
Klein, F. and Sommerfeld, A.
1. Über die Theorie des Kreisels. Teubner, Leipzig. Heft I (1897): Die kinematischen und kinetischen Grundlagen der Theorie. Heft II (1898): Durchführung der Theorie im Falle des schweren symmetrischen Kreisels
Klingbeil, E.
1. Variationsrechnung. Wissenschaftverlag, Mannheim 1977. 2nd edition 1988
Klötzler, R.
1. Untersuchungen über geknickte Extremalen. Wiss. Z. Univ. Leipzig, math. nat. Reihe 1–2, pp. 193–206 (1954–55)
2. Bemerkungen zu einigen Untersuchungen von M.I.Višik im Hinblick auf die Variationsrechnung mehrfacher Integrale. Math. Nachr. **17** 47–56 (1958)
3. Die Konstruktion geodätischer Felder im Grossen in der Variationsrechnung mehrfacher Integrale. Ber. Verh. Sachs. Akad. Wiss. Leipzig **104** 84 pp. (1961)
4. Mehrdimensionale Variationsrechnung. Deutscher Verlag der Wiss., Berlin 1969. Reprint Birkhäuser
5. On Pontryagin's Maximum Principles for multiple integrals. Beitr. Anal., **8** 67–75 (1976)
6. On a general conception of duality in optimal control. Proceedings Equadiff 4, Prague, pp. 189–196 (1977)
7. Starke Dualität in der Steuerungstheorie. Math. Nachr. **95** 253–263 (1980)
8. Adolph Mayer und die Variationsrechnung. Deutscher Verlag der Wiss., Berlin 1981. In: 100 Jahre Mathematisches Seminar der Karl-Marx Universität Leipzig (H. Beckert and H. Schumann, eds.)
9. Dualität bei diskreten Steuerungsproblemen. Optimization **12** 411–420 (1981)
10. Globale Optimierung in der Steuerungstheorie. Z. Angew. Math. Mech., **63** 305–312 (1983)
Kneser, A.
1. Variationsrechnung. Encyk. math. Wiss. **2.1** IIA8, 571–625 B.G. Teubner, Leipzig 1900
2. Zur Variationsrechnung. Math. Ann. **50** 27–50 (1898)
3. Lehrbuch der Variationsrechnung. Vieweg, Braunschweig 1900. 2nd edition 1925
4. Euler und die Variationsrechnung. Abhandl. zur Geschichte der Mathematischen Wissenschaften, Heft 25, pp. 21–60, 1907. In: Festschrift zur Feier des 200. Geburtstages Leonhard Eulers, herausgeg. vom Vorstande der Berliner Mathematischen Gesellschaft
5. Das Prinzip der kleinsten Wirkung von Leibniz bis zur Gegenwart. Teubner, Leipzig 1928. In: Wissenschaftliche Grundfragen der Gegenwart, Bd. 9
Knopp, K. and Schmidt, R.
1. Funktionaldeterminanten und Abhängigkeit von Funktionen. Math. Z., **25** 373–381, 1926
Kobayashi, S. and Nomizu, K.
1. Foundations of differential geometry, vols. 1 and 2. Interscience Publ., New York London Sydney 1963 and 1969

Kolmogorov, A.
1. Théorie générale des systèmes dynamiques et mécanique classique. Proc. Int. Congress Math., Amsterdam 1957 (see also Abraham-Marsden, Appendix)

Koschmieder, L.
1. Variationsrechnung. Sammlung Göschen 1074. W. de Gruyter, Berlin 1933

Kowalewski, G.
1. Einführung in die Determinantentheorie, 4th edn. W. de Gruyter, Berlin 1954
2. Einführung in die Theorie der kontinuierlichen Gruppen. AVG, Leipzig 1931

Kronecker, L.
1. Werke. Edited by K. Hensel et al. 5 vols. Leipzig, Berlin 1895–1930

Krotow, W.F. and Gurman, W.J.
1. Methoden und Aufgaben der optimalen Steuerung. Nauka, Moskau 1973 (Russian)

Krupka, D.
1. A geometric theory of ordinary first order variational problems in fibered manifolds. I: Critical sections. II: Invariance. J. Math. Anal. Appl. **49** 180–206, 469–476 (1975)

Lacroix, S.F.
1. Traité du calcul differentiel et du calcul integral, vol. 2. Courcier, Paris 1797. 2nd edition 1814

Lagrange, J.L.
1. Mécanique analytique, 2nd edition, vol. 1 (1811), vol. 2 (1815). Courcier, Paris. First ed.: Méchanique analitique, La Veuve Desaint, Paris 1788
2. Essai d'une nouvelle méthode pour determiner les maxima et les minima des formules intégrales indéfinies. Miscellanea Taurinensia **2** 173–195 (1760/61) Oeuvres 1, pp. 333–362; Application de la méthode exposée dans le mémoire précedent à la solution de différents problèmes de dynamique. Miscellanea Taurinensia **2**. Oeuvres 1, pp. 363–468
3. Sur la méthode des variations. Miscellanea Taurinensia **4** 163–187 (1766/69, 1771) Oeuvres 2, pp. 36–63
4. Sur l'integration des équations à différences partielles du premier ordre. Nouveaux Mém. Acad. Roy. Sci. Berlin, (1772). Oeuvres 3, pp. 549–577
5. Sur les intégrales particulières des équations différentielles. Noveaux Mém. Acad. Roy. Sci. Berlin, (1774). Oeuvres 4, pp. 5–108
6. Sur l'intégration des équations aux dérivées partielles du premier ordre. Noveaux Mém. Acad. Roy. Sci. Berlin, (1779). Oeuvres 4, pp. 624–634
7. Méthode générale pour intégrer les équations aux différences partielles du premier ordre, lorsque ces différences ne sont que linéaires. Noveaux Mém. Acad. Roy. Sci. Berlin, (1785). Oeuvres 5, pp. 543–562
8. Théorie des fonctions analytiques. L'Imprimerie de la République, Prairial an V, Paris 1797. Nouvelle édition: Paris, Courcier 1813
9. Leçons sur le calcul des fonctions. Courcier, Paris, 1806, second edition. Cf. also Oeuvres, vol. 10
10. Mémoire sur la théorie des variations des éléments des planètes. Mém. Cl. Sci. Inst. France 1–72 (1808)
11. Second mémoire sur la théorie de la variation des constantes arbitraires dans les problèmes de mécanique. Mém. Cl. Sci. Inst. France 343–352 (1809)
12. Œuvres, volume 1–14. Gauthier-Villars, Paris 1867–1892. Edited by Serret et Darboux
13. Lettre de Lagrange à Euler. August 12, 1755. Oeuvre 14, 138–144 (1892) (Euler's answer: loc. cit., pp. 144–146)

Lanczos, C.
1. The variational principles of mechanics. University of Toronto Press, Toronto 1949. Reprinted by Dover Publ. 1970

Landau, L. and Lifschitz, E.
1. Lehrbuch der theoretischen Physik, vol. 1: Mechanik, vol. 2: Feldtheorie. Akademie-Verlag, Berlin 1963

Langer, J. and Singer, D.A.
1. Knotted elastic curves in $\mathbb{R}^3$. J. Lond. Math. Soc. II. Ser. **30** 512–520 (1984)
2. The total squared curvature of closed curves. I. Differ. Geom. **20** 1–22 (1984)

Lavrentiev, M. and Lyusternik, L.
1. Fundamentals of the calculus of variations. Gostechizdat Moscow 1935 (in Russian)

Lebesgue, H.
1. Intégral, longueur, aire. Ann. Mat. Pura Appl. (III), **7** 231–359 (1902)
2. Sur la méthode de Carl Neumann. J. Math. Pures Appl. **16** 205–217 and 421–423 (1937)
3. En marge du calcul des variations. L'enseignement mathématique, Série II, t.9, 1963

Lecat, M.
1. Bibliographie du calcul des variations 1850–1913. Grand Hoste, Paris 1913
2. Bibliographie du calcul des variations depuis les origines jusqu'à 1850. Grand Hoste, Hermann, Paris 1916
3. Calcul des variations. Exposé, d'après articles allemands de A. Kneser, E. Zermelo et H. Hahn. In: Encycl. des sciences math., éd. franc. II, **6** (31) (J. Molk). Gauthier-Villars 1913

Lee, H.-C.
1. The universal integral invariants of Hamiltonian systems and application to the theory of canonical transformations. Proc. Roy. Soc. Edinburgh **A62** 237–246 (1947)

Legendre, A.
1. Sur la maniere de distinguer les maxima des minima dans le calcul des variations. Mémoires de l'Acad. Roy. des Sciences, pages 7–37 (1786) 1788

Lehto, O.
1. Univalent functions and Teichmüller theory. Springer, New York 1987

Leis, R.
1. Initial boundary value problems in mathematical physics. Teubner and John Wiley, New York 1986

Leitman, G.
1. The calculus of variations and optimal control. Plenum Press, New York London 1981

Lepage, J.T.
1. Sur les champs géodesiques du calcul des variations. Bull. Acad. Roy. Belg., Cl. Sci. V. s. **22** 716–729, 1036–1046 (1936)
2. Sur les champs géodesiques des integrales multiples. Bull. Acad. Roy. Belg., Cl. Sci. V s. **27** 27–46 (1941)
3. Champs stationnaires, champs géodesiques et formes integrables. Bull. Acad. Roy. Bel., Cl. Sci. V s. **28** 73–92, 247–265 (1942)

Leray, J.
1. Sur le mouvement d'un liquide visqueux emplissant l'espace. Acta Math. **63** 193–248 (1943)

Levi, E.E.
1. Elementi della teoria delle funzioni e Calcolo delle variazioni. Tip-litografia G.B. Castello, Genova 1915

Levi-Civita, T.
1. Sur la regularisation du problème des trois corps. Acta Math. **42** 99–144 (1920)
2. Fragen der klassischen und relativistischen Mechanik. Springer, Berlin Heidelberg New York 1924

Levi-Civita, T. and Amaldi, U.
1. Lezioni di mechanica razionale, vols. I, II.1, II.2. Zanichelli, Bologna 1923, 1926, 1927

Lévy, P.
1. Leçons d'Analyse fonctionnelles. Gauthier-Villars, Paris 1922

Lewy, H.
1. Aspects of calculus of variations. Univ. California Press, Berkeley 1939

Libermann, P. and Marle, C.

1. Symplectic geometry and analytical mechanics. D. Reidel Publ., Dordrecht 1987

Lichtenstein, L.

1. Untersuchungen über zweidimensionale reguläre Variationsprobleme. I. Das einfachste Problem bei fester Begrenzung. Jacobische Bedingung und die Existenz des Feldes. Verzweigung der Extremalflächen. Monatsh. Math. u. Phys. **28** 3–51 (1912)
2. Über einige Existenzprobleme der Variationsrechnung. Methode der unendlich vielen Variablen. J. Math. **145** 24–85 (1914)
3. Zur Analysis der unendlich vielen Variablen. I. Entwicklungssätze der Theorie gewöhnlicher linearer Differentialgleichungen zweiter Ordnung. Rend. Circ. Mat. Palermo. II. Ser. **38** 113–166 (1914)
4. Die Jacobische Bedingung bei zweidimensionalen regulären Variationsproblemen. Sitzungsber. BMG **14** 119–121 (1915)
5. Untersuchungen über zweidimensionale reguläre Variationsprobleme. I. Monatsh. Math. **28** 3–51 (1917)
6. Untersuchungen über zweidimensionale reguläre Variationsprobleme. 2. Abhandlung: Das einfachste Problem bei fester und bei freier Begrenzung. Math. Z. **5** 26–51 (1919)
7. Zur Variationsrechnung. I. Göttinger Nachr. pp. 161–192 (1919)
8. Zur Analysis der unendlichen vielen Variablen. 2. Abhandlung: Reihenentwicklungen nach Eigenfunktionen linearer partieller Differentialgleichungen von elliptischen Typus. Math. Z. **3** 127–160 (1919/20)
9. Über ein spezielles Problem der Variationsrechnung. Berichte Akad. Leipzig **79** 137–144 (1927)
10. Zur Variationsrechnung. II: Das isoperimetrische Problem. J. Math. **165** 194–216 (1931)

Lie, S.

1. Theorie der Transformationsgruppen I–III. Teubner, Leipzig 1888 (I), 1890 (II), 1893 (III). Unter Mitwirkung von F. Engel. Reprint Chelsea Publ. Comp., 1970
2. Vorlesungen über Differentialgleichungen mit bekannten infinitesimalen Transformationen. Teubner, Leipzig 1891
3. Gesammelte Abhandlungen, vols. 1–7. Teubner, Leipzig and Aschehoug, Oslo 1922–1960

Lie, S. and Scheffers, G.

1. Geometrie der Berührungstransformationen, vol. 1. Teubner, Leipzig 1896

Liebmann, H.

1. Lehrbuch der Differentialgleichungen. Veit and Co., Leipzig 1901
2. Berührungstransformationen. Encyclop. Math. Wiss. III D7, pages 441–502, Teubner, Leipzig

Liebmann, H. and Engel, F.

1. Die Berührungstransformationen. Geschichte und Invariantentheorie. Jahresberichte DMV, Ergänzungsbände: V. Band, pp. 1–79 (1914)

Liesen, A.

1. Feldtheorie in der Variationsrechnung mehrfacher Integrale I, II. Math. Ann. **171** 194–218, 273–392 (1967)

Li-Jost, X.

1. Uniqueness of minimal surfaces in Euclidean and hyperbolic 3-spaces. Math. Z. **217** 275–285 (1994)
2. Bifurcation near solutions of variational problems with degenerate second variation. Manuscr. math. **86** 1–14 (1995)

Lin, F.H.

1. Une remarque sur l'application $\frac{x}{|x|}$. C. R. Acad. Sci. Paris **305** 529–531 (1987)

Lindelöf, E.L.

1. Leçons de calcul des variations. Mallet-Bachelier, Paris 1861. This book also appeared as vol. 4 of F.M. Moigno, Leçons sur le calcul différentiel et intégral, Paris 1840–1861

Lions, P.L.
1. Generalized solutions of Hamilton-Jacobi equations. Pitman, London 1982

Ljusternik, L. and Schnirelman, L.
1. Méthode topologique dans les problèmes variationnels. Hermann, Paris 1934

Lovelock, D. and Rund, H.
1. Tensors, differential forms, and variational principles. Wiley, New York London Sydney Toronto 1975

MacLane, S.
1. Hamiltonian mechanics and geometry. Am. Math. Monthly **77** 570–586 (1970)

MacNeish, H.
1. Concerning the discontinuous solution in the problem of the minimum surface of revolution. Ann. Math. (2) **7** 72–80 (1905)
2. On the determination of a catenary with given directrix and passing through two given points. Ann. Math. (2) **7** 65–71 (1905)

Mammana, G.
1. Calcolo della variazioni. Circolo Matematico di Catania, Catania 1939

Mangoldt, H. von
1. Geodätische Linien auf positiv gekrümmten Flächen. J. Reine Angew. Math. **91** 23–52 (1881)

Maslov, V.P.
1. Théorie des perturbations et méthodes asymptotiques. Dunod, Paris, 1972. Russian original: 1965

Matsumoto, M.
1. Foundations of Finsler geometry and Finsler spaces. Kaiseicha, Otsu 1986

Mawhin, J. and Willem, M.
1. Critical point theory and Hamiltonian systems. Applied Mathematical Sciences, vol. 74. Springer, Berlin Heidelberg New York 1989

Mayer, A.
1. Beiträge zur Theorie der Maxima und Minima der einfachen Integrale. Habilitationsschrift. Leipzig 1866
2. Die Kriterien des Maximums und des Minimums der einfachen Integrale in dem isoperimetrischen Problem. Ber. Verh. Ges. Wiss. Leipzig **29** 114–132 (1877)
3. Über das allgemeinste Problem der Variationsrechnung bei einer einzigen unabhängigen Variablen. Ber. Verh. Ges. Wiss. Leipzig **30** 16–32 (1878)
4. Zur Aufstellung des Kriteriums des Maximums und Minimums der einfachen Integrale bei variablen Grenzwerten. Ber. Verh. Ges. Wiss. Leipzig **36** 99–127 (1884)
5. Begründung der Lagrangeschen Multiplikatorenmethode in der Variationsrechnung. Ber. Verh. Ges. Wiss. Leipzig **37** 7–14 (1885)
6. Zur Theorie des gewöhnlichen Maximums und Minimums. Ber. Verh. Ges. Wiss. Leipzig **41** 122–144 (1889)
7. Die Lagrangesche Multiplikatorenmethode und das allgemeinste Problem der Variationsrechnung bei einer unabhängigen Variablen. Ber. Verh. Ges. Wiss. Leipzig **47** 129–144 (1895)
8. Die Kriterien des Minimums einfacher Integrale bei variablen Grenzwerten. Ber. Verh. Ges. Wiss. Leipzig **48** 436–465 (1896)
9. Über den Hilbertschen Unabhängigkeitssatz der Theorie des Maximums und Minumums der einfachen Integrale. Ber. Verh. Ges. Wiss. Leipzig **55** 131–145 (1903)
10. Über den Hilbertschen Unabhängigkeitssatz in der Theorie des Maximums und Minimums der einfachen Integrale, zweite Mitteilung. Ber. Verh. Ges. Wiss. Leipzig **57**, 49–67 (1905), and: Nachträgliche Bemerkung zu meiner II. Mitteilung, loc. cit., vol. **57** (1905)

McShane, E.
1. On the necessary condition of Weierstrass in the multiple integral problem in the calculus of variations I, II. Ann. Math. **32** 578–590, 723–733 (1931)

2. On the second variation in certain anormal problems of the calculus of variations. Am. J. Math. **63** 516–530 (1941)
3. Sufficient conditions for a weak relative minimum in the problem of Bolza. Trans. Am. Math. Soc. **52** 344–379 (1942)
4. The calculus of variations from the beginning through optimal control theory. Academic Press, New York 1978 (A.B. Schwarzkopf, W.G. Kelley, S.B. Eliason, eds.)

Meusnier, J.
1. Mémoire sur la courbure des surface. Mémoires de Math. et Phys. (de savans etrangers) de l'Acad. **10** 447–550 (1785, lu 1776). Paris

Meyer, A.
1. Nouveaux éléments du calcul des variations. H. Dessain, Leipzig et Liège 1856

Milnor, J.
1. Morse theory. Princeton Univ. Press, Princeton 1963

Minkowski, H.
1. Vorlesungen über Variationsrechnung. Vorlesungsausarbeitung, Göttingen Sommersemester 1907
2. Gesammelte Abhandlungen. Teubner, Leipzig Berlin 1911. 2 vols., edited by D. Hilbert, assisted by A. Speiser and H. Weyl

Mishenko, A., Shatalov, V. and Sternin, B.
1. Lagrangian manifolds and the Maslov operator. Springer, Berlin Heidelberg New York 1990

Misner, C., Thorne, K. and Wheeler, J.
1. Gravitation. W.H. Freeman, San Francisco 1973

Möbius, A.F.
1. Der barycentrische Calcul. Johann Ambrosius Barth, Leipzig 1827

Momsen, P.
1. Elementa calculi variationum ratione ad analysin infinitorum quam proxime accedente tractata. Altona 1833

Monge, G.
1. Mémoire sur le calcul intégral des équations aux différences partielles. Histoire de l'Académie des Sciences, pages 168–185 (1784)
2. Application de l'analyse à la géométrie. Bachelier, Paris 1850. 5th edition

Monna, A.F.
1. Dirichlet's principle. Oosthoek, Scheltema and Holkema, Utrecht 1975

Moreau, J.J.
1. Fonctionnelles convexes. Séminaire Leray, Collège de France, Paris 1966

Morrey, C.B.
1. Multiple integrals in the calculus of variations. Grundlehren der mathematischen Wissenschaften, vol. 130. Springer, Berlin Heidelberg New York 1966

Morse, M.
1. Sufficient conditions in the problem of Lagrange with fixed end points. Ann. Math. **32** 567–577 (1931)
2. Sufficient conditions in the problem of Lagrange with variable end conditions. Am. J. Math. **53** 517–546 (1931)
3. The calculus of variations in the large. Amer. Math. Soc. Colloq. Publ., New York 1934
4. Sufficient conditions in the problem of Lagrange without assumption of normality. Trans. Am. Math. Soc. **37** 147–160 (1935)
5. Variational analysis. Wiley, New York 1973

Moser, J.
1. Lectures on Hamiltonian systems. Mem. Am. Math. Soc. **81** (1968)
2. A sharp form of an inequality of N. Trudinger. Indiana Univ. Math. J. **20** 1077–1092 (1971)
3. On a nonlinear problem in differential geometry. Acad. Press, New York 1973. In: Dynamical systems, ed. by M. Peixoto

4. Stable and random motions in dynamical systems with special emphasis on celestial mechanics. Princeton Univ. Press and Univ. of Tokyo Press, Princeton, N.J. 1973. Hermann Weyl Lectures, Institute for Advanced Study
5. Finitely many mass points on the line under the influence of an exponential potential – An integrable system. Lect. Notes Phys., **38** 467–497 (1975). Springer, Berlin Heidelberg New York
6. Three integrable Hamiltonian systems connected with isospectral deformation. Adv. Math. **16** 197–220 (1975)
7. Various aspects of integrable Hamiltonian systems. Birkhäuser, Boston-Basel-Stuttgart, pp. 233–289 (1980). In: Progress in Mathematics 8, "Dynamical systems", CIME Lectures Bressanone 1978

Moser, J. and Zehnder, E.
1. Lecture notes. Unpublished manuscript

Munkres, J.
1. Elementary differential topology. Princeton Univ. Press, Princeton, N.J. 1966. Annals of Math. Studies Nr. 54

Murnaghan, F.D.
1. The calculus of variations. Spartan Books, Washington 1962

Natani, L.
1. Die Variationsrechnung. Wiegand und Hempel, Berlin 1866

Nevanlinna, R.
1. Prinzipien der Variationsrechnung mit Anwendungen auf die Physik. Lecture Notes T.H. Karlsruhe, Karlsruhe 1964

Newton, I.
1. Philosophiae Naturalis Principia Mathematica. Apud plures Bibliopolas/T. Streater, London 1687. 2nd edition 1713, 3rd edition 1725–26. (English transl.: A. Motte, Sir Isaac Newton Mathematical Principles of Natural Phylosophy and his System of the World, London 1729)
2. The mathematical papers of Isaac Newton, 7 vols. Cambridge University Press, Cambridge, 1967–1976. Edited by T. Whiteside.

Nitsche, J.C.C.
1. Vorlesungen über Minimalflächen. Grundlehren der mathematischen Wissenschaften, vol. 199. Springer, Berlin Heidelberg New York 1975
2. Lectures on minimal surfaces. Vol. 1: Introduction, fundamentals, geometry and basic boundary problems. Cambridge Univ. Press, Cambridge 1989

Noether, E.
1. Invariante Variationsprobleme. Göttinger Nachr., Math.-Phys. Klasse, pages 235–257 (1918)

Nordheim, L.
1. Die Prinzipe der Dynamik. Handbuch der Physik, vol. V, pp. 43–90. Springer, Berlin 1927

Nordheim, L. and Fues, E.
1. Die Hamilton-Jacobische Theorie der Dynamik. Handbuch der Physik, vol. V, pp. 91–130. Springer, Berlin 1927

Ohm, M.
1. Die Lehre von Grössten und Kleinsten. Riemann, Berlin 1825

Olver, P.
1. Applications of Lie groups to differential equations. Springer, New York Berlin Heidelberg 1986

O'Neill, B.
1. Semi-Riemannian geometry with applications to relativity. Academic Press, New York 1983

Ostrowski, A.
1. Funktionaldeterminanten und Abhängigkeit von Funktionen. Jahresbe. Dtsch. Math.-Ver., **36** 129–134 (1927)

Palais, R.
1. Foundations of global non-linear analysis. Benjamin, New York Amsterdam 1968
2. The principle of symmetric criticality. Commun. Math. Phys. **69** 19–30 (1979)

Pars, L.A.
1. An introduction to the calculus of variations. Heinemann, London 1962
2. A treatise on analytical dynamics. Heinemann, London 1965

Pascal, E.
1. Calcolo delle variazioni. Hoepli, Milano 1897. 2nd edition 1918. German transl. by A. Schepp, B.G. Teubner, Leipzig 1899

Pauc, C.
1. La methode métrique en calcul des variations. Hermann, Paris 1941

Pauli, W.
1. Relativitätstheorie. Enzykl. math. Wiss., V. 19, vol. 4, part 2, pages 539–775. Teubner, Leipzig

Pfaff, J.
1. Methodus generalis, aequationes diffentiarum partialium, nec non aequationes differentiales vulgares, utrasque primi ordinis, inter quotcunque variabiles, complete integrandi. Abhandl. Königl. Akad. Wiss. Berlin, pages 76–136 (1814–1815)

Pincherle, S.
1. Mémoire sur le calcul fonctionnel distributif. Math. Ann. **49** 325–382 (1897) (cf. also Opere, vol. 2, note 16)
2. Funktionenoperationen und -gleichungen. Encyklopädie Math. Wiss., II.1.2, 763–817 (1904–1916). B.G. Teubner, Leipzig
3. Sulle operazioni funzionali lineari. Proceedings Congress Toronto, August 1924, pages 129–137 (1928)
4. Opere Scelte, vols. 1 and 2 Ed. Cremonese, Roma 1954

Plücker, J.
1. Über eine neue Art, in der analytische Geometrie Punkte und Curven durch Gleichungen darzustellen. Crelle's Journal **7** 107–146 (1829). Abhandlungen, pp. 178–219
2. System der Geometrie des Raumes in neuer analytischer Behandlungsweise, insbesondere die Theorie der Flächen zweiter Ordnung und Classe enthaltend. Schaub, Düsseldorf 1846. 2nd edition 1852
3. Neue Geometrie des Raumes, gegründet auf die Betrachtung der geraden Linie als Raumelement. B.G. Teubner, Leipzig 1868–69, edited by F. Klein
4. Gesammelte mathematische Abhandlungen. Teubner, Leipzig 1895. Edited by A. Schoenflies

Poincaré, H.
1. Sur le problème des trois corps et les équations de la dynamique. Acta Math., **13** 1–27 (1889). Mémoire couronné du prix de S.M. le Roi Oscar II le 21 Janvier 1889
2. Les méthodes nouvelles de la mécanique céleste, tomes I–III. Gauthier-Villars, Paris 1892, 1893, 1899
3. Oeuvres, vols. I–XI. Gauthier-Villars, Paris 1951–56

Poisson, S.
1. Mémoire sur le calcul des variations. Mem. Acad. Roy. Sic., **12** 223–331 (1833)

Poncelet, J.V.
1. Traité des propriétés projectives des figures. Bachelier, Paris 1822
2. Mémoire sur la théorie génerale des polaires réciproques. Crelle's Journal, **4** 1–71 (1829). Presented 1824 to the Paris Academy

Pontryagin, L.S., Boltyanskii, V.G., Gamkrelidze, R.V. and Mishchenko, E.F.
1. The mathematical theory of optimal process. Interscience, New York 1962

Popoff, A.
1. Elements of the calculus of variations. Kazan 1856 (in Russian)

Prange, G.
1. W.R. Hamilton's Arbeiten zur Strahlenoptik und analytischen Mechanik. Nova Acta Abh. Leopold., Neue Folge **107** 1–35 (1923)
2. Die allgemeinen Integrationsmethoden der analytischen Mechanik. Enzyklopädie math. Wiss., 4.1 II, 505–804. Teubner, Leipzig 1935

Pulte, H.
1. Das Prinzip der kleinsten Wirkung und die Kraftkonzeptionen der rationalen Mechanik. Franz Steiner Verlag, Stuttgart 1989

Quetelet, L.A.J.
1. Resumé d'une nouvelle théorie des caustiques. Nouv. Mémoires de l'Académie de Bruxelles, **4** p. 81

Rabinowitz, P.
1. Periodic solutions of Hamiltonian systems. Commun. Pure Appl. Math. **31** 157–184 (1978)
2. Periodic solutions of a Hamiltonian system on a prescribed energy surface. J. Differ. Equations **33** 336–352 (1979)
3. Periodic solutions of Hamiltonian systems: a survey. SIAM J. Math. Anal. **13** 343–352 (1982)

Rademacher, H.
1. Über partielle und totale Differenzierbarkeit von Funktionen mehrerer Variablen, und über die Transformation der Doppelintegrale. Math. Ann. **79** 340–359 (1918)

Radó, T.
1. On the problem of Plateau. Ergebnisse der Mathematik und ihrer Grenzgebiete, vol. 2. Springer, Berlin 1933

Radon, J.
1. Über das Minimum des Integrals $\int_{s_1}^{s_2} F(x, y, \vartheta, \kappa)\, ds$. Sitzungsber. Kaiserliche Akad. Wiss. Wien. Math.-nat. Kl., **69** 1257–1326 (1910)
2. Die Kettenlinie bei allgemeinster Massenverteilung. Sitzungsber. Kaiserliche Akad. Wiss. Wien. Math.-nat. Kl., **125** 221–240 (1916). Berichtigung: p. 339
3. Über die Oszillationstheoreme der konjugierten Punkte beim Problem von Lagrange. Münchner Berichte, pp. 243–257 (1927)
4. Zum Problem von Lagrange. Abh. Math. Semin. Univ. Hamb., **6** 273–299 (1928)
5. Bewegungsinvariante Variationsprobleme, betreffend Kurvenscharen. Abh. Math. Semin. Univ. Hamb. **12** 70–82 (1937)
6. Singuläre Variationsprobleme. Jahresber. Dtsch. Math.-Ver. **47** 220–232 (1937)
7. Gesammelte Abhandlungen, vols. 1 and 2. Publ. by the Austrian Acad. Sci. Verlag Österreich. Akad. Wiss./Birkhäuser, Wien 1987

Rayleigh, J.
1. The theory of sound. Reprint: Dover Publ., New York 1945. Second revised and enlarged edition 1894 and 1896

Reid, W.T.
1. Analogues of the Jacobi condition for the problem of Mayer in the calculus of variations. Ann. Math. **35** 836–848 (1934)
2. Discontinuous solutions in the non-parametric problem of Mayer in the calculus of variations. Am. J. Math. **57** 69–93 (1935)
3. The theory of the second variation for the non-parametric problem of Bolza. Am. J. Math. **57** 573–586 (1935)
4. A direct expansion proof of sufficient conditions for the non-parametric problem of Bolza. Trans. Am. Math. Soc. **42** 183–190 (1937)
5. Sufficient conditions by expansion methods for the problem of Bolza in the calculus of variations. Ann. Math., **38** 662–678 (1937)
6. Riccati differential equations. Academic Press, New York 1972
7. A historical note on Sturmian theory. J. Differ. Equations, **20** 316–320 (1976)
8. Sturmian theory for ordinary differential equations. Applied Mathematical Sciences, vol. 31. Springer, Berlin Heidelberg New York 1980

Riemann, B.
1. Über die Hypothesen, welche der Geometrie zu Grunde liegen. Habilitationskolloquium Göttingen, Göttinger Abh. 13, (1854). (Cf. also Werke, pp. 254–269 in the first edn., pp. 272–287 in the second edn.)

2. Commentatio mathematica, qua respondere tentatur quaestioni ab Illustrissima Academia Parisiensi propositae (1861). See Werke, pp. 370–399
3. Bernhard Riemann's Gesammelte Mathematische Werke. Teubner, Leipzig, First edition 1876, second edition 1892

Ritz, W.
1. Oeuvres. Gauthier-Villars, Paris 1911
2. Über eine neue Methode zur Lösung gewisser Variationsprobleme der mathematischen Physik. J. Reine Angew. Math. **135** 1–61 (1961)

Roberts, A.W. and Varberg, D.E.
1. Convex functions. Academic Press, New York 1973

Rockafellar, R.
1. Convex analysis. Princeton University Press, Princeton 1970

Routh, E.J.
1. The advanced part of a treatise on the dynamics of a system of rigid bodies. MacMillan, London, 6th edition 1905

Rund, H.
1. Die Hamiltonsche Funktion bei allgemeinen dynamischen Systemen. Arch. Math. **3** 207–215 (1952)
2. The differential geometry of Finsler spaces. Grundlehren der mathematischen Wissenschaften, vol. 101. Springer, Berlin Heidelberg New York 1959
3. On Carathéodory's methods of "equivalent integrals" in the calculus of variations. Nederl. Akad. Wetensch. Proc., Ser. A **62** (Indag. Math. 21), 135–141 (1959)
4. The Hamilton-Jacobi theory in the calculus of variations. Van Nostrand, London 1966
5. A canonical formalism for multiple integral problems in the calculus of variations. Aequations Math. **3** 44–63 (1969)
6. The Hamilton-Jacobi theory of the geodesic fields of Carathéodory in the calculus of variations of multiple integrals. The Greek Math. Soc., C. Carathéodory Symposium, pages 496–536 (1973)
7. Integral formulae associated with the Euler–Lagrange operators of multiple integral problems in the calculus of variations. Aequation Math. **11** 212–229 (1974)
8. Pontryagin functions for multiple integral control problems. J. Optimization Theory and Appl. **18** 511–520 (1976)
9. Invariant theory of variational problems on subspaces of a Riemannian submanifold. Hamburger Math. Einzelschriften Heft 5. Van denhoeck & Ruprecht, Göttingen 1971

Sabinin, G.
1. Treatise of the calculus of variations. Moscow 1893 (in Russian)

Sagan, H.
1. Introduction to calculus of variations. Mc Graw-Hill, New York 1969

Sarrus, M.
1. Recherches sur le calcul des variations. Imprimerie Royal, Paris 1844

Scheeffer, L.
1. Bemerkungen zu dem vorstehenden Aufsatze. Math. Ann. **25** 594–595 (1885)
2. Die Maxima und Minima der einfachen Integrale zwischen festen Grenzen. Math. Ann. **25** 522–593 (1885)
3. Über die Bedeutung der Begriffe "Maximum und Minimum" in der Variationsrechnung. Math. Ann. **26** 197–208 (1886)
4. Theorie der Maxima und Minima einer Funktion von 2 Variablen. Math. Ann. **35** 541–576 (1889/90). (Aus seinen hinterlassenen Papieren mitgeteilt von A. Mayer in Leipzig. Wiederabgedruckt aus den Berichten der Kgl. Sächs. Ges. der Wiss., 1886)

Schell, W.
1. Grundzüge einer neuen Methode der höheren Analysis. Archiv der Mathematik und Physik **25** 1–56 (1855)

Schramm, M.
1. Natur ohne Sinn? Das Ende des teleologischen Weltbildes. Styria, Graz Wien Köln 1985

Schrödinger, E.
1. Vier Vorlesungen über Wellenmechanik. Springer, Berlin 1928

Schwartz, L.
1. Théorie des distributions, vols. 1 and 2. Hermann, Paris 1951. Second edition Paris 1966

Schwarz, H.A.
1. Über ein die Flächen kleinsten Inhalts betreffendes Problem der Variationsrechnung. Acta soc. sci. Fenn. **15** 315–362 (1885). Cf. also Ges. Math. Abh. [1], vol. 1, pp. 223–269
2. Gesammelte Mathematische Abhandlungen, vols. 1 and 2. Springer, Berlin 1890

Schwarz, J. von
1. Das Delaunaysche Problem der Variationsrechnung in kanonischen Koordinaten. Math. Ann. **10** 357–389 (1934)

Seifert, H. and Threlfall, W.
1. Lehrbuch der Topologie. Teubner, Leipzig 1934. Reprint Chelsea, New York
2. Variationsrechnung im Grossen. Hamburger Math. Einzelschriften, Heft 24. Teubner, Leipzig 1938

Siegel, C.L.
1. Gesammelte Abhandlungen, vols. I–III (1966), vol. IV (1979). Springer, Berlin Heidelberg New York
2. Vorlesungen über Himmelsmechanik. Springer, Berlin Göttingen Heidelberg 1956
3. Integralfreie Variationsrechnung. Göttinger Nachrichten **4** 81–86 (1957)

Siegel, C.L. and Moser, J.
1. Lectures on Celestial Mechanics. Springer, Berlin Heidelberg New York 1971

Simon, O.
1. Die Theorie der Variationsrechnung. Berlin 1857

Sinclair, M.E.
1. On the minimum surface of revolution in the case of one variable end point. Ann. Math. (2), **8** 177–188 (1906–1907)
2. The absolute minimum in the problem of the surface of revolution of minimum area. Ann. Math. **9** 151–155 (1907–1908)
3. Concerning a compound discontinuous solution in the problem of the surface of revolution of minimum area. Ann. Math. (2) **10** 55–80 (1908–1909)

Smale, N.
1. A bridge principle for minimal and constant mean curvature submanifolds of R^N. Invent. Math. **90** 505–549 (1987)

Smale, S.
1. Differentiable dynamical systems. Bull. Am. Math. Soc., **73** 747–817 (1967)

Smirnov, V., Krylov, V. and Kantorovich, L.
1. The calculus of variations. Kubuch, 1933 (in Russian)

Sommerfeld, A.
1. Atombau und Spektrallinien, vols. I and II. Vieweg, Braunschweig. (Vol. I: first edition 1919, sixth edition 1944; vol. II: second edition 1944)
2. Mechanik. Akad. Verlagsgesellschaft, Leipzig, 1955. (First edition 1942)

Spivak, M.
1. Differential geometry, vols. 1–5. Publish or Perish, Berkeley 1979

Stäckel, P.
1. Antwort auf die Anfrage 84 über die Legendre'sche Transformation. Btbliotheca mathematica (3. Folge) **1** 517 (1900)
2. Über die Gestalt der Bahnkurven bei einer Klasse dynamischer Probleme. Math. Ann. **54** 86–90 (1901)

Steffen, K.
1. Two-dimensional minimal surfaces and harmonic maps. Technical report, Handwritten Notes, 1993

Stegmann, F.L.
1. Lehrbuch der Variationsrechnung und ihrer Anwendung bei Untersuchungen über das Maximum und Minimum. J.G. Luckardt, Kassel 1854

Steiner, J.
1. Sur le maximum et le minimum de figures dans le plan, sur la sphere et dans l'espace en general I, II. J. Reine Angew. Math. **24** 93–152, 189–250 (1842)
2. Gesammelte Werke, vols. 1, 2. G. Reimer, Berlin 1881–1882. Edited by Weierstrass

Sternberg, S.
1. Celestial mechanics, vols. 1 and 2. W.A. Benjamin, New York 1969
2. On the role of field theories in our physical conception of geometry. Lecture Notes in Mathematics, 676 (ed. by Bleuler/Petry/Reetz), Springer, Berlin Heidelberg New York 1978, 1–80

Strauch, G.W.
1. Theorie und Anwendung des sogenannten Variationscalculs. Meyer and Zeller, Zürich 1849, 2 vols.

Struwe, M.
1. Plateau's problem and the calculus of variations. Ann. Math. Studies nr. 35. Princeton Univ. Press, Princeton 1988

Study, E.
1. Über Hamilton's geometrische Optik und deren Beziehungen zur Geometrie der Berührungstransformationen. Jahresber. Dtsch. Math.-Ver. **14** 424–438 (1905)

Stumpf, K.
1. Himmelsmechanik, volume 1 and 2. Deutscher Verl. Wiss., Berlin 1959, 1965

Sundman, K.
1. Recherches sur le problème des trois corps. Acta Soc. Sci. Fenn. **34** No. 6, 1–43 (1907)
2. Mémoire sur le problème de trois corps. Acta Math. **36** 105–179 (1913)

Synge, J.
1. The absolute optical instrument. Trans. Am. Math. Soc. **44** 32–46 (1938)
2. Classical dynamics. Encyclopedia of Physics, Springer, III/I, 1–225 (1960)

Talenti, G.
1. Calcolo delle variazioni. Quaderni dell'Unione Mat. Italiana. Pitagora Ed., Bologna 1977

Thomson, W.
1. Isoperimetrical problems. Nature, p. 517 (1894)

Thomson, W. and Tait, P.G.
1. Treatise on natural philosophy. Cambridge Univ. Press, Cambridge 1867. (German transl.: H. Helmholtz and G. Wertheim: Handbuch der theoretischen Physik, 2 vols. Vieweg, Braunschweig 1871–1874)

Tichomirov, V.
1. Grundprinzipien der Theorie der Extremalaufgaben. Teubner-Texte zur Mathematik 30. Teubner, Leipzig 1982

Todhunter, I.
1. A history of the progress of the calculus of variations during the nineteenth century. Macmillan, Cambridge and London 1861
2. Researches in the Calculus of Variations, principally on the theory of discontinuous solutions. Macmillan, London Cambridge 1871

Tonelli, L.
1. Fondamenti del calcolo delle variazioni. Zanichelli, Bologna 1921–1923. 2 vols.
2. Opere scelte 4 vols. Edizioni Cremonese, Roma 1960–63

Trèves, F.
1. Applications of distributions to pde theory. Am. Math. Monthly **77** 241–248 (1970)
Tromba, A.
1. Teichmüller theory in Riemannian geometry. Birkhäuser, BaseL 1992
Troutman, J.
1. Variational calculus with elementary convexity. Springer, New York 1983
Truesdell, C.
1. The rational mechanics of flexible or elastic bodies 1638–1788. Appeared in Euler's Opera Omnia, Ser. II, vol. XI.2
2. Essays in the history of mechanics. Springer, New York 1968
Tuckey, C.
1. Nonstandard methods in calculus of variations. Wiley, Chichester 1993
Vainberg, M.M.
1. Variational methods for the study of nonlinear operators, Holden-Day, San Francisco 1964
Valentine, F.A.
1. Convex sets. McGraw-Hill, New York 1964
Vash'chenko-Zakharchenko, M.
1. Calculus of variations. Kiev, 1889 (in Russian)
Velte, W.
1. Bemerkung zu einer Arbeit von H. Rund. Arch. Math., **4** 343–345 (1953)
2. Zur Variationsrechnung mehrfacher Integrale in Parameterdarstellung. Mitt. Math. Semin. Gießen H.45, (1953)
3. Zur Variationsrechnung mehrfacher Integrale. Math. Z. **60** 367–383 (1954)
Venske, O.
1. Behandlung einiger Aufgaben der Variationsrechnung. Thesis, Göttingen 1891, pp. 1–60
Vessoit, E.
1. Sur l'interpretation mécanique des transformations de contact infinitésimales. Bull. Soc. Math. France **34** 230–269 (1906)
2. Essai sur la propagation par ondes. Ann. Ec. Norm. Sup. **26** 405–448 (1909)
Viterbo, C.
1. Capacités symplectiques et applications. Séminaire Bourbaki, June 1989. Astérisque **695**
Vivanti, G.
1. Elementi di calcolo delle variazioni. Principato, Messina 1923
Volterra, V.
1. Opere Matematiche, volume 1 (1954); vol. 2 (1956); vol. 3 (1957); vol. 4 (1960); vol. 5 (1962). Accademia Nazionale dei Lincei, Roma
2. Sopra le funzioni che dipendono da altre funzioni. Rend. R. Accad. Lincei, Ser. *IV* **3** 97–105 (Nota I); pp. 141–146 (Nota II); pp. 153–158 (Nota III), 1887. (Opere Matematiche vol. I, nota XVII, pp. 315–328)
3. Sopra le funzioni dipendenti da linee. Rend. R. Accad. Lincei, Ser. IV **3** 229–230 (Nota I); pp. 274–281 (Nota II), 1887. (Opere mathematiche vol. I, nota XVIII, pp. 319–328)
4. Leçons sur les equations intégrales et les equations integro-differentielles. Gauthier-Villars, Paris 1913
5. Leçons sur les fonctions de lignes. Gauthier-Villars, Paris 1913
6. Theory of functionals and of integral and integro-differential equations. Blaskie, London Glasgow 1930
7. Le calcul des variations, son evolution et ses progrès, son rôle dans la physique mathématiques. Publ. Fac. Sci. Univ. Charles e de l'Université Masaryk, Praha-Brno, 54pp., (1932). (Opere Matematiche, vol. V, note XI, pp. 217–267)
Warner, F.W.
1. Foundations of differentiable manifolds and Lie groups. Graduate Texts in Mathematics, vol. 94, Springer, New York Berlin Heidelberg 1983. (First edn.: Scott, Foresman, Glenview: In. 1971)

Weber, E. von
1. Vorlesungen über das Pfaffsche Problem. Teubner, Leipzig 1900
2. Partielle Differentialgleichungen. Enzykl. Math. Wiss. II **A5** 294–399. Teubner, Leipzig

Weierstrass, K.
1. Mathematische Werke, vols. 1–7. Mayer and Müller, Berlin and Akademische Verlagsgesellschaft Leipzig 1894–1927
2. Vorlesungen über Variationsrechnung, Werke, Bd. 7. Akademische Verlagsgesellschaft, Leipzig 1927

Weinstein, A.
1. Lectures on symplectic manifolds. CBMS regional conference series in Mathematics, vol. 29. AMS, Providence 1977
2. Symplectic geometry. In: The Mathematical Heritage of Henri Cartan. Proc. Symp. Pure Math. 39, 1983, pp. 61–70

Weinstock, R.
1. Calculus of variations. Mc Graw-Hill, New York 1952. Reprinted by Dover Publ., 1974

Weyl, H.
1. Die Idee der Riemannschen Fläche. Teubner, Leipzig Berlin 1913
2. Raum, Zeit und Materie. Springer, Berlin 1918. 5th edition 1923
3. Observations on Hilbert's independence theorem and Born's quantizations of field equations. Phys. Rev. **46** 505–508 (1934)
4. Geodesics fields in the calculus of variations of multiple integrals. Ann. Math. **36** 607–629 (1935)

Whitney, H.
1. A function not constant on a connected set of critical points. Duke Math. J. **1** 514–517 (1935)

Whittaker, E.
1. A treatise on the analytical dynamics of particles and rigid bodies. Cambridge Univ. Press, Cambridge, 1964. German transl: Analytische Dynamik der Punkte und starren Körper, Springer, Berlin 1924

Whittemore, J.
1. Lagrange's equation in the calculus of variations, and the extension of a theorem by Erdmann. Ann. Math. **2** 130–136 (1899–1901)

Wintner, A.
1. The analytical foundations of celestial mechanics. Princeton Univ. Press, Princeton 1947

Woodhouse, R.
1. A treatise on isoperimetrical problems and the calculus of variations. Deighton, Cambridge 1810. (A reprint under the title "A history of the calculus of variations in the eighteenth century" has been published by Chelsea Publ. Comp., New York)

Young, L.
1. Lectures on the calculus of variations and optimal control theory. W.B. Saunders, Philadelphia London Toronto 1968

Zeidan, V.
1. Sufficient conditions for the generalized problem of Bolza. Trans. Am. Math. Soc. **275** 561–586 (1983)
2. Extended Jacobi sufficiency criterion for optimal control. SIAM J. Control Optimization, **22** 294–301 (1984)
3. First- and second-order sufficient conditions for optimal control and calculus of variations. Appl. Math. Optimization **11** 209–226 (1984)

Zeidler, E.
1. Nonlinear fundtional analysis and its applications, volume 1: Fixed-point theorems (1986); vol. 2A: Linear monotone operators (1990); vol. 2B: Nonlinear monotone operators (1990); vol. 3:

Variational methods and optimization (1985); vol. 4: Applications to mathematical physics; vol. 5 to appear. Springer, New York Berlin Heidelberg

Zermelo, E.

1. Untersuchungen zur Variationsrechnung. Thesis, Berlin 1894
2. Zur Theorie der kürzesten Linien. Jahresberichte der Deutsch. Math.-Ver. **11** 184–187 (1902)
3. Über das Navigationsproblem bei ruhender oder veränderlicher Windverteilung. Z. Angew. Math. Mech., **11** 114–124 (1931)

Zermelo, E. and Hahn, H.

1. Weiterentwicklung der Variationsrechnung in den letzten Jahren. Encycl. math. Wiss. II 1,1 pp. 626–641. Teubner, Leipzig 1904

Subject Index

(Page numbers in roman type refer to this volume, those in italics to Volume 311.)

Grundlehren der mathematischen Wissenschaften

A Series of Comprehensive Studies in Mathematics

A Selection

204. Popov: Hyperstability of Control Systems
205. Nikol'skiĭ: Approximation of Functions of Several Variables and Imbedding Theorems
206. André: Homologie des Algébres Commutatives
207. Donoghue: Monotone Matrix Functions and Analytic Continuation
208. Lacey: The Isometric Theory of Classical Banach Spaces
209. Ringel: Map Color Theorem
210. Gihman/Skorohod: The Theory of Stochastic Processes I
211. Comfort/Negrepontis: The Theory of Ultrafilters
212. Switzer: Algebraic Topology – Homotopy and Homology
215. Schaefer: Banach Lattices and Positive Operators
217. Stenström: Rings of Quotients
218. Gihman/Skorohod: The Theory of Stochastic Procrsses II
219. Duvant/Lions: Inequalities in Mechanics and Physics
220. Kirillov: Elements of the Theory of Representations
221. Mumford: Algebraic Geometry I: Complex Projective Varieties
222. Lang: Introduction to Modular Forms
223. Bergh/Löfström: Interpolation Spaces. An Introduction
224. Gilbarg/Trudinger: Elliptic Partial Differential Equations of Second Order
225. Schütte: Proof Theory
226. Karoubi: K-Theory. An Introduction
227. Grauert/Remmert: Theorie der Steinschen Räume
228. Segal/Kunze: Integrals and Operators
229. Hasse: Number Theory
230. Klingenberg: Lectures on Closed Geodesics
231. Lang: Elliptic Curves: Diophantine Analysis
232. Gihman/Skorohod: The Theory of Stochastic Processes III
233. Stroock/Varadhan: Multidimensional Diffusion Processes
234. Aigner: Combinatorial Theory
235. Dynkin/Yushkevich: Controlled Markov Processes
236. Grauert/Remmert: Theory of Stein Spaces
237. Köthe: Topological Vector Spaces II
238. Graham/McGehee: Essays in Commutative Harmonic Analysis
239. Elliott: Proabilistic Number Theory I
240. Elliott: Proabilistic Number Theory II
241. Rudin: Function Theory in the Unit Ball of C^n
242. Huppert/Blackburn: Finite Groups II
243. Huppert/Blackburn: Finite Groups III
244. Kubert/Lang: Modular Units
245. Cornfeld/Fomin/Sinai: Ergodic Theory
246. Naimark/Stern: Theory of Group Representations
247. Suzuki: Group Theory I
248. Suzuki: Group Theory II
249. Chung: Lectures from Markov Processes to Brownian Motion
250. Arnold: Geometrical Methods in the Theory of Ordinary Differential Equations
251. Chow/Hale: Methods of Bifurcation Theory
252. Aubin: Nonlinear Analysis on Manifolds. Monge-Ampère Equations
253. Dwork: Lectures on ρ-adic Differential Equations
254. Freitag: Siegelsche Modulfunktionen
255. Lang: Complex Multiplication
256. Hörmander: The Analysis of Linear Partial Differential Operators I

257. Hörmander: The Analysis of Linear Partial Differential Operators II
258. Smoller: Shock Waves and Reaction-Diffusion Equations
259. Duren: Univalent Functions
260. Freidlin/Wentzell: Random Perturbations of Dynamical Systems
261. Bosch/Güntzer/Remmert: Non Archimedian Analysis – A System Approach to Rigid Analytic Geometry
262. Doob: Classical Potential Theory and Its Probabilistic Counterpart
263. Krasnosel'skiĭ/Zabreĭko: Geometrical Methods of Nonlinear Analysis
264. Aubin/Cellina: Differential Inclusions
265. Grauert/Remmert: Coherent Analytic Sheaves
266. de Rham: Differentiable Manifolds
267. Arbarello/Cornalba/Griffiths/Harris: Geometry of Algebraic Curves, Vol. I
268. Arbarello/Cornalba/Griffiths/Harris: Geometry of Algebraic Curves, Vol. II
269. Schapira: Microdifferential Systems in the Complex Domain
270. Scharlau: Quadratic and Hermitian Forms
271. Ellis: Entropy, Large Deviations, and Statistical Mechanics
272. Elliott: Arithmetic Functions and Integer Products
273. Nikol'skiĭ: Treatise on the Shift Operator
274. Hörmander: The Analysis of Linear Partial Differential Operators III
275. Hörmander: The Analysis of Linear Partial Differential Operators IV
276. Liggett: Interacting Particle Systems
277. Fulton/Lang: Riemann-Roch Algebra
278. Barr/Wells: Toposes, Triples and Theories
279. Bishop/Bridges: Constructive Analysis
280. Neukirch: Class Field Theory
281. Chandrasekharan: Elliptic Functions
282. Lelong/Gruman: Entire Functions of Several Complex Variables
283. Kodaira: Complex Manifolds and Deformation of Complex Structures
284. Finn: Equilibrium Capillary Surfaces
285. Burago/Zalgaller: Geometric Inequalities
286. Andrianov: Quadratic Forms and Hecke Operators
287. Maskit: Kleinian Groups
288. Jacod/Shiryaev: Limit Theorems for Stochastic Processes
289. Manin: Gauge Field Theory and Complex Geometry
290. Conway/Sloane: Sphere Packings, Lattices and Groups
291. Hahn/O'Meara: The Classical Groups and K-Theory
292. Kashiwara/Schapira: Sheaves on Manifolds
293. Revuz/Yor: Continuous Martingales and Brownian Motion
294. Knus: Quadratic and Hermitian Forms over Rings
295. Dierkes/Hildebrandt/Küster/Wohlrab: Minimal Surfaces I
296. Dierkes/Hildebrandt/Küster/Wohlrab: Minimal Surfaces II
297. Pastur/Figotin: Spectra of Random and Almost-Periodic Operators
298. Berline/Getzler/Vergne: Heat Kernels and Dirac Operators
299. Pommerenke: Boundary Behaviour of Conformal Maps
300. Orlik/Terao: Arrangements of Hyperplanes
301. Loday: Cyclic Homology
303. Lange/Birkenhake: Complex Abelian Varieties
303. DeVore/Lorentz: Constructive Approximation
304. Lorentz/v. Golitschek/Makovoz: Constructive Approximation. Advanced Problems
305. Hiriart-Urruty/Lemaréchal: Convex Analysis and Minimization Algorithms I. Fundamentals
306. Hiriart-Urruty/Lemaréchal: Convex Analysis and Minimization Algorithms II. Advanced Theory and Bundle Methods
307. Schwarz: Quantum Field Theory and Topology
308. Schwarz: Topology for Physicists
309. Adem/Milgram: Cohomology of Finite Groups
310. Giaquinta/Hildebrandt: Calculus of Variations, Vol. I: The Lagrangian Formalism
311. Giaquinta/Hildebrandt: Calculus of Variations, Vol. II: The Hamiltonian Formalism

Springer-Verlag and the Environment

DRUCK: STRAUSS OFFSETDRUCK, MÖRLENBACH
BINDEN: SCHÄFFER, GRÜNSTADT